Gonorynchiformes and
Ostariophysan Relationships
A Comprehensive Review

Series on **Teleostean Fish Biology**

Series Editors: Francisco José Poyato-Ariza and Rui Diogo

Gonorynchiformes and Ostariophysan Relationships
A Comprehensive Review

Senior Volume Editor
Terry Grande
Loyola University
Chicago, IL
USA

Volume Coeditors
Francisco José Poyato-Ariza
Universidad Autónoma de Madrid
Madrid
Spain

Rui Diogo
Universitè de Liège
Liège
Belgium

CRC Press
Taylor & Francis Group
Boca Raton London New York

CRC Press is an imprint of the
Taylor & Francis Group, an **informa** business
A SCIENCE PUBLISHERS BOOK

First published 2010 by Science Publishers Inc.

Published 2019 by CRC Press
Taylor & Francis Group
6000 Broken Sound Parkway NW, Suite 300
Boca Raton, FL 33487-2742

First issued in paperback 2019

No claim to original U.S. Government works

ISBN 13: 978-0-367-45239-1 (pbk)
ISBN 13: 978-1-57808-374-9 (hbk)

Visit the Taylor & Francis Web site at
http://www.taylorandfrancis.com

and the CRC Press Web site at
http://www.crcpress.com

Cover Illustration reproduced by courtesy of Shedd Aquarium/Patrice Ceisel

Library of Congress Cataloging-in-Publication Data

Gonorynchiformes and of Ostariophysan relationships : a comprehensive review / senior volume editor, Terry Grande; volume co-editors, Francisco José Poyato-Ariza, Rui Diogo.
    p. cm. -- (Series on teleostean fish biology)
 Includes bibliographical references and index.
 ISBN 978-1-57808-374-9 (hardcover)
1. Gonorynchiformes. I. Grande, Terry. II. Poyato-Ariza,
    Francisco José. III. Diogo, Rui.
 QL637.9.G6G66 2009
 597'.5--dc22

                                                          2009016942

Preface to the Series

Fishes are, by far, the most abundant and diversified group of vertebrates in existence today. They constitute more than half of all living vertebrate species, and their fossil record is equally as impressive. The Teleostei, about 96% of all fish species, demonstrates an endless account of amazing and exciting diversity in terms of morphology, ecology, and evolutionary history. The unprecedented adaptive radiation of the group beginning in the Mesozoic era, has resulted in species colonizing virtually every aquatic habitat throughout the globe. Wherever there is, or has been water available, from the permanent snow-line of a mountain to the abyssal depths of the oceans, teleosts are there. They are indispensable components of our fragile ecosystems, and serve as an important food resource for much of the world's human population.

And yet, what do we know about them? The Teleostean Fish Biology Series will attempt to answer this question. This series will provide comprehensive analyses and updates of major teleostean groups based on cutting edge research by leading ichthyologists focused on morphology, development, palaeontology, systematics, ecology, biogeography, and physiology. Through this series, the wealth of teleost literature often too vast for one person to assimilate, as well as current research, will be synthesized and made available to both the professional and student of ichthyology alike.

The series begins with a collection of peer-reviewed chapters on Gonorynchiformes set within the context of the superorder Ostariophysi. The book "Gonorynchiformes and Ostariophysan Relationships: A Comprehensive Review" is intended to be followed by volumes on other equally important groups. Like in the gonorynchiform book, each volume will be devoted to all possible aspects of a particular telostean fish group and set within a phylogenetic framework. Great effort will be made to offer in the series not only the information itself, but also all the elements of consensus and disagreements, in order to promote discussions and debates that are indispensable in any healthy science. We do expect to provide many volumes for those interested in, or actively working on, this remarkable and astonishing group of creatures.

Francisco José POYATO-ARIZA
Universidad Autónoma de Madrid, Spain

Rui DIOGO
The George Washington University, USA

Preface to the Volume

Gonorynchiform fishes have been of interest to students of fish biology and evolution for many years. This interest stems from the extreme morphological diversities and environmental adaptations seen among some of the gonorynchiform taxa, and from the group's evolutionary position among teleost fishes. An understanding of gonorynchiform morphology and systematic inter- and intrarelationships has proven vital to a better understanding of the evolution of lower teleosts in general, and more specifically of groups such as the clupeiforms (e.g., herrings and anchovies), and otophysans (e.g., carps, minnows and catfishes).

Over the last decade or so, there have been a number of advances in our knowledge about gonorynchiforms. For example, new fossil gonorynchiforms have been described from the Upper Cretaceous of Italy adding biogeographic data for the group. In addition, gonorynchids such as †*Notogoneus osculus* have been redescribed in much more detail based on new and better prepared material, thus leading to a reanalysis of the family Gonorynchidae. The morphologies of paedomorphic groups from freshwater rivers of Africa such as *Cromeria* and *Grasseichthys* have been reexamined using new and better-prepared material, and detailed developmental studies of key species such as *Chanos chanos* have been completed, adding significant character information for phylogenetic reconstruction. Finally, for the first time, extant gonorynchiform taxa have been included in molecular analyses, and fossil taxa were used to calculate divergence times and molecular clocks for Ostariophysi, as well as, Clupeocephala.

This book through a series of peer-reviewed chapters brings together a number of authors to examine the current knowledge of gonorynchiform biology, including comparative osteology, myology, epibranchial morphology and development. Phylogenetic interrelationships among gonorynchiform fishes are reexamined in light of recent publications and a reexamination of all described taxa. A chapter on the gonorynchiform fossil record and paleobiography provides a historical framework. To put Gonorynchiformes into a larger context, Lecointre examines the relationship between gonorynchiforms and the rest of Ostariophysi, as well as the relationship of Ostariophysi within Clupeocephala based on both molecular and morphological data. Additionally, there are chapters written by experts currently working on each of the four otophysan subgroups, Cypriniformes (Simons and Gidmark), Characiformes (Dahdul), Siluriformes (Diogo), and Gymnotiformes (Alves-Gomes). These authors provide a systematic overview of each group and an assessment of the field's current phylogenetic and biogeographic knowledge using fossils, morphology and molecular data. Insights into the future of ostariophysan research are explored. The book ends with Eschmeyer et al. providing a thorough nomenclatural analysis of all gonorynchiform taxa described to date, that will serve as a practical tool to those interested in the group.

Contributions to this book were submitted and strictly reviewed by other specialists. We are grateful to the reviewers for contributing their time and expertise to this effort. Each chapter was greatly enhanced by their critiques and suggestions. These reviewers are: Dominique Adriaens (Belgium), Gloria Arratia (USA), Christopher Braun (USA), Lionel Cavin (Switzerland), Michel Chardon (Belgium), Mario de Pinna (Brazil), Christina Cox Fernandes (USA), Lance Grande (USA), Phillip Haris (USA), Eric Hilton (USA), John Lundberg (USA), Claudia Malabarba (Brazil), Luiz Malabarba (Brazil), Richard Mayden (USA), Janelle Morano (USA), Francisco Poyato-Ariza (Spain), Mark Westneat (USA), and Richard Winterbottom (Canada).

Special thanks to the National Science Foundation for funding (DEB 0128794 and EF 0732589) that helped in the production of this book, and to the Department of Biology, Loyola University, Chicago, for supporting this project by assuming the costs of photocopies, mailing and needed clerical assistance. Finally, thanks to Amanda Burdi for helping to edit and format each chapter and to Joseph Schluep for his technical expertise in improving some of the illustrations presented in this book.

Terry Grande

Contents

Preface to the Series   v

Preface to the Volume   vii

1. Reassessment and Comparative Morphology of the   1
   Gonorynchiform Head Skeleton
   Terry Grande and *Francisco José Poyato-Ariza*

2. Morphological Analysis of the Gonorynchiform   39
   Postcranial Skeleton
   Terry Grande and *Gloria Arratia*

3. Early Ossification and Development of the Cranium and   73
   Paired Girdles of *Chanos chanos* (Teleostei, Gonorynchiformes)
   Gloria Arratia and *Teodora Bagarinao*

4. A Review of the Cranial and Pectoral Musculature of   107
   Gonorynchiform Fishes, with Comments on Their Functional
   Morphology and a Comparison with Other Otocephalans
   Rui Diogo

5. The Epibranchial Organ and Its Anatomical Environment   145
   in the Gonorynchiformes, with Functional Discussions
   Françoise Pasleau, Rui Diogo and *Michel Chardon*

6. The Fossil Record of Gonorynchiformes   173
   Emmanuel Fara, Mireille Gayet and *Louis Taverne*

7. Gonorynchiform Interrelationships: Historic Overview,   227
   Analysis, and Revised Systematics of the Group
   Francisco José Poyato-Ariza, Terry Grande and *Rui Diogo*

8. A New Teleostean Fish from the Early Late Cretaceous   339
   (Cenomanian) of SE Morocco, with a Discussion of Its
   Relationships with Ostariophysans
   Frédéric Pittet, Lionel Cavin and *Francisco José Poyato-Ariza*

9. Gonorynchiformes in the Teleostean Phylogeny: 363
   Molecules and Morphology Used to Investigate
   Interrelationships of the Ostariophysi
   Guillaume Lecointre

10. Systematics and Phylogenetic Relationships of Cypriniformes   409
    Andrew M. Simons and *Nicholas J. Gidmark*

11. Review of the Phylogenetic Relationships and Fossil   441
    Record of Characiformes
    Wasila M. Dahdul
12. State of the Art of Siluriform Higher-level Phylogeny   465
    Rui Diogo and *Zuogang Peng*
13. The Mitochondrial Phylogeny of the South American Electric   517
    Fish (Gymnotiformes) and an Alternative Hypothesis
    for the Otophysan Historical Biogeography
    José A. Alves-Gomes
14. A Nomenclatural Analysis of Gonorynchiform Taxa   567
    William N. Eschmeyer, Terry Grande and *Lance Grande*

Index   589

1

Reassessment and Comparative Morphology of the Gonorynchiform Head Skeleton

Terry Grande[1] and Francisco José Poyato-Ariza[2]

Introduction

For many years the Gonorynchiformes was considered a systematic enigma; its placement within the Teleostei was debated and, as discussed in other chapters of this book, its monophyly questioned. This confusion was probably due to the enormous amount of morphological and behavioral variation within the group. The amount of morphological variation is very clear when examining the skulls of these fishes. *Chanos chanos* represents a more basal gonorynchiform according to most phylogenetic analyses (Fink and Fink 1981, Patterson 1984, Poyato-Ariza 1996a, Grande and Poyato-Ariza 1999), and its skull morphology is usually considered quite generalized for the group. On the other hand, *Cromeria* and *Grasseichthys* show extreme cranial miniaturization (Grande 1994, Britz and Moritz 2007), while *Phractolaemus* exhibits modifications for breathing atmospheric air.

Irrespective of generic variation, the gonorynchiform head skeleton is characterized by the loss of the orbitosphenoids in the interorbital septum, a reduction in the size of the pterosphenoids, and parietals that, when present, are reduced in size and separated from each other by the supraoccipital. Gonorynchiforms are also known for possessing cephalic

[1] Loyola University Chicago, 1032 West Sheridan Road, Chicago, Illinois 60626, U.S.A.
[2] Unidad de Paleontologia, Departamento de Biologia, Universidad Autónoma de Madrid, Cantoblanco, 28049–Madrid, Spain.

ribs that extend from the back of the skull to the shoulder girdle. Cephalic ribs observed in *Chanos* were hypothesized by Rosen and Greenwood (1970) to be involved in transmitting sound to the inner ear.

Gonorynchiforms share with other ostariophysans the loss of the basisphenoid and the dermal component of the palatine. Supramaxillary bones are lost as separate elements. The sacculi and lagenae of the inner ear are positioned posteriorly along the midline. As pointed out by Grande and de Pinna (2003), the posterior midline position of these chambers in ostariophysans is like that found in otophysans and most clupeomorphs, possibly facilitating the reception of far field sound waves transmitted via the swimbladder.

Although previous authors have described the cranial morphology of several gonorynchiform taxa (e.g., Swinnerton 1901, Ridewood 1904, 1905, Gregory 1933, Gréy 1964, Perkins 1970, Blum 1991, Gayet 1993a), in some cases in great detail (e.g., Rabor 1938, Thys van den Audenaerde 1961, Monod 1963, Poll 1965, Langet 1974, Gayet 1986, Britz and Moritz 2007), the descriptive comparative morphology of these fishes has not been examined within a single work. In this chapter, therefore, we examine the comparative skull morphology and associated structures of gonorynchiform fishes within the phylogenetic framework of Poyato-Ariza *et al.* (this volume). Using the Recent *Chanos* as a reference, we examine evolutionary significant modifications and adaptations found within the Chanidae, the Gonorynchidae, and the Kneriidae. Proposed cranial characters diagnosing gonorynchiform subgroups were rechecked against specimens (see Grande and Poyato-Ariza 1999 for a list of specimens), and new morphological analyses (e.g., extrascapulars) are presented.

Types of Skull Bones

The types of actinopterygian skull bones and their origins have recently been discussed in Grande and Bemis (1998) and Arratia (2003) but are worth restating here. Ontogenetically, the skull is formed by the endocranium (= chondrocranium), splanchnocranium and dermatocranium.

The endocranium is chondral in origin (i.e., cartilage precursor) and is often among the first skull structures to form. The endocranial bones surround the brain on the ventral, lateral and posterior sides and, in gonorynchiforms, comprise the mesethmoid, lateral ethmoids, epiotics, prootics, pterotics, sphenotics, basioccipital, and exoccipitals.

The splanchnocranium is also chondral in origin and consists of the structures derived from the visceral arches (mandibular, hyoid and branchial arches). The mandibular arch, plus its associated dermal bones, gives rise to the jaws. The chondral bones are the paired quadrate, metapterygoid, autopalatine, articular and retroarticular. The dermal bone components of the mandibular arch are the paired premaxilla, maxilla, ecto- and endopterygoids in the upper jaw, angular and dentary in the lower jaw.

The hyoid arch consists of the hyomandibula and symplectic dorsally, and anterior and posterior ceratobranchials ventrally. Posteriorly, the branchial arches form the support of the branchial cavity and the gills. Tooth plates may form in this region and join to different elements of the arches.

Lastly, the dermatocranium consists of dermal bones. These bones are the only ones to bear sensory canals, and their formation is actually induced by the canal. Dermal bones basically form the roof, snout and floor of the cranium or braincase (e.g., frontal, nasal, vomer, parasphenoid). Other dermal bones join the mandibular arch to form the jaws, as we just saw above. Finally, the opercular bones and branchiostegal rays close the buccopharyngeal cavity laterally and ventrally.

In the present chapter, the gonorynchiform skull will be described topographically according to its structural and functional regions, regardless of the ontogenetic origin of the different bones. This chapter is also focused on Recent forms, since cranial bones are better observed than in fossils. Comments on fossil taxa are, nonetheless, noted whenever pertinent.

Cranium

Dorsal View

Orienting from anterior to posterior and from medial to lateral, the dorsal surface of the cranium consists of the following bones: mesethmoid, lateral ethmoids, nasals, frontals, sphenotics, pterotics, parietals, epiotics, and supraoccipital (Fig. 1.1A–G). All bones are without ornamentation and without dorsally directed fossae. The comparative size of many of these bones varies largely from genus to genus. In chanids, for example, frontals are long and large, accounting for the elongate and broad aspect of their skull. Gonorynchids demonstrate extreme elongation of the frontal bones yielding a head that is longer than deep, while in *Phractolaemus* (Fig. 1.1C) the frontals are truncate, giving the head a short or squat appearance. This morphometric variation is one of the factors that explains the heterogeneous head appearance of the different gonorynchiform genera.

Variation in mesethmoid size and shape is common among gonorynchiforms. In general the mesethmoid is a compressed bone with two anterolateral processes and two posterolateral processes. Among the chanids where this can be observed, the posterolateral processes, or wings,

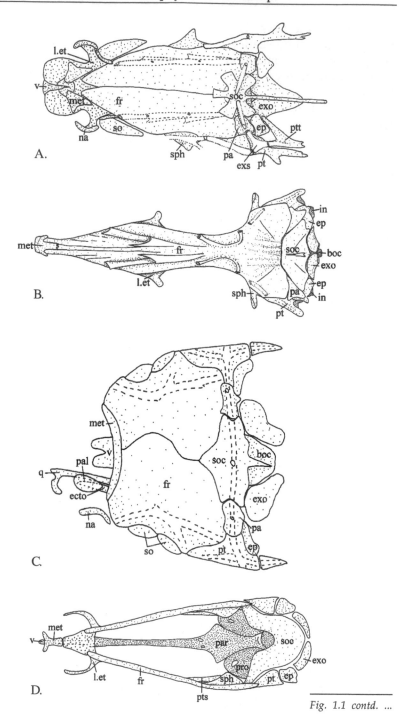

A.

B.

C.

D.

Fig. 1.1 contd. ...

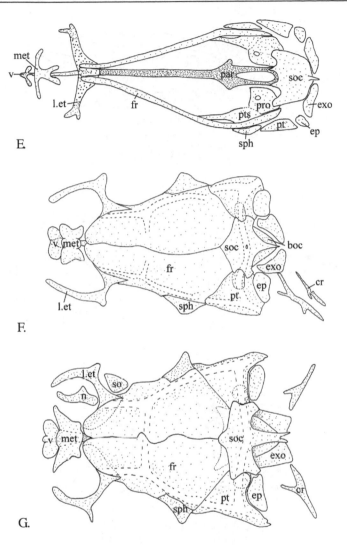

.... *Fig. 1.1 contd.*

Fig. 1.1 Skull roof in dorsal view. All drawings with anterior facing up. (A) *Chanos chanos*, CAS 05022-05075, SL = 103 mm. (B) *Gonorynchus abbreviatus*, SL = 308 mm. (C) *Phractolaemus ansorgei*, FMNH 63938A, SL = 125 mm. (D) *Cromeria occidentalis*, MRAC 141098, SL = 23.2 mm. (E) *Grasseichthys gabonensis*, MRAC 73-02 P-264, SL = 18.9 mm. (F) *Parakneria tanzania*, BMNH 10.21-163, SL = 43.7 mm. (G) *Kneria wittei*, MRAC 79-01-P-516, SL = 54.3 mm. boc, basioccipital; cr, cephalic rib; ds, dermosphenotic; ep, epiotic; exo, exoccipital; exs, extrascapular; fr, frontal bone; in, intercalar; l.et, lateral ethmoid; met, mesethmoid; na, nasal; pa, parietal bone; pal, palatine; par, parasphenoid; pt, pterotic; pts, pterosphenoid; q, quadrate; sph, sphenotic; so, supraorbital bone; soc, supraoccipital; v, vomer.

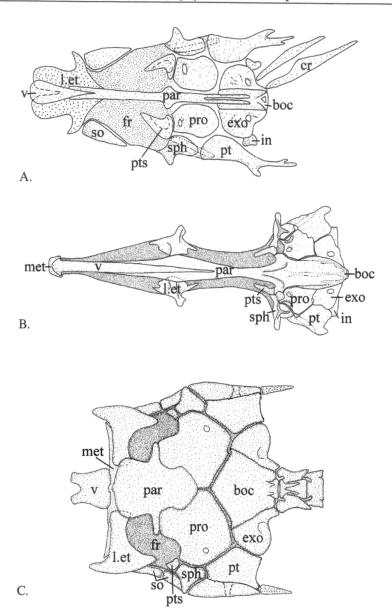

A.

B.

C.

Fig. 1.2 contd. ...

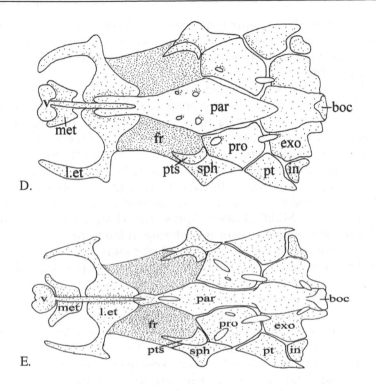

.... *Fig. 1.2 contd.*

Fig. 1.2 Skull in ventral view. All drawings with anterior facing up. (A) *Chanos chanos*, CAS 05022-05075, SL = 103 mm. (B) *Gonorynchus abbreviatus*, FMNH 76746, SL = 308 mm. (C) *Phractolaemus ansorgei*, FMNH 63938A, SL = 125 mm. (D) *Parakneria tanzania*, BMNH 10.21-163, SL = 43.7 mm. (E) *Kneria wittei*, MARC 79-01-P-516, SL = 54.3 mm. boc, basioccipital; cr, cephalic rib; ds, dermosphenotic; ep, epiotic; exo, exoccipital; fr, frontal; in, intercalar; l.et, lateral ethmoid; met, mesethmoid; na, nasal; par, parasphenoid; pro, prootic; pt, pterotic; pts, pterosphenoid; sph, sphenotic; so, suborbital; v, vomer.

are longer than the anterior ones (e.g., †*Tharrhias*), while in the Kneriinae (Fig. 1.1D–G), the anterolateral processes are more prominent and the mesethmoid takes on a butterfly-like shape. The anterolateral processes in *Grasseichthys* (Fig. 1.1E) are delicate and elongated. According to Swinnerton (1901), the anterior part of the cranium in *Cromeria* is greatly reduced and highly cartilaginous, and the minute mesethmoid is fused with two nasals and the suborbitals from either side. We were unable to confirm the mesethmoid condition in our specimens. In Britz and Moritz (2007), autogenous nasal bones are not identified in either species of *Cromeria*. Developmental material is needed to discern whether the ethmoid is a compound structure in *Cromeria*. Based on our material it was difficult to

clearly discern the mesethmoid from the vomer, although the vomer/
parasphenoid connection was quite clear. In *Phractolaemus* (Fig. 1.2C), with
its wide squat skull, the mesethmoid is elongated and spans the length of
the lateral ethmoids.

The position of the mesethmoid relative to the vomer is of particular
interest. The common condition among gonorynchiforms is for the vomer
to extend anteriorly beyond the margin of the mesethmoid. This condition
is found in the chanids, kneriids, †*Hakeliosomus* and, contrary to Gayet
(1993a), †*Ramallichthys* and †*Judeichthys*. In †*Notogoneus* and *Gonorynchus*,
however, the mesethmoid extends anteriorly beyond the vomer (Grande
and Grande 2008). According to Gayet (1993a), this condition is also found
in †*Charitopsis* and †*Charitosomus*, and was used by her as a synapomorphy
to unite the two Middle Eastern forms. This character poses interesting
phylogenetic implications worth exploring in the future.

From the dorsal aspect, the lateral ethmoids appear as wings that
extend from the skull laterally. In the African forms (i.e., kneriids), these
wings have both anterior and posterior extensions. In *Phractolaemus*
(Fig. 1.1C), the extensions are not as pronounced and are of equal length,
while in the Kneriinae the anterior extensions are considerably longer than
the posterior ones. The lateral ethmoids are overlapped by the frontals in
all gonorynchiforms, meaning that most of the bone is better viewed from
the ventral aspect (Fig. 1.2A–E), and are consequently very poorly known in
fossils. The relative size of the lateral ethmoids varies among
gonorynchiforms. They are broad and robust in *Phractolaemus* and exhibit
elongation in *Cromeria occidentalis*. As described in Britz and Moritz (2007),
and illustrated here (Fig. 1.1D), the lateral ethmoids are separated from each
other by a broad cartilaginous region. This cartilage represents the last of
the ethmoid cartilage. This cartilage also extends anteriorly and separates
the lateral ethmoids from the mesethmoid. In *Grasseichthys*
(Fig. 1.1E), the lateral ethmoids are also separated from each other by
cartilage allowing the lateral ethmoids to extend laterally in front of the
orbit (Britz and Moritz 2007). In the gonorynchids and chanids the lateral
ethmoids are reduced in size and do not exhibit extensions comparable to
Cromeria. Because of the anteriorly elongated frontals, in most
gonorynchids the lateral ethmoids appear equidistant between the
sphenotics and the mesethmoid. In other gonorynchiforms the lateral
ethmoids are closer to the mesethmoid. In ventral aspect the lateral
ethmoids articulate with the anterior part of the parasphenoid and the
posterior part of the vomer medially. In the *Chanos* specimens examined,
however (Fig. 1.1A), the ethmoidal region remains cartilaginous and lies
dorsal to the parasphenoid. The examination of very large specimens of
Chanos is necessary to determine how the ethmoid ossifies or whether it
ossifies at all (Arratia and Grande in prep.). In *Phractolaemus* (Fig. 1.2C), the

lateral ethmoids articulate with the parasphenoid along a posterior corner, but they never articulate with the vomer, or meet medially.

The nasals in the Chanidae are very small bones, little more than canal-bearing ossifications, placed just in front of the frontals and lateral to the mesethmoid. Both nasals are well separated from each other medially. They are rather flat bones in smaller forms (e.g., †Rubiesichthyinae), and slightly more tubular in shape in larger forms (e.g., *Chanos*), but always small and delicate, and easily removable, therefore rarely preserved in fossil specimens. The nasals in the Recent Gonorynchoidei resemble that of *Chanos*, and are unknown in fossil forms.

The frontals in *Chanos* (Fig. 1.1A) and other chanids are long and broad, even in their preorbital region, and their orbital curvature is very faintly marked. The supraorbital canal runs within each frontal bone and terminates before the parietals. In the Gonorynchidae, the frontals are long and narrow in the preorbital region with a preorbital length of about 76 percent of the head length (Grande and Poyato-Ariza 1999). The elongation of the frontals in †*Ramallichthys*, †*Hakeliosomus* and †*Judeichthys* is not as extreme as in *Gonorynchus*, †*Notogoneus*, †*Charitosomus* and †*Charitopsis*, but contrary to Gayet (1993a), they still exhibit the diagnostic frontal shape and proportions of the family. *Phractolaemus* (Fig. 1.1C) exhibits a different frontal condition with respect to both the Chanidae and the Gonorynchidae. In *Phractolaemus*, the frontals are as short as they are broad and contribute to the fish's robust-looking head. In most gonorynchiforms the frontal bones are medially sutured tightly together, as in the majority of teleosts. However, in the case of *Grasseichthys* and *Cromeria* (Fig. 1.1D, E), the frontal bones are medially separated from each other by a large space that appears early in the embryonic form. The frontals extend laterally along the skull and are separated by a protruding brain. The brain is covered only by connective tissue and skin. This condition in *Cromeria* and *Grasseichthys* is likely due to paedomorphosis (Grande 1994). In *Cromeria nilotica* and *Grasseichthys gabonensis*, the frontals appear to extend anterior to the level of the lateral ethmoids, where in *C. occidentalis* the frontals terminate anteriorly at the level of the lateral ethmoids (Britz and Moritz 2007). In adult *Gonorynchus* (Fig. 1.1B), the frontals are medially fused to each other. In specimens examined about 75 mm in standard length and smaller (e.g., *G. mosleyi*, BPBM 15437, SL = 74.4 mm), however, two distinct frontal bones were observed. In a specimen of *G. mosleyi* of about 82 mm SL (BPBM 25364), frontals are fused to each other in the ethmoid region, but separate posteriorly. In specimens greater than 87 mm SL (e.g., *G. greyi*, SU 09178, SL = 87.5 mm), only a median frontal was observed. This condition appears to be unique to *Gonorynchus*. Contrary to Perkins (1970), †*Notogoneus* does not have medially fused frontals. Autogenous frontals are found in all adult gonorynchiforms except *Gonorynchus*.

The sphenotics are thought to be composite bones that consist of a fusion between a dermal canal-bearing portion positioned on the dorsal surface (i.e., otic canal), and a chondral portion below. These portions are distinct in †*Tharrhias*, where a small, flat autogenous dermosphenotic bears the sensory canal and forms part of the skull roof, independent of the more massive, triangular autosphenotic below (Brito and Wenz 1990, pers. obs.). In other fossil chanids, the condition is unknown. In all gonorynchiforms the sphenotics are triangular in shape, and with the pterotics form the posterolateral margin of the skull. The sphenotics are predominantly ventrally positioned bones (Fig. 1.2A–E) so in dorsal aspect (Fig. 1.1A–G), they appear smaller than they really are, extending beyond the lateral margin of the frontal bones. In ventral view the fully observable sphenotics articulate with the reduced and separated pterosphenoids and the prootics. In *Gonorynchus* (Figs. 1.1B, 1.2B), the orbital region is greatly enlarged resulting in a more posterior position of the sphenotics. In *Phractolaemus* (Figs. 1.1C, 1.2C), the sphenotics are considerably smaller than in other gonorynchiforms and are completely obscured from view dorsally by supraorbitals 1 and 2.

Like the sphenotics, the pterotics are complex bones consisting of dermal and chondral components. As in all teleosts, these bones form the lateroposterior corners of the skull roof. They articulate with the frontals anterolaterally, the parietals medially, and epiotics posterolaterally. In the case of *Grasseichthys* and *Cromeria* (Fig. 1.1C, D), the pterotics contact the supraoccipital medially. In *Parakneria* and *Kneria* (Fig. 1.1F, G), the pterotics are robust triangularly shaped bones, while in the gonorynchids they are thin elongate elements. In *Chanos* and †*Tharrhias*, the pterotics are highly developed in that they are elongated posteriorly, emitting stout projections that slope backward and downward to the posterior end of the posttemporal bones, above the occipital region and the first vertebra. These robust posterioventral processes are absent in all other chanids. The condition in *Phractolaemus* consists of massive pterotics similar to that found in *Kneria* and *Parakneria*, but, similar to the chanids, extends posteriorly forming a single spine. A comparatively small pterotic posterior expansion, or spine, is found in the gonorynchids or other kneriids. The otic canal housed within the dermal portion of the pterotics leads to the lateral portion of the supratemporal commissure of the extrascapulars.

In gonorynchiforms, the parietals, when present, are either partially (†*Gordichthys*) or completely separated from each other by the supraoccipital. In †*Gordichthys*, the supraoccipital partially separates the parietals along the posterior half while the anterior portion of these bones make contact (Poyato-Ariza 1996a); this is probably a consequence of the relative shortening of the skull of †*Gordichthys* with respect to other fossil chanids and notably to its sister group, †*Rubiesichthys*. As a result of the lateral position of the parietals in gonorynchiforms, the supraorbital canal that runs through the skull terminates at the posterior margin of the frontals instead of the parietals. In other teleosts (e.g., *Elops*, *Albula*), this canal terminates more posteriorly. In gonorynchiforms with autogenous parietals, they are reduced in size and considered diagnostic of the order. In †*Aethalionopsis*, †*Tharrhias*, and other fossil chanids, the parietals are somewhat reduced, but flat and blade-like in shape, while in *Chanos* they are comparatively smaller, but very similar in shape. In the Gonorynchoidei the parietals show the greatest amount of reduction and in some taxa serve as little more than canal bearing bones. In specimens of *Chanos* and *Gonorynchus* examined, the medial portion of the supratemporal commissure that lies superficial to the supraoccipital extends on to the surface of the parietals. Canals do not run through the parietals in these taxa. The parietals in *Parakneria* and *Kneria* (Fig. 1.1F, G) have been reported by several authors to be lost as autogenous elements (Langlet 1974, Grande 1992, 1994). As reported in Grande and Poyato-Ariza (1999), in *Kneria wittei* specimens examined of about 150 mm SL, the "parietals" seem to appear as distinct bones. In specimens examined larger than 300 mm SL, autogenous parietals were no longer observed. In two large specimens reexamined for this study (MRAC 79-01-P-516-826 and CAS 16150), the margins of the "parietals" positioned on either side of the supraoccipital were clearly observed, and a canal was observed running through the elements (see below). Fink and Fink (1981, 1996) also observed parietals in their specimens of *Kneria*. If extrascapular bones are identified as canal-bearing bones and parietals not, it is possible that the elements tentatively identified as "parietals" are actually part of the extrascapular series. Additional developmental study is necessary to better understand this condition. According to d'Aubenton (1954), parietals are absent in *Cromeria*. Swinnerton (1901), on the other hand, reported their presence in *Cromeria*. According to Britz and Moritz (2007), both *Cromeria* and *Grasseichthys* lack parietals. Based on material on hand for this paper, we did not see parietals abutting the large supraoccipital in the specimens of *Cromeria* examined (MRAC 141098). We were also unable to confirm the presence of parietals in *Grasseichthys* based upon our material. These observations differ from those made in Grande and Poyato-Ariza (1999) where the margins of parietals, not epiotics, were observed. We conclude that a more thorough study of the parietal condition in Kneriinae is probably necessary to better assess its ontogenetic and taxonomic variation in these fishes, and to determine what the common parietal condition is. Suffice it to say, within gonorynchiforms, there is a trend towards reduction in the size of the parietals.

Extrascapulars and the supratemporal commissure are worthy of special note. As discussed in Fink and Fink (1981) and then in more detail in Arratia and Gayet (1995), the extrascapulars are plate-like dermal bones that carry the supratemporal commissure. Generally they articulate with the parietals, pterotics, epiotics and possibly the supraoccipital. According to Lecointre and Nelson (1996), the extrascapulars are fused with the parietals in ostariophysans and clupeomorphs. In Arratia and Gayet (1995), however, extrascapulars were illustrated in *Chanos* as distinct elements positioned posterolateral to the parietals and posteromedial to the pterotics. If, as according to Fink and Fink (1981), the key identifying character of an extrascapular is the presence of the supratemporal commissure, it is worthwhile reviewing the posterior lateral line condition observed in our gonorynchiform specimens. We may thus gain a better understanding of the extrascapular condition in these fishes. In our specimens of *Chanos* (Fig. 1.1A, left side of drawing), a long canal identified as the medial portion of the supratemporal commissure runs superficially to the supraoccipital and parietals. This canal does not penetrate these bones and can easily be removed by forceps. The medial canal leads to the intervening and lateral parts of the supratemporal commissure (terminology of Britz and Moritz 2007). The lateral part of the medial canal and the medial part of the lateral canal (Fig. 1.1A) extend over the parietals, but not within them as stated above. Sitting above the pterotics appears to be a thin bone containing the lateral and intervening parts of the supratemporal commissure. This bone seems to have different margins than the pterotics, in that it extends medially beyond the margin of the pterotics, yet laterally does not quite extend to the far margin of the pterotics. We identify this bone as the lateral extrascapular. We also identify the bone carrying the medial part of the supratemporal commissure as the medial extrascapular. Thus in *Chanos*, the extrascapulars are autogenous elements and our observations agree with those of Arratia and Gayet (1995). In †*Tharrhias*, however (Poyato-Ariza 1996: fig. 14b), the supratemporal commissure appears to be embedded in both the supraoccipital and parietals. Our observations parallel those of Britz and Moritz (2007) with respect to *Gonorynchus*. In *Gonorynchus* (Fig. 1.1B), all canals of the supratemporal commissure are superficial tubes lying over the supraoccipital (medial portion), parietals (intervening portion) and pterotics (lateral portion). Five small canals were observed positioned between the lateral portion of the commissure and the post-temporal. In *Gonorynchus* we could find no evidence of extrascapulars, so developmental material is necessary to better understand the condition in this fish. As described in Grande (1999), the lateral line canal system in *Gonorynchus* is unique among gonorynchiforms and worthy of careful study. The condition in †*Notogoneus* appears to be similar to that of *Gonorynchus*. No canal-bearing bone was observed in the posterior skull region (Grande and Grande 2008). We infer from this that the canals of the supratemporal commissure were superficial to the underlying bones like that of *Gonorynchus*. In Grande and Grande (2008), extrascapular bones were not identifiable in †*Notogoneus*. The condition in the Middle Eastern forms is different from *Gonorynchus* and †*Notogoneus*. In †*Ramallichthys*, †*Judeichthys* and †*Hakeliosomus*, canals were observed carved into the supraoccipital, parietals and pterotics. Extrascapulars were not identified in these fish, nor were they identified by Gayet (1985, 1986, 1993). This suggests possible fusion of the extrascapulars with one or more of these bones. Although not observed by us, Gayet (1993a) illustrated †*Charitosomus lineolatus* (MNHN-SHA-281) with a canal-bearing pterotic. Canals were not illustrated in the supraoccipital or parietals. This suggests that the condition in †*C. lineolatus* might be like that of *Chanos*.

The condition in *Phractolaemus* is quite interesting (Figs. 1.1C, 1.2C). Dorsally it appears that all of the elements of the supratemporal commissure run within the supraoccipital (medial portion), parietal (intervening portion) and pterotic (lateral portion) bones. This observation is consistent with that of Thys van den Audenaerde (1961). Ventrally, however, the pterotic situation is complicated (Fig. 1.2C). Upon careful examination, the limits of the pterotics are clearly defined ventrally, but the canal-bearing component seen in dorsal aspect seems to extend beyond the lateral limits of the pterotics. The canal-bearing portion and the underlying pterotics seem inseparable. The canal-bearing component is fused with the pterotics, and the junction between the two bones is visible on the ventral side. From the dorsal aspect only, it appears that the canal runs through the pterotics. An alternative explanation is that the extrascapulars carrying the intervening and lateral supratemporal commissure are fused with the pterotics, and only in ventral aspect is this fusion obvious. As discussed above, elements in the position of the parietals were observed in a few specimens of *Kneria* and *Parakneria*. The common condition, however, in both genera is the loss of parietals. In a further examination, these elements (i.e., "parietals") contain a canal and, although positioned on either side of the supraoccipital, are also positioned within the limits of the large pterotics. It seems likely then that these elements are part of the extrascapular series and that in these specimens the extrascapulars have not completely fused with the pterotics. We were unable to determine the extrascapular condition in *Cromeria* and *Grasseichthys*. Britz and Moritz (2007) do not mention extrascapulars in their descriptions of either genus. So, based on our observations of gonorynchiform fishes, the extrascapular and supratemporal commissure conditions are variable. It is most likely that if the extrascapulars are fused with other bones they are fused with the pterotics, not the parietals. Additional histological studies are necessary to fully understand this situation.

In gonorynchiforms, epiotics (= epioccipitals) (Fig. 1.1A–G) are positioned posterior to the parietals in dorsal view. They contact the exoccipitals ventrally, the supraoccipital medially, and the pterotics anteroventrally, and form the superior margin of the posttemporal fossa in occipital view (e.g., Thys van den Audenaerde 1961: fig. 16). They are of similar shape and proportions in all extant gonorynchiform taxa and are usually quite difficult to observe in fossil taxa; for instance, in †*Tharrhias* they are visible in lateral view only when the intercalars are removed, according to Brito and Wenz (1990: 378).

With the exclusion of the frontal bones, the supraoccipital is the largest bone on the dorsal surface of the skull. The size of the supraoccipital varies, with the †Rubiesichthyinae and †*Dastilbe* having the smallest and the kneriids (e.g., *Kneria*: supraoccipital is expanded laterally) exhibiting some of the largest. In dorsal view, it is bounded laterally by the parietals and anteriorly by the frontals. With the exception of †*Gordichthys* (Poyato-Ariza 1994, 1996a, b), the supraoccipital separates the parietals. Dorsally, the medial part of the supratemporal commissure runs either along the surface of the supraoccipital (i.e., *Chanos*) or within it (i.e., †*Ramallichthys*). In the kneriids the supraoccipital forms the dorsal margin of the foramen magnum. In occipital view, it is bounded by the epiotics laterally and the exoccipitals ventrally or ventrolaterally. In most gonorynchiforms (exceptions include †*Dastilbe* and *Phractolaemus*), the supraoccipital extends posteriorly forming a crest above the foramen magnum that varies in size among gonorynchiforms. *Chanos* (Fig. 1.1C) exhibits a very unique supraoccipital condition where the supraoccipital crest forms a pectinate blade that divides into two sets of 8 to 12 brush-like bony filaments (Rabor 1938). These filaments extend posteriorly to vertebra 3 and are situated between the left and right sets of epaxial muscles. Such a crest is also present in †*Tharrhias*. It is, however, slightly smaller than that of *Chanos* and is devoid of filaments. The lack of filaments in †*Tharrhias* may be an artifact of preservation since the filaments are delicate and easily broken off. The supraoccipital crest is present, but significantly smaller in most other gonorynchiforms, although in *Parakneria* and *Kneria* (Fig. 1.1F, G) the crest is, comparatively, about three times that observed in gonorynchids. It is forked in *Kneria* and not in *Parakneria*. In *Cromeria*, because of the smaller size of other bones, the supraoccipital appears comparatively enlarged.

Ventral View

The bones that constitute the cranium in ventral aspect are the mesethmoid, lateral ethmoids, vomer, parasphenoid, pterosphenoids, sphenotics, pterotics, prootics, intercalar, basioccipital, and exoccipitals. To avoid repetition, bones that have already been described in the section above will not be described again.

In the *Chanos* specimens (Fig. 1.2A) examined for this study (100–115 mm SL), the vomer is reduced in size and is positioned in the center of a mostly cartilaginous ethmoid region. It articulates with the anterior part of the mesethmoid on the ventral surface. The shape and relative position of the vomer and mesethmoid seems very similar to those fossil chanids in which this region is more or less accessible (e.g., †*Tharrhias*). The position of the vomer relative to the mesethmoid has been discussed above. Suffice it to say that the generalized condition among gonorynchiforms is for the vomer to extend beyond the anterior margin of the mesethomid. The opposite condition is found in *Gonorynchus* (Fig. 1.2B), †*Notogoneus* and possibly †*Charitosomus* and †*Charitopsis*. In *Gonorynchus*, the vomer is long and straight, extending about one-third the length of the parasphenoid. In *Phractolaemus* (Fig. 1.2C), the vomer is square, with an anterior notch, and extends beyond the anterior margin of the skull, as does the palatine, quadrate, interopercle and dentary. Posteriorly, the vomer is short and does not articulate with the parasphenoid as it does in all other gonorynchiforms. In *Cromeria occidentalis*, the vomer extends anteriorly from the parasphenoid and is bent ventrally. This condition, ventral inclination of the vomer, is shared with both *Parakneria* and *Kneria* and differs from all other gonorynchiforms, which have a straight, flat vomer. According to Britz and Moritz (2007), the vomer of *Cromeria nilotica* is not bent ventrally as it is in *C. occidentalis*. The vomer is devoid of dentition in all gonorynchiforms except for †*Gordichthys*, which sports a few tiny teeth (Poyato-Ariza 1994).

The long, stout parasphenoid in *Chanos* (Fig. 1.2A) and most gonorynchiforms bears a very strong ascending process that rises high and passes in front of the prootic. This process braces the parasphenoid on the ventral side, thus supporting the very long frontals, and sustains the cranium ventrally. In related teleostean taxa (e.g., otophysans), this ascending process is weak and is situated beneath the prootics, not in front of them as in gonorynchiform fishes. In *Chanos*, †*Dastilbe*, †*Tharrhias*, and probably other chanids, the ascending process of the parasphenoid is located anterior to the hyomandibula and articulates with the prootics; the main body of the parasphenoid is bent upwards posterior to its ascending process. This modification of the parasphenoid seems to be correlated with the elongated frontals, characteristic of gonorynchiforms. The parasphenoid in *Grasseichthys* (Fig. 1.1E) is straight most of its length and forks posteriorly. In most gonorynchiforms, and like many other teleosts, the two arms of the parasphenoid ultimately meet and tightly articulate with the basioccipital. In *Phractolaemus* (Fig. 1.2C), however, the parasphenoid is compact and short and does not extend posteriorly beyond the prootics. Two lateral processes extend from the parasphenoid starting at about its midpoint, abutting the prootics anteriorly. In *Gonorynchus* the parasphenoid expands at its posterior end near the prootics and does not run the length of the skull, but ends anterior to the basioccipital. Lastly, in *Parakneria* and *Kneria* (Fig. 1.2D, E), the parasphenoid is massive and terminates at the anterior end of the basioccipital. Once again, with the exception of †*Gordichthys* (Poyato-Ariza 1994), dentition is absent from the parasphenoid in gonorynchiforms.

In *Chanos*, like in all other observed gonorynchiforms, the pterosphenoid bones are reduced and widely separated from each other (Fink and Fink 1981, Grande and Poyato-Ariza 1999). The amount of reduction varies among the different genera within the order. In fossil chanids the reduction is moderate. In *Chanos* (Fig. 1.2A) the pterosphenoids articulate laterally with the prootics along their posterior margin, and barely abut the sphenotics. In gonorynchoids, the pterosphenoids are comparatively reduced and more broadly separated. In these fishes the pterosphenoids articulate more lengthily with the sphenotics, and in some taxa (e.g., *Phractolaemus*), they are so separated that they hardly contact the prootics. This separation and reduction in pterosphenoid size was used by Grande and Poyato-Ariza (1999) to help diagnose the Gonorynchoidei. Ventrally, near the suture with the prootics, the pterosphenoids present the foramen for the exit of the trigemino-facial nerves (V + part of VII).

The prootics as in all ostariophysans are massive bones and contain a foramen for the exit of the hyomandibular branch of cranial nerve VII (facial), in addition to the more ventral jugular openings. The prootics articulate with the pterotics and sphenotics laterally, and with the parasphenoid medially in all observed gonorynchiforms except *Phractolaemus*. In *Phractolaemus* (Fig. 1.2C), as just explained, the parasphenoid is truncated in length and does not articulate with the basioccipital. As a result, the prootics meet at the midline of the ventral skull posterior to the parasphenoid and anterior to the basioccipital. The utricular otoliths and their chambers are housed within the prootics.

The intercalars, if present, are very small bones placed behind the pterotics, near the pterotic posterior process, and dorsoposterior to the exoccipitals. They form a small part of the posterior border of the cranium, although, as pointed out by Rabor (1938: 354, his "opisthotic") they are firmly connected to the posttemporal through a strong fibrous connection, so intercalars are easily detachable and rare to observe in prepared crania. They are present in at least *Chanos* and †*Tharrhias*, but this region is not accurately observable in any other fossil chanid. Intercalars are reported for *Gonorynchus* by Le Danois (1966) and for *Phractolaemus* by van den Audeaerde (1961) and observed by us here. Intercalars were not reported for *Cromeria* and *Grasseichthys* by Britz and Moritz (2007). They are assumed to be lost in these taxa.

Behind the prootics, the basioccipital is a single posteromedian bone that forms the floor of the braincase as well as the floor of the foramen magnum and most of the posteromedian border of the cranium. It houses both lagenar otoliths in their entirety and the posterior half of the large saccular otolith. As discussed in Rosen and Greenwood (1970), the posteromedian position of the inner ear chambers and corresponding otoliths is common among all ostariophysans, indicating a close phylogenetic relationship. The basioccipital articulates with the parasphenoid anteriorly and with centrum one posteriorly. In most gonorynchiforms, the basioccipital is a small bone, often hard to distinguish from the parasphenoid as their articulation is usually very tight, especially in chanids. Unlike the condition in clupeomorphs and elopomorphs (de Pinna and Grande 2003), anterior vertebral centra have not fused with the basioccipital during development, so its articulation with the vertebral column is direct, forming a monopartite occipital condyle.

The exoccipitals articulate with the basioccipital medially, and with the pterotics and epiotics laterally, forming the posterolateral walls of the cranium. They always frame the foramen magnum laterally, and they frame it dorsally as well in chanids and *Gonorynchus*. They present three paired foramina, the posterior one for the vagus nerve (X) and the two anterior ones, probably for the passage of two glossopharyngeal branches (IX) (Brito and Wenz 1990). In *Chanos* (Figs. 1.1A, 1.2A) the exoccipitals are greatly expanded into plates that slope backward and meet in the midline immediately below the supraoccipital. The exoccipital plates form a roof over the neural arch of vertebra one and provide added protection for the cranial exit of the neural tube. The condition in †*Tharrhias* resembles that of *Chanos* in that the exoccipitals are enlarged and frame the foramen magnum dorsally, although in †*Tharrhias*, they extend posteriorly to a lesser extent, so that they form a roof over the basioccipital, not over the first neural arch (Patterson 1984, pers. obs.). The expanded exoccipitals of *Chanos* and †*Tharrhias* articulate posteriorly with the expanded first neural arch (Poyato-Ariza 1996a: fig. 14, 1996b: fig. 4),

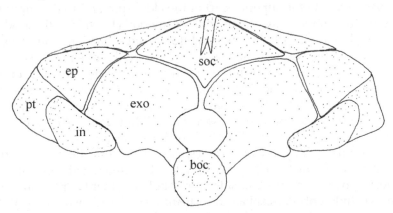

Fig. 1.3 Posterior view of the skull of *Gonorynchus abbreviatus*, FMNH 76746, SL = 308 mm. boc, basioccipital; exo, exoccipital; in, intercalar; pt, pterotic; soc, supraoccipital.

a condition not found in other gonorynchiforms. In the Chanidae in which this region is accessible and in *Gonorynchus*, the large exoccipitals meet each other above the basioccipital (e.g., Rabor 1938: 354, pl. 2A, Brito and Wenz 1990: 377, Grande and Poyato-Ariza 1999: fig. 5A), but in *Phractolaemus* and all other Recent African gonorynchiforms, the right and left exoccipitals do not meet above the basioccipital (Fig. 1.3). This results in an enlarged foramen magnum that is not bounded by the exoccipitals above, but by the supraoccipital. Exoccipitals are moderate in size in *Parakneria* and *Kneria*. The exoccipitals in *Grasseichthys* and *Cromeria* are expanded, thus accommodating the large occipital region of the skull (d'Aubenton 1961). In any case, the exoccipitals of the Gonorynchoidei are not involved in connecting the vertebral column with the skull. The exoccipitals do not participate in the formation of the occipital condyle in any gonorynchiform, resulting in a monopartite occipital condyle.

Cranial Intermuscular Bones

Although intermuscular bones are customarily discussed with the postcranial skeleton, gonorynchiforms exhibit a unique and diagnostic series of intermusculars that are associated with the back of the skull, so it is pertinent to describe them here, in relation to the occipital region of the braincase. Cranial (cephalic) ribs are blade-like intermuscular bones, belonging to the epicentral series (Patterson and Johnson 1995), and positioned between the exoccipitals and the cleithra. The cephalic ribs in *Chanos* (Fig. 1.1A) show the greatest degree of specialization of all observed gonorynchiform fishes (Fig. 1.2A). Anteriorly, the cephalic ribs are joined by ligaments and are in direct contact with the exoccipitals. Posteriorly, these ribs are attached to the first two pleural ribs by strips of connective tissue and by two muscles that are continuous with the epaxial musculature. They are also joined to the superior part of the cleithrum and to the superficial musculature of the body. Rosen and Greenwood (1970) hypothesized that the cephalic (cranial) ribs in gonorynchiform fishes function in an acoustical role. There is no evidence, however, that the cranial ribs in *Chanos* or any other gonorynchiform transmit sound and could be considered to function like the Weberian ossicles tripus and claustrum of, for instance, *Brycon*. A more plausible functional role may be simply to help support the shoulder girdle. With the exception of *Grasseichthys*, all living and observable fossil forms have cephalic ribs. The lack of cephalic ribs, as well as other intermuscular bones in *Grasseichthys*, is likely due to its paedomorphic condition. In *Kneria* there are two cephalic ribs fused together forming one large rib positioned on either side of the skull. In other kneriids and in the gonorynchids, only one cephalic rib articulates with each exoccipital. Within the Gonorynchidae the cephalic ribs have been slightly modified in that they are not blade-like, but look more like delicate rods.

Cheek Bones

The cheek bones are the most superficial bones on the lateral surface of the osteichthyan skull. In gonorynchiforms, they are flat, often support the cephalic lateral line canals, and, as in most osteichthyans, can be ornamented (e.g., †Parachanos). Bones in this region are all paired and dermal in origin, and consist of the circumorbital bones that surround the orbit housing the infraorbital canal (i.e., antorbital, infraorbitals, dermosphenotic, plus the canal-free supraorbital bones) and the opercular series (i.e., opercle, interopercle, preopercle, and subopercle).

Circumorbital Series

Gonorynchiformes lack suborbitals. *Phractolaemus* has two supraorbitals, while *Chanos*, *Kneria* and *Parakneria* have one. *Gonorynchus*, †*Notogoneus* and *Grasseichthys* lack supraorbitals. According to Moritz *et al.* (2006) and Britz and Moritz (2007), *Cromeria nilotica* differs from *C. occidentalis* in not having a supraorbital.

The generalized gonorynchiform condition is a reduction in the number of infraorbitals, from five or six found in other teleosts to four. There are three exceptions: *Gonorynchus* and Cromeriini present a further reduction, with three or fewer autogenous infraorbitals, whereas *Kneria* and *Parakneria* have five. The lacrimal (infraorbital one), as well as the antorbital, is lost altogether in *Grasseichthys*. Among gonorynchiforms, the size and shape of the infraorbitals also varies. Chanids exhibit flat, large, well-developed infraorbitals 2–4, morphology inferred as primitive for gonorynchiforms. *Phractolaemus* (Fig. 1.4C) exhibits the most interesting, unexpected condition of all (Fig. 1.4C). Infraorbitals 2–4 are exceptionally deep laterally and extend to the ventral side of the head. They cover the suspensorium, pterygoid bones, and interopercle, obstructing them from view (Grande and Poyato-Ariza 1999: fig. 11A). In most gonorynchiform taxa the lacrimal, is a small bone (e.g., *Phractolaemus*). Among gonorynchids, however, the lacrimal is enlarged longitudinally, and in *Gonorynchus* (Fig. 1.4B), it exhibits an interesting keel near the lower edge. According to Monod (1963), the lacrimal in *Gonorynchus* consists of a fusion of the antorbital and infraorbitals 1–3. We agree with Monod (1963) that the lacrimal in *Gonorynchus*, and most likely †*Notogoneus* (Grande and Grande 2008), is a composite structure. The last infraorbital or dermosphenotic is described above as a structural part of the cranium.

Opercular Series

Gonorynchiforms exhibit a full complement of bones in the opercular region, plus a rather uncommon suprapreopercular bone (Fig. 1.4A–E). The suprapreopercle is a broad but delicate canal-bearing bone situated above

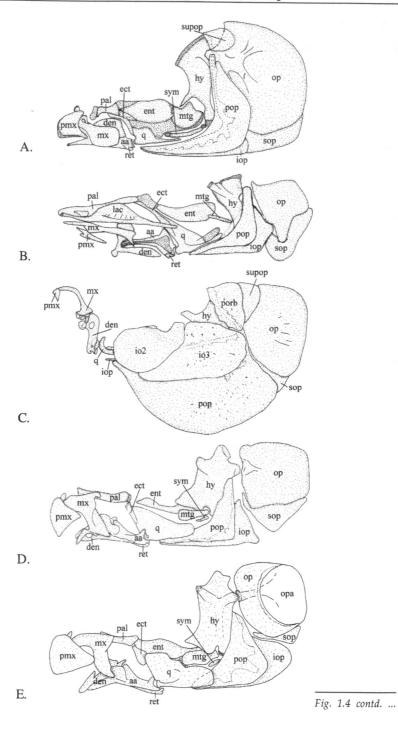

Fig. 1.4 contd. ...

the preopercle and between the hyomandibula and the opercle. In several gonorynchiform taxa (*Chanos*, †*Parachanos*, †*Tharrhias*, †*Dastilbe*, *Phractolaemus*), one large suprapreopercle is present directly anterior to the dorsal part of the opercular bone. In *Gonorynchus*, *Parakneria* and *Kneria* (Fig. 1.4B, D, E), the suprapreopercles are greatly reduced to mearly canals. One suprapreopercular bone is present in *Parakneria*, two are found in *Kneria*. Langet (1974) argued that the condition found in *Parakneria* and *Kneria* is the derived condition for ostariophysans. This is most likely correct since the suprapreopercles found in many otophysans (e.g., *Xenocharax*) are also reduced to canals. Unlike Gayet (1986, 1993a), we did not observe suprapreopercles in our †*Ramallichthys* and †*Judeichthys* material. We were also unable to confirm the presence of suprapreopercles in †*Charitosomus*. If the suprapreopercles in the Middle Eastern fossils are similar in structure to that of *Gonorynchus*, it is highly possible that they were lost in preservation or preparation. Suprapreopercles were not observed in the paedomorphs *Grasseichthys* and *Cromeria*.

The gonorynchiform preopercle is usually as high as it is long, that is, both limbs are equal or subequal in length; the relative elongation of the anteroventral limb is correlated with the general elongation of the suspensorium. The preopercular sensory canal extends dorsally and continues within the suprapreopercle. According to Perkins (1970), unlike other gonorynchiforms, the ventral portion of the preopercular canal is absent in †*Notogoneus*. Gayet (1993b) found that, among gonorynchids, the angle made by the intersection of lines drawn through the preopercular limbs is an obtuse angle (i.e., larger than 90 degrees). Grande and Poyato-Ariza (1999) found that such an obtuse angle is also found in kneriids and was used to diagnose the Gonorynchoidei. In turn, a roughly 90 degree opercular limb angle is found in chanids, with the exception of the †Rubiesichthyinae, where it exhibits an acute angle (Poyato-Ariza 1996a, b). The anteroventral limb of the preopercular bone of fossil Chaninae (e.g., †*Parachanos*, †*Tharrhias*) exhibits a conspicuous flat ridge, distinct from the faint ridge that occasionally houses the preopercular sensory canal. This large, flat preopercular ridge runs along most of the anteroventral limb, sometimes ascending part of the posterodorsal limb of the bone.

.... *Fig. 1.4 contd.*

Fig. 1.4 Skull in lateral view. All drawings with anterior facing the left. (A) *Chanos chanos*, CAS 05022-05075, SL = 103 mm. (B) *Gonorynchus greyi* AMNH 32973, SL = 147 mm. (C) *Phractolaemus ansorgei*, FMNH 63938A, SL = 125 mm. (D) *Parakneria tanzania*, BMNH 10.21-63, SL = 43.7 mm. (E) *Kneria wittei*, MARC 79-01-P-516, SL = 54.3 mm. aa, anguloarticular; den, dentary; ect, ectopterygoid; ent, endopterygoid; hy, hyomandibula; io, infraorbital; iop, interopercle; la, lacrimal (= infraorbital 1); mx, maxilla; mtg, metapterygoid; op, opercle; opa, opercular apparatus; pal, autopalatine; pmx, premaxilla; pop, preopercle; q, quadrate; ret, retroarticular; sop, subopercle; supop, suprapreopercle; sym, symplectic.

The preopercular condition observed in *Phractolaemus* (Figs. 1.4C, 1.5C) deserves special consideration. In this fish, the ventral arms of the right and left preopercles are greatly enlarged to the point that one peropercle overlaps the other along the ventral midline (Thys van den Audenaerde 1961: fig. 14). From our observations, the overlapping side appears to form randomly in development. Sometimes the right preopercle overlaps the left, and in some fish the left overlaps the right. The underlying preopercle, although expanded in size, is never as large as the preopercle that overlaps it. This condition has never been observed in other ostariophysans, although this peculiarity was mistakenly compared with the gular plates of *Polypterus* (Ridewood 1905). The opercular series, along with the expanded infraorbitals, give the *Phractolaemus* skull a remarkable appearance of solidity.

In all chanids (Fig. 1.4A), the opercle is quite broad, at least one-third the head size, and is the most prominent bone on the lateral part of the skull. It is greatly expanded, with a straight anterior border and a more or less curved posterior border. The roughly straight ventral border is usually inclined, although in †*Parachanos* and †*Tharrhias* it is nearly horizontal. Its anterior border is covered by the preopercle, and it possesses a prominent articulating socket on the inner surface for the reception of the corresponding head of the hyomandibula. As discussed by Gayet (1993b), gonorynchids exhibit a triangular opercular shape. In all other gonorynchiforms the opercular is rounded or oval. Grande and Poyato-Ariza (1999) considered the triangular opercular diagnostic of Gonorynchidae. Although *Gonorynchus* (Fig. 1.4B) and †*Notogoneus* exhibit the general triangular opercular shape of gonorynchids, their opercles are semi-divided into two lobes (Fig. 1.4B). The more anterior lobe in *Gonorynchus* is larger and more ventrally inclined. In †*Notogoneus*, each lobe is about the same size (Grande and Grande 2008). †*Charitopsis spinosus* (Gayet 1993b) is diagnosed by the presence of seven opercular spines positioned on its posterior margin.

Kneria (Figs. 1.4E, 1.6) exhibits a unique opercular morphology that is worthy of special discussion. As discussed in Grande and Young (1997), the males of the genus sport an interesting structure on the external lateral surface (Fig. 1.6). This opercular apparatus (term coined by Peters 1967) comprises two structures: an opercular cup positioned directly on the operculum and an opercular flange positioned directly behind the operculum. Both structures are composed largely of connective tissue with superficial keratinized papillae and ridges. It has been hypothesized that this structure functions as an adhesive device helping the male fish anchor to a rock in fast-moving water or to a female during mating. The opercular apparatus is tied closely to modifications of the skeletal system. Those associated with the flange are discussed in the chapter on the postcranial skeleton, but modifications of the opercle are discussed here. As shown in

Figs. 1.5E and 1.6, the opercle is modified to accommodate or support the massive opercular cup. About two-thirds of the opercle is actually reshaped during development to conform to the shape of the external structure. Unlike the keratinized breeding tubercles of ostariophysans that disappear once breeding season is complete, the opercular apparatus plus its associated skeletal elements, once formed, are permanent structures.

Immediately ventral to the opercle is the subopercle, which in *Chanos* (Fig. 1.5A) and other chanids curves slightly upward posteriorly. The subopercle conforms to the ventral margin of the opercle and in lateral aspect almost appears innocuous. This is especially true in †*Rubiesichthys*, where it is practically hidden behind the remarkably hypertrophied opercle. †*Notogoneus* is distinct from all other gonorynchiforms with its diagnostic subopercle. In †*Notogoneus*, the subopercle is divided by deep clefts. Four clefts are found in the type species, †*Notogoneus osculus*, while three were observed in †*Notogoneus montenensis* (Grande and Grande 1999).

The interopercle, mostly covered by the preopercle and subopercle in lateral aspect, is generally a remarkably long and relatively broad bone that extends anteriorly to nearly the level of the retroarticular (Fig. 1.5A–E). The elongation of the interopercle is correlated with the general elongation of the suspensorium. Observed in medial aspect, one interesting variation in the gonorynchiform condition occurs. In *Phractolaemus* (Fig. 1.5C), the interopercle is reduced to a rod or spine that extends beyond the length of the skull. This condition is part of its highly specialized jaw and suspensorial anatomy resulting in a uniquely protrusible jaw.

In most gonorynchiforms, opercular bones are smooth, but in †*Parachanos* they are ornamented with thin but conspicuous radial ridges. Radial ridges are present in †*Parachanos* in addition to the preopercular ridge mentioned above.

Jaws and Suspensorium

Two regions of the skull that show the greatest amount of variation in the greatly variable gonorynchiform skull are the jaws and the suspensorium (Figs. 1.4, 1.5). Elements of these regions, all of which are paired, are those formed from the mandibular and hyoid arches and associated dermal bones such as the premaxilla and the maxilla. Differences in jaw and suspensorium morphology among gonorynchiforms are often correlated to adaptations in jaw protrusibility and feeding mechanisms. A few generalizations, however, can be made about gonorynchiform jaws and suspensoria.

First of all, teeth are missing from the jaws of all gonorynchiforms. Although the suspensorium in chanids is anteriorly more massive than in gonorynchoids, in all gonorynchiforms, the suspensorium is elongated in a parasaggital plane in the region between the anguloarticular condyle of the quadrate and the hyomandibula (Fink and Fink 1981, Poyato-Ariza 1996a).

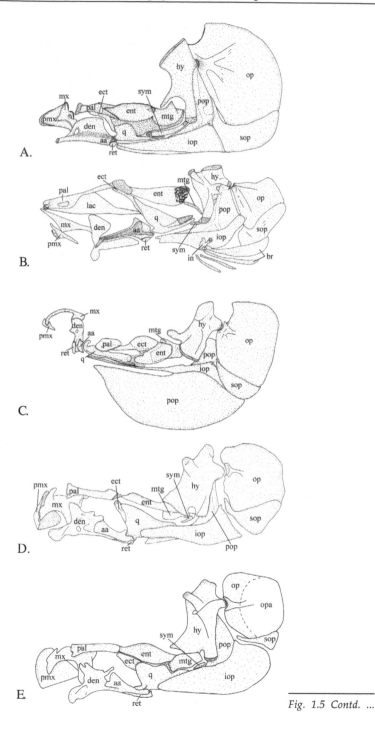

Fig. 1.5 Contd. ...

This results in a loose and flexible suspensorium. The lower limb of the preopercle and the dorsoventral part of the interopercle are equally elongate, as mentioned in the previous section. The elongation of the suspensorium correlates with the elongation of the frontal bones in these fishes, and vice versa, as we saw above.

Upper Jaw and Suspensorium

In chanids (Figs. 1.4A, 1.5A), the premaxilla is a thin, flat, delicate and concave-convex bone and bears a conspicuous posteroventral process that forms part of the oral border. As Gayet (1993a) pointed out, with the exception of †*Charitopsis*, the premaxilla in gonorynchids is thick, with several processes increasing its thickness. We agree with Gayet's (1993a) assessment and consider the thick premaxilla of gonorynchids to be a distinctive character of that family. In Gonorynchiformes, the maxilla is practically excluded from the gape by the premaxilla (in Chanidae largely due to the premaxillary posteroventral process), and the jaws are devoid of dentition. In chanids the posterior region of the maxilla is expanded and, correlatively, the posterior margin of the maxilla is swollen to a bulbous outline. In all other gonorynchiforms, the posterior region of the maxilla is narrower and, correlatively, the posterior margin of the maxilla is narrow and straight. As in all other ostariophysans, supramaxillary bones are lost as separate ossifications (Fink and Fink 1981). The maxilla articulates with the autopalatine; the dermal portion of the palatine is absent in ostariophysans. In *Chanos* (Figs. 1.4A, 1.5A), this articulation is through cartilage, while in other gonorynchiforms examined (e.g., *Kneria* and *Parakneria*, Figs. 1.4D, E, 1.5D, E), the two bones seem to articulate directly with each other. A short but robust process that bears the articulating surface for the autopalatine is present in some gonorynchiforms only (e.g., *Chanos*, †*Charitosomus*, *Kneria*, †*Tharrhias*).

In *Chanos* (Figs. 1.4A, 1.5A), the autopalatine is a thick bone of the mandibular arch bounded by cartilage on its anterior end, where it articulates with the maxilla, and on its posterior end, separating it from the endopterygoid. In *Gonorynchus* (Figs. 1.4B, 1.5B), the autopalatine articulates with the endopterygoid through cartilage as well. In *Gonorynchus*, however,

---

.... *Fig. 1.5 contd.*

Fig. 1.5 Skull in medial view. All drawings with anterior facing the left. (A) *Chanos chanos*, CAS 05022-05075, SL = 103 mm. (B) *Gonorynchus greyi* AMNH 32973, SL = 147 mm. (C) *Phractolaemus ansorgei*, FMNH 63938A, SL = 125 mm. (D) *Parakneria tanzania*, BMNH 10.21-163, SL = 43.7 mm. (E) *Kneria wittei*, MARC 79-01-P-516, SL = 54.3 mm. aa, anguloarticular; br, branchiostegal rays; den, dentary; ect, ectopterygoid; ent, endopterygoid; hy, hyomandibula; int, interhyal; iop, interopercle; mx, maxilla; mtg, metapterygoid; op, opercle;opa, opercular apparatus; pal, autopalatine; pmx, premaxilla; pop, preopercle; q, quadrate; ret, retroarticular; sop, subopercle; sym, symplectic.

the anterior end is enlarged dorsoventrally with a small cartilaginous articulation surface for the maxilla. In *Phractolaemus* (Fig. 1.5C), the autopalatine is oval instead of rectangular and articulates with the ectopterygoid posteriorly and instead of the maxilla, the quadrate anteriorly. As mentioned above, the dermal component of the palatine is lost in gonorynchiforms.

The ectopterygoid is another paired dermal bone that in *Chanos,* †*Tharrhias,* †*Dastilbe,* and other chanids where this bone is observable, is large and overlaps with the autopalatine by over 50 percent, resulting in a comparatively massive suspensorium anteriorly. In the Gonorynchidae and Kneriidae (i.e., Gonorynchoidei), the ectopterygoids are reduced and overlap with the autopalatines by at most 10 percent, resulting in a diagnostic loosely articulated, mobile palatine. In the Kneriinae, the ectopterygoids are reduced even more (Figs. 1.4D, E, 1.5D, E). In *Cromeria* and *Grasseichthys* the ectopterygoids are lost, while in *Kneria* the ectopterygoids are larger and more rounded than those found in *Parakneria*. In *Kneria* and *Phractolaemus,* the ectopterygoids directly articulate with the mesethmoids. In all other gonorynchiforms, there is no contact between these two bones except through cartilage, as in *Chanos* (Figs. 1.4A, 1.5A). Where it is accurately observable, this region is a lacuna of ossification in all fossil chanids, so that a cartilaginous connection can be assumed for them as well.

The endopterygoids are also dermal bones in origin and largely form the floor of the orbit. They vary in shape and size among gonorynchiforms and articulate with the metapterygoids posteriorly and usually with the autopalatines anteriorly, the exception being *Phractolaemus* (Figs. 1.4C, 1.5C), as discussed above. The most interesting endopterygoid modification occurs within the Gonorynchidae (Fig. 1.5B). This family is diagnosed by the presence of conical teeth on the ventral surface of the endopterygoids. There

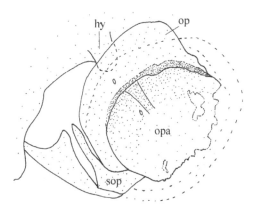

Fig. 1.6 Opercular apparatus in *Kneria wittei*, CAS 16150, SL = 50.5 mm. hy, hyomandibula; opa, opercular apparatus; sop, subopercle.

are about 20 such teeth on each endopterygoid of *Gonorynchus*. The exact number of teeth on the endopterygoids of the fossil forms is uncertain, but it is estimated to approximate that of *Gonorynchus*. For many years, it was thought that teeth on the endopterygoids were lost in all species of †*Notogoneus* (e.g., Woodward 1896, Chabanaud 1931, Perkins 1970, Grande and Poyato-Ariza 1999, Grande and Grande 1999). In a more recent study, however (Grande and Grande 2008), gonorynchid teeth were observed in several well-preserved and better-prepared specimens of †*Notogoneus osculus* from the Green River Formation of southwestern Wyoming. Patches of over 20 conical teeth were found in place in fully articulated specimens (FMNH PF 11955).

The metapterygoid connects the endopterygoid with the hyomandibula. It is a very large and relatively robust ossification in chanids. In contrast, it is reduced to a slender rod in *Gonorynchus*. In spite of Perkins (1970), the metapterygoid morphology is still imperfectly known in †*Notogoneus*. In the other fossil gonorynchids it is a more massive bone, unlike that found in *Gonorynchus*. In *Cromeria* we were not able to discern the division between the endopterygoid and the metapterygoid in the specimens examined. As shown in Britz and Moritz (2007: figs. 4, 10), both bones are distinct.

The hyomandibula is the largest bone in the hyoid arch and in the entire suspensorium. It functions in attaching the jaws of the fish on to the cranium. It is a chondral bone, and one of the first to begin forming during development. Posteriorly the hyomandibula articulates with the opercle, so that this bone can also be considered part of the suspensorium *sensu lato* (but not as part of the splachnocranium, which is always endoskeletal in origin and is formed by the mandibular, hyoid, and branchial arches only). The hyomandibula articulates with the symplectic and the metapterygoid, while anteroventrally it forms a conspicuous, elongate spinous process in the chanids, *Phractolaemus* and the Kneriinae (Fig. 1.5A, C–E). Such a process was not observed in the Gonorynchidae (Fig. 1.5B). As discussed in Arratia (1992: fig. 2A) and Grande and Poyato-Ariza (1999), contrary to the reports of d'Aubenton (1961), Langlet (1974), and Fink and Fink (1981), the articulation between the hyomandibula and the cranium in gonorynchiform fishes, including the Chanidae, is always by means of two distinct articular surfaces. This is evident when examining very large specimens of *Chanos* and well-preserved, three-dimensional specimens of †*Dastilbe* and †*Tharrhias*. In the chanids and *Phractolaemus*, the two articular facets are both positioned on the dorsal surface of the hyomandibula and are not projected, although they are distinct, separate surfaces (Grande and Poyato-Ariza 1999: fig. 10). In the gonorynchids and the Kneriinae, the two articular surfaces do not lie in the same plane, but form projected articular heads that extend from the main body of the bone (Figs. 1.4D, E, 1.5D, E). In either case, the more anterior articular surface articulates with the sphenotic and the posterior articular surface with the pterotic.

The quadrate is a large bone within the suspensorium that articulates with the lower jaw and forms the juncture between the ectopterygoid and anguloarticular, and connects with the hyomandibula via the symplectic (Figs. 1.4, 1.5). In *Chanos* (Figs. 1.4A, 1.5A) and fossil chanids, the quadrate has a distinct posteroventral process that is very long, extending below the symplectic. In chanids the quadrate-mandibular articulation is anterior to the orbit, resulting in a robust appearance of the snout and head. The anatomical modifications associated with this displacement in chanoids include the displacement of the body of the quadrate, the rotation of the longitudinal axis of the quadrate and symplectic to almost horizontal, and the elongation of the posteroventral process of the quadrate, interopercle and anteroventral limb of the preopercle (Poyato-Ariza 1996a, b, Grande and Poyato-Ariza 1999). In *Phractolaemus* (Figs. 1.4C, 1.5C) the quadrate-mandibular articulation is also positioned anteriorly. The quadrate is elongated, with a distinctive upturned hook that articulates with the coronoid beyond the anterior skull margin. The quadrate articulates with the autopalatine dorsolaterally, and with the mesethmoid through cartilage posteriorly. In *Cromeria occidentalis* the posterior margin of quadrate is deeply forked. This condition was not observed in other gonorynchiforms.

The symplectic, if present, commonly articulates directly with quadrate. In *Chanos*, †*Tharrhias* and other fossil chanids, however, there is no direct contact between the symplectic and the quadrate. They contact each other only through cartilage in *Chanos* (Figs. 1.4A, 1.5A). We acknowledge that the condition in *Chanos* is based on smaller specimens, and in larger specimens the condition may be different. The symplectic is absent in *Cromeria, Grasseichthys* and *Phractolaemus*.

The ventral elements of the hyoid arch and its related dermal bones are more intimately associated to the branchial arches series and are therefore discussed in the corresponding section below.

Lower Jaw

The lower jaw of the Gonorynchiformes consists of the dentary, angular, articular, and retroarticular bones; the first two are dermal bones, whereas the other two are endochondral bones of the mandibular arch. The first three of these bones also form what is called the coronoid process, a conspicuous projection for the insertion of relevant mandibular muscles. The morphology and arrangement of the dentary component of the coronoid process in chanids is unique among gonorynchiforms. In these fishes the dentary is anteriorly very thin and posteriorly very deep, forming the anterior part of the high coronoid process. The coronoid process of

Chanos (Fig. 1.5A) is exceptionally high and large. On the ascending part of the coronoid process, a distinctive notch is present in the anterodorsal border of the dentary in the Chanidae. Although this region is not accessible in known specimens of †*Aethalionopsis* and †*Parachanos*, where it is hidden by the maxilla, this notch is present in all other chanids species and used by Poyato-Ariza (1996a) and Grande and Poyato-Ariza (1999) as a diagnostic character of the Chanidae. In all gonorynchiforms, the angular and articular bones of the coronoid process are fused into a single bone forming the anguloarticular (Nelson 1973). Like the chanids, the height of the coronoid process in gonorynchids, *Phractolaemus* and the Kneriini is largely due to the height of the dentary. This component of the coronoid process is at least 50 percent of the mandibular length. In *Grasseichthys* and *Cromeria*, the dentary component is not as high, and the anguloarticular component has more prominence in the coronoid process. Finally, the retroarticular bone is autogenous, very small, placed in the posteroventral corner of the mandible, and excluded (Chanidae) or nearly excluded (Gonorynchoidei) from the articular facet of the lower jaw.

The articulation between the dentary and the anguloarticular in †*Notogoneus* and *Gonorynchus* (Figs. 1.4B, 1.5B) is unique among gononorynchiforms, and probably among ostariophysans. In these two genera, the dentary has a conspicuous V-shape, with one process extending high and dorsally while the other extends posteriorly. The long anterior process of the anguloarticular articulates with the dentary at the junction between the two dentary processes (Figs. 1.4B, 1.5B).

The anguloarticular + retroarticular form an articular facet posteriorly that articulates with the head of the quadrate. As described by Gayet (1993a, b), the lower margin of the articulating surface of the anguloarticular + retroarticular is uniquely elongated in †*Hakeliosomus*, †*Charitosomus* and †*Charitopsis* (= †Charitosomidae *sensu* Gayet 1993a). This retroarticular process as coined by Gayet (1993a) cradles the quadrate ventrally, resulting in a unique type of jaw protrusion. Grande and Grande (2008) report a lower retroarticular process in †*Ramallichthys* and †*Judeichthys* as well, suggesting a close relationship with †*Charitosomus* and its relatives. In addition to the lower retroarticular process, Gayet (1993a) described an upper retroarticular process restricted to the taxa †*Charitosomus* and †*Charitopsis*. The upper and lower processes in these taxa cup the articulating head of the quadrate, thus reducing the potential protrusibility of the lower jaw. An upper retroarticular process was not observed in any other gonorynchiform taxa.

The ensemble of the upper and lower jaws in *Phractolaemus* warrants special attention (Figs. 1.4C, 1.5C). Like other gonorynchiforms, the premaxilla and maxilla are thin and delicate but, together with the coronoid process, are positioned dorsal to the quadrate and autopalatine, not anterior as in all other gonorynchiforms. In addition, the upper and lower jaws plus the quadrate extend beyond the anterior margin of the skull. The irregular position of these jaw bones accommodates the musculature of a small superior protractile mouth. This highly projectile mouth is capable of being thrust forward but, when at rest, folded over and received into a depression on the upper surface of the head. Interestingly, only a few studies have investigated these uniquely protrusible jaws (e.g., Thys van den Audenaerde 1961, Gréy 1962). A study of the jaw mechanics of *Phractolaemus* using modern digital modeling techniques should prove quite interesting and informative. This particular capability for jaws protrusion in *Phractolaemus* has obviously been acquired independently from other teleostean lineages.

Branchial Arches and Associated Structures

Epibranchial Organ Support

The comparative dorsal gill arch morphology in gonorynchiforms is presented in detail by Johnson and Patterson (1997), so we will not repeat it here. Our discussion will concentrate on the gill arch structures that support the epibranchial organ, which are discussed in more detail later in this volume. In *Chanos* the epibranchial organ is supported by the expanded epibranchials 4 and 5. According to Johnson and Patterson (1997: figs. 2B, 4B, 5B), *Gonorynchus* exhibits a fusion of epibranchials 4 and 5 that hypertrophy and form the efferent arterial foramen, or passage for the efferent artery. The origin of an "accessory cartilage" that also supports the epibranchial organ in *Gonorynchus* remains undetermined. In *Phractolaemus* the massive epibranchial organ is supported only by epibranchial 5. In the kneriids, both the epibranchial organ and its support are reduced in structure.

The epibranchial organ itself consists of a pair of muscular pockets. Each pocket is a diverticulum of the roof of the pharynx and has two parts: (1) a canal passage consisting of two rows of gill rakers originating from the lower limb of the fourth and fifth gill arches, which continues into a blind sac, and (2) a blind sac devoid of gill rakers (Kapoor 1954). Bertmar *et al.* (1969) suggest that these structures developed as sites for the accumulation and storage of plankton or other small organisms. Epibranchial organs have been found in most living gonorynchiform fishes. The most advanced, or structurally complex, is that found in *Chanos* and *Phractolaemus*. The presence of an epibranchial organ had been reported in many other lower teleosts such as clupeomorphs, osteoglossids, and salmonids but, according to Nelson (1967), the gill arch structures of these fishes suggest that these organs were independently derived in each of the major teleostean groups in relation to a repeated evolutionary tendency, namely the adaptation of microphagous habits.

Branchial Basket and Branchiostegal Rays

The ventral part of the gill basket most anteriorly consists of the basihyal, an unpaired element that forms the major portion of the "tongue skeleton". In young specimens, the basihyal is predominantly cartilaginous. Extending from the posterior part of the basihyal is the hypohyal, followed by the broadened anterior and posterior ceratohyals, all of which are paired elements (Fig. 1.7A–D).

Primitively in chanids up to six branchiostegal rays are attached to the ceratohyals (e.g., six in †*Dastilbe*, five in the †Rubiesichthyinae, four in *Chanos*). Four rays were observed in †*Notogoneus montanensis* (Grande and Grande 1999). This number is reduced to three in the Kneriinae, and one extending from the posterior ceratohyal in *Phractolaemus*.

The basihyal is posteriorly followed by five basibranchials, all unpaired. *Chanos* (Fig. 1.7A) exhibits an ossified first basibranchial, which can be interpreted as the primitive condition among gonorynchiforms. Basibranchial one is cartilaginous in the Gonorynchoidei (Fink and Fink 1981, Johnson and Patterson 1997). In *Chanos*, basibranchial 3 is considerably larger than basibranchials 1 and 2. Articulating with the posterior end of basibranchial 3 is a narrow bar of cartilage that runs between ceratobranchials 3, 4, and 5; this cartilage represents basibranchials 4 and 5. Basibranchial 5 is ossified in *Parakneria*, *Kneria* (Fig. 1.7D) and *Cromeria*; basibranchials 4 and 5 are ossified in *Parakneria* and *Kneria* (Grande and Poyato-Ariza 1999). According to Johnson and Patterson (1997), basibranchials 4 and 5 are ossified in the Kneriinae. We were not able to support that observation based on our material.

Extending laterally from basibranchials 1 to 3 (unpaired) are hypobranchials 1 to 3 (paired), correspondingly. In *Chanos*, hypobranchial 3 is fused with the sides of basibranchial 3, although the boundary line between them is not obliterated. Extending from hypobranchials 1 to 3 (paired) are ceratobranchials 1 to 3 (paired), and from basibranchials 4 and 5 (unpaired) are ceratobranchials 4 and 5 (paired).

One important adaptation among gonorynchiforms is the expansion and widening of ceratobranchials 4 and 5. Together with part of the dorsal gill arches, they support the epibranchial organ characteristic of the order, as discussed above.

Branchial Teeth

Positioned on basibranchial 2 in *Gonorynchus* (Fig. 1.7B) is a patch of conical teeth. The structure of these teeth is identical to the ones found on the endopterygoids. Teeth have been reported on the branchial arches of the Middle Eastern gonorynchids (Gayet 1985, 1986, 1993a, b) and observed by us. We presume that the tooth patches in Middle Eastern gonorynchids are also positioned on basibranchial 2. Although it was once thought that branchial teeth were lost in the genus †*Notogoneus*, tooth patches on the ventral gill arches of †*Notogoneus osculus* were recently observed by Grande and Grande (2008). Unlike in other otophysans and most primitive teleosts, teeth are absent from ceratobranchial 5 in extant gonorynchiforms. Although the detailed gill arch structure is unknown in fossil forms, teeth are absent on the observable branchial arches remains of all fossil chanids, for instance, from ceratobranchial 5 of †*Tharrhias* (e.g., specimen FMNH PF 10736) or from the numerous observed branchial arch remains of the †Rubiesichthyinae.

Final Remarks

The Gonorynchiformes are characterized by profound modifications of the skull that account for a distinct "gonorynchiform look": a small, edentulous mouth, usually diminished or displaced; a reduced orbital septum; and a more or less large opercular region. Anatomic modifications of the skull can be grouped in several informal categories:

1. Loss of orbitosphenoid, basisphenoid, dermopalatines, supra-maxillae (lost also in otophysans), last two pharyngobranchial tooth plates, teeth on the jaws, and on the fifth ceratobranchial.
2. Reduction in size of pterosphenoids, parietals (which are mostly separated).
3. General elongation of the suspensorium and anterior displacement of the buccal opening, which is usually reduced.
4. Presence of a posterior cartilaginous margin on the exoccipitals and supraoccipital; cephalic ribs; a particular epibranchial organ supported by modified branchial arches.

In addition to the general features shared by all gonorynchiforms, there are also profound modifications of the skull in the different evolutionary lineages within the order, each of which has had a very long, separate history. The Chanidae are characterized by several derived anatomic characters that account for the "chanid look": quadrate and quadrato-mandibular articulation displaced, usually before the level of the orbit, resulting in a very small, anteriorly directed mouth cleft; an expansion of the premaxilla, which bears a long oral process; an expansion of the posterior region of the maxilla; a straight to acute angle between the preopercular limbs; and an expansion or even hypertrophy of the opercular bone. Some of these features are relatively easy to recognize at first sight, even in the field. They appear consistently in all the genera of the family since their first known fossil record, in the Early Cretaceous, showing that the Chanidae are a homogeneous, very conservative evolutionary lineage. *Chanos*, the only

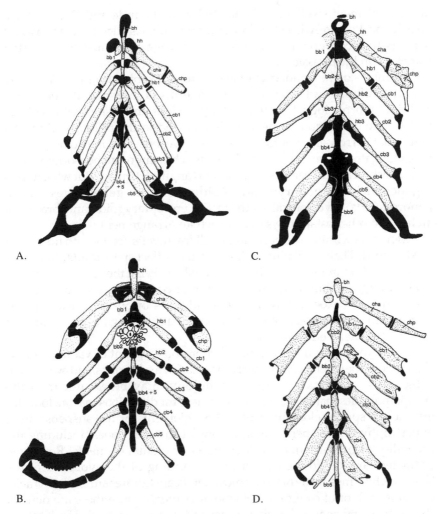

Fig. 1.7 Ventral gill arches. All drawings with anterior facing up. (A) *Chanos chanos*, CAS 05022-05075, SL = 103 mm. (B) *Gonorynchus greyi* AMNH 32973, SL = 147 mm. (C) *Phractolaemus ansorgei*, FMNH 63938A, SL = 125 mm. (D) *Kneria wittei*, MARC 79-01-P-516, SL = 54.3 mm. bb, basibranchial; bh, basihyal; br, branchiostegal rays; cb, certatobranchial; cha, anterior ceratohyal; chp, posterior ceratohyal; hb, hypobranchial; hh, hypohyal.

Recent chanid, shows all of the consistent chanid cranial characters together with a number of differentiations of its own, which are either autopomorphic characters (occasionally shared with †*Tharrhias*) or convergent with other teleostean lineages (mostly those of the caudal endoskeleton). Despite these differentiations, *Chanos chanos* can be considered a "living fossil", versus the hypothesis by Patterson (1984) that it is not; for the phylogenetic support of this statement see Poyato-Ariza (1996a, b), Grande and Poyato-Ariza (1999), and Poyato-Ariza *et al.* (present volume). As such a living fossil, it belongs to a conservative evolutionary lineage that must have started sometime in the Jurassic.

Members of the Gonorynchidae are unmistakably recognizable among gonorynchiforms and ostariophysans in terms of their overall body form and skull morphology. Essentially, they are long thin fishes with skulls to match. The elongation of the skull is reflected in their elongated frontal bones that narrow anterior to their very large orbits. This is also reflected in their elongated suspensorium and loose attachments of suspensorium bones (e.g., attachment between the quadrate and anguloarticular). But what makes a gonorynchid a gonorynchid? Aside from the general gestalt, all gonorynchids have conical teeth on the endopterygoids and branchial arches. No other ostariophysan has such an arrangement.

Members within the Kneriidae are all freshwater fishes with no known fossil record. Their skull morphologies often reflect specific adaptations to the localized environment in which they evolved. What they have in common is a very large foramen magnum that is bounded by the supraoccipital from above and the exoccipitals from the sides, as opposed to exoccipitals surrounding the foramen magnum from all angles (e.g., *Chanos*), and the presence of lateral ethmoid extensions.

Gonorynchiform fishes exhibit an enormous amount of variation in the skull region. The differences among gonorynchiform subgroups, and even within subgroups, in terms of skull morphology, may be greater than the synapomorphies that tie them together. These modifications may include, for instance, a special elongation of certain elements of the suspensorium, hypertrophy of the opercle, adaptations of the oral bones, including the capability of protrusion, or further losses of cranial elements. Reductive characters are key to the correct understanding of the evolution of the Gonorynchiformes, many of whose most significant features are losses of cranial elements and/or of ossification of cranial elements. Since these are derived characters that diagnose the whole order and that are especially significant in the lineage of the Kneriidae, it seems clear that reductive cranial characters, together with the profound modifications of their oral and opercular regions, are indispensable to accurately comprehend the anatomy and the evolutionary history of the Gonorynchiformes. An evolutionary history that is surprisingly long and unexpectedly varied for so small an order.

Acknowledgements

The authors wish to thank Joseph Schlupe and Amanda Burdi (LU) for their assistance in producing the figures for this paper. For permission and assistance in examining specimens we would like to thank Gloria Arratia

(KU), Pierre Bultynk (IRSNB), Sebastian Calzada (MS), Armando Diaz-Romeral (UAM), Darrel Siebert and Petr Forey (NHM), Lance Grande, Mark Westneat and Mary Anne Rogers (FMNH), Jamie Gallemi and Julio Gomez-Alba (MGB), Jesus Madero (Museo de Cuenca), Xavier Martinez-Delclos (IEI), Melanie Stiassny and Barbara Brown (AMNH), Jose Santafe (IPMC), Guy Teugels (MARC), Richard Vari (USNM), and Daniel Goujet (MNHN). Special thanks to Gloria Arratia (KU), Lance Grande (FMNH) and Eric Hilton (VIMS) for their critical and thoughtful comments during the review of this manuscript. This research was supported in part by a National Science Foundation grant to TG (DEB 0128794 and EF 0732589).

References

Arratia, G. 1992. Development and variation of the suspensorium of primitive catfishes (Teleostei: Ostariophysi) and their phylogenetic relationships. Bonner zoologische Monographien 32: 1–149.

Arratia, G. 2003. Catfish Head Skeleton—An Overview, pp. 3– 46. In: G. Arratia, B.G. Kapoor, M. Chardon, and R. Diogo [eds.]. Catfishes. Scientific Publishers, Inc., Enfield, USA.

Arratia, G. and M. Gayet. 1995. Sensory canals and related bones of Tertiary siluriform crania from Bolivia and North America and comparison with Recent forms. J. Vert. Paleontol 15(3): 482–505.

Bemis, W.E. and P.L. Forey. 2001. Occipital structure and the posterior limit of the skull in actinopterygians, pp. 350–369. In: P.E. Ahlberg [ed.]. Major Events in Early Vertebrate Evolution. Paleontology, Phylogeny, Genetics and Development. Taylor and Francis, London.

Bertmar, G., B.G. Kapoor and R.V. Miller. 1969. Epibranchial organs in lower teleostean fishes—an example of structural adaptation. Trop. Atlan. Biol. Lab. Bur. Comm. Fish. Miami 76: 1–49.

Blum, S. 1991. Tharrhias Jordan and Branner, 1908, pp. 286–296. In: J.G. Maisey [ed.]. Santana Fossils, An Illustrated Atlas. TFH Publications, Neptune City, New Jersey.

Brito, P. and S. Wenz. 1990. O endocrânio de Tharrhias (Telcostei, Gonorynchiformes) do Cretáceo inferior da Chapada do Araripe, nordeste do Brasil. Atlas do I Sympósio sobre a Bacia do Araripe e Bacias Interiores do Nordeste Crato: 375–382. DNPM ed., Fortaleza, Brazil.

Britz, R. and T. Moritz. 2007. Reinvestigation of the osteology of the miniature African freshwater fishes Cromeria and Grasseichthys (Teleostei, Gonoirynchiformes, Kneriidae), with comments on kenriid relationships. Mitteilungen aus dem Museum für Naturkunde in Berlin, Zoologie Reihe 83(1): 3–42.

Chabanaud, P. 1931. Notes ichthyologiques II—sur la presence de cotes craniennes chez un Gonorynchus Gronov. Bull. Soc. Zool. Fr. LVI: 115–118.

d'Aubenton, F. 1961. Morphologie du crane de Cromeria nilotica occidentalis Daget 1954. Bull. Inst. Fr. Afrique Noire, sér. A 23: 187–249.

de Pinna, M. and T. Grande. 2003. Ontogeny of the accessory neural arch in pristogasteroid clupeomorphs and its bearing on the homology of the otophysan claustrum (Teleostei). Copeia 2003: 838–845.

Fink, S.V. and W.L. Fink. 1981. Interrelationships of ostariophysan fishes (Teleostei). Zool. J. Linn. Soc. London 72(4): 297–353.

Fink, S.V. and W.L. Fink. 1996. Interrelationships of ostariophysan fishes (Teleostei) pp. 209–249. In: M.L.J. Stiassny, L.R. Parenti and G.D. Johnson [eds.]. Interrelationships of Fishes. Academic Press Inc., San Diego.

Gayet, M. 1982. Cypriniforme ou Gonorhynchiforme? *Ramallichthys*, nouveau genre du Cénomanien inférieur de Ramallah (Monts de Judée). C.R. Acad. Sci. Paris 295(2): 405–407.

Gayet, M. 1985. Gonorhynchiform nouveau du Cénomanien inférieur marin de Ramallah (Monts de Judée): *Judeichthys haasi* nov. gen. nov. sp. (Teleostei, Ostariophysi, Judeichthyidae nov. fam.). Bull. Mus. natl. Hist. nat. Paris, sér. 47C(1): 65–85.

Gayet, M. 1986. *Ramallichthys* Gayet du Cénomanien inférieur marin de Ramallah (Judée) une introduction aux relations phylogénétiques des Ostariophysi. Mém. Muséum natl. Hist. nat. Paris, n.s. C, Sciences de la Terre 51: 1–81.

Gayet, M. 1993a. Gonorhynchoidei du Crétacé Supérieur marin du Liban et relations phylogénétiques des Charitosomidae nov. fam. Documents des laboratoires de Géologie 126: 131 pp., Lyon.

Gayet, M. 1993b. Nouveau genre de Gonorhynchidae du Cénomanien inférieur marin de Hakel (Liban). Implications phylogénétiques. C.R. Acad. Sci. Paris, sér. II 432: 57–163.

Grande, L. and W. Bemis. 1998. A comprehensive phylogenetic study of amid fishes (Amiidae) based on comparative skeletal anatomy. An empirical search for interconnected patterns of natural history. Society of Vertebrate Paleontology, Memoir 4, supplement to J. Vert. Paleontol. 18: 1–690.

Grande, L. and T. Grande. 1999. A new species of †*Notogoneus* (Teleostei: Gonorynchidae) from the Upper Cretaceous Two Medicine Formation of Montana, and the poor Cretaceous Record of freshwater fishes from North America. J. Vert. Paleontol. 19(4): 612–622.

Grande L. and T. Grande. 2008. Rediscription of the type species for the genus †*Notogoneus* (Teleostei: Gonorynchidae) based on new, well-preserved material. J. Paleontol. 82(5): 1–31.

Grande, T. 1994. Phylogeny and paedomorphosis in an African family of freshwater fishes (Gonorynchiformes: Kneriidae). Fieldiana 78: 1–20.

Grande, T. 1999. Revision of the genus *Gonorynchus* Scopoli, 1777 (Teoeostei: Ostariophysi). Copeia 1999(2): 453–469.

Grande, T. and M. de Pinna. 2004. The evolution of the Weberian apparatus: A phylogenetic perspective, pp. 429–448. *In*: G. Arratia and A. Tintori [eds.]. Mesozoic Fishes 3—Systematics, Paleoenvironments and Biodiversity. Verlag Dr. Friedrich Pfeil. Munich.

Grande, T. and L. Grande. 2008. Revaluation of the gonorynchiform genera †*Rammallichthys*, †*Judeichthys* and †*Notogoneus*, with comments on the families †Charitosomidae and Gonorynchidae. *In*: G. Arratia and H.-P. Schultze [eds.]. Mesozoic Fishes 4. Verlag Dr. Friedrich Pfeil. Munich.

Grande, T. and F.J. Poyato-Ariza. 1999. Phylogenetic relationships of fossil and Recent gonorynchiform fishes (Teleostei: Ostariophysi). Zool. J. Linn. Soc. 125: 197–238.

Grande, T. and B. Young. 1997. Morphological development of the opercular apparatus in *Kneria wittei* (Ostariophysi: Gonorynchiformes) with comments on its possible function. Acta Zool. 78(2): 145–162.

Gregory, W.K. 1933. Fish skulls, a study of the evolution of natural mechanisms. Trans. Am. Phil. Soc. 23(2): 75–481.

Gréy, J. 1964. Une nouvelle famille des poissons dulcaquicoles africaina; les Grasseichthyidae. C.R. Acad. Sci. Paris, 259(245): 4805–4807.

Johnson, G.D. and C. Patterson. 1997. The gill-arches of gonorynchiform fishes. S. African J. Sci. 93: 594–600.

Kapoor, B.G. 1954. The pharyngeal organ and its associated structures in the milk-fish *Chanos chanos* (Forskal). J. Zool. Soc. India 6(1): 51–58.

Lecointre, G. and G. Nelson. 1996. *In*: M.L.J. Stiassny, L.R. Parenti and G.D. Johnson [eds.]. Interrelationships of Fishes. Academic Press Inc., San Diego.

Le Danois, Y. 1966. Remarques anatomiques sur le region cephalique de *Gonorynchus gonorynchus* (Linne, 1766). Bull. Inst. Fr. d'Afrique Noire, ser. A 28(1): 283–342.

Lenglet, G. 1974. Contribution à l'étude ostéologique des Kneriidae. Ann. Soc. Roy. Zool. Belg. 104: 51–103.

Monod, T. 1963. Sur quelques points de l'anatomie de *Gonorhynchus gonorhynchus* (Linné 1766). Mém. Inst. Fr. Afr. Noire, mélanges ichthyologiques 66: 255–313.

Moritz, T., R. Britz and E. Linsenmaier. 2006. *Cromeria nilotica* and *C. occidentalis*, two valid species of the African freshwater family Kneriidae (Gonorynchiformes, Teleostei). Ichthyol. Explor. Freshwaters 17: 65–72.

Nelson, G. 1973. Relationships of clupeomorphs with remarks on the structure of the lower jaw in fishes, pp. 333–349. *In*: P.H. Greenwood, R.S. Miles and C. Patterson [eds.]. Interrelationships of Fishes. *Zoological Journal of the Linnean Society of London* 53, suppl. 1. Academic Press, London.

Nelson, G. 1967. Epibranchial organs in lower teleostean fishes. J. Zool. 153: 71–89.

Patterson, C. 1984a. Family Chanidae and other teleostean fishes as living fossils, pp. 132–139. *In*: N. Eldredge and S.M. Stanley, [eds.]. Living Fossils. Springer Verlag, Berlin.

Patterson, C. and G.D. Johnson. 1995. The intermuscular bones and ligaments of teleostean fishes. Smithsonian Contributions, Zoology 559: 1–85.

Perkins, P.L. 1970. *Notogoneus osculus* Cope, an Eocene fish from Wyoming (Gonorynchiformes, Gonorynchidae). Postilla 147: 1–18.

Peters, N. 1967. Opercular- und Postopercularorgan (Occipitalorgan) der Gattung *Kneria* (Kneriidae, Pisces) und ein Vergleich mit verwandten Strukturen. Zeitschrift für Morphologie und Ökologie der Tiere 59: 381–435.

Poll, M. 1965. Contribution a l'etude des Kneriidae et description d'un nouveau genre, le genre *Parakneria* (Pisces, Kneriidae). Mém. Acad. Roy. Belg., Classe des Sci., sér. 8 36(4): 1–28.

Poyato-Ariza, F.J. 1994. A new Early Cretaceous gonorynchiform fish (Teleostei: Ostariophysi) from Las Hoyas (Cuenca, Spain). Occasional Papers of the Museum of Natural History, The University of Kansas, Lawrence (164): 37.

Poyato-Ariza, F.J. 1996a. A revision of the ostariophysan fish family Chanidae, with special reference to the Mesozoic forms. Paleo Ichthyol 6: 1–52.

Poyato-Ariza, F.J. 1996b. The phylogenetic relationships of *Rubiesichthys gregalis* and *Gordichthys conquensis* (Ostariophysi: Channidae) from Early Cretaceous of Spain, pp. 329–348. *In*: G. Arratia and G. Viohl [eds.]. Mesozoic Fishes—Systematics and Paleoecology. Verlag Dr. Friedrich Pfeil, Munich.

Rabor, D.R. 1938. Studies on the anatomy of the bangos *Chanos chanos* (Forskål), 1. The skeletal system. Philipp. J. Sci. 67:351–377.

Ridewood, W.G. 1904. On the cranial osteology of the clupeoid fishes. Proc. Zool. Soc. London 2: 448–493.

Ridewood. W.G. 1905. On the skull of *Gonorhynchus greyi*. Ann. Mag. Nat. Hist. 7(15): 361–372.

Rosen, D.E. and P.H. Greenwood. 1970. Origin of the Weberian apparatus and the relationships of the ostariophysan and gonorynchiform fishes. Am. Mus. Novitates 2428:1–25.

Swinnerton, H.H. 1901. The osteology of *Cromeria nilotica* and *Galaxias attenuatus*. Zoologische Jahrbücher, Abteilung für Anatomie 18: 58–70.

Thys van den Audenaerde, D.F.E. 1961. L'anatomie de *Phractolaemus ansorgei* Blgr. et la position systématique des Phractolæmidae. Ann. Mus. Roy. Afr. Centr., Sci. Zool., sér. 8 103: 101–167.

Woodward, A. 1896. On some extinct fishes of the teleostean Gonorhynchidae. Proc. Zool. Soc. London 500–504.

2

Morphological Analysis of the Gonorynchiform Postcranial Skeleton

Terry Grande[1] and Gloria Arratia[2]

Introduction

The postcranial skeleton forms the framework of the body and functions in support and movement (Liem *et al.* 2001). It can be divided into two components. First, the axial skeleton lies in the longitudinal axis of the body and consists of the notochord, vertebral column, median fins, ribs and intermuscular bones (Arratia *et al.* 2001, Arratia 2003: fig. 1). Second, the appendicular skeleton consists of the pectoral and pelvic girdles and their corresponding fins. They aid the median fins (dorsal, anal and caudal) in stabilizing the body and resisting the tendency to roll, pitch and yaw as the fish swims. The paired fins also aid in steering and turning.

Like the skull, the postcranial skeleton in fishes is morphologically diverse and has been the subject of much examination and debate by researchers for many years. The principal interest lies in three general areas: (1) origin of the vertebrae and role of the notochord, (2) the anterior vertebral region and its modifications, and (3) the caudal fin skeleton. Beginning with the sister-group alignment of the Gonorynchiformes with Otophysi (i.e., taxa with a functioning Weberian apparatus) by Greenwood *et al.* (1966) and Rosen and Greenwood (1970), hypotheses concerning the evolution of the Weberian apparatus emerged with gonorynchiform fishes at the center,

[1] Loyola University Chicago, 1032 West Sheridan Road, Chicago, Illinois 60626, U.S.A.
[2] Biodiversity Research Center, The University of Kansas, Dyche Hall, 1345 Jayhawk Boulevard, Lawrence, Kansas 66045, U.S.A., and Field Museum of Natural History, Roosevelt Road at Lake Shore Drive, Chicago, IL 60605, U.S.A.

either exhibiting a possible intermediate or primitive form, or holding the key to a better understanding of the evolution of this sound transmission device. Modifications specific to the Gonorynchidae, such as expanded neural arches and supraneurals, led many researchers (e.g., Gayet 1986a, Gayet and Chardon 1987, Chardon and Vandewalle 1997) to speculate whether the Weberian apparatus evolved multiple times within Ostariophysi, and whether taxa such as †*Ramallichthys* could constitute an intermediate form.

Gonorychiform fishes share several modifications of the caudal fin skeleton with the Otophysi as described by Fink and Fink (1981), providing further evidence for their sister-group relationship. Yet, similarities found to be in common with the otophysans do not preclude the large amount of variation in caudal fin structure among and within gonorynchiform subgroups, especially with respect to fossil forms. This variation has been realized only recently (Grande and Grande 2008) and warrants further study.

In this chapter, we, in a comparative framework, provide an overview of the gonorynchiform postcranial skeleton. We assess previously published morphological works and add new data that we hope will aid in future phylogenetic reconstruction. For consistency, we discuss the elements of the axial skeleton followed by those of the appendicular skeleton (e.g., pectoral girdles). The pelvic girdle was not examined in detail for this study. Within the axial skeleton, we focus on the anterior vertebral region and caudal fin, including interspecific modifications and adaptations. We also pay particular attention to the postcranial skeleton and its development in *Gonorynchus*, the only living representative of the family Gonorynchidae. Because fossil gonorynchids exhibit the most interesting anterior vertebral column modifications, linked by some to the evolution of the Weberian apparatus, an understanding of the axial skeleton of *Gonorynchus* is crucial to a better understanding of the fossil forms. For additional specimens examined see Grande and Poyato-Ariza (1999).

Vertebral Column

In general, the notochord forms the hydrostatic skeleton for the developing embryo. It extends from the skull to the tip of the tail and serves as the foundation for the developing vertebral column. In gonorynchiforms, as well as in other fishes, the notochord and its obliterations set the path along where each centrum will form (Schultze and Arratia 1988, Arratia 1991, Arratia *et al.* 2001). The centrum is termed chordacentrum, arcocentrum, or autocentrum, depending on its origin. The arcocentra (or ossification of the cartilage extending from the arcualia) are the portions of the arches included in the centrum. They are consistently present in fishes, whereas the chordacentrum and autocentrum are variably present. In ostariophysans such as *Chanos*, chordacentra develop along the vertebral column as

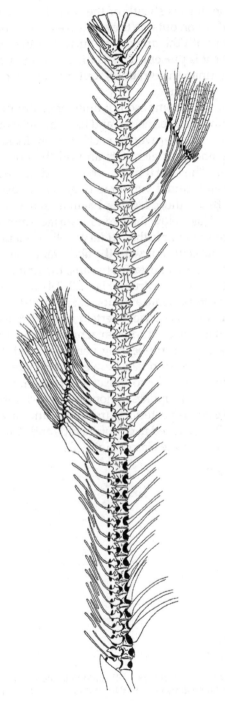

Fig. 2.1 Postcranial skeleton of *Chanos chanos*. Paired girdles and fins are omitted as well as most intermuscular bones. CAS 05022-05075, SL = 103 mm.

mineralizations of the fibrous sheath of the notochord. The autocentrum forms as a direct ossification outside the elastica externa; consequently, the autocentrum together with the arcocentra surround the notochord or the chordacentrum where it is present. This is the condition present in *Chanos* (see Arratia and Bagarinao this volume), but it is unclear how the centra form in other gonorynchiforms.

The vertebral column is composed of units or centra plus the elements directly associated with them, i.e., neural and haemal arches and their spines, parapophyses and ribs. In other words, each vertebra includes all ossified, cartilaginous, and ligamentous elements around the notochord (Schultze and Arratia 1988). The vertebrae are divided into abdominal (= precaudal, trunk) and caudal vertebrae (Figs. 2.1, 2.2A–C). Abdominal vertebrae (Fig. 2.2A–B) are those lacking haemal arches so that the major blood vessels are placed just below their ventral surfaces and are not enclosed by bone. They are placed anterior to the caudal vertebrae. Pleural ribs articulate only with the abdominal centra or with the abdominal centra plus the first caudal ones (e.g., Fig. 2.1). In all ostariophysans, the first pleural rib (or the tripus) articulates with the third centrum (Figs. 2.3, 2.4). Because of the absence of ribs on the first two centra, some authors (e.g., Morin-Kensicki *et al.* 2002) have referred to these units as cervical vertebrae. As discussed in Bird and Mabee (2003), molecular genetic studies in zebrafish have revealed that the Hoxc6 gene is correlated with rib-bearing vertebrae, and that the anterior limit of theHoxc6 gene expression falls between vertebrae 2 and 3. Because the first two centra form anterior to Hoxc6 expression, the amniote term "cervical vertebra" may be appropriate. Among gonorynchiforms, a prominent neural spine (Fig. 2.4) extends posteriorly from centrum 1 in all chanids. This spine (Figs. 2.5, 2.6A–B) is reduced in gonorynchids and *Phractolaemus* and is absent on the first vertebra in the Cromeriini.

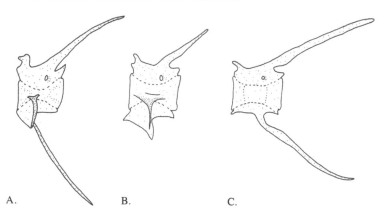

A.   B.   C.

Fig. 2.2 (A) Abdominal centrum 14 and rib, (B) abdominal centrum 45, and (C) caudal centrum 58 from *Gonorynchus abbreviatus*, FMNH 96053, SL = 240 mm.

Caudal vertebrae in ostariophysans including gonorynchiforms are those having haemal arches; thus, the first caudal vertebra is the first vertebra with both halves of the haemal arch fused medially forming an arch, whereas preural centrum 1 is the last vertebra with a haemal arch. The major blood vessels, e.g., dorsal aorta or primary caudal artery, primary caudal vein and secondary arterial and venous trunks, exit the vertebral column posterior to the haemal arch of preural centrum 1 and run laterally to hypurals 1 and 2 (for details see Schultze and Arratia 1989: 222–223, fig. 20).

Finally, the most posterior vertebra in the ostariophysan vertebral column consists of an assumed fusion of preural centrum 1, ural centrum 1, and uroneural 1. The so called ural centrum 2 may be included in the fusion or remains as an independent centrum. This compound terminal centrum, often called the urostyle or complex terminal centrum, was considered a synapomorphy of the Otophysi, but, as argued by Rosen and Greenwood (1970) and Fink and Fink (1981), this series of fusions is also found among gonorynchiforms, indicating a possible sister-group relationship. As discussed by Lecointre and Nelson (1996) and Grande and de Pinna (2004), similar fusions are also common within certain clupeiform subgroups. However, it has not yet been demonstrated whether these structures are homologous (see Schultze and Arratia 1989: 203, fig. 10A), or whether the

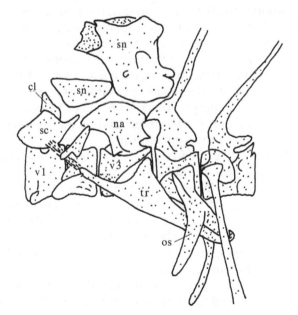

Fig. 2.3 Weberian apparatus of *Opsariichthys uncirostris* (modified from Patterson 1984). cl, claustrum; na, neural arch; os, os suspensorium; sc, scaphium; sn, supraneurals; tr, tripus; v1, vertebra 1; v3, vertebra 3.

complex terminal centrum forms differently in different cypriniform subgroups as well as in basal clupeomorphs (Arratia *et al.* 2007).

Anterior Vertebral Region in Gonorynchiformes

Ostariophysan fishes are known for specializations of the anterior vertebral column. The most well-known specialization, diagnostic of the Otophysi (e.g., carps, minnows, suckers and catfishes), is the Weberian apparatus (Fig. 2.3). This mechanical linkage consists of modified centra, neural arches, supraneurals and pleural ribs that lie in a linear sequence and transmits motion of the swimbladder wall directly to a perilymphatic space, the sinus impar, of the inner ear (Braun and Grande 2008). Although gonorynchiforms do not exhibit a functioning Weberian apparatus, modifications of the gonorynchiform vertebral column similar to those of a Weberian apparatus, or proto-Weberian apparatus, were described (Rosen and Greenwood 1970, Gayet 1985, 1986a). Rosen and Greenwood (1970) examined the vertebral column in the gonorynchiform *Chanos chanos* and compared it to the juvenile characiform *Brycon*. They reported characters in *Chanos* such as a modified first pleural rib and parapophysis, a divided swimbladder, and a swimbladder that is anteriorly covered by a dense silvery peritoneum supported by the first rib to be comparable to *Brycon*. They also reported additional "otophysan-like" specializations in *Gonorynchus*, such as reduction of the first neural spine and expansion of anterior neural arches that, together with similar caudal fin morphologies, gave them reason to include gonorynchiforms within Ostariophysi. Gayet and Chardon (1987) proposed that certain gonorynchiforms exhibit an intermediate or primitive form of the ostariophysan Weberian apparatus. Essentially they argued that the gonorynchiform morphology may be closer to the ancestral proto-otophysic linkage, and the keys to understanding the Weberian apparatus itself may lie in studies of gonorynchiforms.

Fink and Fink (1981), in the first cladistic study of extant ostariophysan fishes, corroborated Rosen and Greenwood's (1970) placement of the Gonorynchiformes within Ostariophysi and cited several anterior vertebral characters to help support their hypothesis. These characters include a dorsomedial expansion of the neural arches, an articulation of the first neural arch with the exoccipitals, absence of the unattached arch anterior to vertebra 1 (counted here as the first independent vertebra), and the absence of the supraneural anterior to neural arch 1.

So, it seems clear that gonorynchiform fishes share several anterior vertebral characters or modifications with otophysans. There are also many adaptations and specialized trends indicative of major gonorynchiform subgroups (i.e., Chanidae, Gonorynchidae, Kneriidae). In this chapter, we will examine some of the major vertebral column modifications found within Gonorynchiformes and discuss trends and modifications of this region

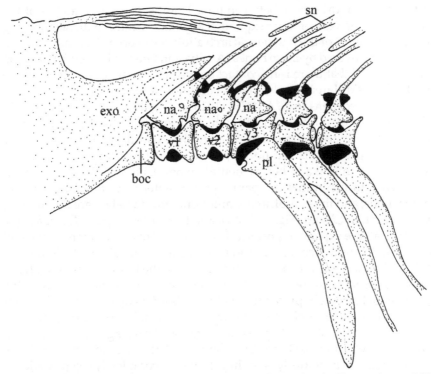

Fig. 2.4 The anterior vertebrae and their relationship to the cranium in *Chanos chanos*, CAS 05022-05075, SL = 103 mm. Tips of neural spines and pleural ribs not complete. Suture between exoccipital and basioccipital not complete. boc, basioccipital; exo, exoccipital; na, neural arch; sn, supraneural; pl, pleural rib; v1–v3, vertebrae 1–3. Cartilaginous elements are represented in black.

within a phylogenetic framework. On the basis of previous work (Fink and Fink 1981, 1996, Poyato-Ariza 1996, Grande and Poyato-Ariza 1999), it can be said that the Chanidae exhibit a more generalized axial skeleton condition than the Gonorynchidae and Kneriidae. We will begin our discussion there.

The Chanidae are represented by one extant form, *Chanos chanos*, and several fossil taxa with a widespread geographic distribution and a fossil record dating back to the Early Cretaceous. The morphology of the anterior vertebral column is often difficult to discern in the fossil chanid taxa because of poor preservation and a large opercle that often masks the first few vertebrae. Nevertheless, we can make some generalizations. According to Poyato-Ariza (1996), the first three centra are shorter than the more posterior ones in all chanids, the anterior neural arches are enlarged but usually do not make contact with neighboring arches, and the first neural arch abuts the exoccipitals. However, in large specimens of *Chanos* (over 300 mm SL) the first 8 to 10 centra are shorter than the rest of the abdominal centra, but their lengths increase progressively caudally, so that it is very difficult to establish a real difference between their lengths; the neural arches of the first four to five vertebrae are expanded and make contact with neighboring arches; and most of the first neural arch abuts the exoccipitals and is almost completely covered by these bones (Arratia and Grande in prep.). Patterson (1984) argued that there is no contact between the first neural arch and the back of the skull in †*Tharrhias*. This condition was observed, however, in †*Tharrhias* specimens (e.g., AMNH 11916 and NHM P. 54331) examined in Grande and Poyato-Ariza (1999), which strengthens the hypothesis that this character is not restricted to extant forms. Also in chanids (Fig. 2.1), the anterior supraneurals are not expanded, but form long rods that are separated from one another by neural spines. Chanids differ from other gonorynchiforms in that the parapophyses anterior to the dorsal fin are autogenous, and the neural arches anterior to the dorsal fin are not fused to their centra. In *Chanos* (Figs. 2.1, 2.4), and to a lesser degree †*Tharrhias*, the exoccipitals are expanded posteriorly and extend over the anterodorsal half of the first neural arch. As seen in *Chanos* specimens of about 120 mm SL (Fig. 2.4), relative to the more posterior arches, the first three neural arches are considerably more expanded, and their dorsal margins partly cover the supradorsal cartilages placed medially. The borders of the first three arches are in contact with each other. Between the dorsal parts of the halves of neural arches 3 to about 32, pairs of supradorsal cartilages are observed in specimens of about 200 mm SL. The size of the supradorsal cartilages decreases going posteriorly, and they disappear completely just posterior to the anal fin in young individuals. In larger individuals the supradorsal cartilages ossify as part of the wall of the neural arch dorsal to the neural cord, at the base of the neural spine. Rosen and Greenwood (1970) referred to these cartilages as supradorsals. For details concerning the supradorsal cartilages and their diversity in otophysans, see Hoffman and Britz (2006).

The Gonorynchidae possess some interesting anterior vertebral modifications that have caused some debate over the years (Gayet 1982, 1986a, b, Fink *et al.* 1984, Grande 1996, Grande and Poyato-Ariza 1999). The Gonorynchidae consist of one extant genus *Gonorynchus*, and four (Grande and Grande 2008) or six (Gayet 1993a, b) fossil genera. These fossil genera include †*Notogoneus* from Upper Cretaceous to Oligocene localities of North America, Europe, Asia and Australia, and a group of predominantly Middle Eastern marine Cretaceous forms (e.g., †*Ramallichthys*, †*Charitosomus*, †*Charitopsis*). In general, the Gonorynchidae exhibit an exaggerated form of the gonorynchiform condition (i.e., the dorsomedial expansion of the anterior neural arches and an expansion of the anterior supraneurals). This condition is unparalleled in the Middle Eastern forms (e.g., †*Ramallichthys* and †*Charitosomus*), where both the arches and supraneurals are expanded to such a degree that they form a lateral wall of bone on each side of the neural tube (Fig. 2.5A–B). Explanations for these modifications have been many. For example, Gayet (1982, 1986a) hypothesized that the modifications of the neural arches in †*Ramallichthys* are homologous with the scaphium, claustrum and intercalarium of otophysans. Although most researchers, including Gayet (1993a, b), now disagree with this hypothesis, the Middle Eastern gonorynchids were also at the center of Gayet and Chardon's (1987) hypothesis for a multiple origin of the Weberian apparatus, suggesting that these fishes might exhibit a primitive or intermediate otophysic condition. Based on new material of both †*Ramallichthys* and the Eocene Green River species †*Notogoneus osculus*, Grande and Grande (2008, in press) reexamined the morphology of these taxa and reassessed previously proposed gonorynchid relationships (e.g., Grande 1996, Grande and Poyato-Ariza 1999).

Like *Chanos* (Fig. 2.4), the first neural arch in *Gonorynchus* articulates with the exoccipitals. The connection between the arch and the back of the skull is, however, made through a thick band of connective tissue. Unfortunately, this condition was not determinable in the fossil gonorynchids examined. If in the fossil taxa the connection between neural arch 1 and the exoccipitals was made through connective tissue as it is in *Gonorynchus*, the connective tissue would not have been preserved, giving the false impression that the arch is naturally disconnected from the back of the skull. Additional well-preserved specimens are necessary if we are to understand the condition in the fossil forms. Relative to the chanids, all gonorynchids exhibit a slight reduction in the length of the first neural spine. This spine is also reduced in the Middle Eastern forms. As shown in *Gonorynchus* (Fig. 2.6A–B), neural arches and parapophyses anterior to the dorsal fin are fused with their centra. This condition was also observed in †*Notogoneus* (Grande and Grande 1999, 2008). Of the Middle Eastern forms, however, the anterior neural arches in †*Ramallichthys*, †*Judeichthys* and †*Hakeliosomus* are not fused with their centra. The parapophyses anterior to the dorsal fin in these taxa appear fused, but additional material is necessary to confirm this observation. The condition is more difficult to assess in †*Charitosomus* and †*Charitopsis*. The anterior vertebrae in the holotype of †*Charitopsis spinosus* (AMNH FF 3895) are obstructed and partly destroyed, and in †*C. spinosus* (AMNH FF 3740) the anterior vertebral column is missing altogether. In our specimens of †*Charitosomus lineolatus* SHA 1454 and 200, the parapophyses appear fused, but the anterior vertebrae are poorly preserved and we can not determine the condition of the neural arches with any degree of certainty. Based on Gayet's (1993a) drawings of †*Charitosomus formosus* and †*C. lineolatus*, both the neural arches and the parapophyses are autogenous. As in Monod (1963), and contrary to Britz and Moritz (2007), the first three expanded neural arches in *Gonorynchus* abut each other. In small specimens (Fig. 2.6A, SL = 147 mm), contact between anterior arches is minimal, and although the supraneurals are expanded, they do not make

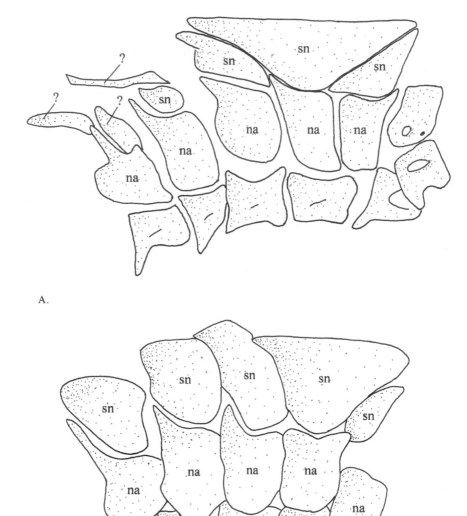

Fig. 2.5 Anterior vertebral region of (A) †*Ramallichthys orientalis* (EY 386, SL = 144 mm) and (B) †*Hakeliosomus hakelensis* (MNHN 130d). Illustration modified from Grande and Grande (2008). Anterior to the right. na, neural arch; sn, supraneural; ?, unknown.

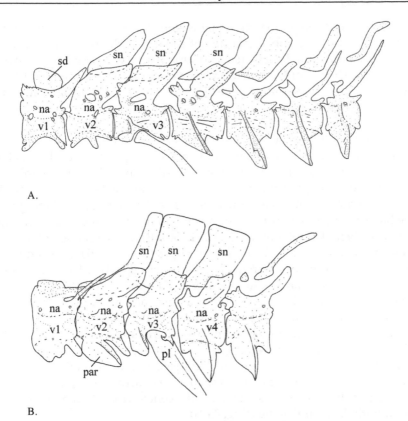

A.

B.

Fig. 2.6 Anterior vertebral region of (A) *Gonorynchus greyi*, AMNH 32973, SL = 147 mm (B) *Gonorynchus abbreviatus*, FMNH 96053, SL = 240 mm. na, neural arch; par, parapophysis; pl, pleural rib; sd, supradorsal cartilage; v1–v4, vertebra 1–4.

contact with each other. In larger specimens, however (Fig. 2.6B, SL = 240 mm), the supraneurals are more expanded and the first two are in contact with each other. Supraneurals beginning at vertebra 5 or 6 are more characteristic in shape (i.e., long and thin) and do not make contact with their neighbors.

Gonorynchus exhibits an interesting condition with respect to vertebra 1 and is best examined developmentally. In the *Gonorynchus* specimens examined, basidorsals are formed in specimens of about 18 mm SL (AMNH 55562). Supraneurals were observed in specimens of about 20 mm SL (AMS I.288400004). The first observable supraneural is positioned between neural arches 1 and 2. There is no supraneural anterior to or above neural arch 1. The supraneurals form and ossify in a linear sequence beginning with the supraneural above vertebra 2. As the anterior neural arches expand, so do the supraneurals. In specimens of about 75 mm SL (AMS I.5478), the centra and most of the neural arches and supraneurals have ossified.

A thin cartilaginous margin remains along the periphery of each anterior arch. The expanded dorsal margins of the first arch approach each other along the midline but do not touch. No supraneural was observed positioned between the halves of the neural arch of vertebra 1. The supraneural above vertebra 2 and those more posterior are flat, expanded laterally, and positioned tightly between the halves of the neural arches (or spines). In specimens of 85 mm SL (AMS I.15048A), the supraneural positioned above vertebra 2 abuts neural arch 1, which has developed a spine (which is unfused medially). Although most of arch 1 has (perichondrally) ossified, the upper and central portion is still cartilaginous. In other words, the ossified arch has a prominent dip in the center, and in between the anterior margin of the arch and the spine is cartilage. The anterodorsal margins of the halves of neural arch 1 do not meet, and no supraneural is positioned between them. By 106 mm SL (AMS I.15048B), the vertebral column has completely ossified. A pair of median supradorsal cartilages is, however, observed extending between the halves of neural arch 1. This supradorsal cartilage is domed, and from lateral view it appears to extend above the neural arch, giving the impression of a supraneural (Fig. 2.6A); it is thus referred to as a supraneural by Poyato-Ariza (1996). The lateral view of the vertebral column and vertebra 1 of a similarly sized fish is illustrated in Monod (1963) but, surprisingly, the morphology of vertebra 1 was not described. In larger specimens (AMNH 96053), it appears that centrum 1 has thinly ossified and the autocentrum has surrounded the base of the neural arch. In even larger specimens, the centra, from the lateral view, are fused with the neural arches (Fig. 2.6B).

†*Notogoneus* and *Gonorynchus* have been considered sister taxa by most researchers (e.g., Perkins 1970, Grande and Poyato-Ariza 1999). The two genera have the same general body shape and share several synapomorphies of the skull. It was therefore assumed that the postcranial morphology of †*Notogoneus* would be similar, if not identical, to that of *Gonorynchus*. As discussed in Grande and Grande (2008), †*Notogoneus* does not exhibit many of the postcranial characters of Gonorynchidae and has been hypothesized to be the sister group to the rest of the gonorynchids (Grande and Grande 2008). For example, the neural arches in †*Notogoneus* are not expanded like those of *Gonorynchus* and the Middle Eastern forms. The supraneurals, although not rod-like as in *Chanos*, are not characteristically expanded. Even in very large specimens of †*Notogoneus*, neither the neural arches nor supraneurals make contact with each other. The first pleural rib in †*Notogoneus* is also more slender and elongate than that of *Gonorynchus*.

The Kneriidae (Figs. 2.7A–D, 2.8A–B), *sensu* Grande and Poyato-Ariza (1999), comprise a group of African freshwater forms with no known fossil record. These fishes are quite interesting in that they not only exhibit their own unique morphologies, but they also exhibit several modifications and trends first seen in the Gonorynchidae. For example, in *Chanos* the spine of neural arch 1 is long and slender. In *Gonorynchus* this spine is reduced but distinctly observable. The presence of a spine on the first neural arch in *Phractolaemus* (Fig. 2.7A) is the object of some debate. Contrary to Britz and Moritz (2007), Grande and Poyato-Ariza (1999) referred to the small pointed structure extending from the posterior part of the first neural arch as a spine. Additional developmental information (Arratia and Grande in prep.) might clarify this point. The posteriorly directed spine is completely lost in the Kneriinae (i.e., *Cromeria*, *Grasseichthys*, *Parakneria* and *Kneria*). If additional data shows that the spine on neural arch 1 is indeed lost in *Phractolaemus* as it is in the Kneriinae, this would lend further support for the monophyly of the more inclusive Kneriidae (Grande and Poyato-Ariza 1999, Poyato-Ariza *et al.* this volume). In *Gonorynchus* and *Phractolaemus* the first neural arch articulates with the exoccipitals, but in the Kneriinae (Figs. 2.7B–D, 2.8A–B), the first arch articulates with the exoccipitals plus the supraoccipital. Interestingly, as illustrated for *Grasseichthys* and *Parakneria* (Figs 2.7B, D), neural arch 1 is attached to the supraoccipital through processes that extend from the neural arch proper. Additional study is necessary to determine whether these processes are neural spines directed anteriorly, or the neural arch proper. As discussed in Moritz *et al.* (2006) and Britz and Moritz (2007), *Cromeria* exhibits two different morphologies of the second vertebra. In *Cromeria nilotica*, the second neural arch sports a rather traditional spine, while in *C. occidentalis* the spine of neural arch 2 is short and slightly expanded. The neural arch of vertebra 3 overlaps with this modified spine.

In our specimen of *Phractolaemus* (Fig. 2.7A: FMNH 63938A), the first neural arch is fused not only to the centrum 1, but also to the basioccipital. In smaller specimens of *Phractolaemus* this fusion is not present. Additional large specimens are needed to investigate whether this fusion is anomalous to this specimen or a result of ontogeny. The Kneriidae also exhibits a reduction in the number of supraneurals. One disk-shaped supraneural was observed in *Phractolaemus*, *Parakneria* and *Kneria*, while supraneurals are completely absent in *Cromeria* and *Grasseichthys*. It should be noted that both *Cromeria* and *Grasseichthys* are paedomorphic forms and the lack of several skeletal elements may be the result of independent paedomorphic histories and not reflective of their phylogenetic relationships (Grande 1994).

Of the kneriids, the most interesting postcranial adaptation is that found in *Kneria* (Fig. 2.8A–B). The opercular apparatus is an external body structure that is highly developed in the adult males and vestigial in the adult females. This structure consists of two parts: a cup-like structure that sits directly on the opercular bone, and a posterior part that originates on the side of the body just posterior to the opercular opening and extends posteriorly past the pectoral fin margin (Grande and Young 1997). The cup consists of thickened uplifted loose connective tissue and is thought to function as an

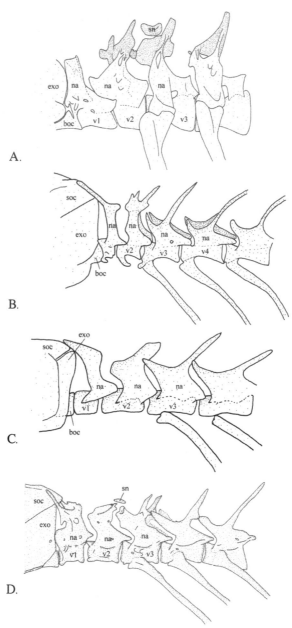

Fig. 2.7 Anterior vertebral region of Kneriidae and its relationship to the posterior part of the braincase. (A) *Phractolaemus ansorgei*, FMNH 63938A, SL = 123.9 mm. Redrawn from Grande and Poyato-Ariza (1999). (B) *Cromeria occidentalis*, CAS SU 54641, SL = 32.6 mm. (C) *Grasseichthys gabonensis,* MRAC 73-02-P-264-905, SL = 18.9 mm. (D) *Parakneria tanzaniae*, BMNH 10.21:163, SL = 43.7. boc, basioccipital; exo, exoccipital; na, neural arch; sn, supraneural; soc, supraoccipital; v1–v4, vertebrae 1–4.

adhesive device anchoring the fish in fast-moving water during courtship and reproduction (Peters 1967, Lenglet 1974, Grande and Young 1997). The opercular apparatus is accompanied by a series of skeletal modifications. Internally, the anterior epicentral intermuscular bones (i.e., intermusculars that lie in the horizontal septum and articulate via ligaments to their corresponding parapophyses or pleural ribs) are enlarged and branched at the distal ends resulting in bones that take on a blade-like appearance. The first three epicentrals show the greatest amount of enlargement, both ventrally and posteriorly. In the adult male, the first six epicentral intermuscular bones are greatly enlarged and are connected to the inner body wall by connective tissue (Fig. 2.8A). The position of these intermusculars corresponds to the posterior portion of the opercular apparatus and may serve to support the structure from the inside. In the adult female (Fig. 2.8B), only the first three epicentrals are expanded. These epicentrals are also about one-third the size of those found in the males, but are still connected to the inner body wall via tendons.

Gonorynchiforms generally have three sets of intermuscular bones. Epicentrals, as discussed above, positioned along the horizontal septum, epineurals found above the horizontal septum, and epipleurals below it. According to Patterson and Johnson (1995), *Chanos* exhibits the most generalized intermuscular condition among gonorynchiforms, while *Gonorynchus* and the kneriids exhibit a more complicated arrangement. The intricate arrangement of epineural, epipleural and epicentral intermuscular bones found in *Gonorynchus* is laboriously described in Patterson and Johnson (1995). To summarize, in *Chanos* (young specimens) and *Gonorynchus*, proximally forked epineurals begin at vertebra 2 and end at about vertebrae 30 or 35 respectively. The anteromedial branch of these epineurals attaches to the neural arches by ligaments. Anterior to these epineurals are additional four epineurals consisting only of the anteroventral branch. These intermusculars have brush-like posterior ends and are subdivided anteriorly. Epicentrals are branched and are attached to the head of the rib through ligaments. The most anterior epicentrals extend from the exoccipitals and basioccipital and are referred to as cranial ribs (Rosen and Greenwood 1970). Cranial ribs or epicentrals attach to the dorsal part of the cleithrum by ligaments. According to Patterson and Johnson (1995), *Parakneria* has the same intermuscular configuration as *Kneria* but without the epicentral modifications. As discussed in Grande and Young (1997), the first two epicentrals connect via ligaments with their corresponding vertebrae. Epicentrals 3–21 articulate with the proximal ends of the pleural ribs, epicentrals 22–26 attach to the parapophyses of their respective vertebrae, and finally epicentrals 27–30 articulate directly to their centra or are attached through ligaments. Epineurals lie above the epicentrals beginning at vertebra 2, attaching to the axial skeleton via ligaments.

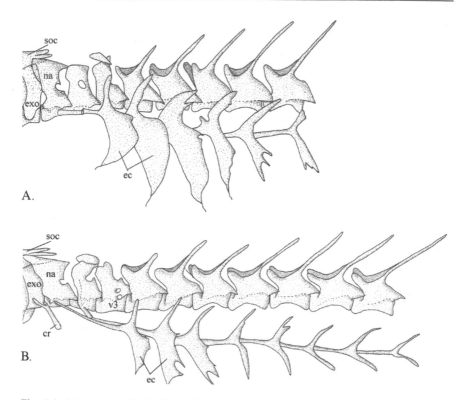

Fig. 2.8 Anterior vertebral region of *Kneria wittei*. (A) MRAC 79-01-P-572, male specimen, SL = 50 mm. (B) MRAC 79-01-P-516, female specimen, SL = 45.9. Redrawn from Grande and Young (1997). Anterior to the left. cr, cranial or cephalic rib; ec, epicentral intermuscular bones; exo, exoccipital; na, neural arch; soc, supraoccipital; v3, vertebra 3.

Epipleurals begin at vertebra 21 and attach to the axial skeleton via parapophyses or haemal arches. Epipleurals terminate at preural centrum 1.

Based on our specimens of *Phractolaemus* (FMNH 63638AB), epineurals begin as small, thin splints at vertebra 7. They are attached to their neural arches by ligaments. The epineurals bifurcate distally at vertebrae 25–27 and fork proximally from vertebrae 28–30. An additional series of six epineurals follow, but these are thin and rod-like and are not directly attached to neural arches. Ventrally, seven epipleurals are associated with the haemal arches. In one specimen, the first epipleural is attached to the last pleural rib. In the caudal fin region lateral to the hypurals, there are two groupings (clumps) of horizontal intermusculars, one above and one below the horizontal body axis. A single intermuscular is positioned lateral to each grouping. The epicentral series is represented in *Phractolaemus* only by one large cranial rib positioned on each side of the skull.

According to Britz and Moritz (2007), two sets of intermuscular bones (i.e., epineural and epipleural intermusculars) are present in the genus *Cromeria*. Epicentrals are represented only by cranial ribs. All intermusculars are absent in the paedomorph *Grasseichthys*.

All fossil gonorynchid taxa were observed with three sets of intermuscular bones. Intermuscular bones were observed in the fossil chanids †*Dastilbe* and †*Tharrhias*. Because of the preservation of these taxa, it is difficult to assess which of these intermusculars are epicentrals, epineurals or epipleurals. The intermuscular condition was not described in Poyato-Ariza (1996), so the condition in the remaining fossil chanids is uncertain.

Dorsal and Anal Fin Skeletons

Dorsal and anal fins, positioned along the midline, help stabilize the fish as it moves through the water column, thus preventing rolling. All gonorynchiform fishes have one dorsal and one anal fin consisting of only soft rays, or lepidotrichia (Table 2.1). The fin rays are supported by deeper bones called pterygiophores. Each pterygiophore is divided into two or three components or radials (i.e., proximal, middle, and distal). According to Grande and Bemis (1998) and Mabee *et al.* (2002), "pterygiophore" is a collective term referring to each set of radials (Fig. 2.9). According to Fink and Fink (1981), *Chanos* has three ossified radials, while *Kneria* and *Grasseichthys* have two. Proximal, middle and distal radials were observed in our specimens of *Chanos*, *Gonorynchus* and *Kneria*. Based on our specimens, it appears that *Phractolaemus* is missing middle radials in the anal fin. One middle radial was observed in one specimen of *Phractolaemus* (FMNH 63938B). This observation is in general agreement with that of Thys van den Audenaerde (1961). We were unable to determine the radial condition in our specimens of *Cromeria* and *Grasseichthys*. Britz and Moritz (2007) use the term proximal-middle and distal radials when referring to *Cromeria* pterygiophores. We assume that proximal-middle means that there is a fusion between the two radials, and that only two radials are present in median fins of these taxa. There is a one-to-one relationship between most principal fin rays and the pterygiophores, with the exception of the first pterygiphore which bears several rays and last pterygiphore which bares a double ray.

As discussed in Mabee *et al.* (2003), symmetrical positioning of the dorsal and anal fins in relation to each other may constitute a modular system (i.e., Dorsal + Anal Fin Positioning Module), and primitively these fins shared specific myomeres and a common anterior boundary (i.e., the anus). Deviation from this pattern would constitute a decoupling of this linkage and an independence of positioning arising from different myomeres. Although the anal fin is constrained in position by the anus,

Table 2.1 Meristic data for gonorynchiform taxa. Data taken from the literature (Gayet 1993, Poyato-Ariza 1994, Britz and Moritz 2007, Dietze 2007) and from specimens examined as indicated by*.

Taxon	Total centra	Abdominal centra	Caudal centra	Dorsal fin rays	Anal fin rays	Pectoral fin rays	Pelvic fin rays
Chanos*	44–51	16–17	26–28	13–17	6–8	15–17	10–11
†Tharrhias	47–52	?	?	13	?	11–12	10
†Dastilbe	38–39	?	15–16	11–12	8–9	10–12	7–9
†Parachanos	38	23	15	10–13	8–10	12?	8–9
†Gordichthys	34–40	21–23	14–15	9–10	7–8	10–11	9–10
†Rubiesichthys	39–40	22–24	16–17	10	9	10–11	?
Gonorynchus*	54–66	43–51	11–15	10–12	7–9	10–11	8–9
†Notogoneus*	46–60	35–37	13–16	8–15	8–11	9–12	7–9
†Ramallichthys*	42	28	14	13	8	14	8
†Judeichthys*	?	?	13 or 14	?	6+	14	8
†Hakeliosomus*	42–43	30–31	12–14	13	8	15	8
†Charitosomus*	51–59	33–37	15–16	10–12	8–12	8	7–9
†Charitopsis*	43–45	29–30	14–15	12–13	7	9	8
Phractolaemus*	37–38	29–30	8	6	6	18	6
Grasseichthys*	36–37	18	18	7–8	9–10	7–8	6
Cromeria*	40–43	27–28	12–16	9–10	7–10	6–9	6–7
Parakneria*	41–43	25	18	8–9	6	16–20	9–10
Kneria*	41–45	24–26	17–19	9	8	12–16	8

the dorsal fin is essentially "free" of any particular position. The same is true for the paired fins. The position of pectoral fins is constrained because the skull serves as their anterior limit, while the pelvic fins often show plasticity in position. Gonorynchiform fishes exhibit a wide array of body types (e.g., *Chanos* has a very hydrodynamic shape, with a torpedo-like body, extremely narrow caudal peduncle, and large homocercal caudal fin, while *Gonorynchus* is very elongate) that upon first inspection might be reflective of fin position. But do gonorynchiforms demonstrate symmetry in median fin position? Calculating predorsal and preanal lengths as a percentage of SL, *Gonorynchus* exhibits a predorsal length of about 70% and a preanal length of about 82%. Although both fins are positioned posteriorly, the dorsal fin is more anterior and is not in symmetry with the anal fin. Among other gonorynchiforms measured (i.e., *Chanos*, †*Ramallichthys*, †*Notogoneus*, *Kneria* and *Cromeria*), all predorsal lengths fall between 51 and 60% of SL, *Chanos* being at the low end. Preanal lengths in these fish range between 72 and 83% of SL, with *Chanos* being at the high end. In other words, the predorsal length is consistently shorter than the preanal length in all gonorynchiforms examined, and the dorsal fin is always anterior to the pelvic fins. For comparison, *Esox lucius* has a predorsal length of about 74% of SL and a preanal length of about 75% of SL. As discussed in Mabee *et al.* (2002), *Esox* exhibits the primitive condition of positioned symmetry of the dorsal and anal fins, implying a modular

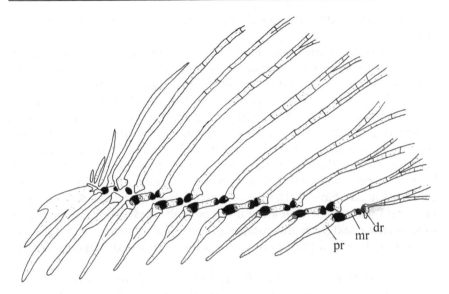

Fig. 2.9 Dorsal fin skeleton of *Gonorynchus forsteri*, AMNH 57120, SL = 226 mm. Anterior to the left. dr, distal radial; mr, middle radial; pr, proximal radial. Cartilage is represented in black.

system in the fish. A modular system, however, can not be argued for gonorynchiforms. Based on our measurements it appears that the dorsal and anal fins form independently in different myomeres, and that the genes regulating fin position are probably decoupled in these fishes. Developmental data in *Gonorynchus* does not indicate that the dorsal and anal fins share myomeres at any time in ontogeny. Developmental information of *Chanos* (Arratia and Grande in prep.) might provide clues to fin patterns in other gonorynchiforms.

Caudal Fin Skeleton

The vertebral column ends with the caudal fin, which delivers a strong propulsive thrust for swimming (Liem *et al.* 2001). Like the anterior vertebral region, the gonorynchiform caudal fin skeleton exhibits interesting interspecific morphological trends, characters diagnostic of the order and of the superorder Ostariophysi, and characters that are highly variable. Among the extant taxa, the caudal skeleton shares several synapomorphies with the Otophysi (i.e., a compound terminal centrum, a pleurostyle). In Recent gonorynchiforms and otophysans, the terminal centrum consists of an assumed fusion of an unknown number of ural centra and the first preural centrum plus the first uroneural. The shared caudal fin similarities between these two groups provided Rosen and Greenwood (1970) and Fink and Fink (1981) additional justification for their sister-group assignment.

However, when adding the fossil taxa into the mix, the story becomes more complicated.

In all fossil chanids (e.g., Fig. 2.10A), preural centrum 1, and the so-called ural centra 1 and 2 and all uroneurals are autogenous. Among the fossil chanids, †Dastilbe and †Gordichthys have two autogenous uroneurals, while †Tharrhias has three. Two epurals are found in all fossil taxa. In addition, all hypurals and the parhypural are free from the caudal centra. Chanos (Fig. 2.10B), on the other hand, differs from the fossil forms and shares with the gonorynchoids a compound terminal centrum. Uroneural 2, however, is still autogenous. One epural is present in Chanos, and all haemal arches anterior to the second preural centrum are fused to their centra (Fink and Fink 1981). Like the fossil forms, Chanos has six hypurals. Hypural 1 is wide and triangular in shape with a reduced proximal region that lacks contact with the compound terminal centrum. The diastema separating the distal parts of hypurals 2 and 3 almost disappears in large specimens.

Interpretations of the caudal fin morphology within the Gonorynchidae (Fig. 2.11A–F) have resulted in considerable debate and some errors. In general, all gonorynchids are assumed to have a compound terminal centrum. The number of hypurals is reduced from six to five, only one epural is present in all forms, and among the fossil taxa (i.e., †Notogoneus, †Ramallichthys, †Judeichthys, †Hakeliosomus, †Charitosomus, †Charitopsis), uroneural 2 is autogenous. Gonorynchus lacks an autogenous uroneural 2. A detailed developmental study is necessary to determine whether uroneural 2 is lost or fused to the compound terminal centrum in Gonorynchus. The observations of Grande and Grande (2008) differ from those of Gayet (1986, 1993a) in that, like Gonorynchus (Fig. 2.11A), the haemal arch of preural centrum 2 is fused to the centrum in †Ramallichthys, †Judeichthys and †Hakeliosomus (Fig. 2.11C, D, F). In addition, not only is hypural 2 fused with the centrum in Gonorynchus and the Middle Eastern taxa, so is hypural 1. Variation in hypural fusion was observed among †Hakeliosomus specimens examined (Fig. 2.11E, F), which might explain discrepancies in the literature (Grande and Grande 2008). Variation in hypural fusion is common among gonorynchiforms and will be addressed later.

Contrary to Grande (1996) and Grande and Poyato-Ariza (1999), better-preserved and prepared material of †Notogoneus osculus has revealed that the caudal skeleton is surprisingly more primitive than previously proposed. Because †Notogoneus was considered the sister group to Gonorynchus, and because of striking superficial resemblances between the two, it was assumed that the caudal skeletons of the two taxa would be similar. In †Notogoneus osculus (type species), however, all of the hypurals are autogenous, as well as the parhypural and the haemal arch on preural centrum 2 (Fig. 2.11B). Haemal arches anterior to preural centrum 2 also appear to be autogenous,

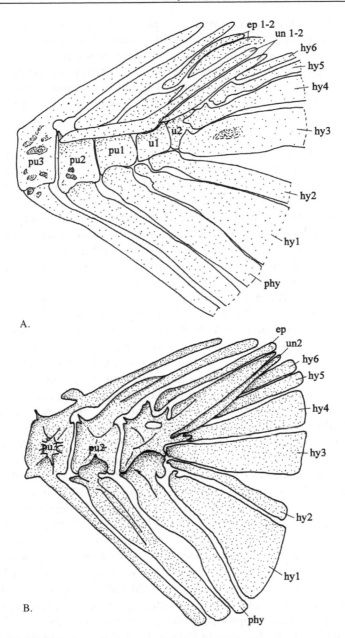

Fig. 2.10 (A) Caudal fin skeleton of †*Dastilbe elongatus* AMNH 12721. Redrawn from Poyato-Ariza (1996). (B) *Chanos chanos* CAS 05022-05075, SL = 103 mm. ep, epural; ep 1–2, epurals 1–2; hy 1–6, hypurals 1–6; phy, parhypural; pu1–3, preural centra 1–3; u1, ural centrum 1; u2, ural centrum 2 (diural terminology); un1–2, uroneurals 1–2.

although additional well-preserved specimens are needed to confirm this observation. Unlike in *Gonorynchus*, hypurals 1 and 2 are not fused distally. Thus, the only caudal fin character †*Notogoneus* shares with the rest of Gonorynchidae is a compound terminal centrum consisting of preural centrum 1, ural centra and uroneural 1, and one epural. The †*Notogoneus* caudal skeleton is similar to the Middle Eastern forms in that it has an autogenous second uroneural.

In *Gonorynchus* (Fig. 2.11A) the terminal centrum shows the greatest number of fusions (i.e., preural centrum 1, ural centra, parhypural, hypurals 1 to 2, uroneural 1 and possibly uroneural 2). Within *Gonorynchus*, variation in hypural fusion was observed (Grande 1999). Hypural 3 is fused to the terminal centrum in *G. abbreviatus* and *G. moseleyi*, but autogenous in *G. gonorynchus*, *G. greyi* and *G. forsteri*. This proposed synapomorphy and close geographic locality between *G. abbreviatus* and *G. moseleyi* (*G. abbreviatus* is found off the coast of Japan and Taiwan, while *G. moseleyi* is found throughout the waters of the Hawaiian Islands) has resulted in their alignment as sister species (Grande 1999).

Within the Kneriidae, considerable variation in caudal fin morphology is observable (Fig. 2.12A–I). As in *Gonorynchus*, the second uroneural is lost as an autogenous element. With the exception of *Kneria* and *Parakneria*, all genera have five hypurals, and the most common condition observed is for the hypurals to articulate, but not fuse with, the terminal centrum. In *Phractolaemus* (Fig. 2.12A), the haemal arch of the parhypural is autogenous. Hypural 1 is enlarged, and the articulating head of the bone is reduced in size or lost. Because of the size of this hypural, there is no space for its articulation with the terminal centrum, so there is a hiatus between the hypural and the compound centrum. The haemal arch on preural centrum 2 is fused in *Phractolaemus*, as well as all haemal arches anteriorly.

The caudal fins of *Cromeria* and *Grasseichthys* are considerably variable (Fig. 2.12B–E). According to Britz and Moritz (2007), the parhypural and hypural 2 are fused with the terminal centrum in *Cromeria nilotica* (BMNH 1949.7.22.1-7) and *C. occidentalis* (BMNH 2005.7.20.34). In *C. nilotica* (BMNH 1949.7.22.1-7), however, the proximal region of hypural 1 is connected to the base of hypural 2. In our specimens of *Cromeria occidentalis* (e.g., 141098-111A), the haemal arch of the parhypural is fused to the terminal centrum, but all hypurals are autogenous (Fig. 2.12B). All haemal arches from preural centrum 1 and anteriorly seem to be fused to their centra, although the spine of preural centrum 2 appears double. In specimen 141098-111B (Fig. 2.12C), hypurals 1 and 2 are fused to the terminal centrum. In no specimen examined was hypural 1 attached to hypural 2.

Variation in hypural fusion was observed in the *Grasseichthys* specimens examined also. In all specimens examined, the haemal arch of parhypural and hypural 1 are fused to the centrum. In one specimen (73-02-P-264-905A),

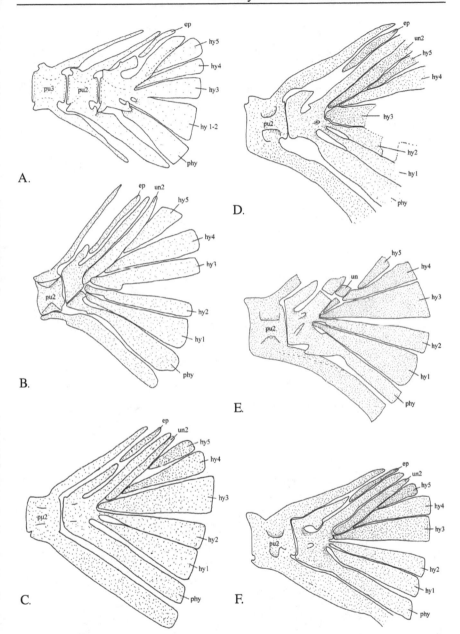

Fig. 2.11 Caudal fin skeleton of Gonorynchidae. (A) *Gonorynchus abbreviatus*, FMNH 96053, SL = 240 mm. (B) †*Notogoneus osculus*, FMNH PF 13043. (C) †*Judeichthys haasi*, AJ 432, SL = 110 mm. (D) †*Ramallichthys orientalis*, EY-25, SL = 129 mm. (E) †*Hakeliosomus hakelensis*, AMNH 5829, SL = 100 mm. (F) †*Hakeliosomus hakeliosomus* MNHN-HAK130d. Illustrations redrawn from Grande and Grande (2008). Anterior to the left. ep, epural; hy1–5, hypurals 1–5; phy, parhypural; pu2–3, preural centrum 2–3; un2, uroneural 2.

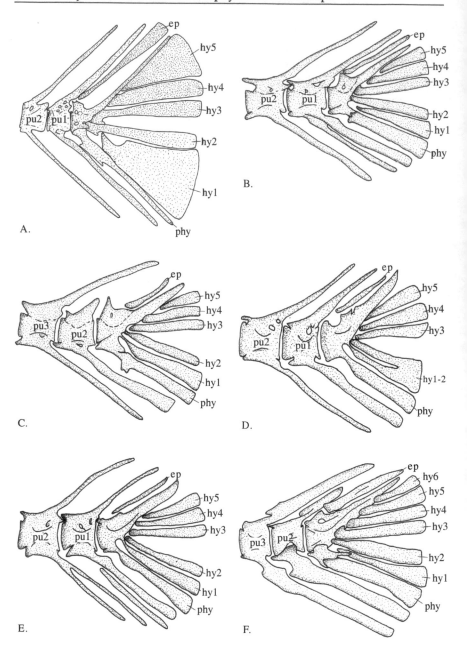

Fig. 2.12 contd. ...

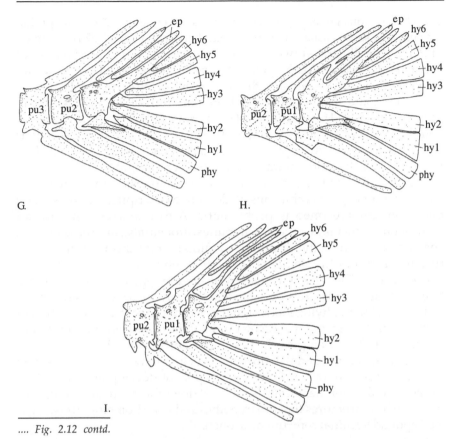

.... *Fig. 2.12 contd.*

Fig. 2.12 Caudal fin skeleton of Kneriidae. (A) *Phractolaemus ansorgei*, FMNH 63938A, SL = 123.9 mm. (B) *Cromeria occidentalis*, MRAC 14198-111A, SL = 25.6 mm. (C) *Cromeria occidentalis*, MRAC 141098-111B, SL = 23.2 mm. (D) *Grasseichthys gabonensis*, MRAC 73-02-P-264-905, SL =18.9. (E) *Grasseichthys gabonensis*, MRAC 73-02-P-264-905, SL = 19.5 mm. (F) *Parakneria tanzaniae*, BMNH 10.21:163-165, SL = 47.3. (G) *Parakneria malaissei* MRAC 164685, SL = 52.3. (H) *Kneria* sp., CAS 16150, SL = 50.5 mm. (I) *Kneria wittei*, MRAC 79-01-P-516-521, SL = 54.3 mm. ep, epural; hy1–6, hypurals 1–6; phy, parhypural; pu1, preural centrum 1; pu2, preural centrum 2.

hypural 2 is distally fused with hypural 1. In all specimens, the haemal arch of preural centrum 2 is fused to the centrum, and in specimen 73-02-P-264-905B, the haemal arch of preural centrum 2 is double. Britz and Moritz (2007) also found some variation in the *Grasseichthys* specimens that they examined. In four out of five specimens examined, the proximal region of hypural 1 is connected to the terminal centrum.

Lastly, unlike the other kneriids, *Parakneria* (Fig. 2.12F, G) and *Kneria* (Fig. 2.12H, I) have six hypurals. In all specimens of *Parakneria* examined, both the arch of the parhypural and the haemal arch of preural centrum 2 are autogenous and no hypural was observed fused to the terminal centrum.

Our observations generally agree with those of Lenglet (1974), except that we observed some variation in epural number. In Lenglet's (1974) specimen of *P. thysi*, he illustrated two epurals. Two epurals were observed in our specimen of *P. malaisse* (paratype), and only one was observed in *P. tanzaniae*. In our specimens of *Kneria wittei*, the haemal arch of the parhypural is always autogenous, and the haemal arch of preural centrum 2 is autogenous in some specimens and fused in others. Hypurals are not fused in our specimens and those described by Lenglet (1974). Like *Parakneria*, some specimens were observed with two epurals, while others had one. As described by Lenglet (1974), the more common condition is one epural.

As described in this section, variation in caudal skeleton structure is common among gonorychiforms. A definitive description of any species based on one specimen is problematic. A caudal skeleton with all autogenous elements (e.g., centra, hypurals, uroneurals), like those of fossil chanids, seems to be the primitive condition. From this primitive state, fusions of caudal fin elements seem to be both synapomorphies and convergences. The trick is to tease out the phylogenetically important fusions from the noise. It is possible that the only consistent fusion on the ordinal level with phylogenetic signal is a terminal centrum consisting of fused preural centrum 1, ural centrum 1 (of the diural terminology), and uroneural 1. That being said, it is important to note that the current interpretation of the caudal endoskeleton of gonorynchiforms is based on numerous assumptions without the support of developmental studies. While those developmental studies are not done, there is really no support for any of the structures that are hypothesized to be homologous, such as a compound terminal centrum or uroneural 1.

Pectoral Girdle and Fin

The pectoral girdle is illustrated in three representative gonorynchiforms, *Chanos*, *Gonorynchus*, and *Phractolaemus* (Figs. 2.13A–C, 2.14A–C). Each fin support is composed of the supracleithrum, cleithrum, scapula, coracoid, and medially the mesocoracoid. Above the cleithrum is the long supracleithrum, and from the supracleithrum the posttemporal attaches the pectoral girdle to the cranium. The cleithrum lies just posterior to the opercle and is a long thin bone that forms the backbone of the pectoral girdle. It broadens ventrally at about half its length and forms a shelf protruding medially. From the cleithrum the other bones of the girdle articulate. In *Chanos* and *Gonorynchus* (Figs. 2.13A–B, 2.14A–B), the anterior projection of the cleithrum is longer and thinner in comparison to the bone's wider posterior projection. In *Chanos*, an elongate flat bone is placed at the posteroventral corner of the cleithrum and lies on an elongated process of the scapula. This bone gives support to the elongate pectoral axillary process.

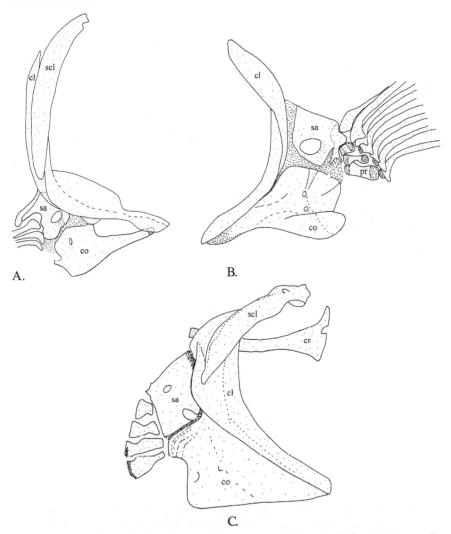

Fig. 2.13 Pectoral girdle, lateral view. (A) *Chanos chanos*, right side, CAS 05022-05075, SL = 103 mm. (B) *Gonorynchus forsteri*, left side, AMNH 57120, SL = 226 mm. (C) *Phractolaemus ansorgei*, right side, FMNH 63938A, SL = 123.9 mm. cl, cleithrum; co, coracoid; cr, cephalic rib; pr, proximal radial; sa, scapula; scl, supracleithrum.

In *Phractolaemus* (Figs. 2.13C, 2.14C), however, the anterior portion of the cleithrum is short and stout, corresponding to its small and robust head. According to Poyato-Ariza (1996), postcleithra are present in the fossil chanids †*Dastilbe*, †*Gordichthys*, †*Parachanos* and †*Rubiesichthys*. Blum (1991) reported postcleithra in †*Tharrhias*. There are no postcleithra (1–3) in *Chanos* or the Gonorynchoidei.

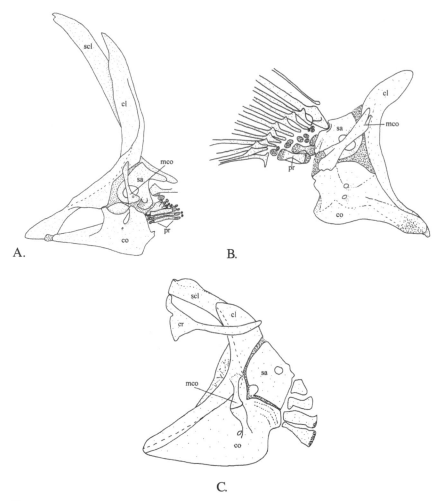

Fig. 2.14 Pectoral girdle, medial view. (A) *Chanos chanos*, right side, CAS 05022-05075, SL = 103 mm. (B) *Gonorynchus forsteri, left side,* AMNH 57120, SL = 226 mm. (C) *Phractolaemus ansorgei*, right side with cephalic rib, FMNH 63938A, SL = 123.9 mm. cl, cleithrum; co, coracoid; cr, cephalic rib; mco, mesocoracoid; pr, proximal radial; sa, scapula; scl, supracleithrum.

Articulating with the posteroventral margin of the cleithrum is the coracoid. The coracoid is a large flat bone, and in *Chanos* it loosely articulates with the cleithrum in two spots. A large foramen is found between the two articulation points. In *Gonorynchus* and *Phractolaemus,* the anteroventral extension of the coracoid firmly articulates with the cleithrum. As seen in lateral aspect, in all three taxa the ventral tip of the coracoid articulates medially to the cleithrum.

The scapula is found dorsal to the coracoid and posterior to the cleithrum. The large scapular foramen is positioned in the anteroventral portion of the scapula, surrounded by a ring of bone. In all three taxa, the scapula does not articulate directly with the coracoid or cleithrum but through cartilage. Three (*Gonorynchus*) to four (*Chanos, Phractolaemus*) proximal radials are placed posteriorly to the primary girdle. In *Chanos*, the proximal radials are positioned closer to the coracoid. In *Gonorynchus*, they are more dorsal and articulate with the scapula. In *Phractolaemus*, two radials extend from the scapula and two from the coracoid. The position of the proximal and distal radials seems to set the position of the fin proper. In *Chanos* and *Gonorynchus*, the pectoral fin shape is more tapered and streamlined. In *Phractolaemus*, the pectoral fin is fan-like (pectoral rays = 18, Table 2.1), and positioned more towards the center of the pectoral girdle.

In all species of *Gonorynchus*, the paired fins have a fleshy lobe positioned dorsally to the base of each rayed fin (Grande 1999). Within each fleshy lobe is a single unbranched and unsegmented ray. This ray is not connected to either the pectoral or the pelvic girdle. Its function is unknown except to support the fleshy lobe.

Final Remarks

The gonorynchiform postcranial skeleton is an amazingly variable yet strong phylogenetic signal in terms of anterior vertebral and caudal fin anatomy that binds these taxa into one monophyletic group. This paper serves to reanalyze the gonorynchiform postcranial skeleton and make comparisons not only between subgroups, but also between fossil and extant forms. New morphological information presented here has led to a reinterpretation of gonorycnhiform interrelationships. Although one might argue that at this point all is known about the gonorynchiform skeleton, this paper has in actuality led the authors down new paths of investigation and raised several new questions about gonorynchiform morphology, many of which might be best answered in light of a comprehensive developmental study (Arratia and Grande in prep.). For example, the association between the first vertebra and the back of the skull is quite interesting, yet still poorly understood. In most gonorynchiforms examined, the first neural arch is anteriorly expanded and articulates with the exoccipitals and/or the supraoccipital. But how and for what purpose did this anterior expansion develop? Neural arches and associated elements surround and protect the nervous system. Maybe the development and modification of the anterior neural arch in these fishes is associated with, or is a response to, developmental modifications of the anterior neural canal. Recent gonorynchoids share a loss of all autogenous uroneurals (Figs. 2.11A, 2.12A–I). Autogenous uoneurals are, however, present in *Chanos* (Fig. 2.10B) and all fossil forms (Figs. 2.10A, 2.11B–F). Because of the lack of ontogenetic information, it was assumed in Fink and Fink (1981) that these uroneurals were lost in the caudal fin skeleton instead of fused to the compound terminal centrum (S. Fink, pers. comm., 2008). The formation of the caudal skeleton among gonorynchiforms is thus not without questions and is an important avenue of future study. Gonorynchiforms hold a pivotal phylogenetic position among otocephalans or ostarioclupeomorphs. A better understanding of gonorynchiform developmental morphology is essential to a better understanding of the otophysan and possibly otocephalan morphology. We thus argue that this paper does not necessarily constitute the ultimate work on gonorynchiform postcranial morphology, but serves as a basis for new avenues of investigation. There is still much to be learned and more work to be done to better understand these amazing fishes.

Acknowledgements

The authors wish to thank Joseph Schlupe and Amanda Burdi (LU) for their help in preparing the figures for this manuscript. For permission and assistance in examining specimens we would like to thank Darrel Siebert and Peter Forey (NHM), Lance Grande, Mark Westneat and Mary Anne Rogers (FMNH), Melanie Stiassny and Barbara Brown (AMNH), Guy Teugels (MARC), Richard Vari (USNM), and Daniel Goujet (MNHN). This research was supported in part by National Science Foundation grants to TG (EF 0128794 and EF 0732589) and GA (EF 037870).

References

Arratia, G. 1991. The caudal skeleton of Jurassic teleosts; a phylogenetic analysis, pp. 249–340. *In*: M.-M. Chang, Y.-H. Liu and G.-R. Zhang [eds.]. Early Vertebrates and Related Problems in Evolutionary Biology. Science Press, Beijing.

Arratia, G. 2003. Catfish Head Skeleton—An Overview, pp. 3– 46. *In*: G. Arratia, B.G. Kapoor, M. Chardon and R. Diogo [eds.]. Catfishes. Scientific Publishers, Inc., Enfield, USA.

Arratia, G., M. Coburn and P. Mabee. 2007. Caudal skeleton of ostariophysans: issues of homology and new characters supporting the monophyly of cypriniforms. XII European Congress of Ichthyology: Dubrounik, Croatia: 256.

Arratia, G., H.-P. Schultze and J. Casciotta. 2001. Vertebral column and associated elements in Dipnoans and comparison with other fishes: Development and homology. J. Morphol 250: 101–172.

Bird, N. and P.M. Mabee. 2003. Development of the axial skeleton of the zebrafish, *Danio rerio* (Ostariophysi: Cyprinidae). Developmental Dynamics 228: 33– 357.

Blum, S. 1991. *Tharrhias* Jordan and Branner, 1908, pp. 286–296. *In*: J.G. Maisey [ed.] Santana Fossils, An Illustrated Atlas. TFH Publications, Neptune City, New Jersey.

Braun, C. and T. Grande. 2008. Evolution of peripheral mechanisms for the enhancement of sound reception, pp. 99–144. *In*: J. Webb, R. Fay and A. Popper [eds.]. Springer Handbook of Auditory Research, vol. 32, Fish Bioacoustics. Springer, New York.

Britz, R. and T. Moritz. 2007. Reinvestigation of the osteology of the miniature African freshwater fishes *Cromeria* and *Grasseichthys* (Teleostei, Gonorynchiformes, Kneriidae), with comments on kneriid relationships. Mitteilungen aus dem Museum für Naturkunde in Berlin, Zoologie Reihe 83(1): 3–42.

Chardon, M. and P. Vandewalle. 1997. Evolutionary trends and possible origin of the Weberian apparatus. Neth. J. Zool. 47(4): 383–403.

Fink, S.V. and W.L. Fink. 1981. Interrelationships of ostariophysan fishes (Teleostei). Zool. J. Linn. Soc. London 72(4): 297–353.

Fink, S.V. and W.L. Fink. 1996. Interrelationships of ostariophysan fishes (Teleostei), pp. 209–249. In: M.L.J. Stiassny, L.R. Parenti and G.D. Johnson [eds.]. Interrelationships of Fishes. Academic Press, San Diego.

Fink, S.V., P.H. Greenwood and W.L. Fink. 1984. A critique of recent work on fossil ostariophysan fishes. Copeia 1984(4): 1033–1041.

Gayet, M. 1982. Cypriniforme ou Gonorhynchiforme? Ramallichthys, nouveau genre du Cénomanien inférieur de Ramallah (Monts de Judée). C.R. Acad. Sci. 295(2): 405–407.

Gayet, M. 1985. Gonorhynchiform nouveau du Cénomanien inférieur marin de Ramallah (Monts de Judée): Judeichthys haasi nov. gen. nov. sp. (Teleostei, Ostariophysi, Judeichthyidae nov. fam.). Bull. Mus. natl. Hist. nat. Paris, sér. 4 7C(1): 65–85.

Gayet, M. 1986a. Ramallichthys Gayet du Cénomanien inférieur marin de Ramallah (Judée) une introduction aux relations phylogénétiques des Ostariophysi. Mém. Mus. natl. Hist. nat. Paris, n.s. C, Sciences de la Terre 51: 1–81.

Gayet, M. 1986b. About ostariophysan fishes: a reply to S.V. Fink, P.H. Greenwood and W.L. Fink's criticisms. Bull. Mus. natl. Hist. nat. Paris, sér. 4, C(3): 393–409.

Gayet, M. 1993a. Gonorhynchoidei du Crétacé Supérieur marin du Liban et relations phylogénétiques des Charitosomidae nov. fam. Documents des laboratoires de Géologie, 126: 131 pp., Lyon, France.

Gayet, M. 1993b. Nouveau genre de Gonorhynchidae du Cénomanien inférieur marin de Hakel (Liban). Implications phylogénétiques. C.R. Acad. Sci. de Paris, série II, 432: 57–163.

Gayet, M. and M. Chardon. 1987. Possible otophysic connections in some fossil and living ostariophysan fishes. Proceedings of the European Ichthyological. Congress. Stockholm 1985: 31–42.

Grande, L. and W. Bemis. 1998. A comprehensive phylogenetic study of amid fishes (Amiidae) based on comparative skeletal anatomy. An empirical search for interconnected patterns of natural history. Society of Vertebrate Paleontology, Memoir 4, supplement to Journal of Vertebrate Paleontology 18: 1–690.

Grande, L. and T. Grande 1999. A new species of Notogoneus (Teleostei: Gonorynchidae) from the Upper Cretaceous Two Medicine Formation of Montana, and the poor Cretaceous Record of freshwater fishes from North America. J. Vert. Paleontol. 19(4): 612–622.

Grande, L. and T. Grande. 2008. Redescription of the type species for the genus †Notogoneus (Teleostei: Gonorynchidae) based on new, well-preserved material. J. Paleontol. 82(5): 1–31.

Grande, T. 1994. Phylogeny and paedomorphosis in an African family of freshwater fishes (Gonorynchiformes: Kneriidae). Fieldiana 78: 1–20.

Grande, T. 1996. The interrelationships of fossil and Recent gonorynchid fishes with comments on two Cretaceous taxa from Israel, pp. 299–318. In: G. Arratia and G. Viohl [eds.]. Mesozoic fishes—Systematics and Paleoecology. Verlag Dr. Friedrich Pfeil, Munich.

Grande, T. 1999. Revision of the genus Gonorynchus Scopoli, 1777 (Teleostei, Ostariophysi). Copeia 1999(2): 453–469.

Grande, T. and M. de Pinna. 2004. The evolution of the Weberian apparatus: A phylogenetic perspective, pp. 429–448. In: G. Arratia and A.Tintori. [eds.]. Mesozoic Fishes 3—Systematics, Paleoenvironments and Biodiversity. Verlag Dr. Friedrich Pfeil, Munich.

Grande, T. and L. Grande. 2008. Revaluation of the gonorynchiform genera †Rammallichthys, †Judeichthys and †Notogoneus, with comments on the families

†Charitosomidae and Gonorynchidae. *In*: G. Arratia, H.-P. Schultze and M.V.H. Wilson [eds.]. Mesozoic Fishes 4—Homology and Phylogeny. Verlag Dr. Friedrich Pfeil, Munich.

Grande, T. and F.J. Poyato-Ariza. 1999. Phylogenetic relationships of fossil and Recent gonorynchiform fishes (Teleostei: Ostariophysi). Zool. J. Linn. Soc. 125: 197–238.

Grande, T. and B. Young. 1997. Morphological development of the opercular apparatus in *Kneria wittei* (Ostariophysi: Gonorynchiformes) with comments on its possible function. Acta Zoologica 78(2): 145–162.

Greenwood, P.H., D.E. Rosen, S.H. Weitzman and G.S. Myers. 1966. Phyletic studies of teleostean fishes, with a provisional classification of living forms. Bull. Am. Mus. Nat. Hist. 131(4): 339–456.

Hoffmann, M. and R. Britz. 2006. Ontogeny and homology of the neural complex of otophysan Ostariophysi. Zool. J. Linn. Soc. 147: 301–330.

Lecointre, G. and G. Nelson. 1996. Clupeomorpha, sister-group of Ostariophysi, pp. 193–207. *In*: M.L.J. Stiassny, L.R. Parenti and G.D. Johnson [eds.]. Interrelationships of Fishes. Academic Press, San Diego.

Lenglet, G. 1974. Contribution à l'étude ostéologique des Kneriidae. Ann. Soc. Roy. Zool. Belg. 104: 51–103.

Liem, K.F., W.E. Bemis, W.F. Walker and L. Grande. 2001. Functional Anatomy of the Vertebrates. An Evolutionary Perspective (3rd ed.). Harcourt College Publishers, New York.

Mabee, P.M., P.L. Crotwell, N.C. Bird and A.C. Burke. 2002. Evolution of median fin modules in the axial skeleton of fishes. J. Exp. Zool. 294: 77–90.

Monod, T. 1963. Sur quelques points de l'anatomie de *Gonorhynchus gonorhynchus* (Linné 1766). Mém. Inst. Fr. Afr. Noire, mélanges ichthyologiques 66: 255–313.

Morin-Kensick, E.M., E. Melancon and J.S. Eisen. 2002. Segmental relationship between somites and vertebral column in zebrafish. Development 129: 3851–3860.

Moritz, T., R. Britz and K.E. Linsenmair. 2006. *Cromeria nilotica* and *C. occidentalis*, two valid species of the African freshwater fish family Kneriidae (Teleostei: Gonorynchiformes). Ichthyol. Explor. Freshwaters 17(1): 65–72.

Patterson, C. 1984. Family Chanidae and other teleostean fishes as living fossils, pp. 132–139. *In*: N. Eldredge and S.M. Stanley, [eds.]. Living Fossils, Springer Verlag, Berlin.

Patterson, C. and G.D. Johnson. 1995. The intermuscular bones and ligaments of teleostean fishes. Smithsonian Contributions, Zoology 559: 1–85.

Perkins, P.L. 1970. *Notogoneus osculus* Cope, an Eocene fish from Wyoming (Gonorynchiformes, Gonorynchidae). Postilla 147: 1–18.

Peters, N. 1967. Opercular- und Postopercularorgan (Occipitalorgan) der Gattung *Kneria* (Kneriidae, Pisces) und ein Vergleich mit verwandten Strukturen. Zeitschrift für Morphologie und Ökologie der Tiere 59: 381–435.

Poyato-Ariza, F.J. 1994. A new Early Cretaceous gonorynchiform fish (Teleostei: Ostariophysi) from Las Hoyas (Cuenca, Spain). Occasional papers of the Museum of Natural History, The University of Kansas, Lawrence (164): 1–37.

Poyato-Ariza, F.J. 1996. A revision of the ostariophysan fish family Chanidae, with special reference to the Mesozoic forms. Paleo Ichthyologica 6: 1–52.

Rosen, D.E. and P.H. Greenwood. 1970. Origin of the Weberian apparatus and the relationships of the ostariophysan and gonorynchiform fishes. Am. Mus. Novitates 2428: 1–25.

Schultze, H.-P. and G. Arratia. 1988. Reevaluation of the caudal skeleton of some actinopterygian fishes. II. *Hiodon, Elops* and *Albula*. J. Morphol. 195: 257–303.

Schultze, H.-P. and G. Arratia. 1989. The composition of the caudal skeleton of teleosts (Actinopterygii: Osteichthyes). Zool. J. Linn. Soc. 97: 189–231.

Thys van den Audenaerde, D.F.E. 1961. L'anatomie de *Phractolaemus ansorgei* Blgr. et la position systématique des Phractolæmidae. Ann. Mus. Roy. Afr. Centr., Sci. Zool., sér. 8 103: 101–167.

Appendix

Institutional abbreviations used in text.

AJ or EY, Hebrew University, Jerusalem; AMNH, American Museum of Natural History; AMS, Australian Museum, Sydney; BMNH, British Museum of Natural History=The Natural History Museum, London; CAS, California Academy of Sciences, San Francisco; FMNH, Field Museum of Natural History, Chicago; MNHN, Muséum national d'Histoire naturelle, Paris; MRAC, Museum Royal d'Afrique Centrale, Tervuren.

3

Early Ossification and Development of the Cranium and Paired Girdles of *Chanos chanos* (Teleostei, Gonorynchiformes)

Gloria Arratia[1,*] and Teodora Bagarinao[2]

Introduction

Numerous studies have been dedicated to larval development, growth, and larval behavior of commercially important fishes (some of the results can be found in Moser 1984) such as *Chanos chanos* (e.g., Delsman 1926, 1929, Rabanal *et al.* 1953, Liao *et al.* 1977, Chaudhuri *et al.* 1978, Villaluz 1979, Buri 1980, Buri *et al.* 1981, Villaluz and Unggui 1983, Kamawura 1984, Kumagai *et al.* 1985, Bagarinao 1986, 1994, Juliano and Hirano 1986, Bagarinao and Kumagai 1987). In general, many authors have investigated the embryonic development and larval developmental osteology of numerous teleosts in approximately the last 150 years (e.g., Parker 1873, Gegenbauer 1878, Stöhr 1882, Tischomiroff 1885, Gaupp 1903). Some authors described the development of the chondrocranium (e.g., Wells 1923, Bhargava 1958, Bertmar 1959, Ristovska *et al.* 2006, Adrians and Verraes 1997a, b), others described only the osteocranium (e.g., Morris and Gaudin 1982, Jollie 1984, Pottoff *et al.* 1988,

[1] Biodiversity Research Center, University of Kansas, Dyche Hall, Lawrence, Kansas 66045-7561, U.S.A. and Department of Geology, Field Museum of Natural History, Chicago, Illinois, U.S.A. e-mail: garratia@ku.edu
[2] Southeast Asian Fisheries Development Center (SEAFDEC), Aquaculture Department, 5021 Iloilo, Philippines.
* author for correspondence.

Adriaens and Verraes 1998), and others described a sequential combination of chondrification and ossification of skull elements (e.g., Langille and Hall 1987, Vandewalle *et al.* 1992, Mabee and Trendler 1996, Cubbage and Mabee 1996, Faustino and Power 2001). Other papers have been dedicated to the development of certain specialized areas of the skull, such as the urohyal (Arratia and Schultze 1990), the palatoquadrate (Arratia and Schultze 1991), the ceratohyal (Verraes 1974), the suspensorium of ostariophysans, especially catfishes (Arratia 1990, 1992), the posterior end of the lower jaw (e.g., Haines 1937, Francillon 1974), skull canal bones (e.g., Lekander 1949, Kapoor 1970, Adriaens *et al.* 1997), and skull chondral bones (e.g., Bertmar 1959).

Among head bones, the development of bones associated with feeding and respiration has attracted the attention of a few authors searching for a relationship between development and functional demands (e.g., Vandewalle *et al.* 1992, Taki *et al.* 1987, Kohno *et al.* 1996a, b). Only a few studies concerning the developmental osteology of *Chanos chanos* are available. For instance, research has been done on the elements of the palatoquadrate and hyosymplectic (e.g., Arratia 1992), certain cranial elements comprising the bucco-pharyngeal cavity, especially the branchial arches (Taki *et al.* 1987, Kohno *et al.* 1996a), a few elements of the paired girdles, for example, and the fin rays and their supports (Taki *et al.* 1986, 1987, Kohno *et al.* 1996a), and the anterior vertebrae (Coburn and Chai 2003).

In this chapter, we provide new data on the timing of ossification of cranial and paired girdle elements, and compare our results, when it is possible, with previous work on early ossification development of *Chanos chanos*. This is particularly important because of the basal phylogenetic position of *Chanos* among living ostariophysans and among extant gonorynchiforms as well (see Fig. 3.1). We describe the normal cranial and girdle patterns of ossification in *Chanos chanos* and evaluate to what extent the cranial development is consistent, or whether some intraspecific differences occur in comparison to previous results by Taki *et al.* (1987) and Kohno *et al.* (1996a) based on ontogenetic series grown in the same Aquaculture Department as the specimens used in this study. We assess previously published developmental work in a few ostariophysans and test some previous hypotheses on heretochrony and patterns of diversification. [A study of chondrification versus ossification processes in *Chanos chanos* is outside the scope of this paper, but it is the subject of a separate paper (Arratia and Grande, in preparation).]

Material and Methods

The specimens used in this study originated from naturally spawned eggs, hatched in October 2004 and reared in facilities of the Southeast Asian Fisheries Development Center (SEAFDEC), Aquaculture Department, Iloilo, Philippines. The water temperature ranged between 26 and 30°C. Samples (10 to 20 larvae)

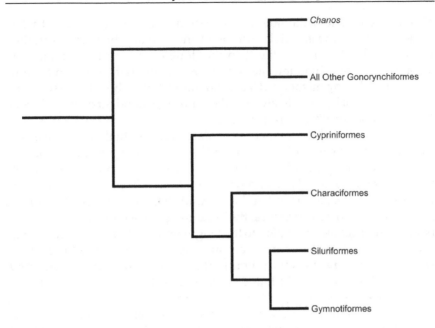

Fig. 3.1 Hypothesis of phylogenetic relationships of extant ostariophysans showing the basal position of *Chanos* (slightly modified from Fink and Fink, 1981).

were taken every day from day 1 to day 45, and then irregularly from day 46 to day 65, with sizes up to 80 mm standard length (SL). Samples were preserved in 70% ethanol. All these specimens are catalogued at the collection of the Division of Ichthyology of the Natural History Museum, the University of Kansas, Lawrence, Kansas (KUNHM 39848 to 39889, 39890, 39891 to 39894). The specimens were measured under a Leica MZ9 stereomicroscope with ocular micrometer and measurements were double-checked with a precision caliper reading to the nearest 0.1. Two hundred and seven specimens were cleared. Most cleared specimens were double-stained for both cartilage and bone following a technique described in Arratia and Schultze (1992: 190) and the rest were stained only for bone. In addition, seven young cleared and stained specimens belonging to the Scripps Institution of Oceanography, La Jolla, California (SIO 80–199), and one specimen belonging to the Museum of Zoology, University of Michigan, Ann Arbor, Michigan (UMMZ 196864), were studied.

Small larvae were studied under an Olympus microscope with normal optic, face contrast and polarized light with a Nikon digital camera attachment. Larger specimens were studied under a Leica MZ9 stereomicroscope with a Leica digital camera attachment.

The absence and beginning of ossification of each skeletal element in the head and paired fins was recorded (bilaterally for paired elements) for all specimens. The beginning of ossification was scored as present when alizarin red stain was visible. Although the size at which an ossification begins can be detected earlier than the double staining by staining with alizarin red alone (Vandewalle *et al.* 1998, Arratia pers. obs.) and can be detected earlier by histology (Clark and Smith 1993), the relative sequence of ossification of bones is likely to be the same, independent of the technique (Clark and Smith 1993, Arratia pers. obs.).

Prenotochordal flexion and early flexion individuals were measured from the anterior end of the head to the posterior tip of the notochord (notochordal length, NL; Ahlstrom *et al.* 1976, Leis and Trnski 1989). Specimens with a flexed notochord were measured from the anterior end of the head to the posterior end of the hypurals (standard length, SL). In this study, all references concerning the beginning of ossification of particular bones are expressed in relation to the length of the fish. In contrast, in some previous studies of *Chanos* and a few other fishes, all references to time were standardized and expressed as hours after initial mouth opening (HAMO) (e.g., for *Chanos chanos*, 54 h after hatching, Kohno *et al.* 1996a; for *Lates calcifer*, 40 h after hatching, Kohno *et al.* 1996b). The opening of the mouth at 54 h = 5.44±0.14 mm NL = about 2.2 days after hatching (Kohno *et al.* 1996a). According to Bagarinao (1994) and Kohno *et al.* (1996a), the initial mouth opening in *Chanos chanos* occurs about 54 to 55 h after hatching.

Bones are classified as dermal, chondral and membrane bones following Patterson (1977). Bones formed in tendons are identified here as tendon bones (e.g., urohyal and sesamoid coronoid or coronomeckelian at the medial side of the lower jaw) and not as dermal bones as in Mabee and Trendler (1996), Cubbage and Mabee (1996), and others, because their origin is different. A list of all cartilage, dermal, membrane and tendon bones of *Chanos chanos* is shown in Table 3.1. To facilitate comparisons, data from two other ostariophysans are added, the zebra fish *Danio rerio* and the barbel *Barbus barbus*.

Terminology

Because of contradictory opinions, some of the terminology used is explained below.

The notochord is not considered as a bone herein, in contrast to Cubbage and Mabee (1996), who interpreted it as a bone resulting from a perichordal ossification. The anterior cranial end of the notochord mineralizes, and its mineralization is a different process than an ossification. In addition, the formation of the chordacentra is a different process than the formation of the bony autocentra (Arratia *et al.* 2001).

There is a conflict over names between traditional terminology and terminology based on homology. For example, there is some confusion over the frontal and parietal bones of the skull roof. The frontal of traditional

Table 3.1 Types of bones present in nine cranial regions and paired girdles and fins of adult *Chanos chanos*. Bones are arranged alphabetically within regions. Ch, chondral bone; Dm, dermal bone; iorb, infraorbital sensory canal; sorb.c, supraorbital sensory canal; Tb, tendon bone; *, bone of compound origin. Cartilages are excluded. Bones that are absent in *Chanos*, but present in other teleosts are indicated in bold type.

Region and bone	Type of bones	Bone bearing sensory canal	Number of bones left/right
Olfactory			
Latheral ethmoid	Ch	—	1/1
Mesethmoid (= ethmoid)	Ch	—	1
Nasal	Dm	sorb.c	1/1
Vomer	Dm	—	1
Orbital			
Antorbital	Dm	—	1/1
Basisphenoid	**Absent**		
Dermosphenotic	Dm	iorb.c	1/1
Frontal (= parietal) bone	Dm	sorb.c	1/1
Infraorbital	Dm	iorb.c	4/4
Orbitosphenoid	**Absent**		
Pterosphenoid	Ch	—	1/1
Parasphenoid	Dm	—	1
Sclerotic, anterior	**Absent**		
Sclerotic, posterior	**Absent**		
Supraorbital	Dm	—	1/1
Otic			
Autosphenotic	Ch	—	1/1
Epiotic (epioccipital)	Ch	—	1/1
Extrascapular (= supratemporal)	Dm	Exc.com	2/2
Intercalar	Mb	—	1/1
Parietal (= postparietal) bone	Dm	—	1/1
Posttemporal	Dm	ll.c	1/1
Prootic	Ch	—	1/1
Pterotic*	Ch	ot.c	1/1
Occipital			
Basioccipital	Ch	—	1
Exoccipital	Ch	—	1/1
Supraoccipital	Ch	—	1
Mandibular arch			
Anguloarticular*	Dm/Ch	ma.c	1/1
Coronomeckelian	Tb	—	1/1
Dentary	Dm	ma.c	1/1
Maxilla	Dm	—	1/1
Premaxilla	Dm	—	1/1
Retroarticular	Ch	—	1/1
Palatoquadrate arch			
Autopalatine	Ch	—	1/1
Dermopalatine	**Absent**		
Ectopterygoid	Dm	—	1/1
Entopterygoid	Dm	—	1/1
Metapterygoid	Ch	—	1/1
Quadrate	Ch	—	1/1

Table 6.1 contd....

... *Table 6.1 contd.*

Region and bone	Type of bones	Bone bearing sensory canal	Number of bones left/right
Hyoid arch			
Basihyal	Ch	—	1
Branchiostegal ray	Dm	—	4/4
Ceratohyal, anterior	Ch	—	1/1
Ceratohyal, posterior	Ch	—	1/1
Hyomandibula	Ch	—	1/1
Hypohyal, dorsal	Ch	—	1/1
Hypohyal, ventral	Ch	—	1/1
Interhyal	**Absent**		
Symplectic	Ch	—	1/1
Urohyal	Tb	—	1
Branchial arches			
Basibranchial	Ch	—	3
Ceratobranchial	Ch	—	5/5
Epibranchial	Ch	—	5/5
Hypobranchial (1 to 3)	Ch	—	3/3
Pharyngobranchial (1 to 3)	Ch	—	2/2
Dermal tooth plates	**Absent**		
Opercular			
Interopercle	Dm	—	1/1
Opercle	Dm	—	1/1
Preopercle	Dm	pre.c	1/1
Subopercle	Dm	—	1/1
Suprapreopercle	Dm	pre.c	1/1
Pectoral girdle			
Cleithrum	Dm	—	1/1
Coracoid	Ch	—	1/1
Mesocoracoid	Ch	—	1/1
Pectoral axillary process*			
Postcleithrum (1 to 3)	**Absent**		
Propterygium	Ch	—	1/1
Proximal radial	Ch	—	4/4
Rays	Dm	—	15–16/15–16
Scapula	Ch	—	1/1
Supracleithrum	Dm	ll.c	1/1
Pelvic girdle			
Basipterygium	Ch	—	1/1
Pelvic axillary process*			
Pelvic splint	Dm	—	1/1
Radials	Ch	—	4/4
Rays	Dm	—	12–13/12–13

terminology corresponds to the parietal bone, while the parietal of traditional terminology corresponds to the postparietal bone (e.g., Jollie 1962; see Schultze 2008 for literature concerning this subject; Wiley 2008). We keep the traditional terminology because it is followed in this book, but at each reference we add the corresponding name based on homologization of structures.

One synapomorphy of clupeocephalans is the presence of an anguloarticular bone in the lower jaw of large individuals. However, as the present study confirms (see below), the bone is the result of a fusion between the dermal angular and the chondral articular. Both bones arise from two ossification centers that ossify at different times; consequently, both bones are treated separately herein. This observation is contrary to Cubbage and Mabee (1996: 143, figs. 1–3, table 1), who incorrectly interpreted the anguloarticular as a dermal bone (p. 143), apparently arisen from only one ossification center.

According to the available information, the first element of the infrapharyngobranchial series in *Chanos chanos* has been traditionally named pharyngobranchial 1 (= infrapharyngobranchial 1), following Nelson (1969) and Patterson and Johnson (1997). We follow this terminology here, but new evidence concerning the development and relationships of the elements named pharyngobranchials will be presented elsewhere.

To understand part of the section on comparisons, a few names that are frequently used in milkfish aquaculture need to be explained here. For instance, milkfish "fry" are large average larvae (13 to 14 mm total length, about 20 days old and without yolk) approaching metamorphosis and the end of the pelagic interval. Following metamorphosis, the larvae become juveniles (Bagarinao 1994: 33–34). "Juveniles" are milkfish larger than 17 mm SL that have the characteristic shape and structures of the adult (Taki *et al.* 1987, Bagarinao 1994).

Results

Observation of Certain Major Events in Early Ontogeny

There are some major events in the development of a fish, such as the opening of the mouth (see Material and Methods), notochordal flexion, and changes in the notochord sheaths, that are specifically mentioned here because only a few authors have referred their results in connection to the mouth opening and notochordal changes of the fishes.

Appearance of Cartilaginous Gill Rakers

The first one to three cartilaginous gill rakers appear on the anterior margin or above the first cartilaginous ceratobranchial in our specimens of 7.3 mm SL (day 25); then they begin to develop on the anterior margin or above the cartilaginous ceratobranchial 2 (7.8 mm SL), and a few days later, they become visible on cartilaginous ceratobranchials 3 and 4 (8 to 9 mm SL; day 28). A few cartilaginous gill rakers are present on the anterior margin of cartilaginous ceratobranchial 5 at about 10 mm SL. This is a continuous process for which it is difficult to set limits, because while the gill rakers are appearing on cartilaginous ceratobranchials 3 and 4, gill rakers are slightly increasing in number and length on cartilaginous ceratobranchials 1 and 2. The appearance of cartilaginous gill rakers is from cartilaginous ceratobranchial 1 to the next ceratobranchial and so on. However, the quantity of gill rakers is independent of the sequence of ceratobranchials, because in a short time the number of gill rakers on ceratobranchials 3 and 4 becomes larger than those on the other ceratobranchials.

Gill rakers were first observed at 10.0 mm SL in *Chanos chanos* studied by Taki *et al.* (1987) and grown at 25.5 to 30°C at the Igang Substation of the SEAFDEC Aquaculture Department, Iloilo, Philippines in 1983. Two cartilaginous gill rakers were recorded on ceratobranchial 1, three on ceratobranchial 2, and another three on ceratobranchial 3.

Notochordal Flexion

The notochordal flexion in *Chanos chanos* begins in our specimens at about 6 mm NL, but the appearance of the flexion varies from about 6 to 10 mm NL (= days 7 to 10 after hatching). The chondrification of elements of the caudal skeleton surrounding the caudal tip of the notochord varies intraspecifically. For instance, the notochordal flexion begins when: (1) hypurals 1, 2 and 3 are chondrified and the chondrification of the parhypural is beginning at its hemal arch; (2) parhypural and hypurals 1, 2 and 3 are chondrified; (3) hypurals 1, 2 and 3 are chondrified, but the chondrifications of the parhypural and hypural 4 are beginning to appear. According to our results, the beginning of the notochordal flexion in *Chanos chanos* occurs slightly later than the chondrification of hypurals 1, 2 and 3.

Information concerning the timing of the notochordal flexion of *Chanos chanos* was not reported by Kohno *et al.* (1996a), although the authors described the early chondrification and ossification of hypurals and the parhypural. According to Taki *et al.* (1987), the initial phase—which corresponds to the preflexion larva of Kendall *et al.* (1984)—is up to 6.5 mm SL; consequently, the notochordal flexion would be about 6.6 mm SL. Bagarinao (1994: 33) described notochordal postflexion larvae as larvae of 10 mm NL and larger. Thus, the notochordal flexion occurs in larvae under 10 mm NL, similar to the results obtained here. Notochordal flexion occurred at about 5.5 mm NL (= 9 to 10 days after hatching) in the cypriniform *Danio rerio* as reported by Cubbage and Mabee (1996).

Notochordal Changes

The notochord, which in small specimens shows a smooth appearance, begins to show changes in its density, so that a regular pattern (Fig. 3.2) of alternate dense (darker) versus non-dense (paler) regions is observed in specimens of about 7 to 13 mm NL. The notochord begins to constrict at the level of the dense regions (Fig. 3.3; about 7 to 11 mm SL), setting the path where an intervertebral region will be formed (Fig. 3.3), in turn setting the region where a chordacentrum will be formed in the middle sheath of the

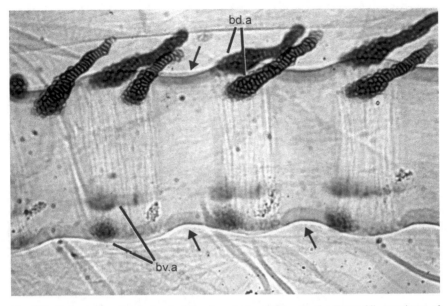

Fig. 3.2 Abdominal region in front of the dorsal pterygiophores of a 7 mm NL specimen of *Chanos chanos* (KU 39857). Early notochordal changes and setting of the pathway for the vertebral centra formation. Note the modifications of the notochord where chordacentra (framed by basidorsal and basiventral arcualia) and intervertebral spaces (indicated by arrows) will form. bd.a, basidorsal arcualia; bv.a, basiventral arcualia.

notochord in *Chanos chanos*. The changes of the notochord observed in *Chanos chanos* confirm previous observations in other fishes that the notochord sets the path where the vertebral centra will develop (see Arratia 1991, 2003, Schultze and Arratia 1988, Arratia and Schultze 1992, Arratia *et al.* 2001).

The anterior cranial end of the notochord begins to mineralize in specimens of about 7.6 mm SL, almost at the same time that the rest of the notochord shows the changes described above.

Timing and Sequence of Bone Ossification

There are 52 different bones present in the cranium of large individuals of *Chanos chanos* (Table 3.1). However, when all bones, including left and right sides of the head, are included, the total number is 134. From these, 51% are chondral and 77% are dermal bones. The remaining 6% correspond to compound (e.g., anguloarticular, pterotic), membrane, and tendon bones. Seven bones form each pectoral girdle, two of them dermal in origin (supracleithrum and cleithrum), and five are chondral (e.g., scapula, proximal radials) (see Table 3.1). If all left and right bones are counted, the pectoral girdle has 20 bones (the count does not include the cartilaginous distal and proximal radials). Only two chondral bones form each pelvic

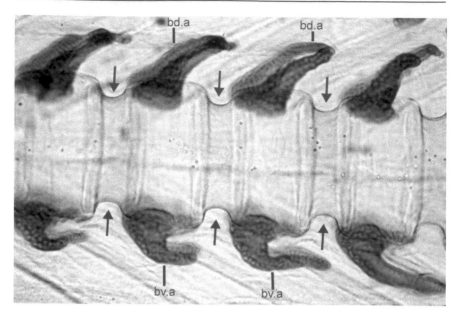

Fig. 3.3 Abdominal region in front of the dorsal pterygiophores of a 7.8 mm SL specimen of *Chanos chanos* (KU 39875). Notochordal changes and setting of the pathway for the vertebral centra formation. Note the obliterations of the notochord enclosing the regions where chordacentra (framed by basidorsal and basiventral arcualia) will form and constrictions where intervertebral spaces (indicated by arrows) will form. bd.a, basidorsal arcualia; bv.a, basiventral arcualia.

girdle: the basipterygium and radials (Table 3.1). If all left and right bones are counted, the pelvic girdle has 10 bones.

The sequence of ossification of the cranial bones of *Chanos chanos* is represented in Fig. 3.4, and the same sequence is represented in bar graph form in Fig. 3.5 for easy visualizing, following Taki *et al.* (1987 for *C. chanos*), Kohno *et al.* (1996a for *C. chanos*, and 1996b for *Lates calcifer*), and Cubbage and Mabee (1996 for *Danio rerio*). There is individual variation between the ossification of bilateral bones; usually one side is complete before the other begins. Since this is a common pattern, the beginning of ossification of any paired bone is recorded as present when one side begins to ossify.

Four bones begin to ossify simultaneously in specimens of 10.3 mm SL, two from the cranium and two from the pectoral girdle. Three of these bones are dermal (**cleithrum, opercle,** and **supracleithrum**) and one is chondral (**hyomandibula**). The opercle starts at the region of the articulatory facet for the hyomandibula, while the hyomandibula ossifies perichondrally at the posterodorsal process bearing the condyle to articulate with the facet of the opercle. From the hyomandibula-opercle articulation, the ossification continues first at the dorsal region of both bones, and then extends ventrally. The opercle and hyomandibula are almost completely ossified at 12.9 mm

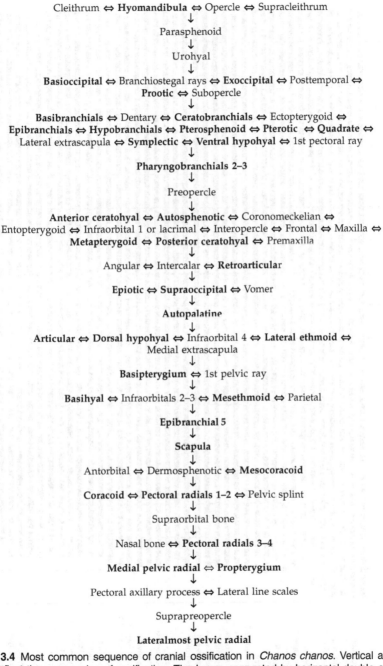

Cleithrum ⇔ **Hyomandibula** ⇔ Opercle ⇔ Supracleithrum
↓
Parasphenoid
↓
Urohyal
↓
Basioccipital ⇔ Branchiostegal rays ⇔ **Exoccipital** ⇔ Posttemporal ⇔
Prootic ⇔ Subopercle
↓
Basibranchials ⇔ Dentary ⇔ **Ceratobranchials** ⇔ Ectopterygoid ⇔
Epibranchials ⇔ **Hypobranchials** ⇔ Pterosphenoid ⇔ **Pterotic** ⇔ **Quadrate** ⇔
Lateral extrascapula ⇔ **Symplectic** ⇔ **Ventral hypohyal** ⇔ 1st pectoral ray
↓
Pharyngobranchials 2–3
↓
Preopercle
↓
Anterior ceratohyal ⇔ **Autosphenotic** ⇔ **Coronomeckelian** ⇔
Entopterygoid ⇔ Infraorbital 1 or lacrimal ⇔ Interopercle ⇔ Frontal ⇔ Maxilla ⇔
Metapterygoid ⇔ **Posterior ceratohyal** ⇔ Premaxilla
↓
Angular ⇔ Intercalar ⇔ **Retroarticular**
↓
Epiotic ⇔ **Supraoccipital** ⇔ Vomer
↓
Autopalatine
↓
Articular ⇔ **Dorsal hypohyal** ⇔ Infraorbital 4 ⇔ **Lateral ethmoid** ⇔
Medial extrascapula
↓
Basipterygium ⇔ 1st pelvic ray
↓
Basihyal ⇔ Infraorbitals 2–3 ⇔ **Mesethmoid** ⇔ Parietal
↓
Epibranchial 5
↓
Scapula
↓
Antorbital ⇔ Dermosphenotic ⇔ **Mesocoracoid**
↓
Coracoid ⇔ **Pectoral radials 1–2** ⇔ Pelvic splint
↓
Supraorbital bone
↓
Nasal bone ⇔ **Pectoral radials 3–4**
↓
Medial pelvic radial ⇔ **Propterygium**
↓
Pectoral axillary process ⇔ Lateral line scales
↓
Suprapreopercle
↓
Lateralmost pelvic radial

Fig. 3.4 Most common sequence of cranial ossification in *Chanos chanos*. Vertical arrows (↓) reflect the progression of ossification. The bones connected by horizontal double arrows (⇔) develop simultaneously or one may develop slightly before the other and they are listed alphabetically in the different ossification levels. Chondral bones are represented in bold type.

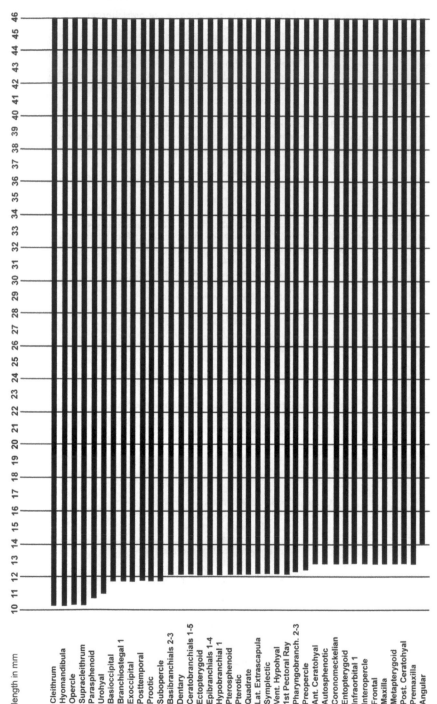

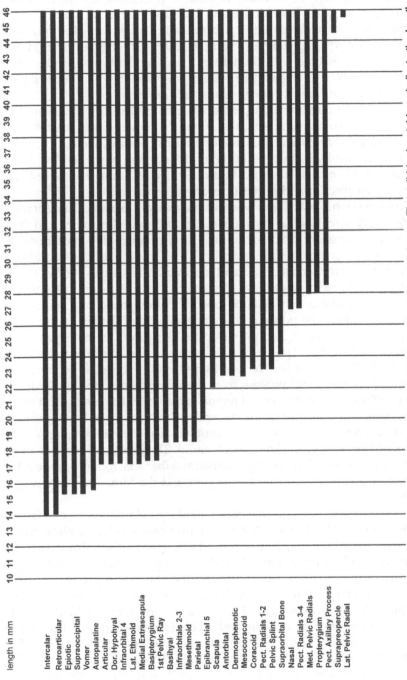

Fig. 3.5 Ossification sequence of the entire cranial and paired girdles bones of *Chanos chanos*. The solid horizontal bars indicate the length at which ossification begins to appear. See text for explanation on the sequence of ossification and characteristic of certain bones.

and are totally ossified at about 14 mm SL (including the membranous anterior process of the hyomandibula). The supracleithrum, which in adult specimens of *Chanos chanos* is a large bone and comparatively the longest of the pectoral girdle, appears as a long, slightly fusiform ossification, not associated at this time with the lateral line. The cleithrum, on the other hand, appears first as an elongate bone represented first by its dorsal part; the ventral part develops later as a long, fine ossification.

The ossification of the **parasphenoid** begins at 10.6 mm, with the middle region, and is complete at about 12.9 mm SL, when it has acquired the sharp, ventral curvature that characterizes this bone in *Chanos*.

The tiny **urohyal** emerges in the middle region of the tendon of the sternohyoideus muscle at about 11 mm SL. By 11.7 mm, the entire bone, including its anterior and posterior processes, has ossified.

The next step is at 11.7 mm when six bones begin to ossify. Three of these are dermal (**branchiostegal ray** 1, **posttemporal**, and **subopercle**), and three are chondral (**basioccipital**, **exoccipital**, and **prootic**). The posteriormost branchiostegal ray, or branchiostegal ray 1, ossifies first, followed almost immediately by the second, third and fourth branchiostegals, so that at 12.9 mm the four branchiostegal rays characteristic of *Chanos* are present. The appearance, as well as the numbering, of branchiostegals is from posterior to anterior direction. The posttemporal appears around the lateral line canal and then ossifies in a thin, slightly oval, plate-like pattern. In some specimens, however, the long dorsal process ossifies almost at the same time as the small, plate-like body of the bone. The ossification of the subopercle begins at the region of its anterodorsal process; from there, it spreads to the whole dorsal region to finally reach the ventral region of the bone. The ossification of the basioccipital encloses part of the mineralized cranial tip of the notochord, and at 12.6 mm SL the process is complete and the basioccipital totally surrounds the anterior end of the notochord. The prootic appears as a thin ossification between the auditory foramen and the facial foramen, as does the exoccipital that begins to thinly ossify around the foramen for the vagus nerve.

At 12.1 mm SL, three dermal bones (**dentary, ectopterygoid**, and **lateral extrascapula**), eight chondral bones (**basibranchials 2–3, ceratobranchials 1–5, epibranchials 1–4, hypobranchial 1, pterosphenoid, quadrate, symplectic**, and **ventral hypohyal**), and one compound bone (**pterotic**) begin to ossify in the cranium, along with the **1st ray** in the pectoral fin. The ossification of the dentary starts at the region of the mandibular symphysis and the anteroventral margin of the bone around the mandibular canal, and from there it spreads posteriorly. This is a slow process, because in some specimens of 18.5 mm SL the bone is still not completely ossified. The lateral extrascapula develops as a thin, irregular ossification surrounding the confluence of the otic and lateral line canals and extrascapular or supratemporal commissure, whereas the ectopterygoid appears as a thin, slightly boomerang-shaped tiny bone at the anterior margin of *pars quadrata*. Basibranchials 2 and 3 are the first to appear as thin elongate ossifications, basibranchial 3 being very long (in one specimen a tiny, elongate basibranchial 4 was observed). The last element to appear is basibranchial 1 at 12.9 mm SL. The five ceratobranchials and the four epibranchials appear simultaneously as thin, perichondral ossifications at their middle regions. Hypobranchial 1 is the first to ossify, followed almost immediately by hypobranchial 2 (12.2 mm SL), and finally hypobranchial 3 (12.3 mm SL). A tiny, slightly elongate ossification, the pterosphenoid, emerges medial to the authosphenotic and prootic. The cartilaginous quadrate starts its ossification process as one unit at the quadrate condyle and the membranous posteroventral process, confirming that the posteroventral process is not a separate ossification, as it has been shown previously for *Esox* and salmonids, respectively (Jollie 1984, Arratia and Schultze 1991). A few authors have proposed that the posteroventral process of the quadrate is the quadratojugal [of primitive actinopterygians] that fuses to the body of the quadrate during ontogeny. However, the presence of the posteroventral process is a unique synapomorphy of Teleostei (Arratia and Schultze 1991, Arratia 1999). The perichondral ossification of the body of the quadrate starts much later, when the fish is almost 18 mm SL. The perichondral ossification of the symplectic begins at its middle region, closer to its articulation with the hyomandibula. The ventral hypohyal begins to ossify at the point of insertion of the tendon of the sternohyoideus muscle. The dermal component (dermopterotic) of the pterotic appears around the otic canal. Next, the lateral wall of the autopterotic starts to ossify and encloses the dermal component that has developed around the canal forming the pterotic. At 17 mm SL, the wall of the pterotic that forms part of the hyomandibular fossa is still unossified. The dorso-medial part of the pterotic is thinly ossified at 22 mm SL, and the whole bone is ossified in specimens of about 22 to 23 mm SL.

At 12.2 mm SL, **pharyngobranchials 2** and **3 (or infrapharyngobranchials 2** and **3)** are ossifying. At 28 mm SL, the last element of the series of branchial arches to begin the process is the so-called pharyngobranchial 1 (see comments in terminology section). Pharyngobranchial 4 remains as cartilage.

At 12.3 mm SL, the **preopercle** emerges as a thin, irregular ossification around the preopercular canal.

At 12.9 mm SL, several other bones begin to appear. These include dermal bones (**entopterygoid, frontal** of traditional terminology [= parietal, based on homologization of bones], **infraorbital 1** or **lacrimal, interopercle, maxilla,** and **premaxilla**), chondral bones (**anterior ceratohyal, autosphenotic, metapterygoid,** and **posterior ceratohyal)**, and a tendon-bone, the **coronomeckelian** or sesamoid coronoid. The entopterygoid develops as a thin, elongate ossification below the eyeball, whereas the interopercle arises as a thin, elongate ossification ventral to the anterior limb of the preopercle. The thin, elongate, irregular ossification of the frontal begins around the supraorbital canal. The first element of the circumorbital series to develop is the infraorbital 1 or lacrimal that starts as a thin, elongate, irregular ossification surrounding the anterior portion of the infraorbital canal. The premaxilla begins to ossify at its lateral portion; later, the medial portion of the bone appears as an irregular, thin bony plate that grows medially to meet its counterpart at the midline. The ossification of the maxilla begins at the anteromedial region and progresses laterally and posteriorly to form the blade. The anterior ceratohyal ossifies perichondrally at the middle section of the bone, whereas the posterior ceratohyal ossifies at the posterior corner of the arch. (*Chanos chanos* lacks an interhyal from early ontogeny on.) The autosphenotic starts ossifying at the lateral corner that will become the ventrolateral process of the bone, whereas the metapterygoid begins at its dorsal curved margin, where a notch is produced between the sharp, small cartilaginous *processus basalis* and the *processus metapterygoideus lateralis* (see Arratia 1992: fig. 4B–D). The conoromeckelian begins to ossify at the insertion of an adductor tendon on the Meckel cartilage, at the medial side of the lower jaw.

At 14 mm SL, the ossification of the **angular**, **intercalar**, and **retroarticular** begins in the cranial region, while the process continues in pectoral rays 2 to 4. Two distinct ossifications are present at the posterior region of the lower jaw: the thin, irregular dermal ossification of the angular that surrounds the mandibular canal lateral to the Meckel cartilage, and the chondral retroarticular that develops at the posteroventral corner of the Meckel cartilage where the retroarticular-interopercular ligament inserts. The intercalar membrane bone emerges as a small ossification at the posterolateral corner of the braincase between the pterotic and exoccipital regions.

At 15.3 mm SL, the dermal **vomer** and the chondral **epiotic** and **supraoccipital** begin to ossify. The epiotic appears as a small, cup-like ossification at the posterior end of the cranium (there is no evidence that this bone is a result of a fusion between two elements, the epiotic and an occipital element). The vomer emerges as a tiny ossification placed below and anterior to the anteromedian region of the large mass of ethmoidal cartilage. At the opposite median end of the cranium, the supraoccipital is ossifying at the occipital crest; later, when most of the crest is present, the bone begins to ossify anteriorly as a thin plate.

At 15.5 mm SL, the **autopalatine** appears as a thin perichondral ossification around *pars autopalatina*.

At 17.2 mm SL, the dermal **infraorbital 4** and **medial extrascapula** and the chondral **articular**, **dorsal hypohyal** and **lateral ethmoid** start to ossify. The infraorbital bone 4, at the posterodorsal corner of the orbit, develops as a thin, elongate, irregular ossification around the infraorbital canal. A second tube-like extrascapular bone begins to ossify around the extrascapular commissure, medial to the lateral extrascapula. This ossification is positioned partially above the level of the epiotic. During growth of the juvenile milkfish, it is displaced slightly anteriorly onto the posterior part of the parietal bone of traditional terminology and part of the supraoccipital. The medial extrascapula has a large, round foramen where the supraorbital canal and extrascapular commissure meet. This seems to be a weak point, because in many large specimens the bone breaks at this level and it looks like the fish has three extrascapular bones. When three pairs of extrascapular bones are found in large specimens, the second of the series is the smallest one. The articular begins to develop as an endochrondral ossification of the Meckel cartilage at its articular facet, in a similar fashion as described for *Esox* by Jollie (1984: 77); however, it ossifies perichondrally at the medial wall of the cartilage. The angular and articular develop from two independent ossification centers that appear at different times (see Figs. 3.4, 3.5). The fusion between the two bones occurs later in the ontogeny of the juvenile milkfish. The lateral ethmoid begins as a tiny ossification at the posterolateral border of the large ethmoid cartilage.

At 17.4 mm SL, the **basipterygium** and **1st pelvic ray** begin to ossify. The basipterygium, or pelvic plate, ossifies perichondrally at its middle region. As in the pectoral fin, the first ray—or the most lateral ray of the pelvic fin—ossifies first, and then the rays develop one by one.

At 18.5 mm SL, the **basihyal, mesethmoid, parietal** bone of traditional terminology, infraorbitals 2 and 3, last two pectoral rays, and pelvic rays 2 to 4 or 5 begin to ossify. At the ventromedian region of the branchial arch series, a tiny, elongated ossification appears at the posteroventral part of the basihyal cartilage, and from this point the ossification progresses anteriorly. The ossification of the mesethmoid begins as a tiny formation at its anteromedian margin and from there expands posteriorly and laterally. The parietal develops from a plate-like ossification that grows rapidly to reach an almost squarish shape. At about 45 mm SL, the bone is still well exposed (Fig. 3.6), and it is possible to observe the posterior part of the supraorbital canal running in a tube along the bone and joining the extrascapular commissure running in the medial extrascapula at the level of a foramen placed along the mid-section of the medial extrascapula. During growth, the anterior part of the parietal is progressively covered by the posterior part of the frontal bone, so that only a narrow posterior portion of the parietal bone is exposed dorsally in larger specimens of *Chanos*.

At about 20 mm SL, the last chondral element of the gill arches, **epibranchial 5**, begins to ossify perichondrally.

At 22 mm SL, one chondral element of the pectoral girdle, the **scapula**, starts to ossify.

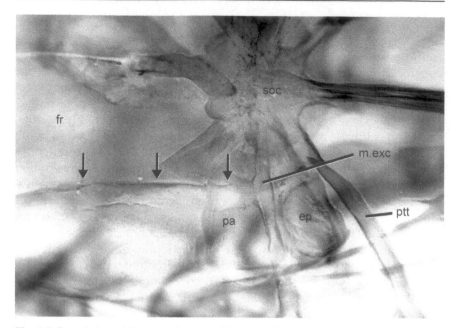

Fig. 3.6 Dorsal view of the posterior part of the skull of *Chanos chanos* (45.8 mm SL; KU 39894). Small arrows point to the path of the supraorbital canal. fr[=pa], frontal bone [= parietal bone based on homology]; ep, epiotic; m.exc, medial extrascapula; pa[=ppa], parietal bone [= postparietal bone based on homology]; ptt, posttemporal; soc, supraoccipital.

At 22.8 mm SL, two other elements of the circumorbital ring appear anteriorly and posteriorly: the **antorbital** and the **dermosphenotic,** respectively. The ossification of the antorbital begins at its ventral margin and extends dorsally from there, whereas the dermosphenotic emerges as a thin ossification surrounding the infraorbital canal and then the ossification spreads. At this point, the bone begins to acquire its triangular shape, with an anterior elongate process. The infraorbital canal curves posteriorly within the dermosphenotic and exits the bone to continue as the otic canal. In the pectoral girdle, the **mesocoracoid** starts to ossify medial to the cartilaginous scapulo-coracoid cartilage.

At 23.2 mm SL, two other chondral elements of the pectoral girdle, the **coracoid** and the **proximal radials 1** and **2,** start to ossify, as well as the dermal **pelvic splint** in the pelvic fin. In the pectoral girdle, two other elements appear later: the proximal radials 3 and 4. Thus, at about 27 mm SL, the four proximal pectoral radials are present.

At 24.1 mm SL, the **supraorbital** bone appears at the anterodorsal corner of the orbit. This is the last element of the circumorbital ring to ossify.

At 27 mm SL, the **nasal** bone starts as a thin ossification surrounding the anterior section of the supraorbital canal. At about 20 mm SL, there are three cartilaginous pelvic radials placed posteriorly to the basipterygium; the medial one is elongated caudally, whereas the most lateral element of the series is slightly larger than the middle element. The lateralmost element divides into two pieces of cartilage, so that four cartilaginous radials are present. At 28 mm SL, the **medial pelvic radial,** the largest of the series, ossifies. The lateralmost pelvic radial follows at 45.5 mm SL. At 28 mm SL, the **propterygium** starts to ossify between the hemilepidotrichia of the first pectoral ray. At about 40 mm SL, the propterygium fuses to the base of the dorsal hemilepidotrichium of the 1st pectoral ray.

The scapula begins to ossify at the scapular process where the developing **pectoral axillary process** will sit (at 28.6 mm). The pectoral axillary process is formed by a combination of a thin bony plate and modified scales. The first three scales of the lateral line, just posterior to the supracleithrum, also appear at this time. Enclosed in bone, the lateral line runs near the dorsal margin of the supracleithrum; thus, the position of the first scales carrying the lateral line canal is above the middle flank.

At 44.5 mm SL, long after all other cranial bones are present and the cranium is almost fully ossified, the **suprapreopercle** becomes visible as a thin ossification surrounding the preopercular canal, above the dorsal border of the preopercle. From this point, the ossification spreads posteriorly as a thin, expanded plate.

At 87.8 mm SL, the anterior and posterior sclerotic bones are not present, nor were they found in specimens of 420–500 mm SL examined here. The sclerotic ring, which is well developed in small larvae, begins to get thinner and thinner through growth, and in specimens of about 44 mm SL the ring has almost completely disappeared. In the available specimens of about 80 mm SL, no remnants of the cartilaginous sclerotic ring were present.

Comparisons and Discussion

Chanos chanos

The results obtained in this study of *Chanos chanos* are different from those from studies also based on specimens rised in the same hatchery, specifically in the Aquaculture Department of the SEAFDEC, Iloilo, the Philippines, but in different years. For instance, the goals of previous studies differ from the present one, because the investigation of early ossification of *C. chanos* was restricted to certain bones, for example, those associated with feeding and swimming abilities. Additionally, it is difficult to compare the results concerning the beginning of ossification of bones associated with feeding and swimming, because the recorded timings of ossification are different, as is shown below.

Kohno *et al.* (1996a) observed the initial opening of the mouth of *Chanos chanos* at 54 h after hatching (5.44±0.14 mm and about 2.2 days) for larvae grown at 29°C. Different results were previously reported by Bagarinao (1986), who found that at hatching, the "yolk-sac larvae" are of 3.5 mm total length and that, 3 days later at 27–30°C, the larvae begin to feed when the eyes become fully pigmented, the mouth has opened, and some yolk sac is still present. "Egg size, size at hatching, amount of yolk, and initial mouth size are greater in milkfish than in many other tropical marine fishes, a size advantage that probably determines in part the survival of larvae in plankton and in the hatchery (Bagarinao 1986)" (in Bagarinao 1994: 33).

There are major differences between our results and those results of Taki *et al.* (1987) and Kohno *et al.* (1996a) regarding the timing of ossification of certain bones. One major difference is the size of the specimens at certain events. For instance, the cleithrum is the first bone to ossify at 4.65 mm NL in Taki *et al.* (1987). The cleithrum begins to ossify at about 5 mm NL (about day 1.8) and the next bone is the maxilla at 5.44±0.14 mm (about day 2.2) in the study by Kohno *et al.* (1996a: fig. 6). According to these authors, the maxilla is ossifying at the same time as the initial mouth opening. In contrast, the present results (Figs. 3.2, 3.3, 3.4) show that the process of ossification begins much later, when the larvae are almost double the length at which they were previously observed, e.g., 10.3 mm SL for the cleithrum and 12.9 mm SL for the maxilla. The supracleithrum, which in the present study ossified along with the cleithrum, was not mentioned at all in Kohno *et al.* (1996a). The supracleithrum began to ossify at 15 mm SL in specimens studied by Taki *et al.* (1987). [The process of ossification of the jaws was not studied by Taki *et al.* (1987).]

The opercle, as well as ceratobranchials 1–3, started to ossify at 6.52±0.35 mm in the specimens studied by Kohno *et al.* (1996a), unlike the results presented here that show that the opercle begins to ossify at the same time as the cleithrum at 10.3 mm SL, and the ceratobranchials at about 12.1 mm SL. The hyomandibula, which in the specimens studied herein appeared at the same time as the cleithrum, opercle and supracleithrum, began to ossify at 6.83±0.42 mm NL in specimens studied by Kohno *et al.* (1996a), along with the symplectic, dentary, angular, and quadrate. [Bones of the opercular apparatus were not studied by Taki *et al.* (1987).]

According to Taki *et al.* (1987), the first signs of ossification were seen at 13.8 mm SL in ceratobranchials 1–5 and epibranchials 1–4. All these elements began to ossify at 12.1 mm in the specimens studied here. Independent of differences in the sizes of the specimens, most bones of the branchial arches begin to ossify simultaneously. However, there are some differences between the timing of ossification of branchial arch elements and sequence of ossification of certain bones between Taki *et al.* (1987) and the present study:

Branchial arches (Taki et al. 1987)	Branchial arches (present paper)
13.8 mm SL	12.1 mm SL
Basibranchials 1–4 ⇔ Hypobranchial 1 ⇔	Basibranchials 2–3 ⇔ Hypobranchial 1
Ceratobranchials 1–5 ⇔	Ceratobranchials 1–5 ⇔
Epibranchials 1–4 ⇔	Epibranchials 1–4 ⇔
Pharyngobranchials 2–4	↓
↓	12.2 mm SL
15.7 mm SL	Hypobranchial 2 ⇔ Pharyngobranchials 2–3
Hypobranchials 2–3	↓
↓	12.3 mm SL
17.2 mm SL	Hypobranchial 3
Epibranchial 5	↓
↓	12.9 mm SL
18.5 mm SL	Basibranchial 1
Phraryngobranchial 1	↓
	20 mm SL
	Epibranchial 5
	↓
	28 mm SL
	Pharyngobranchial 1

Basibranchials 1–4 and pharyngobranchials 2–4 were mistakenly reported as beginning to ossify at 13.8 mm SL (Taki *et al.* 1987). However, according to the present results and the available literature on larger specimens of *Chanos chanos*, basibranchial 4 and pharyngobranchial 4 remain as cartilage in this species.

According to Taki *et al.* (1987), the cleithrum is the first element to ossify at 4.65 mm NL, and the (first) pectoral fin rays are next at 13.8 mm SL. There is a large gap between the first element to ossify and the next in Taki *et al.* (1987), in contrast to the results obtained here that show a more continuous ossification process. The 1st pectoral ray ossifies first in the present study and is followed by the 2nd, 3rd, etc. In contrast, Taki *et al.* (1987) reported several rays ossifying simultaneously. The results obtained here are different, not only in timing but also in sequence of appearance of ossification, as shown below:

Differences are also registered concerning the early ossification of the pelvic girdle. For instance, the pelvic rays begin to ossify at 13.8 mm SL and the basipterygium at 17.15 mm SL in specimens studied by Taki *et al.* (1987), whereas the first pelvic ray and basipterygium begin to ossify at 17.4 mm SL in specimens in the present study. Thus, the pectoral and pelvic rays ossify at the same time in Taki *et al.* (1987), whereas the pectoral rays ossify earlier than the pelvic rays in the specimens studied here. [Kohno *et al.* (1996a) did not include the pelvic girdle and fin in their study.]

There are differences not only in the size of the specimens versus beginning of ossification of certain bones, but also in the sequence of ossification of the bones. These can be compared, as reported by Taki *et al.* (1987), Kohno *et al.*

Pectoral girdle and fin (Taki et al. 1987)	Pectoral girdle and fin (present paper)
4.65 mm NL Cleithrum ↓	10.3 mm SL Cleithrum ⇔ Supracleithrum ↓
13.8 mm SL First pectoral rays ↓	11.7 mm SL Posttemporal ↓
14.9 mm SL Posttemporal ⇔supracleithrum ⇔ Scapula ⇔ Coracoid ↓	12.1 mm SL 1st pectoral ray ↓
18.55 mm SL All pectoral rays ossified ↓	22 mm SL Scapula ↓
24.6 mm SL Pectoral proximal radial 4	22.8 mm SL Mesocoracoid ↓
	23.2 mm SL Coracoid ⇔ Proximal radials 1 and 2 ↓
	27 mm SL Pectoral proximal radials 3 and 4

(1996a), and herein. Possible interpretations for the differences may in part be explained by differences in temperature, food, and other possible variables of the environment where the fishes developed. However, differences in observations of the studied material cannot be discarded. It is already known that changes in the temperature and diet have effects on growth, development, behavior, and survival of young milkfish (see, for instance, Buri 1980, Kumagai and Bagarinao 1980, Villaluz and Unggui 1983, Kumagai et al. 1985, Taki et al. 1987). The available studies and differences reported in *Chanos chanos* grown in different years and in slightly different environmental conditions (in the same Aquacultural Department in the Philippines) confirm that the larvae are very sensitive to changes in the environment, as demonstrated by variations in the development of the skeletal structures and their time of ossification.

Ossification Sequence of Cranial and Paired Girdle Regions

Cranial Regions

Cranial bones are separated in nine regions to facilitate comparison with previous studies (see Table 3.2). The analysis of each region shows that there is no rule concerning which type of bone (dermal or chondral) ossifies first in each region. For instance, a dermal bone (vomer) ossifies first in the olfactory region of *Chanos chanos*, as well as in the orbital (parasphenoid) and mandibular (e.g., dentary) regions, whereas a chondral bone ossifies first in the hyoid arch (e.g., hyomandibula). Both types of bones may also ossify at the same time in other places, e.g., otic region (prootic and

Table 3.2   Sequence of ossification of bones of nine cranial regions in *Chanos chanos* and comparison with the cypriniform *Danio rerio* (after Cubbage and Mabee, 1996). Names written in bold represent chondral bones. The vertical arrangement follows the sequence of ossification of bones in each cranial region. ** denotes a wrong interpretation, e.g., the bone results from the fusion of two elements that ossify at different times. °°, a name given according to current literature, but see Nomenclature on page 79.

Chanos chanos	*Danio rerio*
Olfactory region	
Vomer	**Lateral ethmoid**
Lateral ethmoid	Vomer
Mesethmoid	Nasal bone
Nasal bone	**Mesethmoid**
Orbital region	
Parasphenoid	Parasphenoid
Pterosphenoid	**Pterosphenoid**
Infraorbital 1 or lacrimal and Frontal	Infraorbital 1 or lacrimal
Infraorbital 4	Frontal bone
Infraorbitals 2–3	Supraorbital bone
Dermosphenotic	Dermosphenotic
Antorbital	Infraorbitals 2–3
Supraorbital bone	Infraorbital 4
Otic region	
Prootic and **Posttemporal**	**Prootic**
Lateral extrascapula and **Pterotic**	Posttemporal
Autosphenotic	**Pterotic**
Intercalar	**Autosphenotic**
Epiotic	Parietal
Medial extrascapula	**Epiotic**
Parietal	Lateral extrascapula and Intercalar
Occipital region	
Basi- and **exoccipital**	**Exoccipital**
Supraoccipital	**Basioccipital**
	Exoccipital
Mandibular arch	
Dentary	Dentary and Maxilla
Premaxilla, Maxilla and	Anguloarticular** and
Coronomeckelian	**Retroarticular**
Angular and **Retroarticular**	Premaxilla
Articular	Coronomeckelian
Palatoquadrate arch	
Quadrate and Ectopterygoid	Entopterygoid
Metapterygoid and Entopterygoid	**Quadrate**
Autopalatine	Ectopterygoid
	Metapterygoid
	Autopalatine
Hyoid arch	
Hyomandibula	Branchiostegal rays
Urohyal and Branchiostegal rays	**Hyomandibula**
Symplectic and **Ventral Hypohyal**	Urohyal
Anterior and **posterior ceratohyals**	**Anterior ceratohyal**
Dorsal hypohyal	**Ventral hypohyal**

Table 3.2 contd....

... Table 3.2 contd.

Chanos chanos	Danio rerio
Basihyal	Posterior ceratohyal
	Symplectic
	Basihyal
	Dorsal hypohyal
Branchial arches	
Ceratobranchials 1–5 and	Ceratobranchial 5
Epibranchials 1–4 and	Ceratobranchial 4
Basibranchials 2–3 and	Epibranchial 4
Hypobranchial 1	Ceratobranchials 1–3
Hypobranchial 2	Epibranchials 1–3
Pharyngobranchials 2–3	Pharyngobranchials 2–3
Hypobranchial 3	Basibranchials 1–3
Basibranchial 1	Hypobranchial 3
Epibranchial 5	Pharyngobranchial 1
Pharyngobranchial 1°°	Hypobranchial 2
Opercular region	
Opercle	Opercle
Subopercle	Interopercle and Subopercle
Preopercle	Preopercle
Interopercle	
Suprapreopercle	

posttemporal) (see Table 3.2 for details). Comparisons with the cypriniform *Danio rerio* show that there are major differences in all regions. For instance, in the olfactory region of *Chanos*, the first bone to ossify is the dermal vomer, whereas in *Danio* it is the chondral lateral ethmoid. The dermal parasphenoid and the chondral pterosphenoid ossify first in the orbital region of *Chanos* and *Danio*, but apart from this similarity all other bones of this region ossify in different sequences in both taxa (see Table 3.2).

The sequence of ossification of the cranial bones in each region is represented in bar graph form in Fig. 3.7. The first two cranial regions to begin to ossify in *Chanos* are the hyoid arch, with the hyomandibula, and the opercular region, with the opercle (corresponding with two dermal bones of the pectoral girdle). The second is the orbital region with the parasphenoid, followed by the otic region (prootic) and occipital region (basi- and exoccipitals). The palatoquadrate, mandibular arch and branchial arches follow. The last cranial region to start to ossify is the olfactory one. However, the initiation of the ossification of a cranial region does not ensure that all bones belonging to this specific region will appear earlier than others. For instance, in the opercular region, the opercle ossifies first at 10.3 mm SL, whereas the suprapreopercle is the last cranial bone to begin to ossify (at 45.5 mm SL), at which point all other bones are almost totally ossified.

As shown in Figs. 3.4 and 3.7, the first elements to ossify are simultaneously dermal (opercle) and chondral (hyomandibula) and form part of the cover and protection of the bucco-pharyngeal cavity and the branchial arches of the fish. The opercle, which in *Chanos* forms most of the lateral cover of the opercular apparatus, does not ossify alone; it needs the support of the hyomandibula, a chondral bone that provides the facet with which the opercle articulates. Apparently, the synchronic ossification of the two bones responds to their functions in *Chanos*. This correlation is not usually observed in a variety of teleosts, where both bones ossify independently (first the opercle and later the hyomandibula), for example, in the esociform *Esox lucius* (Jollie 1984), the salmoniform *Oncorhynchus mykiss* (Verraes 1977, Arratia pers. obs.), the clupeomorph *Dorosoma cepedianum* (Arratia pers. obs.), and the ostariophysans *Clarias gariepinus* (Surlemont *et al.* 1989), *Barbus barbus* (Vandewalle *et al.* 1992), and *Danio rerio* (Cubbage and Mabee 1996).

The dermal elements of the jaws, such as the premaxilla, maxilla and dentary, are not among the earliest bones to form in *Chanos chanos*, unlike in other fishes, such as the cyprinid *Barbus barbus* (Vandewalle at al. 1992), the catfish *Ictalurus punctatus* (Arratia 1992), the clupeomorph *Dorosoma cepedianum* (Arratia pers. obs.), the esociform *Esox lucius* (Pehrson 1944, Jollie 1984), the salmonid *Oncorhynchus mykiss* (Arratia pers. obs.), the lophiiform *Lophius gastrophysis* (Matsuuru and Yoneda 1987), the gasterosteiform *Nerophis aequoreus* (Kadan 1961), and the percomorph *Sparus aurata* (Faustino and Power 2001). However, this could also be a response to certain demands of the feeding mechanism. *Chanos* lacks dentition in its jaws and palate and, based on the morphology and developmental patterns, the feeding mode of early larval milkfish was interpreted as "straining" by Kohno *et al.* (1996a), whereas Bagarinao (1986) stated that the milkfish larvae feed on plankton, and that following metamorphosis (after 10–17 mm total length and 14–29 days), the zooplankton-feeding larvae become benthic-feeding juveniles, opportunistically herbivorous, detrivorous or omnivorous.

Splanchnocranium versus Neurocranium

The bones of the nine cranial regions discussed above (Table 3.2, Fig. 3.7) can be re-grouped into two major regions, the splanchnocranium and viscerocranium. Despite the variation in the patterns of development and timing of ossification of cranial bones described in the available literature (see for instance, de Beer 1937), it is believed that the first bones to appear are always dermal and, additionally, belong to the splanchnocranium. This is confirmed in *Chanos chanos*, with the opercle as the first bone of the splanchnocranium to ossify, although in *Chanos* a chondral bone, the hyomandibula, ossifies simultaneously with the opercle.

The parasphenoid, which in *Chanos* is the first bone (at 10.6 mm SL) of the neurocranium that ossifies, has the same characteristic as other teleosts where the development is known, for example, the esociform *Esox lucius* (Pehrson 1944, Jollie 1984) and the cypriniforms *Barbus barbus* (Vandewalle

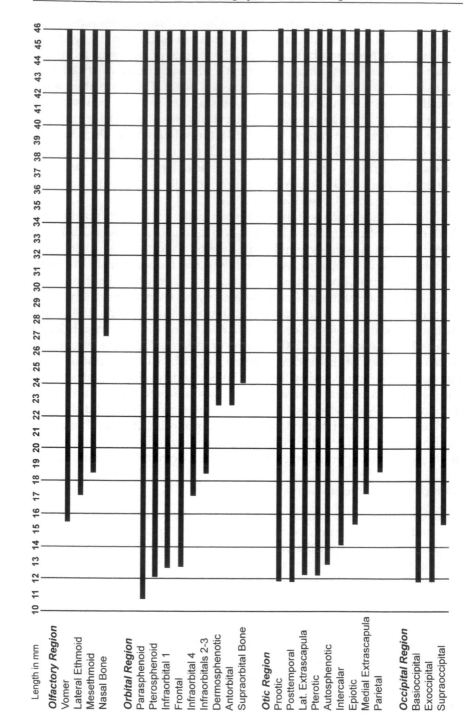

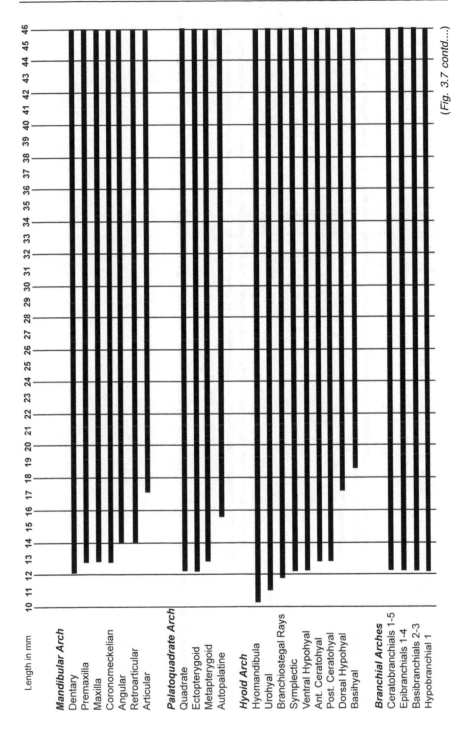

(Fig. 3.7 contd....)

(Fig. 3.7 contd....)

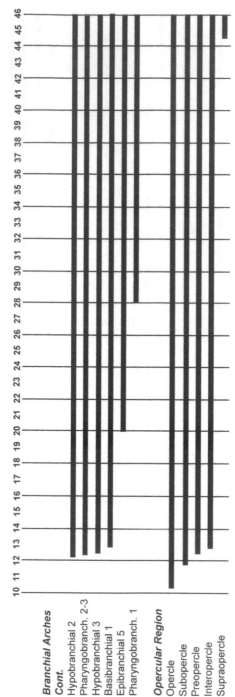

Branchial Arches Cont.
Hypobranchial 2
Pharyngobranch. 2-3
Hypobranchial 3
Basibranchial 1
Epibranchial 5
Pharyngobranch. 1

Opercular Region
Opercle
Subopercle
Preopercle
Interopercle
Supraopercle

Fig. 3.7 Regional ossification sequences in the cranium of *Chanos chanos*. The solid, horizontal lines indicate the length at which ossification begins to appear.

et al. 1992) and *Danio rerio* (Cubbage and Mabee (1996). Bones of the skull roof (e.g., lateral extrascapula, frontal bone) initiate their ossification later than those forming the base of the cranium (parasphenoid) or its lateral walls (e.g., prootic). Among them, the last bone to begin to ossify is the mesethmoid at 18.5 mm SL.

Paired Girdles and Fins

The dermal bones of the pectoral girdle, the cleithrum and supracleithrum, are the first bones to ossify, even earlier than the pectoral rays. The appearance of the chondral bones occurs much later during development, when most of the pectoral rays are ossified. The pattern observed in the pelvic fin is different, because the 1st pelvic ray and the basipterygium ossify simultaneously. There are a few differences between both girdles: the pectoral rays ossify before the pelvic rays, and the chondral elements of the pelvic girdle ossify before those of the pectoral girdle (see Figs. 3.4 and 3.5).

The sequence of ossification of bones belonging to different regions does not clearly confirm that dermal bones ossify before chondral bones or vice versa (Figs. 3.4, 3.5, 3.7, Table 3.2). Regions where both dermal and chondral elements form simultaneously show ambiguous results. For instance, in the otic region a chondral bone (prootic) ossifies before the dermal posttemporal and lateral extrascapula, whereas in the olfactory region the dermal vomer ossifies before the chondral lateral ethmoid and mesethmoid. In the pectoral girdle, the dermal supracleithrum and cleithrum ossify long before any of the chondral elements of the girdle (see Fig. 3.7, Table 3.2).

Ossification Sequences

The beginning of ossification of bones of the cranium and paired girdles and fins in *Chanos chanos* shows two very distinct paths.

Cranial bones in *Chanos chanos*, with a few exceptions, ossify in groups (see Fig. 3.4). For instance, at 11.7 mm SL, the basioccipital, branchiostegal rays, exoccipital, posttemporal, prootic, and subopercle all begin. At 12.1 mm SL, 13 cranial bones and at 12.9 mm SL, 11 bones initiate their ossification. Thus, before the larvae reach 13 mm SL, 37 bones are ossifying; this corresponds to 68% of the total number of cranial bones. Additionally, the process is rapid, because from the moment that the first cranial bone begins to ossify until the larva reaches 13 mm SL, in a growth of 2.7 mm, 37 bones have begun to ossify. Bones ossifying in groups have been described for other teleosts, e.g., *Barbus barbus* (Vandewalle *et al.* 1992) and *Esox lucius* (Jollie 1984). In contrast, the results of a study based on many specimens of *Danio rerio* showed that the process does not involve many bones ossifying simultaneously (see Cubbage and Mabee 1996: fig. 1).

The sequence differs in the paired girdles and fins, where the ossification process spreads from one bone to the next. For instance, the pectoral girdle begins to ossify at 10.3 mm SL with both the cleithrum and supracleithrum. Then, at 12.1 mm SL, the 1st pectoral ray appears, at 22 mm SL the scapula, at 22.8 the mesocoracoid, and at 23.2 the coracoid. The last elements to appear are the proximal pectoral radials 3 and 4 and propterygium at 27 mm SL. The ossification of the paired girdles and fins is a longer process. For instance, the pectoral girdle and fin begin at 10.3 mm SL and end at about 27 mm SL.

Phases of Development

Based on the scheme of development of swimming and feeding-related characters, Taki *et al.* (1987) proposed that the development of *Chanos chanos* can be divided into four phases:

(1) The initial phase that corresponds to the preflexion larval stage of Kendall (1984). During this phase, the supports of the fins and branchial arches have not yet developed. This phase ends at about 6.5 mm SL in Taki *et al.* (1987). However, because of the descriptions of the phase and of larvae, we suppose that this so-called standard length is indeed the notochordal length. The initial phase varies slightly in the specimens studied here, when the preflexion stage may extend from 5.9 to 10 mm NL.

(2) The phase in which principal structures of the unpaired fins, vertebrae and branchial arches are developed; up to 10.5 mm SL. As far as the available information permits a comparison, the branchial arches in the present study begin to ossify at 12.1 mm SL, much later than in the larvae studied by Taki *et al.* (1987).

(3) The phase in which the mechanical support of the paired fins, gill rakers and epibranchial organ are developed and the body form attains juvenile proportions; up to 17 mm SL. The mechanical supports of the paired girdles, such as the scapula, coracoid and basipterygium, begin to ossify after the larvae reach 17 mm SL in the specimens studied herein, a size that corresponds to the next phase (see below).

(4) The juvenile phase; beyond 17 mm SL.

The results obtained here conflict with the last three phases of Taki *et al.* This issue will be extensively analyzed in a study of cranial chondrification versus ossification of *Chanos chanos* by Arratia and Grande (in preparation).

Final Comments

The available information on the development of cranial and paired girdles of *Chanos chanos* reveals that the larvae are sensitive to changes in the environment, and consequently there are major differences concerning the timing of ossification, and also the sequence of ossification between studies. Further studies on the osteological development of *Chanos* should include exhaustive analyses of the environmental conditions to find an explanation for the observed differences in bone development. The understanding of these differences will make it possible to compare the osteological development of other extant gonorynchiforms, most of which are freshwater forms.

Acknowledgements

G. Arratia is grateful to R. Rosemblat (SIO) and W.L. Fink and D. Nelson (UMMZ) for permission to study material housed at their institutions and to K. Myck and M. Davis (Lawrence, K.S.) for their help with the illustrations. This research was supported by the National Science Foundation's Tree of Life of Cypriniformes Project (EF 037870).

References

Adriaens, D. and W. Verraes. 1997a. The ontogeny of the chondrocranium in *Clarias gariepinus*: trends in siluroids. J. Fish Biol. 50: 1221–1257.

Adriaens, D. and W. Verraes. 1997b. Some of the consequences in transformations in siluriform chondrocrania: a case study of *Clarias gariepinus* (Burchell, 1822) (Siluroidei: Clariidae). Netherland J. Zool. 47: 1–15.

Adriaens, D. and W. Verraes. 1998. Ontogeny of the osteocranium in the African Catfish *Clarias gariepinus* Burchell (1822) (Siluriformes: Clariidae): Ossification sequence as a response to functional demands. J. Morphol. 235: 183–237.

Adriaens, D., W. Verraes and L. Taverne. 1997. The cranial lateral-line system in *Clarias gariepinus* (Burchell, 1822) (Siluroidei: Clariidae). Morphology and development of canal related bones. Eur. J. Morphol. 35: 181–208.

Ahlstrom, E.H., J.L. Butler and B.Y. Sumisa. 1976. Pelagic stromateoid fishes (Pisces, Perciformes) of the eastern Pacific: kinds, distribution, and early life histories and observations on five of these from the northwest Atlantic. Bull. Mar. Sci. 26: 285–402.

Arratia, G. 1990. Development and diversity of the suspensorium of the trichomycterids and comparison with loricarioids (Teleostei: Siluriformes). J. Morphol. 205: 193–218.

Arratia, G. 1991. The caudal skeleton of Jurassic teleosts; a phylogenetic analysis. pp. 249–340. *In*: M.-M. Chang, Y.-H. Liu, and G.-R. Zhang [eds.]. Early Vertebrates and Related Problems in Evolutionary Biology. Science Press, Beijing.

Arratia, G. 1992. Development and variation of the suspensorium of primitive Catfishes (Teleostei: Ostariophysi) and their phylogenetic relationships. Bonner zoologische Monographien 32: 1–149.

Arratia, G. 1999. The monophyly of Teleostei and stem-group teleosts. Consensus and disagreements. pp. 265–334. *In*: G. Arratia and H.-P. Schultze [eds.]. Mesozoic Fishes 2-Systematics and Fossil Record. Verlag Dr. F. Pfeil, Munich.

Arratia, G. 2003. The siluriform postcranial skeleton—An overview. pp. 121–157. *In*: G. Arratia, B.G. Kapoor, M. Chardon and R. Diogo [eds.]. Catfishes, Vol. 1. Science Publishers, Inc., Enfield, NH, and Plymouth.

Arratia, G. and H.-P. Schultze. 1990. The urohyal: development and homology within osteichthyans. J. Morphol. 203: 247–282.

Arratia, G. and H.-P. Schultze. 1991. Development and homology of the palatoquadrate within osteichthyans. J. Morphol. 208: 1–81.

Arratia, G. and H.-P. Schultze. 1992. Reevaluation of the caudal skeleton of certain actinopterygian fishes. III. Salmonidae. Homologization of caudal skeletal structures. J. Morphol. 214: 187–249.

Arratia, G., H.-P. Schultze and J. Casciotta. 2001. Vertebral column and associated elements in Dipnoans and comparison with other fishes: Development and homology. J. Morphol. 250: 101–172.

Bagarinao, T. 1986. Yolk resorption, onset of feeding and survival potential of larvae of three tropical marine fish species reared in hatchery. Mar. Biol. 91: 449–459.

Bagarinao. T. 1994. Systematics, distribution, genetics and life history of milkfish, *Chanos chanos*. Environ. Biol. Fishes 39: 23–41.

Bagarinao, T. and S. Kumagai. 1987. Occurrence and distribution of milkfish, *Chanos chanos*, larvae off the western coast of Panay island, Philippines. Environ. Biol. Fishes 19:155–160.

Bertmar, G. 1959. On the ontogeny of the chondral skull in Characidae, with a discussion on the chondrocranial base and the visceral chondrocranium in fishes. Acta Zool., Stockholm 40: 203–364.

Bhargava, H.N. 1958. The development of the chondrocranium of *Mastacembelus armatus* (Cuv. et Val.). J. Morphol. 102: 401–426.

Buri, P. 1980. Ecology on the feeding of milkfish fry and juveniles, *Chanos chanos* (Forsskål) in the Philippines. Mem. Kagoshima University Research Center, South Pacific 1: 25–42.

Buri, P., V. Bañada and A. Triño. 1981. Developmental and ecological stages in the life history of milkfish *Chanos chanos* (Forsskål). Fish. Res. J., Philippines 6(2): 33–58.

Chaudhuri, H., J.V. Juario, J.H. Primavera, R. Samson and R. Mateo. 1978. Observations on the artificial fertilization of eggs and the embryonic development of milkfish *Chanos chanos* (Forsskål). Aquaculture 13: 95–113.

Clark, C. and K. Smith. 1993. Cranial osteogenesis in *Monodelphis domesticata* (Didelphidae) and *Macropus eugenii* (Macropodidae). J. Morphol. 215: 119–149.

Coburn, M. and P. Chai. 2003. Development of the anterior vertebrae of *Chanos chanos* (Ostariophysi: Gonorynchiformes). Copeia 1: 175–180.

Cubbage, C.C. and P.M. Mabee. 1996. Development of the cranium and paires fins in the zebrafish *Danio rerio* (Ostariophysi, Cyprinidae). J. Morphol. 229: 121–160.

de Beer, G.R. 1937. The Development of the Vertebrate Skull. Clarendon Press, Oxford.

Delsman, H.C. 1926. Fish eggs and larvae from the Java Sea 10. On a few larvae of empang fishes. Treubia 8: 400–412.

Delsman, H.C. 1929. Fish eggs and larvae from the Java Sea 13. *Chanos chanos* (Forsskål). Treubia 11: 281–286.

Faustino. M. and D.M. Power. 2001. Osteological development of the viscerocranial skeleton in sea bream: alternative ossification strategies in teleost fish. J. Fish Biol. 58: 537–572.

Fink, S.V. and W.L. Fink. 1981. Interrelationships of the Ostariophysan fishes (Teleostei). Zool. J. Linn. Soc. 72(4): 297–353.

Francillon, H. 1974. Développement de la partie postérieure de la mandibule de *Salmo truta fario* L. (Pisces, Teleostei, Salmonidae). Zool. Scripta 3: 41–51.

Gaupp, E. 1903. Zur Entwicklung der Schädelknochen bei den Teleostiern. Verhandlungen der anatomischen Gesellschaft, Jena 1903: 113–123.

Gegenbauer, C. 1878. Über das Kopfskelet von *Alepocephalus rostratus*. Morphologisches Jarhbücher 4: 1–41.

Haines, R.W. 1937. The posterior end of the Meckel's cartilage and related ossifications in bony fishes. Q.J. Microscop. Sci. 80: 1–38.

Jollie, M. 1962. Chordate Morphology. Reinhold Publishing Coorporation, New York, Chapman & Hall, Ltd., London, 478 pp.

Jollie, M. 1984. Development of the head skeleton and pectoral girdle in *Esox*. J. Morphol. 147: 61–88.

Juliano, R.O. and R. Hirano. 1986. The growth rate of milkfish, *Chanos chanos*, in brackishwater ponds of the Philippines. pp. 63–66. *In*: J.L. MacLean, L.B. Dizon, and L.V. Hosillos [eds.]. The First Asian Fisheries Forum. Asian Fisheries Society, Manila.

Kadan, K.M. 1961. The development of the skull in *Nerophis* (Lephobranchii). Acta Zool., Stockholm 42: 1–42.

Kamawura, G. 1984. The sense organs and behavior of milkfish fry in relation to collection techniques. pp 69–84. *In*: J.V. Juario, R.P. Ferraris, and L.V. Benitez [eds.]. Advances in Milkfish Biology and Culture. Island Publishing House, Manila.

Kapoor, A.S. 1970. Development of dermal bones related to sensory canals of the head in the fishes *Ophiocephalus punctatus* Bloch (Ophiocephalidae) and *Wallago attu* Bl. & Sch. (Siluridae). Zool. J. Linn. Soc. London 49: 69–97.

Kendall, A.W. Jr., E.H. Ahlstrom and H.G. Moser. 1984. Early life history stages of fishes and their characters. pp. 11–22. *In*: Ontogeny and Systematics of Fishes. American Society of Ichthyologists and Herpetologists, Special Publication 1.

Kohno, H., R. Ordonio-Aguilar, A. Ohno and Y. Taki. 1996a. Morphological aspects of feeding and improvement in feeding ability in early stage larvae of the milkfish, *Chanos chanos*. Ichthyol. Res. 43: 133–140.

Kohno, H., R. Ordonio-Aguilar, A. Ohno and Y. Taki. 1996b. Osteological development of the feeding apparatus in early stage larvae of the seabass, *Lates calcarifer*. Ichthyol. Res. 43: 1–9.

Kumagai, S. and T. Bagarinao. 1980. Studies on the habitats and food of juvenile milkfish in the wild. Fish. Res. J., Philippines 6: 1–10.

Kumagai, S., T. Bagarinao and A. Unggui. 1985. Growth of juvenile milkfish, *Chanos chanos*, in a natural habitat. Mar. Ecol. Progr. Ser. 22: 1–6.

Langille, R.M. and B.K. Hall. 1987. Development of the head skeleton of the Japanesse Medaka, *Oryzias latipes* (Teleostei). J. Morphol. 193: 135–158.

Leis. J.M. and T. Trnski. 1989. The Larvae of Indo-Pacific Shore-fishes. New South Wales University Press and University Press of Hawai Press, Honolulu 371 pp.

Lekander, B. 1949. The sensory line system and the canal bones in the head of some Ostariophysi. Acta Zool., Stockholm 30: 1–131.

Liao, I.C., H.Y. Yan and M.S. Su. 1977. Studies on milkfish fry-I. On the morphology and its related problems of milkfish fry from the coast of Tungkank. J. Fish Soc. Taiwan 6: 73–83.

Mabee, P.M. and T.A. Trendler. 1996. Development of the cranium and paired fins in *Betta splendens* (Teleostei: Percomorpha): Intraspecific variation and interspecific comparisons. J. Morphol. 227: 249–287.

Matsuuru, Y. and N.T. Yoneda. 1987. Osteological development of the lophiid anglerfish, *Lophius gastrophysus*. Japan. J. Ichthyol. 33: 360–367.

Morris, S.L. and A.J. Gaudin. 1982. Osteocranial development in the viviparous surfperch *Amphistoichus argenteus* (Pisces: Embiotocidae). J. Morphol. 174: 95–120.

Moser, H.G. [ed.]. 1984. Ontogeny and Systematics of Fishes. American Society of Ichthyologists and Herpetologists, Special Publication 1.

Nelson, G. 1969. Gill arches and the phylogeny of fishes, with notes on the classification of vertebrates. Bull. Amer. Mus. Nat. Hist. 141, article 4: 477–552.

Parker, W.K. 1873. On the structure and development of the skull in salmon (*Salmo salar*). Phil. Trans. R. Soc. London 163: 95–145.

Patterson, C. 1977. Cartilage bones, dermal bones and membrane bones or the exoskeleton versus the endoskeleton. pp. 77–121. *In*: S.M. Andrew, R.S. Miles, and A.D. Walker [eds.]. Problems in Vertebrate Evolution. Academic Press, London.

Patterson, C. and G.D. Johnson. 1997. The gill-arches of gonorynchiform fishes. S. Afr. J. Sci. 93: 594–600.

Pehrson, T. 1944. Development of the latero-sensory canal in the skull of *Esox lucius*. Acta Zool., Stockholm 25: 135–157.

Pottoff, T., S. Kelly and L.A. Collins. 1988. Osteological development of the red snapper, *Lutjanus campechanus* (Lutjanidae). Bull. Mar. Sci. 43: 1–40.

Rabanal, H.R., R.S. Esquerra and M.M. Nepomuceno. 1953. Studies on the rate of growth of milkfish, *Chanos chanos* Forsskål, under cultivation. Rate of growth of fry and fingerlings in fishpond nurseries. Proc. Indo-Pacific Fish. Council 4(2): 171–180.

Ristovska, M., B. Karamn, W. Verraes and D. Adriaens. 2006. Early development of the chondrocranium in *Salmo letnica* (Karamn, 1924) (Teleostei: Salmonidae). J. Fish Biol. 68: 458–480.

Schultze, H.-P. 2008. Nomenclature and homologization of cranial bones in actinopterygians. pp. 23–48. *In*: G. Arratia, H.-P. Schultze, and M.V.H. Wilson [eds.]. Mesozoic Fishes 4—Homology and Phylogeny. Verlag Dr. F. Pfeil, Munich.

Schultze, H.-P. and G. Arratia. 1988. Reevaluation of the caudal skeleton of some actinopterygian fishes. II. *Hiodon, Elops* and *Albula*. J. Morphol. 195: 257–303.

Stöhr, P. 1882. Zur Entwicklungsgeschichte des Kopfskelettes der Teleostier. *In* Festschrift 3 Säkulafeier Alma Julia Maximiliana Universitäts Würzburg, Leipzig 2: 1–23.

Surlemont, C., M. Chardon and M. Vandewalle. 1989. Skeleton, muscles and movements of the head of a 5.2 mm fry of *Clarias gariepinus* (Burchell) (Pisces Siluriformes). Fortschritte der Zoologie 35: 459–562.

Taki, Y., H. Kohno and S. Hara. 1986. Early development of fin-supports and fin-rays in the milkfish *Chanos chanos*. Japan. J. Ichthyol. 32: 413–420.

Taki, Y., H. Kohno and S. Hara. 1987. Morphological aspects of the development of swimming and feeding functions in the milkfish *Chanos chanos*. Japan. J. Ichthyol. 34: 198–208.

Tischomiroff, A. 1885. Zür Entwicklung des Schädels bei den Teleostieren. Zoologischer Anzeiger 8: 533–537.

Vandewalle, P., B. Focant, F. Huriaux and M. Chardon. 1992. Early development of the cephalic skeleton of *Barbus barbus* (Teleotei, Cyprinidae). J. Fish Biol. 41: 43–62.

Vandewalle, P., I. Gluckman and F. Wagemans. 1998. A critical assessment of the Alcian blue/Alizarine double staining in fish larvae and fry. Belg. J. Zool. 128: 93–95.

Verraes, W. 1974. Some functional aspects of ossification in the cartilaginous ceratohyal during postembryonic development in *Salmo gairdneri* Richardson, 1836 (Teleostei: Salmonidae). Forma et Function 8: 27–32.

Verraes, W. 1977. Postembryonic ontogeny and functional anatomy of the ligamentum mandibulo-hyoideum and the ligamentum interoperculo-mandibulare, with notes on the opercular bones and some other cranial elements in *Salmo gairdneri* Richardson, 1836 (Teleostei: Salmonidae). J. Morphol. 151: 11–120.

Villaluz, A.C. 1979. Early ontogeny and meristic variability of milkfish *Chanos chanos* (Forsskål) in Philippine waters. M Sc. Thesis, University of Guelph, Guelph, 110 pp.

Villaluz, A.C. and A.S. Unggui. 1983. Effects of temperature on behavior, growth, development and survival of young milkfish, *Chanos chanos* (Forsskål). Aquaculture 35: 321–330.

Wells, F.R. 1923. On the morphology of the chondrocranium of the larval herring (*Clupea harengus*). Proc. Zool. Soc. London 92: 1213–1229.

Wiley, E.O. 2008. Homology, identity and transformation. pp. 9–21. *In*: G. Arratia, H.-P. Schultze, and M.V.H. Wilson [eds.]. Mesozoic Fishes 4—Homology and Phylogeny. Verlag Dr. F. Pfeil, Munich.

4

A Review of the Cranial and Pectoral Musculature of Gonorynchiform Fishes, with Comments on Their Functional Morphology and a Comparison with Other Otocephalans

Rui Diogo[1]

Abstract

Very few studies focused on the anatomy of gonorynchiforms deal in detail with their musculature. In this paper I provide an overview of the cranial musculature (excluding branchial and extrinsic eye musculature) and pectoral girdle musculature of extant gonorynchiforms. Some comments on the functional morphology of these fishes, as well as a comparison with other otocephalans, are also given. The paper is based on a review of the available literature and on the author's own analysis of the cephalic and pectoral girdle musculature of members of the various major otocephalan groups, including representatives of the seven extant gonorynchiform genera. The overview provided here points out that there is a considerable morphological diversity of the cranial and pectoral girdle muscles within the Gonorynchiformes. Interestingly, with the exception

[1] Rui Diogo (corresponding author): Department of Anthropology, The George Washington University, 2110 G St. NW, Washington, DC 20052, USA.
e-mail: Rui_Diogo@hotmail.com

of the adductor mandibulae, hyohyoideus inferioris, adductor profundus, and eventually adductor hyomandibulae and arrector dorsalis, the plesiomorphic gonorynchiform configuration for each of the muscles discussed seems to represent the plesiomorphic configuration for the Otocephala as a whole. Because of their apparent basal position within otocephalans and the rather plesiomorphic configuration of their myological structures, the gonorynchiforms can thus play an important role in studies of the comparative anatomy, functional morphology and evolution of not only otocephalans but also teleosts in general.

Introduction

The study of myological structures of a certain group can not only reveal useful information to pave the way for subsequent functional and/or ecomorphological studies, but also help to clarify the inter and/or intrarelationships of that group. However, only very few of the several studies focused on gonorynchiform anatomy deal in detail with the musculature of these fishes (Le Danois 1966, Pasleau 1974, Howes 1985). Howes (1985) provided a comparison between some cranial muscles of the genera *Chanos, Gonorynchus, Phractolaemus, Cromeria, Grasseichthys, Parakneria* and *Kneria*, this being the only paper to compare the seven extant gonorynchiform genera. However, Howes' (1985) study did not include the muscles of the pectoral girdle. Pasleau's unpublished Bachelor's thesis (1974) did provide information on the pectoral girdle muscles, as well as on the cranial ones, but the myological descriptions pertain exclusively to the genera *Chanos* and *Gonorynchus*. With respect to Le Danois (1966), one should, as stressed by Howes (1985), praise the effort of that author to provide an extensive, ambitious work dealing with many aspects of the anatomy of the species *Gonorynchus gonorynchus*. However, as also noted by Howes (1985: 274), Le Danois' (1996) descriptions on the cranial and pectoral girdle musculature have "numerous inaccuracies, misidentifications and misinterpretations that render this work virtually useless".

   The aim of the present paper is to provide an overview of the cranial musculature (excluding branchial and extrinsic eye musculature) and pectoral girdle musculature of extant gonorynchiforms. Some comments on the functional morphology of these fishes, as well as a comparison with other otocephalans, are also given. The paper is based on a review of the available literature and on the author's own analysis of the cephalic and pectoral girdle musculature of members of the various major otocephalan groups, including representatives of the seven gonorynchiform extant genera. A list of these species is given below (*MNCN*, Museo Nacional de Ciencias Naturales de Madrid; *LFEM*, Laboratory of Functional and Evolutionary Morphology of the University of Liège; *CAS*, California Academy of Sciences; *MRAC*, Musée Royal de l'Afrique Centrale; *FMNH*, Field Museum of Natural History; *USNM*, National Museum of Natural History of Washington, DC; *UNB*, Université Nationale du Bénin; *alc*, alcohol-preserved specimens):

Alestes macrophthalmus LFEM, 3 (alc). *Arius heudelotii* LFEM, 4 (alc). *Bagrus bayad* LFEM, 2 (alc). *Bagrus docmak* MRAC 86-07-P-512, 1 (alc). *Barbus guiraonis* MNCN 245730, 3 (alc). *Brycon guatemalensis* MNCN180536, 3 (alc). *Brycon henni* CAS 39499, 1 (alc). *Callichthys callichthys* USNM 226210, 2 (alc). *Cetopsis coecutiens* USNM 265628, 2 (alc). *Chanos chanos* USNM 347536, 1 (alc); LFEM, 3 (alc). *Chrysichthys auratus* UNB, 2 (alc). *Chrysichthys nigrodigitatus* UNB, 2 (alc). *Clarias anguillaris* LFEM, 2 (alc). *Clarias batrachus* LFEM, 2 (alc). *Clupea pallasii* MNCN 48869, 3 (alc); LFEM, 1 (alc). *Cromeria nilotica* MRAC P.141098-99, 2 (alc). *Danio rerio* LFEM, 5 (alc). *Denticeps clupeoides* MRAC 76-032-P-1, 1 (alc); LFEM, 3 (alc). *Diplomystes chilensis* LFEM, 3 (alc). *Engraulis anchoita* LFEM, 2 (alc). *Gonorynchus gonorynchus* LFEM, 2 (alc). *Gonorynchus greyi* FMNH 103977, 1 (alc); LFEM, 2 (alc). *Grasseichthys gabonensis* MRAC 73-002-P-264-266, 3 (alc). *Gymnotus carapo* MNCN 115675, 2 (alc). *Hoplerythrinus unitaeniatus* LFEM, 4 (alc). *Ictalurus furcatus* LFEM, 2 (alc). *Ictalurus punctatus* USNM 244950, 2 (alc). *Kneria auriculata* LFEM, 3 (alc). *Kneria wittei* MRAC P-33512-513, 2 (alc). *Nematogenys inermis* USNM 084346, 2 (alc); LFEM, 2 (alc). *Opsariichthys uncirostris* MNCN 56668, 3 (alc). *Parakneria abbreviata* MRAC 99-090-P-703-705, 3 (alc). *Parakneria tanzaniae* LFEM, 1 (alc). *Phractolaemus ansorgii* MRAC P.137982-84, 3 (alc). *Pimelodus blochii* LFEM, 2 (alc). *Pristigater cayana* LFEM, 2 (alc). *Rhamphichthys rostratus* LFEM, 3 (alc). *Silurus aristotelis* LFEM, 2 (alc). *Silurus glanis* LFEM, 2 (alc). *Sternopygus macrurus* CSA 48241, 1 (alc); LFEM, 3 (alc). *Trichomycterus areolatus* LFEM, 2 (alc). *Trichomycterus banneaui* LFEM, 2 (alc). *Zacco platypus* MNCN 35712, 3 (alc).

Morphological Diversity of the Cranial and Pectoral Musculature of Gonorynchiforms and Comparison with Other Otocephalans

In this section, I not only refer to the morphological diversity of the different cranial and pectoral muscles of gonorynchiforms, but also compare these muscles with those of other otocephalan fishes. It is important to emphasize that, unless stated otherwise, the anatomical descriptions given below refer to adult fishes. The nomenclature of the myological and osteological structures mentioned in this work follows that of Diogo (in press).

Cheek Musculature

Adductor mandibulae

In the members of all the seven extant gonorynchiform genera the adductor mandibulae is divided into various sections (Figs. 4.1, 4.2, 4.3, 4.4, 4.5, 4.6, 4.7). In *Chanos* the most lateral of these sections, the A1-OST (*sensu* Diogo and Chardon 2000a), is subdivided into lateral (A1-OST-L) and mesial (A1-OST-M) bundles (Figs. 4.1 and 4.2). The A1-OST-L runs from the preopercle and quadrate to the maxilla (Figs. 4.1 and 4.2). The A1-OST-M originates on the preopercle and quadrate and inserts on the mesial surface of the mandible, being mixed anteriorly with fibers of the adductor mandibulae Aω (Figs. 4.1 and 4.2). In *Gonorynchus* the A1-OST also originates on the preopercle and quadrate, but is subdivided into three, and not two, bundles. The most mesial, which seems to correspond to the A1-OST-M of *Chanos*, attaches anteriorly into the thick connective tissue surrounding the coronoid process of the dentary

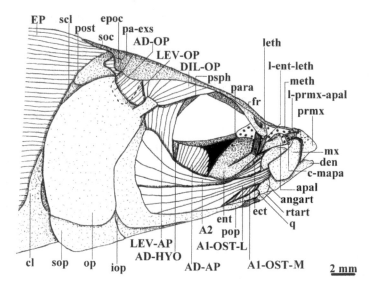

Fig. 4.1 Lateral view of the cephalic musculature of *Chanos chanos*. The pectoral girdle muscles are not illustrated; the nasals and infraorbitals were removed. A1-OST-L, A1-OST-M, A2, sections of adductor mandibulae; AD-AP, adductor arcus palatini; AD-OP, adductor operculi; AD-HYO, adductor hyomandibulae; angart, angulo-articular; apal, autopalatine; c-mapa, small cartilage between maxilla and autopalatine; cl, cleithrum; den, dentary bone; DIL-OP, dilatator operculi; ect, ectopterygoid; ent, entopterygoid; EP, epaxialis; epoc, epioccipital; fr, frontal; iop, interopercle; l-ent-leth, ligament between entopterygoid and lateral-ethmoid; l-prmx-apal, ligament between premaxilla and autopalatine; leth, lateral-ethmoid; LEV-AP, levator arcus palatini; LEV-OP, levator operculi; meth, mesethmoid; mx, maxilla; op, opercle; pa-exs, parieto-extrascapular; para, parasphenoid; pop, preopercle; post, post-temporal; prmx, premaxilla; psph, pterosphenoid; q, quadrate; rtart, retroarticular; scl, supracleithrum; soc, supraoccipital; sop, subopercle.

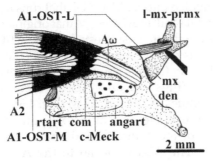

Fig. 4.2 Mesial view of the left mandible, adductor mandibulae and maxilla of *Chanos chanos*. A1-OST-L, A1-OST-M, A2, Aω, sections of adductor mandibulae; angart, angulo-articular; c-Meck, Meckel's cartilage; com, coronomeckelian bone; den, dentary bone; l-mx-prmx, ligament between maxilla and premaxilla; mx, maxilla; rtart, retroarticular.

bone (see Howes 1985, fig. 5: a1i). Another bundle, which seems to correspond to the A1-L of *Chanos*, attaches anteriorly on the maxilla (see Howes 1985, fig. 5: a10). The other bundle attaches anteriorly on the lacrimal (see Howes 1985, fig. 5: A2L). It should be noted that Howes (1985) considered this latter bundle to be part of the A2. However, as shown in Howes' (1985) fig. 5, this is the most lateral bundle of the adductor mandibulae complex and thus seems to correspond to part of the A1-OST of other otocephalans (*sensu* Diogo and Chardon 2000a: see below). In *Phractolaemus* (Fig. 4.4) the A1-OST narrows into a thin tendon as it approaches the jaw articulation, passing beneath a bridge formed by a short, thick ligament attached from the outer quadrate spine to the lateral surface of this bone. Exiting from the bridge the tendon turns through 90°, whereupon it joins a broad aponeurosis from which extends the A1-OST-L. This bundle A1-OST-L (Fig. 4.4) inserts on the distal cavity of the maxilla and on the connective tissue extending between the maxilla and the small premaxilla, covering the rictal cartilages that lie ventrally and between the dentary bones. The A1-OST-M (Fig. 4.4) originates ventrally from the quadrate spine, covers the outer face of the lower jaw, and inserts on the posterior process of the curved maxilla. In *Parakneria* and *Kneria* the A1-OST is also subdivided into an A1-OST-L and an A1-OST-M: A1-OST-L connects the preopercle to the maxilla; A1-OST-M originates on the quadrate and attaches anteriorly on the maxilla, mesially to the A1-OST-L (see figs. 11 and 12 of Howes 1985, and also Fig. 4.6 in this chapter). In *Grasseichthys* (Fig. 4.5C) the muscle adductor mandibulae is quite reduced in size. Only the A1-OST seems to be present, and, although there is some differentiation between the more lateral and more mesial fibers of this A1-OST, there seems to be no complete differentiation between the A1-OST-L and the A1-OST-M. The A1-OST runs from the inner face of the infraorbital series and the preopercle to the connective tissue investing the rictus of the jaws, the majority of the fibers appearing to attach into the tissue covering the maxilla. With respect to *Cromeria* (Fig. 4.5A,

B), the A1-OST is subdivided into an A1-OST-L that runs from the preopercle and quadrate to the maxilla and an A1-OST-M that originates on the quadrate and attaches anteriorly on the maxilla, mesially to the A1-OST-L.

The section A2 (*sensu* Diogo and Chardon 2000a) of the adductor mandibulae complex in *Chanos* runs from the preopercle to the coronomeckelian bone (Figs. 4.1 and 4.2). In *Gonorynchus* the A2 is subdivided into two bundles, A2-L and A2-M (Fig. 4.3). The A2-L attaches posteriorly on the preopercle and anteriorly on the posterior and posterolateral surfaces of the mandible and on the connective tissue of the coronoid process. The A2-M, mesial to the A2-L, runs from the preopercle to the mesial surface of the dentary bone. A different configuration is found in *Phractolaemus*. In the members of this genus the A2 (Fig. 4.4) originates on the preopercle and hyomandibula. Anteriorly, this section is subdivided into a dorsal bundle (A2-D) inserting on the connective tissue of the coronoid process and on the antorbital, and a ventral bundle (A2-V) inserted on the coronomeckelian bone and on the tendon of the A1-OST. In *Parakneria* and *Kneria* the A2 is constituted by a single mass of fibers running from the preopercle, quadrate and hyomandibula to the coronomeckelian bone (see figs. 11 and 12 of Howes 1985, and also Fig. 4.6 in this chapter). In *Cromeria* (Fig. 4.5A, B) the A2 is also constituted by a single mass of fibers, which originates from the preopercle. Below the center of the orbit the A2 becomes aponeurotically constricted and linked to the quadrate by a tendinous band that runs mesially from the aponeurosis (Fig. 4.5B). From the point of its constriction to its insertion, the muscle is represented by a tendinous sheet that fans out to a tripartite attachment to the autopalatine, the coronomeckelian bone, and the maxilla.

Apart from the A1-OST and A2, in *Chanos* there is another section of the adductor mandibulae, the Aω. This is a small section lodged on the mesial surface of the mandible. Posteriorly it is mixed with the fibers of the A2 and attached to the anguloarticular, and anteriorly it is attached to the dentary (Fig. 4.2). In the other six extant gonorynchiform genera there is no Aω (e.g., Fig. 4.3). The adductor mandibulae sections A0 and A3 (*sensu* Diogo and Chardon 2000a) are absent in gonorynchiforms.

Gosline (1989) and Diogo and Chardon (2000a) noted that there are two basically different pathways of differentiation in the cheek part of the adductor mandibulae in teleostean fishes. The plesiomorphic teleostean pattern is that in which an undivided cheek muscle attaches to the mesial face of the mandible and the Aω is present (the presence of an Aω is seemingly an actinopterygian plesiomorphy; it was probably the first adductor mandibulae bundle to separate as a distinct entity: Edgeworth 1935, Winterbottom 1974, Lauder 1980a). This configuration is found, for example, in the otocephalan clupeiform *Denticeps* (Fig. 4.8), as well as in

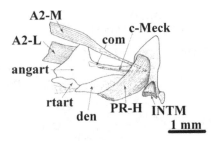

Fig. 4.3 Mesial view of the left mandible, adductor mandibulae, protractor hyoidei and intermandibularis of *Gonorynchus gonorynchus*. Modified from Howes (1985). A2-L, A2-M, sections of adductor mandibulae; angart, angulo-articular; c-Meck, Meckel's cartilage; com, coronomeckelian bone; den, dentary bone; INTM, intermandibularis; PR-H, protractor hyoidei; rtart, retroarticular.

other, more apomorphic clupeiforms such as *Clupea* and in non-otocephalan basal teleosts such as *Elops* (Gosline 1989). It should, however, be noted that according to Wu and Shen (2004: 722) some derived clupeiforms (e.g., *Ilisha*, *Amblygaster*, *Nematalosa*) may also exhibit, apart from the sections A2 and Aω, a separate section A3 of the adductor mandibulae (for a discussion on the homologies of the A3 within teleosts, see Diogo and Chardon 2000a).

From the rather simple configuration found in fishes such as *Denticeps* or *Elops*, two types of differentiation occur. In acanthopterygians, an *anterodorsolateral* part of the cheek muscle (A1 *sensu* Diogo and Chardon 2000a) has seemingly developed an attachment to the maxilla *via* the primordial ligament. This situation seems to be plesiomorphic for acanthopterygians and is present, for example, in *Aulopus* (Lauder and Liem 1983, Gosline 1989) and *Neoscopelus* (Winterbottom 1974). A secondary differentiation of the adductor mandibulae (A3 *sensu* Diogo and Chardon 2000a), mesial to all the others, is found in many acanthopterygian fishes.

In ostariophysans, i.e., Gonorynchiformes, Cypriniformes, Characiformes, Gymnotiformes and Siluriformes, an *anteroventrolateral* part (A1-OST *sensu* Diogo and Chardon 2000a) of the cheek muscle separates and attaches to the posterodorsolateral face of the mandible. This section is effectively found in all ostariophysans examined in the present work (Figs. 4.1, 4.2, 4.3, 4.4, 4.5, 4.9, 4.10). Since an A1-OST *sensu* Diogo and Chardon (2000a) is absent in *Denticeps* (Fig. 4.8) and in other clupeiforms examined, as well as in non-otocephalan basal teleosts such as elopomorphs and osteoglossomorphs (e.g., Gosline 1989, Diogo and Chardon 2000a), the presence of this section may constitute an ostariophysan synapomorphy. According to Gosline (1989), the A1-OST seems to have developed as a supplementary system for raising the mandible. It should be noted that Wu and Shen (2004) have seemingly misinterpreted the identity of the A1-OST. They state that this bundle is distributed in many acanthopterygian teleosts

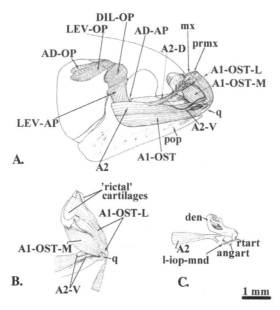

Fig. 4.4 Cephalic musculature of *Phractolaemus ansorgei*. Modified from Howes (1985). (A) Lateral view of the cephalic muscles. (B) Dorsolateral view of the upper and lower jaws, the quadrate and the adductor mandibulae; the A2-D has been removed and A1-OST-L has been cut posteriorly and moved laterally to the quadrate to expose the lower portion of the A1-OST-M. (C) Lateral view of the lower jaw and the section A2 of the adductor mandibulae. A1-OST, A1-OST-L, A1-OST-M, A2, A2-D, A2-D, sections of adductor mandibulae; AD-AP, adductor arcus palatini; AD-OP, adductor operculi; angart, angulo-articular; den, dentary bone; DIL-OP, dilatator operculi; I-iop-mnd, ligament between interopercle and mandible; LEV-AP, levator arcus palatini; LEV-OP, levator operculi; mx, maxilla; pop, preopercle; prmx, premaxilla; q, quadrate; rtart, retroarticular.

and is absent in some otocephalans such as the catfish *Parasilurus*. However, according to Gosline (1989) the A1-OST (his "ostariophysine external division") is missing in most, if not all, the acanthopterygians. Moreover, an A1-OST is in fact present in the members of the catfish genus *Parasilurus*, as well as in all the other numerous catfishes examined by Gosline (1989), Diogo and Chardon (2000a) and Diogo (2004).

Apart from the A1-OST, another section of the adductor mandibulae (A3 *sensu* Diogo and Chardon 2000a), mesial to the A1 and A2, may also be present in ostariophysans. It is interesting to note that a well-differentiated, separated A3 *sensu* Diogo and Chardon (2000a) is absent in all seven gonorynchiform extant genera (see above) as well as in the cypriniforms examined in the present work (Fig. 4.9). However, such a section is found in the characiforms, siluriforms and gymnotiforms examined (Fig. 4.10). The consistent presence of a well-differentiated A3 in these three latter groups may constitute a synapomorphy to support their close relationship, as

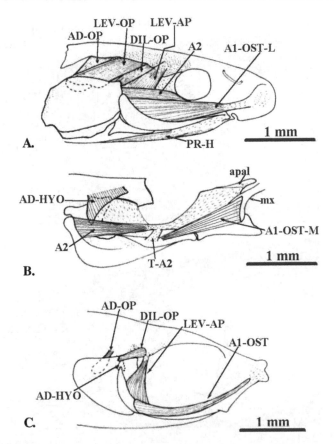

Fig. 4.5 Cephalic musculature of *Cromeria nilotica* and *Grasseichthys gabonensis*. Modified from Howes (1985). (A) *Cromeria nilotica*, lateral view of the cephalic muscles. (B) *Cromeria nilotica*, most lateral muscles were removed to show details of the adductor hyomandibulae and of the sections A2 and A1-OST-M of the adductor mandibulae. (C) *Grasseichthys gabonensis*, lateral view of the cephalic muscles. A1-OST-L, A1-OST-M, A2, sections of adductor mandibulae; AD-AP, adductor arcus palatini; AD-HYO, adductor hyomandibulae; AD-OP, adductor operculi; apal, autopalatine; DIL-OP, dilatator operculi; LEV-AP, levator arcus palatini; LEV-OP, levator operculi; mx, maxilla; PR-H, protractor hyoidei; T-A2, tendon of adductor mandibulae A2.

suggested by, for example, Fink and Fink (1981, 1996). However, it should be noted that, although eventually constituting a synapomorphy of the clade constituted by these three groups, an A3 *sensu* Diogo and Chardon (2000a: fig. 8) was also acquired in certain other teleosts such as many acanthopterygians (very likely by parallel evolution: for a recent discussion on this subject, see Diogo 2004).

In many ostariophysan fishes, that is, some characiforms, a large number of gymnotiforms, and all gonorynchiforms (Figs. 4.1, 4.2, 4.3, 4.4, 4.5), the

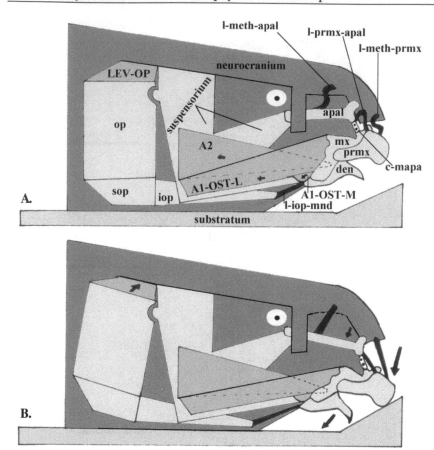

Fig. 4.6 Diagram of the author's hypothesis concerning the mechanisms of mouth closure by the action of the adductor mandibulae and mouth opening by the action of the levator operculi in *Parakneria abbreviata* (note that only some cephalic structures are illustrated and that the ventral cephalic muscles and the mechanisms of mouth opening and mouth closure associated with these muscles are not represented; also note that the movements shown are exaggerated, in order to facilitate the understanding of the diagram). (A) The premaxilla is retracted and the mandible is raised because of contraction of the adductor mandibulae. (B) The premaxilla is protracted and the mandible is lowered because of contraction of the levator operculi (for more specific details, see text). A1-OST-L, A1-OST-M, A2, sections of adductor mandibulae; c-mapa, small cartilage between maxilla and autopalatine; den, dentary bone; iop, interopercle; l-iop-mnd, ligament between interopercle and mandible; l-meth-apal, ligament between mesethmoid and autopalatine; l-meth-prmx, ligament between mesethmoid and premaxilla; l-prmx-apal, ligament between premaxilla and autopalatine; LEV-OP, levator operculi; mx, maxilla; op, opercle; prmx, premaxilla; sop, subopercle.

most external bundles of the adductor mandibulae attach directly to the upper jaw and/or the anterior infraorbital bones (Figs. 4.1, 4.2, 4.3, 4.4, 4.5, 4.9; see also Takahasi 1925, Alexander 1964, Chardon and De la Hoz 1973, Winterbottom 1974, De la Hoz 1974, Vandewalle 1975, 1977, Howes 1976,

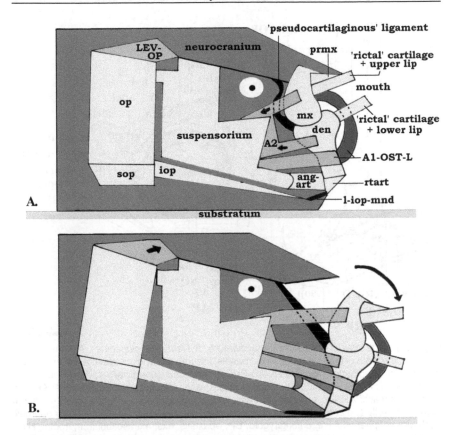

Fig. 4.7 Diagram of the author's hypothesis concerning the mechanisms of mouth closure by the action of the adductor mandibulae and mouth opening by the action of the levator operculi in *Phractolaemus ansorgei* (note that only some cephalic structures are illustrated and that the ventral cephalic muscles and the mechanisms of mouth opening and mouth closure associated with these muscles are not represented; also note that the movements shown are exaggerated, in order to facilitate the understanding of the diagram). (A) The mouth is closed because of contraction of the adductor mandibulae. (B) The mouth is opened and projected anteroventrally because of contraction of the levator operculi (for more specific details, see text). A1-OST-L, A2, sections of adductor mandibulae; ang-art, angulo-articular; den, dentary bone; iop, interopercle; l-iop-mnd, ligament between interopercle and mandible; LEV-OP, levator operculi; mx, maxilla; op, opercle; prmx, premaxilla; rtart, retroarticular; sop, subopercle.

Vari 1979, Fink and Fink 1981, De la Hoz and Chardon 1984, Aguilera 1986, Gosline 1989, Diogo 2004). [Note: as explained by Howes (1983), Gosline (1989) and Diogo and Chardon (2000a) the retractor tentaculi muscle attached directly on the maxilla in certain derived catfishes is the result of a differentiation of the most internal, and not of the most external, sections of the adductor mandibulae.] The well-developed and rather independent

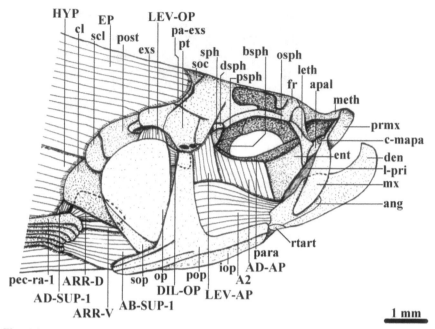

Fig. 4.8 Lateral view of the cephalic musculature of *Denticeps clupeoides*. All muscles are exposed; the teeth of the jaws, onodontes, nasals, infraorbitals and postcleithra were removed. A2, section of adductor mandibulae; AB-SUP-1, abductor superficialis 1; AD-AP, adductor arcus palatini; AD-SUP-1, adductor superficialis 1; ang, angular; apal, autopalatine; ARR-D, arrector dorsalis; ARR-V, arrector ventralis; bsph, basisphenoid; c-mapa, small cartilage between maxilla and autopalatine; cl, cleithrum; den, dentary bone; DIL-OP, dilatator operculi; dsph, dermosphenotic; ent, entopterygoid; EP, epaxialis; exs, extrascapular; fr, frontal; HYP, hypaxialis; iop, interopercle; l-pri, primordial ligament; leth, lateral-ethmoid; LEV-AP, levator arcus palatini; LEV-OP, levator operculi; meth, mesethmoid; mx, maxilla; op, opercle; osph, orbitosphenoid; pa-exs, parieto-extrascapular; para, parasphenoid; pec-ra-1, pectoral ray 1; pop, preopercle; post, post-temporal; prmx, premaxilla; psph, pterosphenoid; pt, pterotic; rtart, retroarticular; scl, supracleithrum; soc, supraoccipital; sop, subopercle; sph, sphenotic.

lateral section of the adductor mandibulae attaching directly on the maxilla in cypriniform fishes was named "adductor mandibulae A0" by Diogo and Chardon (2000a) (see fig. 9).

It has been widely discussed whether a direct insertion of the adductor mandibulae on the upper jaw represents a plesiomorphic or an apomorphic condition for ostariophysans. Takahasi (1925), Fink and Fink (1981) and other authors defended the view that the attachment of the most external bundles of the adductor mandibulae on the upper jaw/infraorbitals is the basal ostariophysan configuration. However, Alexander (1964), Vari (1979), and Gosline (1989) argued that such an attachment of the adductor mandible on the upper jaw is a derived character within the Ostariophysi, which was acquired in different groups exhibiting a small mouth and/or

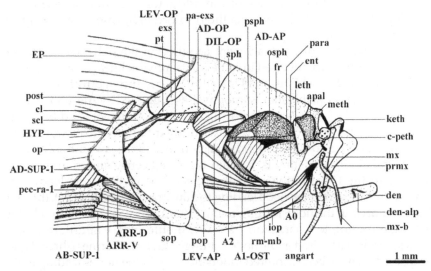

Fig. 4.9 Lateral view of the cephalic musculature of *Danio rerio*. All muscles are exposed, the mesial branch of the ramus mandibularis is also illustrated; the nasals, infraorbitals and postcleithra were removed. A0, A1-OST, A2, sections of adductor mandibulae; AB-SUP-1, abductor superficialis 1; AD-AP, adductor arcus palatini; AD-OP, adductor operculi; AD-SUP-1, adductor superficialis 1; angart, angulo-articular; apal, autopalatine; ARR-D, arrector dorsalis; ARR-V, arrector ventralis; c-peth, pre-ethmoid cartilage; cl, cleithrum; den, dentary bone; den-alp, anterolateral process of dentary bone; DIL-OP, dilatator operculi; ent, entopterygoid; EP, epaxialis; exs, extrascapular; fr, frontal; HYP, hypaxialis; iop, interopercle; keth, kinethmoid; leth, lateral-ethmoid; LEV-AP, levator arcus palatini; LEV-OP, levator operculi; meth, mesethmoid; mx, maxilla; mx-b, maxillary barbel; op, opercle; osph, orbitosphenoid; pa-exs, parieto-extrascapular; para, parasphenoid; pec-ra-1, pectoral ray 1; pop, preopercle; post, post-temporal; prmx, premaxilla; psph, pterosphenoid; pt, pterotic; rm-mb, mesial branch of ramus mandibularis; scl, supracleithrum; sop, subopercle; sph, sphenotic.

a protrusile upper jaw. It should be noted, however, that the vast majority of these discussions did not place emphasis on the situation found in the clupeiforms, since in the 1960s, 1970s and 1980s most authors did not accept a monophyletic otocephalan clade formed by the Clupeomorpha and the Ostariophysi (see, e.g., Stiassny *et al.* 2004).

When one takes the clupeiforms into account, it seems that, at least for otocephalans, the plesiomorphic condition is without a direct, external attachment of the adductor mandibulae on the upper jaw and/or infraorbitals. In fact, in all clupeiforms examined, including *Denticeps* (Fig. 4.8), as well as in basal teleosts such as *Elops*, there is no such direct attachment. What is found in clupeiforms and in fishes such as *Elops*, as well as in many ostariophysans without a direct adductor mandibulae attachment on the maxilla, is an association of some lateral fibers of this muscle with the posteroventral portion of the primordial ligament (e.g., Figs. 4.8 and 4.10). Thus, at a certain moment in otocephalan evolutionary history the outermost bundles of the adductor mandibulae seemingly developed a direct attachment, via the primordial ligament, on the upper jaw/infraorbitals.

But was such a direct attachment acquired in the node leading to the Ostariophysi and then reversed in some ostariophysan lineages, as suggested by Takahasi (1925) and Fink and Fink (1981)? Or is such a direct attachment a derived character within the Ostariophysi, as suggested by Alexander (1964) or Gosline (1989)? This is a rather difficult question. As stressed by Diogo (2004), in a strictly phylogenetic context, the distribution of a direct, external attachment of the adductor mandibulae on the upper jaw/infraorbitals is somewhat ambiguous, that is, it could (a) have been acquired in the node leading to the Ostariophysi and then been lost in the node leading to Characiformes and in the node leading to Siluriformes (three steps), (b) have been acquired in the node leading to the Ostariophysi and then been lost in the node leading to Characiformes + Gymnotiformes + Siluriformes and subsequently reacquired in the node leading to Gymnotiformes (three steps), or instead (c) have been acquired independently in the nodes leading to the Gonorynchiformes, to the Cypriniformes and to the Gymnotiformes, respectively (three steps). This, of course, if we assume that the plesiomorphic condition for Characiformes is the absence of a direct attachment of the adductor mandibulae on the upper jaw/infraorbitals, which seems indeed to be the case (e.g., Alexander 1964, Vari 1979, Gosline 1989), and that the plesiomorphic condition for Gymnotiformes is to have such an attachment, which is less clear. In fact, in the members of two genera that are often considered examples of plesiomorphic extant gymnotiforms, *Gymnotus* and *Electrophorus* (e.g., Albert and Campos-da-Paz 1998, Albert 2001), there is no direct attachment of the external bundles of the adductor mandibulae to the upper jaw/infraorbitals. There is, however, an indirect attachment of these bundles to these latter structures by means of the primordial ligament (e.g., De la Hoz 1974, Aguilera 1986). Thus, it may be that such a direct attachment is only plesiomorphic for two of the six otocephalan orders, namely for the Cypriniformes and Gonorynchiformes. Moreover, the situation found in the plesiomorphic gonorynchiform genera *Chanos* and *Gonorynchus* (see Grande and Poyato-Ariza 1999, Lavoué *et al.* 2005) is rather different from that found in Cypriniformes: the outermost adductor mandibulae bundle (A1-OST) inserts on both the mandible (A1-OST-M) and the maxilla (A1-OST-L) (Fig. 4.1) and, thus, there is no well-differentiated bundle attaching exclusively on the maxilla, as in most cypriniforms (Fig. 4.9: A0). Taking into account the points mentioned above, and in face of the data available so far, I am inclined to agree with Alexander (1964), Vari (1979) and Gosline (1989) that the plesiomorphic ostariophysan condition is a lack of a direct attachment of the adductor mandibulae on the upper jaw and/or infraorbitals.

As mentioned above, these latter authors argued that the homoplasic acquisition of such a direct attachment of the adductor mandibulae on the upper jaw in some ostariophysan taxa is probably related with the fact that the members of those taxa exhibit a small mouth and/or a protrusile upper jaw. An analysis of the functional morphology of gonorynchiform fishes such as *Parakneria* and *Kneria* reveals that these fishes effectively have a protrusile upper jaw, which in certain aspects is strikingly similar to that found in most cypriniforms. As can be seen in Fig. 4.6, in *Parakneria* and *Kneria* the premaxilla and maxilla are relatively free from the neurocranium and the suspensorium, as is the case in cypriniforms (Fig. 4.9). When the mouth is opened (e.g., by the contraction of the levator operculi, as shown in Fig. 4.6), the upper jaw is protruded anteroventrally, because of the long ligaments existing between the mesethmoid and autopalatine, the premaxilla and mesethmoid, and the premaxilla and autopalatine (Fig. 4.6B). When the adductor mandibulae contracts, the direct attachment of the A1–OST on the upper jaw, together with the attachment of the A2 on the lower jaw and the firm connection between the coronoid process of the mandible and the maxilla, causes the retraction of the upper jaw (Fig. 4.6A). A particularly peculiar mechanism of mouth opening/closure is found in the gonorynchiform *Phractolaemus*. In this genus, the only connection between the upper jaw and the suspensorium/neurocranium is by the peculiar, strong "pseudocartilaginous ligament" connecting the prevomers, the autopalatines and the upper and lower jaws. The upper jaw is thus extremely mobile. As described by Thys van den Audenaerde (1961), when *Phractolaemus* is at rest and its mouth is in a retracted position, the mouth is somewhat dorsally directed (Fig. 4.7A). Then, when the mouth is protracted (e.g., by the contraction of the levator operculi), the "oral cavity does not move to the front but rather describes a circular arch", in a way that, when the mouth is totally protracted, the oral cavity is now completely directed anteriorly, or even anteroventrally (Fig. 4.7B) (Thys van den Audenaerde 1961: 109). It is interesting to note that if the highly peculiar section A1-OST-L (Fig. 4.4A, B) of the adductor mandibulae of *Phractolaemus* could be contracted independently, it would seem to contribute to the anteroventral circular movement of the upper jaw (Fig. 4.4B). That is, it would have an antagonist function to that of the adductor mandibulae A2, the contraction of which contributes to the posterodorsal circular movement of the upper jaw (Fig. 4.7A). However, only a detailed electromyographic study can show if such an independent contraction and an antagonist function of the A1-OST-L and A2 sections of the adductor mandibulae is effectively possible in *Phractolaemus*. Regarding the situation in *Chanos*, it is somewhat intermediate between that present in derived gonorynchiforms such as *Parakneria* and *Kneria*, and that exhibited by otocephalan fishes with non-protrusile mouths (Figs. 4.8 and 4.10). In fact, although in *Chanos* the upper jaws are not as mobile as in these latter genera, the dorsal portions of the maxilla and premaxilla do have a certain mobility in relation to the anterior region of the neurocranium, which allows them to undertake a small anteroventral protraction when the mouth is opened. The mechanism is somewhat similar to that described above for *Parakneria* and *Kneria* (Fig. 4.6), with the main difference being that the movements of the upper jaw are more limited in *Chanos* than in these two genera. This seems to support the view expressed by Vari (1979), Howes (1985) and Gosline (1989), that the highly protrusile mouths in gonorynchiforms such as *Parakneria* and *Kneria*, in most cypriniforms, and in certain characiforms were independently acquired.

The adductor mandibulae Aω is plesiomorphically present in the Teleostei, as explained above. The presence of an Aω clearly seems to also constitute a plesiomorphic condition for otocephalan teleosts (Diogo in press). Within the evolutionary history of the Otocephala, the Aω was independently lost in some lineages, for example, in extant non-chanid gonorynchiforms, and in many catfishes (Howes 1985, Diogo 2004, this work).

In summary, it can thus be said that a rather simple adductor mandibulae subdivided into a section A2 and a section Aω, with the tendon of the former attaching on the latter, for example, in *Denticeps* (Fig. 4.9), very likely represents the plesiomorphic condition for otocephalans. Regarding the Gonorynchiformes, the plesiomorphic condition seems to be that in which there is an A2, an Aω and an A1-OST, the latter section being attached to both the upper and lower jaws as in *Chanos* and *Gonorynchus* (Figs. 4.1 and 4.2). Thus, an A1-OST almost exclusively associated with the upper jaw, as found in the remaining extant gonorynchiform genera, seems to represent an apomorphic condition within the Gonorynchiformes.

Levator arcus palatini

In *Chanos, Kneria* and *Cromeria* the levator arcus palatini is a thick muscle originating from a lateral process of the sphenotic and inserting on the hyomandibula and metapterygoid (Figs. 4.1 and 4.5A). In *Gonorynchus*, this muscle (see Howes 1985: fig. 5) inserts on the hyomandibula and metapterygoid, as well as on the preopercle, its fibers being extremely mixed with those of the dilatator operculi. The fibers of the levator arcus palatini and the dilatator operculi are also intermingled in *Phractolaemus* (Fig. 4.4A), but in this genus the former muscle inserts only on the hyomandibula and metapterygoid. In *Parakneria* the levator arcus palatini inserts exclusively on the hyomandibula, while in *Grasseichthys* this muscle inserts on both the hyomandibula and preopercle (Fig. 4.5C).

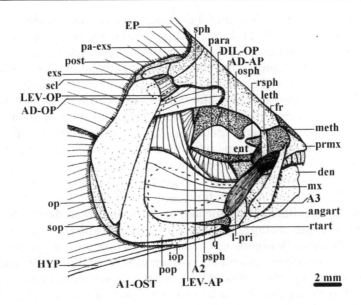

Fig. 4.10 Lateral view of the cephalic musculature of *Brycon guatemalensis*. The pectoral girdle muscles are not illustrated; the postcleithra and the most ventral elements of the pectoral girdle, as well as the nasals and infraorbitals, were removed. A1-OST, A2, A3, sections of adductor mandibulae; AD-AP, adductor arcus palatini; AD-OP, adductor operculi; angart, angulo-articular; den, dentary bone; DIL-OP, dilatator operculi; ent, entopterygoid; EP, epaxialis; exs, extrascapular; fr, frontal; HYP, hypaxialis; iop, interopercle; l-pri, primordial ligament; leth, lateral-ethmoid; LEV-AP, levator arcus palatini; LEV-OP, levator operculi; meth, mesethmoid; mx, maxilla; op, opercle; osph, orbitosphenoid; pa-exs, parieto-extrascapular; para, parasphenoid; pop, preopercle; post, post-temporal; prmx, premaxilla; psph, pterosphenoid; q, quadrate; rsph, rhinosphenoid; rtart, retroarticular; scl, supracleithrum; sop, subopercle; sph, sphenotic.

The configuration of the levator arcus palatini is rather conservative within otocephalans. However, as noted by Greenwood (1968), in non-denticipitoid extant clupeiforms such as *Clupea*, the levator arcus palatini is divided into two well-differentiated bundles. The upper, larger bundle originates on the frontal and sphenotic and inserts on the hyomandibula; the lower, smaller bundle runs from the sphenotic to the hyomandibula and metapterygoid. Greenwood (1968) stated that the undivided levator arcus palatini of *Denticeps* (Fig. 4.8) corresponds, very likely, to the plesiomorphic condition for clupeiforms. Taking into account the configuration found in ostariophysans and in basal teleosts such as elopomorphs and osteoglossomorphs, the statement of Greenwood seems well founded. In fact, in the great majority of the members of these groups the levator arcus palatini is undivided (Figs. 4.1, 4.4, 4.5, 4.9, 4.10). An exception is the divided levator arcus palatini of the *Opsariichthys* specimens examined, which is morphologically very similar to that of *Clupea*.

However, the other cypriniform fishes dissected have an undivided adductor arcus palatini (Fig. 4.9), and the most parsimonious interpretation is that the configuration found in the cypriniform *Opsariichthys* and in non-denticipitoid extant clupeiforms such as *Clupea* was acquired independently.

Another interesting variation of the levator arcus palatini within otocephalans concerns the attachment of this muscle on the metapterygoid. In many clupeiforms, including *Denticeps* (Fig. 4.8), as well as in a great number of ostariophysans, including most cypriniforms and the great majority of gonorynchiforms examined (see above), the levator arcus palatini inserts on both the hyomandibula and the metapterygoid. As a partial attachment on the metapterygoid is also found in basal teleosts such as *Elops* (e.g., Vrba 1968), this seems to represent the plesiomorphic otocephalan condition. Thus, the loss of contact between the levator arcus palatini and the metapterygoid, found in otocephalans such as the gonorynchiforms *Parakneria* and *Cromeria* as well as in most gymnotiforms and most characiforms examined, was seemingly acquired independently within the Otocephala. It should be noted that in some otocephalans, such as some loricariid catfishes, the levator arcus palatini is absent (e.g., Howes 1983, Diogo and Vandewalle 2003).

In summary, the plesiomorphic otocephalan situation seems to be that in which the levator arcus palatini is well separated from the dilatator operculi, is constituted by a single bundle, and is inserted on both the hyomandibula and the metapterygoid. From this situation, different apomorphic configurations were acquired in the evolutionary history of the Otocephala: in some cases the muscle became extremely mixed with the dilatator operculi (e.g., *Phractolaemus*, *Gonorynchus*), differentiated into different bundles (e.g., *Clupea*, *Opsariichthys*), inserted in bones such as the preopercle (e.g., *Gonorynchus*, *Grasseichthys*) and/or lost contact with structures such as the metapterygoid (e.g., *Parakneria*, *Grasseichthys*).

Adductor arcus palatini

In *Chanos* the adductor arcus palatini (Fig. 4.1) is a large muscle situated mesial to the adductor mandibulae and the levator arcus palatini and extending from the lateral surface of the parasphenoid to the mesial surface of the metapterygoid and entopterygoid. A similar overall configuration is found in *Gonorynchus* (see Howes 1985: figs. 11 and 12), but the origins of the muscle are on both the parasphenoid and the prootic, and the insertions are on the metapterygoid, entopterygoid, quadrate and preopercle. In *Phractolaemus* this muscle (Fig. 4.4) originates on the parasphenoid and prootic, and inserts on the mesial and posterodorsal surfaces of the metapterygoid and on the mesial and anterodorsal surfaces of the hyomandibula. In *Parakneria* the adductor arcus palatini (Howes 1985: figs. 11 and 12) connects the parasphenoid to the hyomandibula, while in *Kneria* it runs from the parasphenoid to both the hyomandibula and metapterygoid. In *Cromeria* this muscle runs from the prootic and pterotic to the hyomandibula. In *Grasseichthys* it originates on the parasphenoid and prootic and inserts on the hyomandibula.

As noted by Winterbottom (1974), many authors have used the name "adductor hyomandibulae" to designate the large muscle connecting the neurocranium to the mesial surface of the suspensorium (i.e. the adductor arcus palatini of the present work). This is due to the fact that in many teleosts the mesial insertion of this muscle is exclusively on the hyomandibula. However, as noted by that author, this name becomes inappropriate when the muscle is expanded anteriorly along the floor of the orbit and attaches also on more anterior elements of the suspensorium such as the metapterygoid and/or entopterygoid. As described above, this is precisely the case in the gonorynchiforms *Chanos, Gonorynchus, Phractolaemus* and *Kneria* (Fig. 4.1) and in most otocephalans examined in the present work (Figs. 4.8, 4.9, 4.10). Winterbottom (1974) therefore opted to use the name "adductor arcus palatine" to designate the "adductor hyomandibulae" of Greenwood (1968) and Vrba (1968). The nomenclature of Winterbottom (1974) is followed here.

Some otocephalans, such as catfishes, exhibit a peculiar muscle that is usually named "extensor tentaculi" and that, as stated by, for example, Takahasi (1925), Alexander (1965), Gosline (1975), and Diogo *et al.* (2000), has arisen from the differentiation of the anterior portion of the adductor arcus palatini. The extensor tentaculi, plesiomorphically present in catfishes (e.g., Diogo 2004, in press), inserts on the autopalatine and is related with the movements of the maxillary barbels of these fishes. Another derived configuration of the adductor arcus palatini that was acquired within otocephalan evolutionary history is that found in *Gonorynchus*, in which part of this muscle attaches on the preopercle (see above). Such an attachment was not found in any other otocephalan examined.

It can thus be said that the adductor arcus palatini of otocephalans usually inserts in bones such as the hyomandibula, quadrate, metapterygoid and/or entopterygoid, although it may eventually also insert on other structures such as the preopercle (e.g., *Gonorynchus*) or the autopalatine (e.g., catfishes, in which the differentiation of the portion attaching to the autopalatine has given rise to the extensor tentaculi).

Adductor hyomandibulae

As stated by Winterbottom (1974: 239), in some teleosts, apart from the adductor arcus palatini, there are other muscles connecting the neurocranium to the mesial surface of the suspensorium and promoting the adduction of this structure, which "apparently separate 1) either from the posterior region

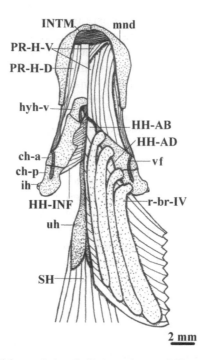

Fig. 4.11 Ventral view of the ventral cephalic musculature of *Chanos chanos*. The hypaxialis and the pectoral girdle musculature, as well as the pectoral girdle, interopercle, opercle, preopercle and subopercle, are not illustrated. On the right side all the hyoid muscles are exposed; on the left side the dorsal section of the protractor hyoidei, the hyohyoideus abductor and the hyohyoidei adductores were removed. ch-a, ch-p, anterior and posterior ceratohyals; HH-AB, hyohyoideus abductor; HH-AD, hyohyoidei adductores; HH-INF, hyohyoideus inferior; hyh-v, ventral hypohyal; ih, interhyal; INTM, intermandibularis; mnd, mandible; PR-H-D, PR-H-V, sections of protractor hyoidei; r-br-IV, branchiostegal ray IV; SH, sternohyoideus; uh, urohyal; vf, dorsal fossa of anterior and posterior ceratohyals.

of the adductor arcus palatini or 2) from the anterior fibers of the adductor operculi". Winterbottom described both these two latter muscles under the name "adductor hyomandibulae", but I prefer to designate them as *adductor hyomandibulae 1* and *adductor hyomandibulae 2*, since they are not homologous. In many cases, the position and attachments of the muscle give good cues about whether it is an adductor hyomandibulae 1 or an adductor hyomandibulae 2. This is the case, for example, in catfishes, in which the adductor hyomandibulae, when present, is a thin muscle just situated posteriorly to, and usually mixed with, the adductor operculi, indicating that this is an adductor hyomandibulae 2 (see Diogo 2004). However, the situation is more complicated in other otocephalan groups, for example, the gonorynchiforms. In *Parakneria*, *Kneria* and *Grasseichthys* the situation is quite similar to that found in catfishes with an adductor hyomandibulae 2, with a

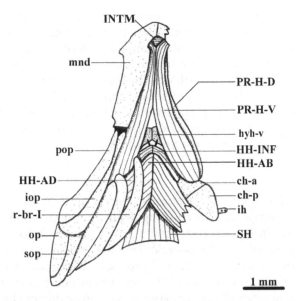

Fig. 4.12 Ventral view of the ventral cephalic musculature of *Danio rerio*. The hypaxialis and the pectoral girdle musculature, as well as the pectoral girdle, are not illustrated. On the left side all the hyoid muscles are exposed; a portion of the hyohyoidei adductores, as well as of the mandible, was cut, and the opercle, interopercle, subopercle and preopercle are not represented. ch-a, ch-p, anterior and posterior ceratohyals; HH-AB, hyohyoideus abductor; HH-AD, hyohyoidei adductores; HH-INF, hyohyoideus inferior; hyh-v, ventral hypohyal; ih, interhyal; INTM, intermandibularis; iop, interopercle; mnd, mandible; op, opercle; pop, preopercle; PR-H-D, PR-H-V, sections of protactor hyoidei; r-br-I, branchiostegal ray I; SH, sternohyoideus; sop, subopercle.

small, thin muscle running from the neurocranium (pterotic and/or prootic) to the posterodorsal surface of the hyomandibula (Fig. 4.5C). This suggests that this small muscle is an adductor hyomandibulae 2 derived from the adductor operculi. However, in *Chanos* the adductor hyomandibulae is situated in a much more anterior position, being much larger than the muscle found in *Parakneria*, *Kneria* and *Grasseichthys* and, it should be noted, being intermingled with fibers of the adductor arcus palatini (Fig. 4.1; see fig. 3 of Howes 1985). In *Gonorynchus*, both types of adductor hyomandibulae seem to be present: one is a thin, small muscle situated just anteriorly to the adductor operculi and attaching on the posterodorsal surface of the hyomandibula; the other is a broad muscle situated just posterior to, and mixed with, the adductor arcus palatini and attaching on the ventromesial surface of the hyomandibula (see e.g., Howes 1985: fig. 8). The situation is still more complex in the examined specimens of *Phractolaemus* and *Cromeria*, in which it is particularly difficult to appraise whether the adductor hyomandibulae 1 and/or the adductor hyomandibulae 2 are present or not. There seems to be a great variability and homoplasy concerning the presence of an "adductor hyomandibulae" *sensu* Winterbottom in teleosts (Winterbottom 1974). The high taxonomic diversity and probably homoplasy associated with the presence/shape of the adductor hyomandibulae 1 and adductor hyomandibulae 2 seems to follow this rule. However, the scarce data available on the configuration and diversity of these poorly studied muscles in other teleosts, and especially on their ontogenetic development in both gonorynchiform and non-gonorynchiform fishes, makes it difficult to provide a well-grounded discussion on this subject at present. I plan to address this subject in a future work.

Levator operculi

In *Chanos*, *Gonorynchus*, *Parakneria*, *Kneria* and *Cromeria*, the levator operculi is a large muscle running from the ventrolateral margin of the pterotic and posterodorsal surface of the hyomandibula to the dorsomesial edge of the opercle (Fig. 4.1). In *Phractolaemus* the levator operculi passes from the pterotic and prootic to the opercle, with the place of insertion on the mesial surface of this latter bone being much more ventral than in the five genera listed just above. As in the study of Howes (1985), it was not possible to discern, in the present work, whether or not a levator operculi is present in *Grasseichthys*. In the great majority of the non-gonorynchiform otocephalans dissected the levator operculi displays a configuration basically similar to that of these five genera (Figs. 4.8, 4.9, 4.10).

It can thus be said that this configuration seemingly represents the plesiomorphic condition for gonorynchiforms, the configuration found in *Phractolaemus* being apomorphic. Another notable, derived configuration of the levator operculi within otocephalans is that found in fishes such as the siluriforms *Nematogenys* and *Trichomycterus*, in which the levator operculi covers a significant part of the lateral surface of the opercle (e.g., Howes 1983, Diogo 2004). It should also be noted that, as stressed by these authors and Diogo and Vandewalle (2003), the levator operculi may be missing in some otocephalan fishes, for example, some loricariid siluriforms.

Dilatator operculi

In *Chanos* and *Gonorynchus* the dilatator operculi is a thick muscle originating on the ventrolateral surfaces of the sphenotic, pterotic and frontal and the dorsolateral surface of the hyomandibula and inserting on the anterodorsal margin of the opercle (Fig. 4.1). In *Phractolaemus* and *Parakneria* the levator operculi is somewhat similar to that of *Chanos* and *Gonorynchus* (Fig. 4.4A), but there is no attachment of the muscle on the frontal. In *Kneria* the dilatator operculi originates on the sphenotic, pterotic and hyomandibula, as in

Phractolaemus and *Parakneria*, but the attachments on the sphenotic are on both its ventrolateral and dorsolateral surfaces. In *Grasseichthys* and *Cromeria* the levator operculi is a thin muscle (Fig. 4.5A, C) running from the ventrolateral margin of the sphenotic and the dorsolateral margin of the hyomandibula to the anterodorsal surface of the opercle.

Greenwood (1968) stated that the dilatator operculi is missing in the clupeiform *Denticeps*. However, this muscle is present in the *Denticeps* specimens examined in the present work, although it is rather small and a significant part of it is covered in lateral view by the preopercle (Fig. 4.8). In the vast majority of the other otocephalans, including non-denticipitoid clupeiforms such as *Clupea*, as well as in basal teleosts such as elopomorphs and osteoglossomorphs, the dilatator operculi is usually well developed and widely exposed in lateral view (Figs. 4.1, 4.4, 4.9, 4.10; e.g., Greenwood 1968, Vrba 1968, Winterbottom 1974, Howes 1983, 1985, Aguilera 1986). Thus, the configuration of the dilatator operculi in *Denticeps* clearly seems to be derived.

In major lines, it can thus be said that the plesiomorphic gonorynchiform condition seems to be that found in *Chanos, Gonorynchus, Phractolaemus, Parakneria, Cromeria* and *Grasseichthys*, in which this muscle runs from the ventrolateral surface of the neurocranium and dorsolateral surface of the hyomandibula to the anterodorsal surface of the opercle. The configuration found in *Kneria*, in which part of the dilatator operculi originates on the dorsal margin of the cranial roof, is thus apomorphic. Other types of derived configurations of this muscle were acquired within otocephalan evolutionary history: for example, they may be subdivided into different bundles (e.g., some trichomycterid and aspredinid catfishes) or be completely absent (e.g., in some loricariid siluriforms) (Winterbottom 1974, Howes 1983, Diogo 2004, in press).

Adductor operculi

The configuration of this muscle is rather conservative within gonorynchiforms: it originates on the ventrolateral surface of the pterotic, mesially to the levator operculi, and inserts on the posterodorsal surface of the opercle (Figs. 4.1, 4.4A, 4.5A, C). In fact, the differences between the members of the order are essentially related to differences in size, the muscle being better developed in the specimens examined of *Cromeria* (Fig. 4.5A) than in those of *Grasseichthys* (Fig. 4.5C).

The configuration of the adductor operculi is also rather conservative within the non-gonorynchiform otocephalans analyzed and within teleosts in general (Winterbottom 1974). An undivided adductor operculi connecting the neurocranium to the dorsomesial surface of the opercle, such as that found in gonorynchiforms, is effectively found in the vast majority of otocephalans (Figs. 4.9 and 4.10).

It can thus be said that the configuration found in gonorynchiforms seems to represent the plesiomorphic otocephalan condition. A rather peculiar configuration among otocephalans is that found in the clariid siluriforms, in which this muscle inserts on the anterodorsal, and not on the dorsomedial, surface of the opercle (e.g., Adriaens and Verraes 1997, Diogo 2004).

Ventral Cephalic Musculature

Intermandibularis

This muscle is present in all extant gonorynchiforms, connecting the two halves of the lower jaws (Figs. 4.3 and 4.11). However, as stated by Howes (1985), in *Phractolaemus* the insertions of the muscle intermandibularis on the lower jaws are on the angulo-articulars (see Howes 1985: fig. 20), while in the remaining six genera they are on the dentary bones (Fig. 4.3).

Winterbottom (1974) noted that in notopterid and mormyrid osteoglossomorphs, the intermandibularis is divided into an "intermandibularis anterior" and an "intermandibularis posterior". The "intermandibularis anterior" corresponds to the intermandibularis of the present work; the "intermandibularis posterior" is a portion of the intermandibularis that embryologically becomes incorporated in the formation of the protractor hyoidei of most other teleosts, including otocephalans (Figs. 4.3, 4.11, 4.12) (see Edgeworth 1935, Winterbottom 1974, Diogo in press). In the vast majority of otocephalans, as well as of teleosts, the muscle intermandibularis (*sensu* this work) connects the dentary bones of the two lower jaws, as is the case in the gonorynchiforms *Chanos*, *Grasseichthys*, *Gonorynchus*, *Kneria*, *Parakneria* and *Cromeria*.

Briefly, it can thus be said that the configuration found in these latter gonorynchiform genera is seemingly plesiomorphic for otocephalans, the configuration found in *Phractolaemus* (in which this muscle connects the angulo-articulars and not the dentary bones) being therefore apomorphic. Apart from this apomorphic configuration, one other example of a peculiar, derived otocephalan feature is that found in members of some groups such as doumein catfishes, in which this muscle is absent (e.g., Diogo and Vandewalle 2003, Diogo 2004).

Protractor hyoidei

The protractor hyoidei of *Chanos* has ventral and dorsal sections, which connect the anterior ceratohyal to the ventromesial surface of the dentary bone (Fig. 4.11). In the other six extant gonorynchiform genera this muscle has a single section (Figs. 4.3 and 4.5A). Insertions on the hyoid arch may vary: on the anterior and posterior ceratohyals in *Gonorynchus* and *Phractolaemus*; on the anterior ceratohyal and ventral hypohyal in *Parakneria*,

Cromeria and *Grasseichthys*; on the anterior ceratohyal, posterior ceratohyal and ventral hypohyal in *Kneria*. Insertions on the mandible are always on the dentary bone, except in *Phractolaemus*, in which the muscle inserts on the angulo-articular. However, as shown by Howes (1985), the configuration of the protractor hyoidei of *Gonorynchus* is also peculiar in the sense that this muscle attaches on the dorsomesial, and not the ventromesial, surface of the dentary bone (Fig. 4.3).

Within non-gonorynchiform otocephalans, the characiform, gymnotiform and clupeiform fishes examined have a protractor hyoidei divided into dorsal and ventral sections (e.g., Fig. 4.12), similar to those of *Chanos*. This indicates that such a configuration may represent the plesiomorphic condition for otocephalans and for gonorynchiforms and, thus, that the single division found in *Phractolaemus, Gonorynchus, Kneria, Parakneria, Grasseichthys* and *Cromeria* is apomorphic. An example of another derived, peculiar configuration of this muscle within otocephalans is found in siluriforms. In the Diplomystidae and Loricarioidei, the most basal siluriform groups according to Diogo (2004, in press) and Sullivan *et al.* (2006), the protractor hyoidei consists of a single mass of fibers (e.g., Diogo and Chardon 2000b, Diogo and Vandewalle 2003, Diogo 2004). However, in most other catfishes, this muscle is divided into a pars dorsalis, a pars lateralis and a pars ventralis, the latter usually being associated with the cartilages of the characteristic mandibular barbels of these fishes (see Diogo and Chardon 2000b). In addition, various small, apomorphic ventral cephalic muscles associated with the movements of these barbels are often present in these fishes; those muscles were seemingly derived, at least partly, from the protractor hyoidei (e.g., Diogo and Chardon 2000b, Diogo and Vandewalle 2003).

Hyohyoideus inferioris

In *Chanos, Parakneria, Kneria, Gonorynchus, Cromeria* and *Grasseichthys* this muscle lies dorsal to the hyohyoideus abductor and hyohyoidei adductores (Fig. 4.11), some fibers of these muscles being mixed. It originates from the anterior and posterior ceratohyals and runs obliquely towards a marked median aponeurosis (or myocommata) where it fuses with its counterpart. The configuration of the hyohyoideus inferior in *Phractolaemus* is similar to that of these six genera, but the anterior fibers of the muscle lie ventral, and not dorsal, to the hyohyoideus abductor and hyohyoidei adductores.

In the cypriniforms and siluriforms examined the left and right sides of this muscle fuse to each other at the midline, as in gonorynchiforms (Fig. 4.12). However, in the clupeiforms, gymnotiforms and characiforms examined the left and right sides of the hyohyoideus inferior do not fuse to each other at the midline, running directly from one hyoid arch to the hyoid arch of the opposite side. The plesiomorphic condition for actinopterygians seems to be that in which the left and right sides of the hyohyoideus inferior fuse mesially to each other (see e.g., Winterbottom 1974, Lauder 1980b). However, the fact that this is not the case in the clupeiforms examined as well as in basal teleosts such as *Elops* or *Megalops* (Vrba 1968) seems to indicate that the configuration found in these latter fishes could represent the plesiomorphic situation for otocephalans. In that case, the configuration found in gonorynchiforms would be apomorphic within the Otocephala. More detailed data on the distribution of this character in the basal teleost groups Elopomorpha and Osteoglossomorpha is needed to test this hypothesis.

Hyohyoideus abductor

The configuration of the muscle hyohyoideus abductor is rather conservative within gonorynchiforms, the muscle running from the first branchiostegal ray to a median aponeurosis (Fig. 4.11). However, in *Kneria* and *Parakneria* some fibers of this muscle also attach to the cleithrum.

In the cypriniforms, siluriforms and gymnotiforms examined the left and right sides of the hyohyoideus abductor fuse in a median aponeurosis, as in gonorynchiforms (Fig. 4.12). In the clupeiforms and characiforms dissected (but not in the gymnotiforms, as was the case with the hyohyoideus inferior), the left and right sides of the hyohyoideus abductor do not fuse at the midline. The plesiomorphic condition of the hyohyoideus abductor in actinopterygian fishes is not clear. This is because the hyohyoideus abductor exhibits a much greater variation among the actinopterygians than the hyohyoideus inferior (e.g., Winterbottom 1974). In the Otocephala, the absence of a mesial fusion between the left and right parts of the hyohyoideus abductor in the clupeiforms examined and in basal teleosts (e.g., *Elops*: Vrba 1968) indicates that the configuration found in these latter fishes may represent the plesiomorphic otocephalan configuration. Again, more precise data on the distribution of this character in elopomorphs and osteoglossomorphs is needed to test such a hypothesis. Among the otocephalans dissected, the attachment of some fibers of the hyohyoideus abductor on the pectoral girdle occurs, apart from *Parakneria* and *Kneria*, only in a few catfishes (see e.g., Diogo 2004). Therefore, these clearly seem to represent non-homologous apomorphic configurations within the Otocephala.

Hyohyoidei adductores

In extant gonorynchiforms this muscle (e.g., Fig. 4.11) connects the branchiostegal rays, the opercle, the interopercle and the subopercle of the same size of the fish. According to Pasleau (1974), the hyohyoidei adductores also attaches on the preopercle in *Chanos*. However, in the *Chanos* specimens I dissected, this muscle does not attach on the preopercle. As is the case with the hyohyoideus abductor, some fibers of the hyohyoidei adductores attach to the cleithrum in *Parakneria* and *Kneria*.

The configuration of the hyohyoidei adductores is rather conservative in the non-gonorynchiform otocephalans examined (Fig. 4.12), being similar to that found in *Chanos*, *Gonorynchus*, *Cromeria*, *Grasseichthys* and *Phractolaemus*.

Sternohyoideus

In *Chanos*, *Gonorynchus*, *Grasseichthys* and *Cromeria*, this is a well-developed muscle (Fig. 4.11) running from the cleithrum to the urohyal and partly mixing posteriorly with fibers of the hypaxialis. In *Phractolaemus*, *Parakneria* and *Kneria*, the muscle also connects the cleithrum to the urohyal, but does not mix posteriorly with the fibers of the hypaxialis.

A sternohyoideus connecting the anterior surface of the pectoral girdle to the urohyal and eventually to the anteromesial margin of the hyoid arch and/or the branchial apparatus is a feature invariably present in all the otocephalans examined. This thus constitutes the plesiomorphic condition for the Otocephala. The mixing of the posterior fibers of this muscle with the anterior fibers of the hypaxialis is a generalized feature within teleosts (e.g., Winterbottom 1974). It also seems to constitute the plesiomorphic condition for otocephalans, being found, apart from the gonorynchiforms *Chanos*, *Gonorynchus*, *Grasseichthys* and *Cromeria*, in non-denticipitoid clupeiforms such as *Clupea*, and in most otophysans examined.

Pectoral Girdle Musculature

Arrector dorsalis

In *Chanos* and *Gonorynchus* the arrector dorsalis connects the coracoid and cleithrum to the anterolateral surface of the first pectoral ray (Figs. 4.13, 4.14, 4.15). As can be seen in Fig. 4.14B, a portion of this muscle originates on the mesial surface of the pectoral girdle, running posterolaterally through the "coracoid-cleithrum foramen" (*sensu* Diogo 2004) to meet the lateral portion of the muscle (Fig. 4.13) and then attaching to the first pectoral ray. In *Phractolaemus*, *Parakneria*, *Kneria*, *Cromeria* and *Grasseichthys*, the arrector dorsalis originates and lies exclusively on the lateral side of the pectoral girdle (Fig. 4.16B), passing from the cleithrum to the anterolateral margin of the first pectoral ray.

In the vast majority of the non-gonorynchiforms otocephalans examined, including clupeiforms, as well in many other teleosts (e.g., Winterbottom 1974), the arrector dorsalis is not divided into two large portions lying on either side of the pectoral girdle. It is, instead, a single mass of fibers that are not visible (Fig. 4.17), or are only just visible (Fig. 4.18), in a mesial view of the pectoral girdle. In fact, among the non-gonorynchiform otocephalans examined, a subdivided arrector dorsalis is found only in some characiforms (e.g., *Brycon*) and in many catfishes (see above). Moreover, as stressed by Diogo (2004), although this feature is present in many Siluriformes, it is plesiomorphically absent in that order, being absent in the most plesiomorphic extant catfish groups (diplomystids and loricarioids). Therefore, the plesiomorphic condition for the Otocephala as a whole seems to be that in which the arrector dorsalis is divided into two large portions. However, the situation in the Gonorynchiformes is less clear. This is because in the two phylogenetically more plesiomorphic extant genera of the order, *Chanos* and *Gonorynchus* (e.g., Grande and Poyato-Ariza 1999, Lavoué *et al.* 2005), the arrector dorsalis is, as stated above, divided into two large portions lying respectively on the lateral and the mesial surfaces of the pectoral girdle (Figs. 4.13 and 4.14).

It can thus be said that this feature may have been acquired in the lineage leading to all extant gonorynchiform genera and then been lost in the lineage leading to *Phractolaemus*, *Parakneria*, *Kneria*, *Grasseichthys* and *Cromeria*, or it could have been independently acquired in *Gonorynchus* and *Chanos*. [Note: if *Gonorynchus* and *Chanos* were sister groups the most parsimonious option would be, of course, to consider that this feature was only acquired in the lineage leading to *Gonorynchus* + *Chanos*, as this would imply a single evolutionary step.]

Arrector ventralis

The configuration of the arrector ventralis is rather conservative within gonorynchiforms: in the members of all extant genera it is a thin muscle situated lateral to the arrector ventralis and running from the cleithrum to the anteroventral surface of the first pectoral ray (Figs. 4.13, 4.15, 4.16B).

The configuration of this muscle is also rather conservative within the other otocephalans examined, being quite similar to that found in gonorynchiforms (Figs. 4.8 and 4.9). A rare example of apomorphic configuration within otocephalans is that found in catfishes of the family Amphiliidae: in amphiliines and leptoglanidines the arrector ventralis is markedly bifurcated mesially; in doumeines the most posterior fibers of this muscle are fully differentiated into an "additional muscle" (see Diogo 2004).

Adductor superficialis

This muscle is differentiated into two sections in gonorynchiforms. The most mesial section (adductor superficialis 1) (Figs. 4.13, 4.14A, 4.16) originates on the cleithrum, coracoid, scapula and mesocoracoid arch and

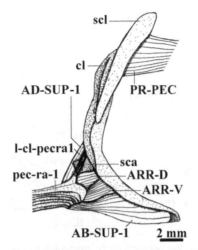

Fig. 4.13 Lateral view of the pectoral girdle musculature of *Chanos chanos*; all the musculature is exposed. AB-SUP-1, section of abductor superficialis; AD-SUP-1, section of adductor superficialis; ARR-D, arrector dorsalis; ARR-V, arrector ventralis; cl, cleithrum; l-cl-pecra1, ligament between cleithrum and pectoral ray 1; pec-ra-1, pectoral ray 1; PR-PEC, protractor pectoralis; sca, scapula; scl, supracleithrum.

inserts on the anterodorsal margin of the dorsal part of the pectoral fin rays. The most lateral section (adductor superficialis 2) (Fig. 4.16A) runs from the coracoid, scapula, mesocoracoid arch and the dorsal surface of the proximal radials to the anteroventral margin of the dorsal part of the pectoral fin rays.

An adductor superficialis divided into two bundles such as those found in gonorynchiforms is found in all the non-gonorynchiform otocephalans examined (Figs. 4.17 and 4.18), as well as in numerous other teleosts (Winterbottom 1974, Diogo 2004), and, thus, represents the plesiomorphic condition for both the Otocephala and the Gonorynchiformes.

Adductor profundus

Before describing the configuration of this muscle in gonorynchiforms (see Figs. 4.1 and 4.16: AD-PRO-1, AD-PRO-2), it is important to note that its identity has been, and still is, controversial (Diogo *et al.* 2001). Diogo *et al.* (2001) attempted to clarify this subject and have tentatively named this muscle "abductor profundus". However, in the present work I adopt the name "adductor profundus". Diogo *et al.* (2001), and subsequently Diogo (2004), were in fact somewhat reluctant to use the name "abductor profundus" to designate a muscle that clearly adducts the pectoral rays. Despite this, they decided on that name because in catfishes this muscle "inserts on the pectoral spine (first pectoral ray), and both the adductor

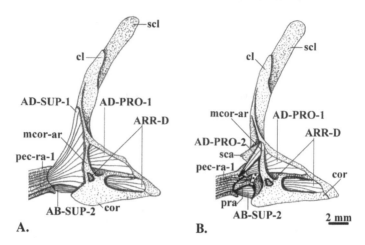

Fig. 4.14 Mesial view of the pectoral girdle musculature of *Chanos chanos*. (A) All the musculature is exposed. (B) The adductor superficialis 1 and 2 were removed. AB-SUP-2, section of abductor superficialis, AD-PRO-1, AD-PRO-2, sections of adductor profundus; AD-SUP-1, section of adductor superficialis; ARR-D, arrector dorsalis; cl, cleithrum; cor, coracoid; mcor-ar, mesocoracoid arch; pec-ra-1, pectoral ray 1; pra, proximal radials; sca, scapula; scl, supracleithrum.

superficialis and the adductor profundus (of Winterbottom 1974) never attach to the first pectoral ray in teleosts" (Diogo *et al.* 2001: 122). However, although the muscle I am referring to (Figs. 4.14, 4.16, 4.17, 4.18: AD-PRO, AD-PRO-1, AD-PRO-2) does not seem to correspond to the "adductor profundus" of Winterbottom (1974), I prefer to follow here the nomenclature of Allis (1903), Dubale and Rao (1961) and Bornbusch (1995), and thus to use the name "adductor profundus".

In *Chanos* and *Gonorynchus* the adductor profundus is divided into two sections. One, named here "adductor profundus 1" (Figs. 4.14 and 4.15), originates on the mesial surfaces of the cleithrum, scapula and coracoid, passes laterally to the mesocoracoid arch and adductor superficialis, and attaches on the anterior margin of the first pectoral ray. The other, smaller section, named here "adductor profundus 2" (Fig. 4.14), runs from the mesial surfaces of the cleithrum, scapula and coracoid to the anterior margin of the second pectoral ray. In the remaining five extant gonorynchiform genera, the adductor profundus is also divided into two sections, but the adductor profundus 2 is much larger than in *Chanos* and *Gonorynchus* (Fig. 4.16A). In these five genera, the adductor profundus 2 (Fig. 4.16A) originates on the mesial surfaces of the cleithrum and scapula and inserts on the anterior margin of the second pectoral ray. The adductor profundus 1 (Fig. 4.16A) inserts on the anterior margin of the first pectoral ray; it originates on the mesial margins of the cleithrum, of the coracoid, and, in the case of *Phractolaemus*, also of the scapula.

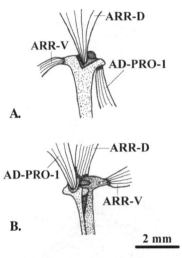

Fig. 4.15 Lateral (A) and mesial (B) views of the anterior portion of the first pectoral ray and the insertions of the arrector dorsalis, the arrector ventralis and the adductor profundus 1 in *Chanos chanos*. AD-PRO-1, section of adductor profundus; ARR-D, arrector dorsalis; ARR-V, arrector ventralis.

In the characiforms and gymnotiforms dissected the adductor profundus is divided into two well-differentiated, separate sections, as in gonorynchiforms (Figs. 4.14 and 4.16; e.g., Gijsen 1974, De la Hoz and Chardon 1984). However, in the clupeiforms, cypriniforms and siluriforms examined the adductor profundus is undivided (Figs. 4.17 and 4.18; e.g., Brosseau 1978, Diogo 2004). By analyzing figs. 32–34 of Winterbottom (1974), it seems that in basal teleosts such as the elopomorph *Elops* and the osteoglossomorph *Osteoglossum* the muscle named "adductor profundus" in the present work is undivided.

As this is also the case in *Denticeps* and the other clupeiforms examined, as well as in the cypriniforms and siluriforms analyzed, the plesiomorphic condition for the Otocephala is probably that in which the adductor profundus is undivided. In fact, it should be noted that the configuration of the two adductor profundus sections found in the gymnotiforms examined differs from that found in the gonorynchiform and characiform fishes dissected: in gymnotiforms both sections attach on the first pectoral ray, while in characiforms and gonorynchiforms the sections attach respectively on the first and the second pectoral rays (Fig. 4.14).

Abductor superficialis

In gonorynchiforms the abductor superficialis is differentiated into two bundles. The most lateral bundle, abductor superficialis 1 (Figs. 4.13 and 4.16B), runs from the coracoid and cleithrum to the anteroventral margin of

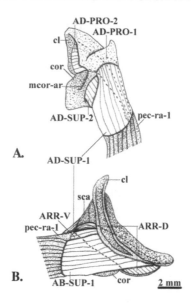

Fig. 4.16 Dorsal (A) and lateral (B) views of the pectoral girdle musculature of *Phractolaemus ansorgei*; all the musculature is exposed. AB-SUP-1, section of abductor superficialis; AD-PRO-1, AD-PRO-2, sections of adductor profundus; AD-SUP-1, AD-SUP-2, sections of adductor superficialis; ARR-D, arrector dorsalis; ARR-V, arrector ventralis; cl, cleithrum; cor, coracoid; mcor-ar, mesocoracoid arch; pec-ra-1, pectoral ray 1; sca, scapula.

the ventral part of the pectoral fin rays. The most mesial section, abductor superficialis 2 (Fig. 4.14), runs from the cleithrum and ventral surfaces of the proximal radials to the anterodorsal margin of the ventral part of the pectoral fin rays.

Such a division into two bundles is found in the vast majority of the non-gonorynchiform otocephalans examined (Figs. 4.8, 4.9, 4.17, 4.18), as well as in numerous other teleosts (e.g., Winterbottom 1974). This therefore seems to represent the plesiomorphic condition for otocephalans. It should be noted that the abductor superficialis is absent in a few otocephalans, for example, the members of the catfish *Chaca* (Diogo 2004).

Protractor pectoralis

In all gonorynchiforms examined the protractor pectoralis (Fig. 4.13) is a thick, large muscle connecting the posterior region of the pterotic to the anterodorsal region of the pectoral girdle. However, the attachments on the pectoral girdle may vary: on the cleithrum and supracleithrum in *Chanos*, *Parakneria*, *Cromeria* and *Grasseichthys* (Fig. 4.13); on the post-temporo-supracleithrum in *Phractolaemus*; and on the supracleithrum in *Gonorynchus*.

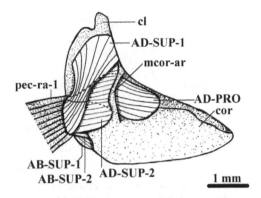

Fig. 4.17 Mesial view of the pectoral girdle musculature of *Denticeps clupeoides*; all the musculature is exposed. AB-SUP-1, AB-SUP-2, sections of abductor superficialis; AD-PRO, adductor profundus; AD-SUP-1, AD-SUP-2, sections of adductor superficialis; cl, cleithrum; cor, coracoid; mcor-ar, mesocoracoid arch; pec-ra-1, pectoral ray 1.

In the clupeiform *Denticeps* the protractor pectoralis is missing (Greenwood and Lauder 1981). However, a muscle protractor pectoralis connecting the posterior region of the neurocranium to the anterodorsal margin of the pectoral girdle is found in many non-denticipitoid clupeiforms, as well as in the vast majority of the other otocephalans and in many non-otocephalan teleosts. Thus, the presence of this muscle probably represents the plesiomorphic condition for otocephalans.

Conclusions

As we have seen, there is a considerable morphological diversity of the cranial and pectoral girdle muscles within the Gonorynchiformes, most of these muscles exhibiting different configurations within the order. Therefore, as shown by Diogo (2004) for the Siluriformes, the study of the cranial and pectoral girdle musculature may provide useful data not only for an analysis of the intra-relationships of gonorynchiforms, but also on the phylogenetic position of these fishes within the Otocephala and the Teleostei. It is interesting to note, for example, that except for the adductor mandibulae, hyohyoideus inferioris, adductor profundus and eventually adductor hyomandibulae and arrector dorsalis, the plesiomorphic gonorynchiform configuration for each of the muscles discussed here seemingly represents the plesiomorphic configuration for the Otocephala as a whole. Because of their apparent basal position within otocephalans and the rather plesiomorphic configuration of their myological structures, the gonorynchiforms can thus play an important role in the study of the functional morphology, comparative anatomy, and evolution of not only otocephalans, but also teleosts in general. For instance, the functional analysis of the opening/closure of the mouth in

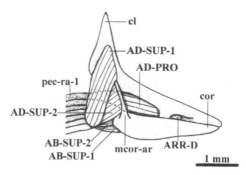

Fig. 4.18 Mesial view of the pectoral girdle musculature of *Danio rerio*; all the musculature is exposed. AB-SUP-1, AB-SUP-2, sections of abductor superficialis; AD-PRO, adductor profundus; AD-SUP-1, AD-SUP-2, sections of adductor superficialis; ARR-D, arrector dorsalis; cl, cleithrum; cor, coracoid; mcor-ar, mesocoracoid arch; pec-ra-1, pectoral ray 1.

gonorynchiforms such as *Kneria* and *Parakneria* seems to support the hypothesis of, for example, Alexander (1964), Vari (1979), and Gosline (1989), according to which the presence of a direct attachment of the adductor mandibulae on the upper jaw in certain teleostean taxa is probably associated with the characteristic small mouth and/or a protrusile upper jaw of the members of those taxa (see above). Numerous opportunities for further research include the following questions: (1) Do the plesiomorphic configuration of muscles of other parts of the body of gonorynchiforms correspond to the plesiomorphic configuration for the otocephalans as a whole? (2) Is the morphological diversity of those muscles within otocephalans and within teleosts greater, or less, than that of the cephalic and pectoral girdle muscles? (3) Would a detailed, extensive study of the functional morphology of gonorynchiforms shed light on the evolution of structural complexes such as those related with the feeding mechanisms or with the movements of the fins, within this order and within the Otocephala? It is hoped that this paper will not only contribute to the knowledge on the anatomy of the gonorynchiform and otocephalan teleosts, but also pave the way for future works concerning the comparative anatomy, functional morphology, phylogeny and evolution of these fishes, and of teleosts in general.

Acknowledgements

I specially thank J. Snoeks, E. Vreven and the late G.G. Teugels (Musée Royal de l'Afrique Centrale), P. Laleyé (Université Nationale du Bénin), R. Vari, J. Williams and S. Jewett (National Museum of Natural History), T. Grande (Field Museum of Natural History) and D. Catania (California Academy of Sciences) for kindly providing a large part of the specimens analyzed for this study. I would also like to acknowledge P. Vandewalle, M. Chardon, I. Peng, G.G. Teugels, R.P. Vari, S. Weitzman, T. Abreu, A.

Zanata, B.G. Kapoor, F. Meunier, S. He, D. Adriaens, F. Wagemans, C. Oliveira, E. Parmentier, M.M. de Pinna, P. Skelton, M.J.L. Stiassny, F.J. Poyato-Ariza, G. Arratia, T. Grande, H. Gebhardt, M. Ebach, A. Wyss, J. Waters, B. Perez-Moreno, G. Cuny, A. Choudhury, M. Vences, S.H. Weitzman, L. Cavin, F. Santini, J.C. Briggs, L.M. Gahagan, Philiphe J.G. Maisey, S. Hughes, M. Gayet, J. Alves-Gomes, G. Lecointre, C. Borden and L. Taverne for their helpful advice and assistance and for their discussions on teleost anatomy, functional moprhology, phylogeny and evolution. This project received financial support from a postdoctoral grant of the Fondation Duesberg of the University of Liège and from the European Community's Programme "Structuring the European Research Area" under SYNTHESIS at the Museo Nacional de Ciencias Naturales de Madrid (MNCN).

References

Adriaens, D. and W. Verraes. 1997. Ontogeny of suspensorial and opercular muscles in *Clarias gariepinus* (Siluroidei: Clariidae), and the consequences for respiratory movements. Neth. J. Zool. 47: 1–29.
Aguilera, O. 1986. La musculatura estriada en los peces Gymnotiformes (Teleostei-Ostariophysi): musculatura facial. Acta. Biol. Venez. 12: 13–23.
Albert, J. 2001. Species diversity and phylogenetic relationships of American knifefishes (Gymnotiformes, Teleostei). Misc. Publ. Mus. Zool., Univ. Mich. 190: 1–127.
Albert, J.S. and R. Campoz-da-Paz. 1998. Phylogenetic systematics of Gymnotiformes with diagnoses of 58 clades: a review of available data, pp. 410–446. In: L.R. Malabarba, R.E. Reis, R.P. Vari, Z.M. Lucena and C.A.S. Lucena [eds.]. Phylogeny and Classification of Neotropical Fishes. Edipucrs, Porto Alegre, Brazil.
Alexander, R. McN. 1964. Adaptation in the skulls and cranial muscles of South American characinoid fish. Zool. J. Linn. Soc. 45: 169–190.
Alexander, R. McN. 1965. Structure and function in catfish. J. Zool. (Lond.) 148: 88–152.
Allis, E.P. 1903. The skull and cranial and first spinal muscles and nerves of *Scomber scomber*. J. Morphol. 18: 45–328.
Bornbusch, A.H. 1995. Phylogenetic relationships within the Eurasian catfish family Siluridae (Pisces: Siluriformes), with comments on generic validities and biogeography. Zool. J. Linn. Soc. 115: 1–46.
Brosseau, A.R. 1978. The pectoral anatomy of selected Ostariophysi. II. The Cypriniformes and Siluriformes. J. Morphol. 140: 79–115.
Chardon, M. and E. De la Hoz. 1973. Notes sur le squelette, les muscles, les tendons et le cerveau des Gymnotoidei. Ann. Soc. Nat. Zool. Paris., 12 Ser. 15: 1–10.
De la Hoz, E. 1974. Definition et classification des poissons Gymnotoidei sur la base de la morphologie comparée et fonctionnelle du squelette et des muscles. Ph.D. thesis, University of Liège, Liege, Belgium.
De la Hoz, E. and M. Chardon. 1984. Skeleton, muscles, ligaments and swim-bladder of a gymnotid fish, *Sternopygus macrurus* Bloch and Schneider (Ostariophysi: Gymnotoidei). Bull. Soc. R. Sci. Liège 53: 9–53.
Diogo, R. 2004. Morphological Evolution, Adaptations, Homoplasies, Constraints, and Evolutionary Trends: Catfishes as a Case Study on General Phylogeny and Macroevolution. Science Publishers, Enfield, USA.
Diogo, R. In press. On the Origin and Evolution of Higher-Clades: Osteology, Myology, Phylogeny and Macroevolution of Bony Fishes and the Rise of Tetrapods. Science Publishers, Enfield, USA.

Diogo, R. and M. Chardon. 2000a. Homologies between different adductor mandibulae sections of teleostean fishes, with a special regard to catfishes (Teleostei: Siluriformes). J. Morphol. 243: 193–208.

Diogo, R. and M. Chardon. 2000b. The structures associated with catfish (Teleostei: Siluriformes) mandibular barbels: Origin, Anatomy, Function, Taxonomic Distribution, Nomenclature and Synonymy. Neth. J. Zool. 50: 455–478.

Diogo, R. and P. Vandewalle. 2003. Review of superficial cranial musculature of catfishes, with comments on plesiomorphic states, pp. 47–69. In: B.G. Kapoor, G. Arratia, M. Chardon and R. Diogo [eds.]. Catfishes. Science Publishers, Enfield, USA.

Diogo, R., C. Oliveira and M. Chardon. 2000. The origin and transformation of catfish palatine-maxillary system: an example of adaptive macroevolution. Neth. J. Zool. 50: 373–388.

Diogo, R., C. Oliveira and M. Chardon. 2001. On the osteology and myology of catfish pectoral girdle, with a reflection on catfish (Teleostei: Siluriformes) plesiomorphies. J. Morphol. 249: 100–125.

Dubale, M.S. and B.V.S. Rao. 1961. Pectoral fin musculature in certain siluroid fishes. J. Univ. Bombay 29: 89–96.

Edgeworth, F.H. 1935. The Cranial Muscles of Vertebrates. University Press, Cambridge, UK.

Fink, S.V. and W. Fink. 1981. Interrelationships of the ostariophysan fishes. Zool. J. Linn. Soc. 72: 297–353.

Fink, S.V. and W. Fink. 1996. Interrelationships of ostariophysan fishes (Teleostei), pp. 209–249. In: M.L.J. Stiassny, L.R. Parenti and G.D. Johnson [eds.]. Interrelationship of Fishes. Academic Press, New York.

Gijsen, L. 1974. Étude comparée du squelette, des muscles et des ligaments de la tête de quatre espéces de poissons téléostéens Characoidei a cáractéres archaiques. Bachelor's thesis, University of Liège, Liège, Belgium.

Gosline, W.A. 1975. The palatine-maxillary mechanism in catfishes with comments on the evolution and zoogeography of modern siluroids. Occas. Pap. Calif. Acad. Sci. 120: 1–31.

Gosline, W.A. 1989. Two patterns of differentiation in the jaw musculature of teleostean fishes. J. Zool. (Lond.) 218: 649–661.

Grande, T. and F.J. Poyato-Ariza. 1999. Phylogenetic relationships of fossil and recent gonorynchiform fishes (Teleostei: Ostariophysi). Zool. J. Linn. Soc. 125: 197–238.

Greenwood, P.H. 1968. The osteology and relationships of the Denticipitidae, a family of clupeomorph fishes. Bull. Br. Mus. Nat. Hist. (Zool.) 16: 215–273.

Greenwood, P.H. and G.V. Lauder. 1981. The protractor pectoralis muscle and the classification of teleost fishes. Bull. Br. Mus. Nat. Hist. (Zool.) 41: 213–234.

Howes, G.J. 1976. The cranial musculature and taxonomy of characoid fishes of the tribes Cynodontini and Characini. Bull. Br. Mus. Nat. Hist. (Zool.) 29: 203–248.

Howes, G.J. 1983. The cranial muscles of the loricarioid catfishes, their homologies and value as taxonomic characters. Bull. Br. Mus. Nat. Hist. (Zool.) 45: 309–345.

Howes, G.J. 1985. Cranial muscles of gonorynchiform fishes, with comments on generic relationships. Bull. Br. Mus. Nat. Hist. (Zool.) 49: 273–303.

Lauder, G.V. 1980a. On the evolution of the jaw adductor musculature in primitive gnathostome fishes. Breviora 460: 1–10.

Lauder, G.V. 1980b. Evolution of the feeding mechanisms in primitive actinopterygian fishes: a functional anatomical analysis of Polypterus, Lepisosteus, and Amia. J. Morphol. 163: 283–317.

Lauder, G.V. and K.F. Liem. 1983. The evolution and interrelationships of the actinopterygian fishes. Bull. Mus. Comp. Zool. 150: 95–197.

Lavoué, S., M. Miya, J.G. Inoue, K. Saitoh, N.B. Ishiguro and M. Nishida M. 2005. Molecular systematics of the gonorynchiforms fishes (Teleostei) based on whole mitogenome sequences: implications for higher-level relationships within the Otocephala. Mol. Phylogenet. Evol. 37: 165–177.

Le Danois, Y. 1966. Remarques anatomiques sur la règion cèphalique de de *Gonorynchus gonorynchus* (Linné, 1766). Bull. Inst. Fond. Afr. Noire 28: 283–342.

Pasleau, F. 1974. Recherches sur la position phylétique des téleostéens Gonorynchiformes, basées sur l'étude de l'osteologie et de myologie cephalique. Bachelor's thesis, University of Liège, Liège, Belgium.

Stiassny, M.L.J., E.O. Wiley, G.D. Johnson and M.R. Carvalho. 2004. Gnathostome fishes, pp. 410–429. *In*: M.J. Donaghue and J. Cracraft [eds.]. 2004. Assembling the Three of Life. Oxford University Press, New York.

Sullivan, J.P., J.G. Lundberg and M. Hardman. 2006. A phylogenetic analysis of the major groups of catfishes (Teleostei: Siluriformes) using rag1 and rag2 nuclear gene sequences. Mol. Phylogenet. Evol. 41: 636–662.

Takahasi, N. 1925. On the homology of the cranial muscles of the cypriniform fishes. J. Morphol. Physiol. 40: 1–109.

Thys van den Audenaerde, D.F.E. 1961. L'anatomie de *Phractolaemus ansorgei* Blgr. et la position systématique des Phractolaemidae. Ann. Mus. Roy. Afr. Centr. 103: 101–167.

Vandewalle, P. 1975. Des formes aux fonctions: une étude de morphologie fonctionnelle et comparée chez trois poissons cyprinidés. Ph.D. thesis, Université de Liège, Liège, Belgium.

Vandewalle, P. 1977. Particularités anatomiques de la tête de deux Poissons Cyprinidés *Barbus barbus* (L.) et *Leuciscus leuciscus* (L). Bull. Acad. Roy. Belg. 5: 469–479.

Vari, R.P. 1979. Anatomy, relationships and classification of the families Citharinidae and Distichodontidae (Pisces, Characoidea). Bull. Br. Mus. Nat. Hist. (Zool.) 36: 261–344.

Vrba, E.S. 1968. Contributions to the functional morphology of fishes, part V, the feeding mechanism of *Elops saurus* Linnaeus. Zool. Afr. 3: 211–236.

Winterbottom, R. 1974. A descriptive synonymy of the striated muscles of the Teleostei. Proc. Acad. Nat. Sci. Philadelphia 125: 225–317.

Wu, K.-Y. and S.-C. Shen. 2004. Review of the teleostean adductor mandibulae and its significance to systematic positions of the Polymixiiformes, Lampridiformes, and Triacanthoidei. Zool. Stud. 43: 712–736.

5

The Epibranchial Organ and Its Anatomical Environment in the Gonorynchiformes, with Functional Discussions

Françoise Pasleau[1], Rui Diogo[1] and Michel Chardon[1]

Abstract

All extant gonorynchiforms except for species within the genus *Gonorynchus* are provided with an epibranchial organ (EBO), which consists in a blind sac opening through a canal into the buccopharyngeal cavity just above and behind the last branchial slit. The proximal part of the EBO is supported by specialized elements of the last two branchial arches. The gill rakers rows of the same arches continue in the canal part of the EBO, but not in the sac. The epithelium and tunicae of the EBO are similar to those of the surrounding buccopharynx; it is plicate and rich in mucous cells. The gill rakers of all the branchial arches are particularly long and bear microbranchiospines (except in *Gonorynchus* and *Grasseichthys*). They constitute a very efficient filter complicated by bridges between the branchial arches and between the opposite gill rakers rows. All gonorynchiforms are considered to be efficient filter feeders and the EBO plays an important part in that function. A hypothesis is presented as to the water currents in the buccopharyngeal, opercular and EBO cavities, and as to the trapping of food particles by mucus and their transport into the esophagus. In

[1] Laboratory of Functional and Evolutionary Morphology. Institut de Chimie, Bat. B6, Universitè de Liège, B-4000 Sart-Tilman (Liège), Belgium.

Gonorynchus, the EBO is functionally replaced by a dorsally elongated fifth branchial slit provided with very long gill rakers. This configuration is puzzling, regarding the phylogenetic position in which the genus has been placed in recent cladistic studies.

Introduction

All extant gonorynchiforms except species within the genus *Gonorynchus* (provided with a differently specialized structure in the same region) possess a so-called epibranchial (= "crumenal") organ (EBO), which is a paired dorsal sac-like expansion of the pharyngeal cavity opening at the top of the fifth branchial slit. A substantial part of the EBO is supported by dorsal branchial arch elements. The branchial basket skeleton and/or muscles have been described by Johnson and Patterson (1997) and the dorsal musculature of the branchial basket by Springer and Johnson (2004). Some good accounts were previously published for particular taxa (Thys van den Audenaerde 1961: *Phractolaemus*; Lenglet 1973: *Kneria*; Monod 1963: *Gonorynchus*; d'Aubenton 1961: *Cromeria*) and also a detailed, well-illustrated unpublished study by Pasleau (1974: essentially *Chanos* and *Gonorynchus*). However, in general the gonorynchiform EBO was seldom thoroughly described, except by Kapoor (1954), Takahasi (1957), Bertmar *et al.* (1969) and Pasleau (1974).

The EBO (or a dorsal extension of the fifth slit and fourth and fifth epibranchials in *Gonorynchus*: see below) is present in all the lineages of extant gonorynchiforms. It is associated with the loss of jaw teeth in extant and fossil members of this order (note: the "premaxillary" teeth that Monod 1963 described in *Gonorynchus* are not truly jaw teeth; e.g., Taverne 1981, Gayet 1993, Grande and Poyato-Ariza 1999). It is important to stress that the branchial basket could never be observed in detail in the fossil gonorynchiforms that have been discovered so far. The absence of a "true" EBO in *Gonorynchus*, a feature considered by Pasleau (1974) an important character separating this taxon from the other extant gonorynchiform genera, remains puzzling and deserves discussion.

An EBO is present in many other teleosts (e.g., Hyrtl 1863, Boulenger 1901, Vialli 1926, Heim 1935, d'Aubenton 1955, 1961, Takahasi 1957, Bertin 1958, Miller 1964, Nelson 1967, Bertmar 1961, 1973, Bertmar and Kapoor 1969, Bertmar *et al.* 1969) such as certain osteoglossiforms, cypriniforms, characiforms, salmoniforms, perciforms *sensu lato* and particularly several clupeomorphs, which are now usually considered the sister group of the Ostariophysi (Stiassny *et al.* 2004, Diogo 2007, Diogo *et al.* 2008). But as an EBO was not described in all clupeiforms (e.g., Takahasi 1957), and particularly not in the plesiomorphic (e.g., Diogo 2007, Diogo *et al.* 2008) clupeiform *Denticeps clupeoides*, the EBO is seemingly not a synapomorphy of a potential clupeiform + gonorynchiform clade such as that proposed in the recent molecular mitochondrial study of Peng *et al.* (2006). A characteristic, distinctive particularity of the EBO of gonorynchiforms relative to that observed in other teleosts is its more dorsal position. However, although the EBO very likely appeared polyphyletically in different unrelated teleostean groups, the histological structure of its epithelium, and the relations to the surrounding pharynx, in gonorynchiforms such as *Chanos* (Kapoor 1954) is rather similar to that found in those other teleosts in which this organ has been studied (see, e.g., Kapoor 1954, d'Aubenton 1961, Thys van den Audenaerde 1961, Bertmar *et al.* 1969, Lenglet 1973, 1974).

There is unfortunately no direct functional study of the EBO in a living fish. Bertmar *et al.* (1969) provided a summary of the different opinions regarding the function of the EBO. The possibility of a respiratory function is now abandoned and most authors hypothesize that it is involved in filter feeding, but they do not propose a precise mechanism. The filter-feeding hypothesis is, for instance, supported by the high number, arrangement and development of the gill rakers. In all genera but *Gonorynchus* the anterior gill rakers rows extend dorsally on the epi- and pharyngobranchials, and ventrally on basibranchials. They are "almost meeting their fellow on the opposite side", according to Johnson and Patterson (1997); however, according to the detailed description of Monod (1949), they are imbedded together with those of the opposite arch in a fleshy cushion ("bourrelet"), leaving only restricted passages for water.

Since the EBO is anatomically and functionally linked to the dorsal part of the last two branchial arches, it is important to discuss the homologies of their epi- and ceratobranchials, and thus to briefly refer, in this introduction, to a much debated and highly controversial issue: the identity of the so-called fifth epibranchials (EBR5). The fifth arch and the fourth epibranchial (EBR4) are specialized in different ways in gonorynchiforms and their interpretation remains controversial (see, e.g., Johnson and Patterson 1996). The ceratobranchial 5 (CBR5) is large in *Phractolaemus* (Thys van den Audenaerde 1961), elongated in *Gonorynchus* (Monod 1963), and triangular in *Grasseichthys* and *Cromeria* (d'Aubenton 1961, Britz and Moritz 2007). Contrary to Fink and Fink (1996), EBR4 is enlarged in *Chanos* and *Grasseichthys*. The much widened EBR4 of *Gonorynchus* is pierced by the enlarged foramen of the efferent artery (the mesial margin of the foramen is called "cartilage semi-lunaire" by Monod 1963), as in the clupeid *Sierrathrissa* (Whitehead and Teugels 1985). It is joined to the EBR4 by an enigmatic cartilage, which is interpreted as a fifth epibranchial by Monod (1963), Pasleau (1974), and Johnson and Patterson (1997). As suggested by Fink and Fink (1996), Johnson and Patterson (1997) propose that the so-called EBR5 of *Gonorynchus* is a neoformation, but that there is a true EBR5 fused with the posterior process of EBR4 (= "cartilage semi-lunaire" of Monod). Johnson and Patterson (1997) interpret the cartilage borne by EBR4 as an EBR5.

An EBR5 is present in the wall of the EBO in all gonorynchiform genera ("épibranchial accessoire" of d'Aubenton 1961: fig.10, in *Cromeria*) except in *Kneria* (Lenglet 1974: fig.16). The photographs of Britz and Moritz (2007) do not clearly demonstrate the presence versus absence of an EBR5 in *Cromeria* and *Grasseichthys* (i.e., the presumed cartilage is not colored in *Grasseichthys*). We could not find any description of an EBR5 in basal teleosts such as *Elops* (Taverne 1974) or osteoglossiforms (Taverne 1977), or in *Amia* (Grande and Bemis 1998). Authors such as Jollie (1975) stated that there is no fifth epibranchial in teleosts or in any other actinopterygians. However, Johnson and Patterson (1997) refer to an EBR5 in the teleosts *Elops* and *Hiodon*. In any case, the cartilaginous fifth EBR described in gonorynchiforms, clupeiforms and other teleosts is probably a polyphyletic neoformation, eventually linked to the development of an EBO. In their synthetic discussion of the problem, Springer and Johnson (2004) consider that "it has not been established" that the element is an epibranchial; they suggest that it is present (autogenous or fused) in some clupeomorphs, e.g., *Denticeps clupeoides*, some Albulidae, some Osmeridae, and inclusively in some salmoniforms. They contest Fink and Fink's (1996: 231) description of an EBR5 distinct primordium at an early stage of *Gonorynchus* and consider that the EBR5 originates from EBR4 and remains continuous with it, or becomes distinct and articulated on it, or is completely free or even sometimes fused with the CBR4. The term EBR5 will be used here in a purely descriptive view, that is, without implying that all the so-called EBR5 are necessarily homologous structures.

The present contribution is based principally on the Ph.D. study of *Gonorynchus* and *Chanos* by the first author and on observations of *Grasseichthys* by the last one. The aim is to summarize the data available on the EBO, to provide more detailed descriptions of the EBO in three gonorynchiform genera, *Gonorynchus, Chanos* and *Grasseichthys* (including its muscles, gill rakers and epithelium), and to propose a functional hypothesis based on this summary and on these descriptions. The EBO is deliberately interpreted in its anatomical environment, since this helps in the understanding of its functional and evolutionary signification. As a consequence, the text and figures presented here also provide aspects of the branchial basket and associated muscles, and the role of the EBO is envisaged in the general context of filter feeding.

Material and Methods

All the gonorynchiform specimens that were analyzed for this work are now part of the collection of the Laboratory of Functional and Evolutionary Morphology (LFEM, ULG, uncatalogued): two *Gonorynchus gonorynchus* given

by the South African Museum and one *G. gonorynchus* offered by the British
Museum of Natural History; two *Chanos chanos* offered by F. Coumans and
five *C. chanos* offered by H.K. Mok; two *Phractolaemus ansorgei* and three
Grasseichthys gabonensis offered by D.E.F. Thys van den Audenaerde and by
M. Poll. In addition, two *G. gabonensis* specimens from the LFEM were stained
with alizarine and toluidine, while two others were serially cut and stained.
Dissections and anatomical drawings of the dissected specimens were made
using a Wild M5 dissecting microscope equipped with a camera lucida.
The phylogenetic framework for the present study is essentially based on
the results of Grande and Poyato-Ariza (1999).

Results and Discussion

Gonorynchus gonorynchus (Figs. 5.1–5.6)

The phylogenetic position of the genus *Gonorynchus* and the family
Gonorynchidae among the Gonorynchiformes is a matter of debate (see,
e.g., Gayet 1993, 1995, Fink and Fink 1996, Johnson and Patterson 1996).
The peculiarity of the fifth slit of *Gonorynchus* is that it is enlarged and
specialized while there is no true EBO (Monod 1963 considers that it is
"something similar (to an EBO), but still very rumentary.. a case study of
interest for the study of more derived types (of EBO)", an opinion that is in
agreement with the position of the genus according to Gayet 1993 and Lavoué
et al. 2005, but not according to Grande and Poyato-Ariza 1999).

The gill-arch skeleton is best described in Monod (1963) and Pasleau
(1974) (for a detailed description of the cranium of *Gonorynchus*, see
Ridewood 1905, Monod 1963, Pasleau 1974, and Howes 1985). The fifth
slit is elongated and curved dorsally and posteriorly (at the place occupied
by the EBO in other gonorynchiforms). The EBR4 is much transformed,
much broadened and pierced by an unusual foramen for the fifth efferent
branchial artery. The posterior cartilaginous edge of the foramen is called
semi-lunar cartilage by Monod (1963). EBR5 is cartilaginous and articulated
on both CBR4 and 5. It dorsally limits the upper part of the fifth slit. Well-
developed branchictenies are borne by EBR4 and 5 and CBR4 and 5. As in
the other gonorynchiforms, the whole branchial basket and especially its
posterior part is perfectly continuous, its elements being united by cartilage
(or by short ligaments between basibranchials) that results in global
elasticity.

The muscles associated with the last two branchial arches and
the EBO are highly specialized and best described by Pasleau (1974:
G. gonorynchus) and Springer and Johnson (2004: *G. moseleyi*). Both
descriptions are in general agreement (see Fig. 5.2). The region of the fifth
slit is coated dorsally by thick circular muscles continuous with the sphincter
oesophagi and the transverse dorsals 4 (joining both EBR4) in *G. moseleyi*

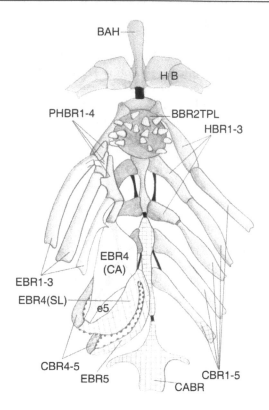

Fig. 5.1 *Gonorynchus gonorynchus.* Dorsal view of the branchial basket. The upper limb of the right arches and the lower part of the left ones have been removed. Cartilages are squared and ligaments are in bold black.

but distinct from them in *G. gonorynchus*. A distinct broad transverse muscle (m.) runs from the esophageal tunica to EBR5. A m. transversus pharyngobranchialis 4 not found in *G. gonorynchus* is figured in the other species. A m. adductor 4 joins EBR4 to CBR4 and a m. elevator 4 attaches the cartilaginous part of EBR4 to the prootic (Figs. 5.2–5.4).

The ventral muscles attaching on CBR5 insert mesially on the cardiobranchial cartilage and those of CBR4 on the last basibranchial cartilage (basibranchials 4+5?) (Fig. 5.3). Paired basicleithrales muscles join the posterior basibranchial to the mesial aspect of the cleithrum and pharyngocleithrales muscles (seemingly homologous to the pharyngocleithralis internus of *Chanos*) run from the cleithrum to CBR4.

The gill filaments conserve a plesiomorphic teleostean configuration (Le Danois 1966); they are supported by an unbranched cartilaginous rod (Johnson and Patterson 1997). Two rows of toothless gill rakers of equal length are borne on cerato- and epibranchials (Johnson and Patterson 1997).

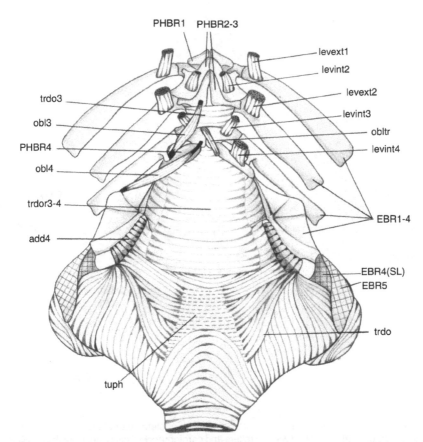

Fig. 5.2 *Gonorynchus gonorynchus*. Dorsal view of the branchial basket and associated musculature. Cartilages are squared, stippled lines indicate the dorsal surface of the pharynx.

The gill rakers are particularly long on CBR and EBR 5 up to the superior end of the slit (Fig. 5.6, after Monod 1963, Le Danois 1966: fig. 17). Monod (1963) gives an accurate description of the fifth slit and figures a papillate pad on the fourth arch and a velum partly covering the unique gill rakers row of the fifth arch. This velum probably plays an important role in controlling the water current through the branchiospinal filter.

Grande and Poyato-Ariza (1999) consider that *Gonorynchus* species are "most likely" adapted to deep water and bottom-dwelling. The absence of teeth on the lower jaw and the long gill rakers are probably associated with filter feeding. The numerous small "premaxillary teeth" *sensu* Monod (1963), which are not true jaw teeth (see above), may help in holding and processing food collected by the protrusive mouth. A large toothed plate with blunt teeth on the second basibranchial faces a similar double one on the entopterygoids (Monod 1963, Pasleau 1974): Pasleau (1974) considers

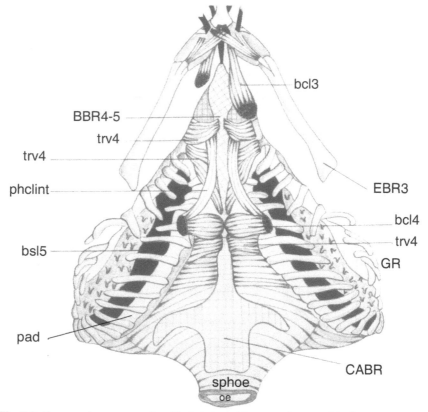

Fig. 5.3 *Gonorynchus gonorynchus*. Ventral view of the posterior part of the branchial basket, associated musculature, fifth slit with gill rakers and cardiobranchial cartilage. Note the papillate pad. Cartilage is squared.

that the ventral tooth plate is mobile longitudinally and transversally relative to the entopterygoideal plate because of the well-developed m. transversus ventralis. At the same time, the anterior and posterior parts of the suspensorium are reciprocally freed by the reduction of the metapterygoid to a thin bar loosely attached to the entopterygoid (Pasleau 1974: fig. 34). So, a chewing mechanism seems likely, in a way somewhat comparable to that of the pharyngeal bones of most cypriniforms. Therefore, there are probably two main feeding mechanisms in *Gonorynchus*: (1) sucking in of large items (e.g., crabs or mollusks) by the inferior mouth and chewing by the toothed plates, and (2) filter feeding (probably of food sucked from the bottom surface; the orientation of the mouth gape does not allow engulfing of water while swimming). In this paper we mainly focus on the second mechanism. Unfortunately, the sole stomach dissected by the first author did not contain any recognizable food.

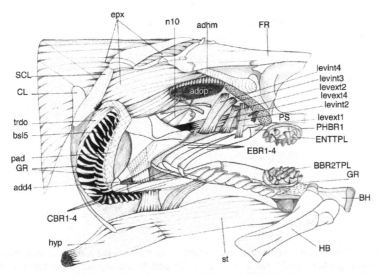

Fig. 5.4 *Gonorynchus moseleyi*. Lateral view of the dorsal skeleton of the last gill arches. Cartilages are marked with small circles (modified from Springer and Johnson 2004).

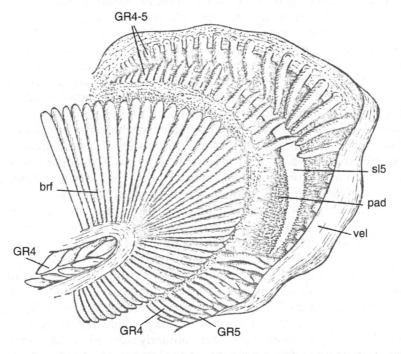

Fig. 5.5 *Gonorynchus gonorynchus*. Lateral aspect of a dissection showing the branchial basket, the dorsal and ventral toothed plates, the gill rakers of the fifth slit and muscles. Cartilage is squared.

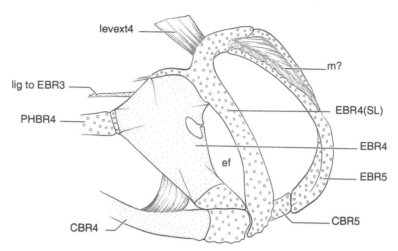

Fig. 5.6 *Gonorynchus gonorynchus*. Lateral aspect of the right branchial arches 5 and 5 and fifth slit with gill rakers. Gill rakers of the middle part of the arches are omitted. Note the papillate pad and the posterior velum on the last arch (modified from Monod 1963).

The well-developed m. sternohyoideus, m. levator arcus palatini, m. levator operculi and m. dilatator operculi, seemingly helped by the well-developed m. transverse dorsales muscles of the fourth and fifth arches, probably allow a powerful dilation of the mouth and opercular cavities (for a description of the " plesiomorphic" breathing movements in teleosts, see, e.g., Hughes 1960). The inhalant water current easily and preferentially passes through the largest slit, the fifth one, while the large m. adductor mandibulae II begins to adduct the suspensorium (for a updated overview of the cephalic and pectoral muscles of gonorynchiforms, see Diogo, this volume). The levatores interni and externi muscles of the gill arches, and especially the well-developed m. levator externus 4, also participate in the dilation of the oro-branchial cavity. During the exhalant phase, the elevation of the large hyoid bars by the m. sternohyoideus, as well as the contraction of the thick transverse muscle of the last two arches and of the sphincter oesophagi inserting on the accessory cartilage of the fifth arch, probably favors sieving on the interlocking gill rakers and food transport to the esophagus. The Adductores four muscles (= attractores 4 *sensu* Pasleau 1974) probably also play a role in this mechanism; the other constrictor and dilatator muscles of the branchial basket seemingly play a lesser part. The function of the "gill filaments muscle" figured by Springer and Johnson (2004: plate 31) is not clear. It is possible that the larger food items chewed by the toothed plates are similarly stopped by the gill rakers. Moreover, Monod (1963) described, on the last branchial arch, a "velum" that is probably able to shut the last slit.

The so-called EBR5 is important in sustaining the dorsal extension of the fifth slit as does the cartilaginous upper part of the EBR4 anteriorly. The specialization of the dorsal part of the last two arches is a clear adaptation to the filtering mechanism. The abundance of cartilage in the branchial basket, and especially around the upper part of the last slit, gives it an elasticity favorable for fast dilation. The particular disposition of the fourth pharyngobranchial and its anterior articulation on the third one allows the EBR4 (and consequently all the skeletal support of the last slit) to move more easily.

Chanos chanos (Figs. 5.7–5.13)

Good descriptions of the EBO of *Chanos* are provided by, for example, Kapoor (1954: soft parts), Pasleau (1974: skeleton and muscles) and Springer and Johnson (2004: muscles). *Chanos* is an exclusive and very efficient filter feeder that is able to trap blue algae; it is provided with a sophisticated disposition of the gill rakers leaving only small openings for water flow helped by numerous mucous cells and sensory organs (Monod 1949: figs. 15–17). Like all other gonorynchiforms except *Gonorynchus*, *Chanos* possesses a sac-like EBO, which is a cul-de-sac opening dorsally in the pharynx just above the fifth slit. The epibranchial sac and the skeleton supporting it are, however, simpler and smaller in the smaller non-*Gonorynchus* extant

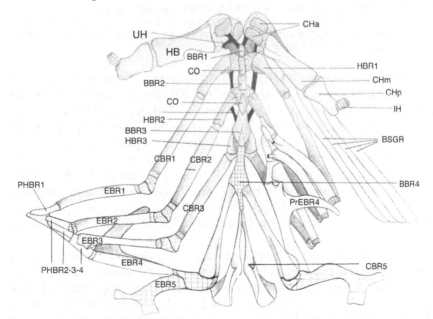

Fig. 5.7 *Chanos chanos.* Dorsal view of the branchial basket. The upper part of the right arches has been extended laterally. Cartilages are squared and ligaments are in dark black.

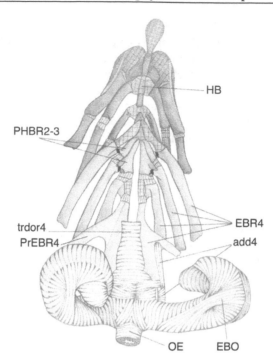

Fig. 5.8 *Chanos chanos.* Dorsal view of the branchial basket with the epibranchial organ and associated musculature. The upper limb of the arches has been removed. The right EBO is partly unrolled. In their normal position the left and right organs almost meet on the midline. Cartilages are squared. HB means BBR here.

gonorynchiform species, and especially in the miniature ones (*sensu* Conway and Moritz 2006 and Britz and Moritz 2007; the miniature species are considered to be paedomorphic by Grande 1994). A plausible explanation of that morphological allometry is that the feeding efficiency is proportional to (1) the length (unidimensional) of the slits, (2) the gill rakers filter surface (bidimensional), and (3) the volume of the sac (tridimensional), while the fish weight to be fed is proportional to the product of the three body dimensions. So a greater size probably results in longer slits and gill rakers and/or in a proportionally longer EBO sac. It would be interesting to observe and measure the growth of the branchial slits, branchictenies and EBO in *Chanos*.

The EBO of *Chanos* is large and rather complicated. It consists in an entrance canal followed by a blind sac. It occupies the space left between the floor of the neurocranium, the supracleithrum and the long EBR1 to 4. That volume seems to be much increased by the contraction of the big oticobranchial muscle whose ends insert respectively on the prootic and on a membrane joining the EBR4 to the pectoral girdle (Pasleau 1974). Coiling

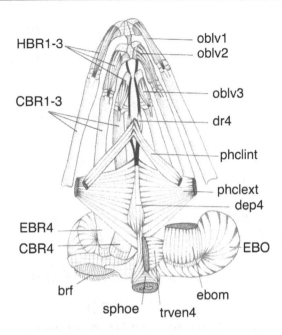

HBR1-3

oblv1
oblv2

CBR1-3

oblv3

dr4

phclint

phclext
dep4

EBR4
CBR4

EBO

brf

ebom

sphoe   trven4

Fig. 5.9 *Chanos chanos*. Ventral view of the branchial basket with the muscles and the EBO. The upper limb of the arches has been removed. Cartilages are squared. HBR 1–3 are part ly hidden by muscles.

allows lodging of a long blind sac in that restricted space. The epithelium of the sac is similar to that of the neighbouring pharynx. It is rather thin, generally made of only two layers: basal cells and flattened cells over them, with many mucous caliciform and club cells. It is coated by a submucosa principally made of striated muscles (the inner are mostly circular and the outer longitudinal) (for details, see Kapoor 1954: figs. 1, 2; see also fig. 16). The sac is almost completely coated by muscles. The entrance canal is supported posteriorly by the cartilaginous EBR5 (Springer and Johnson 2004: fig. 30B) and anteriorly by EBR4, which also supports and attaches to the proximal part of the sac (Kapoor 1954, Pasleau 1974). One of the two triradiate gill rakers rows of the fourth gill arch and the sole row of the fifth arch continue in the canal but taper progressively, up to the mouth of the sac. The gill filaments also become smaller in the canal and disappear in the blind sac. The internal wall of the canal bears papillae similar to those in the esophagus (Kapoor 1954: figs. 1, 2, Bertin 1958).

The EBO of *Chanos* (and Kneriidae *sensu* Grande and Poyato-Ariza 1999) plays an important part in their filter-feeding mechanism, as previously suggested by Kapoor (1954) and Takahasi (1957). Pasleau (1974) found, in the EBO of *Chanos*, an accumulation of unicellular organisms and much

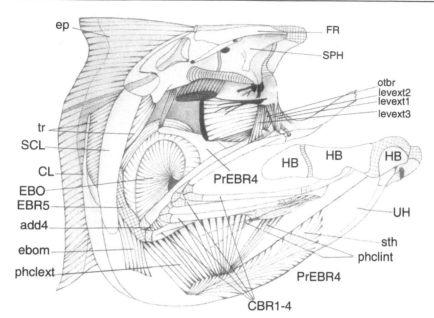

Fig. 5.10 *Chanos chanos.* Lateral view of the EBO in its skelet al-muscular environment. Cartilages are squared.

mucus, supporting the diet described by Takahasi (1957) and Huet and Timmermans (1970): blue and green algae, bacteria, protozoans, entomostraceans, small malacostraceans, worms and detritus mixed with inorganic material (e.g., sand) and imbedded in abundant mucus. Filtering efficiency rests among other things on the high complexity of the filtering gill rakers apparatus (see Monod 1949: figs. 13–17), on the long and specialized branchictenies of all gill arches, and on the volume of water passing through the sieve and especially that treated by the EBO.

Filter feeding and respiration are realized simultaneously and mainly by the same musculo-skeletal apparatus; this probably implies some compromise in morphology and central general pattern. As in teleosts in general (e.g., Ballintijn and Hughes 1965), maintaining an almost continuous rearward water current requires the pressure in the mouth cavity to remain higher than that in the opercular cavities almost throughout the respiratory cycle. The pressure drop in the opercular cavity must be faster than in the mouth cavity; this is probably a first explanation for the powerful levator and dilatator operculi muscles seen in *Chanos*.

Bucco-pharyngeal dilation principally results from contraction of the big m. sternohyoideus (and hypaxial muscles; see, e.g., Ballintijn and Hughes 1965). It is further enhanced by the height of the suspensorium and the non-alignment of its anterior and posterior cranial articulations, which oblige it

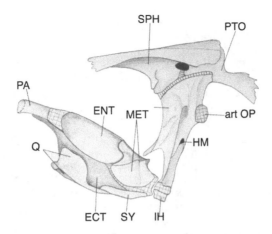

Fig. 5.11 *Chanos chanos.* Lateral view of the suspensorium to show the proportions and the separation of the metapterygoid from the hyomandibula. Cartilages are squared.

to bend between the non -fused hyomandibula and metapterygoid and lets the interhyal (drawing the hyoid bar) protrude laterally (Fig. 5.11). For creating an efficient inhalant current in the EBO, the dilation of the opercular cavities must follow with some delay that of the buccopharyngeal cavity to prevent water from flowing directly through the slits. It is also probably advantageous that the dilation of the branchial basket follows with a short but real delay that of the oral cavity to enhance the rearward current directed to the EBO. The water current in the bucco-pharyngeal cavity will also be accelerated toward the EBO if the protractores hyoidei muscles lowering the mandible contract somewhat before the m. sternohyoideus and the m. levatores arcus palatini (Fig. 5.13).

The branchial basket is dilated by (1) its own elasticity (abundance of cartilage in the branchial basket is evident in Figs. 7 and 8) and the levatores externi 1 to 4 and interni muscles 2 to 4, (2) by the pharyngocleithral muscles (which pull the last two ceratobranchials downward when the powerful m. sternohyoideus is contracted); and (3) by the well-developed characteristic "muscle of the EBO" (which pulls downward EBR4) of *Chanos* (Pasleau 1974). At the same time, or a little later, the blind sac is dilated by the relaxation of the constrictor oesophagi and by contraction of the controversial m. adductor 5 attaching on to the long process of EBR5 (see, e.g., Springer and Johnson 2004 : fig. 30, and Pasleau 1974).

Filling and emptying the EBO is probably not integrated in all respiratory cycles, but only when enough food in suspension is present and detected by taste buds and/or mechanoreceptors. The decoupling of the last branchial arch (also in kneriids, but not in *Gonorynchus*) by the separation of CBR5 from the fourth pharyngobranchial allows a partly independent control of

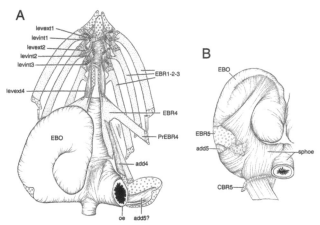

Fig. 5.12 *Chanos chanos.* Dorsal (A) and ventral (B) views of the epibranchial organ; in (A) the right organ has been cut on the right side. Cartilages are squared (modified from Springer and Johnson 2004).

the water currents in the EBO. The relative freedom of pharyngobranchial 4 also facilitates movements of EBR4 at the base of the EBO. Food particles are trapped and packed in mucous strings inside the sac of the EBO; that process is enhanced by the papillae and longitudinal grooves of the sac wall. A breathing and feeding cycle can thus be hypothesized as follows, in two main phases , the first one mainly of dilations, the second one of contractions.

1. As in a usual breathing cycle, the buccopharyngeal cavity dilation begins during the end of opercular adduction (at that moment the fact that opercular pressure is temporarily higher than buccopharyngeal pressure cannot be avoided).
2. Just after mouth opening and buccal region dilation, the branchial basket dilation begins.
3. EBO sac dilates and water enters it. Food items are trapped there in mucous strings.
4. Branchial basket contraction and rapid dilation of the opercular cavities occur while EBO and buccopharyngeal cavity begin to contract. Water flows rapidly from the buccopharyngeal cavity and from the EBO through the slits. Small food is stopped by the gill rakers filter of the branchial arches brought nearer by the elastic recoil of the basket helped by the adductor (functionally a depressor?) of the fourth branchial arch (Pasleau 1974, Springer and Johnson 2004). The enigmatic adductor 5 of the latter authors possibly controls the entry of the canal of the EBO by pulling the cartilaginous EBR5.

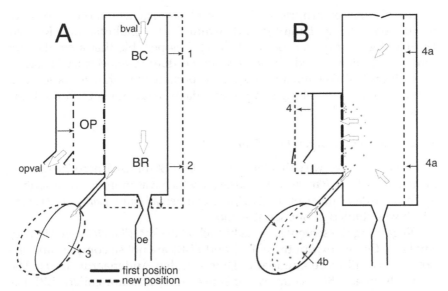

Fig. 5.13 *Chanos chanos.* Diagram of our hypothesis about water flow in the EBO and respiratory cavities: (A) dilations phase; (B) contraction phase. For more details, see text.

5. Food from the EBO and from the branchial sieve is sucked and processed rearward by alternate contraction and relaxation of the sphincter oesophagi (note that the esophagus of *Chanos* is helicoidally folded according to Bertin 1958); that latter mechanism is difficult to understand completely because pharyngeal jaws that often realize it (e.g., Vandewalle *et al.* 2000) are lacking. However, the mechanism probably resembles that observed by cineradiographic studies in several teleosts, such as cypriniforms (e.g., Sibbing 1982), cichlids (e.g., Aerts *et al.* 1986), labrids (e.g., Liem and Sanderson 1986), serranids (e.g., Vandewalle *et al.* 1992), and sparids (Vandewalle *et al.* 2000): food items (in this case, small food trapped in mucous balls) undergo a longitudinal see-saw movement in the branchial cavity until they enter the esophagus. The normal antero-posterior water flow can indeed be reversed, as is the case, for instance, in coughing in other teleosts (see Osse 1969). Two different central pattern generators may be activated, one for normal breathing and one for integrating filtering by the EBO and food transport.

The EBO is thus probably not the sole filtering part of the branchial basket, but it is likely the most efficient one. Its activity is seemingly controlled by the central nervous system, namely by information (registered by the taste buds) regarding the abundance of food particles in the water current and by hunger. The possibility of disengaging its mechanism saves energy.

It probably rests on proprioceptive control (as in general motor control of respiration: see, e.g., Ballintijn and Bamford 1975, Ballintijn and Roberts 1976) and gustatory control (see below). It is possible that a similar but simpler water flow is produced in *Gonorynchus*, due to the specialized postero-dorsal branchial skeleton and muscles, especially the sphincter oesophagi (Figs. 5.8 and 5.9), resulting in a preferential passage of water through the dorsal part of the last slit.

Phractolaemus ansorgei (Figs. 5.14–5.15)

In the present paper, we will briefly discuss the EBO of two Kneriidae species: the relatively large *Phractolaemus ansorgei*, which was thoroughly described by Thys van den Audenaerde (1961), and of the small *Grasseichthys gabonensis*, which was studied in our serial sections.

Regarding the structure of its branchial basket, *Phractolaemus* is very similar to *Chanos*, except that CBR5 and EBR5 are not so complex and the cartilage is still more abundant (Thys van den Audenaerde 1961). The relatively large EBO occupies the same place; it is not coiled. EBR4 is V-shaped: one ossified branch fuses with the cartilaginous part of the fourth pharyngobranchial; the other, more flattened, cartilaginous branch lies near (but does not really reach) the flat cartilaginous EBR5 at the upper end of the fifth branchial slit, at the level of the aperture of the EBO.

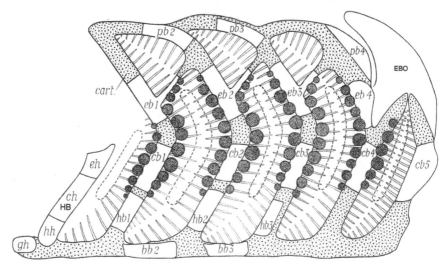

Fig. 5.14 *Phractolaemus ansorgei.* Diagram of lateral view of the branchial basket to show the EBO and the position of the gill rakers; note also the skeletal bridges between the first three arches and the following ones (modified from Thys van den Audenaerde 1961; most of that author's abbreviations are conserved: pb = pharyngobranchial, eb = epibranchial, cb = ceratobranchial, bb = basibranchial, eh+ch+hh = hyoid bar). Cartilage is stippled. Gill filaments are represented by circles. Gill slits are in dotted lines.

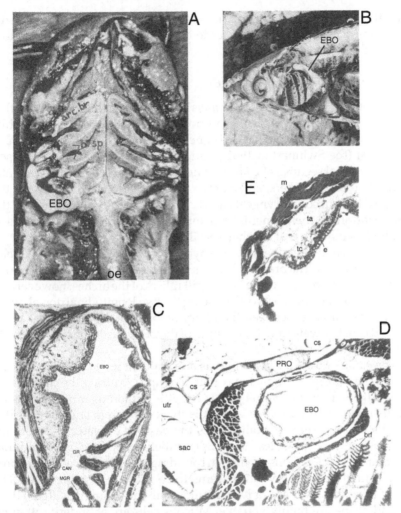

Fig. 5.15 *Phractolaemus ansorgei*. Lateral (A) and ventral (B) photographs of a dissection showing the position of the EBO. (C) Longitudinal section of the EBO to show the opening of the organ into the pharynx, the end of the gill rakers row and the disposition of the microbranchiospines. (D) Cross-section through the head of a young specimen to show the position of the EBO. (E) Magnified detail of the wall of the sac (modified from Thys van den Audenaerde 1961).

The wall of the latter is rich in mucous cells and presents longitudinal grooves. It is anteriorly coated by muscles running parallel to its length. The gill rakers taper progressively up to that level. They bear unicellular blunt microbranchiospines complicating the filter (Thys van den Audenaerde 1961: plate III). Different and complicated movement patterns of the splanchnocranium are probably used to combine typical aquatic respiratory movements with occasional filter feeding by the EBO and aerial respiration (Thys van den Audenaerde 1959) by the specialized pulmonoid swimbladder.

Grasseichthys gabonensi (Fig. 5.16)

The anatomy of the EBO of *Grasseichthys* is, in general, mainly similar to that of *Phractolaemus*, except that the proportion of cartilage in the branchial basket is still higher. The last two branchial arches are built as in *Cromeria* and *Kneria* (see Swinnerton 1901 , Lenglet 1974 , Johnson and Patterson 1997). Our serial sections of *Grasseichthys* mainly provide histological data, information on gill rakers, and a better representation of the shape of the ceiling of the bucco-pharyngeal cavity. They allow us to confirm that the configuration of the branchial basket and of its principal muscles is essentially similar to that of other kneriids.

The branchial filter does not display the specializations found in *Chanos* and *Phractolaemus*. The branchial arches remain separate and the gill rakers do not bear microbranchiospines. The gill rakers of the arches, however, are efficiently interlocking. It is worth noticing the two longitudinal dorsal canal-like grooves situated between the pharyngobranchials; they are likely to taste and canalize water into the opening of the EBO. Taste buds are regularly disposed on their margins.

The canal of the EBO continues as a prominent incomplete ring in the entry of the sac (Fig. 5.16) that may help to empty it during the contraction phase. The wall of the ring is provided with numerous mucous cells on both its internal and external faces; it is thick enough to suggest that it is rigid. Be that as it may, it complicates the passage of water and perhaps helps to stop food particles. The EBO epithelium is similar, but thinner than that of most regions of the surrounding pharyngeal wall; in some places it might be so thin that it seems to be made of a single layer. The cells of the basal layer are rather flat and their basal lamina is thick; their size and shape are highly variable. Mucous cells are abundant (often on the ridges) and some cells seem to present long cilia (difficult and uncertain observation). Taste buds are lacking in the EBO epithelia. The striated muscular sheath is rather thick and comprises an internal circular layer and an outer longitudinal one. The muscular coating is thicker on the lateral wall.

Conclusions

The genus *Gonorynchus* clearly differs from all the other gonorynchiforms in not having an EBO, but instead a dorsally elongated fifth branchial slit. Its diet could be twofold: filtered small items and larger food collected from the bottom and crushed between the toothed plates (Fig. 5.4). The dorsal parts

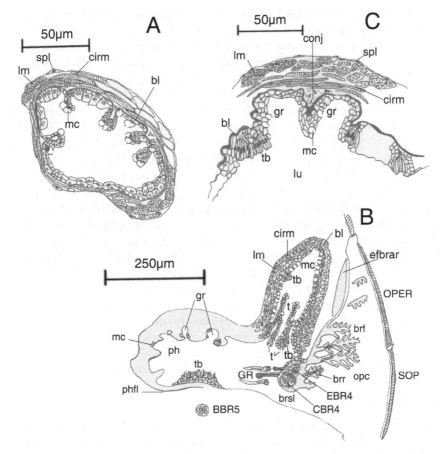

Fig. 5.16 *Grasseichthys gabonensis*. Details of a cross- section through the posterior part of the branchial basket to show the epithelium, gill rakers, taste buds and epithelium of the pharynx and EBO.

of the fourth and fifth branchial arches are highly specialized in order to support the dorsal elongation of the last gill slit.

The EBO and the last two gill arches are similar in the other gonorynchiforms, displaying a so-called EBR5 that is probably a specialized cartilaginous outgrowth of EBR4. The EBO of *Chanos* is proportionally the most elongated; it is coiled helicoidally. The EBO is a long blind pocket opening in the pharynx just above the last branchial slit. It consists in a proximal canal in which two rows of gill rakers are prolonged, followed by an elastic sac , the wall of which is plicate, rich in mucous cells and coated by striated muscles. The skeletal support of the EBO by EBR4 and/or EBR5 is variable and more substantial in *Chanos*. A functional hypothesis is proposed for this genus (Fig. 5.13): small food is sucked with water into the sac of the EBO and is trapped in mucus produced by the wall. It is then expelled with water back to the pharynx and stopped by the branchictenies. It is finally pushed into the esophagus. The inflation-compression cycle of the sac is eventually integrated in the respiratory cycle when enough food is present in the water. Processing of food trapped on the gill rakers probably requires see-saw movements of water into the pharynx and thus repeated reversals of the water current in the bucco-pharyngeal cavity interrupting the breathing cycle. Anatomically, the filter -feeding system and particularly the EBO appear to be more derived in *Chanos*.

A phylogenetic issue remains problematic. Certain cladograms propose that kneriids are more closely related to *Gonorynchus* than to *Chanos* (e.g., Grande and Poyato-Ariza 1999, which, as stated above, constitutes the basic phylogenetic framework for our discussions). It is, however, difficult to accept that the branchial region structure of *Gonorynchus* is derived in relation to that of *Chanos*, because that would imply a secondary shortening of the EBO, followed by the opening of its lateral portion and the dorsal extension of the gill rakers rows. Of course, it is possible that despite evident similarities the EBO in *Chanos* and kneriids is convergent. That would be acceptable because an EBO (although less dorsally situated) exists also in many unrelated teleosts. The existence of a relatively unoccupied space above the end of the branchial basket, over the last epibranchials, thus seems to be a "positive constraint" allowing the formation of an EBO in different lineages by parallel evolution (see, e.g., Diogo 2004, 2007, Diogo *et al.* 2008).

Acknowledgements

We thank Louis Taverne (IRSNB, Brussels), Pierre Vandewalle, Bruno Frederich and Eric Parmentier (ULG, Liège) and other colleagues for their helpful advice and assistance and for their discussions on teleost anatomy, functional morphology, and evolution. We are especially grateful to Nicole Decloux and André Delvaux (UlG, Liège) for preparing and staining the serial sections of *Grasseichthys* and for helping prepare the figures provided in this paper, respectively.

References

Aerts, P., F. De Vree and P. Vandewalle. 1986. Pharyngeal jaw movements in *Oreochromis niloticus* (Teleostei: Cichlidae): preliminary results of a cineradiographic analysis. Ann. Soc. R. Zool. Belg. 116: 75–82.

D'Aubenton, F. 1955. III. Etude de l'appareil branchiospinal et de l'organe suprabranchial d' *Heterotis niloticus* Ehrenberg, 1827. Bull. Inst. Fr. Afr. Noire 17: 1179–1201.

D'Aubenton, F. 1961. Morphologie du crâne de *Cromeria nilotica occidentalis* Daget 1954. Bull. Inst. Fr. Afr. Noire 23, sér. A 1: 129–164.

Ballintijn, C.M. and G.M Hughes. 1965. The muscular basis of the respiratory pumps in the trout. J. Exp. Biol. 43: 349–362.

Ballintijn, C.M and O.S. Bamford. 1975. Proprioceptive motor control in fish respiration. J. Exp. Biol. 62: 99–114.

Ballintijn, C.M. and J.L. Roberts. 1976. Neural control and proprioceptive load matching in reflex respiratory movements of fishes. Federation Proc. 35, 9: 1983–1991.

Bertin, L. 1958. Appareil digestif, pp. 1248–1302. In: P.P. Grassé [ed.]. Traité de Zoologie, anatomie, systématique, biologie, XIII, fasc. 2. Masson, Paris.

Bertmar, G. 1961. Are the accessory branchial organs in characidean fishes modified fifth gills or rudimentary ultimobranchial organs? Acta Zool. (Stockholm) 42: 151–162.

Bertmar, G. 1973. Epibranchial organ en anpassning till planktonupptagning hos benfiskar. Zool. Rev. 35: 5–10.

Bertmar, G. and C. Strömberg. 1969. The feeding mechanisms in plankton eaters. I. The epibranchial organ in whitefish. Mar. Biol. 3: 107–109.

Bertmar, G., B. Kapoor and R. Miller. 1969. Epibranchial organs in lower teleostean fishes— An example of structural adaptation. Int. Rev. Gen. Exp. Zool. 4: 1–48.

Boulenger, G.A. 1901. On the presence of a superbranchial organ in the cyprinoid fish Hypophthalmichthys. Ann. Mag. Nat. Hist. 8: 186–188.

Britz, R. and T. Moritz. 2007. Reinvestigation of the miniature African freshwater fishes Cromeria and Grasseichthys (Teleostei, Gonorynchiformes, Kneriidae), with comments on kneriid relationships. Mitt . Mus. Nat. kd. Berl. Zool. Reihe 83: 3–42.

Conway, K and R. Moritz. 2006. Barboides britzi, a new species of miniature cyprinid from Bénin (Ostariophysi, Cyprinidae) with a neotype designation for B. gracilis. Ichthyol. Explor. Fresh waters 17: 73–84.

Diogo, R. 2004. Evolutionary convergences and parallelisms: their theoretical differences and the difficulty of discriminating them in a practical phylogenetic context. Biol. Phil. 20: 735–744.

Diogo, R. 2007. On the Origin and Evolution of Higher- Clades: Osteology, Myology, Phylogeny and Macroevolution of Bony Fishes and the Rise of Tetrapods. Science Publishers, Enfield.

Diogo, R., I. Doadrio and P. Vandewalle. 2008. Teleostean phylogeny based on osteological and myological characters. Int. J. Morphol. 26: 463–522.

Fink, S.V. and W. Fink. 1996. Interrelationships of ostariophysan fishes (Teleostei), pp. 209–249. In: M.L.J. Stiassny, L.R. Parenti and G.D. Johnson [eds.]. 1996. Interrelationship of Fishes. Academic Press, New York .

Gayet, M. 1993. Relations phylogénétiques des Gonorynchiformes (Ostariophysi). Belg. J. Zool. 123: 165–192.

Gayet, M. 1995. A propos des Ostariophysaires (Poissons, Téléostéens). Bull. Soc. Zool. Fr. 120: 347–360.

Grande, L. and W. Bemis. 1998 . A comprehensive phylogenetic study of amiid fishes (Amiidae) based on comparative skeletal anatomy. An empirical search for interconnected patterns of natural history. J. Vert. Paleontol. 18, suppl.: 1–690.

Grande, T. 1994. Phylogeny and paedomorphosis in an African family of freshwater fishes (Gonorynchiformes: Kneriidae). Fieldiana (Zool.) N. Ser. 78: 1–20.

Grande, T. and F. Poyato-Ariza. 1999. Phylogenetic relationships of fossil and recent gonorynchiform fishes (Teleostei: Ostariophysi). Zool. J. Linn. Soc. 125: 197–238.

Heim, W. 1935. Ueber die Rachensäcke der Characiniden und über verwandte Akzessorische Organe bei andern Teleosteern. Zool. Jahrb. Abt. Anat. Ontog. Tiere. 60: 61–106.

Howes, G.J. 1985. Cranial muscles of gonorynchiform fishes, with comments on generic relationships. Bull. Br. Mus. Nat. Hist. (Zool.) 49: 273–303.

Huet, M. and J.A. Timmermans. 1970. Traité de Pisciculture. Wyngaert, Brussels.

Hughes, G.M. 1960. A comparative study of gill ventilation in marine teleosts. J. Exp. Biol. 37: 28–45.

Hyrtl, C.J. 1863. Ueber besondere Eigenthumlichkeiten der Kiemen und der Skeletes, und über das epigonale Kiemenorgan von *Lutodeira chanos*. Denkschrift K. Ak. Wiss. Wien (Math. Naturw.) 21: 1–8.

Johnson, G. and C. Patterson . 1996. Relationships of lower euteleostean fishes, pp. 251–332. In: M. Stiassny, L. Parenti and G. Johnson [eds.]. 1996. Interrelationships of Fishes. Academic Press, San Diego.

Johnson, G. and C. Patterson. 1997. The gill-arches of gonorynchiform fishes. S. Afr. J. Sci. 93: 594–600.

Jollie, M. 1975. Development of the head skeleton and pectoral girdle in *Esox*. J. Morphol. 147: 61– 88.

Kapoor, B.G. 1954. The pharyngeal organ and its associated structures in the milk-fish *Chanos chanos* (Forskal). J. Zool. Soc. India 6: 51–58.

Lavoué, S., M. Miya, J.G. Inoue, K. Saitoh, N.B. Ishiguro and M. Nishida. 2005. Molecular systematics of the gonorynchiforms fishes (Teleostei) based on whole mitogenome sequences: implications for higher-level relationships within the Otocephala. Mol. Phylogenet. Evol. 37: 165–177.

Le Danois, Y. 1966. Remarques anatomiques sur la région cephalique de *Gonorhynchus gonorynchus* (Linné, 1766). Bull. Inst. Fond. Afr. Noire 28 ser. A 1: 283–342.

Lenglet, G. 1973. Contribution à l'étude de l'anatomie viscérale des Kneriidae. Ann. Soc. Roy. Zool. Belg. 103: 239–270.

Lenglet, G. 1974. Contribution à l'étude ostéologique des Kneriidae. Ann. Soc. Roy. Zool. Belg. 104: 51–103.

Liem, K. and L. Sanderson. 1986. The pharyngeal jaw apparatus of labrid fishes: a functional morphological perspective. J. Morphol. 187: 143–158.

Miller, R.V. 1964. The morphology and function of the pharyngeal organs in the clupeid *Dorosoma petenense* (Günther). Chesepeake Sci. 5: 194–199.

Monod, T. 1949. Sur l'appareil branchiospinal de quelques téléostéens tropicaux. Bull. Inst. Fr. Afr. Noire 11: 36–76.

Monod, T. 1963. Sur quelques points de l'anatomie de *Gonorhynchus gonorhynchus* (Linné 1766). Mém. Inst. Fr. Afr. Noire Dakar 68: 255–310.

Nelson, G.J. 1967. Epibranchial organs in lower teleostean fishes. J. Zool. Lond. 153: 71–89.

Osse, J.W. 1969. Functional morphology of the head of the perch (*Perca fluviatilis* L.): an electromyographic study. Neth. J. Zool. 19: 289–392.

Pasleau, F. 1974. Recherches sur la position phylétique des téléostéens Gonorynchiformes, basées sur l'étude de l'osteologie et de la myologie cephalique. Bachelor's thesis, University of Liège, Liège, Belgium.

Peng, Z., S. He, J. Wang, W. Wang and R. Diogo. 2006. Mitochondrial molecular clocks and the origin of the major Otocephalan clades (Pisces: Teleostei): A new insight. Gene 370: 113–124.

Ridewood, W.G. 1905. On the skull of *Gonorhynchus greyi*. Ann. Mag. Nat. Hist. 7: 367–424.

Sibbing, F.A. 1982. Pharyngeal mastication and food transport in the carp (*Cyprinus carpio*): a cineradiographic and electromyographic study. J. Morphol. 172: 223–268.

Springer, V. and G. Johnson . 2004. Study of the dorsal gill-arch musculature of teleostome fishes, with special reference to the Actinopterygii. Bull. Biol. Soc. Wash. 11: 1–260.

Stiassny, M.L.J., E.O. Wiley, G.D. Johnson and M.R. Carvalho. 2004. Gnathostome fishes, pp. 410–429. In: M.J. Donaghue and J. Cracraft [eds.]. 2004. Assembling the Three of Life. Oxford University Press, New York.

Swinnerton, H.H. 1901. The osteology of *Cromeria nilotica* and *Galaxias attenuatus*. Ann. Mag. Nat. Hist. 8: 444–446.

Takahasi, N. 1957. On the so-called accessory respiratory organ "gill-helix" found in some clupeiform fishes, with special reference to its function and genealogy. Japan. J. Ichthyol. 5: 71–77.

Taverne, L. 1974. L'ostéologie d'*Elops*, Linné, C., 1766 (Pisces Elopiformes) et son intérêt phylogénétique Acad. Roy. Belg., Mém. Cl. Sci. 41: 1–96.

Taverne, L. 1977. Ostéologie, phylogénèse et systématique des Téléostéens fossiles et actuels du super-ordre des Ostéoglossomorphes. Première partie. Ostéologie des genres *Hiodon, Eohiodon, Lycoptera, Osteoglossum, Scleropages* et *Arapaima*. Acad. Roy. Belg., Mém. Cl. Sci. 42: 1–235.

Taverne, L. 1981.Ostéologie et position systématique d'*Aethalionopsis robustus* (Pisces Teleostei) du Crétacé inférieur de Bernissart (Belgique) et considérations sur les affinités des Gonorhynchiformes. Bull Cl. Sci. Acad Roy. Belg. 5ᵉ sér. 47: 958–982.

Thys van den Audenaerde, D.F.E. 1959. Existence d'une vessie natatoire pulmonoïde chez *Phractolaemus ansorgei* Blgr. (Actinopterygii). Rev. Zool. Bot. Afr. 59: 364–366.

Thys van den Audenaerde, D.F.E. 1961. L'anatomie de *Phractolaemus ansorgei* Blgr. et la position systématique des Phractolaemidae. Ann. Mus. Roy. Afr. Centr. 103: 101–167.

Vandewalle. P., M. Havard, G. Claes and F. De Vree. 1992. Mouvements des mâchoires pharyngiennes pendant la prise de nourriture chez le *Serranus scriba* (Linné, 1758) (Pisces Serranidae). Can. J. Zool. 70: 145–160.

Vandewalle, P., P. Saintin and M. Chardon. 1995. Structure and movements of the buccal and pharyngeal jaws in relation to feeding in *Diplodus sargus*. J. Fish Biol. 46: 623–656.

Vandewalle, P., E. Parmentier and M. Chardon. 2000. The branchial basket in teleost feeding. Cybium 24: 319–342.

Vialli, M. 1926. L'organo epibranchiale dei clupeiddi. Monitore Zool. Ital. 37: 174–185.

Whitehead P. and G. Teugels. 1985. The West African pygmy herring *Sierrathrissa leonensis*: General features, visceral anatomy, and osteology. Am. Mus. Novitates 2835: 1–44.

List of Abbreviations

add4,5: adductores muscles of fourth and fifth arches

adhm: adductor hyomandibulae muscle

Arc br or **arc br**: branchial arch

artOP: condyle for opercular

BAH: basihyal.

BBR 1,2,3,4: basibranchials 1,2,3,4

BBR2 TPL: toothed plate on BBR

BC: buccal cavity

bcl3,4: basicleithrales muscles of third and fourth arches

bl: basal lamina

BR: branchial cavity

brf: branchial filaments

brsp: gill rakers

bsl5: fifth branchial slit

BSGR: branchiostegal rays

bval: buccal valve

CABR: cardiobranchial cartilage

conj: conjonctive tissue

cc: ciliated cells

CL: cleithrum

CBR 1,2,3,4,5: ceratobranchials 1,2,3,4,5

Cirm: circular muscles coating

CL: cleithrum

CO: copula

cs: semi-circular canal

dep4: depressor muscle of the fourth arch

dg: dorsal groove

dr4: rectus muscle of the fourth arch

e5: foramen for the fifth efferent branchial artery

EBO: epibranchial organ

ebom: muscle of the EBO

EBR 1,2,3,4,5: epibranchials 1,2,3,4,5

EBR4CAR: accessory cartilage on EBR4

EBR4SL: semilunar cartilage of EBR4

ECT: ectopterygoid

ENT: entopterygoid

ENTTPL: toothed plate on entopterygoids

epx: epaxial musles

FR: frontal

GR: gill rakers

GR4,5: gill rakers of fourth, fifth arches

HB: hyoid bar

HBR1,2,3: hypobranchials1,2,3

HM: hyomandibula

hyp: hypaxial muscles

IH: interhyal

levext 1,2,3,4: levatores externi muscles 1,2,3,4

levint 2,3,4: levatores interni muscles 2,3,4

lig to EBR3: ligament reaching EBR3

lm: longitudinal muscles coating

m: muscles

m?: enigmatic muscle

MBRSP: micdrobranchiospines

mc: mucous cells

MET: metapterygoid

MGR: microbranchiospines

n10: nervus vagus

obl3,4: obliqui dorsales muscles 3,4

oblv1,2,3: obliqui ventrales muscles 1,2,3

obltr: obliquus transversus muscle

oe: esophagus

OP: opercular cavity

opval: opercular valve

otbr: oticobranchialis muscle

PA: palatine
pad: papillate pad on fourth arch
PBR1,2,3,4: pharyngobranchials
    1,2,3,4
phclext: pharyngocleithralis
    externus muscle
phclint: pharyngocleithralis
    internus muscle
PHBR 1,2,3,4: pharyngobranchials
    1,2,3,4
phclint: pharyngocleithralis
    internus muscle
PrEBR4 or PE4: process of EBR4
PRO: prootic
PTO: pterotic
PS: parasphenoid
Q: quadrate
r: ring in the distal part of the EBO

sac: sacculus
SCL: supracleithrum
SPH: sphenotic
sphoe: sphincter oesophagi
sth: sternohyoideus muscle
SY: symplectic
ta, tc: tunicae of EBO
tb: taste bud
tr: trapezius muscle
trdo: transversus dorsalis muscle
trdor3,4: transversi dorsales 3,4
    muscle
trven: transversus ventralis muscle
tuph: tunica of pharynx
UH: urohyal
utr: utriculus
vel: velum on the fifth arch
v pulm: pulmonary vein

6

The Fossil Record of Gonorynchiformes

Emmanuel Fara[1], Mireille Gayet[2] and
Louis Taverne[3]

Abstract

The fossil record of gonorynchiform fishes provides key information on the
diversity, palaeobiogeography, and phylogeny of the group. The first mention
of fossil Gonorynchiformes dates back to Cuvier in the early 19th century,
and there is still a need for a critical review of the earliest descriptions and
of some key taxa today.

Fossil gonorynchiform fishes are known from the earliest Cretaceous
(Berriasian-Valanginian) to the earliest Miocene, and the clade has several
extant representatives. To date, the fossil record has yielded only about 18
genera and 35 species of Gonorynchiformes. With only 46 known localities,
their fossil record is relatively poor compared to that of other groups of
Ostariophysi. The distribution of these localities is heterogeneous in both
space and time.

Debates on the phylogenetic status of Gonorynchiformes have mainly
focused on the identity of the basal-most members of the clade and on its
sister group. Unfortunately, very few large-scale phylogenetic studies have
included the fossil representatives of the clade.

[1] Biogéosciences (CNRS/uB), Université de Bourgogne, 6 Boulevard Gabriel, 21000 Dijon,
France. e-mail: emmanuel.fara@u-bourgogne.fr
[2] 18 rue Vauban, 69006 Lyon, France. e-mail: gayet.mireille@free.fr
[3] Résidence "Les Platanes", Bd du Souverain, 142 (Boîte 8), B-1170 Bruxelles, Belgique.
e-mail: louis.taverne@gmail.com

Using known fossil occurrences and several phylogenetic proposals, we conducted an exploratory diversity analysis. A traditional taxic approach shows that gonorynchiform diversity rose steadily during the Early Cretaceous and reached a peak in the Aptian-Cenomanian interval. It then declined slightly towards the end of the Cretaceous and it decreased further at the dawn of the Cenozoic. This apparent low diversity level is only interrupted by relative diversity peaks in the first half of the Eocene and in the Oligocene. In the absence of fossils after the earliest Miocene, diversity estimates are conjectural for most of the Neogene.

We found a close similarity of the estimates obtained with alternative phylogenetic hypotheses, meaning that the differences among these phylogenies have virtually no impact on inferred diversity patterns. Our diversity analysis points to some major gaps in the known fossil record, and it calls for the integration of most (if not all) fossil taxa in phylogenetic analyses.

Introduction

Fossil Gonorynchiformes have been known for a long time. Cuvier (in Cuvier and Brongniard 1822) first recognized the similarity between some fossil bones from the Paris Basin (Tertiary of France) and the living *Gonorynchus*, described by Gronovius in 1763 (non-available name after Nelson 1994). After the recognition of Gonorynchiformes as a taxonomic entity (Gosline 1960, Greenwood *et al.* 1966), one of the most important published works was certainly that of Rosen and Greenwood (1970), who moved Gonorynchiformes from Clupeiformes to Otophysi.

The monophyly of the order Gonorynchiformes is now well established (Fink and Fink 1981, 1996, Blum 1991a, Poyato-Ariza 1996a, Grande and Poyato-Ariza 1999, Lavoué *et al.* 2005), but the phylogenetic position of many fossil forms remains unclear. This is because some recently described taxa have not been included in a cladistic analysis yet, and also because a critical review of the earliest descriptions is still awaited.

Here we review all taxa that have been assigned to Gonorynchiformes, and we discuss their occurrence and validity. Because there is no consensus about the systematic position of several gonorynchiform species, we have chosen to present the type genus first, and then other genera in alphabetical order. The same choice was made for the species within each genus. Recent species are just listed and briefly commented on. This review is not exhaustive and it does not include all published accounts on fossil Gonorynchiformes. Instead, we selected some major studies dealing with the anatomy and/or phylogeny of the group. In the same way, synonymy lists are limited to contrasting opinions.

In a second part, we present the spatio-temporal distribution of fossil gonorynchiforms, we review some major phylogenetic proposals, and we provide an exploratory diversity analysis.

Systematic Review

GONORYNCHUS Scopoli, 1777

type genus

There is no fossil record of this genus. According to Grande (1999b), five extant species of the marine genus *Gonorynchus* are valid. They are *Gonorynchus gonorynchus* (Linnaeus 1766), Indo-Pacific; *Gonorynchus forsteri* (Ogilby 1911), New Zealand; *Gonorynchus greyi* (Richardson 1845), Hawaii; *Gonorynchus abbreviatus* (Temminck and Schlegel 1846), Japan and Taiwan; and *Gonorynchus moseleyi* (Jordan and Snyder 1923), Hawaii.

†*AETHALIONOPSIS* Gaudant, 1966

†*Anaethalion robustus* Traquair, 1911

type and only species

1911   †*Anaethalion robustus*: R.H. Traquair, p. 50; figs. 19 and 20; table XI.

1966   †*Aethalionopsis robustus* (Traquair, 1911): J. Gaudant, p. 309, fig. 2.

A single species belongs to this genus, †*Aethalionopsis robustus* (Traquair 1911). It comes from the famous dinosaur-bearing locality of Bernissart (Belgium). The age of the fossils is middle Barremian to earliest Aptian, as estimated from angiosperm pollens (Yans *et al.* 2005, 2006).

Gaudant (1966) erected the genus †*Aethalionopsis*, which he considered close to †*Anaethalion* (within Anaethalionidae), the taxon to which the Bernissart specimens were first assigned. †*Aethalionopsis robustus* is known from numerous complete specimens, reaching 10 to 40 cm in standard length.

According to Gaudant (1968), this genus is represented by at least two other species. Indeed, Bassani and Erasmo (1912) and Erasmo (1915) assigned to †*Anaethalion robustus* several specimens from the Aptian-Albian (Early Cretaceous) of Castellamare and Pietraroia (Italy). However, these fossils differ from the type-species by the composition of their dorsal and anal fins, and they represent another, yet unnamed, species of †*Aethalionopsis* (Gaudant 1968). The third species, †*A. valdensis*, was described by Woodward (1907) from the Berriasian-Hauterivian of Sussex, England.

Taverne (1981) revised the anatomy of †*Aethalionopsis robustus* and he designated a lectotype. This author first emphasized its similarities with some Gonorynchiformes, such as †*Dastilbe*, †*Tharrhias* and †*Parachanos*. However, he did not mention the Italian and English occurrences. For Taverne (1981), †*Aethalionopsis* differs from all other gonorynchiforms by some plesiomorphic characters, such as the shortened mandible and the lack of fusion of the third and fourth infraorbitals.

The palaeoenvironment in Bernissart was most probably fresh water with episodic connections to the sea (Marliere and Robaszynski 1975, Martin and Bultynck 1990, Grande and Bemis 1998). Grande (1999a) suggested that †*Aethalionopsis* may have been an euryhaline taxon with a wide salinity tolerance. Because †*Aethalionopsis* was found within a mixed marine and freshwater fauna, isotopic analyses would be necessary to infer the habitat of this taxon.

†*APULICHTHYS* Taverne, 1997

†*Apulichthys gayeti* Taverne, 1997

type and only species

1980   Elopiforme: F. Medizza and L. Sorbini, p. 133, fig.
1997   †*Apulichthys gayeti*: L. Taverne, p. 403; figs. 1 to 13, table.

†*Apulichthys gayeti* is a small (less than 12 cm in total length) marine gonorynchiform from the late Campanian–early Maastrichtian of Porto Selvaggio, near Nardo, Apulia, southern Italy. The age of the fossil locality was determined by nannofossils, and the ichthyofauna suggests a shallow platform environment (Sorbini 1978, Medizza and Sorbini 1980). Only four more or less complete specimens are currently known. The family †Apulichthyidae was erected for this monospecific genus, which is regarded as the sister taxon of all other fossil and modern gonorynchoids (Taverne 1997).

†*CHANOIDES* Woodward, 1901

†*Chanoides* was named for †*Clupea macropoma* Agassiz, 1839/44, a fossil found in the famous Eocene site of Monte Bolca, Italy (Woodward 1901). Because of some anatomical similarities with the extant *Chanos*, Woodward (1901) placed both †*Chanoides* and *Chanos* in the family Albulidae. Patterson (1984a) showed that †*Chanoides macropoma* is neither an albulid nor a chanid, but an ostariophysan fish. A second species of †*Chanoides* has been described recently as †*Chanoides chardoni* (Taverne 2005). It comes from the late Campanian–early Maastrichtian of Nardo, Apulia, southern Italy. The species †*Chanoides striata*, originally described by Weiler (1920), was moved to †*Neohalecopsis* by Weiler (1928), whereas †*Chanoides leptostea* was transferred to †*Coelogaster* (see below). A yet older species of †*Chanoides* from the Santonian of southern Italy (Apricena) is currently under study by Taverne and De Cosmo.

†*CHANOPSIS* Casier, 1961

†*Chanopsis lombardi* Casier, 1961

type and only species

1961   †*Chanopsis lombardi*: E. Casier, p. 60; figs. 17 and 18; table 10, figs. 1 and 2; table 11, figs. 1 to 4; table 12, figs. 1 to 6.

Casier (1961) described †*Chanopsis lombardi* based on some isolated bones of a single individual that was about one metre long. The fossils were found in the Wealdian (perhaps Albian according to Taverne 1984) Loia strata, Democratic Republic of Congo. This species was tentatively assigned to the family Chanidae by Casier (1961), but Taverne (1984) showed that it belongs in Osteoglossidae, on the basis of the peculiar enlarged shape of the frontals, the large opercle, the small or absent subopercle, and the anterior position of the autosphenotic, among other characters. Poyato-Ariza (1996a) agreed with the conclusions of Taverne (1984).

CHANOS Lacepède, 1803

Chanos is the only Recent gonorynchiform genus known also from the fossil record. It lives in tropical and sub-tropical areas of the Indian and Pacific oceans, mainly in coastal and brackish waters that enter estuaries and rivers (Riede 2004). Adults spawn in fully saline water, whereas larvae occasionally enter freshwater lakes (Bagarinao 1994). A single extant species is known, and it is the type species: *Chanos chanos* Forsskål, 1775.

†*Chanos brevis* (Heckel, 1854)

1854   †*Albula brevis*: J.J. Heckel, p. 132.
1863   †*Chanos brevis* (Heckel): R. Kner and F. Steindachner, p. 19; table I.

Several articulated specimens of †*Chanos brevis* are known from the Oligocene beds of Chiavon (Vicentin, Italy). With its 40 vertebrae, †*C. brevis* has an intermediate number of vertebrae compared to †*C. zignoi* (35 vertebrae) and †*C. forcipatus* (45 vertebrae). According to Arambourg and Schneegans (1935a), †*C. brevis*, †*C. zignoi* and †*C. forcipatus* are so close to each other that they may simply represent the geographical variation of a single species. These three species, for which the latest description dates back to the 19th century, clearly need a systematic revision. The palaeoenvironment of †*Chanos brevis* was probably a lagoon with more or less brackish waters (Sorbini 1980).

†*Chanos compressus* Stinton, 1977

1977   †*Chanos compressus*: F.C. Stinton, p. 69; table 5, figs. 5, 6.

Gonorynchiform otoliths are rare. Five specimens from the Wittering Formation (late Ypresian, Early Eocene) in Hampshire, England, have been assigned to *Chanos* by Stinton (1977).

†*Chanos forcipatus* Kner and Steindachner, 1863

1854   †*Megalops forcipatus*: J.J. Heckel, p. 132. listed.
1863   †*Chanos forcipatus*: R. Kner and F. Steindachner, p. 21; table 3.

This species comes from the marine beds of the Monte Postale, Bolca, in Italy. The age is late Lutetian-early Bartonian (Middle Eocene). Kner and Steindachner (1863) moved this species to the genus *Chanos*. According to Blot (1980), another new, but unnamed, species is also present in the same locality. Unfortunately, there is no recent study on this material.

†*Caesus leopoldi* (Costa, 1860)

1857   †*Caesus* sp.: O.G. Costa, p. 235; table.
1860   †*Caesus Leopoldi*: O.G. Costa, p. 45; table 4.
1879   †*Prochanos rectifrons*: F. Bassani, p. 163, listed.
1882   †*Prochanos rectifrons* Bassani: F. Bassani, p. 218; table 13; table 14, fig. 1; table 15.
1915   †*Chanos Leopoldi* (Costa): G. d'Erasmo, p. 35; figs. 32 and 33; table 6, fig. 1.

†*Chanos leopoldi* is known from a single specimen found in Pietraroia, Benevento, Italy. Catenacci and Manfredini (1963) gave a Barremian-Albian age for the "calcari a ittioliti" of Pietraroia, which could be Aptian (Argenio 1963, Scorziello 1980). Erasmo (1915) regarded †*Prochanos rectifrons* (Bassani 1882) as a possible synonym of †*Chanos leopoldi*.

†*Chanos torosus* Danil'chenko, 1968

1968   †*Chanos torosus*: P.G. Danil'chenko, p. 119, fig. 4, table 25, figs. 1–4.

This species was attributed to the genus *Chanos* by Danil'chenko (1968) after comparison with other fossil and extant species of *Chanos*. The fossils come from the Danatinsk Formation, Kopetdag, Turkmenia. Danil'chenko (1968) noticed that this Thanetian (Late Paleocene) fish fauna is similar to several Eocene marine faunas, especially that of Monte Bolca. The palaeoenvironment probably corresponds to an open marine basin with a normal salinity under a warm climate (Danil'chenko 1980, A.F. Bannikov, personal communication to M. Gayet, 2005).

†*Chanos zignoi* Kner and F. Steindachner, 1863

1854   †*Albula de zignii*: J.J. Heckel, p. 129, listed
1854   †*Albula lata*: J.J. Heckel, p. 131.
1863   †*Chanos Zignii* (Heckel): R. Kner and F. Steindachner, p. 20; table 2.

This species is from the Oligocene of Chiavon, Vicenza, Italy. It has not been revised since the works by Kner and Steindachner (1863) and Bassani (1889a, b). According to Bassani (1889a), †*C. zignoi* differs from †*C. forcipatus* mostly by its lower number of vertebrae (only 35 vertebrae *versus* 45 or 46 for †*C. forcipatus*).

†*CHARITOPSIS* Gayet, 1993a

†*Charitopsis spinosus* Gayet, 1993a

type and only species

1993a   †*Charitopsis spinosus*: M. Gayet, p. 259; figs. 1 to 4.

†*Charitopsis* was erected on the basis of several fossils found in early Cenomanian marine deposits at Haqil, Lebanon (Gayet 1993a, b). It is close to †*Charitosomus*, another gonorynchiform fish from the same locality (Gayet 1993a), and Grande and Grande (1999) suggested that †*Charitopsis spinosus* may belong to this later genus. However, †*Charitopsis* shares with †*Notogoneus* and *Gonorynchus* the lengthening of the preopercular branch (and/or the shortening of the upper branch), as well as the separation between the dentary and the angular, two characters absent in †*Charitosomus*. In addition, it is the only gonorynchiform with a flat premaxilla and an opercle whose posterior border is strongly spinous.

In the Late Cretaceous, the Haqil site was probably located near the shore, as attested by the associated macro- and microfauna. Marine algae (Basson and Edgell 1971, Basson 1981), varanoid dolichosaurs and ophiomorph reptiles (Dal Sasso and Renesto 1999) are found, together with terrestrial plant remains (Hückel 1970).

†*CHARITOSOMUS* von der Marck, 1885

†*Charitosomus formosus* von der Marck, 1885

type species

1885   †*Charitosomus formosus*: W. von der Marck, p. 257; table 24, fig. 1.
1954   †*Charitosomus formosus* Marck: P. Siegfried, p. 14; table 5, fig. 3.
1954   †*Charitosomus* (?) cf. *formosus*: P. Siegfried, p. 14; table 5, figs. 4 and 5.

†*Charitosomus formosus*, the type species of the genus, was originally created by von der Mark (1885) for German fossils from the Late Senonian (late Campanian–Maastrichtian) of Baumberg, Westphalia. Only two large specimens are known, and the total length of each is about 26 cm. The anatomy was described by von der Marck (1885) and Woodward (1901). Gayet (1993b) noticed that the subopercle of †*C. formosus* presents a large dorsal spine (as all other †*Charitosomus* species), a character that is absent in †*Hakeliosomus*. Gayet (1993b) proposed the new family †Charitosomidae to accommodate †*Charitosomus*, †*Hakeliosomus* and †*Charitopsis*.

As for all other Lebanese fossil species, the palaeoenvironment was marine (Hückel 1970, Cappetta 1980).

†*Charitosomus* aff. †*C. formosus*

1885   †*Mesogaster cretaceus*: W. von der Marck, p. 247; table 22, fig. 2.
1894   †*Spaniodon lepturus*: W. von der Marck, p. 46; table 5, fig. 5.
1954   †*Charitosomus* (?) cf. *formosus*: P. Siegfried, p. 14, pl. 5, figs. 4 and 5.

Siegfried (1954) tentatively assigned to †*Charitosomus* aff. †*C. formosus* two specimens from the Late Senonian (late Campanian–Maastrichtian) of Baumberg, Westphalia, which were originally described as †*Mesogaster cretaceus* (von der Marck 1885) and †*Spaniodon lepturus* (von der Marck 1894).

†*Charitosomus hermani* Taverne, 1976a

This species was created on the basis of late Albian to middle Cenomanian fossils from Kipala, Kwango, Democratic Republic of Congo. The material is now identified as belonging to an undetermined species of Gonorynchidae.

†*Solenognathus lineolatus* Pictet and Humbert, 1866

1866   †*Solenognathus lineolatus*: F.J. Pictet and A. Humbert, p. 56; table 4, figs. 4 to 7.
1901   †*Charitosomus lineolatus* (Pictet and Humbert): A.S. Woodward, p. 274; table 15, fig. 4.

This eel-shaped species has been reported from Santonian marine strata at Sahel Alma, Lebanon. It is the smallest known †*Charitosomus*, with a maximal total length of about 8 cm. As for all fish from Sahel Alma, anatomical study is rather difficult due to their state of preservation (Pictet and Humbert 1866, Woodward 1901), but this species was redescribed recently (Gayet 1993b) on the basis of numerous specimens and acid preparation. This author noticed the possible presence of a supraneural 1 in †*Charitosomus lineolatus*, as observed in †*Hakeliosomus* and †*Chanoides chardoni* (Gayet 1993b, Taverne 2005). This could be interesting to investigate because the absence of supraneural 1 is the rule among all other Ostariophysi (Fink and Fink 1981).

†*Charitosomus major* Woodward, 1901

1901   †*Charitosomus major*: A.S. Woodward, p. 272; table 15, fig. 3.

This large †*Charitosomus* species, also from the Santonian Sahel Alma locality in Lebanon, can reach a total length of 20 cm. Only five specimens are presently known and described (Gayet 1993b). Like †*Charitosomus lineolatus*, two body shapes are present and they could represent a case of sexual dimorphism.

Most Middle East †*Charitosomus* species (including †*Hakeliosomus hakelensis*) are known from Lebanon only, although an occurrence from Syria was mentioned by Grande and Grande (1999).

†*COELOGASTER* Eastman, 1905

†*Chanoides leptostea* Eastman, 1905

type species

1835   †*Clupea leptostea*: L. Agassiz, p. 306, listed.
1835   †*Coelogaster analis*: L. Agassiz, p. 304, listed.
1844   †*Clupea leptostea*: L. Agassiz, p. 170, listed.
1844   †*Coelogaster analis*: L. Agassiz, t. V, pt II, p. 106, listed.
1905   †*Chanoides leptostea*: C.R. Eastman, p. 11; table 1, fig. 1.
1905   †*Coelogaster analis*: C.R. Eastman, p. 12; table 1, fig. 2.
1984b   ? †*Coelogaster leptostea* (Eastman): C. Patterson, p. 448.

This species from Monte Bolca (late Ypresian–early Lutetian marine deposit, Italy) was based on a rather poorly preserved holotype, the only known specimen. According to Patterson (1984b), †*Coelogaster leptostea* may be a chanid. He discussed the taxonomic status of †*Coelogaster leptostea* and rejected the possibility that this species could belong to †*Chanoides*. Patterson (1984b) also suggested the possible conspecificity with †*Coelogaster analis* Eastman, 1905, from the same locality. We agree with this interpretation and the chanoid affinities based on the similarity of the caudal skeleton with that of †*Chanos*. Consequently, we regard †*Coelogaster leptostea* (Eastman, 1905) as the valid name for the species.

CROMERIA Boulenger, 1901

No fossil has been assigned to this genus. The two Recent subspecies, *Cromeria nilotica nilotica* Boulenger, 1901, and *Cromeria nilotica occidentalis* Daget, 1954, occur in the Nile, Niger, Volta and Chad basins.

†*DASTILBE* Jordan, 1910

Several authors have studied †*Dastilbe* and its various species, but opinions differ about possible synonymies. Four species have been described so far.

†*Dastilbe crandalli* Jordan, 1910

type species

1910   †*Dastilbe crandalli*: D.S. Jordan, p. 30; table 9, figs. 9 to 13.

Type species for the genus and the only valid species of †*Dastilbe* according to Davis and Martill (1999) and Brito and Amaral (2008). It has been reported from black shales in the Aptian Muribeca Formation of Riacho Doce, Sergipe-Alagoas Basin, northeastern Brazil. Santos (1990) cited †*Dastilbe crandalli* in the Aptian Cabo Formation (Brazil). Because he differentiated †*D. crandalli* from †*D. elongatus* by the number of pectoral fin rays, Blum (1991a) reported the former species from the Crato Formation, Araripe Basin, Brazil. Davis and Martill (1999) analysed the very abundant specimens of †*Dastilbe* from the Nova Olinda Member of this formation. Berthou (1990) suggested a latest Aptian to early Albian age for the sequence containing the Nova Olinda Member.

Numerous specimens are known, and they range up to 7 cm in total length. Otoliths rarely occur *in situ* but can be abundant as isolated microfossils (Davis and Martill 1999). In contrast with Blum (1991a) and Martill (1993), who suggested that †*Dastilbe* was a non-marine fish, Davis and Martill (1999) speculated that †*D. crandalli* was an anadromous fish tolerant to hypersalinity.

†*Dastilbe batai* Gayet, 1989

1989   *Dastilbe batai*: M. Gayet, p. 22; table 1, fig. 1.

This species was erected by Gayet (1989) on the basis of a single, complete specimen from the Aptian-Albian beds of Río Benito, south of Bata, Equatorial Guinea. It was distinguished from †*D. crandalli* by the anterior position of the pelvics, with only seven lepidotriches, and by anal fins with eight lepidotriches. Poyato-Ariza (1996a), followed by Davis and Martill (1999), considered the diagnostic characters of this species not convincing, and the later authors considered it synonymous with †*D. crandalli*. The position of the pelvic fins relative to the dorsal fin is a variable feature of little diagnostic value, and the fin ray counts overlap with the variation observed among the Brazilian specimens of †*D. crandalli*. Some of the numerous isolated bones surrounding the holotype may belong to this species, but the poor state of preservation may hamper any robust taxonomic decisions (Brito and Amaral 2008). In any case, the specimen from Bata currently represents the only occurrence of †*Dastilbe* in Africa (but see Dietze 2007).

†*Dastilbe batai* was found in black shales. The same level yielded two clupeomorphs, †*Ellimma goodi* (referred to the marine †*Ellimmichthys* by Chang and Grande 1997) and †*Clupavichthys dufouri* (Gayet 1989). Clupavidae seem to be marine [for example, *Clupavus* is known in marine deposits in Portugal and Morocco (Gayet 1981, Taverne 1977)]. Other fossil fish, including †*Parachanos*, also occur in bituminous shales in several fossil localities close to Rio Benito (Casier and Taverne 1971). They are mostly marine, and the palaeoenvironment corresponds to a calm bay temporarily lagoonal (Weiler 1922).

†*Dastilbe elongatus* Santos, 1947

1947   †*Dastilbe elongatus*: R. da Silva Santos, p. 2; table I, figs. 1 and 2; table 2, fig. 1.
1968   †*Dastilbe elongatus* Santos: R. da Silva Santos and L. G. Valença, p. 349, fig. 6.

†*Dastilbe elongatus* was first reported in an oral communication about some specimens from the Codó Formation of the Parnaíba Basin, Maranhão State, northeastern Brazil (Santos 1947, Duarte and Santos 1993), and in a manuscript that remained unpublished. Santos (1947) described and figured this species on the basis of five specimens from the laminated limestone of the Aptian Crato Member, Santana Formation, Araripe Basin, northeastern Brazil. However, he did not provide any diagnostic character. These specimens seem to be lost (Davis and Martill 1999). †*Dastilbe elongatus* was also listed in a review of the Araripe fish fauna (Santos and Valença 1968).

Blum (1991a) distinguished †*D. elongatus* and †*D. crandalli* on the basis of the count of pectoral fin rays and of total length (respectively 13 rays and TL up to 20 cm for †*D. elongatus*, and 10 rays and TL up to 6 cm for †*D. crandalli*). These characters were criticized by Poyato-Ariza (1996a) and Davis and Martill (1999). Taverne (1981) hypothesized that †*Dastilbe crandalli* and †*D. elongatus* may represent distinct populations of a single species. Davis and Martill (1999), Dietze (2007) and Brito and Amaral (2008) found no skeletal criterion for distinguishing these two species and they regard †*D. elongatus* as a junior synonym of †*D. crandalli*, an opinion that we follow here. Poyato-Ariza (1996a) considered the specimens of †*Dastilbe crandalli* from Riacho Doce and specimens from the Crato member as a distinct species, based on their absence of distema in the caudal skeleton.

†"*Dastilbe minor*" Santos, 1975

1975   †*Dastilbe minor*: R. da Silva Santos, unpublished doctoral thesis.
1990   †*Dastilbe minor* Santos, 1975: R. da Silva Santos, p. 267, listed
1996a   †*Dastilbe minor*?: F. J. Poyato-Ariza, p. 43, listed.

In his unpublished doctoral thesis, Santos (1975) proposed under the name †*Dastilbe minor* a new species from the Marizal Formation (Lower Cretaceous), Bahia State, Brazil. He quoted this name again many years later in another paper (Santos 1990). Unfortunately, that species has never been diagnosed or described, only listed, so that it can be regarded as *nomen nudum* (Brito and Amaral 2008).

†*Dastilbe moraesi* Santos, 1955

1955   †*Dastilbe moraesi*: R. da Silva Santos, p. 19; tables 1 and 2.

Santos (1955) reported this species from the Aptian (or late Barremian–early Aptian, see Brito and Amaral, 2008) of the Areado Formation, Presidente Olegario, State of Minas Gerais, Brazil. Some cranial characters seem to relate this species more closely to †*Tharrhias* than to †*Dastilbe* (Poyato-Ariza 1996a), but the count of vertebrae differs between these two genera (respectively 50 and 36 for the single specimen examined by this author). Davis and Martill (1999) agreed with Poyato-Ariza (1996a) by considering this taxon valid, awaiting a more complete study, whereas Brito and Amaral (2008) consider it as a junior synonym of †*D. crandalli*.

†*Dastilbe* sp.

Malabarba *et al.* (2002) reported †*Dastilbe* from the Aptian/Albian outcrops of the Maceió Formation, Alagoas Basin, northeastern Brazil, and Soares and Calheiros (1991) reported †*Dastilbe* sp. from Aptian-Albian bituminous shales at Rio Largo, Muribeca Formation, in levels younger than those in which †*D. crandalli* was found. According to Soares and Calheiros (1991), these bituminous shales were formed in a paralic flysch depositional environment.

†*ECTASIS* Jordan and Gilbert, 1919

†*Ectasis proriger* Jordan and Gilbert, 1919

type and only species

1919   †*Ectasis proriger*: Jordan and Gilbert (*in* Jordan), p. 62; pl. 22.

A single specimen from the Pliocene of the Los Angeles clay shale, at the Third Street Tunnel, was first assigned with doubts to the family

Gonorynchidae by Jordan and Gilbert (*in* Jordan 1919). Later on, Jordan (1921a) placed †*Ectasis* in Elopidae, close to *Elops*.

Gonorynchiformes? incertae sedis

A single articulated skeleton from the Early Cretaceous freshwater deposits of Kyushu, Japan, was reported and figured by Yabumoto (1994) as "Gonorynchiformes? *incertae sedis*" (probably meaning "Gonorynchiformes? indet."). According to Grande and Grande (1999), based on the specimen figured by Yabumoto, the form of the body is gonorynchid-like.

Gonorynchidae indet.

1965   *Incertae sedis*: E. Casier, p. 47; table 15, fig. 7.
1976a   †*Charitosomus hermani*: L. Taverne, p. 36; fig. 20.
1976b   †*Charitosomus hermani*: L. Taverne, p. 762, listed.

The posterior part of an individual with 23 vertebrae and the caudal skeleton, from the late Albian to middle Cenomanian of Kipala, Kwango, Democratic Republic of Congo, was assigned to a new species of †*Charitosomus*, †*C. hermani* by Taverne (1976a). However, it seems more appropriate to attribute this caudal skeleton to an indeterminate gonorynchid fish. Even if close to fossil genera from the Middle East, it differs by the fusion of the hypural 1 and 2, both attached to the terminal centrum. In all studied specimens of †*Hakeliosomus*, †*Charitosomus* and †*Charitopsis*, the first hypural is free from the second, and it articulates with the terminal centrum. The whole fauna from Kipala suggests a lagoonal palaeoenvironment (Taverne 1976a).

†*GORDICHTHYS* Poyato-Ariza, 1994

†*Gordichthys conquensis* Poyato-Ariza, 1994

type and only species

1988   Undetermined teleost: J.L. Sanz *et al.*, p. 622.
1990   Teleost *incertae sedis* "Tipo B": F.J. Poyato-Ariza and S. Wenz, p. 307, fig. 4C.
1994   †*Gordichthys conquensis*: F.J. Poyato-Ariza, p. 5; figs. 1 and 3; table 1.

First mentioned as an undetermined teleost close to †*Ascalabos* (*Leptolepis*) *voithi* (Sanz *et al.* 1988), and as a teleost *incertae sedis* (Poyato-Ariza and Wenz 1990), this small fish was recognized as a chanid and described in detail by Poyato-Ariza (1994). It comes from the freshwater late

Hauterivian–early Barremian beds of Las Hoyas, Cuenca Province, Spain (Poyato-Ariza 1994, Poyato-Ariza *et al.* 1998). After his cladistic analysis of chanids, Poyato-Ariza (1996a, b) arranged †*Gordichthys* and †*Rubiesichthys* in the new subfamily †Rubiesichthyinae. This result is based on five synapomorphies, including two unique derived characters (acute angle between the preopercular limbs in adults and presence of the posterior process of the first supraneural). The numerous specimens currently known are all less than 4 cm in standard length.

GRASSEICHTHYS Géry, 1964

Only one Recent species, *Grasseichthys gabonensis* Géry, 1964, is reported from the Cuvette Centrale, Democratic Republic of Congo and from the Ivindo Basin, Gabon (Roberts 1972). No fossil species is known for this genus.

†*HAKELIOSOMUS* Gayet, 1993b

†*Spaniodon hakelensis* Davis, 1887

type and only species

1887   †*Spaniodon hakelensis*: J. W. Davis, p. 591; table 34, fig. 4.
1898   †*Charitosomus hakelensis* (Davis): A. S. Woodward, p. 412.
1993b   †*Hakeliosomus hakelensis* (Davis): M. Gayet, p. 19; figs. 1 to18; tables 1 to 3.

†*Hakeliosomus hakelensis* has been reported from the marine early Cenomanian strata of Haqil, Lebanon. The species has long been considered as belonging to †*Charitosomus* (Davis 1887, Woodward 1898, 1901, Patterson 1970). However, acid preparation of numerous specimens allowed a re-description of the species and its transfer into the monospecific genus †*Hakeliosomus* (Gayet 1993b). Grande and Poyato-Ariza (1999) found no justification for this taxonomic choice. Phylogenetically, †*Hakeliosomus* is close to †*Charitosomus* and †*Charitopsis*. However, it differs from all species of †*Charitosomus* and from †*Charitopsis* by the smooth border of its subopercle (that is spinous in these two genera). Also, a supraneural 1 seems to be present, as in some species of †*Charitosomus* (e.g., †*C. lineolatus*). For these reasons, we believe that it represents a valid genus, although Grande and Grande (2008) argued that it is a synonym of †*Ramallichthys*.

†*HALECOPSIS* Woodward, 1901

†*Osmeroides insignis* Delvaux and Ortlieb, 1888

type and only species

1844   †*Halecopsis laevis*: L. Agassiz, p. 139 (*nomen nudum*).
1887   †*Osmeroides insignis*: E. Delvaux, p. 74; table 3, figs. 2 to 7, 10 and 11 (not described).
1888   †*Osmeroides insignis*: E. Delvaux and J. Ortlieb, p. 60; tables 1 and 2.
1901   †*Halecopsis insignis* (Delvaux and Ortlieb): A.S. Woodward, p. 134.

This species was reported from the marine Ypresian London Clay (southeast England), from the "Argile des Flandres" (Belgium), from the Clay of Hemmoor (northwest Germany), and from northern France. Its osteology was studied by Delvaux and Ortlieb (1888), Woodward (1901), and especially Casier (1946, 1966). Jordan (1910) suspected some relationship with the chanid †*Dastilbe*, but Schaeffer (1947) questioned this affinity. Casier (1966) was the first to suggest gonorynchid affinities. For Patterson (1984b), †*Halecopsis* and †*Neohalecopsis* (see below) are chanids rather than members of the family †Halecopsidae, which is not well defined. Because the specimens of these two genera are poorly preserved, they were not included into the phylogenetic analysis by Poyato-Ariza (1996a). However, because they lack two chanid synapomorphies (wide frontals and expanded operculum), that author suspected that they may be closer to *Gonorynchus* than to *Chanos*. Neither Grande (1999a) nor Grande and Grande (1999) mentioned this teleost. Taverne and Gayet (2006) confirmed the gonorynchoid status of †*Halecopsis insignis*, and they placed it between †*Apulichthys* and the other families of the suborder.

†*JUDEICHTHYS* Gayet, 1985

†*Judeichthys haasi* Gayet, 1985

type and only species

1985   †*Judeichthys haasi*: M. Gayet, p. 67; figs. 1 to 8; table 1, figs. 1 to 3; table 2, fig. 1.

Only one specimen from the marine early Cenomanian of Ramallah, Judea Mounts, Israel, is presently known. A new family, †Judeichthidae, was erected for this taxon (Gayet 1985). Grande (1996) rejected the validity of both this family and this genus. This author included both †*Judeichthys* and †*Ramallichthys* into †*Charitosomus*. More recently, Grande and Grande (2008) argued that †*Judeichthys* should be synonymized with †*Ramallichthys*.

There is a controversy for the identity of the two posterior pharyngeal patches of conical teeth. For Grande (1996), these teeth belong to the entopterygoid and to the basibranchials, as in all gonorynchiforms. However, the specimen of †*Judeichthys* was apparently not subject to post-mortem displacement because the two mandibles are nearly superposed. It is therefore difficult to imagine how the right endopterygoid only could have been turned out in a movement placing the patch of teeth at a level as low as the second basibranchial. In addition, two teeth belonging to the upper patch are placed outside the entopterygoid (Gayet 1985).

Gonorynchus and †*Notogoneus* have elongated skull, and Grande (1996) argued that the shortness of the anterior part of the skull of †*Judeichthys* is similar to the 'derived' *Phractolaemus*. However, there was no direct evidence of a relationship between these two taxa.

Although it would be out of place to discuss all the characters of †*Judeichthys* here, we propose not to maintain the family †Judeichthyidae, but we do maintain the generic status of †*Judeichthys*.

KNERIA Steindachner, 1866

This freshwater taxon comprises 13 extant species distributed in East, West and Central Africa. These are: *K. angolensis* (Steindachner 1866), *K. ansorgii* (Boulenger 1910), *K. auriculata* (Pellegrin 1905), *K. katangae* (Poll 1976), *K. maydelli* (Ladiges and Voelker 1961), *K. paucisquamata* (Poll and Stewart 1975), *K. polli* (Trewavas 1936), *K. ruaha* (Seegers 1995), *K. rukwaensis* (Seegers 1995), *K. sjolandersi* (Poll 1967), *K. stappersii* (Boulenger 1915), *K. uluguru* (Seegers 1995), and *K. wittei* (Poll 1944). See Poll (1965) and Roberts (1975) for a detailed account on the distribution of these species.

Kneria mashkovae (Karatajute-Talimaa 1997) from Spain is a chondrichthyan and it was renamed †*Knerialepis mashkovae* (Hanke and Karatajute-Talimaa 2002).

†*LECCEICHTHYS* Taverne, 1998

†*Lecceichthys wautyi* Taverne, 1998

type and only species

1998   †*Lecceichthys wautyi*: L. Taverne, p. 292; figs. 1 to 6; table.

One near complete specimen from late Campanian–early Maastrichtian marine deposits in Nardo, Apulia, Southern Italy, was named †*Lecceichthys wautyi* by Taverne (1998). Close to †*Notogoneus* and *Gonorynchus*, it was placed in the Gonorynchidae, as the basal sister group of these two genera.

†NEOHALECOPSIS Weiler, 1928.

†Neohalecopsis striatus (Weiler, 1920)

type and only species

1920 †Chanoides striata: W. Weiler, p. 4.
1928 †Neohalecopsis striatus (Weiler): W. Weiler, p. 14; table 4, figs. 3 to 5; table 6, fig. 2.

This taxon is based on a single specimen from the Oligocene Septarientones, Flörsheim, Rheinhessen (Germany). The fossil was first described as †Chanoides striata by Weiler (1920), and it was later assigned to a new monospecific genus, †Neohalecopsis, supposed close to †Halecopsis. Casier (1946) erected the family †Halecopsidae for these two genera (see †Halecopsis). However, the skull and the general morphology of †Neohalecopsis greatly differ from †Halecopsis, and they seem to indicate chanid affinities instead (Taverne and Gayet 2006).

†NOTOGONEUS Cope, 1885

†Notogoneus osculus Cope, 1886

type species

1885 †Notogoneus osculus: E.D. Cope, p. 1091, listed.
1886 †Notogoneus osculus Cope: E.D. Cope, p. 163; figs. 4 and 5.
1890 †Protocatostomus constablei: R.P. Whitfield, p. 117; pl. 4.
1984 †Notogoneus osculus Cope 1885: L. Grande, p. 104.

Type species of the genus, †Notogoneus osculus was first reported from the early Eocene of Wyoming (Green River Formation and Bridger Formation). †Notogoneus osculus was apparently not common in Eocene lake faunas, and it was not a near-shore species (Grande 1999a). Hundreds of nearly complete specimens are known. This species has been described by several authors, including Cope (1885), Whitfield (1890), Perkins (1970) and Grande (1984). The known representatives of †Notogoneus osculus are between 20 and 80 cm long.

Isolated scale fragments from the early Eocene freshwater deposits of the Coalmon Formation, northern Colorado, have been assigned to †Notogoneus sp. cf. N. osculus (Wilson 1981).

†*Notogoneus alsheimensis* (Weiler, 1942)

1942   † Otolithus (? *Coregonidarum*) *alsheimensis:* W. Weiler, p. 18, pl. 1, fig. 14–19.
1963   † *Notogoneus alsheimensis* Weiler: W. Weiler, p. 19.

This species was found in the Hydrobia marl (early Miocene, Aquitanian) of Alsheim, middle Rhein Valley, Germany. According to W. Schwarzhans (personal communication to M. Gayet, 2005), it lived in a brackish to freshwater environment.

†*Notogoneus brevirostris* Schwarzhans, 1974

1974   †*Notogoneus brevirostris:* W. Schwarzhans, p. 95; figs. 77 and 78; pl. 3, fig. 20.

Otoliths from the late Oligocene of Niederrhein, Germany, have been reported as a species of †*Notogoneus* by Schwarzhans (1974). Nolf (1985) first agreed with this determination, but further research indicates that these otoliths now appear to belong to a perciform fish from the family Acropomatidae (D. Nolf, personal communication to L. Taverne, 2004).

†*Sphenolepis cuvieri* Agassiz, 1844

1818   †*Anormurus macrolepidotus:* H.D. de Blainville, p. 374.
1822   †"Gonorynque": G. Cuvier, p. 346; table 77, figs. 9, 11, 12, 13 and 15.
1844   †*Sphenolepis cuvieri:* L. Agassiz, part 1, p. 13; part 2, p. 89; table 44, figs. 1, 2, 4 to 9 and 11.
1895   †*Notogoneus cuvieri* (Agassiz): A.S. Woodward, p. 503.
1900   †*Notogoneus* sp.: F Priem, p. 849; table 15, figs. 2 to 5.
1908   †*Notogoneus janeti:* F. Priem, p. 133; pl. 3, figs. 2 and 3.
1911   †*Notogoneus* aff. *squammosseus* (Blainville): F. Priem, p. 34.
1931   †*Phalacropholis centumnucesianus* (Chabanaud): P. Chabanaud, p. 502; table 22, fig. 2.
1934   †*Colcopholis centumnucesianus* (Chabanaud): P. Chabanaud, p. 9.
1981a   †*Notogoneus cuvieri* (Agassiz): J. Gaudant, p. 63; pl. 3.

†*Notogoneus cuvieri* was reported from the late Oligocene gypsum deposits of Montmartre, Paris, France. According to Gaudant (1981a), these layers were deposited in the slightly brackish waters of a lagoon that had some connection to the sea. Only two imperfect articulated skeletons and some isolated bones are known.

Gaudant and Burkhardt (1984) also reported †*Notogoneus* cf. *cuvieri* from the "marnes grises" (grey marls) at Altkirch (Haut-Rhin, France), a site dated as early Oligocene (Sittler 1965, 1972). These †*Notogoneus* specimens represent about two thirds of the ichthyofauna at this locality. The skeletons are more or less fragmentary, but the bones are usually found in connection. The fish fauna, the invertebrates, the plant remains, as well as some isotopic analyses suggest a low salinity for the palaeowater at Altkirch (Gaudant and Burkhardt 1984).

†*Notogoneus fusiformis* Schwarzhans, 1994

1994   †*Notogoneus fusiformis*: W. Schwarzhans, p. 57; fig. 40.

Schwarzhans (1994) described this species from the late Oligocene (Chattian) of Niederrhein (Germany), based on isolated otoliths. The generic determination seems correct (D. Nolf, personal communication to L. Taverne, 2004). Interestingly, Schwarzhans (1994) described the palaeoenvironment as "a shallow marine setting".

†*Notogoneus gracilis*, Sytchevskaya, 1986

1986   †*Notogoneus gracilis*: E.K. Sytchevskaya, p. 51; figs. 15 and 16; tables 8 and 9.

Sytchevskaya (1986) described a new species of a small †*Notogoneus* from the late Paleocene or early Eocene Boltyshka Basin, Ukraine. This taxon is represented by numerous partial or complete specimens of small size, between 5 and 9 cm (Grande and Grande 1999). †*Notogoneus gracilis* is the oldest record of the genus in Europe.

†*Notogoneus janeti* Priem, 1908

1908   †*Notogoneus janeti*: F. Priem, p. 133; pl. 3, figs. 2 and 3.

This species was erected for a single specimen from the late Eocene or early Oligocene "Marnes bleues supragypseuses" of the Paris Basin, France. It was described and figured by Signeux (1961). †*Notogoneus janeti* was regarded as a junior synonym of †*Notogoneus cuvieri* by Gaudant (1981a).

†*Cobitis longiceps* (von Meyer, 1848)

1844   †*Cobitis longiceps*: L. Agassiz, part I, p. 10 (listed)
1848   †*Cobitis longiceps*: H. von Meyer, p. 151; table 20, fig. 2.
1901   †*Notogoneus longiceps* (von Meyer): A.S. Woodward, p. 278; table 20, fig. 2.

Numerous complete specimens from the early Miocene (Aquitanian) freshwater "Hydrobienschichten" deposits in Germany were assigned to †*Notogoneus longiceps* by Weiler (1963). This species is also present at Godrastein-bei-Landau (Palatinat, Germany), in the lower part of the Hydrobienschichten layers (Brelie *et al.* 1973). According to Gaudant (1981b), isolated otoliths from the same areas were also attributed to this species by these authors and by Malz (1978a, b).

Despite the presence of the amphibians *Rana*, †*Palaeobatrachus* and *Salamandra*, as well as of freshwater gastropods (Planorbidae and Lymnaeidae), Gaudant (1981b) argued for oligohaline conditions during the formation of the Hydrobienschichten deposits, because strictly marine fishes are absent.

†*Notogoneus montanensis* Grande and Grande, 1999

1999   †*Notogoneus montanensis*: L. Grande and T. Grande, p. 614; figs. 1 to 4; table 1.

Grande and Grande (1999) described this new species of †*Notogoneus* from the Campanian Two Medecine Formation, northwestern Montana (USA). It is based primarily on a single skeleton that is less than 5 cm in total length and that is missing much of the skull. It represents the earliest ascertained occurrence of †*Notogoneus* in North America, and the earliest known freshwater gonorynchiform from this continent. The palaeoenvironment is described as a small lacustrine or waterhole environment by Varrichio and Horner (1993).

†*Notogoneus parvus* Hills, 1934

1934   †*Notogoneus parvus*: E.S. Hills, p. 164; figs. 8 and 9; pl. 20.

This species, about 20 cm in total length, is known from late Eocene or early Oligocene freshwater deposits of Redbank Plains, southern Queensland, Australia. It was first described by Hills (1934) on the basis on five nearly complete specimens preserved as moulds in a limonitic mudstone. However, the absence of denticles on the scales suggests that the Australian specimens may belong to another genus.

†*Cyprinus squammosseus* Blainville, 1818

1818   †*Cyprinus squammosseus*: H.D. de Blainville, p. 371.
1844   †*Sphenolepis squammosseus*: L. Agassiz, part 1, p. 13; part 2, p. 87; table 65.
1896   †*Notogoneus squammosseus* (Blainville): A.S. Woodward, p. 502; table 18, figs. 3 and 4.

†*Notogoneus squammosseus*, from the late Oligocene gypsum quarry of Aix-en-Provence (southern France), is known from several complete specimens up to 77 cm in total length. They were re-described by Gaudant (1981c), who suggested a brackish palaeoenvironment for the deposits, based on the fish assemblage.

†*Notogoneus* sp.

Numerous specimens, lacking specific diagnostic characters, have been assigned to †*Notogoneus* sp. These specimens come from the late Paleocene freshwater deposits of Alberta (Paskapoo Formation), Canada (Wilson 1980); from the middle Eocene (Lutetian) of Brasles, Aisne Departement, France (subopercle and scales, Gaudant 1981b); from the early Oligocene potassic basin of Alsace, France (caudal skeleton associated with the cyprinodontid †*Prolebias*, Gaudant 1981d); from the early Oligocene of the Isle of Wight, England (Gaudant 1981b); and from the early Oligocene of Hoeleden, Brabant, in Belgium (Nolf 1977, Gaudant 1981b).

? † *Notogoneus*

1996   Undescribed gonorynchidae: S.P. Applegate, p. 534; fig. 5.

A new, undescribed genus, based on two specimens, was reported by Applegate (1996) from the middle-late Albian Tlayúa quarry, near Tepexi de Rodríguez, Puebla, Mexico. There is currently no information about the morphology of the fossil specimens, and only size and body shape were used to tentatively assign it to †*Notogoneus*. The co-occurrence of coral reef, freshwater and open ocean taxa suggests that the paleoenvironment was a lagoon protected behind a reef barrier (Espinosa-Arrubarrena and Applegate 1996). If the specimens really belong to †*Notogoneus*, they would represent the earliest occurrence of the genus.

? † *Notogoneus*

1999   Gonorynchidae: L. Grande and T. Grande, p. 620.
1999a   ? †*Notogoneus*: T. Grande, p. 437.

Some isolated bones from late Paleocene to middle Eocene freshwater oilshale deposits in Longkou county (Shandong Province, China) have been attributed to a gonorynchiform by Grande and Grande (1999). The material, still to be studied in detail, probably represents a species of †*Notogoneus* (Grande 1999a). Such an occurrence is not unexpected, given the wide geographic and temporal distribution of †*Notogoneus*.

†*PARACHANOS* Arambourg and Schneegans, 1935a.

†*Leptosomus aethiopicus* Weiler, 1922

type and only species

1922   †*Leptosomus aethiopicus*: W. Weiler, p. 154; figs. 3 to 5.
1935a   †*Parachanos aethiopicus* (Weiler): C. Arambourg and D. Schneegans, p. 141; table 1, figs. 1 and 2; table 2, figs. 1, 2 and 4; table 3, figs. 1, 5 and 7.

1935b   †*Parachanos aethiopicus* (Weiler): C. Arambourg and D. Schneegans,
          p. 1934, listed.

Weiler (1922) described four small complete specimens from the Aptian/
Albian bituminous shales of Rio San Benito, south of Bata, Equatorial
Guinea. This fossil material led Weiler (1922) to erect a new species:
†*Leptosomus aethiopicus*. Arambourg and Schneegans (1935a, b) later removed
this species from the genus †*Leptosomus* because they regarded the Rio Benito
specimens as juveniles of a new genus they just discovered in Gabon:
†*Parachanos*. This taxon was found in the sub-littoral sandstones of the
Cocobeach series, Gabon. Like *Chanos*, †*Parachanos* was placed among the
clupeiforms at that time (Arambourg and Schneegans 1935a). Later
recognized as a chanid, †*Parachanos* then became the focus of another debate:
its similarity with †*Dastilbe*.

In contrast with Arambourg and Schneegans (1935a), Santos (1947)
suggested a synonymy between †*Parachanos* and †*Dastilbe* because the
anatomical differences (vertebral counts and persistence of a notochordal
canal in †*Dastilbe*) were not regarded as taxonomically significant. Taverne
(1974) first questioned this synonymy, but he later suggested that †*Dastilbe*
(†*D. crandalli* and †*D. elongatus*) and †*Parachanos* represent separate
populations of a single species, †*D. crandalli* (Taverne 1981). After a study of
the type species of †*Parachanos* (†*P. aethiopicus*) and of †*Dastilbe* species,
Gayet (1989) disagreed with this synonymy. Blum (1991a), who regarded
both †*D. crandalli* and †*D. elongatus* as valid species (although he did not
mention †*D. batai*), considered that the validity of †*Parachanos* remains to be
tested. Finally, recent cladistic analyses have considered †*Parachanos* and
†*Dastilbe* (on the basis of †*D. crandalli* and †*D. elongatus*) distinct genera
(Poyato-Ariza 1996a, Grande and Poyato-Ariza 1999).

†*Parachanos* sp.

Erasmo (1952) figured six specimens from the Late Cretaceous at Redipuglia,
Carso Triestino, Komen, in Croatia and he attributed them to †*Parachanos*
sp. According to Radovcic (1975), the beds these fish come from may range
from Cenomanian to Maastrichtian in age. Gayet (1993a) placed it
erroneously in synonymy with †*Dastilbe*, as noted by Poyato-Ariza (1996a).

Nardon (1990) also reported on †*Parachanos* sp. from the Coniacian-
Santonian of Palazzo, Carso Goriziano, Italy.

PARAKNERIA Poll, 1965

No fossil is known for this extant freshwater genus. Fourteen stream-
dwelling species of *Parakneria* are known from East, West and Central Africa
(see Poll 1965 and Roberts 1975 for details). These are: *Parakneria abbreviata*

(Pellegrin, 1931), *P. cameronensis* (Boulenger, 1909), *P. damasi* Poll, 1965, *P. fortuita* Penrith, 1973, *P. kissi* Poll, 1969, *P. ladigesi* Poll, 1967, *P. lufirae* Poll, 1965, *P. malaissei* Poll, 1969, *P. marmorata* (Norman, 1923), *P. mossambica* Jubb and Bell-Cross, 1974, *P. spekii* (Günther, 1868), *P. tanzaniae* Poll, 1984, *P. thysi* Poll, 1965, and *P. vilhenae* Poll, 1965.

PHRACTOLAEMUS Boulenger, 1901

A single Recent species, *Phractolaemus ansorgii*, is known for this genus. There are two subspecies, *P. ansorgii ansorgei* Boulenger, 1901, and *P. ansorgii spinosus* Pellegrin, 1925, that are found respectively in Niger and in the central basin of the Democratic Republic of Congo (Thys van den Audenaerde 1961). No fossil is known for this genus.

†*PROCHANOS* Bassani, 1882

†*Prochanos rectifrons* Bassani, 1882

type and only species

1879   †*Prochanos rectifrons*: F. Bassani, p. 163, listed.
1882   †*Prochanos rectifrons* Bassani: F. Bassani, p. 218; table 13; table 14, fig. 1; table 15.

This large fish (54 cm in total length) was described from the Late Cretaceous of Lesina Island, Hvar, Dalmatia (Croatia). It is probably a junior synonym of †*Chanos leopoldi* (Erasmo, 1915). The limestones from this area are regarded as Turonian to Maastrichtian in age (Radovcic 1975).

†*Prochanos* ? sp.

Bassani (1882) listed †*Prochanos* ? sp. in the early Cenomanian of Haqil, Lebanon. Later, Woodward (1901) reidentified the specimen as †*Chirocentrites libanicus* Pictet and Humbert, 1866, an ichthyodectid fish found in the same beds. In fact, Woodward (1901) transferred that species into the genus †*Ichthyodectes*. Two years later, Hay (1903) erected the new genus †*Eubiodectes* for that species. Clearly, †*Eubiodectes libanicus* does not belong to the Gonorynchiformes. Thus, there is currently no evidence for the presence of †*Prochanos* in the Cretaceous of Lebanon.

†*RAMALLICHTHYS* Gayet, 1982

†*Ramallichthys orientalis* Gayet, 1982

type and only species

1982   †*Ramallichthys orientalis*: M. Gayet, p. 405; fig. p. 406.

Gayet (1982) created this taxon for seven specimens found in the marine Beit-Mer Formation (early Cenomanian) at Ein-Yabrud, near Ramallah, Judea Mounts, Israel. As for †*Judeichthys*, Grande (1996) rejected the validity of the genus and suggested that †*Ramallichthys orientalis* may be a junior synonym of †*Charitosomus hakelensis*. In fact, the study of the type species †*Charitosomus formosus* led Gayet (1993b) to argue that †*Charitosomus hakelensis* is certainly not a †*Charitosomus* (see †*Hakeliosomus hakelensis* above). In addition, the validity of †*Ramallichthys orientalis* is supported by several characters, such as the parhypural and hypural 1 being unfused to the centrum complex, and the strong modification of the anteriormost vertebrae, as figured by Gayet (1986). As for †*Judeichthys*, the detailed controversy about †*Ramallichthys* is well beyond the scope of this paper. However, it is certain that the phylogenetic status of this taxon poses an interesting problem of character distribution among Ostariophysi. Recently, Grande and Grande (2008) proposed to synonymize †*Judeichthys* and †*Hakeliosomus* with †*Ramallichthys*.

†*RUBIESICHTHYS* Wenz, 1984

†*Rubiesichthys gregalis* Wenz, 1984

type and only species

1984   †*Rubiesichthys gregalis*: S. Wenz, p. 276; figs. 1 and 2.

Wenz (1984, 1991) described this taxon on the basis of 10 complete specimens (each less than 5 cm in standard length) from the Berriasian-Valanginian (Early Cretaceous) lithographic limestones of Serra del Montsec, Lérida Province, Spain. Abundant specimens (more than 400) of †*Rubiesichthys gregalis* have been reported from both this locality and the late Hauterivian–early Barremian site of Las Hoyas, Cuenca Province, Spain (Sanz *et al.* 1988, Poyato-Ariza 1996c). Poyato-Ariza (1996a, c) provided a detailed account on this chanid species and found many similarities with †*Gordichthys conquensis*.

†*THARRHIAS* Jordan and Branner, 1908

From Arambourg and Schneegans (1935a, b) onwards, most authors have included †*Tharrhias* species within Chanidae, but Patterson (1984b), followed by Brito and Wenz (1990), Blum (1991b) and Gayet (1993c), treated †*Aethalionopsis*, †*Tharrhias*, †*Dastilbe*, †*Parachanos*, and occasionally †*Rubiesichthys* as stem group gonorynchiforms (or gonorynchoids).

†*Tharrhias araripis* Jordan and Branner, 1908

type species

1908   †*Tharrhias araripis*: D. S. Jordan and J. C. Branner, p. 14; pl. II

Type species of the genus, †*Tharrhias araripis* was erected based on specimens from the Aptian/Albian Romualdo Member of the Santana Formation, Araripe Basin, northeastern Brazil. †*Tharrhias* is the most abundant taxon in the fish assemblages found in the lower part of these deposits, where it typically occurs in early diagenetic carbonate concretions (Fara *et al.* 2005). According to Poyato-Ariza (1996a), †*T. araripis* is currently the only valid species of the genus.

†*Tharrhias castellanoi* Santos and Duarte, 1962

1962   †*Tharrhias castellanoi*: R. da Silva Santos and L. Duarte, pl. 3, figs. 3 and 4.

Santos and Duarte (1962) created this species based on imprints of isolated scales found in the Turonian Açu Sandstones in Rio Grande do Norte, Brazil. Brito and Wenz (1990) first pointed out that the very poor material does not support the creation of a new taxon, and †*T. castellanoi* is certainly not a valid species.

†*Tharrhias feruglioi* (Bordas, 1942)

1942   †*Tharrhias feruglioi*: A. Bordas, p. 316; fig. 1.
1949   †*Tharrias shamani*: M. Dolgopol de Sáez, p. 445; fig. 1.
1949   †*Leptolepis leanzai*: M. Dolgopol de Sáez, p. 447; fig. 2.
1978   †*Leptolepis feruglioi* Bordas, 1942: A. Bocchino, p. 303; figs. 1 and 2, pl. 1.
1987   "†*Tharrhias*" *feruglioi* Bordas, 1942: A.L. Cione and S.M. Pereira, p. 290; table 3 and table 5, fig. C.
2001   †new genus *feruglioi* (Bordas): G. Arratia and A. Lopez-Arbarello, p. 5.

Bordas (1942) erected the species †*Tharrhias feruglioi* for specimens found in the vicinity of Cerro Cóndor, Rio Chubut, Argentina. This taxonomic choice may have been influenced by both the age of the sediments (thought to be Early Cretaceous at that time, but actually Jurassic [Arratia and Lopez-Arbarello, 2001]) and by the overall similarity with †*Tharrhias araripis*. Bocchino (1978) moved this species to genus †*Leptolepis* (*sensu* Nybelin 1974). Cione and Pereira (1987) later suggested that †*T. shamani* Dolgopol de Sáez, 1949 and †*Leptolepis leanzai* Dolgopol de Sáez, 1949 are synonyms of †"*Tharrhias*" *feruglioi*. Poyato-Ariza (1996a: 42) first suggested that this species can be removed from the genus †*Tharrhias* and from the clade Ostariophysi. More recently, Arratia and Lopez-Arbarello (2001) provided a wealth of anatomical detail based on abundant new material for this taxon. Because of its generalized morphology, these authors also concluded that †"*Tharrhias*" *feruglioi* can be removed from the genus †*Tharrhias* as a yet unnamed species of Teleostei *incertae sedis*.

†*Cearana rochae* Jordan and Branner, 1908

1908   †*Cearana rochae*: D.S. Jordan and J.C. Branner, p. 27; pl. VIII, fig. 2.
1921b   †*Tharrhias rochae* (Jordan and Branner): D.S. Jordan, p. 28; pl. VII, fig. 4.
1938   †*Tharrhias Rochai* (Jordan and Branner): G. d'Erasmo, p. 19; pl. III, figs. 1–3; pl. IV, fig. 1.
1991b   †*Tharrhias araripis* Jordan and Branner: S. Blum, p. 287; figs. and pl. pp. 286–293.

Based on specimens from the Aptian/Albian Romualdo Member of the Santana Formation (northeastern Brazil), Jordan and Branner (1908) erected two genera, †*Tharrhias* (†*T. araripis*) and †*Cearana* (†*C. rochae*). They assigned them to the families †Leptolepidae and Osteoglossidae, respectively. Later, Jordan (1921b) moved the two type species to the genus †*Tharrhias* (considered to be a †Leptolepidae at that time). This author regarded †*T. araripis* and †*T. rochae* as valid species that could be distinguished by the body shape, and Erasmo (1938) acknowledged the taxonomic status of the two forms. Taverne (1975) did not discuss the synonymy of the genera †*Tharrhias* and †*Cearana*, but he assigned both taxa to the Gonorynchiformes. The study by Oliveira (1978) provided six anatomical features that discriminate †*T. araripis* and †*T. rochae*. Patterson (1984a) listed the two species, but he only discussed and figured †*T. araripis*. Finally, Blum (1991b) was the first to check for a possible synonymy between †*T. araripis* and †*T. rochae*. He concluded that †*T. rochae* must be regarded as a junior synonym of †*T. araripis*, a suggestion with which Poyato-Ariza (1996a) agreed.

†*Tharrhias shamani* Dolgopol de Sáez, 1949

1949   †*Tharrias shamani*: M. Dolgopol de Sáez, p. 445; fig. 1.
1949   †*Leptolepis leanzai*: M. Dolgopol de Sáez, p. 447; fig. 2.
1987   "†*Tharrhias" feruglioi* Bordas, 1942: A.L. Cione and S.M. Pereira, p. 290; tables 3 and 5, fig. C.

Junior synonym of †*Tharrhias feruglioi*.

Discussion

From the review above, and considering the conflicts among taxonomic opinions, the following list of valid gonorynchiform fossil species is proposed:
  Aethalionopsis robustus (Traquair, 1911)
  Apulichthys gayeti Taverne, 1997
  Chanos brevis Heckel, 1854
  Chanos compressus Stinton, 1977
  Chanos forcipatus Kner and Steindachner, 1863
  Chanos leopoldi (Costa, 1860)
  Chanos torosus Danil'chenko, 1968
  Chanos zignoi Kner and Steindachner, 1863
  Charitopsis spinosus Gayet, 1993a
  Charitosomus formosus von der Marck, 1885
  Charitosomus lineolatus (Pictet and Humbert, 1866)
  Charitosomus major Woodward, 1901
  Coelogaster leptostea (Eastman, 1905)
  Dastilbe crandalli Jordan, 1910
  Dastilbe moraesi Santos, 1955
  Gordichthys conquensis Poyato-Ariza, 1994
  Hakeliosomus hakelensis (Davis, 1887)
  Halecopsis insignis (Delvaux and Ortlieb, 1888)
  Judeichthys haasi Gayet, 1985
  Lecceichthys wautyi Taverne, 1998
  Neohalecopsis striatus (Weiler, 1920)
  Notogoneus alsheimensis (Weiler, 1942)
  Notogoneus cuvieri (Agassiz, 1844)
  Notogoneus fusiformis Schwarzhans, 1994
  Notogoneus gracilis Sytchevskaya, 1986
  Notogoneus longiceps (Meyer, 1848)
  Notogoneus montanensis Grande and Grande, 1999
  Notogoneus osculus Cope, 1886
  Notogoneus parvus Hills, 1934
  Notogoneus squammosseus (Blainville, 1818)

Parachanos aethiopicus (Weiler, 1922)
Ramallichthys orientalis Gayet, 1982
Rubiesichthys gregalis Wenz, 1984
Tharrhias araripis Jordan and Branner, 1908

Fossil Distribution

Fossil gonorynchiform fishes are known from the earliest Cretaceous (Berriasian-Valanginian) to the earliest Miocene, and the clade has several extant representatives. With only 46 known localities, their fossil record is relatively poor compared to that of other groups of Ostariophysi. For example, Siluriformes are present in more than 500 fossil sites (Gayet and Meunier 2003), despite their first appearance in the Late Cretaceous. The paucity of the fossil record of Gonorynchiformes might reflect the actual low diversity of the group through geological time, but this apparent poor diversity is certainly accentuated by some biasing factors, such as the low frequency of diagnostic specimens.

The distribution of gonorynchiform-bearing localities is heterogeneous in both space and time. More than half the localities are Cretaceous in age (16 in the Early Cretaceous and nine in the Late Cretaceous), and there are only three sites known from the Palaeocene, seven from the Eocene, 10 from the Oligocene, and one from the earliest Miocene (Aquitanian). There are no fossil sites with gonorynchiform fossils between the Miocene and the Recent, and the clade could therefore be qualified as a "Lazarus taxon" for that time interval.

The spatial distribution of localities is also unbalanced (see also Grande 1999a and Fara et al. 2007 for summary maps). More than half the sites are located in Europe, and this is accompanied by a higher taxonomic diversity, because the majority of fossil genera and species are European (Figs. 6.1 and 6.2). The oldest representative of the group (†Rubiesichthys) is known from Europe, as well as the most basal taxa. The Italian taxon †Chanos leopoldi is currently the oldest known record of Chanos, which is the only extant gonorynchiform genus with fossil representatives. All fossil Chanos species are restricted to Europe (although the extant Chanos is Indo-Pacific), and they range apparently from the Early Cretaceous to the Oligocene.

Only three fossil genera and three species have been reported from Africa (Fig. 6.3), but their precise taxonomic status remains unresolved (†Parachanos, †Dastilbe, and †Charitosomus). Therefore, gonorynchiform fish are undoubtedly present in Africa since Aptian-Albian times, but the exact nature of the taxa is open to debate. Nothing is known about the origin of the two living African families, Kneriidae and Phractolaemidae, which lack a fossil record.

Occurrences in the Middle East (Fig. 6.3) suggest that gonorynchiform diversity was apparently higher than in Africa, and they provide important information on the Tethyan biogeography in the early Late Cretaceous.

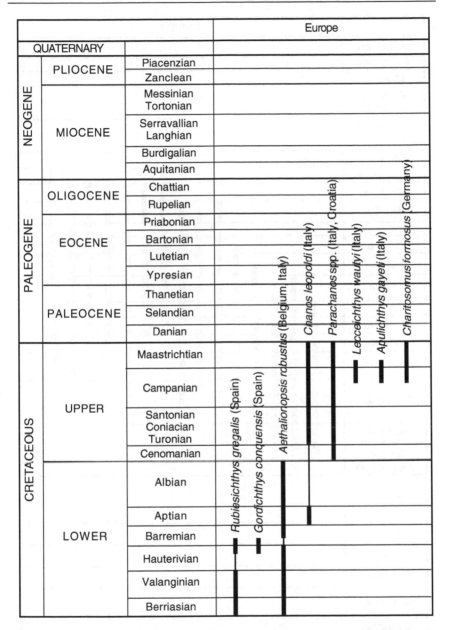

Fig. 6.1 Stratigraphic distribution of Gonorynchiformes in the Cretaceous of Europe. Because there is some uncertainty about the precise age of several localities, the stratigraphic ranges given in Figs. 6.1 to 6.6 provide the maximum possible age for each taxon. Thin bars represent the inferred presence of taxa in the fossil record.

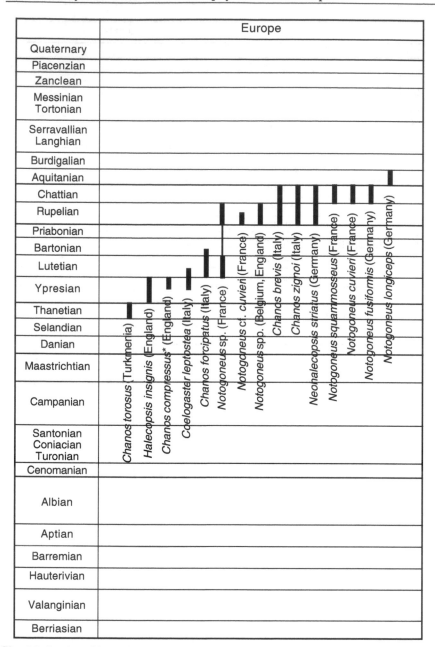

Fig. 6.2 Stratigraphic distribution of post-Cretaceous Gonorynchiformes in Europe.

The most remarkable feature of the gonorynchiform fossil record in North and Central America is that it documents only a few species of a single genus, †*Notogoneus* (Fig. 6.4), unless the yet undescribed *Notogoneus*-like form from Mexico represents another genus. Given the size of the continent, the well-known geology, the intensive sampling and the numerous fossil fish localities, this very low gonorynchiform diversity is puzzling, but it certainly represents a real biological pattern.

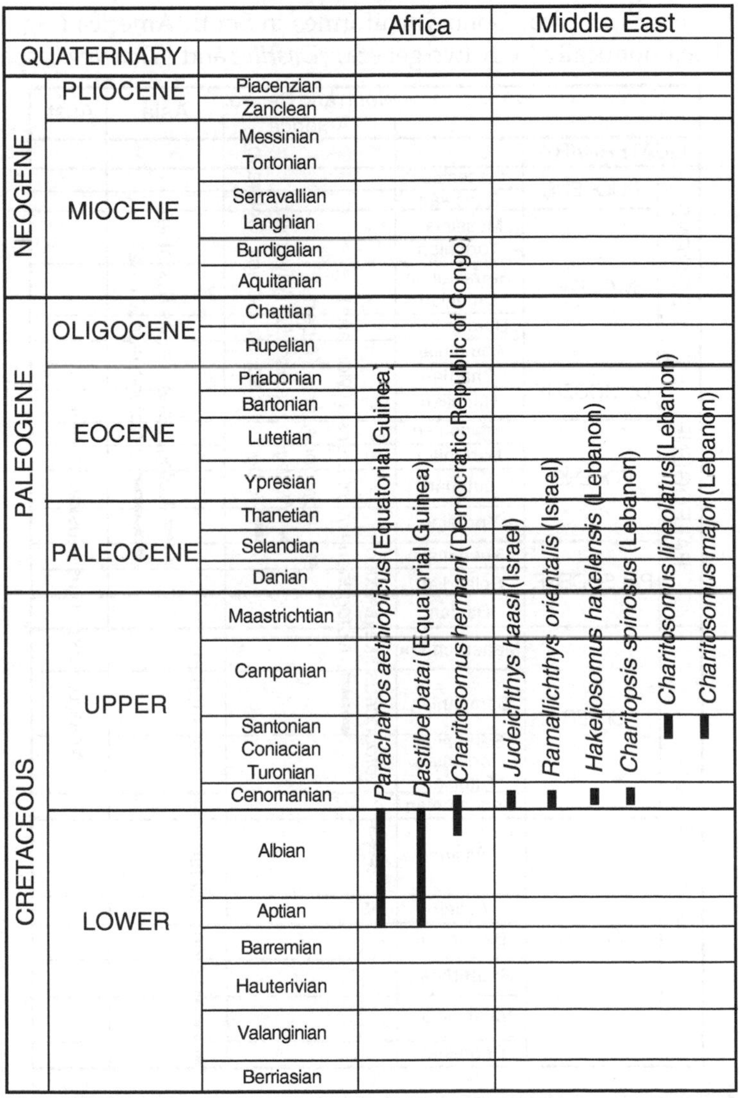

			Africa			Middle East					
			Parachanos aethiopicus (Equatorial Guinea)	*Dastilbe batai* (Equatorial Guinea)	*Charitosomus hermani* (Democratic Republic of Congo)	*Judeichthys haasi* (Israel)	*Ramallichthys orientalis* (Israel)	*Hakeliosomus hakelensis* (Lebanon)	*Charitopsis spinosus* (Lebanon)	*Charitosomus lineolatus* (Lebanon)	*Charitosomus major* (Lebanon)
QUATERNARY											
NEOGENE	PLIOCENE	Piacenzian									
		Zanclean									
	MIOCENE	Messinian									
		Tortonian									
		Serravallian									
		Langhian									
		Burdigalian									
		Aquitanian									
PALEOGENE	OLIGOCENE	Chattian									
		Rupelian									
	EOCENE	Priabonian									
		Bartonian									
		Lutetian									
		Ypresian									
	PALEOCENE	Thanetian									
		Selandian									
		Danian									
CRETACEOUS	UPPER	Maastrichtian									
		Campanian								▮	▮
		Santonian									
		Coniacian									
		Turonian									
		Cenomanian			▮	▮	▮	▮	▮		
	LOWER	Albian	▮	▮							
		Aptian	▮	▮							
		Barremian									
		Hauterivian									
		Valanginian									
		Berriasian									

Fig. 6.3 Stratigraphic distribution of Gonorynchiformes in Africa and in the Middle East.

Asia and Australia have very poor but interesting gonorynchiform fossil records (Fig. 6.4). For example, the oldest known occurrence of the group may be a gonorynchid-like specimen from the earliest Cretaceous (Berriasian-Valanginian) of Japan. Also, there is currently no palaeobiogeographic model to explain the presence of an Eocene/Oligocene freshwater †*Notogoneus* in Australia. In turn, there is no doubt that more Asian and Australian fossils would provide key phylogenetic and palaeobiogeographic information for the group.

The fossil record of Gonorynchiformes in South America (Fig. 6.5) is limited taxonomically (only two genera, †*Dastilbe* and †*Tharrhias*), spatially

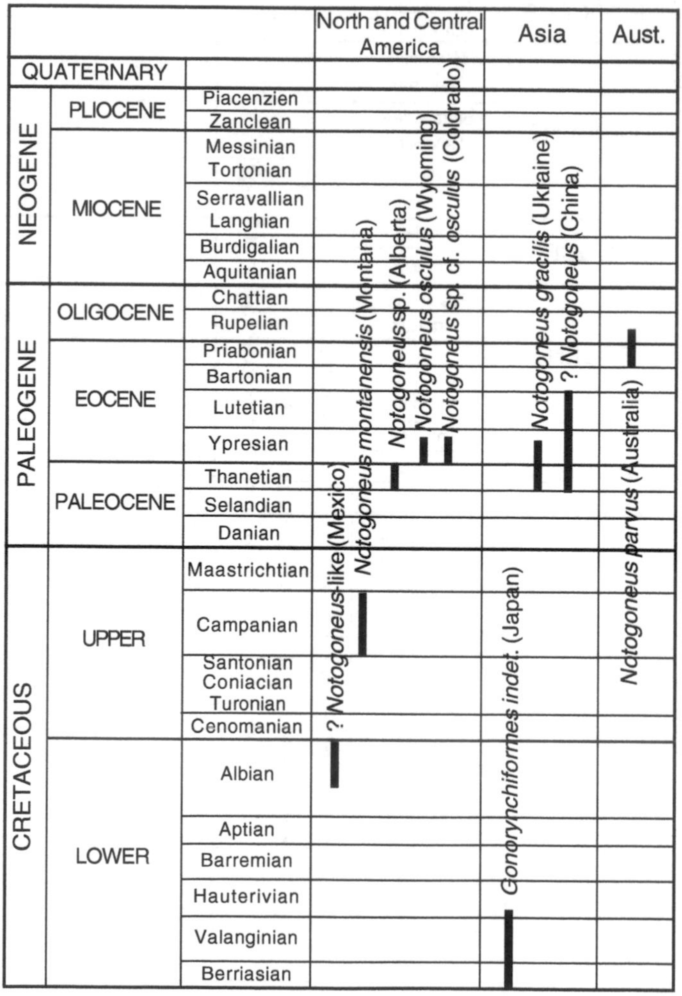

			North and Central America	Asia	Aust.
QUATERNARY					
NEOGENE	PLIOCENE	Piacenzien			
		Zanclean			
	MIOCENE	Messinian			
		Tortonian			
		Serravallian			
		Langhian			
		Burdigalian			
		Aquitanian			
PALEOGENE	OLIGOCENE	Chattian			
		Rupelian			
	EOCENE	Priabonian			
		Bartonian			
		Lutetian			
		Ypresian			
	PALEOCENE	Thanetian			
		Selandian			
		Danian			
CRETACEOUS	UPPER	Maastrichtian			
		Campanian			
		Santonian			
		Coniacian			
		Turonian			
		Cenomanian			
	LOWER	Albian			
		Aptian			
		Barremian			
		Hauterivian			
		Valanginian			
		Berriasian			

North and Central America taxa: *Notogoneus* sp. cf. *osculus* (Colorado); *Notogoneus osculus* (Wyoming); *Notogoneus* sp. (Alberta); *Notogoneus montanensis* (Montana); *Notogoneus-like* (Mexico); *Notogoneus* sp.

Asia taxa: *Notogoneus gracilis* (Ukraine); ? *Notogoneus* (China); *Gonorynchiformes indet.* (Japan).

Aust. taxa: *Notogoneus parvus* (Australia).

Fig. 6.4 Stratigraphic distribution of Gonorynchiformes in North and Central America, Asia and Australia (Aust.).

(most records are from northeastern Brazil), and temporally (Aptian-Albian). The distribution of †*Dastilbe* seems to be linked with regional events during the opening of the South Atlantic Ocean. However, the precise palaeobiogeographical history of †*Dastilbe* awaits a better understanding of the relationships among its various species (Maisey 2000).

The synoptic range-chart in Fig. 6.6 summarizes the temporal distribution of all fossil genera considered valid in this work.

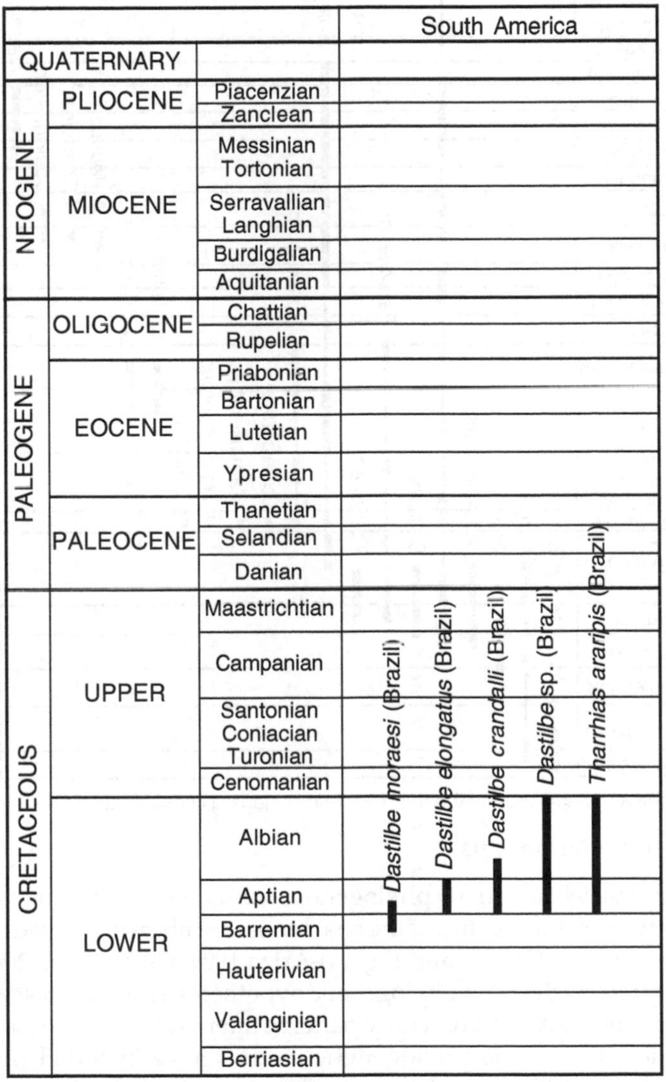

			South America
QUATERNARY			
NEOGENE	PLIOCENE	Piacenzian	
		Zanclean	
	MIOCENE	Messinian	
		Tortonian	
		Serravallian	
		Langhian	
		Burdigalian	
		Aquitanian	
PALEOGENE	OLIGOCENE	Chattian	
		Rupelian	
	EOCENE	Priabonian	
		Bartonian	
		Lutetian	
		Ypresian	
	PALEOCENE	Thanetian	
		Selandian	
		Danian	
CRETACEOUS	UPPER	Maastrichtian	*Dastilbe moraesi* (Brazil) · *Dastilbe elongatus* (Brazil) · *Dastilbe crandalli* (Brazil) · *Dastilbe* sp. (Brazil) · *Tharrhias araripis* (Brazil)
		Campanian	
		Santonian / Coniacian / Turonian	
		Cenomanian	
	LOWER	Albian	
		Aptian	
		Barremian	
		Hauterivian	
		Valanginian	
		Berriasian	

Fig. 6.5 Stratigraphic distribution of Gonorynchiformes in South America.

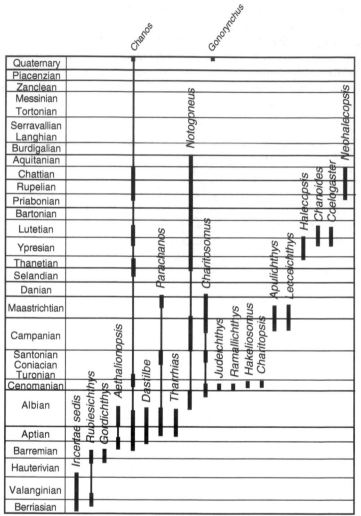

Fig. 6.6 Summary range-chart of the gonorynchiform genera considered valid in this chapter.

Phylogenetic Relationships

In recent years, debates on the phylogenetic status of gonorynchiforms have mainly focused on the identity of the basal-most members of the clade and on its sister group (e.g. Grande and Poyato-Ariza 1999, Lavoué *et al.* 2005). We shall briefly review the main phylogenetic hypotheses in order to test whether conflicting proposals affect diversity patterns inferred from the fossil record.

Very few large-scale phylogenetic studies have included the fossil representatives of Gonorynchiformes. The hypothesis by Gayet (1993c) needed to be updated after the discovery of new taxa such as †*Gordichthys*

(Poyato-Ariza 1994), †*Apulichthys* and †*Lecceichthys* (Taverne 1997, 1998). Grande and Poyato-Ariza (1999) provided the most comprehensive phylogenetic analysis of fossil and extant Gonorynchiformes, although †*Apulichthys* and †*Lecceichthys* could not be included at that time. We summarize below these three proposals as indented lists, and we present the corresponding phylogenetic trees (phylogenetic hypotheses projected on a time scale) in Figs. 6.7–6.9. Poyato-Ariza (1996a) first provided a cladistic analysis of extant and fossil Gonorynchiformes, but it is not included here because it focused primarily on Chanidae and was later updated in Grande and Poyato-Ariza (1999).

Gayet (1993c) first proposed a sister-group relationship between *Gonorynchus* and †*Notogoneus*, and she was followed by subsequent workers

Superorder Ostariophysi *sensu* Rosen and Greenwood 1970
Series Anatophysi *sensu* Rosen and Greenwood 1970
Order Gonorynchiformes *sensu* Rosen and Greenwood 1970

Gayet 1993c (Fig. 7)	Taverne 1997 and 1998 (Fig. 8)	Grande and Poyato-Ariza 1999 (Fig. 9)
†*Aethalionopsis*	Suborder Chanoidei	Suborder Chanoidei
"†*Tharrhias* group"	Family Chanidae	†*Aethalionopsis*
†*Rubiesichthys*	Subfamily	Family Chanidae
†*Tharrhias*	Rubiesichthyinae	Subfamily Chaninae
†*Dastilbe*	†*Rubiesichthys*	†*Dastilbe*
†*Parachanos*	†*Gordichthys*	†*Parachanos*
unnamed clade	Subfamily Chaninae	*Chanos*
Chanos	†*Aethalionopsis*	†*Tharrhias*
Kneriidae	*Chanos*	Subfamily
Phractolaemidae	†*Dastilbe*	†Rubiesichthyinae
Suborder	†*Parachanos*	†*Gordichthys*
Gonorynchoidei	†*Prochanos*	†*Rubiesichthys*
†*Ramallichthys*	†*Tharrhias*	Suborder Gonorynchoidei
Family	Suborder Gonorynchoidei	Family Gonorynchidae
Judeichthyidae	Family Apulichthyidae	Subfamily
†*Judeichthys*	†*Apulichthys*	Gonorynchinae
Family	Family Halecopsidae	*Gonorynchus*
Charitosomidae	†*Halecopsis*	†*Notogoneus*
†*Hakeliosomus*	Family	†*Charitosomus*
†*Charitosomus*	Phractolaemidae	†*Ramallichthys*
†*Charitopsis*	Family Kneriidae	†*Judeichthys*
Family	Family	†*Charitopsis*
Gonorynchidae	Gonorynchidae	Family Kneriidae
†*Notogoneus*	†*Ramallichthys*	Subfamily
Gonorynchus	†*Judeichthys*	Phractolaeminae
	†*Hakeliosomus*	*Phractolaemus*
	†*Charitosomus*	Subfamily Kneriinae
	†*Charitopsis*	Tribe Cromerini
	†*Lecceichthys*	*Cromeria*
	†*Notogoneus*	*Grasseichthys*
	Gonorynchus	Tribe Kneriini
		Kneria
		Parakneria

(but see Grande and Grande 2008 for an alternative opinion). The family Gonorynchidae, as defined by Grande and Poyato-Ariza (1999), includes *Gonorynchus* and †*Notogoneus* plus the Middle East Cretaceous taxa †*Ramallichthys*, †*Judeichthys*, †*Charitosomus*, and †*Charitopsis*. In turn, this family is similar in content to what Gayet (1993b) named the Gonorynchoidei in her taxonomic scheme, with the exception of †*Hakeliosomus*, which was not considered a valid taxon by Grande and Poyato-Ariza (1999). The family names †Judeichthyidae and †Charitosomidae have no more reason to exist, as observed by Grande and Grande (1999). Another similarity between the hypotheses of Gayet (1993c) and Grande and Poyato-Ariza (1999) is the sister-group relationship between *Gonorynchus*, †*Notogoneus* and the Middle East taxa on the one hand and the extant Kneriinae and Phractolaeminae on the other hand.

The monophyly of the clade including all these extant African freshwater taxa is well supported by osteological, myological, and molecular data (Howes 1985, Gayet 1993c, Fink and Fink 1996, Grande and Poyato-Ariza 1999, Lavoué *et al.* 2005). Within this clade, however, there is some disagreement between the results of osteological studies (Grande 1996, Grande and Poyato-Ariza 1999) and those of a recent study based on whole mitogenome sequences (Lavoué *et al.* 2005), in which *Grasseichthys* and *Cromeria* are regarded as sequenced sister groups to the clade (*Kneria+Parakneria*). Such a conflict has a great heuristic value in the systematics of Gonorynchiformes.

Taverne's (1997, 1998) definition of Gonorynchoidei was similar to that of Gayet (1993c) and Grande and Poyato-Ariza (1999), but he added †*Lecceichthys*, †*Apulichthys* and †*Halecopsis* to that group. His definition of Chanoidei made this clade paraphyletic.

The identity of the basal-most members of Gonorynchiformes is a recurrent debate. Poyato-Ariza (1996a) and Taverne (1997, 1998) placed †*Aethalionopsis* among the Chaninae, but the analysis of Grande and Poyato-Ariza (1999) excluded this taxon from the Chanidae, a result in accordance with Gayet's (1993c) early hypothesis. The relationships of †*Tharrhias*, †*Dastilbe*, and †*Parachanos* are similar in the three phylogenetic schemes reviewed here. Although Gayet (1993c) placed *Chanos* outside the "†*Tharrhias* group", this genus certainly belongs to the clade composed of †*Dastilbe*, †*Parachanos*, †*Prochanos*, and †*Tharrhias* (Taverne 1997, 1998, Grande and Poyato-Ariza 1999). A vast majority of osteological studies (Fink and Fink 1981, Gayet 1993c, Poyato-Ariza 1996a, Taverne 1997, 1998, Grande and Poyato-Ariza 1999) agree on the relationships of extant forms, with the structure {*Chanos*[*Gonorynchus*(Phractolaemidae+Kneriidae)]}. However, this configuration has been challenged recently by the mitogenomic study of Lavoué *et al.* (2005), in which the arrangement {*Gonorynchus*[*Chanos*(Phractolaemidae+Kneriidae)]} was found.

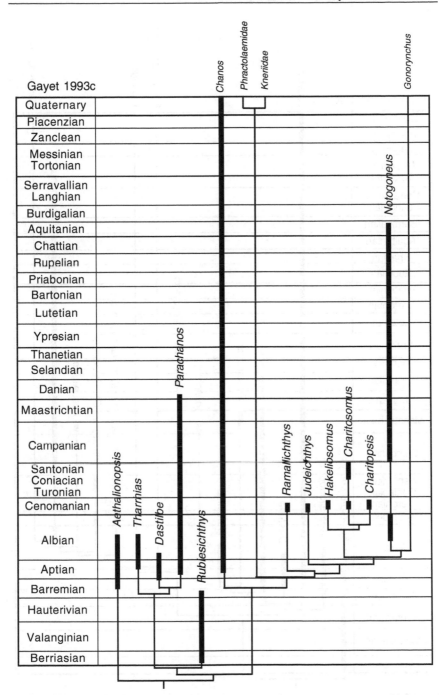

Fig. 6.7 Phylogenetic tree obtained according to the phylogenetic hypothesis by Gayet (1993c).

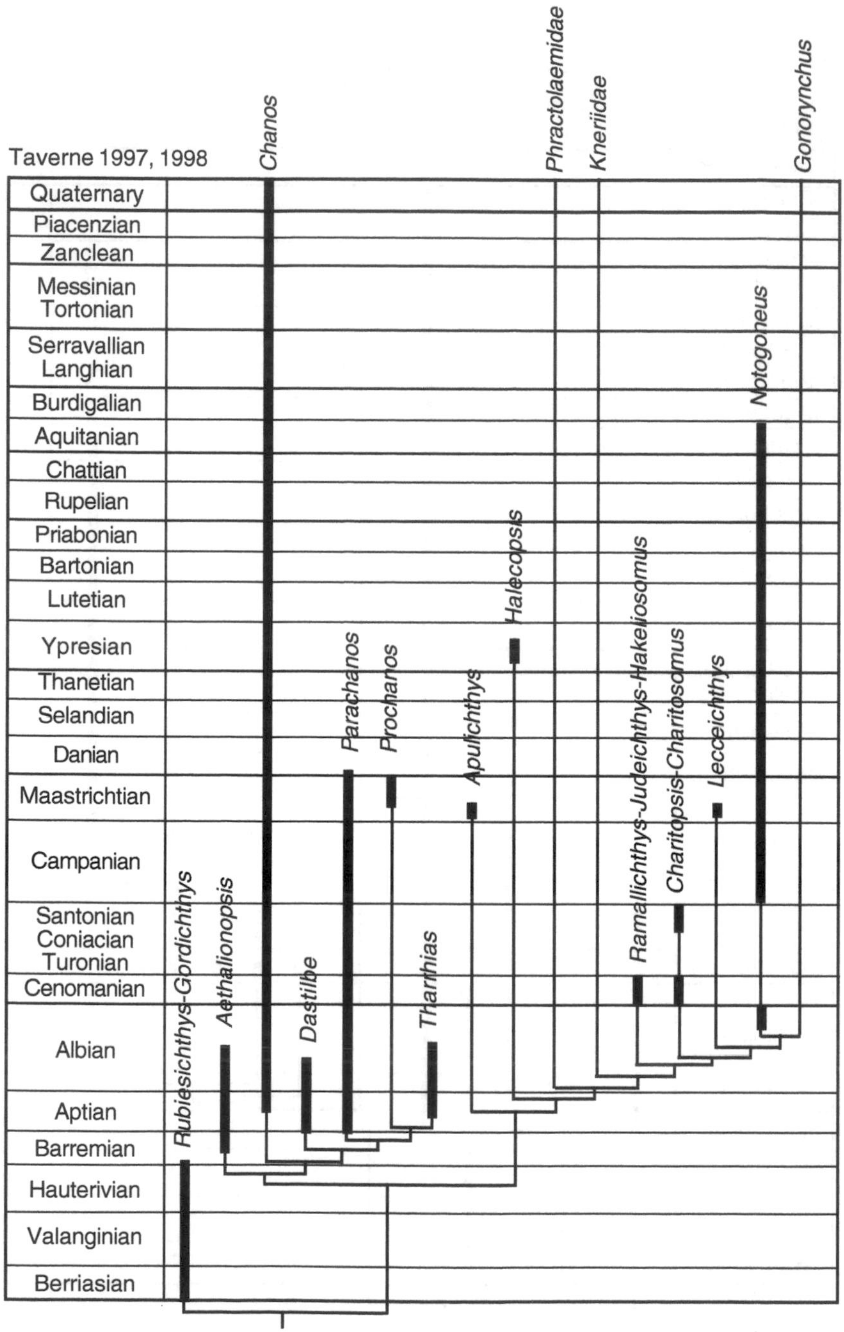

Fig. 6.8 Phylogenetic tree derived from the phylogenetic hypotheses by Taverne (1997, 1998).

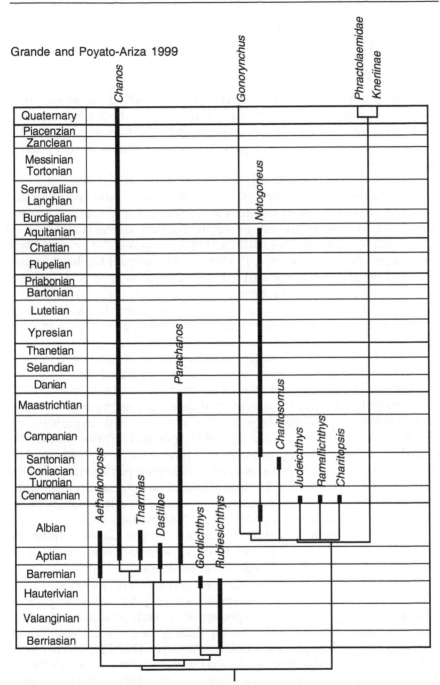

Fig. 6.9 Phylogenetic tree obtained according to the phylogenetic hypothesis by Grande and Poyato-Ariza (1999).

Interestingly, such a structure is in accordance with the early osteology-based study by Greenwood *et al.* (1966).

Diversity Dynamics

From the problems outlined above, it is clear that both the fossil record of Gonorynchiformes and our understanding of their phylogenetic relationships are not yet ripe for carrying out a robust diversity study. The following attempt should be regarded as an exploratory analysis and a base for future work.

In this review, we count 18 genera and 35 valid species of fossil gonorynchiforms between the Early Cretaceous and the Miocene. Diversity dynamics was estimated by different techniques at the species and generic levels. First, we assessed diversity by simply counting observed ranges within each stratigraphic stage. This traditional "taxic approach" (Levinton 1988) yields a raw diversity estimate. The latter was corrected for the "Lazarus effect" (Jablonski 1986, Fara 2001), that is, a taxon is counted as present between its first and last occurrence even if no fossil record is actually known in the intervening stages. Taxa whose estimated ages cross a boundary between two time intervals were counted as present in both intervals.

Second, we used the "phylogenetic approach" (Smith 1994). This method extends the observed stratigraphic ranges of taxa with the "ghost lineages" in order to confirm the predictions of a phylogenetic hypothesis (Smith 1988, 1994, Norell 1993). It supposes that sister taxa originate at the same time. We have applied this approach to the phylogenetic hypotheses by Gayet (1993c), Taverne (1997, 1998), and Grande and Poyato-Ariza (1999) to see how these different proposals affect inferred diversity patterns. For comparison purposes, only taxa common to these three phylogenetic schemes were retained in the analysis. The lack of resolution in some parts of Grande and Poyato-Ariza's (1999) cladogram did not affect the calculations because it concerns a series of either Aptian or mostly Cenomanian taxa.

Third, we computed minimal species diversity of Gonorynchiformes with the method described by Fara (2004). This approach is an intermediate between the taxic and the phylogenetic approaches.

Figure 6.10 shows the raw species and genus diversity estimated with the taxic approach. These two estimates are very similar because most genera have only one species represented in each stratigraphic stage. Gonorynchiform diversity rose steadily during the Early Cretaceous and reached a peak in the Aptian–Cenomanian interval. The high diversity level in the Aptian occurred both in Gondwana (†*Dastilbe*, †*Parachanos*, †*Tharrhias*) and in southern Europe (†*Rubiesichthys*, †*Gordichthys*, †*Aethalionopsis*, *Chanos*), whereas the Cenomanian diversity maximum is dominated by

Tethyan forms (†*Ramallichthys*, †*Judeichthys*, †*Hakeliosomus*, †*Charitopsis*, †*Charitosomus*). Gonorynchiform diversity then declined slightly towards the end of the Cretaceous and dropped at the beginning of the Cenozoic. This apparent low diversity level is only interrupted by relative peaks in the first half of the Eocene (*Chanos*, †*Notogoneus*, †*Halecopsis*, †*Coelogaster*) and, to a lesser extent, in the Oligocene (*Chanos*, †*Notogoneus*, †*Neohalecopsis*). This uneven Cenozoic diversity pattern occurred in Europe, North America, and Asia. After the last record of †*Notogoneus* in the Aquitanian (earliest Miocene) gonorynchiform diversity dropped to a minimum for the rest of the Neogene (Fig. 6.10).

This literal reading of the fossil record cannot be taken at face value, however. It is always possible that gonorynchiform diversity stabilized or decreased in a more or less regular pattern since the Cenomanian, and that new taxa wait to be discovered in Palaeocene, late Eocene, and Mio-Pleistocene deposits.

Figure 6.11 shows the diversity estimates computed with the phylogenetic approach based on the three phylogenetic hypotheses mentioned above (curves), together with the raw diversity of the genera common to these phylogenetic proposals (histogram). The diversity level is of course higher in the curves derived from the phylogenetic approach because of the addition of range extensions. The most remarkable feature is the similarity of the three estimates (Fig. 6.11), meaning that the differences across the studied phylogenies have virtually no impact on inferred diversity patterns. The same is true for the estimates of minimal species diversity (not shown here).

Figure 6.11 also shows a limit of the phylogenetic approach in the case of the Gonorynchiformes. The post-Cenomanian diversity level is remarkably low because several taxa were not included in most phylogenetic analyses (e.g., †*Coelogaster*, †*Apulichthys*, †*Halecopsis*, †*Chanoides*). Updated and comprehensive phylogenetic analyses will certainly help to resolve this issue (see Poyato-Ariza *et al.*, this volume).

This brief overview of diversity is limited by the low number of fossil taxa currently known, and it calls for caution when analysing fossil diversity patterns for Gonorynchiformes. These fossil diversity patterns strongly contrast with the relatively high diversity observed among modern gonorynchiform fishes. Unless most of these extant forms have originated in a sub-Recent radiation, this pattern is suggestive of a bias in the fossil record, especially for the Neogene (see also Fara *et al.* 2007).

Historical Biogeography

The historical biogeography and associated palaeoenvironments of Gonorynchiformes have been recently studied in detail by Grande (1999a),

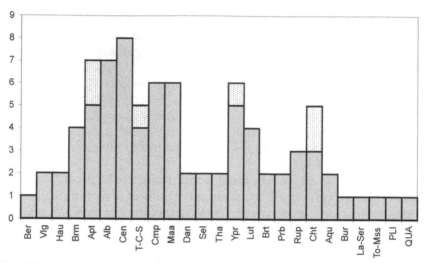

Fig. 6.10 Diversity dynamics of Gonorynchiformes during geological times as assessed by the taxic approach. The histogram represents the total raw diversity at the genus level (gray bars) and species level (dotted bars). The minimal species diversity patterns (not shown here) are virtually identical. Abbreviations of international stratigraphic stages (x-axis) are from Harland *et al.* (1990), except T-C-S, Turonian-Coniacian-Santonian; La-Ser, Langhian-Serravallian; To-Mss, Tortonian-Messinian; PLI, Pliocene; QUA, Quaternary.

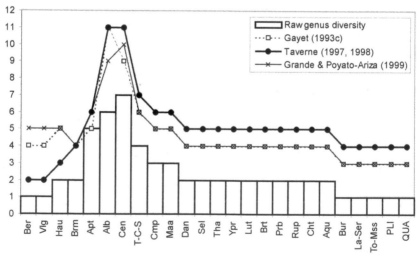

Fig. 6.11 Phylogenetic hypotheses and diversity estimates. The histogram represents the raw diversity of the genera common to the phylogenetic studies of Gayet (1993c), Taverne (1997, 1998), and Grande and Poyato-Ariza (1999). The curves correspond to the diversity estimated from these three phylogenetic proposals using the phylogenetic approach. Note the similarity of these estimates, regardless of the selected phylogenetic scheme. Abbreviations as in Fig. 6.10.

and we shall not duplicate this work here. She used cladistic vicariance biogeography on available fossil data based on the phylogenetic hypothesis by Grande and Poyato-Ariza (1999). She found that the historical biogeography of Gonorynchiformes is complex and probably results from various episodes of vicariance and dispersal (Grande 1999a).

Conclusions

Although they are known since Cuvier, gonorynchiform fishes have still many secrets to unveil. Their fossil record is crucial to this endeavour, and it certainly provides key information that complements recent advances in molecular and developmental biology. However, our current knowledge of Gonorynchiformes makes it difficult to quantify precisely several aspects of their evolutionary history. Whether for biogeography or diversity analysis, the known fossil record and the understanding of their phylogenetic relationships are still insufficient to draw robust inferences and models. In particular, our diversity analysis calls for a special effort in sampling several parts of the fossil record, and it pleads for the integration of most (if not all) fossil gonorynchiform taxa in the phylogenetic analyses of the order.

Acknowledgements

We wish to express our gratitude to the editors for inviting us to contribute to this volume. The quality of the manuscript has greatly been improved by constructive comments from L. Grande and F.J. Poyato-Ariza. We thank our colleagues A.F. Bannikov, D. Nolf, and W. Schwarzhans, who generously shared some unpublished information. EF acknowledges grants from the foundations Singer-Polignac (France) and FUNCAP (Brazil), as well as a CNRS postdoctoral fellowship that helped in the early phase of this work. This study is a contribution of the team "Forme-Evolution-Diversité" of the Laboratoire Biogéosciences (CNRS/uB), Université de Bourgogne.

References

Agassiz, L. 1833–1844. Recherches sur les poissons fossiles. Imprimerie Petitpierre, Neuchâtel, Switzerland.
Applegate, S.P. 1996. An overview of the Cretaceous fishes of the quarries near Tepexi de Rodríguez, Puebla, Mexico, pp. 529–538. In: G. Arratia and G. Viohl [eds.]. Mesozoic Fishes—Systematics and Paleoecology. Verlag Dr. F. Pfeil, Munich.
Arambourg, C. and D. Schneegans. 1935a. Poissons fossiles du bassin sédimentaire du Gabon. Ann. Paléontol. 24: 139–160.
Arambourg, C. and D. Schneegans. 1935b. Les Poissons fossiles du Bassin sédimentaire du Gabon. C. r. somm. Soc. Géol. Fr. 12: 170–171.
Argenio, B.d'. 1963. I calcari ad ittioliti del Cretacico Inferiore del Matese. Mem. Atti. Acc. Sc. Fis. Mat. Napoli, ser 4, 4: 5–63.
Arratia G. and A. Lopez-Arbarello. 2001. Morphology and relationships of "Tharrias" feruglioi from the Jurassic of Argentina, 5. In: A.Tintori [ed.]. III International

Meeting on Mesozoic Fishes: systematics, paleoenvironments and biodiversity, Serpiano-Monte San Giorgio (IT-CH), 26–31 August 2001, UNIMI, Italy.

Bagarinao, T. 1994. Systematics, distribution, genetics and life history of milkfish, *Chanos chanos*. Environ. Biol. Fish. 39: 23–41.

Barale, G., J. Martinell, X. Martínez-Delclòs, F.J. Poyato-Ariza and S. Wenz. 1994. Les gisements de calcaires lithographiques du Crétacé inférieur du Montsec (province de Lérida, Espagne): rapports récents à la paléobiologie. Geobios, Mém. Sp. 16: 177–184.

Bassani, F. 1879. Vorläufige Mittheilungen über die Fischfauna der Insel Lesina. Verhandl. K.K. Geol. Reichs. 8: 161–168.

Bassani, F. 1882. Descrizione dei Pesci fossili di Lesina. Denkschrift Akad. Wiss. Math.-naturwiss. 45: 195–288.

Bassani, F. 1889a. Ricerce sui pesci fossili di Chiavòn. Atti Reale Accad. Sci. Fis. Mat. Napoli 6: 1–104.

Bassani, F. 1889b. Notes of some Researches on the Fossil Fishes of Chiavòn, Vicentino (Stratum of Sotzka, Lower Miocene). Ann. Mag. Nat. Hist. Rep. 1888: 675–677.

Bassani, F. and G. d'Erasmo. 1912. La ittiofauna del Calcare cretacico di Capo d'Orlando presso Castellamare (Napoli). Mem. Soc. Ital. Sci., ser. 3a, 17: 185–243.

Basson, P.W. 1981. Late Cretaceous alga, *Delesserites lebanesis* sp. nov. Rev. Palaeobot. Palynol. 33: 363–370.

Basson, P.W. and H.S. Edgell. 1971. Calcareous algae from the Jurassic and Cretaceous of Lebanon. Am. Mid. Nat. 88: 506–511.

Berthou, P.Y. 1990. Le bassin de l'Araripe et les petits bassins intracontinentaux voisins (N.E. du Brésil): formation et évolution dans le cadre de l'ouverture de l'Atlantique équatoriale. Comparaison avec les bassins ouest-africains situés dans le même contexte, pp. 113–134. *In*: D.A. Campos, M.S.S. Viana, P.M. Brito and G. Beurlen [eds.]. Atas do I Simposio sobre a Bacia do Araripe e Bacias Interiores do Nordeste, URCA, Crato, Brazil.

Blainville, H.D. de. 1818. Sur les ichthyolites, ou, les poissons fossiles. Nouveau Dictionnaire d'Histoire naturelle 28: 1–91.

Blot, J. 1980. La faune ichthyologique des gisements du Monte Bolca (Province de Vérone, Italie). Catalogue systématique présentant l'état actuel des recherches concernant cette faune. Bull. Mus. Natn. Hist. Nat., Paris, 4e sér., 2, sect. C (4): 339–396.

Blum, S. 1991a. *Dastilbe* Jordan, 1910, pp. 274–285. *In*: J.G. Maisey [ed.]. Santana Fossils. An Illustrated Atlas. T.F.H. Publ., Neptune City, USA.

Blum, S. 1991b. *Tharrhias* Jordan and Branner, 1908, pp. 286–295. *In*: J.G. Maisey [ed.]. Santana Fossils. An Illustrated Atlas. T.F.H. Publ., Neptune City, USA.

Bocchino, A. 1978. Revisión de los Osteichthyes fósiles de la República Argentina I. Identidad de *Tharrias feruglioio* Bordas 1943 y *Oligopleurus groeberi* Bordas 1943. Ameghiniana 15: 301–320.

Bordas, A. 1942. Peces del Cretáceo del río Chubut (Patagonia). Physis 19: 313–318.

Boulenger, G.A. 1901. Diagnoses of new fishes discovered by Mr. W.L.S. Loat in the Nile. Ann. Mag. Nat. Hist., ser. 7, 8: 444–446.

Boulenger, G.A. 1909. Catalogue of the fresh-water fishes of Africa in the British Museum (Natural History). I. Trustees of the Brit. Mus. (Nat. Hist.), London.

Boulenger, G.A. 1910. On a large collection of fishes made by Dr. W.J. Ansorge in the Quanza and Bengo rivers, Angola. Ann. Mag. Nat. Hist., ser. 8, 6: 537–561.

Boulenger, G.A. 1915. Diagnoses de poissons nouveaux. II. Mormyrides, Kneriides, Characinides, Cyprinides, Silurides. (Mission Stappers au Tanganika-Moero.). Rev. Zool. Afr. 4: 162–171.

Brelie, G. von der, F. Doebl, F. Geissert, H. Weiler and W. Weiler. 1973. Ein Aufschluss im unteren Bereich der Hydrobien-Schichten (Aquitan) beim Bau der Gruppen-

Kläranlage "Queichtal", Godramstein bei Landau (Pfalz). Oberrhein. Geol. Abh., Karlsruhe 22: 13–44.

Brito, P.M. and C.R.L. Amaral. 2008. An overview of the specific problems of *Dastilbe* Jordan, 1910 (Gonorynchiformes: Chanidae) from the Lower Cretaceous of western Gondwana, pp. 279–294. *In*: G. Arratia, H.-P. Schultze and M.V.H. Wilson [eds.]. Mesozoic Fishes 4—Homology and Phylogeny, Verlag Dr. F. Pfeil, Munich.

Brito, P.M. and S. Wenz. 1990. O endocrânio de *Tharrhias* (Teleostei, Gonorhynchiformes) do Cretáceo inferior da Chapada do Araripe, Nordeste do Brasil, pp. 375–382. *In*: D.A. Campos, M.S.S. Viana, P.M. Brito and G. Beurlen [eds.]. Atas do I Simposio sobre a Bacia do Araripe e Bacias Interiores do Nordeste, URCA, Crato, Brazil.

Cappetta, H. 1980. Les sélaciens du Crétacé supérieur du Liban. I: Requins. Palaeontographica, Abt. A 168: 69–148.

Casier, E. 1946. La faune ichthyologique de l'Yprésien de Belgique. Mém. Mus. Roy. Hist. Nat. Belg. 104: 1–267.

Casier, E. 1961. Matériaux pour la faune ichthyologique éocrétacique du Congo. Ann. Mus. Roy. Afr. Centr., Tervuren, in –8°, Sci. Géol. 39: 1–96.

Casier, E. 1965. Poissons fossiles de la série du Kwango (Congo). Ann. Mus. Roy. Afr. Centr., sér. In –8°, Sci. Géol. 50: 1–64.

Casier, E. 1966. Faune ichthyologique du London Clay. Trustees of the Brit. Mus. (Nat. Hist.), London.

Casier, E. and L. Taverne. 1971. Note préliminaire sur le matériel paléoichthyologique éocrétacique récolté par la Spanish Gulf Oil Company en Guinée Equatoriale et au Gabon. Rev. Zool. Bot. Afr. 83: 16–20.

Catenacci, V. and M. Manfredini. 1963. Osservazioni stratigrafiche sulla Civita di Pietraroia (Benevento). Boll. Soc. Geol. Ital. 82: 1–30.

Chabanaud, P. 1931. Affinités morphologiques, répartitions stratigraphiques et géographiques des Poissons fossiles et actuels de la famille des Gonorhynchidae, avec la description de deux genres nouveaux et d'une espèce nouvelle. Bull. Soc. Géol. Fr. 5: 497–517.

Chabanaud, P. 1934. Les Gonorhynchidae fossiles. Ann. Mus. Hist. Nat. Marseille 25: 5–16.

Chang, M.-M. and L. Grande. 1997. Redescription of *Paraclupea chetungensi*, an early clupeomorph from the Lower Cretaceous of south-eastern China. Fieldiana, Geol. n.s. 37: 1–19.

Cione, A. L. and S.M. Pereira. 1987. Los peces del Jurásico de Argentina. El Jurásico anterior a los movimientos intermálmicos, pp. 287–298. *In*: W. Volkheimer [ed.]. Biostratigrafía de los Sistemas Regionales del Jurásico y Cretácico de América del Sur 1, Mendoza, Editorial Inca, Argentina.

Cope, E.D. 1885. Eocene Paddle-fish and Gonorhynchidae. Am. Nat. 19: 1090–1091.

Cope, E.D. 1886. On two new forms of Polyodont and Gonorhynchid fishes from the Eocene of the Rocky Mountains. Mem. Nat. Acad. Sci. 3: 161–165.

Costa, O.G. 1857. Decripzione di alcuni pesci fossili del Libano. Mem. R. Acad. Sci. Mat. Nat. 2: 97–112.

Costa, O.G. 1860. Ittiologia fossile italiana. Edizione simile all'opera dell'Agassiz sui Pesci fossile alla quale è estinata o sopplimento, Napoli, Italy.

Cuvier, G. and A. Brongniard. 1822. Description géologique des environs de Paris. Paris Dufour et d'Occagne, Paris.

Daget, J. 1954. Les poissons du Niger Supérieur. Mem. Inst. Franc. Afr. Noire 36: 1–391.

Dal Sasso, C. and S. Renesto. 1999. Aquatic varanoid reptiles from the Cenomanian (Upper Cretaceous) lithographic limestones of Lebanon. Riv. Mus. Sci. nat. "Enrico Caffi", Third Intern. Symp. Lithographic Limestones, Bergamo, Italy, suppl. 20: 63–69.

Danil'chenko, P.G. 1968. Ryby verkhnego paleotsena Turkmenii (Upper Paleocene Fishes from Turkmenia, pp. 113–156. *In*: D.V. Obruchev [ed.]. Ocherki po filogenii i sistematicke iskopaemykh ryb i bezcheliustnykh. Nauka Press, Moscow.

Danil'chenko, P.G. 1980. Osnovnye kompleksy ikhtiofauny kainozoiskikh morey Tetisa [Principal assemblages of the ichthyofauna of the Cenozoic seas of Tethys], pp. 175–183. *In*: L.I. Novitskaya [ed.]. Iskopayemye kostistye ryby SSSR [Fossil teleost fishes of the USSR]. Trudy Paleontol. Inst. Akad. Nauk SSSR, 178, Moscow.

Davis, J.W. 1887. The fossil fishes of Chalk of Mount Lebanon in Syria. Sci. Trans. r. Dublin Soc. 4, 2: 457–636.

Davis, S.P. and D.M. Martill. 1999. The gonorynchiform fish *Dastilbe* from the Lower Cretaceous of Brazil. Palaeontology 42: 715–740.

Delvaux, E. 1887. Documents stratigraphiques et paléontologiques pour l'étude monographique de l'étage Yprésien. Ann. Soc. Géol. Belg., Mém. 14: 57–72.

Delvaux, E. and J. Ortlieb. 1888. Les poissons fossiles de l'argile yprésienne de Belgique. Description paléontologique accompagnée de documents stratigraphiques pour servir à l'étude monographique de cet étage. Ann. Soc. Géol. Nord 15: 50–66.

Dietze, K. 2007. Redescription of *Dastilbe crandalli* (Chanidae, Euteleostei) from the Early Cretaceous Crato Formation of north-eastern Brazil. J. Vert. Paleont. 27: 8–16.

Dolgopol de Sáez, M. 1949. Noticias sobre peces fósiles argentinos. I. Peces Cretácicos de Mendoza y Chubut. Notas Mus. La Plata 14: 443–453.

Duarte, L. and R. da Silva Santos. 1993. Plant and fish megafossils of the Codó Formation, Parnaiba Basin, NE Brazil. Cret. Res. 14: 725–747.

Eastman, C.R. 1905. Les types de poisons fossiles du Monte-Bolca au Muséum d'Histoire Naturelle de Paris. Mém. Soc. Géol. Fr., Paléont. 13, fasc. 1: 5–31.

Erasmo, G.d' 1915. La fauna e l'età dei calcari a ittioliti di Pietraroia (Prov. di Benevento). Palaeontogr. Ital. 21: 1–53.

Erasmo, G.d' 1938. Ittioliti Cretacei del Brasile. Atti Acad. Sci. Fis. Mat., Napoli 3, 3: 1–44.

Erasmo, G.d' 1952. Nuovi ittioliti cretacei del Carso triestino. Atti Museo Civ. St. Nat. Trieste 18: 100–122.

Espinosa-Arrubarrena, L. and S.P. Applegate. 1996. A paleoecological model of the vertebrate bearing beds in the Tlayúa Quarries near Tepexi de Rodríguez, Puebla, Mexico, pp. 539–550. *In*: G. Arratia and G. Viohl [eds.]. Mesozoic Fishes— Systematics and Paleoecology. Verlag Dr. F. Pfeil, Munich.

Fara, E. 2001. What are Lazarus taxa ? Geol. J. 36: 291–303.

Fara, E. 2004. Estimating minimum global species diversity for groups with a poor fossil record: a case study of Late Jurassic–Eocene lissamphibians. Palaeogeogr. Palaeoclim. Palaeoecol. 207: 59–82.

Fara, E., A. Saraiva, D.A. Campos, J.K.R. Moreira, D.C. Siebra and A.W.A. Kellner. 2005. Controlled excavations in the Romualdo Member of the Santana Formation (Araripe Basin, Lower Cretaceous, northeastern Brazil): stratigraphic, palaeoenvironmental and palaeoecological implications. Palaeogeogr. Palaeoclim. Palaeoecol. 218: 145–160.

Fara, E., M. Gayet and L. Taverne. 2007. Les Gonorynchiformes fossiles: distribution et diversité. Cybium 31(2): 125–132.

Fink, S.V. and W.L. Fink. 1981. Interrelationships of the Ostariophysan fishes (Teleostei). J. Linn. Soc. London 72: 297–353.

Fink, S. V. and W. L. Fink. 1996. Interrelationships of Ostariophysan fishes (Teleostei), pp. 209–245. *In*: M.L.J. Stiassny, L.R. Parenti and G.D. Johnson [eds.]. Interrelationships of Fishes. Academic Press, San Diego, California.

Forsskål, P. 1775. Descriptiones animalium avium, amphibiorum, piscium, insectorum, vermium; quae in itinere orientali observavit. Post mortem auctoris edidit Carsten Niebuhr. Hauniae.

Gaudant, J. 1966. Sur la nécessité d'une subdivision du genre *Anaethalion* White (Poisson Téléostéen). C.R. Somm. Scéan. Soc. Géol. Fr. 8: 308–309.

Gaudant, J. 1968. Contribution à une révision des *Anaethalion* de Cérin (Ain). Bull. B.R.G.M., 2e sér., Géol. Gén., Sect. 4: 95–107.

Gaudant, J. 1981a. Nouvelles recherches sur l'ichthyofaune des gypses et des marnes supragypseuses (Eocène supérieur) des environs de Paris. Bull. B.R.G.M., 2e sér., sect. 4: 57–75.

Gaudant, J. 1981b. Contribution de la paléoichthyologie continentale à la reconstitution des paléoenvironnements cénozoïques d'Europe occidentale: approche systématique, paléoécologique, paléogéographique et paléoclimatologique. Thèse Dc d'Etat (Ph.D. thesis), Univ. Paris VI, Paris.

Gaudant, J. 1981c. Mise au point sur l'ichtyofaune oligocène des anciennes plâtrières d'Aix-en-Provence (Bouches-du-Rhône). C.R. Acad. Sci. Paris 292, sér. 3: 1109–1112.

Gaudant, J. 1981d. Nouvelles recherches sur l'ichthyofaune des zones salifères moyenne et supérieure (Oligocène inférieur) du Bassin potassique alsacien. Sci. Géol. Bull. 34: 209–218.

Gaudant, J. and T. Burkhardt. 1984. Sur la découverte de poissons fossiles dans les marnes grises rayées de la zone fossilifère (Oligocène basal) d'Altkirch (Haut-Rhin). Sci. Géol. Bull. 37: 153–171.

Gayet, M. 1981. Considérations relatives à la paléoécologie du gisement cénomanien de Laveiras (Portugal). Bull. Mus. Natn. Hist. Nat., Paris, 4e sér., 3, sect. C., 4(1,2): 405–407.

Gayet, M. 1982. Cypriniformes ou Gonorhynchiformes? *Ramallichthys* nouveau genre du Cénomanien inférieur de Ramallah (Monts de Judée). C.R. Acad. Sci. Paris, 295, sér. 2: 405–407.

Gayet, M. 1985. Gonorhynchiformes nouveau du Cénomanien inférieur marin de Ramallah (Monts de Judée): *Judeichthys haasi* nov. gen. nov. sp. (Teleostei, Ostariophysi, Judeichthyidae nov. fam.). Bull. Mus. Natn. Hist. Nat., 4e sér., 7, sect. C(1): 65–85.

Gayet, M. 1986. *Ramallichthys* Gayet du Cénomanien inférieur marin de Ramallah (Judée), une introduction aux relations phylogénétiques des Ostariophysi. Mém. Mus. Natn. Hist. Nat., n.s. sér. C., Sci. Terre 51: 1–81.

Gayet, M. 1989. Note préliminaire sur le matériel paléoichthyologique éocrétacique du Rio Benito (sud de Bata, Guinée Equatoriale). Bull. Mus. Natn. Hist. Nat., 4e sér., 11, sect. C(1): 21–31.

Gayet, M. 1993a. Nouveau genre de Gonorhynchidae du Cénomanien inférieur marin de Hakel (Liban). Implications phylogénétiques. C.R. Acad. Sci. Paris 316, sér. 2: 257–263.

Gayet, M. 1993b. Gonorhynchoidei du Crétacé supérieur marin du Liban et relations phylogénétiques des Charitosomidae nov. Fam. Doc. Lab. Géol. Lyon 126: 1–128.

Gayet, M. 1993c. Relations phylogénétiques des Gonorhynchiformes (Ostariophysi). Belg. J. Zool. 123: 165–192.

Gayet, M. and F. Meunier. 2003. Paleontology and Palaeobiogeography of Catfishes, pp. 491–524. In: G. Arratia, B.G. Kapoor, M. Chardon and R. Diogo [eds.]. Catfishes. Sci. Publ., Inc, Plymouth, UK.

Géry, J. 1964. Une nouvelle famille de poissons dulçaquicoles africains: les Grasseichthyidae. C.R. Acad. Sci. Paris 259(25): 4805–4807.

Gosline, W.A. 1960. Contributions toward a classification of modern isospondylous fishes. Bull. Br. Mus. (Nat. Hist.) Zool. 6: 325–365.

Grande, L. 1984. Paleontology of the Green River Formation, with a review of the fish fauna. Bull. Geol. Surv. Wyoming 63: 1–333.

Grande, L. and W.E. Bemis. 1998. A comprehensive phylogenetic study of amiid fishes (Amiidae) based on comparative skeletal anatomy. An empirical search for interconnected patterns in natural history. J. Vert. Paleontol. 18(suppl. 1), Mem. 4: 1–690.

Grande, L. and T. Grande. 1999. A new species of †*Notogoneus* (Teleostei: Gonorynchidae) from the Upper Cretaceous Two Medicine Formation of Montana, and the poor Cretaceous records of freshwater fishes from North America. J. Vert. Paleontol. 19: 612–622.

Grande, T. 1996. The interrelationships of fossil and Recent gonorynchid fishes with comments on two Cretaceous taxa from Israel, pp. 299–318. *In*: G. Arratia and G. Viohl [eds.]. Mesozoic Fishes—Systematics and Paleoecology. Verlag Dr. F. Pfeil, Munich.

Grande, T. 1999a. Distribution patterns and historical biogeography of gonorhynchiform fishes (Teleostei: Ostariophysi), pp. 425–444. *In*: G. Arratia and H.P. Schultze [eds.]. Mesozoic Fishes 2—Systematics and Fossil Record. Verlag Dr. F. Pfeil, Munich.

Grande, T. 1999b. Revision of the genus *Gonorhynchus* Scopoli, 1777 (Teleostei, Ostariophysi), Copeia 1999(2): 453–469.

Grande, T. and L. Grande. 2008. Reevaluation of the gonorynchiform genera †*Ramallichthys*, †*Judeichthys* and †*Notogoneus*, with comments on the families †Charitosomidae and Gonorynchidae, pp. 295–310. *In*: G. Arratia, H.-P. Schultze and M.V.H. Wilson [eds.] Mesozoic Fishes 4—Homology and Phylogeny, Verlag Dr. F. Pfeil, Munich.

Grande, T. and F.J. Poyato-Ariza. 1999. Phylogenetic relationships of fossil and Recent gonorynchiform fishes (Teleostei: Ostariophysi). Zool. J. Linn. Soc. 125: 197–238.

Greenwood P.H., D.E. Rosen, S.H. Weitzman and G.S. Myers. 1966. Phyletic studies of teleostean Fishes, with a provisional classification of living forms. Bull. Am. Mus. Nat. Hist. 131: 339–456.

Gronovius, L.T. 1763. Zoophylacium Gronovianum, exhibens animalia quadrupeda, amphibia, pisces, insecta, vermes, mollusca, testacea et zoophyta, quae in Museo suo adservavit, examini subjecit, systematice disposuit atque descripsit. Fasciculus primus exhibens animalia quadrupeda, amphibia atque pisces, quae in Museo suo adservat, rite examinavit, systematice disposuit, descripsit, atque iconibus illustravit. Lugduni Batavorum, Leyde, The Netherlands, pp. 1–136.

Günther, A. 1868. Catalogue of the fishes in the British Museum. Catalogue of the Physostomi, containing the families Heteropygii, Cyprinidae, Gonorhynchidae, Hyodontidae, Osteoglossidae, Clupeidae,... [thru]... Halosauridae, in the collection of the British Museum. Cat. Fishes vol. 7, London, pp. 1–512.

Hanke, G. and V. Karatajute-Talimaa. 2002. *Knerialepis*, a new generic name to replace *Kneria* Karatajute-Talimaa. J. Vert. Paleontol. 22: 703.

Harland, W.B., R.L. Armstrong, A.V. Cox, L.E. Craig, A.G. Smith and D.G. Smith. 1990. A Geologic Time Scale 1989. Cambridge University Press, Cambridge, UK.

Hay, O.P. 1903. On a collection of Upper Cretaceous fishes from Mount Lebanon, Syria, with descriptions of four new genera and nineteen new species. Bull. Am. Mus. Nat. Hist. 19: 395–452.

Heckel, J.J. 1854. Bericht über die vom Herrn Cavaliere Achille de Zigno hier Angelangte Sammlung fossiler Fische. Sitzung. Akad. Wiss., Mat-Nat. Wiss. Class, 1853, 11: 122–138.

Hills, E.S. 1934. Tertiary Fresh Water Fishes from Southern Queensland. Mem. Queensl. Mus. 10: 157–174.

Howes, G.J. 1985. Cranial muscles of gonorhynchiform fishes, with comments on generic relationships. Bull. Br. Mus. Nat. Hist. (Zool.) 49: 273–303.

Hückel, U. 1970. Dis Fischschiefer von Haqel und Hjoula in der Oberkreide des Libanons. N. Jb. Geol. Paläont. Abh. 135: 113–149.

Jablonski, D. 1986. Causes and consequences of mass extinctions: a comparative approach, pp. 183–230. *In*: D.K. Elliott [ed.]. Dynamics of Extinction. Wiley and Sons, New York.

Jordan, D.S. 1910. Description of a collection of fossil fishes from the bituminous shales at Riaco Doce, State of Alagôas, Brazil. Ann. Carnegie Mus. 7: 23–34.

Jordan, D.S. 1919. Fossil fishes from the Pliocene formations of southern California. Stand. Univ. Publ., Univ. Ser., Biol. Sci. 61–98.

Jordan, D.S. 1921a. The fish fauna of the Califormia Tertiary. Stand. Univ. Publ., Univ. Ser., Biol. Sci. 1(4): 233–300.

Jordan, D.S. 1921b. Peixes cretaceos do Ceará e Piauhy. Monogr. Serv. Geol. Miner. Brasil, Rio de Janeiro 3: 1–97.

Jordan, D.S. and J.C. Branner. 1908. The Cretaceous fishes of Ceará, Brazil. Smith. Misc. Coll., Wash. 52: 1–29.

Jordan, D.S. and J.O. Snyder. 1923. Gonorhynchus moseleyi, a new species of herring-like fish from Honolulu. J. Wash. Acad. Sci. 347–350.

Jubb, R.A. and G. Bell-Cross. 1974. A new species of Parakneria Poll 1965 (Pisces, Kneriidae) from Mozambique. Arnoldia, ser. Misc. Publ. 6(29): 1–3.

Karatajute-Talimaa, V. 1997. Taxonomy of loganiid thelodonts. Modern Geol. 21: 1–15.

Kner, R. and F. Steindachner. 1863. Neue Beiträge zur Kenntnis der Fossilen Fische österreichs. Denkschr. K. Akad. Wiss Math-naturwiss., cl. 21: 17–36.

Lacepède, B.G.E. 1803. Histoire Naturelle des Poissons, vol 5. Plassan, Paris, pp. 1–803.

Ladiges, W. and J. Voelker. 1961. Untersuchungen über die Fischfauna in Gebirgsgewässern des Wasserscheidenhochlands in Angola. Mitt. Hamb. Zool. Mus. Inst. 117–140.

Lavoué, S., M. Miya, J.G. Inoue, K. Saitoh, N.B. Ishiguro and M. Nishida. 2005. Molecular systematics of the gonorynchiform fishes (Teleostei) based on whole mitogenome sequences: implications for higher-level relationships within the Otocephala. Mol. Phyl. Evol. 37: 165–77.

Levinton, J. 1988. Genetics, Paleontology and Macroevolution. Cambridge University Press, Cambridge, UK.

Linnaeus, C. 1766. Systema naturae sive regna tria naturae, secundum classes, ordines, genera, species, cum characteribus, differentiis, synonymis, locis. Laurentii Salvii, Holmiae.

Maisey, J. G. 2000. Continental break up and the distribution of fishes of Western Gondwana during the Early Cretaceous. Cret. Res. 21: 281–314.

Malabarba, M.C., D.A. Carmo and I.G. Perez. 2002. New fossil fishes from the Maceió Formation, Alagoas Basin, pp. 303–306. In: J.C Castro, D. Dias-Brito, E.A. Musacchio and R. Rohn [eds.]. Bol. 6° Simp. sobre o Cretáceo do Brasil, UNESP, São Paulo, Brazil.

Malz, H. 1978a. Aquitane Otolithen-Horizonte im Untergrund von Frankfurt am Main. Senckenb. Lethaea 58: 451–171.

Malz, H. 1978b. Vergleichend-morphologische Untersuchungen an aquitanien Fisch-otolithen aus dem Untergrund von Frankfurt am Main. Senckenb. Lethaea 59: 441–481.

Marliere, R. and F. Robaszynski. 1975. Crétacé. Conseil Géologique, Commissions Nationales de Stratigraphie (Service Géologique de Belgique) 9: 1–53.

Martill, D.M. 1993. Fossils of the Santana and Crato formations, Brazil. Palaeontological Association, Field Guide to Fossils, 5, London.

Martin, F and P. Bultynck. 1990. The iguanodons of Bernissart. Inst. Roy. Sci. Nat. Belg.: 1–51.

Medizza, F. and L. Sorbini. 1980. Il giacimento del Salento (Lecce), pp. 131–134. In: I Vertebrati Fossili Italiani. 1980. Catalogo della Mostra, Verona.

Nardon, S. 1990. Il giacimento di Polazzo (Carso Goriziano), pp. 81–84. In: A. Tintori, G. Musciio and F. Bizzarini [eds.]. Pesci Fossili Italiani. Scoperte e riscoperte. New Interlitho, Milano.

Nelson, J.S. 1994. Fishes of the World, 3rd ed. John Wiley and Sons, New York.

Nolf, D. 1977. Les otolithes des Téléostéens de l'Oligo-Miocène belge. Ann. Soc. Roy. Zool. Belg. 106 [1976]: 3–119.

Nolf, D. 1985. Otolithi piscium, pp. 1–145. In: H.-P. Schultze [ed.]. Handbook of Paleoichthyology, vol. 10. G. Fischer-Verlag, Stuttgart.

Norell, M.A. 1993. Tree-based approaches to understanding history: comments on ranks, rules, and the quality of the fossil record. Am. J. Sci. 293: 407–417.

Norman, J.R. 1923. A new cyprinoid fish from Tanganyika Territory, and two new fishes from Angola. Ann. Mag. Nat. Hist. ser. 9: 694–696

Nybelin, O. 1974. A revision of the Leptolepid fishes. Acta Reg. Soc. Sient. Litt. Gothoburgensis, Zool. 9: 5–202.

Ogilby, J.D. 1911. On the genus Gonorynchus (Gronovius). Ann. Queensl. Mus. 30–35.

Oliveira, A.F. de. 1978. O gênero Tharrhias no Cretáceo da Chapada do Araripe. Anal. Acad. Brasil. Ciênc. 50: 537–552.

Patterson, C. 1970. Two upper Cretaceous salmoniform fishes from the Lebanon. Bull. Brit. Mus. (Nat. Hist.), Geol. 19: 207–296.

Patterson, C. 1984a. Chanoides, a marine Eocene otophysan fish (Teleostei, Ostariophysi). J. Vert. Paleontol. 4: 430–456.

Patterson, C. 1984b. Family Chanidae and other Teleostean Fishes as living fossils, pp. 132–139. In: N. Eldredge and S.M. Stanley [eds.] Living fossils. Springer-Verlag, New York.

Pellegrin, J. 1905. Poisson nouveau du Mozambique. Bull. Mus. natl. Hist. Nat.: 145–146

Pellegrin, J. 1925. Poissons du nord du Gabon et de la Sangha recueillis par M. Baudon. Description de deux espèces et d'une variété nouvelle. Bull. Soc. Zool. Fr. 50: 97–106.

Pellegrin, J. 1931. Poissons du Kouilou et de la Nyanga recueillis par M. A. Baudon. Bull. Soc. Zool. Fr. 205–211

Penrith, M.J. 1973. A new species of Parakneria from Angola (Pisces: Kneriidae). Cimbebasia ser. A: 131–135

Perkins, P.L. 1970. Notogoneus osculus Cope, an Eocene fish from Wyoming (Gonorhynchiformes, Gonorhynchidae). Postilla 147: 1–18.

Pictet, F.J. and A. Humbert. 1866. Nouvelles recherches sur les poissons fossiles du Mont Liban, George éd., Geneva.

Poll, M. 1944. Descriptions de poissons nouveaux recueillis dans la région d'Albertville (Congo belge) par le Dr. G. Pojer. Bull. Mus. Roy. Hist. nat. Belg. 20: 1–12.

Poll, M. 1965. Contribution à l'étude des Kneriidae et description d'un nouveau genre, le genre Parakneria (Pisces, Kneriidae). Mém. Acad. Roy. Belg. 36: 1–28.

Poll, M. 1967. Contribution à la faune ichthyologique de l'Angola. Publ. Cult. Cia. Diamantes Angola 1–381.

Poll, M. 1969. Contribution à la connaissance des Parakneria. Rev. Zool. Bot. Afr. 80: 359–368.

Poll, M. 1976. Poissons. Exploration Parc National Upemba Mission G. F. de Witte. Explor. Parc Upemba 1–127.

Poll, M. 1984. Parakneria tanzaniae, espèce nouvelle des chutes de la rivière Kimani, Tanzanie (Pisces, Kneriidae). Rev. Zool. Afr. 98: 1–8.

Poll, M. and D.J. Stewart. 1975. Un Mochocidae et un Kneriidae nouveaux de la rivière Luongo (Zambia), affluent du bassin du Congo (Pisces). Rev. Zool. Afr. 89: 151–158.

Poyato-Ariza, F.J. 1994. A new Early Cretaceous gonorynchiform fish (Teleostei: Ostariophysi) from Las Hoyas (Cuenca, Spain). Occ. Pap. Mus. Nat. Hist., Univ. Kansas 164: 1–37.

Poyato-Ariza, F.J. 1996a. A revision of the ostariophysan fish family Chanidae, with special reference to the Mesozoic forms. Palaeo Ichthyologica 6: 5–52.

Poyato-Ariza, F.J. 1996b. The phylogenetic relationships of Rubiesichthys gregalis and Gordichthys conquensis (Teleostei, Ostariophysi), from the Early Cretaceous of Spain, pp. 329–348. *In*: G. Arratia and G. Viohl [eds.]. Mesozoic Fishes: Systematics and Paleoecology. Verlag Dr. Friedrich Pfeil, Munich.

Poyato-Ariza, F.J. 1996c. A revision of *Rubiesichthys gregalis* Wenz 1984 (Ostariophysi, Gonorynchiformes), from the Early Cretaceous of Spain, pp. 319–328. *In*: G. Arratia and G. Viohl [eds.]. Mesozoic Fishes—Systematics and Paleoecology. Verlag Dr. Friedrich Pfeil, Munich.

Poyato-Ariza, F.J. and S. Wenz. 1990. La ictiofauna española del Cretácico inferior, pp. 299–311. *In*: J. Civis Llovera and J.A Flores [eds]. Acta de Paleont. 68, Salamanca, Spain.

Poyato-Ariza, F.J., M.R. Talbot, M.A. Fregenal-Martínez, N. Meléndez and S. Wenz. 1998. First isotopic and multidisciplinary evidence for nonmarine coelacanths and pycnodontiform fishes: palaeoenvironmental implications. Palaeogeogr. Palaeoclim. Palaeoecol. 144: 65–84.

Priem, F. 1900. Sur les poissons fossiles du gypse de Paris. Bull. Soc. Géol. Fr. (3)28: 841–860.

Priem, F. 1908. Étude des poissons fossiles du Bassin parisien. Ann. Paléont. 1–144.

Priem, F. 1911. Étude des poissons fossiles du Bassin parisien (supplément). Ann. Paléontol. 6: 1–44.

Radovcic, J. 1975. Some new Upper Cretaceous teleosts from Yugoslavia with special reference to localities, geology and palaeoenvironment. Paleontol. Jugoslav. 17: 7–55.

Richardson, J. 1844-48. Ichthyology of the voyage of H.M.S. Erebus and Terror, pp. 1–139. *In*: J. Richardson and J.E. Gray [eds.]. The zoology of the voyage of H. H. S. "Erebus and Terror," under the command of Captain Sir J.C. Ross during the years 1839–43. Ichthy. Erebus and Terror, London.

Riede, K. 2004. Global register of migratory species—from global to regional scales. Final Report of the R&D-Projekt 808 05 081. Fed. Agency Nat. Cons. Bonn.

Roberts, T.R. 1972. Ecology of fishes in the Amazon and Congo basins. Bull. Mus. Comp. Zool. Harv. 143: 117–147.

Roberts, T.R. 1975. Geographical distribution of African freshwater fishes. Zool. J. Linn. Soc. 57: 249–319.

Rosen, D.E. and P.H. Greenwood. 1970. Origin of the Weberian apparatus and the relationships of the Ostariophysan and Gonorhynchiform fishes. Am. Mus. Novitates. 2428: 1–25.

Santos, R. da Silva. 1947. Uma redescrição de *Dastilbe elongatus*, com algumas considerações sobre o gênero *Dastilbe*. DNPM, Div. Geol. Min. Bras., Notas Prelim. Est. 42: 1–7.

Santos, R. da Silva. 1955. Descrição dos peixes fósseis. Boll. Div. Geol. Miner. Bras. 155: 1–27.

Santos, R. da Silva. 1975. Peixes da Formação Marizal, Estado da Bahia. Ph.D. thesis, Univ. do Estado de Sâo Paulo, Brazil.

Santos, R. da Silva. 1990. Clupeiformes e Gonorhynchiformes do Cretáceo Inferior (Aptiano) da Formaçâo Cabo, Nordeste do Brasil. An. Acad. Brasil. Ciênc. 62: 261–268.

Santos, R. da Silva and L. Duarte. 1962. Fósseis do Arenito Açu. Col. Mossoroense, sér. B, 62: 156–174.

Santos, R. da Silva and J.G. Valença. 1968. A Formação Santana e sua paleoictiofauna. An. Acad. Brasil. Ciênc. 40: 339–360.

Sanz, J.L., S. Wenz, A. Yebenes, R. Estes, X. Matinez-Delclos, E. Jimenez-Fuentes, C. Diéguez, A.D. Buscalioni, L.J. Barbadillo and L. Via. 1988. An Early Cretaceous faunal and floral continental assemblage: Las Hoyas fossil site (Cuenca, Spain). Geobios 21: 611–635.

Schaeffer, B. 1947. Cretaceous and Tertiary actinopterygian fishes from Brazil. Bull. Am. Mus. Nat. Hist. 89: 1–39.

Schwarzhans, W. 1974. Die Otolithen-Fauna des Chatt A und B (Oberoligozän, Tertiär) vom Niederrhein, unter Einbeziehung weiterer Fundstellen. Decheniana 126: 91–132.

Schwarzhans, W. 1994. Die Fisch-Otolithen aus dem Oberoligozän der Niederrheinischen Bucht. Systematik, Paläoökologie, Paläobiogeographie, Biostratigraphie und Otolithen-Zonierung. Geol. Jahrb., A 140: 3–248.

Scopoli, G.A. 1777. Introductio ad historiam naturalem, sistens genera lapidum, plantarum et animalium hactenus detecta, caracteribus essentialibus donata, in tribus divisa, subinde ad leges naturae. Prague, pp. 1–506.

Scorziello, R. 1980. L'Ittiofauna di Pietraroia (Benevento), pp. 111–114. In: I vertebrati fossili italiani. Catalogo della Mostra, Verona.

Seegers, L. 1995. Revision of the Kneriidae of Tanzania with description of three new Kneria-species (Teleostei: Gonorhynchiformes). Ichthyol. Explor. Freshwater 6: 97–128.

Siegfried, P. 1954. Die Fisch-Fauna des Westfälischen Ober-Senons. Palaeontographica, A 106: 1–36.

Signeux, J. 1961. Sur quelques poissons fossiles du Bassin Parisien. Bull. Soc. Géol. Fr., Paris, 7ᵉ sér. 8: 417–423.

Sittler, C. 1965. Le Paléogène des fossés rhénan et rhodanien. Études sédimentologiques et paléoclimatiques. Mém. Service Carte Géol. Alsace-Lorraine 24: 1–392.

Sittler, C. 1972. Le Sundgau: aspect géologique et structural. Sci. Géol. Bull. 25: 93–118.

Smith, A.B. 1988. Patterns of diversification and extinction in Early Palaeozoic echinoderms. Palaeontology 31: 799–828.

Smith, A.B. 1994. Systematics and the fossil record: documenting evolutionary patterns. Blackwell Science, Oxford, UK.

Soares, R.M.C. and M.E.V. Calheiros. 1991. Notas sobre a Bacia Sergipe-Alagoas: Cretáceo Inferior da porção Alagoana. Geociências, São Paulo 10: 211–229.

Sorbini, L. 1978. New fish bed localities of latest Campanian age (Lecce-South Italy). A preliminary paper. Boll. Mus. Civ. St. Nat. Verona 5: 607–608.

Sorbini, L. 1980. Il giacimento di Chiavon (Vicenza), pp. 177–179. In I vertebrati fossili italiani. Catalogo della Mostra, Verona.

Steindachner, F. 1866. Ichthyologische Mittheilungen (IX). Verh. K.-K. Zool.-Bot. Ges. Wien 761–796

Stinton, F.C. 1977. Fish otolith from the English Eocene, II. Palaeontogr. Soc. (Monogr.) 57–126.

Sytchevskaya, E.K. 1986. Presnovodnaya paleogenovaya ikhtiofauna SSSR i Mongolii [Paleogene freshwater fish fauna of the USSR and Mongolia). Trudy-Sovmestanaya Sovetsko-Mongo'skaya Nauchno-Issledovatel'skaya Nauchno-Issleddovatel (stov "Nauka") 29: 1–157.

Taverne, L. 1974. Parachanos Arambourg et Schneegans (Pisces, Gonorhynchiformes) du Crétacé inférieur du Gabon et de Guinée Equatoriale et l'origine des Téléostéens Ostariophysi. Rev. Zool. Afr. 88: 683–688.

Taverne, L. 1975. À propos de trois téléostéens fossiles déterminés erronément comme ostéoglossides, Cearana Jordan, D. S. et Branner, J. C. 1908, Eurychir Jordan, D. S. 1924, et Genartina Frizzell, D. L. et Dante, J. H. 1965. Ann. Soc. Roy. Zool. Belg. 105: 15–30.

Taverne, L. 1976a. Les téléostéens fossiles du Crétacé moyen de Kipala (Kwango, Zaïre). Ann. Mus. Roy. Af. Centr., Tervuren, sér. In –8°, Sci. Géol. 79: 1–50.

Taverne, L. 1976b. La faune paléoichthyologique du Crétacé moyen de Kipala (Kwango, Zaïre) et son intérêt géochronologique et paléobiogéographique. Rev. Trav. Inst. Pêches Marit. 40: 762–763.

The Fossil Record of Gonorynchiformes   225

Taverne, L. 1977. Ostéologie de *Clupavus maroccanus* (Crétacé supérieur du Maroc) et considérations sur la position systématique et les relations des *Clupavidae* au sein de l'ordre des Clupéiformes *sensu stricto* (Pisces, Teleostei). Geobios 10(5): 697–722

Taverne, L. 1981. Ostéologie et position systématique d'*Aethalionopsis robustus* (Pisces, Teleostei) du Crétacé inférieur de Bernissart (Belgique) et considération sur les affinités des Gonorhynchiformes. Acad. Roy. Belg., Bull. Cl. Sci., 5ᵉ sér. 68: 958–982.

Taverne, L. 1984. À propos de *Chanopsis lombardi* du Crétacé inférieur du Zaïre. Rev. Zool. Afr. 98: 578–590.

Taverne, L. 1997. Les poissons crétacés de Nardo. 4°. *Apulichthys gayeti* gen. nov., sp. nov. (Teleostei, Ostariophysi, Gonorhynchiformes). Boll. Mus. Civ. St. Nat. Verona 21: 401–436.

Taverne, L. 1998. Les poissons crétacés de Nardo. 7°. *Lecceichthys wautyi* gen. nov., sp. nov. (Teleostei, Ostariophysi, Gonorhynchiformes) et considération sur la phylogénie des Gonorhynchidae. Boll. Mus. Civ. St. Nat. Verona 22: 291–316.

Taverne, L. 2005. Les poissons crétacés de Nardo. 20°. *Chanoides chardoni* sp. nov. (Teleostei, Ostariophysi, Otophysi). Boll. Mus. Civ. St. Nat. Verona, Geol. Paleont. Preist. 29: 39–54.

Taverne, L. and M. Gayet. 2006. Nouvelle description d'†*Halecopsis insignis* de l'Éocène marin de l'Europe et les relations de ce taxon avec les Gonorhynchiformes (Teleostei, Ostariophysi). Cybium 30: 109–114.

Temminck, C.J. and H. Schlegel. 1846. Pisces. *In*: Fauna Japonica, sive descriptio animalium quae in itinere per Japoniam suscepto annis 1823–30 collegit, notis observationibus et adumbrationibus illustravit. P.F. de Siebold, Part 10–14, pp. 173–269.

Thys van den Audenaerde, D.F.E. 1961. L'anatomie de *Phractolaemus ansorgi* Blgr. et la position systématique des Phractolaemidae. Ann. Mus. Roy. Afr. Centr., Tervuren, in –8°, Sci. Zool. 103: 101–167.

Traquair, R.H. 1911. Les poissons wealdiens de Bernissart. Mém. Mus. Roy. Hist. Nat. Belg. 6: 1–65.

Trewavas, E. 1936. Dr. Karl Jordan's expedition to South-West Africa and Angola: The fresh-water fishes. Novit. Zool. 63–74.

Varrichio, D. and J.R. Horner. 1993. Hadrosaurid and lambeosaurid bone beds from the Upper Two Medecine Formation of Montana: taphonomic and biologic implications. Canad. J. Earth Sci. 30: 997–1006.

von der Marck, W. 1885. Fische der Oberen Kreide Westfalen. Palaeontographica 31: 233–267.

von der Marck, W. 1894. Vierter Nachtrag zu: Die fossilen Fische des Westphälischen Kreide. Palaeontographica 41: 41–48.

von der Meyer, H. 1848. Die fossilen Fische aus dem Tertiär-Thone von Unter-Kirchberg. N. Jb. Min. Geogr. Geol. 1848: 781–784.

Weiler, W. 1920. Die Septarientonfische des Mainzer Beckens. Eine vorläufige Mitteilung. Jahr. Nass. Ver. Natur. 72: 2–15.

Weiler, W. 1922. Die Fischreste aus den bituminösen Schiefern v. Ibando bei Bata (Spanish Guinea). Palaont. Zeitschr. 5: 148–160.

Weiler, W. 1928. Beiträge zur Kenntnis der tertiären Fische des mainzer Beckens II. 3. Teil. Die Fische des Septarientones. Abh. Hess. Geol. Landesanst. Darmstadt 8, 3: 1–63.

Weiler, W. 1942. Die Otolithen des rheinischen und nordwestdeutschen Tertiärs. Abh. Reichs. Bodenforsch. Neue Folge 206: 1–140.

Weiler, W. 1963. Die Fischfauna der Tertiäres im oberrheinischen Graben des Mainzer Beckens, des unteren Maintals und der Wetterau, unter besonderer Berücksichtigung des Untermiozänes. Abh. Senckenb. Naturf. Gesel. 504: 1–75.

Wenz, S. 1984. *Rubiesichthys gregalis* n. g., n. sp., Pisces Gonorhynchiformes, du Crétacé inférieur du Montsech (Province de Lérida, Espagne). Bull. Mus. Natn. Hist. Nat., 4ᵉ sér., 6, sect. C(3): 275–285.

Wenz, S. 1991. Lower Cretaceous fishes from the Serra del Montsec (Spain). The Lower Cretaceous lithographic limestones of Montsec. Ten years of paleontological expeditions pp. 73–84.

Whitfield, R.P. 1890. Observations on a fossil fish from the Eocene beds of Wyoming, Bull. Am. Mus. Nat. Hist. III(1): 117–120.

Wilson, M.V.H. 1980. Oldest known *Esox* (Pisces: Esocidae), part of a new Paleocene teleost fauna from western Canada. Canad. Earth Sci. 17: 307–312.

Wilson, M.V.H. 1981. Eocene freshwater fishes from the Coalmont Formation, Colorado. J. Paleont. 55: 671–674.

Woodward, A.S. 1895. Catalogue of the fossil fishes in the British Museum (Natural History), Part. III. Trust. Brit. Mus. (Nat. Hist.), London.

Woodward, A.S. 1896. On some extinct fishes of the Teleostean Family Gonorhynchidae. Proc. Zool. Soc. London 500–504.

Woodward, A.S. 1898. Notes on some type specimens of Cretaceous fishes from Mount Lebanon in the Edinburgh Museum of Science and Art. Ann. Mag. Nat. Hist. ser 7, 2: 405–414.

Woodward, A.S. 1901. Catalogue of the fossil fishes in the British Museum (Natural History), Part IV. Trust. Brit. Mus. (Nat. Hist.), London.

Woodward, A.S. 1907. On a new Leptolepid fish from the Weald Clay of Southwater, Sussex. Ann. Mag. Nat. Hist. (7), 20: 93–95.

Yabumoto, Y. 1994. Early Cretaceous freshwater fish fauna in Kyushu, Japan. Bull. Kitak. Mus. Nat. Hist. 13: 107–254.

Yans, J., J. Dejax, D. Pons, C. Dupuis and P. Taquet. 2005. Implications paléontologiques et géodynamiques de la datation palynologique des sédiments à faciès wealdien de Bernissart (bassin de Mons, Belgique). C.R. Palevol. 4: 135–150.

Yans, J., J. Dejax, D. Pons, L. Taverne and P. Bultinck. 2006. The iguanodons of Bernissart (Belgium) are middle Barremian to earliest Aptian in age. Bull. Inst. Roy. Sci. Nat. Belg. 76: 91–95.

7

Gonorynchiform Interrelationships: Historic Overview, Analysis, and Revised Systematics of the Group

Francisco José Poyato-Ariza[1], Terry Grande[2] and Rui Diogo[3]

Abstract

The interrelationships of the gonorynchiform fishes, including fossil and Recent genera, are revised and updated. A historic summary of the different hypotheses of their phylogenetic relationships is presented. The morphological disparity of the Gonorynchiformes, as known today, prevented its two longest-known genera, *Chanos* and *Gonorynchus*, from being gathered into the same order until 1960, although their kinship had already been proposed in 1846. Discoveries of more Recent forms, and of their impressive fossil record, have added both abundant data and new challenges to the study of their phylogenetic relationships.

In addition to the historic outline, we present herein a new, entirely revised cladistic analysis of the Gonorynchiformes, including a total of 24 nominal fossil and Recent genera. Only seven of the gonorynchiform genera are living taxa, evidencing that this is a group with a diverse fossil

[1] Unidad de Paleontología, Departamento de Biología, Universidad Autónoma de Madrid, Cantoblanco, 28049-Madrid, Spain. e-mail: francisco.poyato@uam.es
[2] Department of Biology, Loyola University of Chicago, 1032 West Sheridan Road, Chicago, Illinois 60626, USA. e-mail: tgrande@luc.edu
[3] Center for the Advanced Study of Hominid Paleobiology, Department of Anthropology, The George Washington University, 2110 G St. NW, Washington DC 20052, USA. e-mail: rui_diogo@hotmail.com

record. The fossil record of the Gonorynchiformes, as an ensemble, is quite geographically widespread for such a small group, extending to Europe, North and South America, Africa, Australia, and Asia; it dates back to the Early Cretaceous, so that gonorynchiforms may be considered a relict or "living fossil" group.

Previous evidence indicates that the Gonorynchiformes are sister group to the Otophysi, (i.e., fishes with a functioning Weberian apparatus), together forming the clade Ostariophysi. The results of the present analysis, consisting of 130 revised characters, including osteological and myological ones, confirm that the Gonorynchiformes comprise a monophyletic group that consists of two monophyletic, sister-group suborders, the Chanoidei and the Gonorynchoidei. The relationships of the three families {Chanidae + [Gonorynchidae + Kneriidae]} are confirmed. Each family has at least one Recent genus, and the Kneriidae have no known fossil record.

An evaluation of the evolutionary history of the Gonorynchiformes and a systematic revision are also provided in this chapter. Future research on this group should include an evaluation of some problematic taxa, mostly †*Dastilbe* and the Middle Eastern fossil gonorynchids. A separate approach to the interrelationships of the Chanoidei and the Gonorynchoidei, the latter including molecular studies, may help unravel further elements of phylogenetic uncertainty.

Introduction: A Historic Review

The phylogenetic interrelationships, and consequently the evolutionary history, of the Recent and fossil genera that are nowadays included in the order Gonorynchiformes have been a matter of debate for over 150 years. The type species has been known since the Systema Naturae of Linnaeus (1766), yet its alignment with other forms was not widely accepted until the 1960s. The main reason probably lies in that their synapomorphies, or derived anatomic similarities, are far from obvious or evident. Their varied external aspect, wide-ranging behavior, distinct distribution, sparse fossil record, even their overall anatomic features are very diverse, so that this group is superficially very heterogeneous, even though there are about 24 nominal genera included at present (most likely estimation, including problematic genera and possible synonyms, pending revision of challenging forms). Their resemblance, and therefore their kinship, are not easily seen and have eluded researchers for a very long time. Gonorynchiforms do share an evolutionary history as a group, but this has been masked by the distinct evolutionary history of each particular taxon, which resulted in profound particular modifications and adaptations. In addition, gonorynchiform fossils were quite scarce until a few decades ago, were not always introduced in the discussion of the interrelationships together with the Recent taxa, or were misinterpreted when considered. For a long time, the interrelationships of the gonorynchiforms were thus understood almost exclusively in terms of living taxa, and therefore an understanding of the evolutionary history of the group was not as complete as it could be.

Much uncertainty has surrounded gonorynchiforms, and the hypotheses about their relationships were, for a long time, very diverse. The approach to their study can be confusing, even overwhelming. Before getting into the new cladistic analyses performed herein, it is the purpose of the first part of this chapter to summarily review the main hypotheses of the phylogenetic relationships of the Gonorynchiformes from a historical perspective. One can roughly distinguish three historical phases, which largely overlap each other: the first, when hypotheses were mostly or exclusively based on the known Recent taxa; the second, when fossils were slowly incorporated in the research; and the third, when cladistic methodology is consistently applied.

The Kinship of Recent Gonorynchiforms

Historically, most genera now recognized as gonorynchiforms were aligned with other taxa; this is especially true among the Recent genera. Possibly because of the apparent morphological variation among groups, most researchers could not acknowledge that taxa such as *Chanos*, *Gonorynchus*, *Grasseichthys*, *Kneria*, *Parakneria*, and *Phractolaemus* could belong to one and the same evolutionary lineage.

The two more prominent genera in the systematic literature are *Chanos* and *Gonorynchus*. They were related one way or another to the Clupeiformes, but not necessarily to one another. Historically, clupeiforms formed a considerably heterogeneous group that gathered a large and diverse array of teleosts, and these included *Chanos* and *Gonorynchus*. Surprisingly enough, though, the "modern" idea of a close relationship between *Chanos* and *Gonorynchus* dates back to Cuvier and Valenciennes (1846), who first thought that they were more closely related to each other than to any other fish. They considered that these two genera "se rapprochent, en effet, par la grandeur de leur membrane branchiostège qui enveloppe le dessous du cou. Ces deux genres n'ont point des dents" ("they are close, effectively, by the largeness of the membrane that covers the lower part of the neck. These two genera lack teeth") (Cuvier and Valenciennes 1846: 179). They were both distinctly gathered in a family called Lutodeires, and, together with a variety of other families (namely "Mormyres, Hyodontes, Butirins, Élopiens, Amies, Chirocentres, Alépocéphales"; p. 145), placed among those "Familles de Malacoptérygiens, intermediaires entre les Brochets (= Lucioïdes) et les Clupes" ("Families of malacopterygians, intermediates between the pikes (=Lucioids) and the sardines") (p. 145). Furthermore, and more interestingly, the French naturalists gave a discussion of this taxonomic arrangement in terms of what were considered "natural groups" more than 160 years ago: by taking *Chanos* and *Gonorynchus* (and those other taxa) out of the "clupéoïdes", the resulting new, rearranged families and groups "restent alors mieux circonscrits et deviennent des véritables familles naturelles" ("remain therefore better defined, and become true natural families") (p. 148).

Unfortunately, this idea of a close relationship between *Chanos* and *Gonorynchus* was not followed at all after Cuvier and Valenciennes (1846). For instance, according to Günther (1868), *Chanos* was a clupeid fish. Other authors separated *Chanos* and *Gonorynchus* from the evolutionary lineage of the clupeids and their allies, but a close relationship between the two genera was not considered. For instance, Gill (1872), simply mentioned distinct Chanoidae (sic) and Gonorhynchidae (sic) in his arrangement of families of fishes. The Chanoidae was his family number 168, between Elopidae and Dussumeridae; the Gonorhynchidae, number 164, between Alepocephalidae and Hyodontidae. The two families are not far from each other numerically, but are nonetheless distinct and separated by other families resulting in no special relationships between the two. The same relationship is maintained in his subsequent arrangement of fish families (Gill 1893).

A few years later, Günther (1880) placed *Chanos* among the Clupeidae (22nd family of his large order Physostomi), together with, and therefore related to, genera such as *Albula*, *Elops*, *Megalops*, and *Clupea*, while *Gonorhynchus* (sic) was again placed in its own family, the monospecific Gonorhynchidae (sic; 18th family of the Physostomi). In turn, the then recently described *Kneria* was placed in a monospecific family of its own as well, the Kneriidae (fourth family of the Physostomi). As we can see, there was no special relationship acknowledged among these three genera, as they were simply part of a huge array of highly varied fishes forming a very heterogeneous group.

Subsequently, and for a long time, *Chanos* was considered a clupeiform fish by many authors (i.e., Perrier 1903, Gregory 1933, Rabor 1938, and Schuster 1960). But the different hypotheses (often simply inferred) of the relationships concerning *Chanos* and *Gonorynchus* in the literature of the early 20th century are complicated to follow up and summarize coherently, because these hypotheses are anything but consistent. For instance, Jordan (1923), among others, considered both genera somehow related, but only because they were thought to be distinct families within the Clupeoidei. *Gonorynchus* is, nonetheless, so peculiarly different from other fishes that it has also been occasionally assigned to the Cypriniformes (together with the Kneriidae: Perrier 1903) or to a group of its own, although considered of uncertain affinities, or "isolated" (Gregory 1933: 175).

The relationships of other Recent gonorynchiform genera were even more diversely treated in the literature prior to the 1960s. For instance, Ridewood (1905) already suggested a kinship between *Gonorynchus* and *Phractolaemus*, but explicitly objected to the kinship between *Gonorynchus* and *Chanos* proposed by Cuvier and Valenciennes (1846), whereas *Cromeria* was thought by this author not to have any special relationship with the other genera. As in this case, *Cromeria* has usually been placed far from *Gonorynchus* (e.g., Jordan 1923) and even allied to the "salmonoids", with uncertain affinities, by Gregory (1933), for instance.

The old hypothesis of Cuvier and Valenciennes (1846) was not really reconsidered until Gosline (1960). This author proposed, for the first time in more than 100 years, a close relationship between *Chanos* and *Gonorynchus*, and also of both genera with *Cromeria*, *Kneria*, and *Phractolaemus*. They still formed five separate superfamilies of the suborder Gonorhynchoidei (sic), but, not surprisingly at the time, still within the order Clupeiformes. This hypothesis of phylogenetic relationships among these five genera (plus the subsequently described *Grasseichthys*) was later developed and discussed at length by Greenwood *et al.* (1966), who gathered them in their newly defined order Gonorynchiformes. The ordinal taxon had been formally erected by Berg (1940), but these authors gave it a new sense. Greenwood *et al.* (1966), breaking a long tradition, clearly separated Recent gonorynchiforms from the evolutionary lineage Clupeomorpha (Fig. 7.1). They related the Gonorynchiformes with the Protacanthopterygii, both sharing a salmonoid ancestral group with myctophoids, †Ctenothrissiformes, and neoscopelid-like fishes. There was also a hint of a vague relationship with the Ostariophysi, which were nonetheless still kept as a distinct lineage (Fig. 7.1). It must be noted that the common descent of gonorynchiforms and ostariophysans from a shared stem was explicitly acknowledged for the first time, with both groups related by descent from a common salmoniform ancestry.

This hypothesis of kinship among the Recent gonorynchiform genera was broadly accepted and followed later on. For instance, according to Monod (1968), "Dans l'ensemble *Phractolaemus* et *Gonorhynchus* (sic) sont très voisins et les deux familles doivent être rapprochées" ("In overall, *Phractolaemus* and *Gonorhynchus* (sic) are very close and the two families must be related") (p. 241); in fact, *Chanos*, *Cromeria*, *Phractolaemus*, *Grasseichthys*, and *Gonorhynchus* (sic) are presented as monotypic and closely related families. This was supported by additional studies of Rosen and Greenwood (1970), as well as the sister-group relationship between the Gonorynchiformes (Anotophysi) and the Otophysi (those fish with a functioning Weberian apparatus). This kinship of the gonorynchiforms with the otophysans was, however, a matter of intense debate for some time, and the classification by Gosline (1971) still relating the Gonorhynchoidei (sic)

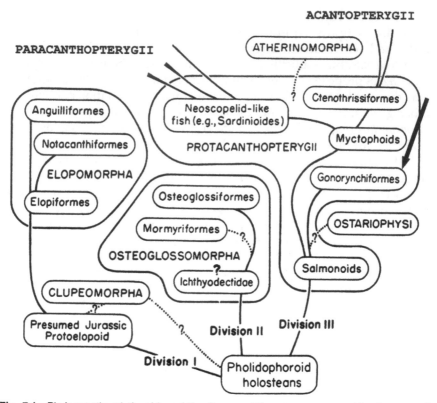

Fig. 7.1 Phylogenetic relationships of the Gonorynchiformes as proposed by Greenwood *et al.* (1966). The arrow indicates this order, which is closer to, but clearly separated from, the Ostariophysi. Modified from Greenwood *et al.* (1966: fig. 1).

with the Clupeoidei and the Elopoidei within the order Clupeiformes remained. Roberts (1973) pointed out that "placing the Gonorynchiformes in Ostariophysi could make the Ostariophysi (otherwise the most clearly defined higher teleostean taxon) an unnatural group" (p. 374). According to this author, these relationships should be considered with reservations, so the gonorynchiforms were placed in a distinct order, well separated from the ostariophysans, which in turn were grouped in an "order Cypriniformes" (p. 377). The precise phylogenetic relationships of the gonorynchiforms were, nonetheless, not dealt with. This issue would not be satisfactorily solved until cladistic methodology was applied later. In these debates, fossil gonorynchiforms, when correctly interpreted, made a major contribution to the understanding of the evolutionary history and phylogenetic relationships of this group.

Phylogenetic History of Fossil Forms

Fossils have not always been taken into account in the study of gonorynchiform relationships. Whenever included and interpreted correctly, a better understanding of the phylogenetic relationships among Recent forms, including *Chanos* and *Gonorynchus*, often resulted. Historically though, fossils have not always been interpreted the way they are nowadays and, like Recent taxa, have a complex history. For instance, Zittel (1893) gathered †*Caeus* (later to be replaced by †*Parachanos*), †*Prochanos* and †*Hypsospondylus* (*nomen nudum*), together with the fossil and Recent species of the genus *Chanos*, in the "subfamily Chanina". This subfamily was included in the very large family Clupeidae, the eighth one of the "order Physostomi", placed between the Chirocentridae and the Salmonidae. In turn, †*Charitosomus* was considered vaguely related to them, as part of the subfamily Thrissopina within the same, large, heterogeneous Clupeidae. Finally, †*Notogoneus* was considered very closely related to *Gonorhynchus* (sic), forming the Gonorhynchidae (sic), the 14th family of the same order Physostomi, placed between the Cyprinoidae and the Murenidae. As a consequence, their relationships with *Chanos* and allies were very vague and distant again.

The relationship of †*Notogoneus* and *Gonorynchus* was confirmed by Woodward (1896), following the hints provided by Cuvier (1822 *fide* Woodward 1896). Later on, Woodward (1901) included †*Chanoides* and †*Prochanos*, together with *Chanos*, in the family Albulidae, quite distinct from the family Gonorhynchidae (sic), which included †*Charitosomus* and †*Notogoneus* together with *Gonorhynchus* (sic). Both families were distantly related within the very large suborder Isospondyli; the Albulidae were placed between the Elopidae and the Osteoglossidae, and the Gonorhynchidae (sic), between the Scopelidae and the Chirothricidae. *Chanos* was also considered an Albulidae by Eastman (1905), although this albulid kinship had been contested by Ridewood (1904a, b), who related this genus and its fossil allies to the clupeoids again. Also nearer to more traditional views, the huge order of the Clupeiformes *sensu* Goodrich (1909) contained the more or less related but distinct families Gonorhynchidae (sic) with *Gonorhynchus* (sic), †*Notogoneus*, and †*Charitosomus*; Cromeridae (sic); and Phractolaemidae. *Chanos* was placed, together with †*Prochanos*, in the subfamily Chaninae, within the family Clupeidae. And the Kneriidae was placed with the Esociformes. No close kinship was recognized among these gonorynchiform taxa.

The whole fossil picture is, unfortunately, quite complex. Traditionally, many small to medium-sized teleosts found in Mesozoic beds were in the early 20th century consistently included in the vast, heterogeneous group of the †"leptolepids", thus misleading those who studied their evolutionary history. Following this scheme, the fossil chanid †*Tharrhias* was a †"Leptolepidae" according to its original description by Jordan and Branner (1908) and later assigned to the Chanidae within the Clupeiformes (e.g., de Oliveira 1978). Later on, †*Charitosomus* and †*Notogoneus* were related to the Gonorynchidae, but in a distinct †Notogoneidae family listed just afterwards, within the Clupeoidei, by Jordan (1923). In contrast, these two fossil genera were subsequently included in the Gonorynchoidea by Gregory (1933).

In turn, another chanid, †*Dastilbe*, was initially described as a Clupeidae (Jordan 1910), not surprisingly. It is notable that the relationship of †*Dastilbe* with *Chanos* was already acknowledged by Arambourg (1935), who included it in the "Chanoïnés" (followed by Silva-Santos 1947, 1979, and others), together with, and related to, †*Parachanos* and *Chanos*. According to Arambourg (1935), the Cretaceous fossils from Africa (†*Parachanos* and others) are a link between †"Leptolépidés" and "Clupéidés", allowing "de montrer le polyphylétisme (…) de ces derniers" ("to show the polyphyletism (...) of the latter") (p. 1). In contrast with this view, the subsequent classification by Bertin and Arambourg (1958) included the Chanoidei (just after the Clupeoidei) in the very large order Clupeiformes. According to these authors, the Chanoidei gathered three families: the monospecific Kneriidae and Phractolaemidae plus the Chanidae, consisting of *Chanos*, †*Dastilbe*, †*Parachanos*, and †*Prochanos*. *Cromeria*, although closely related, formed the next, monospecific suborder. The Gonorhynchidae (sic), forming a monotypic suborder, appeared in their classification after the Salmonoidei and the Cromerioidei, gathering †*Charitosomus*, *Gonorhynchus* (sic), and †*Notogoneus*. This arrangement seems to follow Berg (1940) rather than Arambourg (1935). Berg (1940) again related the Chanoidei (*Chanos*, †*Dastilbe*, and †*Parachanos*), Kneriidae, Phractolaemoidei, and Cromerioidei among them and to the subsequent Salmonoidei, quite far from the Gonorhynchoidei (sic), formed by Gonorhynchidae (sic) plus †Notogoneidae. All of them were, once again, just vaguely related within the heterogeneous Clupeiformes.

This clupeomorph kinship was, not surprisingly, maintained in some palaeontological works for some time (e.g., Lehman 1966), whereas a salmoniform kinship was proposed for the Chanoidei plus Gonorhynchoidei (sic) by Taverne (1974a). The ostariophysan kinship was explicitly rejected by Taverne (1974b), while the close affinity of Chanoidei and Gonorynchoidei was recognized.

Later on, Patterson (1975) also acknowledged the close relationship between chanoids and gonorynchoids, proposing a "sister-group relationship" between them (p. 168). This author indicated also that, if this hypothesis was correct, the fusions in the caudal endoskeleton in chanoids, gonorynchoids, and ostariophysans (which he used in the sense of otophysans) were acquired independently. Subsequently, Taverne (1981) indicated, in the paper erecting and describing the Cretaceous genus †*Aethalionopsis*, that nothing justified the kinship of Gonorynchiformes and Ostariophysi. He placed the former in an evolutionary position more derived than Osteoglossomorpha and Elopomorpha, and more primitive than Clupeomorpha and Euteleostei. Gonorynchiforms were therefore closer to clupeomorphs than to ostariophysans. Within this arrangement, †*Aethalionopsis* was considered very close to, but slightly more primitive than, †*Tharrhias*, †*Dastilbe*, and †*Parachanos*.

The study of the phylogenetic hypotheses of the Gonorynchiformes with cladistic methodology was also to start with the Recent forms, having the fossils added later on.

Cladistic Approaches

The history of the phylogenetic hypotheses of the Gonorynchiformes and the use of cladistic methodology were not without controversy. Focusing on extant forms, the work by Fink and Fink (1981) was the first cladistic approach to better understand the phylogenetic relationships of the Ostariophysi. In that paper, the Gonorynchiformes were hypothesized to be the basalmost ostariophysans, sister group to the Otophysi (Fig. 7.2); *Chanos* "appears to be the sister group to all other Recent gonorynchiforms" (Fink and Fink 1981: 304). Although some character distribution within the order Gonorynchiformes was discussed, a more thorough study that would include fossil taxa was left for future investigations.

Following this, Lauder and Liem (1983) placed the sister groups [Chanoidei + Gonorynchoidei] as the basalmost ostariophysan clade (Anotophysi). Although no formal cladistic analysis was performed, Patterson (1984a) dealt with some fossil gonorynchiforms in a cladistic framework for the first time. This author concluded that fossil forms like †*Tharrhias*, †*Dastilbe*, †*Parachanos*, and †*Aethalionopsis* should be considered

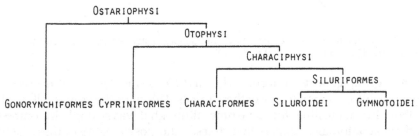

Fig. 7.2 Phylogenetic relationships of the Gonorynchiformes within the Ostariophysi as proposed by Fink and Fink (1981). The Gonorynchiformes appears as the sister group to the Otophysi. Modified from Fink and Fink (1981: fig. 1).

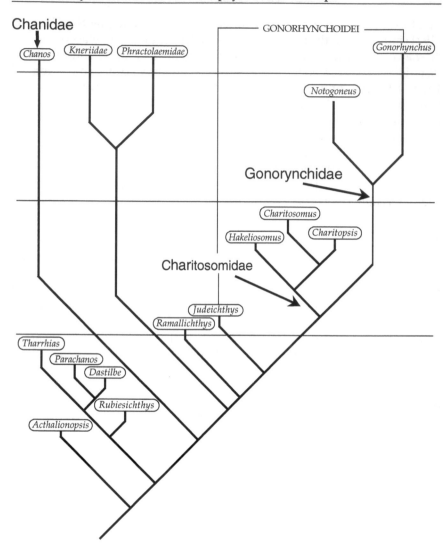

Fig. 7.3 Phylogenetic interrelationships of the Gonorynchiformes as proposed by Gayet (1993a). Modified from Gayet (1993a: fig. 10).

as stem-group gonorynchiforms, whereas *Chanos* was closer to other Recent gonorynchiforms than to any of these fossils. This hypothesis was followed, without further testing, for some time, although it was strongly in contrast with palaeontological papers that stressed the close kinship of these fossil genera with *Chanos* (e.g., Taverne 1974b, 1981, Wenz 1984 in her description and erection of †*Rubiesichthys*).

In a phylogenetic analysis of fossil and extant gonorynchiforms, Gayet (1993a) also considered *Chanos* to be more closely related to other gonorynchiforms than to †*Tharrhias* and its related fossil genera (Fig. 7.3). In this study the Kneriidae and Phractolaemidae were removed from the Gonorynchoidei (contrary to Grande 1992) and more closely aligned with *Chanos*. In this and other studies, the Middle Eastern taxa †*Charitosomus*, †*Charitopsis* and †*Hakeliosomus* were placed within their own family †Charitosomidae as the sister group to the Gonorhynchidae (sic) (*Gonorhynchus* (sic) + †*Notogoneus*); and †*Judeichthys* and †*Ramallichthys* were placed as plesions within the Gonorhynchoidei (sic) (Gayet 1980, 1982a, b, 1985, 1986a, b, 1989, 1993b, c).

The first cladistic analyses to include both fossil and Recent gonorynchiforms were separately performed as unpublished doctoral theses by Poyato-Ariza (1991) and Grande (1992). The first published paper that provided a cladistic analysis with both fossil and Recent gonorynchiforms was by Poyato-Ariza (1996a). This study provided evidence that, contrary to Patterson (1984a) and to Gayet (1993a), chanids are a monophyletic group that includes *Chanos* and all fossils like †*Tharrhias* and related genera. The analyses by Grande (1992) and Poyato-Ariza (1991, 1996a) were independently accomplished, and both were incomplete, since the former was principally focused on the Gonorynchidae and the latter on the Chanidae, although both included representative genera of all the families of the order. The first paper to actually include all fossil and Recent gonorynchiform genera known at the time was the collaborative effort by Grande and Poyato-Ariza (1999), joining and updating the data previously published separately. This work confirmed the monophyly of the Chanidae, including fossil forms. Contrary to Gayet (1993a and previous papers listed above), †*Ramallichthys* and other genera from the Middle East (considering †*Hakeliosomus* as an invalid genus, its only species assessed to †*Charitosomus*) do not form a distinct †Ramallichthyidae family, but are included in the family Gonorynchidae. In this chapter, the new cladistic analyses will update and clarify these and other issues concerning gonorynchiform interrelationships; the reader is referred to the Discussion section below.

In the meantime, Fink and Fink (1996) published some comments on the conclusions by Poyato-Ariza (1996a) concerning the diagnosis of the Chanidae and the polarization of the corresponding characters (see Discussion below). Also, new gonorynchiform genera were described by Taverne (1997: †*Apulichthys*; 1998: †*Lecceichthys*) while the cladistic analysis by Grande and Poyato-Ariza (1999) was going through the revision and publication process, so they were not included in it. The results of the cladistic analyses of the present chapter will also be compared with the corresponding phylogenetic hypotheses by Taverne (1997, 1998).

It is worth noting here that the recent molecular phylogenies, which obviously do not include discussions of interrelationships of fossil forms, have in some instances challenged the sister-group relationship between *Chanos* and *Gonorynchus* (e.g., Lavoué *et al.* 2005, hypothesizing *Gonorynchus* and *Chanos* as consecutive basal groups within Gonorynchiformes) and in other instances corroborated it (e.g., Peng *et al.* 2006, although excluding the other Recent gonorynchiforms). The phylogeny by Lavoué *et al.* (2005) also differed in the kneriid interrelationships, with *Grasseichthys* as sister group of the Kneriini.

New Cladistic Analysis on Gonorynchiform Interrelationships

For this updated study of gonorynchiform interrelationships, we would like to present a new data set that is formed by: (1) osteological characters revised and updated from Grande and Poyato-Ariza (1999); (2) new osteological characters; and (3) myological characters adapted from Diogo (2007, present volume). These characters are presented and discussed in the next section.

As for the taxa in this new analysis, we include all gonorynchiform genera, both fossil and Recent, from Grande and Poyato-Ariza (1999), plus, in a separate analysis, those that have been described since then (Taverne 1997, 1998, 1999). We have also updated additional information that has been published concerning gonorynchiform taxa (e.g., Britz and Moritz 2007, Davis and Martill 1999, Dietze 2007, Taverne and Gayet 2006). Thus, the 27 genera included in the data matrix are as follows. Three of them are used as outgroup, formed by *Brycon*, †*Diplomystus* and *Opsariichthys*, as in Grande and Poyato-Ariza (1999). As ingroup, we include a total of 24 genera. They are, first of all, those in Grande and Poyato-Ariza (1999): †*Aethalionopsis*, *Chanos*, †*Charitopsis*, †*Charitosomus*, *Cromeria*, †*Dastilbe*, †*Halecopsis*, *Gonorynchus*, †*Gordichthys*, *Grasseichthys*, †*Judeichthys*, *Kneria*, †*Notogoneus*, †*Parachanos*, †*Parakneria*, *Phractolaemus*, †*Ramallichthys*, †*Rubiesichthys*, and †*Tharrhias*. To these, we added those published since that paper entered the revision process, that is: †*Apulichthys* Taverne 1997; †*Lecceichthys* Taverne 1998; †*Sorbininardus* Taverne 1999. We could not examine the new Taverne material personally, so we coded according to their original descriptions. These taxa from the literature are added to the analysis to round it out and to represent all proposed or nominal gonorynchiform or gonorynchiform-related taxa. Based on the papers it is difficult to confirm character coding in a number of cases, so we complemented their coding with personal communications with Taverne (2007). The new genus by Pittet *et al.* (present volume), henceforth called "new genus", is also included, coded according to its original description plus personal communications by Cavin (2006, 2007), who suggested new characters for this analysis.

Some material has been re-examined for the present paper. This includes: the lectotype (IRSNB P.1244a,b) and paralectotypes (P.2405a,b, P.2402, and P.3388) of †*Aethalionopsis robustus* (Traquair did not indicate a holotype in his description of the new species in 1911, so Taverne chose a lectotype in his revision of 1981); the type series of †*Gordichthys conquensis*, currently housed at the MCCM (see complete list of material in Poyato-Ariza 1994, 1996a); the most significant specimens of †*Rubiesichthys gregalis* at the MCCM collection (see complete list of material in Poyato-Ariza 1996a, b); †*Charitosomus spinosus*: AMNH 3895 (holotype), 3746; HAK 133; †*Charitosomus lineolatus*: SHA 136, 200, 206, 1454; †*Hakeliosomus hakelensis*: AMNH 3639, 3757, 3770, 3856, 5859, 19449; MNHN HAK 340, HAK 130d, 136d, 111g, 112, 113g; †*Ramallichthys orientalis*: FMNH PF 79, 14367–14369, FMNH uncatalogued. Lists of material examined are provided in Grande (1994, 1996), Poyato-Ariza (1994, 1996a–c) and Grande and Poyato-Ariza (1999). Institutional abbreviations for these and other specimens mentioned in this chapter: AMNH, American Museum of Natural History, New York; FMNH, Field Museum of Natural History, Chicago; IRSNB, Institut Royal des Sciences Naturelles de Belgique (Département de Paléontologie), Brussels; MCCM, Museo de las Ciencias de Castilla-La Mancha, Cuenca, Spain; NHM, Natural History Museum, London.

Three papers have revised the genus †*Dastilbe* after the publication of the cladistic analysis by Grande and Poyato-Ariza (1999). Davis and Martill (1999), Dietze (2007), and Brito and Amaral (2008) all provided further insight concerning the variation of some relevant characters of this genus. The first concluded that all the Brazilian specimens of this genus belong to the type species, †*Dastilbe crandalli*. The second agrees that †*Dastilbe elongatus* is a junior synonym of the type species, and that the other Brazilian species, †*Dastilbe moraesi*, is too poorly known to make any conclusion in this respect. Finally, Brito and Amaral (2008) conclude that, due to the high plasticity of the characters, it is impossible to distinguish any nominal species from the type species. However, neither of these studies conducted a re-examination of type material of any nominal species of this genus, and this is, in our opinion, an indispensable requirement to achieve more definitive taxonomic conclusions. This problem still demands further confirmation from broader revisions of all nominal species by locating and examining all type series. The variations in several key characters are also examined in all of the papers, but possible sources of variation (e.g., ontogenetic versus taxonomic and/or geographic) are not explored in depth, nor are satisfactory explanations offered for these variations. For the present analysis, we have consistently coded for †*Dastilbe* characters as observed from large adult specimens. In case of variation, we code the state more commonly present among larger individuals (see also comments on character 97 below).

†*Halecopsis* was incompletely known for a long time. The revision of Taverne and Gayet (2006) has provided more information on this genus, so characters have been coded according to this revision (see comments on characters list below). Unfortunately, the few known specimens of this genus are rather incomplete and not very well preserved, so many character states still remain unkown for this taxon. The genera †*Coelogaster*, †*Neohalecopsis*, and †*Prochanos* are still poorly known and in need of revision. They are represented by very few available specimens, and those that are known are incompletely and badly preserved. Only a detailed revision based on new material would likely provide enough information for these three genera to be properly included in a cladistic analysis. Finally, a brief mention of †*Chanoides* and †*Chanopsis* seems pertinent here, since these generic names might suggest some kind of chanid affinities. However, the former was shown to be an otophysian fish by Patterson (1984b), and the latter a member of the Osteoglossidae by Taverne (1984), both followed by Poyato-Ariza (1996c). Consequently, they are not included in this study of gonorynchiform interrelationships.

Characters

Osteological characters listed below as part 1 of this section form most of the data matrix. The myological characters are presented as part 2, after the osteological list. An abbreviated list of characters is included as Appendix 1. The complete data matrix is presented in Appendix 2.

In this section we take the opportunity to update characters and correct the occasional errors and polymorphic characters that were proposed in Grande and Poyato-Ariza (1999). To avoid misunderstanding of character coding and character distribution (as in Britz and Moritz 2007), in the present section we discuss character coding only; character distribution is presented later on, according to the results of the cladistic analysis (see Discussion of Results and Appendix 3). This way, we hope to avoid further mixing of character coding and character distribution, which are different issues both methodologically and logically.

List of Characters, Part 1 (Osteological)

The largest part of the osteological character section consists of a revision and updating of characters used in Grande and Poyato-Ariza (1999). The original character numbers are given in brackets. In the list below, we include new, specific comments only for those characters whose states and/or coding have been revised in the present analysis. In addition, modifications to character descriptions have been made in many cases to improve their wording and make them more precise, or clear, thus avoiding potential misunderstandings (the original numbers of those characters are: 5, 7, 8, 10, 12, 13, 16, 17, 20, 21, 23, 26–28, 33, 34, 40, 42, 43, 45, 48, 49, 54, 56, 58, 59, 61, 62, 67–70, 76, 84, 88, and 93 in Grande and Poyato-Ariza 1999). Modifications to characters originally numbered 26, 27, 40, and 59 were done in consultation with Cavin (pers. comm. 2006), and are based on his work with the new genus by Pittet *et al.* (present volume). The remainder of the characters from Grande and Poyato-Ariza (1999) (i.e., without any comments), are coded as in Grande and Poyato-Ariza (1999), so they are merely listed here, and we refer the reader to the original publication for a full discussion of each one. Based upon further study, some characters from Grande and Poyato-Ariza (1999) were omitted from this analysis; they are briefly commented on at the end of the list of characters.

Characters in the list below that do not have a number in brackets after the number in this list are those newly added to this study. We discuss them for all the taxa included in the analysis, and they are intercalated with the rest of the characters in order to preserve the anatomical sequence.

Braincase

1 (1).   Orbitosphenoid: present [0]; absent [1]. Among the ingroup genera, the orbitosphenoid is present in †*Sorbininardus* only (Taverne 1999: 82–83, fig. 3; primitive state). This character is difficult to assess with certainty in most fossils, but no traces of orbitosphenoid have been observed in any re-examined skulls of †*Charitopsis*, †*Gordichthys* or †*Rubiesichthys*. Gayet (1993a–c) does not mention the presence of an orbitosphenoid in †*Charitopsis*.

2.   Basisphenoid: present [0]; absent [1]. The loss of basisphenoids (Fink and Fink 1981) is diagnostic of ostariophysans. However, a very small, peculiarly shaped basisphenoid is present in †*Apulichthys* according to Taverne (1997: 408–409, fig. 4). This character is difficult to assess with certainty in other fossils, but no traces of basisphenoid have been observed in any skull of †*Charitopsis*, †*Charitosomus*, †*Dastilbe*, †*Gordichthys*, †*Judeichthys*, †*Ramallichthys*, †*Rubiesichthys*, and †*Tharrhias*. In †*Sorbininardus* there is no basisphenoid, according to Taverne (1999: 82; coded as 1 as well).

3 (2).   Pterosphenoids: well developed and articulating with each other [0]; slightly reduced, not articulating anteroventrally but approaching each other anterodorsally [1]; greatly reduced and broadly separated both anteroventrally and anterodorsally [2]. Among the ingroup genera, large, articulating pterosphenoids are present in †*Sorbininardus* only (Taverne 1999: 82–83, 96, fig. 3; primitive state).

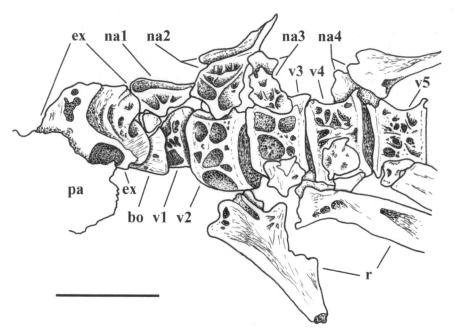

Fig. 7.4 Camera lucida drawing of posterior region of skull and anteriomost vertebrae as preserved in specimen NHM P.54331 of †*Tharrhias araripis* (Natural History Museum, London). bo, basioccipital; ex, left exoccipital; na1–4, neural arches one to four (note the autogenous neurapophyses anteroventral to neural arches one and two); pa, left parietal; r, ribs; v1–5, vertebral centra one to five (note the autogenous paraphophyses preserved on centra three to five). Scale bar represents 5 mm.

4 (3).   Posterolateral expansion of exoccipitals: absent [0]; present [1]. †*Halecopsis* was coded as [0] by Grande and Poyato-Ariza (1999) as result of a typing error, but this region is not accessible in the holotype and only specimen of this genus, so it is now corrected and coded as unknown [?]. See Fig. 7.4 for the exoccipitals of †*Tharrhias*.

5.   Exoccipitals: posteriorly smooth, with no projection above the basioccipital [0]; with a posterior concave-convex border, and a projection above basioccipital [1]. In lateral view, the exoccipitals are normally flat to slightly convex bones forming part of a smooth occipital region. In *Chanos* and †*Tharrhias* they are modified in that, in lateral view, they are ventrally concave and dorsally convex projecting above the supraoccipital (e.g., Fink and Fink 1981: fig. 6; Poyato-Ariza 1996a: fig. 14, 1996c: fig. 4; Fig. 7.4). This region, unfortunately, is not accurately observable in most fossil genera, although the exoccipitals seem to be posteriorly smooth in, at least, †*Dastilbe* (e.g., Dietze 2007: fig. 4).

6 (4).   Cephalic ribs: absent [0]; present and all articulating with the exoccipitals [1]; present and articulating with both the exoccipitals and basioccipital [2]. Cephalic ribs are present in †*Apulichthys* according to Taverne (1997: fig. 8), but their mode of articulation with the skull is unknown (Taverne pers. comm. 2007), so it is coded as [1 and 2] simultaneously for this genus.

7 (5).   Brush-like cranial intermuscular bones (*sensu* Patterson and Johnson 1995): absent [0]; present [1]. They are present in †*Apulichthys* according to Taverne (1997: 416, fig. 8, labelled as "EPN", epineurals; pers. comm. 2007).

8.   Nasal bone: small but flat [0]; just a tubular ossification around the canal [1]; absent as independent ossification [2]. Suggested by Cavin (pers. comm. 2006). The primitive state is present in †*Diplomystus* within the outgroup; *Brycon* and *Opsariichthys* present state [1]. The nasal is a small bone, but with small laminar portions lateral to the ossified tube of the sensory canal, in †*Gordichthys*, *Kneria*, *Parakneria*, †*Rubiesichthys*, and the new genus by Pittet *et al.* (present volume) (state 0). The nasal bone is just a tubular ossification surrounding the anterior part of the supraorbitary sensory canal in †*Apulichthys*, *Chanos*, *Gonorynchus*, †*Parachanos* (e.g., Poyato-Ariza 1996a: fig. 14, unlabelled ossification dorsal to premaxilla and maxilla), *Phractolaemus*, and †*Tharrhias* among the ingroup (state 1). It is absent as an independent ossification in, at least, †*Charitosomus*, *Cromeria*, and *Grasseichthys* (state 2). Although Blum (1991a) and Dietze (2007) restore †*Dastilbe* without a nasal bone, it is not unlikely that such small, delicate bone has been lost during fossilization and/or preparation; to be conservative, we coded this character as [?] for this genus until its ethmoid region is well known. For the same reason, it was coded as unknown in †*Charitosomus*, †*Charitopsis*, †*Hakeliosomus*, †*Judeichthys*, †*Notogoneus*, and †*Ramallichthys* as well. We prefer this conservative coding, because, according to Grande and Grande (2008), the presence of nasals could not be confirmed in †*Notogoneus*, although an elongate bone in about the correct position was observed. For the same reason, we conservatively code as [?] the other fossil forms, where the state of preservation prevents certainty. Similarly, in †*Sorbininardus* the nasals are not conserved (Taverne 1999: 81; conservatively coded as unknown for the same reason).

9 (6).   Frontals: wide through most of their length, narrowing anteriorly to form a triangular anterior border [0]; elongate and narrow except in postorbital region [1]; wide, anteriorly shortened,

anterior border roughly straight [2]; roughly rectangular in outline, narrow throughout their length [3]. Character state [3] is added in this analysis for the new genus; according to its description by Pittet *et al.* (present volume), it does not fit in any of the previous character states.

10 (7).   Interfrontal fontanelle: absent [0]; present [1]. The interfrontal fontalle is a lack of ossification between the frontals, separating one from the other. We code this character as [0] in *Gonorynchus* for this analysis because we reworded the character as presence or absence of a fontanelle between the frontals, regardless of the presence or absence of fusion between them in adult specimens (see next character; this one was coded as [N], non-applicable, for *Gonorynchus* by Grande and Poyato-Ariza 1999). We code this character as 1 for *Brycon* in the outgroup after Weitzman (1962: 21, fig. 65). Other than that, the interfrontal fontanelle is only observed in *Cromeria* and *Grasseichthys* (state 1).

11 (8).   Frontal bones: paired in adult [0]; co-ossified, with no median suture [1]. Contrary to Perkins (1970), †*Notogoneus osculus*, like all other species of †*Notogoneus*, have paired frontals, with no co-ossification (Grande and Grande 2008; coded as 0). Both frontals are fused in all adult specimens of *Gonorynchus* observed, although younger specimens do present at least a partial suture between them (e.g., American Museum of Natural History 96053; Poyato-Ariza 1996a: fig. 6B). We have, nonetheless, coded for the condition in adults, as for all other characters (coded as 1).

12 (9).   Foramen for olfactory nerve in frontal bones: absent [0]; present [1].

13 (10).   Relative position of the parietals: medioparietal (in full contact with each other along their midline) [0]; mesoparietal (*sensu* Poyato-Ariza 1994; partly separated by the supraoccipital, posteriorly, and partly in contact with each other, anteriorly) [1]; lateroparietal (completely separated from each other by the supraoccipital) [2].

14 (11).   Parietal portion of the supraorbital canal: absent [0]; present [1]. The posterior portion of the supraorbital canal does not pierce the parietals in any genus, except in †*Aethalionopsis*, where the canal enters the parietals after exiting the frontals.

15 (12).   Parietals: large [0]; reduced but flat and blade-like in shape [1]; highly reduced [2]; absent as independent ossifications [3]. This is a character whose states are difficult to define. It is not easy to establish degrees of reduction, since we practically find a

continuum among gonorynchiforms. However, the reduction of the parietals is evident in this group, so we do think it is worth keeping this character, traditionally used to diagnose the Gonorynchiformes (e.g., Fink and Fink 1981: 302, 313). We will try to refer parietal reduction with recognizable references. We code the primitive state in the three outgroup genera because their parietals are nearly as broad as the frontals and therefore are not considered to be reduced (e.g., Weitzman 1962: fig. 9; Chang and Maisey 2003: fig. 10B). We therefore consider the primitive state that in which parietals are nearly as wide as frontals are posteriorly. If this same criterion is applied to †*Aethalionopsis*, and although the parietal borders are never clearly visible in any specimen, these bones are considerably narrower than the frontals (Taverne 1981: fig. 3; pers. obs.) and therefore reduced in size with respect to the genera of the outgroup. Therefore, although the precise borders of the parietals are not clear, it seems better to code it as state [1] for this genus. By applying the same criterion to †*Halecopsis*, and according to the restorations by Taverne and Gayet (2006: figs. 1, 2) we code also this character as [1] for this genus. After re-examination, we also code †*Charitopsis*, †*Charitosomus*, *Gonorynchus*, †*Hakeliosomus*, †*Judeichthys*, †*Notogoneus*, and †*Ramallichthys* as state 1, because their parietals still have a flat portion and are more similar in size to, for example, those of †*Tharrhias* (state 1 as well). *Chanos* (state 2) exemplifies the maximum reduction in parietals, which are little more than canal-supporting bones, with little else, if any, beside the bone surrounding the canal itself. Even so, this character remains problematic, since the relative reduction of the parietals can differ very subtly when different genera are compared to each other, not to mention its possible fusion with other bones in *Kneria* and *Parakneria*, where they are usually absent as independent ossifications in most adult specimens, and consequently coded as 3, like *Cromeria* and *Grasseichthys*. It is necessary to evaluate the detailed ontogenetic and individual variation, especially in *Kneria* and *Parakneria*, where there might be a great variation of possible fusions with supraoccipital or pterotics, which is not yet fully understood or accounted for. Much work is still necessary concerning the ontogenetic origin of the parietals and their possible fusions with other bones in kneriids. We nonetheless keep this character, despite all its difficulties, because there is a clear reduction of the parietals in gonorynchiforms which is better to be accounted for.

16 (13).   Supraoccipital crest: small, short in lateral view [0]; long and enlarged, projecting above occipital region and first vertebrae, forming a vertical, posteriorly deeply pectinated blade [1]. New evidence shows that a relatively large, pectinated supraoccipital crest is present in †*Dastilbe* (Dietze 2007: figs. 3, 8); this character state was previously unknown in this genus. Despite Britz and Moritz's (2007: 37) interpretation that the supraoccipital crest is absent in *Parakneria*, the flat posterior projection of the supraoccipital in this genus is clearly the supraoccipital crest, homologous to that of *Kneria* and other genera, despite its more or less flat shape (e.g., Lenglet 1974: figs. 1, 2, 3, 4, 5, 11; pers. obs.). In none of these genera, though, is the crest comparable to that of the mentioned chanids (plus there is a certain individual variation in size), so we reworded this character and code it as [0] for *Kneria* and *Parakneria*.

17 (15).   Foramen magnum: dorsally bounded by exoccipitals [0]; enlarged and dorsally bounded by supraoccipital [1]. As discussed in Fink and Fink (1981), in *Chanos* the exoccipital bones extend more medially over the dorsal part of the foramen magnum; this is also the case of, at least, †*Charitosomus*, †*Dastilbe*, *Gonorynchus*, †*Notogoneus*, and †*Tharrhias* (state 0). The exoccipital bones in *Cromeria*, *Grasseichthys*, *Kneria*, *Parakneria*, and *Phractolaemus* are separated and remain lateral to the foramen magnum; only the supraoccipital bone was observed dorsal to the foramen magnum in these taxa (state 1). This character apparently was misunderstood by Britz and Moritz (2007: 28), for it has nothing to do with cartilages surrounding the foramen magnum.

18 (16).   Mesethmoid: wide and short [0]; long and slender, with anterior elongate lateral extensions [1]; large, with broad posterolateral wing-like expansions [2]; elongated and thin [3]. We coded [0] for †*Lecceichthys* according to the restoration and description by Taverne (1999: 294, fig. 2), although this bone is but partly preserved. Derived state [3] is added for the present analysis, as it is present in †*Apulichthys* (after Taverne 1997), †*Halecopsis* (after Taverne and Gayet 2006) and the new genus.

19 (17).   Wings (extensions) on lateral ethmoids: absent [0]; present [1]. It is coded as primitive state for †*Lecceichthys* according to the description by Taverne (1999: 294), although these bones are not represented or restored. It is coded as [0] for †*Halecopsis* as well, following Taverne and Gayet (2006).

Jaws

20. Teeth in premaxilla, maxilla, and dentary: present [0]; absent [1]. Among the outgroup, these bones bear teeth in *Brycon* and †*Diplomystus*, but not in *Opsariichthys* (teeth are independently absent in Gonorynchiformes and Cypriniformes according to Fink and Fink 1981: 322). True teeth are absent on these oral bones in all the genera of our ingroup (but see character 22, which refers to the gingival "teeth" of soft tissue in *Gonorynchus* (Monod 1963)).

21 (18). Premaxilla: consisting of one solid element [0]; premaxilla consisting of two distinct elements, with a shorter, non-osseous element lying ventral to a much longer osseous portion, which in turn articulates with the maxilla [1].

22 (19). Premaxillary "gingival teeth" (Monod 1963): absent [0]; present [1].

23 (24). Premaxilla: small, flat and roughly triangular [0]; large, very broad, concave-convex, with long oral process [1]; narrow and elongated, its length more than one half of the length of the maxilla [2]. In Grande and Poyato-Ariza (1999) this character was coded as unknown [?] for †*Aethalionopsis* (at the reviewer's request), although it had been discussed and coded as derived [1] by Poyato-Ariza (1996a). Subsequent study of the material has verified that the premaxillary condition is derived in †*Aethalionopsis*. The restoration of †*Aethalionopsis* by Taverne (1981: fig. 2) shows the whole anterior border of the premaxilla in dotted line, thus indicating that the real anterior border, and therefore the actual morphology of the bone, cannot be precisely restored. This is totally correct: no specimen of †*Aethalionopsis* shows the premaxilla in its entirety; the anterior border is never observed in its entirety either. However, in some specimens (all housed at the IRSNB, Département de Paléontologie de l'Institut Royal des Sciences Naturelles de Belgique), enough of the premaxilla remains, or its impressions are clear enough, to be reasonably sure that this bone was very broad and concave-convex (in addition to presenting the long oral border with an acute posterior process, as accurately restored by Taverne 1981: fig. 2). The lectotype of †*Aethalionopsis* (P.1244a/b) shows impressions that can be attributed to the premaxilla, and they are impressions of a very large bone. Paralectotype P.2405a,b shows remains of a very large plate where the premaxilla should be. They are unclear impressions in the part (a), but actual fragments of bone in the counterpart (b). Finally, paralectotype

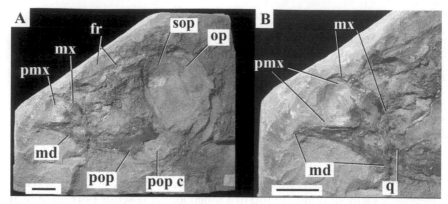

Fig. 7.5 Skull and oral region of †*Aethalionopsis robustus.* Photos of paralectotype P.2402 in lateral view (Institut Royal des Sciences Naturelles de Belgique, Brussels). Most bones are those of the right side in medial view (pop is from left side in lateral view). **A**, complete skull (note clear impressions of large premaxilla, suprapreoperculum, and preopercular crest). **B**, detail of oral region of specimen in A; compare with Fig. 7.6. fr, frontals; md, mandible (mostly actual bone, part impressions); mx, maxilla (part impression, part actual bone); pmx, premaxilla (impression; note its large, rounded shape and its concave surface; see character number 23 in text for further explanations); op, opercular bone (impression); pop, preopercular bone (part impression, part actual bone); pop c, preopercular crest (partly preserved as impression); q, quadrate; sop, suprapreopercular bone (impression). Scale bars represent 10 mm. Photos courtesy E. Bultynk.

P.2402 is the one that provides the clearest evidence (Fig. 7.5). In this specimen, there are remains of the left maxilla in lateral view, nearly complete but displaced, leaving room for observation of the elements from the right side. There are remains of the right maxilla in medial view, roughly in place but very incomplete, leaving room for the observation of clear impressions of the right premaxilla, in medial view as well. These impressions of the premaxilla (Fig. 7.5A, B) are those of a very large, concave bone (in medial view; therefore convex in lateral view). This is the typical morphology of the main body of the premaxilla of *Chanos* and of any other chanid fish, confirming the original coding by Poyato-Ariza (1996a). Therefore, we have changed this character back to the derived condition [1] in †*Aethalionopsis* for the present analysis. State [2] is for †*Apulichthys* and †*Sorbininardus* only, according to the restorations and descriptions by Taverne (1997: fig. 6, 1999: fig. 2) of a premaxilla with an oral border "qui s'étire en une longue et fine tigelle osseuse qui exclut la plus grande partie du maxillaire du bord oral de la mâchoire" ("which forms a long and thin osseous rod that excludes most of the maxilla from the oral border of the mouth") (1997: 411) and a maxilla that "est totalement exclu du bord buccal" ("is totally excluded from the border of the mouth") by the elongated premaxilla (1999: 83).

24 (25). Premaxillary ascending process: present [0]; absent [1]. As in the previous character, the premaxilla of †*Apulichthys* is very different from any other ingroup taxon, as it exhibits a small but conspicuous ascending process (Taverne 1997: 411, fig. 6; coded as primitive state). Among the ingroup, †*Hakeliosomus* is the only taxon with a reported process in the premaxilla (Gayet 1993b: 26, fig. 6). However, we could not confirm its presence in the observed specimens of †*Hakeliosomus*, and, in consequence, we code it as absent [1] in this genus.

25 (22). Dorsal and ventral borders of the maxillary articular process: straight or slightly curved [0]; very curved, almost describing an angle [1]. Britz and Moritz (2007: 37) seem to be "uncertain" to what this structure refers. The "articular process of the maxilla" or "maxillary articular process", as per, for instance, Rojo (1988: 303), is a standard term in actinopterygian anatomy that refers to the anterior projection of the maxilla for its articulation with the premaxilla (e.g., Grande and Bemis 1998: 84). An abrupt curvature in the articular process of the maxilla occurs in †*Gordichthys* and †*Rubiesichthys* only among gonorynchiforms (Poyato-Ariza 1994, 1996 a–c).

26. Maxillary process for articulation with autopalatine: absent [0]; present [1]. A special process on the anterodorsal border of the maxilla for its articulation with the autopalatine is apparently absent in all members of the outgroup, plus *Cromeria*, *Gonorynchus*, †*Gordichthys*, *Grasseichthys*, †*Notogoneus*, *Phractolaemus*, †*Rubiesichthys*, †*Sorbininardus*, and the new genus. A short, but robust process with a flat distal facet for articulation with the autopalatine is present on the anterodorsal border of the maxilla in: †*Aethalionopsis* (Taverne 1981: 964; pers. obs. on paralectotype IRSNB P.3388), †*Apulichthys* (Taverne 1997: 411, fig. 6), *Chanos* (e.g., Poyato-Ariza 1996b: fig. 5), †*Charitopsis* (e.g., Gayet 1993b: fig. 38), †*Charitosomus* (e.g., Gayet 1993b: fig. 23), †*Dastilbe* (e.g., Dietze 2007: fig. 3; Fig. 7.6A), †*Hakeliosomus* (e.g., Gayet 1993b: 28, fig. 6; specimen HAK340), †*Judeichthys* (e.g., Gayet 1985: fig. 3), *Kneria*, †*Parachanos* (e.g., Poyato-Ariza 1996b: fig. 9; Fig. 7.6B), *Parakneria*, †*Ramallichthys* (e.g., Gayet 1986a: fig. 12), and †*Tharrhias* (see Fig. 7.6).

27 (23). Posterior region of the maxilla: slightly and progressively expanded to form a thin blade, with roughly straight posterior border [0]; very enlarged, swollen to a bulbous outline, with curved posterior border [1]. See Fig. 7.6.

28.   Supramaxilla(e): present [0]; absent [1]. Supramaxilla(e) are present only in †*Diplomystus* and in the new genus. The rest of the taxa are devoid of supramaxilla(e), bones that are absent as separate ossifications in ostariophysans with a few exceptions, according to Fink and Fink (1981).

29.   Notch between the dentary and the angulo-articular bones: absent [0]; present [1]. The derived state of this character refers to the presence of a deep dorsal notch in the dorsal border of the mandible; this notch is formed between the posterior part of the dentary and the anterior part of the angulo-articular. Such a notch is independent from the occurrence of a posteriorly V-shaped dentary (see next character), and is present in, at least †*Charitopsis* (e.g., Gayet 1993b: figs. 39, 40), †*Charitosomus* (e.g., Gayet 1993b: figs. 23, 24, 27), *Gonorynchus* (e.g., Grande 1996: figs. 5A, 6; Britz and Moritz 2007: fig. 18), †*Judeichthys* (e.g., Gayet 1985: fig. 3), *Kneria* (e.g., Grande 1994: fig. 17b), †*Notogoneus* (e.g., Woodward 1896: pl. 18; Grande and Poyato-Ariza 1999: fig. 9A), *Parakneria* (with an additional complexity due to an anterior projection of the angulo-articular; e.g., Grande 1994: fig. 15b), and †*Ramallichthys* (e.g., Gayet 1986a: figs. 14, 25). This notch is filled with cartilage in Recent forms (e.g., Grande 1994: fig. 11; Britz and Moritz 2007: figs. 11a, b, 15a, 18), so it is consistent to assume that this was also the case in fossil forms.

30 (28).   Articulation between dentary and angulo-articular: strong, dentary not V-shaped posteriorly [0]; loose, with a posteriorly V- shaped dentary [1]. A dentary bone that is V-shaped posteriorly, with one long posterodorsal and one long posteroventral process, is present in *Gonorynchus* and †*Notogoneus* only (see comments on previous character).

31 (29).   Notch in the anterodorsal border of the dentary: absent [0]; present [1]. A notch in this position is often called "leptolepid" notch (e.g., Arratia 1997), but, due to the high homoplasy of this character among phylogenetically distant teleostean groups (in this case, †Leptolepidae and Gonorynchiformes), we prefer not to use that name herein. This notch differs from the notch of character 29 both in position (it is more anterior) and in the elements involved (only the dentary). In †*Dastilbe*, "a notch in the anterodorsal border of the lower jaw was not observed", according to Dietze (2007: 15). Although not discussed by the author, this is due to the fact that the anterior border of the coronoid process, where this notch occurs, is seldom exposed in the specimens of this genus, and apparently never in her material (e.g., Dietze 2007: figs. 3, 8). When normally articulated,

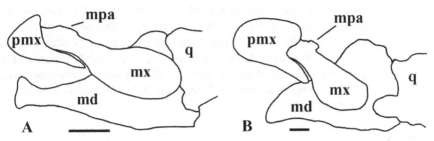

Fig. 7.6 Oral region of two chanid fishes; outlines as restored from camera lucida drawings. **A,** †*Dastilbe elongatus*, restored mostly from specimen AMNH 31 (American Museum of Natural History, New York). **B,** †*Parachanos aethiopicus*, restored mostly from the lectotype, MNHN GAB 1 (Muséum national d'Histoire naturelle de Paris). Left side, lateral view, anterior to the left. md, mandible; mpa, maxillary process for articulation with autopalatine; mx, maxilla; pmx, premaxilla; q, quadrate. Scale bars represent 2 mm.

the posterior, bulbous expansion of the maxilla of †*Dastilbe* always conceals this part of the mandible (Fig. 7.6A), as is the case of all articulated chanid material (e.g., Fig. 7.6B). Consequently, if this region is not accessible, the character is not verifiable, but that does not imply absence, of course. Partial disarticulation allowing observation of the anterior border of the coronoid process occurs very rarely in †*Dastilbe*; however, whenever this happens and that zone of the mandible is accessible, a notch is clearly observable (e.g., American Museum of Natural History, New York, specimen 19432, Natural History Museum, London, specimen 4300; Silva-Santos 1947: pl. 1, fig. 3; Poyato-Ariza 1996a: 20, fig. 10B). Therefore, we continue to code this character as present [1] for †*Dastilbe*, as in Poyato-Ariza (1996a) and Grande and Poyato-Ariza (1999).

32 (30). Mandibular sensory canal: present [0]; absent [1]. Previous illustrations report a mandibulary sensory canal in †*Charitopsis* (e.g., Gayet 1993b: figs. 39, 40; pl. 7, fig. 3), †*Charitosomus* (e.g., Gayet 1993b: fig. 29), †*Judeichthys* (e.g., Gayet 1985: pl. 1, fig. 2), and †*Ramallichthys* (e.g., Gayet 1986a: figs. 14, 25). However, our observations could not confirm this presence, so we conservatively coded these genera as unknown [?]. This character was also coded as unknown for †*Halecopsis* by Grande and Poyato-Ariza (1999); for the present analysis, it is coded as [1] according to the new restoration by Taverne and Gayet (2006: fig. 2).

33 (31). Inferior and superior enlarged retroarticular processes of mandible: both absent [0]; inferior retroarticular process present, superior retroarticular process absent [1]; both inferior and superior retroarticular processes present [2]. As discussed

in Grande and Grande (2008) and Grande and Poyato-Ariza (this volume), a posterior retroarticular process, although absent in *Gonorynchus* and †*Notogoneus*, is present in †*Hakeliosomus*, †*Judeichthys*, and †*Ramallichthys*. A superior retroarticular process was reported by Gayet (1993b) for †*Charitopsis* and †*Charitosomus*, but only confirmed for †*Charitosomus* (Grande and Grande 2008).

Suspensorium

34 (32).   Quadrate with: posterior margin smooth [0]; elongated forked posterior process [1].

35 (33).   Quadrate-mandibular articulation: below or posterior to orbit, no elongation or displacement of quadrate [0]; anterior to orbit, quadrate displaced but not elongate [1]; anterior to orbit, with elongation of the body of quadrate instead of displacement [2]. In some cases (e.g., †*Charitosomus*, †*Hakeliosomus*) the quadrate-mandibular articulation is placed anterior to the orbit as marked by the lateral ethmoids; since there is no displacement or elongation of the quadrate, a modification inherent to the derived state, this character is coded as [0] in these cases. The articulation in †*Lecceichthys* is clearly quite anterior to the orbit; as restored by Taverne (1999: fig. 2), the precise limits of the quadrate are not known, but there is no room for an elongation of this bone, which is small, according to Taverne's description (p. 296), and therefore displaced. This character is consequently coded as [1] for this genus. Although the anterior limits of the orbit are not precise in †*Halecopsis*, the articulation lies at approximately the level of the lateral ethmoid and very anterior to the posterior limit of the orbit, according to the restoration by Taverne and Gayet (2006: fig. 2), so it is reasonable to admit that the articulation "est située en avant de l'aplomb de l'orbite" ("is placed before the middle part of the orbit") (p. 111). According to their restoration and description, the quadrate is elongated, so the character state coded for this genus is [2].

36 (34).   Symplectic: elongated in shape but relatively short [0]; very long, about twice the length of the ingroup [1]; absent as an independent ossification [2]. Recent new ontogenetic evidence provided by Britz and Moritz (2007: 31, fig. 15) indicates that the structure identified by Howes (1985) as the symplectic in *Grasseichthys*, interpretation followed by Grande (1994) and Grande and Poyato-Ariza (1999), might actually be the metapterygoid. However, we find their interpretation

inconclusive (see comments on character 38), and therefore, this character is conservatively coded as [1] for this genus, according to the traditional interpretation (Howes 1985, Grande 1994, Grande and Poyato-Ariza 1999; see detailed comments on character 38). The symplectic is absent in *Cromeria* and *Phractolaemus* (coded as 2).

37 (35).   Symplectic and quadrate: articulating directly with each other [0]; separated through cartilage [1]; no contact due to absence of symplectic [2]. This character is coded as [2] in *Cromeria* and *Phractolaemus*.

38 (36).   Metapterygoid: large, broad and in contact with quadrate and symplectic through cartilage [0]; reduced to a thin rod [1]. The metapterygoid is absent as an independent ossification in *Grasseichthys* according to Howes (1985) and the phylogenetic hypotheses by Grande (1994) and Grande and Poyato-Ariza (1999). The puzzling claim that, in *Grasseichthys*, "without any explanation this character was omitted in Grande and Poyato-Ariza (1999)" (Britz and Moritz 2007: 35) is in clear contradiction with the statement that "a metapterygoid is absent in *Grasseichthys*" (Grande and Poyato-Ariza 1999: 216). Character 36 in that paper deals with the morphology and arrangement of the metapterygoid, so it was coded as non-applicable [N] for this genus, as explained in that text. Furthermore, additional confusion has arisen from the interpretation of the osseous bar between the posterior part of the quadrate and the lower anterior part of the hyomandibular bone in *Grasseichthys* as the symplectic (Howes 1985, Grande 1994, Grande and Poyato-Ariza 1999) or the metapterygoid (Britz and Moritz 2007). The latter interpretation is supported with the argument that "our illustration of the hyopalatine arch of *Grasseichthys* (Fig. 7.15a) clearly shows that the bone in question is an ossification in the posterior part of the palatoquadrate cartilage and therefore represents the metapterygoid" (Britz and Moritz 2007: 34). No further arguments are provided. The mentioned figure illustrates a nearly fully ossified bar between the quadrate-endopterygoid anteriorly and the hyomandibular bone posteriorly. Let us note that this is the relative position that the symplectic occupies in other gonorynchoids (e.g., *Gonorynchus*: Britz and Moritz 2007: fig. 19a). We reckon that this problem is complex, because, in some cases, both the metapterygoid and the symplectic occupy this position, as in *Kneria* (e.g., Britz and Moritz 2007: fig. 18c). So, the connectivity criteria could support either interpretation of the osseous bar of *Grasseichthys* as the symplectic

or the metapterygoid, since one of them is absent. However, the evidence presented by Britz and Moritz (2007: fig. 15a) is, in our opinion, inconclusive, because their figure shows a nearly fully ossified bone, but does not show whether the original cartilage comes from the *pars metapterygoidea* of the palatoquadrate cartilage or not. In consequence, pending further ontogenetic evidence to support a change in the traditional interpretation, we still code this character as non-applicable [N] for *Grasseichthys*.

39.   Dermopalatine: present [0]; absent [1]. The Ostariophysi are partly diagnosed by the loss of the dermal portion of the palatine (Fink and Fink 1981). In all genera examined but †*Diplomystus* (outgroup), the dermopalatines are absent. Admittedly, this character is difficult to assess in fossil taxa, but we feel that we have examined a good number of fossils of each genus where we can feel confident that the lack of dermopalatine in these specimens is not a result of poor preservation, at least in †*Apulichthys* (Taverne 1997: 412), †*Charitosomus* (e.g., Gayet 1993b, fig. 23), †*Dastilbe*, †*Gordichthys*, †*Judeichthys*, †*Ramallichthys*, †*Rubiesichthys*, and †*Tharrhias*. Although Taverne (1999) reports the loss of dermopalatine in †*Lecceichthys* (p. 305), this region seems too incompletely preserved in the holotype and only specimen of this genus, but the shape of the preserved palatine resembles very much that of other gonorynchids, and no traces of any dermal component are observed (Taverne pers. comm. 2007; coded as 1).

40 (37).   A patch of about 20 conical teeth on endopterygoids and basibranchial 2: absent [0]; present [1]. According to the description of †*Lecceichthys*, the 27 teeth observed at the orbital level belong to the left endopterygoid (Taverne 1999: 296, fig. 2), although the bone itself is not visible (Taverne pers. comm. 2007) and the level of the teeth is rather high, apparently closer to the parasphenoid level than to the endopterygoid level. Even though nothing is known about the branchial skeleton (Taverne 1999: 298), some teeth are identified as teeth associated with the second basibranchial (p. 314, fig. 2). We code this character as [1] for this genus, pending further confirmation. A patch of endopterygoid teeth comparable to those of *Gonorynchus* is present in the new genus; basibranchial teeth are observed as well, according to Pittet *et al.* (present volume).

41 (38).   Ectopterygoids: well developed, ectopterygoid overlapping with the ventral surface of the autopalatine by at least 50% [0]; well developed, with three branches in lateral view, reduced but direct contact with autopalatine [1]; reduced, articulating

with the ventral surface of the autopalatine by at most 10% through cartilage, resulting in a loosely articulated suspensorium [2]; absent as distinct ossifications [3]. State 1 is for †*Apulichthys*, according to the description and restoration by Taverne (1997: 412, fig. 7; pers. comm. 2007); this is very different from any other taxon in our analysis. We coded [2] for †*Sorbininardus* as the most likely character state after the description and restoration by Taverne (1999: 83: "The ectopterygoid is a short and very fine osseous strip"; fig. 3; pers. comm. 2007), although the precise spatial relationships of the ectopterygoid and the palatine are not known. The ectopterygoids are absent as independent ossifications in *Cromeria* and *Grasseichthys* (state 3).

42 (39). Teeth on vomer and parasphenoid: absent [0]; present [1]. †*Lecceichthys* is coded [0], but see character 40 for comments on this genus.

43 (40). Anterior portion of vomer: horizontal [0]; anteroventrally inclined, nearly vertical [1]; dorsally curved [2]. According to Britz and Moritz (2007), this character is polymorphic in *Cromeria*, since the type species, *C. nilotica*, presents a straight, horizontal vomer (state 0), and the other species, *C. occidentalis*, has an inclined vomer (state 1, the only one coded by Grande and Poyato-Ariza 1999). We therefore code this character as [0 and 1] simultaneously for this genus. The anterior portion of the vomer appears anteroventrally inclined in the restoration of †*Sorbininardus* by Taverne (1999: figs. 2, 3; state 3).

44.   Spatial relationship between vomer and mesethmoid anteriorly: vomer and mesethmoid ending at about the same anterior level [0]; mesethmoid extending anteriorly beyond the level of anterior margin of vomer [1]; vomer extending anteriorly beyond the level of anterior margin of mesethmoid [2]. In †*Charitopsis*, †*Charitosomus*, *Gonorynchus*, †*Notogoneus* (Grande and Grande 2008), and †*Sorbininardus* (Taverne 1999: fig. 3) the mesethmoid extends anterior to the level of the anterior margin of the vomer (state 1). In †*Apulichthys* (Taverne 1997: fig. 4), *Chanos* (e.g., Arratia 1992: fig. 5A; pers. obs.), *Cromeria nilotica* (e.g., Grande 1994, Britz and Moritz 2007, showing that in *C. occidentalis* both bones end at about the same level, so we simultaneously coded as polymorphic [0 and 2] for this genus), *Grasseichthys* (e.g., Britz and Moritz 2007), †*Hakeliosomus*, †*Judeichthys*, *Kneria*, *Parakneria*, *Phractolaemus* (e.g., Thys van den Audenaerde 1961: fig. 15), †*Ramallichthys*, and †*Tharrhias* (e.g., Blum 1991b: 296; pers. obs.) the vomer extends anterior to the level of the anterior margin of the mesethmoid (state 2).

45 (41).   Articular head of hyomandibular bone: double, with both articular surfaces placed on the dorsal border of the main body of the bone [0]; double, with the anterior articular surface forming a separate head from the posterior articular surface [1]. According to the new restoration by Taverne and Gayet (2006: fig. 2), the articular and posterior surfaces apparently form separate heads, at least in lateral view, in †*Halecopsis* (coded as 1).

46 (42).   Metapterygoid process of hyomandibular bone: absent [0]; present, anterior [1]; present, ventral [2]. In their criticisms of this character, Britz and Moritz (2007: 42) seem to confuse this process with the anterior membrane of the hyomandibula, broadly present in osteichthyans. This character refers to a distinct process (therefore a projection, not a membranous outgrowth) of the border of the hyomandibular bone that articulates the hyomandibula with the metapterygoid (and sometimes the endopterygoid as well). In the primitive state, this process is absent, and the articulation with the metapterygoid is simply made through part of the anterior border of the bone, without any special process, in, for instance, *Brycon* and †*Diplomystus* in the outgroup (Weitzman 1962: fig. 10; Chang and Maisey 2003: fig. 10), †*Charitopsis* (e.g., Gayet 1993b: 78–79), †*Charitosomus* (e.g., Gayet 1986a: 32; 1993b: fig. 29), *Gonorynchus* (e.g., Grande 1996: fig. 5A), and †*Notogoneus* (Grande and Poyato-Ariza 1999: fig. 9) among the ingroup. There is no process on the hyomandibular bone in †*Halecopsis* either, as coded by Grande and Poyato-Ariza (1999: table 1) and confirmed by Taverne and Gayet (2006: 111, fig. 2). The metapterygoid process of the hyomandibular bone is pointed to blunt in shape and may or may not be placed on the anterior membrane, but always represents a distinct projection of the bone that abuts against the metapterygoid. The presence of this process was the unique derived state in Grande and Poyato-Ariza (1999), but our character revision has shown that there are two different processes, or at least two different positions for this process. It is placed on the anterior border of the hyomandibula in, for instance, *Chanos* (e.g., Grande and Poyato-Ariza 1999: fig. 10A), †*Dastilbe* (e.g., Poyato-Ariza 1996a: fig. 2A, Dietze 2007: figs. 4, 5, 8), †*Parachanos* (e.g., Poyato-Ariza 1996a: fig. 2B), and †*Tharrhias* (e.g., Poyato-Ariza 1996a: fig. 2C), coded as [1]; and on the ventral border of the hyomandibula in *Cromeria* (Britz and Moritz 2007: fig. 11a), *Grasseichthys* (Britz and Moritz 2007: fig. 15a), *Kneria* (e.g., Britz and Moritz 2007: fig. 18c), *Parakneria* (e.g., Grande 1994:

fig. 15), *Phractolaemus* (Thys van den Audenaerde 1961: fig. 17), and †*Sorbininardus* (Taverne 1999: fig. 2); coded as [2]. We could not corroborate Gayet's observations of a spine on the hyomandibula in †*Ramallichthys* (e.g., Gayet 1986a: fig. 26) and certainly not in †*Judeichthys* (e.g., Gayet 1985: 72: fig. 5). New specimens seem to favor absence, but we conservatively code these two taxa as unknown [?].

47 (43).   Ossified interhyal: present [0]; absent as an independent ossification [1]. An ossification of †*Sorbininardus*, partly hidden under fragments of infraorbitals, is identified as a ceratobranchial plus hypobranchial by Taverne (1999: fig. 2; confirmed by pers. comm. 2007). However, this position is too anterior for elements of a branchial arch and, moreover, represents the approximate position and spatial relationships that correspond to the interhyal: anteroventrally articulating with the quadrate and posterodorsally directed towards the hyomandibular bone (although the distal end is not preserved, it is directed towards this bone). It seems, though, that the relative size and stoutness of this bony element are excessive for a standard interhyal (Taverne pers. comm. 2007). Therefore, we code this character as [?] for this genus, due to uncertainty in the interpretation of that structure. An ossified interhyal is absent in *Cromeria* and *Grasseichthys* (derived state).

Branchial Arches

48 (44).   Teeth on fifth ceratobranchial: present [0]; absent [1].
49 (45).   First basibranchial in adult specimens: ossified [0]; unossified [1]. As a word of caution concerning fossil forms, in that it is never possible to know whether an absence of ossified bone is also an absence of its corresponding unossified cartilage, we changed the description of this character from presence/absence to ossified/unossified.
50 (46).   Fifth basibranchial in adult specimens: cartilaginous [0]; ossified [1]. Following the revision of this character by Britz and Moritz (2007), it is coded as [1] for *Grasseichthys* in the present analysis (as for *Cromeria*, *Kneria*, and *Parakneria*).
51 (47).   First pharyngobranchial in adult specimens: ossified [0]; unossified [1]. See comment on character 49 above.

Cheek Bones

52 (48).   Size of opercular bone: normal, about one quarter of the head length [0]; expanded, at least one third of the head length [1].

53 (49).   Shape of opercular bone in lateral view: rounded/oval [0]; triangular [1]. This character was previously coded as [0] for †*Halecopsis*, but it is now coded as [1] following the new restoration by Taverne and Gayet (2006: fig. 2).

54 (50).   Opercular spines: absent [0]; present [1]. Among the genera newly included for the present study, spines on the posterior border of the opercular bone are present in †*Sorbininardus* only (Taverne 1999: 84, figs. 2, 4).

55 (51).   Opercular apparatus on external surface of operculum: absent [0]; present [1]. Males of the genus *Kneria* exhibit an external structure called an opercular apparatus, consisting of two parts: a cup-like structure positioned on the opercle and a posterior component identified as a flange (Grande and Young 1997). Contrary to Britz and Moritz (2007), the posterior component or flange is indeed part of the opercular apparatus and its identification is not a misnomer. The entire structure works as an adhesive device to secure the fish in fast-moving water, possibly during courtship and mating (pers obs.).

56 (52).   Opercular borders: free from side of head [0]; partly or almost completely connected to side of head with skin [1]. As stated, the derived character is present in *Cromeria*, *Gonorynchus*, *Kneria*, *Grasseichthys*, *Parakneria*, and *Phractolaemus*. The differences in opening size among these genera commented on by Britz and Moritz (2007: 30) are difficult to delimit with precision as distinct character states, since they are part of a progressive process of reduction; and they are all contained in this derived state as defined here and also as described by these authors (p. 30).

57 (53).   Angle formed by preopercular limbs: obtuse [0]; approximately straight [1]; acute [2]. The angle is clearly obtuse in †*Apulichthys* (Taverne 1997: 412, fig. 3; coded as primitive state). That is also unmistakably the case of †*Lecceichthys* according to its restoration by Taverne (1999: fig. 2). Although this angle is mentioned to be slightly greater than 45° in the text (p. 298), we assume this to be an error (it was meant to read "90°", as confirmed by Taverne pers. comm. 2007). We therefore code this character as [0] for this genus. The angle formed by the anterior border of the preopercular limbs is obtuse in †*Sorbininardus* (Taverne 1999: fig. 2), and so is the angle formed by the two corresponding portions of the preopercular canal, despite the shape of the posterior expansion of the bone (coded as 0).

58 (54).   Posterodorsal limb of preopercular bone: well developed [0]; reduced, correlated with expansion of anteroventral limb that meets its fellow along the ventral midline [1].

59.   Ridge on anteroventral limb of preopercular bone: absent [0]; present [1]. Character suggested by Cavin (pers. comm. 2006). There is a conspicuous ridge running along most of the anteroventral limb of the preopercular bone, often ascending part of the posterodorsal limb as well, in †*Aethalionopsis* (Fig. 7.5A), †*Dastilbe*, †*Parachanos* (e.g., Poyato-Ariza 1996a: fig. 2B), †*Tharrhias*, and the new genus by Pittet *et al.* (present volume). This ridge is not directly related with the main preopercular sensory canal but with the distal openings of its branches, so that it also runs more or less parallel to it, in the direction of the long axis of the preopercular limb, but it is always placed more ventrally (distally) on the bone than the main canal, following the distal openings of the branches. The inconspicuous ridges on the preopercular of †*Apulichthys* and †*Sorbininardus* correspond to the preopercular canal, as labelled by Taverne (1997: fig. 3; 1999: fig. 2; state 0).

60.   Preopercular expansion distal to the terminal openings of the preopercular canal branches: absent, preopercule not enlarged [0]; present, restricted to the posteroventral corner [1]; present in posteroventral corner and part of the posterodorsal limb [2]; present in anteroventral limb only [3]. Revised from Poyato-Ariza (1996a, character 33). The preopercular bone lacks any notorious enlargement in the outgroup genera and in †*Apulichthys*, †*Charitopsis*, †*Charitosomus*, *Cromeria*, †*Dastilbe*, *Gonorynchus*, *Grasseichthys*, †*Hakeliosomus*, †*Judeichthys*, *Kneria*, †*Lecceichthys*, †*Notogoneus*, *Parakneria*, and †*Ramallichthys* (primitive state). We consider that the preopercular has an expansion when it extends well beyond the terminal openings of the preopercular canal branches. An expansion is present in the posteroventral corner of the preopercular bone in †*Aethalionopsis*, †*Gordichthys*, †*Parachanos*, and †*Rubiesichthys* (state 1). A larger expansion, occupying the posteroventral corner plus a good part of the posterodorsal limb, is present in *Chanos*, †*Halecopsis*, †*Sorbininardus*, and †*Tharrhias* (state 2). A large expansion on the anteroventral limb is present only in *Phractolaemus* (state 3). In the new genus the posterior border of the preopercular bone is not preserved and the posterior part of the bone is broken and imperfectly preserved, so this character is conservatively coded as unknown. For further details, see Poyato-Ariza (1996a: 23–24, character 33, fig. 12).

61.   Suprapreopercular bone: absent [0]; present as a relatively large, flat bone [1]; present as tubular ossification(s) [2]. Suggested by Cavin (pers. comm. 2006) and adapted and expanded from

Poyato-Ariza (1996a: 22, character 31). The presence of a suprapreopercular bone is a relatively common occurrence among ostariophysan fishes, and gonorynchiforms are no exception. There is absence in *Brycon*, †*Diplomystus*, and *Opsariichthys* (the whole outgroup). Among the ingroup, a suprapreopercular bone is also absent (state 0) in *Cromeria*, *Gonorynchus*, *Grasseichthys*, †*Halecopsis*, †*Lecceichthys*, †*Sorbininardus*, and the new genus. Contrary to Gayet (1986a), we could not confirm the presence of a suprapreopercular bone or its impressions in †*Ramallichthys*, †*Hakeliosomus*, †*Judeichthys*, †*Charitopsis*, and †*Charitosomus*. In addition, Grande and Grande (2008) did not observe a suprapreopercle in †*Notogoneus osculus*. The suprapreopercle is a very delicate bone, often broken and lost during fossilization, although its presence can be deduced from the occurrence of a small concave surface for its accommodation on the opercular bone (e.g., holotype of †*Rubiesichthys gregalis*, Wenz 1984, Poyato-Ariza 1996a: fig. 3C). A large suprapreopercular bone directly anterior to the dorsal part of the opercular bone is present in †*Aethalionopsis* (re-examination of specimens of †*Aethalionopsis* at the IRSNB, such as the impressions clearly visible on paralectotype P.2402, provides additional support for the presence of a broad suprapreoperculum in this genus; see Fig. 7.5A), *Chanos*, †*Dastilbe*, †*Gordichthys*, †*Parachanos*, *Phractolaemus*, †*Rubiesichthys*, and †*Tharrhias* (state 1; see also Poyato-Ariza 1996a: 22, character 31). In †*Apulichthys*, *Gonorynchus* (e.g., AMNH 57120), *Kneria*, and *Parakneria* they are present and reduced to narrow tubes that run alongside the operculum (state 2); there is one small ossification in †*Apulichthys* and *Parakneria*, and two in *Kneria*. Although the otophysans examined for this study as part of the outgroup lacked suprapreoperculi, Fink and Fink (1981) reported many other otophysans with suprapreoperculi; the character may be a case of mosaic evolution, with suprapreoperculi evolving more than once within the Ostariophysi.

62.   Spine on posterior border of subopercular bone: absent [0]; present [1]. A conspicuous spine on the posterior border of the subopercular bone is present only in †*Charitopsis*, †*Charitosomus*, and †*Sorbininardus* (e.g., Gayet 1993b, Taverne 1999; state 1).

63.   Major axis of subopercular bone in lateral view: inclined [0]; subhorizontal [1]; subvertical [2]. Regardless of the particular extent, shape, and ornamentation, the major (longest) axis of the subopercular bone, as seen in lateral view, is usually

inclined, forming an angle of 30° to 60° with the saggital plane of the body (primitive state). It is nearly horizontal in *Chanos*, †*Dastilbe*, the new genus (although this region is imperfectly preserved), †*Parachanos*, and †*Tharrhias* (state 1; e.g., Poyato-Ariza 1996a: figs. 1B, 2; Grande and Poyato-Ariza 1999: fig. 10; Dietze 2007: figs. 3, 4, 8; pers. obs.), and nearly vertical in †*Sorbininardus* only (Taverne 1999: figs. 2, 4).

64 (55).   Subopercular clefts: absent [0]; present [1]. Deep clefts on the posterior margin of the subopercular bone are present in †*Notogoneus* only; their number varies at specific level (for further details see Grande and Poyato-Ariza present volume).

65 (56).   Interopercular bone: relatively broad and positioned medioventral to preopercular bone [0]; reduced to a long thin spine and positioned mediodorsal to preopercular bone [1]; reduced to a long thin spine and positioned lateroventral to preopercular bone [2]. The primitive state of this character is the usual elongated, but flat and relatively large, interopercular bone that is placed medial and ventral to the anteroventral limb of the preopercular bone, as, for instance, in *Chanos* (e.g., Grande and Poyato-Ariza 1999: fig. 10A, B). In *Phractolaemus* the interopercular bone is mostly a thin rod that is placed medial and dorsal to the preopercular bone (e.g., Grande and Poyato-Ariza 1999: fig. 11B), and thus coded as [1]. As pointed out by Britz and Moritz (2007), the morphology of the interopercular bone in *Cromeria* and *Grasseichthys* is similar to that of *Phractolaemus*, but not its position (state 1 is coded for *Phractolaemus* only, state 2 for *Cromeria* and *Grasseichthys*).

66.   Spine on posterior border of interopercular bone: absent [0]; present [1]. There is a spine on the posteroventral corner of the interopercular bone only in †*Lecceichthys* (Taverne 1998: 298, fig. 2; not to be confused with the little posterodorsal ascending process of the next character).

67.   Posterodorsal ascending process of interopercular bone: absent [0]; present [1]. A small ascending process is present on the posterodorsal corner of the interopercular bone (better visible on medial view) in *Cromeria*, *Grasseichthys*, *Kneria*, *Parakneria*, and *Phractolaemus* (e.g., Grande 1996, Britz and Moritz 2007). The interpretation is difficult in *Grasseichthys*; it seems present in our observed material and is shown on the photo by Britz and Moritz (2007: fig. 15a), although it is not mentioned in their text.

68 (57).   Number of infraorbitals: five or more [0]; four [1]; three or fewer [2]. Britz and Moritz (2007: 36) reported polymorphism in the

number of infraorbitals in *Cromeria*: four for *C. occidentalis* (five minus the dermosphenotic, which is not considered an infraorbital) and none in *C. nilotica*. We therefore code this character simultaneously as [1 and 2] for this genus in the present analysis. It is, however, difficult to interpret the number of infraorbitals in *Grasseichthys* according to these authors. In their osteological description, they state only that "…the antorbital and lacrimal are lacking" (p. 20). Nothing else is mentioned about the infraorbitals in their osteological description of *Grasseichthys*. Later on, in their discussion, they state that "We found four infraorbitals in *Grasseichthys*, but they are also poorly ossified and difficult to see except in transmitted light" (p. 36). We assume that these four infraorbitals do not include the lacrimal since the authors state that the lacrimal (first infraorbital) is missing. In addition, they do not specify whether they include the dermosphenotic in their count of the infraorbital bones. Lacking any graphics showing the infraorbitals of *Grasseichthys*, the exact number of infraorbitals in this genus is difficult to discern. Pending further evidence, however, we simultaneously code this character as [1 and 2] for *Grasseichthys*. Britz and Moritz (2007: 36) also corrected the erroneous coding of this character for *Kneria* and *Parakneria* in Grande and Poyato-Ariza (1999); these genera show four infraorbitals (including lacrimal, but not dermosphenotic; e.g., Lenglet 1974: figs. 1, 2) and are therefore coded [1] for this character in the present analysis. We have consistently included the lacrimal, but not the dermosphenotic, when counting the number of infraorbitals. There are four infraorbitals (state 1) in †*Apulichthys* according to Taverne (1997: 411, fig. 3) and in †*Halecopsis*, according to Taverne and Gayet (2006: 111, fig. 2). This is also the case of †*Lecceichthys* (Taverne 1999: 297, fig. 2). A space is present in the the posterior corner of the orbit, between the fourth infraorbital and the dermosphenotic. A putative fifth infraorbital might have been present (Taverne pers. comm. 2007) but lost in the specimen (easily done since the posterior infraorbitals are tubular). Therefore, this character is coded as [0 and 1] simultaneously for this genus.

69 (58).   Infraorbital bones not including lacrimal: well developed [0]; reduced to small, tubular ossifications [1]; hypertrophied [2]. The infraorbital bones are reduced in *Kneria* and *Parakneria* (state 1; coded as 0 by typing error in Grande and Poyato-Ariza 1999, table 1). In †*Lecceichthys* (Taverne 1999: 297–298, fig. 2) infraorbital two is large and infraorbitals three and four are

tubular, so this character is coded [0 and 1] simultaneously for this genus. The observable remains of infraorbitals in †*Sorbininardus* (Taverne 1999: fig. 2) indicate that they were large, well developed (coded as 0).

70 (59).   Lacrimal: flat and comparable in length to subsequent infraorbitals [0]; tube-like and extremely long, without keel [1]; flat, long and large, with keel near lower edge [2]; long and large, with spines and crests [3]. †*Apulichthys* and †*Lecceichthys* are coded as [0] according to the respective restorations and descriptions by Taverne (1997, 411, fig. 3; 1999, 297, fig. 2), and so is the new genus according to the description by Pittet *et al.* (present volume). State 3 is new in this analysis, added for †*Sorbininardus* according to the restoration and description by Taverne (1999: 83, fig. 2). The condition in †*Notogoneus* is in need of revision (Grande and Grande 2008), so this character is coded as unknown for this analysis, pending further confirmation.

71 (60).   Supraorbital: present [0]; absent [1]. This character was coded as unknown in *Grasseichthys* by Grande and Poyato-Ariza (1999) because of the difficulty in observing the infraorbitals in the specimens examined. A supraorbital is present in *Grasseichthys* (as in *Cromeria*) according to Britz and Moritz (2007); we therefore code it as [0] for this genus in the present analysis. In †*Sorbininardus* there is no supraorbital (Taverne 1999: 83; coded as 1). It is absent in †*Charitopsis* (coded as 1).

Vertebrae

72.   Two anteriormost vertebrae: as long as posterior ones [0]; shorter than posterior ones [1]. In ostariophysans the two anteriormost vertebrae are shorter than the more posterior ones (Fink and Fink 1981), and this is the case of all the taxa in which this character was accessible, except the outgroup genus †*Diplomystus*.

73.   Second abdominal centrum: as long as first [0]; shorter than first [1]. In *Phractolaemus* the second abdominal centrum is shorter than the first. This condition seems unique to *Phractolaemus*, and is found in all specimens examined regardless of size and age. In nearly all other fishes examined where this region is accessible, the second abdominal is always larger than the first. It must be mentioned that †*Apulichthys* is restored as having a second vertebral centrum possibly shorter than the first (Taverne 1997: fig. 8) but the first vertebra of this genus is difficult to distinguish because it is covered by opercular debris (p. 414). However, its contour can be outlined from its relief under the bony pieces,

showing that this vertebral centrum is longer than the second one (Taverne pers. comm. 2007). Consequently, the character is coded as [1] for this genus as well, with precautions.

74.   Autogenous neural arch anterior to arch of first vertebra: present [0]; absent [1]. This condition is unobservable in the specimens of †*Aethalionopsis*, †*Charitopsis*, †*Gordichthys*, and †*Halecopsis* examined. In all other fossil gonorynchiforms examined, the first neural arch either directly abuts the back of the skull (i.e., †*Tharrhias*, specimens AMNH 11916, NHM P. 54331: Poyato-Ariza 1996a: fig. 14; 1996c: fig. 4; Fig. 7.4) or is in such close proximity to the back of the skull that an additional neural arch element fitting between the two would not be possible (state 1).

75 (62).   Neural arch of first vertebra and exoccipitals: separate [0]; in contact [1].

76 (63).   Neural arch of first vertebra and supraoccipital: separate [0]; in contact [1].

77 (64).   Spine on the neural arch of first vertebra: present, well developed [0]; present but reduced [1]; absent [2]. Contrary to Britz and Moritz (2007: 33), Grande and Poyato Ariza (1999) identified a small spine-like structure extending from the posterior margin of neural arch 1 of *Phractolaemus* as a neural spine. Pending additional developmental data, we have coded *Phractolaemus* as [1]. If additional data shows that the first neural spine is indeed absent in *Phractolaemus*, as it is in the Kneriinae [2], this would lend further support for the monophyly of the more inclusive Kneriidae.

78. (61)   Anterior neural arches: no contact with adjoining arches [0]; abutting contact laterally with adjoining arches, no overlapping [1]; overlapping contact with adjoining arches laterally [2]. Our new observations indicate that there is no contact in †*Notogoneus* (coded as 0); slight contact, but no strictly overlapping, in *Gonorynchus* and Recent African forms (coded as 1); and contact plus overlapping in fossil Middle Eastern forms (coded as 2). In †*Apulichthys*, according to Taverne (1997: 411, figs. 1, 8), there is a very slight contact between arch 1 and arch 2 only, whereas all the others are well separated (coded as 0). In turn, †*Sorbininardus* shows laterally contacting arches wherever accesible (Taverne 1999: 86, fig. 4; coded as 1).

79.   Neural arches 5–10 in adults: fused to centra [0]; autogenous, at least laterally [1]. For comparison with the otophysan outgroups that have fused anterior neural arches as part of the Weberian apparatus, we restrict this character to those before

the dorsal fin but posterior to the Weberian apparatus. Within the ingroup, fusion is exhibited by †*Apulichthys* (Taverne 1997: 416), *Cromeria, Gonorynchus, Grasseichthys, Kneria,* †*Notogoneus, Parakneria, Phractolaemus,* and †*Sorbininardus* (state 0). Although this region is seldom accurately exposed in fossil forms (e.g., Grande 1996: fig. 11A), especially when specimens of a given taxa are scarce, our observations for the present analysis indicate that the neural arches anterior to the dorsal fin are autogenous in †*Aethalionopsis, Chanos,* †*Dastilbe* (e.g., AMNH 12721R, 12736, 19432), †*Gordichthys,* †*Parachanos,* †*Rubiesichthys,* †*Tharrhias* (Fig. 7.4), and the new genus. Although Gayet (e.g., 1993a-c) has illustrated free neural arches in this region for †*Charitopsis,* †*Charitosomus,* †*Hakeliosomus,* †*Judeichthys,* and †*Ramallichthys,* we have conservatively coded these taxa with a [?] because of the uncertainty in this character due to conditions of preservation, or obstructions such as intermuscular bones overlying the area.

80.   Neural arches of vertebrae posterior to the dorsal fin in adults: fused to centrum [0]; autogenous, at least laterally [1]. As for the previous character, fusions are primitive in this particular phylogenetic context. The common condition observed in all taxa in our analysis is a fusion of the neural arches of vertebrae posterior to the dorsal fin. Only in †*Aethalionopsis* are these neural arches autogenous (Taverne 1981: fig. 5; pers. obs.).

81 (65).   First two anterior parapophyses: autogenous [0]; fused to centra [1].

82 (66).   Rib on third vertebral centrum: similar to posterior ones [0]; widened and shortened [1]; modified into Weberian apparatus [2]. We still code †*Aethalionopsis* as unknown [?], although the condition of this character in this genus deserves some comment. This region is never well preserved, always concealed by the operculum in the material, but fragments of an enlarged anterior rib are visible in the type series whenever this region is partly accessible, although it is not possible to be certain whether or not this rib is attached to the third vertebral centrum (Grande and Poyato-Ariza 1999: 223). The description of †*Aethalionopsis* ribs by Taverne (1981) only states that "Les vertèbres abdominales portent de longues et fines côtes ventrales enserrant la cavité viscérale" ("the abdominal vertebrae bear long and thin ventral ribs sorrounding the visceral cavity") (p. 968). This refers to the general shape of ribs, and no specific comments are made about the rib on the third vertebral centrum; as a matter of fact, Taverne's restoration of the body of this genus (Taverne 1981: fig. 1) does

depict an anterior enlarged rib, which has been confirmed by the additional revision of the type series material for this new analysis. We conservatively code †*Aethalionopsis* as [?] because it is not possible to be certain that the expanded rib is attached to the third centrum, but we do note the expansion of the anteriormost rib in this genus. We code the second derived state for the further modifications involved in the formation of the Weberian apparatus (*Brycon* and *Opsariichthys*, part of the outgroup).

Intermuscular Bones

83 (67).   Paired intermuscular bones consisting of three series: epipleurals, epicentrals, and epineurals: absent [0]; present [1]. Epicentrals are absent in †*Apulichthys* according to Taverne (1997: 416, figs. 1, 9), so this character is coded as primitive for this genus. The three series are present in †*Lecceichthys* and †*Sorbininardus* after Taverne (1998: 301, 1999: 87; state 1). Cephalic ribs are considered modified epicentrals (Patterson and Johnson 1995), and they are specifically dealt with by character 6.

84 (68).   Anterior (first six) epicentral bones: unmodified, no differences in size from others [0]; highly modified, much larger than posterior ones [1]; epicentrals in anterior vertebrae absent [2]. See previous character for comments on the coding in †*Apulichthys* (state 2).

85 (70).   Size of anterior supraneurals (whatever the number present): narrow [0]; large [1]. This character refers to the size of the anterior supraneurals, whatever the number present (a feature that is treated by character 87 below). When enlarged, supraneurals usually make contact with their neighbors. After revision (Grande and Grande 2008), †*Notogoneus* is coded as [0], since the supraneurals are narrow and do not make contact with each other (unlike *Gonorynchus*, where they are larger and make a slight contact with each other, state 1). The accessible regions of the enlarged supraneurals of †*Sorbininardus* indicate that they are enlarged (Taverne 1999: fig. 4; derived state). Although †*Dastilbe* has been restored without supraneurals by Blum (1991a: fig. on p. 274) and Dietze (2007: fig. 8b, in contrast with supraneurals present on her fig. 9a), this genus does exhibit supraneurals whenever this region is accessible and well preserved; they do not seem to be enlarged (e.g., AMNH 12736, state 0); however, this character requires further confirmation

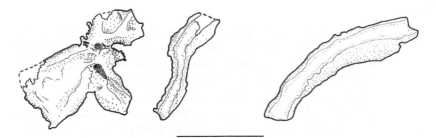

Fig. 7.7 †*Gordichthys conquensis*, anterior supraneurals one to three (left to right) as preserved in paratype MCCM LH-4989-R (Museo de las Ciencias de Castilla-La Mancha, Cuenca, Spain). Modified from Poyato-Ariza (1996a: fig. 15), in order to show the actual spatial relationship of these elements. Scale bar represents 0.5 mm.

of its ontogenetic and specific variability within the genus. The remains of the anterior supraneurals preserved in †*Lecceichthys* suffice to indicate that they are very broad and in contact (Taverne 1999: 300–301, fig. 3; state 1). In *Phractolaemus*, the only supraneural present is very large, although it cannot obviously be in contact with any neighbors (as implied by the unfortunate wording in Grande and Poyato-Ariza 1999: 224, and pointed out by Britz and Moritz 2007: 31), but it is comparable in relative size to the supraneurals of, for instance, *Gonorynchus* (coded as 1), and larger than those of *Chanos*, for instance (which is coded as 0). Therefore, we maintain the coding of this character as [1] in *Phractolaemus*. In *Grasseichthys*, *Kneria*, and *Parakneria* the supraneurals are smaller than those in other genera, regardless of the number of supraneurals present (state 0; see character 87 for the reduction in the number of supraneurals). This character is not applicable in *Cromeria*, which has lost all supraneurals (coded as N).

86 (71).   Posterior process on the posterior border of first supraneural: absent [0]; present [1]. See Fig. 7.7.

87 (72).   Number of supraneurals: several supraneurals in a long series [0]; two or fewer supraneurals [1]. The outgroup genera plus most of the ingroup genera present a more or less long series of several supraneural bones, in variable number, extending beyond the fourth vertebra. In *Cromeria*, *Grasseichthys*, *Kneria*, *Parakneria*, and *Phractolaemus* the number of supraneurals is reduced; there are two (*Grasseichthys*) to none (*Cromeria*) (e.g., Grande 1994: figs. 5, 6, 9).

Fins

88 (73).   Postcleithra: present [0]; absent [1]. The character is coded as primitive for †*Apulichthys* after the restoration by Taverne (1997: fig. 3), and for †*Lecceichthys* as well, since the fragmentary "hypercleithrum" in Taverne (1999: 299, figs. 2, 3) seems better interpreted as a fragmentary postcleithrum, due to its relative position with respect to the cleithrum (confirmed by Taverne pers. comm. 2007). This character was previously coded as unknown for †*Halecopsis*, but it is now coded as [1] following the description by Taverne and Gayet (2006: 111; "there is no postcleithrum"). In †*Sorbininardus* there is no postcleithrum (Taverne 1999: 86, fig. 4) or any other bone that could be interpreted as such (coded 1).

89 (74).   Lateral line and supracleithrum: supracleithrum pierced through dorsal region [0]; supracleithrum pierced all through its length [1]; lateral line does not pierce supracleithrum [2].

90 (75).   Fleshy lobe of paired fins: absent [0]; present [1].

91 (76).   Caudal fin morphology: elongated, posteriorly forked [0]; higher than long, slightly incurved posteriorly [1]; crescent-shaped [2].

92.   Fringing fulcra in dorsal lobe of caudal fin: present [0]; absent [1]. Among the ingroup, a few small fringing fulcra are present in †*Aethalionopsis* only (Taverne 1981: fig. 7; pers. obs. on specimen IRSNB 1249). Fringing fulcra are completely absent in all other genera from the ingroup.

93 (77).   Caudal scutes: absent [0]; present [1].

Caudal Endoskeleton

94 (78).   Ural centra (u1, u2), preural centrum one (pu1), and uroneural one (un1): autogenous [0]; fused [1]; fused except for ural centrum two, which is autogenous [2]. Recent revision has shown that the caudal endoskeleton is not well preserved in †*Charitopsis*, so we conservatively code all caudal endoskeleton characters as unknown in this genus, pending further confirmation. Taverne (1999: 303, fig. 5) describes and restores a compound terminal centrum in †*Lecceichthys*; although this particular region of the caudal endoskeleton is not completely preserved in the holotype and only specimen of this genus, this restoration is made on the basis of the remains (as imprints) that are observable. This character is coded as [1] for this genus for the present analysis (confirmed by Taverne pers. comm. 2007).

State [2] is coded for †*Apulichthys*, which, unlike any other taxon, presents a fusion of preural centrum one, ural centrum one and uroneural one, but ural centrum two remains autogenous (Taverne 1997: 419, fig. 11; pers. comm. 2007).

95 (79).   Neural arch and spine of preural centrum one: both well developed, spine about half as long as preceding ones [0]; arch complete and closed, spine rudimentary [1]; arch open, no spine [2]. The primitive state, a neural arch with a well-developed spine on preural centrum one, usually about half as long as preceding ones, is found in †*Diplomystus* (outgroup), †*Apulichthys* (Taverne 1997: 420, fig. 11), and †*Tharrhias* (primitive state). A well-developed arch with a reduced spine, less than half as long as preceding neural spines and usually rudimentary, is exhibited by *Brycon* and *Opsariichthys* (outgroup), †*Aethalionopsis*, †*Charitosomus*, †*Dastilbe*, *Gonorynchus*, †*Gordichthys*, *Grasseichthys*, †*Judeichthys*, †*Hakeliosomus*, *Kneria*, †*Notogoneus*, †*Parachanos*, †*Ramallichthys*, †*Rubiesichthys*, and †*Sorbininardus* (state 1). The coding in †*Judeichthys* must be taken with caution; the spine seems slightly longer than that of, for instance, *Gonorynchus*, †*Notogoneus*, and †*Ramallichthys*, but is still notably shorter than in, for instance, †*Tharrhias*; it is comparable to that of †*Hakeliosomus*, and this is also shorter than in †*Tharrhias*. In any case, †*Judeichthys* is known from only one specimen, so variation is not accounted for, and from our personal observations, we find there is considerable variation in the caudal skeleton of gonorynchoids. See Grande and Grande (2008) and Grande and Arratia (present volume) for further information on variation. A reduced, open neural arch (occasionally barely closed dorsally) with total absence of spine is found in *Chanos*, *Cromeria*, *Parakneria*, and *Phractolaemus* (state 2). As many other characters involving progressively reduced structures, it may be difficult to establish the precise limit between states in some cases (e.g., †*Charitosomus*, †*Diplomystus*, and †*Judeichthys*), but the limit seems better defined with more than one derived state. This character is unknown in †*Lecceichthys* (Taverne 1999: 303, fig. 5; pers. comm. 2007). For illustrations of caudal endoskeletons of the genera mentioned in this and other caudal endoskeleton characters, see Britz and Moritz (2007), Fink and Fink (1981), Gayet (1985, 1986a, 1993b), Grande (1996), Lenglet (1974), Poyato-Ariza (1996a, b), Poyato-Ariza *et al.* (2000), Taverne (1981), Thys van den Audenaerde (1961), Weitzman (1962), and Wenz (1984).

96 (80).   Uroneurals: arranged in a linear series [0]; arranged in a double series [1]. The distal part of the uroneural series is not known in †*Lecceichthys* (Taverne 1999: fig. 5) and, although it is more likely to interpret a linear series, the character is conservatively coded as unknown in this genus.

97 (81).   Number of uroneurals: three [0]; two or one [1]; none [2]. We count the total number of uroneurals, and not only the autogenous ones, in order to keep this character independent from the next one, where uroneural fusion is dealt with. Within the ingroup, three uroneurals (state 0) are present in †*Aethalionopsis* (Taverne 1981: fig. 8; pers. obs.) and †*Tharrhias* (e.g., Poyato-Ariza 1996a: fig. 18). Two uroneurals (state 1) are present in †*Apulichthys*, *Chanos*, †*Charitosomus*, †*Gordichthys*, †*Judeichthys*, †*Notogoneus*, †*Ramallichthys*, †*Rubiesichthys*, and †*Sorbininardus*. One uroneural, fused to the caudal complex, is exhibited by *Cromeria*, *Gonorynchus*, *Grasseichthys*, *Kneria*, *Parakneria*, and *Phractolaemus* (state 2). There are three uroneurals in †*Hakeliosomus* according to Gayet (1993b: 42, fig. 17), but only two according to Patterson (1970: fig. 45) and to what can be interpreted on the photo by Gayet (1993b: pl. 3, fig. 4); this is confirmed by revision of the material (Grande and Grande 2008). Therefore, we code this character as [1] for †*Hakeliosomus*. There are at least two uroneurals in †*Lecceichthys* (Taverne 1999: 303, fig. 5), but the distal region of the uroneural series is not preserved in the specimen, so we conservatively code this character as [?] in this genus, because a third uroneural may or may not be present (Taverne pers. comm. 2007). The number of uroneurals in †*Dastilbe* is reportedly variable; although most specimens show two, exceptional cases show three (Poyato-Ariza 1996a) or even one only (Davis and Martill 1999). The variation in the number of uroneurals in the caudal endoskeleton, particularly when this number is so low, seems too important to be disregarded as a simple case of individual variation, as in Davis and Martill (1999) and Dietze (2007), without a detailed study. However, pending further studies of the ontogenetic, geographic, and possible taxonomic variation of this character in †*Dastilbe*, we code it as [1] (two uroneurals) for this genus, as this is the usual number in most observed specimens, especially in those that are larger and better preserved (Blum 1991a, Poyato-Ariza 1996a, Davis and Martill 1999, Dietze 2007, pers. obs.). The case of †*Parachanos* is also complex. The caudal endoskeleton is accurately preserved in the holotype only, and is partly broken and fractured (Fig. 7.8). In this

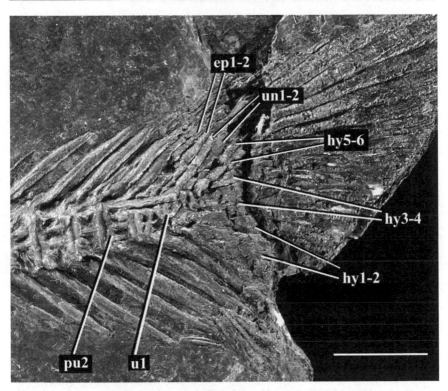

Fig. 7.8 Caudal endoskeleton of †*Parachanos aethiopicus*, lectotype, MNHN GAB 1 (Muséum national d'Histoire naturelle de Paris). ep, epurals; hy, hypurals; u1, ural centrum one; un, uroneurals; pu2, preural centrum two. New photo by P. Loubry, courtesy M. Véran. Scale bar represents 10 mm.

specimen, Taverne (1974b: 686, fig. 3) interpreted the occurrence of three uroneurals, whereas Poyato-Ariza (1996a) interpreted two. Additional revision of the holotype for the present analysis confirms the interpretation that uroneural 3 in Taverne (1974) is probably "a broken distal fragment of uroneural 2" (Poyato-Ariza 1996a: 28), so we code it as [1] (two uroneurals) for this genus (Fig. 7.8).

98 (82). Anterior extent of first uroneural: to anterior end of first preural [0]; to anterior end of second preural [1]; to anterior end of third preural [2]; uroneural fused to caudal fin complex [3]. In †*Diplomystus*, the autogenous first uroneural reaches the anterior end of the first preural centrum (primitive state). The autogenous un1 reaches the anterior end of the second preural centrum in †*Dastilbe*, †*Gordichthys*, †*Rubiesichthys*, and †*Tharrhias* (state 1), and of the third preural centrum in †*Parachanos* (paratype MNHN

GAB 4; the holotype, as shown by Fig. 7.8, has an anteriorly broken un1; the third preural centrum shows the groove for the articulation of the missing part of un1) (state 2). In †*Aethalionopsis* un1 is also autogenous and long, but the anterior region is never accurately preserved (coded ?). The first uroneural is fused to the caudal complex, which is formed by the ural centra and preural centrum one, in *Brycon* and *Opsariichthys* among the genera of the outgroup, and in †*Apulichthys* (with ural centrum two autogenous, see character 94), *Chanos*, †*Charitosomus*, *Cromeria*, *Gonorynchus*, *Grasseichthys*, †*Judeichthys*, *Kneria*, †*Notogoneus*, *Parakneria*, *Phractolaemus*, †*Ramallichthys*, and †*Sorbininardus*, among the ingroup. See character 94 for comments about †*Lecceichthys* (coded as 3).

99.   Uroneural two and second ural centrum: in contact [0]; separated [1]; uroneural two absent as an autogenous ossification [2]. Uroneural two and the second ural centrum articulate (are in contact) in †*Diplomystus* among the ingroup (e.g., Grande 1982: figs. 10, 12; Poyato-Ariza *et al.* 2000: fig. 9), and in †*Apulichthys* (Taverne 1997: 420, fig. 11), †*Dastilbe* (e.g., Poyato-Ariza 1996a: fig. 16; Dietze 2007: fig. 10), †*Gordichthys* (Poyato-Ariza 1994: figs. 15, 16A), †*Lecceichthys* (Taverne 1999: 303, fig. 5), †*Notogoneus* (e.g., Grande 1996: fig. 4B), †*Parachanos* (e.g., Poyato-Ariza 1999a: fig. 17), †*Rubiesichthys* (e.g., Poyato-Ariza 1996b: fig. 4), and †*Tharrhias* (e.g., Poyato-Ariza 1996b: fig. 4B; 1996a: fig. 18) among the outgroup (state 0). Uroneural two is separated from the second ural centrum in *Brycon* and *Opsariichthys* among the outgroup (e.g., Weitzman 1962: fig. 15; Fink and Fink 1981: fig. 23B) and in †*Aethalionopsis* (Taverne 1981: fig. 8; pers. obs.), *Chanos* (e.g., Fink and Fink 1981: fig. 23A; Poyato-Ariza 1996a: fig. 19), †*Charitopsis* (e.g., Gayet 1993b: fig. 42), †*Charitosomus* (e.g., Gayet 1993b: fig. 21; Grande 1996: fig. 4D), †*Ramallichthys* (e.g., Gayet 1986: fig. 37, pl. 2, fig. 5; Grande 1996: fig. 4E), and †*Sorbininardus* (as restored by Taverne 1999: fig. 7) among the ingroup (state 1). In †*Judeichthys* uroneural 2 does not contact the second ural centrum according to the interpretation by Gayet (1985: fig. 8); however, it does make contact according to Grande (1996: fig. 4C). Our new observations on the only known specimen of this genus for the present analysis confirm Gayet's (1985) interpretation; uroneural two appears a bit too long in Grande (1996: fig. 4C). Consequently, we code this character as [1] for †*Judeichthys*, pending further re-evaluation if and when additional material is found. In *Cromeria*, *Gonorynchus*, *Grasseichthys*, *Kneria*,

Parakneria, and *Phractolaemus* (e.g., Britz and Moritz 2007: figs. 6f, 12f, 17f; Grande 1996: fig. 4A; Lenglet 1974: fig. 24; Poyato-Ariza 1996a: figs. 20, 21; Thys van den Audenaerde 1961: fig. 24), uroneural two is absent as an autogenous ossification (state 2).

100 (83).   Parahypural and preural centrum 1: independent in adults [0]; fused only in large adults [1]; fused since early ontogenetic stages [2]. Revision by Grande and Grande (2008) has shown that the parahypural and preural centrum 1 are autogenous in †*Notogoneus* (state 0). In *Chanos*, there is a fusion of the parahypural with preural centrum 1 in large adults, but not in juvenile and subadult specimens (state 1). In other cases of taxa with ontogenetic series available for study, such fusion occurs since early ontogenetic stages. In the case of fossil taxa with no ontogenetic series available for study, we have consistently coded for the character state as observed and/or reported in the known adult specimens. The region between the parahypural and its centrum is broken in †*Lecceichthys* (Taverne 1999: fig. 5), but the ventral region of the preural one component is preserved so that one can make sure that the parahypural is fused (Taverne pers. comm. 2007; coded as 2 for this genus).

101 (84).   Reduction in the number of hypurals: six [0]; fewer than six [1]. The putative variation in the number of hypurals in †*Dastilbe* reported by Davis and Martill (1999: 728) is not backed up by photographs or accurate observational accounts. Their "bifid fifth hypural" seems better explained as a partial fusion between hypurals 5 and 6. Their ambiguous interpretations fit better with the usual arrangement of these elements, in which hypural 5 presents a small dorsal concavity, just posterior to its proximal articular head, where hypural 6 comfortably fits (e.g., Poyato-Ariza 1996a: fig. 16). Even their own illustration (Davis and Martill 1999: pl. 3, fig. 1) clearly shows six hypurals (they do not label the bones, so we ignore their precise interpretation of the elements of this caudal endoskeleton). Their reported variability in this crucial character is not tested to be either individual or specific. According to Dietze (2007: 14), "In a few of the smaller specimens, five hypurals instead of six are present in the caudal skeleton" and, admittedly, "This could be related to the degree of ossification and/or preservation. It is also possible that the sixth hypural is concealed beneath the fin rays." Our personal observations confirm that, whenever this region is well preserved, and not hidden by the proximal ends of the caudal fin rays, the

number of hypurals shown by this genus is six, so we maintain this coding for this genus, pending a much needed detailed revision of its individual-versus-specific variability. The distal region of the hypural series is not accessible in †*Lecceichthys* (Taverne 1999: fig. 5), where there are at least five hypurals, but there might be more (confirmed by Taverne pers. comm. 2007; coded as unknown). As pointed out above, comments by Britz and Moritz (2007) on this and other characters implicitly confuse character coding with character distribution; the latter will be commented on later in this chapter.

102 (85).   Hypurals 1 and 2 more commonly: independent [0]; partly fused to each other [1]; totally fused to each other [2]. We use the criterion of the prevalent condition because of the high individual variability of fusions between hy1 and hy2 with each other (and with the ural centrum): Grande and Grande (2008), Grande and Arratia (present volume). Revision by Grande and Grande (2008) has shown that the hypurals 1 and 2 are not fused to each other in †*Notogoneus* (state 0). Dietze (2007) describes and illustrates partly fused hypurals 1 and 2 in †*Dastilbe*; such fusion has not been cited in any paper dealing with this genus (e.g., Poyato-Ariza 1996a, Davis and Martill 1999), and never observed personally by us. Therefore, pending thorough study of caudal endoskeleton variation in †*Dastilbe* (which does occur, but is far from being accurately accounted for at present), we code for the state observed in most large adult specimens personally observed and in those large adult specimens cited and/or figured in the literature, as for all other caudal endoskeleton characters concerning this genus (in this case, coded as primitive). In most taxa with fusion between hypurals 1 and 2, this fusion is partial, since part of the corresponding contiguous borders is usually preserved and normally there is an open fenestra that corresponds to the narrowest part of each hypural (the "neck", just distal to each articular head; e.g., Grande 1996: fig. 4). Among the taxa newly included for this analysis, hypurals 1 and 2 are partly fused to each other in †*Lecceichthys* (Taverne 1999: 303, fig. 5). Hypurals 1 and 2 are independent in †*Hakeliosomus* according to Gayet (1993b: fig. 17), but the situation is not clear, as shown by the photo in Gayet (1993b: pl. 3, fig. 4), from which a fusion could be interpreted. In the restoration by Patterson (1970: fig. 45), hypurals 1 and 2 appear partly fused to each other. Personal observations reveal that, although there is individual variation in this character, the prevalent condition is the absence of fusion. Therefore, we code this character as [0] for

†*Hakeliosomus*, pending further re-evaluation. Both hypurals appear totally fused in †*Apulichthys*, where even the above-mentioned fenestra is closed, according to the interpretation by Taverne (1997: 420, fig. 11). The second derived state, [2], is coded for †*Apulichthys* only, and newly added to this character for the present analysis.

103 (86). Hypural 1 and terminal centrum most commonly: articulating [0]; separated by a hiatus [1]; fused [2]. In †*Diplomystus* and *Opsariichthys* (outgroup) and †*Aethalionopsis*, †*Apulichthys* (with autogenous ural centrum two, see character 94), †*Dastilbe*, †*Gordichthys*, †*Notogoneus* (Grande and Grande 2008), †*Parachanos*, †*Rubiesichthys*, and †*Tharrhias*, hypural 1 articulates with the terminal ural centrum but is not fused to it, so we code it as state [0]. It is separated by a hiatus in *Brycon* (outgroup), and in *Chanos*, *Cromeria*, *Kneria*, *Parakneria*, and *Phractolaemus* (state 1). It is fused in †*Charitosomus*, *Gonorynchus*, *Grasseichthys*, †*Judeichthys*, †*Ramallichthys*, and †*Sorbininardus* (state 2). They are separated by a hiatus in †*Hakeliosomus* according to Gayet (1993b: fig. 17), but the zone is obscure in the photo in Gayet (1993b: pl. 3, fig. 4), and hypural 1 is fused to the terminal centrum according to Patterson (1970: fig. 45). There seems to be individual variation in this character; as indicated in Grande and Grande (2008), †*Hakeliosomus hakelensis* (AMNH 5829) shows fusion, while MNHN-HAK130d does not. †*Hakeliosomus* is polymorphic concerning the fusion of hypural 1 to the compound centrum; personal observations confirm that the prevalent condition is a fusion of hypurals 1 and 2 (see Grande and Arratia present volume for comments on variation of caudal endoskeleton). Therefore, we code this character as [2] for †*Hakeliosomus*. In †*Lecceichthys*, the proximal region of hypural one seems imperfectly known from impression on the matrix only (Taverne 1999: fig. 5), but this impression clearly shows that hypural 1 is anteriorly fused to hypural 2 and to the terminal centrum as well (Taverne pers. comm. 2007). We therefore code this character as [2] for this genus.

104 (87). Hypural 2 and terminal centrum: fused [0]; articulating [1]. In the three outgroup genera, hypural 2 is fused to the terminal complex, so it turns out to be the primitive state in this particular phylogenetic context. State [0] is coded for †*Charitosomus*, *Gonorynchus*, †*Judeichthys*, *Phractolaemus*, †*Ramallichthys*, and †*Sorbininardus* within the ingroup. In †*Apulichthys*, †*Aethalionopsis*, *Chanos*, *Cromeria*, †*Dastilbe*, †*Gordichthys*, *Grasseichthys*, *Kneria*, †*Notogoneus* (Grande and Grande 2008), †*Parachanos*, *Parakneria*, †*Rubiesichthys*,

and †*Tharrhias*, hypural 2 articulates with the terminal ural centrum or complex, but is not fused, so we code it as the state [1]. The character state is unknown in †*Lecceichthys*, because the proximal region of hypural two is preserved as impression only (see previous character). See Grande and Arratia (present volume) for comments on individual variation of fusions in the caudal endoskeleton.

105 (88).   Hypural 5: of comparable size to preceding ones [0]; considerably larger [1]. The hypural 5 is not preserved in †*Lecceichthys* (see comments on character 101 above), but there is no room in its caudal endoskeleton for any hypertrophied hypural after the last one preserved (hypural 3), so the character is coded [0] for this genus.

106 (89).   Hypural 5 (plus 6 if present) and second ural centrum: separate [0]; articulating [1].

107 (90).   Haemal arch and preural centrum 2: fused [0]; independent [1]. They are fused in †*Hakeliosomus* according to Patterson (1970: fig. 45), Gayet (1993b: fig. 17; state 0), and Grande and Grande (2008). They are independent in †*Notogoneus* as well according to recent revision (Grande and Grande 2008).

108 (91).   Posterolateral process of caudal endoskeleton: absent [0]; present [1].

Scales and Lateral Line

109 (93).   Scales on body: present [0]; absent [1].
110 (92).   Type of scales: cycloid [0]; modified ctenoid [1]. This character is coded as non-applicable [N] for *Cromeria* and *Grasseichthys*, since scales are absent (see previous character).
111 (94).   Lateral line: not extending to posterior margin of hypurals [0]; extending to posterior margin of hypurals [1].

List of Characters, Part 2 (Myological)

The following muscular characters are mostly adapted from Diogo (2007) for the present analysis. Character 120, however, is new. A number of characters are inspired from Howes (1985), they and are indicated accordingly. Myological characters are coded exclusively for Recent taxa because muscles are not preserved in fossils. All fossils are coded as unknown [?] in these characters for this analysis.

Ventral Cephalic Musculature

112.   Intermandibularis: mainly attaching on the dentary [0]; exclusively attaching to angulo-articular [1] (inspired from Howes 1985). In the

vast majority of otocephalans, as well as of other teleosts, the intermandibularis muscle connects the two lower jaws by attaching mostly on the dentary bones (state 0). This configuration is found in *Brycon* and *Opsariichthys* (outgroup) and in *Chanos, Cromeria, Gonorynchus, Grasseichthys, Kneria,* and *Parakneria* among the ingroup. In *Phractolaemus* the intermandibularis exclusively connects the angulo-articulars of each side of the jaw (state 1; Howes 1985: figs. 20, 21).

113. Protractor hyoidei: not inserting on coronoid process [0]; inserting on coronoid process [1] (inspired from Howes 1985). Unlike *Brycon, Opsariichthys,* and the other extant gonorynchiforms, in which the protractor hyoidei inserts below the coronoid process of the mandible (state 0), in *Gonorynchus* the insertion of this muscle extends dorsally in order to attach on the mesial surface of the coronoid process (Diogo present volume, Fig. 3; state 1).

114. Hyohyoideus inferioris of both sides: mostly overlapping each other [0]; mostly mixing mesially with each other [1]. In many teleosts, as, for instance, *Brycon* (outgroup), the hyohyoideus inferioris from both sides mainly overlap each other and run from the anteromesial surface of the hyoid arch to the hyoid arch (and eventually the branchiostegal rays) of the opposite side of the animal (state 0); in *Opsariichthys* (outgroup), *Chanos, Cromeria, Gonorynchus, Grasseichthys, Kneria, Parakneria,* and *Phractolaemus* (state 1), each side of the hyohyoideus inferioris mainly attaches anteriorly on a median aponeurosis and/or on the anterior region of the hyoid arch side from which it originates (Diogo 2007).

115. Hyohyoideus abductor: not attaching on pectoral girdle [0]; with significant part of its fibers attaching on pectoral girdle [1]. As in most other teleosts, in *Brycon* and *Opsariichthys* (outgroup), and in *Chanos, Cromeria, Gonorynchus, Grasseichthys,* and *Phractolaemus,* the hyohyoideus abductor runs from the first branchiostegal ray to a median aponeurosis (Diogo present volume, Fig. 7.11; state 0). However, in *Kneria* and *Parakneria* a significant part of the fibers of this muscle also attach to the pectoral girdle (see Diogo present volume; state 1).

Musculature of Pectoral Girdle and of Pectoral Fins/Forelimbs

116. Adductor profundus: not subdivided into different sections [0]; subdivided into different sections [1]. As noted by Diogo (present volume), the plesiomorphic condition of this character for the Otocephala is seemingly that in which the adductor profundus is undivided. Within the extant genera included in our matrix, that

configuration is found in *Opsariichthys* (Diogo present volume, fig. 17; state 0), whereas in the other genera the adductor profundus is divided into two well-differentiated, distinct sections (Diogo present volume, figs. 13–16; state 1).

117.   Attachment of adductor profundus: on first pectoral ray only [0]; on first and second pectoral rays [1]. Within otocephalans, the plesiomorphic condition seems to be that in which the adductor profundus attaches exclusively on the first ray of the pectoral fin (*Opsariichthys*; Diogo present volume, fig. 17; state 0). In *Brycon, Chanos, Cromeria, Gonorynchus, Grasseichthys, Kneria, Parakneria,* and *Phractolaemus*, this muscle attaches on both the first and the second pectoral rays (Diogo present volume, Figs. 13–16; state 1).

Lateral Cephalic Musculature

118.   Most lateral bundles of adductor mandibulae: inserting on mandible and/or primordial ligament [0]; attaching also, or even exclusively, on other bones such as the maxilla or the premaxilla [1]. The plesiomorphic condition for otocephalans is that in which the most external bundles of the adductor mandibulae are exclusively attached to the mandible (and/or eventually to the primordial ligament, near the insertion of this ligament on the mandible) and, thus, do not insert directly on other bony structures (*Brycon*; Diogo present volume, Fig. 10; state 0). In *Opsariichthys* (outgroup), *Chanos, Cromeria, Gonorynchus, Grasseichthys, Kneria, Parakneria,* and *Phractolaemus*, part of the external bundles of the adductor mandibulae attaches also, or even exclusively, on other bones (mostly the maxilla and the premaxilla, but also other bones such as the lacrimal; Diogo present volume, Figs. 1, 2, 4, 5; state 1).

119.   Position of adductor mandibulae A1-OST: mostly horizontal [0]; with a peculiar anterior portion almost perpendicular to its posterior portion [1] (inspired from Howes 1985). As noted by Howes (1985) and by Diogo (present volume, Figs. 7.1, 7.5, 7.11), an adductor mandibulae A1-OST arranged in a mostly horizontal position is found in *Brycon* and *Opsariichthys* (outgroup) and in *Chanos, Cromeria, Gonorynchus, Grasseichthys, Kneria,* and *Parakneria* (state 0). In turn, *Phractolaemus* exhibits a quite peculiar adductor mandibulae A1-OST in which the anterior portion of this section is arranged almost perpendicular to its posterior portion (Diogo present volume, Fig. 7.4; state 1).

120.   Section A2 of adductor mandibulae: present [0]; absent [1]. All the Recent genera in the present analysis possess an adductor mandibulae A2 (state 0), except *Grasseichthys*, where, as described

by Diogo (present volume, Fig. 7.5C), the adductor mandibulae of its adult members is quite reduced in size, being formed mostly by a single bundle. That is, the A2 and A1-OST-M are not present as independent structures in *Grasseichthys* (state 1). For this reason, this genus was coded as [N] (non-applicable) for characters 121, 123, 124, and 125.

121.   Several small tendons branching off from adductor mandibulae A2: absent [0]; present [1] (based on Howes 1985). The only taxon with the derived state (*Phractolaemus*; Diogo present volume, Fig. 7.4A; state 1) exhibits a peculiar configuration in which several small tendons branch off from the dorsal portion of the adductor mandibulae A2.

122.   Peculiar adductor mandibulae A1-OST-M: absent [0]; present [1] (adapted from Howes 1985). Recent genera from the outgroup, as well as *Chanos* and *Gonorynchus* among the ingroup (Diogo present volume, Figs. 7.1, 7.2, 7.10; state 0), lack the peculiar, distinct adductor mandibulae A1-OST-M that connects the anteroventral surface of the quadrate with the maxilla in *Cromeria*, *Kneria*, *Parakneria*, and *Phractolaemus* (Diogo present volume, Figs. 7.4A,B, 7.5A,B; state 1). *Grasseichthys* is coded as [?] (unknown) for this character, because it was not possible to discern whether the configuration shown by its members corresponds to character state 0 or to character state 1.

123.   Direct insertion of adductor mandibulae A2 far anteriorly on the anteromesial surface of dentary: absent [0]; present [1]. The derived state, present only in *Gonorynchus* among the extant taxa analyzed, consists of the insertion of the adductor mandibulae A2 directly on the anteromesial surface of the dentary (Diogo present volume, Fig. 7.3; state 1), instead of the broadly present primitive state, where this section mainly inserts on the posteromesial and/or posterolateral surface of the mandible (Diogo present volume, Figs. 7.1, 7.2, 7.4, 7.5, 7.10; state 0).

124.   Dilatator operculi: mainly mesial and/or dorsal to adductor mandibulae A2 [0]; markedly lateral to A2 [1]. In the primitive state, the dilatator operculi is mainly mesial and/or dorsal to the adductor mandibulae A2 (*Opsariichthys*, *Chanos*, *Cromeria*, *Gonorynchus*, *Kneria*, *Parakneria*, and *Phractolaemus*; Diogo present volume, Figs. 7.1, 7.4, 7.5; state 0), but in the derived state, the former is clearly lateral to the latter (*Brycon*; Diogo present volume, Fig. 7.10; state 1).

125.   Peculiar tendon of adductor mandibulae A2 running perpendicular to the main body of this section and connecting it to the anteroventral surface of the quadrate: absent [0]; present [1] (adapted from Howes 1985). Among the extant genera analyzed, the configuration of the

derived condition is found only in *Cromeria* (Diogo present volume, Fig. 7.5B, compare with, for instance, Figs. 7.2, 7.3, 7.4; state 1).

126.   Distinct section A3 of adductor mandibulae: absent [0]; present [1]. As explained by Diogo (present volume, Figs. 7.1–7.5), the plesiomorphic condition for ostariophysans is seemingly that in which the adductor mandibulae A3 is not present as an independent structure (state 0). This condition is found in all the extant genera analyzed but *Brycon* (Diogo present volume, Fig. 7.10; state 1).

127.   Adductor mandibulae Aω: present [0]; absent [1]. As explained by Diogo (present volume), plesiomorphically in ostariophysans the adductor mandibulae Aω is present as an independent structure. Among the extant genera analyzed, Aω is present in *Brycon*, *Opsariichthys*, and *Chanos* (Diogo present volume, Fig. 7.2; state 0) and absent in *Cromeria*, *Gonorynchus*, *Grasseichthys*, *Kneria*, *Parakneria*, and *Phractolaemus* (Diogo present volume, Figs. 7.3, 7.4, 7.5; state 1).

128.   Adductor arcus palatini: not inserting on preopercle [0]; inserting also on preopercle [1] (adapted from Howes 1985). The adductor arcus palatini of ostariophysans usually inserts in bones such as the hyomandibula, quadrate, metapterygoid and/or endopterygoid (Diogo present volume, Figs. 7.1, 7.4, 7.8–7.10). This occurs in all extant genera analyzed (state 0) except *Gonorynchus*, which presents a peculiar configuration in which this muscle also inserts on the preopercle (see Howes 1985, Diogo present volume; state 1).

129.   Levator arcus palatini: not divided [0]; divided into two well-differentiated bundles [1]. Plesiomorphically in ostariophysans the levator arcus palatini is undivided (Diogo present volume, Figs. 7.1, 7.4, 7.5, 7.8–7.10); this plesiomorphic configuration is present in all extant genera analyzed (state 0) except *Opsariichthys*, in which this muscle is divided into two well-differentiated bundles (state 1).

130.   Origin of dilatator operculi: on ventrolateral surface of neurocranium [0]; on dorsal margin of cranial roof [1]. The primitive state for ostariophysans is that in which the dilatator operculi runs from the ventrolateral surface of the neurocranium and/or eventually from the dorsolateral surface of the hyomandibula to the anterodorsal surface of the opercle (Diogo present volume, Figs. 7.1, 7.4, 7.5, 7.8, 7.9). Among the Recent taxa included in the analysis, this configuration is found in *Opsariichthys*, *Chanos*, *Cromeria*, *Gonorynchus*, *Grasseichthys*, *Parakneria*, and *Phractolaemus* (state 0). The derived state, in which a part of this muscle originates on the dorsal margin of the cranial roof, is found in *Brycon* and *Kneria* only (Diogo present volume, Fig. 7.10; state 1).

Withdrawn Characters

Character 14 in Grande and Poyato-Ariza (1999; morphology of supraoccipital bone in dorsal view) is not used in the present analysis because, after additional revision, it is found to present too great individual variation, and because of the consequent difficulty of establishing clear criteria to distinguish between the character states.

This study has shown variation in maxillary process morphology between the Chanoidei, a group in which the morphology of this process significantly varies but is easily accounted for (Poyato-Ariza 1996a, c), and the Gonorynchoidei, in which the general morphology, orientation, and extension of the maxilla, and of this process in particular, is highly variable and significantly different from the Chanoidei, to the point that, at present, we cannot find a way for them to be adequately compared. In addition, it introduces strong noise in the interrelationships of the Gonorynchoidei. Therefore, we do not use characters 20 and 21 from Grande and Poyato-Ariza (1999) for the present analysis, although we reckon their relevance for the Chanidae, and maybe further revision of the ontogenetic and individual variation of the very particular maxilla of the Gonorynchoids may provide in the future sound grounds to be used again in this group.

Characters 26 and 27 in Grande and Poyato-Ariza (1999), concerning the relative height of the dentary and its position, have proven problematic during revision. Not because this character is based on a highest point that is not homologous, as claimed by Britz and Moritz (2007: 28); these characters described the morphology (shape according to the point where the maximum height is placed) of an element (the dentary bone) that is homologous in all taxa involved. However, additional problems did arise during revision. In the first place, the mandible of the Gonorynchoids is very different from that of the Chanoidei, for which these characters were initially coded (see comments in previous paragraph). In the second place, there is great ontogenetic and individual variation, and the variations are not well understood at present. For instance, claims by Britz and Moritz (2007: 28, fig. 18) that the highest point of the dentary in *Kneria* is not at the symphysis are based on evidence from juvenile specimens, 8.1 to 19.9 mm in standard length, which do present the maximum height of the dentary at its anterior half, but not at the symphysis. However, adult specimens seem to present the maximum height of the dentary at the symphysis (e.g., Grande 1994: fig. 3B, 54.3 mm in standard length). The claim by Britz and Moritz (2007) that the maximum height of the mandible of *Phractolaemus* is placed "equally distant from its posterior terminus and the anterior symphysis" (p. 28) is contradicted by their own illustration of this genus (Britz and Moritz 2007: fig. 19B), which clearly shows a dentary whose highest point is just beside the symphysis (see also Thys van den Audenaerde 1961: fig. 17; height measured with longitudinal axis of lower jaw in horizontal position, as in all other taxa). This shows that there is yet another problem with these characters, involving accurate orientation of the mandible for comparison of its general shape, a problem we did find while revising these characters for the present analysis. Consequently, we prefer not to use them herein. We reckon that general shape of the dentary is still a good character for establishing chanid interrelationships, but their application to gonorynchoids (where the lower jaw is as derived and particular as the upper jaw) must be re-evaluated in the future on the basis of an ontogenetic revision, an extensive study of their individual variations, and a possible use of the Chanidae as an additional outgroup for establishing a precise definition and polarization of these characters in the Gonorynchoidei.

Character 61 from Grande and Poyato-Ariza about the contact of the anterior neural arches received strong criticism by Britz and Moritz (2007: 30, 31, 34). It is true that this character is difficult to establish in gonorynchids and kneriids, but it is also true that it significantly varies throughout ontogeny by increasing the contact of initially separated arches; in addition, it may also present sexual and/or other individual variation. We find the arguments by Britz and Moritz (2007) inconclusive, because they do not discuss the ontogenetic and individual variation in detail, and do not present fully adult specimens to justify their interpretation. Therefore, pending detailed study of ontogenetic, taxonomic, and individual variation of the character in kneriids, we conservatively decided not to use it in the present analysis.

Character 69 from Grande and Poyato-Ariza (1999), concerning contact of anterior supraneurals, is not used because our revision has shown that it is somehow linked to their relative enlargement (see character 85 above).

First Results

To facilitate the discussion of our results, an abbreviated list of characters can be found in Appendix 1. The complete data matrix showing the coding of all characters as discussed above is presented in Appendix 2. This data matrix was analyzed using PAUP program, version 4.10. The analyses were carried out using an iMac computer at the Unidad de Paleontología, Universidad Autónoma de Madrid. Characters 3, 8, 13, 15, 33, 57, 77, 78, 82, 95, 97, 99, 100, and 102 were processed as ordered. For a justification, with references, of why certain multistate characters can sensibly be treated as ordered, see Poyato-Ariza (2005: 544).

The data matrix of all Recent and fossil taxa was processed with the DELTRAN option, although the character distribution in Appendix 3 also presents the results with the ACCTRAN option. The heuristic search, general option, resulted in 92 most parsimonious trees of 297 steps each. The strict consensus tree is shown by Fig. 7.9. General configuration is as follows. The Gonorynchiformes do not appear defined because of the noise introduced by †*Apulichthys* and the new genus (see details below). The Chanoidei is a well-supported clade, sister group to the Gonorynchoidei, which in turn consist of the Gonorynchidae plus the Kneriidae. Relationships within the Chanoidei are not fully solved, although there are two sets of sister groups in all trees: [*Chanos* + †*Tharrhias*] and [†*Gordichthys* + †*Rubiesichthys*]. Relationships within the Kneriidae are fully solved (see below for details). Relationships within the Gonorynchidae are not fully solved, although †*Notogoneus* always appear as the most basal gonorynchid; the clades formed by [†*Halecopsis* + †*Sorbininardus*] and by †*Hakeliosomus*, †*Judeichthys*, and †*Ramallichthys* appear in all trees. The consistency index of each most parsimonious tree is 0.613, the homoplasy index is 0.411, and the retention index is 0.743.

We computed the 50% majority rule consensus tree in order to check the improvement of the indeterminations of the strict consensus tree. This 50% majority rule consensus tree (Fig. 7.10) did show higher resolution, so its configuration is better solved than that of the strict consensus tree (Fig. 7.9). Each node in the 50% consensus tree (Fig. 7.10) shows the percentage of individual most parsimonious trees in which the particular node is present. The clade of the Gonorynchiformes distinctly appears now, with the new genus outside of them (and outside the Ostariophysi), although †*Apulichthys* is still introducing noise in this definition. Relationships within the Chanoidei are fully solved now, except for the relative position of †*Dastilbe* and †*Parachanos* (see details below). Relationships within the Gonorynchidae are nearly fully solved, except for the indetermination among some Middle Eastern forms (see below).

The Systematics section at the end of the main text contains updated diagnoses of all clades above generic level based on the results of the present analysis (Figs. 7.9–7.12, Appendix 3). The list of autapomorphies for each genus at the end of Appendix 3 provides the respective diagnostic characters according to the present hypothesis of phylogentic relationships, but, since additional work is needed for a good number of genera in order to solve species composition, variation, and nomenclatural problems, we do not provide revised generic diagnoses at present.

In this section, we first discuss the relationships of some problematic genera according to both the strict and the 50% consensus trees. After that, we discuss in detail the character distribution and nodes support after those problematic genera are removed from the data matrix in order to perform a second cladistic analysis.

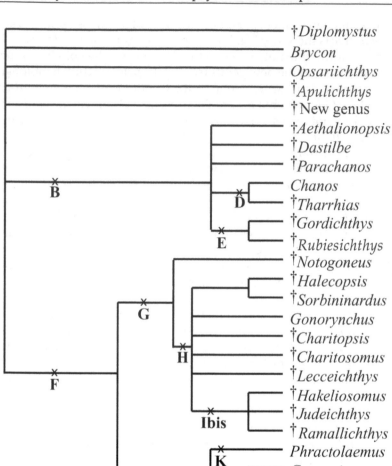

Fig. 7.9 Strict consensus tree including all fossil and Recent genera in the data matrix. Nodes: **B**, Chanoidei = Chanidae; **D**, Chanini; **E**, †Rubiesichthyinae; **F**, Gonorynchoidei; **G**, Gonorynchidae; **H**, Gonorynchidae minus †*Notogoneus*; **J**, Kneriidae; **K**, Phractolaeminae = *Phractolaemus*; **L**, Kneriinae; **M**, Cromeriini; **N**, Kneriini. Nodes without lettering involve problematic genera and are discussed in the main text. For information on character distribution, see discussion of results in the main text and list of apomorphies in Appendix 3.

Comments on Some Problematic Genera

Some genera included in the analyses show largely indeterminate position or other particular problems that are commented on in this section. They are the new genus by Pittet *et al.* (present volume), †*Halecopsis*, and the fishes from Nardò (†*Apulichthys*, †*Lecceichthys*, and †*Sorbininardus*).

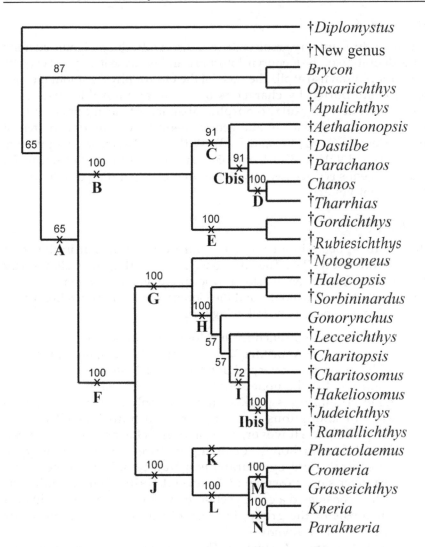

Fig. 7.10 Fifty percent majority rule consensus tree including all fossil and Recent genera in the data matrix. Nodes: **A,** Gonorynchiformes; **B,** Chanoidei = Chanidae; **C,** Chaninae; **D,** Chanini; **E,** †Rubiesichthyinae; **F,** Gonorynchoidei; **G,** Gonorynchidae; **H,** Gonorynchidae minus †*Notogoneus*; **I,** Middle Eastern genera; **J,** Kneriidae; **K,** Phractolaeminae = *Phractolaemus*; **L,** Kneriinae; **M,** Cromeriini; **N,** Kneriini. Nodes without lettering correspond to problematic genera and are discussed in the main text. For information on character distribution, see discussion of results in the main text and list of apomorphies in Appendix 3. Numbers indicate the percentage of most parsimonious trees where each corresponding node appears.

New Genus

Its relationships are very problematic, because it introduces significant noise in the definition of the Gonorynchiformes, and because it may not even be an ostariophysan fish at all. The clade of the Ostariophysi without the new genus is supported by characters 8, 28 (autapomorphic absence of supramaxillae), 39 (autapomorphic absence of dermopalatine), 74 (autapomorphic absence of autogenous neural arch anterior to arch of first vertebra), 77, 92, 97, and 98(3). With ACCTRAN it also presents characters 2, 13(2), 81, 82, 94, and 95, and does not present character 77. This new genus is discussed in detail in the corresponding chapter (Pittet *et al.* present volume).

†Apulichthys

This genus appears in a completely unresolved position in the strict consensus tree (Fig. 7.9). The 50% consensus tree (Fig. 7.10) shows only slightly better resolution. It actually appears in an indeterminate position with respect to the Chanoidei and the Gonorynchoidei. Its position varies widely. It may appear in the following forms:

(1) As a basal gonorynchiform. Most traditional gonorynchiform characters (see Appendix 3), and notably the autapomorphic ones, still support the Gonorynchiformes with *†Apulichthys* inside them. Additionally, the node separating *†Apulichthys* from the other gonorynchiforms is weakly supported: only characters 24 and 83 are not ambiguous (i.e., appear with both DELTRAN and ACCTRAN). However, one of these characters (83, paired intermuscular bones consisting of three series: epipleurals, epicentrals, and epineurals) is a traditional gonorynchiform character since Rosen and Greenwood (1970). Consequently, this fish, if considered a gonorynchiform, would change the traditional diagnosis of the group, because epicentrals are absent in its description by Taverne (1997).

(2) As a stem-group Chanoidei. The characters supporting this node are: 6(2: cephalic ribs present and articulating with both the exoccipitals and basioccipital), 35(1: quadrate-mandibular articulation placed anterior to orbit, quadrate displaced but not elongated), 46(1: metapterygoid process of hyomandibular bone present, anterior), and 104(1: hypural 2 and terminal centrum articulating). None of these characters is autapomorphic. With ACCTRAN optimization it also presents characters 23, 26, 31, 61(reversion), 81(reversion), 94(reversion) and 99(reversion). The position of this genus as stem-group Chanoidei affects the support of

this clade in character 35 (displacement of quadrate and quadrate-mandibular articulation), which is a typical chanid diagnostic character that is also present in †*Apulichthys*. This position is quite unexpected, since it does not show any other typical chanid character, and Taverne (1997) considered it the most basal Gonorynchoidei (in a hand-made tree without cladistic analysis). However, it does not appear in that position in any of the 92 most parsimonious trees obtained in this analysis. †*Apulichthys* appears in this stem-group Chanoidei position only in those trees in which the new genus by Pittet *et al.* (present volume) is not the stem-group Chanoidei. Interestingly, this is a position therefore incompatible between these two forms: see Pittet *et al.* (present volume) for further details. This possible position of †*Apulichthys* in this analysis is also quite unexpected and rather problematic.

On top of this, the terminal branch of †*Apulichthys* presents numerous reversions (see Appendix 3), especially when it appears as a stem-group Chanoidei. It is clear that its phylogenetic relationships are impossible to solve at present, and there is even the possibility that this genus is not a Gonorynchiform and/or would force us to alter the definition of the group if considered one. This genus shows a mosaic of characters that cannot be accurately interpreted at present. As a matter of fact, the premaxilla, the ectopterygoid, and the caudal endoskeleton of †*Apulichthys* (Taverne 1997: 411, 412, figs. 7, 11) are very unlike those present in any gonorynchiform, adding more confusion to this problem.

In consequence, in this moment we think it is wiser to consider †*Apulichthys* as a teleostean *incertae sedis*, pending further confirmation of its relationships by means of additional cladistic analyses of this genus together with non-gonorynchiform teleosteans.

†*Lecceichthys*

This genus appears in an indeterminate position in the strict consensus tree together with all gonorynchids except †*Notogoneus* (Fig. 7.9); the 50% consensus tree (Fig. 7.10) shows better resolution, as it appears as the stem-group to the clade formed by the Middle Eastern forms (†*Charitopsis*, †*Charitosomus*, †*Hakeliosomus*, †*Judeichthys*, and †*Ramallichthys*) in 57% of the trees. This node is supported only by one non-ambiguous character, number 71 (supraorbital present, reversion). It also presents characters 26, 77(2), and 78(2) with ACCTRAN. This support is consequently very weak, and the characters involved are conflicting.

Other trees show †*Lecceichthys* in a more basal position, as sister group to all gonorynchids but †*Notogoneus*. The node separating †*Lecceichthys* from the rest of the gonorynchids is, in this case, supported by characters 45, 71,

78, 99, and 104 (reversion) with DELTRAN, but the only character that also appears with ACCTRAN, and is therefore unambiguous, is number 99: uroneural two and second ural centrum separated from each other. This support is, consequently, also very weak, with equally conflicting characters. This is not the only other possibility; †*Lecceichthys* can also appear within the clade formed by the Middle Eastern forms in other trees, thus altering their interrelationships and the support of this node.

As we can see, the results of our analysis concerning †*Lecceichthys* are clearly inconclusive. Moreover, the interrelationships within the Gonorynchidae and the character distribution inside the group are affected by †*Lecceichthys* in a conflicting manner. In any case, our results essentially agree with Taverne (1998), who placed †*Lecceichthys* within the Gonorynchidae (in a fully solved hand-made tree without cladistic analysis). †*Lecceichthys* does have some typical gonorynchid characters, such as the presence of conical endopterygoid and basibranchial teeth (Taverne 1998: 296, fig. 2). The characters of this genus are incomplete because, unfortunately, nothing is known of most of its oral bones. It is a valid genus, since it has at least one autapomorphic character, the presence of an interopercular spine, and very few vertebrae. However, a number of its characters, as terminal branch, are reversions (see Appendix 3). At present, its precise relationships cannot be fully solved by the data of the present analysis, so we propose to consider this genus a Gonorynchidae *incertae sedis*. We removed it from the data matrix for the additional analyses, in order to test whether the relationships of the Gonorynchidae remain the same, and especially if the interrelationships of the Middle Eastern genera (†*Charitopsis*, †*Charitosomus*, †*Hakeliosomus*, †*Judeichthys*, and †*Ramallichthys*) could be better solved without its influence.

†*Halecopsis* and †*Sorbininardus*

These genera appear within the Gonorynchidae, as sister group to each other (Figs. 7.9, 7.10). The node supporting their sister-group relationship is rather puzzling, supported only by two unambiguous characters: 60(2: preopercular expansion present in posteroventral corner and part of the posterodorsal limb, convergent with some chanids) and 69(reversion: infraorbital bones not including lacrimal well developed); they are unambiguous, but obviously homoplastic. These are the only two characters with DELTRAN, but with ACCTRAN they also present characters 23(2), 24(reversion), 29(reversion), 43, 62, 84(2), and 101(reversion). In turn, the node separating them from the basal gonorynchid †*Notogoneus* is even more conflicting, as it is supported by four characters, curiously enough all of them concerning the caudal endoskeleton, which is unknown in †*Halecopsis*. Additionally, [†*Halecopsis* + †*Sorbininardus*] can appear in other positions within the Gonorynchidae as well; for instance, as sister group to the Middle Eastern forms; these additional characters with ACCTRAN do vary according to the alternative positions of [†*Halecopsis* + †*Sorbininardus*] within the Gonorynchidae, which do change the distribution of characters within the family.

Taverne and Gayet (2006) considered †*Halecopsis* a stem-group Gonorynchoidei in a tree without cladistic analysis. The problem with †*Halecopsis* is that it presents a very high number of unknown character states because of the significant incompleteness of the known specimens. It does not show diagnostic gonorynchoid characters, but it exhibits a "gonorynchid look" and two gonorynchid synapomorphies, such as elongate and narrow frontals (character 9), and triangular shape of the operculum (character 53). In addition, it does appear within the gonorynchidae in our analysis, so we consider it a gonorynchid fish *incertae sedis*; its precise relationships would be better defined only with more complete specimens.

†*Halecopsis* was removed from the data matrix in order to check the influence of this high percentage of missing data (over 70%) on it. When the datra matrix is run without †*Halecopsis*, the number of most parsimonious trees is 48, nearly half as many: it is clear that this genus introduces a great noise in the analysis. As we saw above, a good part of the conflict in character interpretation arises from its unkown characters. Surprisingly enough, its removal resulted in placing †*Apulichthys*, †*Lecceichthys*, and †*Sorbininardus* in a multiple indetermination, together with the Chanoidei and the Gonorynchoidei, in the strict consensus tree. These three genera are successive stem-groups Gonorynchidae in the 50% majority rule consensus tree. Therefore, deletion of †*Halecopsis* adds more confusion to the position of these genera, but it also implies, quite curiously, that a putative revision of †*Halecopsis* based on additional, more complete material may add some light to the relationships of the Nardò fishes.

Although †*Neohalecopsis* is not included in the analysis because of incompleteness and poor preservation (see presentation of genera included in the analysis above), we note herein that the general proportions of the head are not sufficient to place †*Neohalecopsis* within the Chanidae, as tentatively suggested by Taverne and Gayet (2006), although we agree with these authors that there is no reason to relate it closely to †*Halecopsis* either. This genus seems better considered a teleostean *incertae sedis* until a thorough revision is carried out.

In addition to this confusion, †*Sorbininardus* has an unexpected combination of characters because it lacks two gonorynchiform autapomorphies (its orbitosphenoid is present, and its pterosphenoids are of primitive, non-gonorynchiform type). Many of its autapomorphic characters, as terminal branch, are reversions (see Appendix 3). The premaxilla, maxilla, lacrimal, and subopercular of this genus are very different from those of any gonorynchiform. The spines on the opercular bones and the fusion in the caudal endoskeleton that seem to link it to some gonorynchids could be convergent. Taverne (1999) actually placed †*Sorbininardus* as a stem-group Gonorynchiformes. Therefore, the relationships of this genus are at present highly problematic, and it is necessary to test them by means of further cladistic analyses that include non-gonorynchiform teleosts. It may not be a gonorynchiform fish. Therefore, we also deleted it from the data matrix in order to test its influence in the support of the different clades, and we consider it a teleostean *incertae sedis*, pending futher analyses with non-gonorynchiform fishes.

As a consequence, for the time being, we propose to consider †*Halecopsis* as a Gonorynchidae *incertae sedis* because of the presence of two gonorynchid synapomorphies, together with its incompleteness and its problematic, poorly defined relationships with other gonorynchids. †*Sorbininardus* might be a gonorynchid, a stem-group gonorynchid, or even a stem-group Gonorynchiformes. Therefore, it is not considered a gonorynchiform fish, although it might be an Anotophysi (as in Taverne 1999). Its unexpected combination of primitive and derived characters might grant it quite different phylogentic relationships when compared with non-gonorynchiform teleosteans in future analyses.

Second Analysis

The problematic genera discussed above were removed from the data matrix for a second analysis; the corresponding consensus tree is shown by Fig. 7.11. The relationships among the major clades remain the same: [Chanoidei + (Gonorynchidae + Kneriidae)]. The Kneriidae inter-relationships appear, once again, fully solved; those of the Chanoidei are not completely solved; and those of the Gonorynchidae are rather unsolved. The situation is better with the 50% majority rule consensus tree, as shown by Fig. 7.12. The indetermination within the Chanoidei remains, but the interrelationships within the Gonorynchidae are nearly fully solved. The results concerning the support of the main clades remained unaltered in the essential characters, and the noise introduced by the problematic forms logically dissapears. This second analysis finds a slightly smaller number of most parsimonious trees, 90, each of which is obviously shorter, 235 steps long; the consistency index is significantly higher than in the first analysis, 0.716; the homoplasy index is correspondingly lower, 0.302; and the retention index is higher, 0.821.

Curiously enough, the interrelationships of the remaining genera in the new 50% consensus tree (Fig. 7.12) are the same as the relative

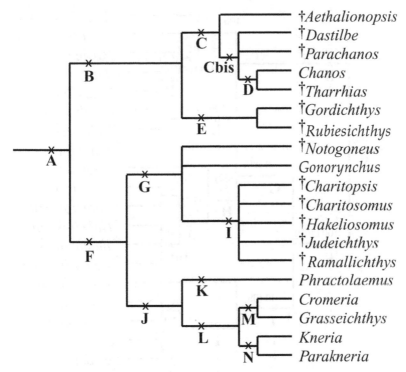

†*Aethalionopsis*
†*Dastilbe*
†*Parachanos*
Chanos
†*Tharrhias*
†*Gordichthys*
†*Rubiesichthys*
†*Notogoneus*
Gonorynchus
†*Charitopsis*
†*Charitosomus*
†*Hakeliosomus*
†*Judeichthys*
†*Ramallichthys*
Phractolaemus
Cromeria
Grasseichthys
Kneria
Parakneria

Fig. 7.11 Strict consensus tree without the problematic fossil genera commented on in the text; outgroup not depicted. Nodes: **A,** Gonorynchiformes; **B,** Chanoidei = Chanidae; **C,** Chaninae; **D,** Chanini; **E,** †Rubiesichthyinae; **F,** Gonorynchoidei; **G,** Gonorynchidae; **I,** Middle Eastern genera; **J,** Kneriidae; **K,** Phractolaeminae = *Phractolaemus*; **L,** Kneriinae; **M,** Cromeriini; **N,** Kneriini. For information on character distribution, see discussion of results in the main text and list of apomorphies in Appendix 3.

interrelationships indicated in the 50% consensus tree with all taxa (Fig. 7.10). In other words, removal of the problematic genera improves the support of the nodes but does not alter the interrelationships of the remaining genera.

Bremer Support and Bootstrap Analysis

An additional check of the nodes strength was done by means of the Bremer support test. A heuristic search was programmed to compute all trees one step longer than the most parsimonious ones (i.e., 236 steps). The result was a total of 450 trees saved, whose consensus trees are shown in Figs. 7.13 and 7.14. Most nodes in the original analyses stand the Bremer support test. All nodes corresponding to higher clades stand the test, confirming that they are strongly supported. Only three minor clades (C, Cbis, H, and Ibis, see discussion below) do not appear in the strict consensus tree (Fig. 7.13); in

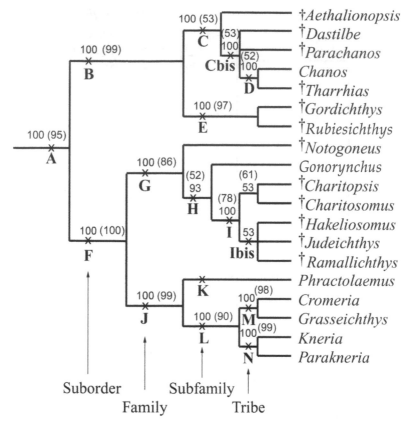

Fig. 7.12 Fifty percent majority consensus tree without the problematic fossil genera commented on in the text; outgroup not depicted. Nodes: **A,** Gonorynchiformes; **B,** Chanoidei = Chanidae; **C,** Chaninae; **D,** Chanini; **E,** †Rubiesichthyinae; **F,** Gonorynchoidei; **G,** Gonorynchidae; **H,** Gonorynchidae minus †*Notogoneus*; **I,** Middle Eastern genera; **J,** Kneriidae; **K,** Phractolaeminae = *Phractolaemus*; **L,** Kneriinae; **M,** Cromeriini; **N,** Kneriini. For information on character distribution, see discussion of results in the main text and list of apomorphies in Appendix 3. Numbers indicate the percentage of most parsimonious trees where the corresponding node appears; numbers in brackets indicate the percentage of most parsimonious trees of the bootstrap analysis where the corresponding node appears. The names at the bottom indicate the levels of the higher taxa established according to the present hypothesis of phylogenetic relationships by following the principle of subordination, as discussed in the text.

turn, only one of the minor nodes (Ibis) does not appear in the 50% consensus tree (Fig. 7.14). This confirms the strength of most minor nodes in the original analyses as well.

An additional test of strength was provided by a bootstrap analysis. With 100 bootstrap replicates, the results confirmed those of the heuristic search. As for the Bremer support, most nodes appear in the 50% majority rule consensus tree provided by the bootstrap, except node Ibis (part of the Middle Eastern forms). The percentages of the boostrap analysis for each node are provided by Fig. 7.12; note the very high values of the Gonorynchiformes, the Chanoidei and Gonorynchoidei, the three families, the †Rubiesichthyinae, and all the nodes within the Kneriidae.

Discussion of General Relationships and Clades Support

The consensus trees (Figs. 7.9–7.12) support the general interrelationships of the Gonorynchiformes as proposed by Grande and Poyato-Ariza (1999). The Gonorynchiformes is a clade formed by two sister groups, the Chanoidei and the Gonorynchoidei. The Chanoidei contains the family Chanidae only, and the Gonorynchoidei is formed by two sister families, the Gonorynchidae

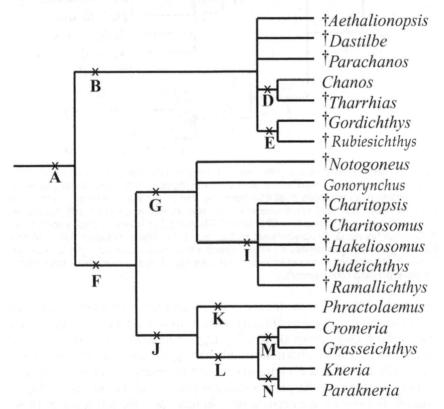

Fig. 7.13 Bremer support test. Strict consensus tree without the problematic fossil genera commented on in the text; outgroup not depicted. It represents the strict consensus of all 450 trees one step longer than the most parsimonious ones. Most nodes stand unaltered (see text for further explanations). Nodes: **A**, Gonorynchiformes; **B**, Chanoidei = Chanidae; **D**, Chanini; **E**, †Rubiesichthyinae; **F**, Gonorynchoidei; **G**, Gonorynchidae; **I**, Middle Eastern genera; **J**, Kneriidae; **K**, Phractolaeminae = *Phractolaemus*; **L**, Kneriinae **M**, Cromeriini; **N**, Kneriini.

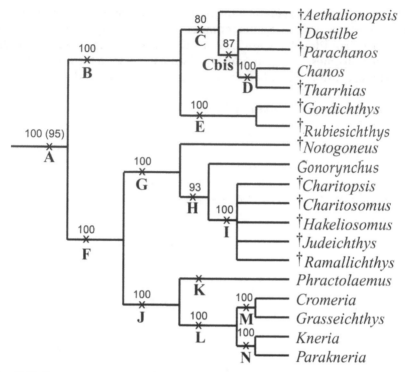

Fig. 7.14 Bremer support test. Fifty percent majority rule consensus tree without the problematic fossil genera commented on in the text; outgroup not depicted. It represents the 50% majority rule consensus of all 450 trees one step longer than the most parsimonious ones. All nodes (but Ibis) stand unaltered (see text for further explanations). Nodes: **A,** Gonorynchiformes; **B,** Chanoidei = Chanidae; **C,** Chaninae; **D,** Chanini; **E,** †Rubiesichthyinae; **F,** Gonorynchoidei; **G,** Gonorynchidae; **H,** Gonorynchidae minus †*Notogoneus*; **I,** Middle Eastern genera; **J,** Kneriidae; **K,** Phractolaeminae = *Phractolaemus*; **L,** Kneriinae; **M,** Cromeriini; **N,** Kneriini. Numbers indicate the percentage of most parsimonious trees where the corresponding node appears.

and the Kneriidae. Except for the cases specifically mentioned, the nodes discussed below are present in all consensus tress of all the analyses carried out for the present paper (Figs. 7.9–7.12,). Their diagnostic characters are taken from the trees without the problematic genera discussed above (Figs. 7.11, 7.12). We think it is better to characterize major nodes and diagnose higher taxa without the genera that are poorly known, *incertae sedis*, or possibly not even a gonorynchiform; in any case, the influence of these problematic genera was already presented in detail above.

We base the discussion and diagnoses below on unambiguous characters, although some ambiguous characters are occasionally used whenever considered relevant or "traditional" for a particular taxon. In general, we prefer the DELTRAN optimization option; we particularly interpret as convergent the characters of the caudal endoskeleton, whose fusions are traditionally considered to be independently acquired by different teleostean lineages.

In the discussion below, (A) indicates autapomorphic character.

Order Gonorynchiformes

The Gonorynchiformes (Figs. 7.10–7.14, node A) is a strongly supported clade in the present analysis. This node is supported by eight unambiguous characters, five of which are autapomorphic. They include: loss of the orbitosphenoid (A); reduction and separation of pterosphenoids (A); presence of cephalic ribs; reduction of parietals; separation of parietals by supraoccipital; absence of teeth in premaxilla, maxilla, and dentary; loss of premaxillary ascending process (A); absence of teeth on fifth ceratobranchial (A); neural arch of first vertebra and exoccipitals in contact (A); rib on third vertebral centrum wide and short; and paired intermuscular bones consisting of three series (epipleurals, epicentrals, and epineurals). These results are totally congruent with the sense of the monophyletic Gonorynchiformes since Greenwood *et al.* (1966) and Rosen and Greenwood (1970), to recent studies (e.g., Diogo 2007), and the characters are consistent with the usual gonorynchiform diagnostic characters, although expanded and revised. Therefore, the diagnosis of the Gonorynchiformes in this chapter is essentially as in Grande and Poyato-Ariza (1999).

Suborder Chanoidei and Family Chanidae

The Chanoidei is a monophyletic group that includes the family Chanidae only (Figs. 7.9–7.14, node B). This result differs from Grande and Poyato-Ariza (1999) in the present inclusion of †*Aethalionopsis* within the Chanidae, whereas in that paper it was a stem-group Chanidae. This is due to re-examination of the premaxilla morphology and the reinterpretation of this character in this genus, which was originally coded as unknown in the 1999 paper (see discussion of character 23 above). This apparently trivial change had an important consequence that resulted in a rediagnosis of the Chanoidei to equal that of the Chanidae, as in Poyato-Ariza (1996a, c). Comparison of the number of unambiguous characters and of autapomorphies indicates that the node at the base of Chanoidei (= Chanidae) is well supported even in comparison to Gonorynchiformes. The node of the chanids is supported by 13 unambiguous characters, seven of which are autapomorphic. They include: a large, very broad, concave-convex premaxilla with long oral process (A); a greatly enlarged maxilla, posteriorly swollen to a bulbous outline, with a clearly curved posterior border (A); presence of a notch in the anterodorsal border of the dentary ("leptolepid" notch, see discussion of character 31) (A); a quadrate-mandibular articulation anteriorly displaced (A), due to displacement of the quadrate bone, a bone that is not elongated and maintains the usual triangular shape of its main body (the articulation is anterior to the orbit except in †*Aethalionopsis*, where the orbit is very large; see Poyato-Ariza 1996c: 21); the symplectic correlatively elongated, at least twice the normal length; a metapterygoid process of hyomandibular bone present and placed on the anterior border of the bone (A); opercular bone expanded, at least one third the head length (A); angle formed by preopercular limbs straight to acute; preopercular expansion present (except in †*Dastilbe*); suprapreopercular bone present as a relatively large, flat bone; and unmodified neural arches anterior to dorsal fin in adults autogenous, at least laterally (A).

The diagnosis and taxonomic composition of the Chanidae in the present paper are essentially as those in Poyato-Ariza (1996a, c). Fink and Fink (1996) criticised some chanid apomorphies in that paper, mostly because of homoplasy and objections of character polarization. However, the analysis by Grande and Poyato-Ariza (1999) confirmed the characters supporting the chanid clade, in terms of both character polarization and distribution, and so does the present analysis. Therefore, we consider that the family Chanidae is a strongly supported clade, indistinguishable from the suborder Chanoidei, that is, both the suborder and the family contain the same genera: *Chanos* plus all the fossil chanids (see Systematics section below).

Subfamilies Chaninae and †Rubiesichthyinae

The two subfamilies that form the Chanidae are sister groups, as in Poyato-Ariza (1996a) and Grande and Poyato-Ariza (1999). According to the relationships in the present analysis (without the problematic forms such as †*Apulichthys*), there are two alternatives for defining the subfamily Chaninae: with or without †*Aethalionopsis*. Both alternatives are supported by the strict consensus tree (Fig. 7.11) and appear in the 50% majority rule consensus tree with 100% (Fig. 7.12).

The node formed by [(*Chanos* + †*Tharrhias*), †*Dastilbe*, †*Parachanos*] (Cbis in Figs. 7.9–7.14) is supported by two autapomorphic characters: supraoccipital crest long and enlarged, projecting above occipital region and first vertebrae forming a vertical, posteriorly deeply pectinated blade (not accessible in some fossils) and major axis of subopercular bone in subhorizontal lateral view. Alternatively, the clade including †*Aethalionopsis* (Figs. 7.9–7.14, node C) is supported by two characters that are easily accessible in all fossils. In addition, placing the Chaninae here follows the principle of subordination (that, whenever possible, sister groups should have the same taxonomic rank; see Fig. 7.12 for further details). Therefore, we prefer to define the subfamily Chaninae including †*Aethalionopsis* (node C) and, consequently, we consider that there are two sister-group subfamilies within the Chanidae.

Thus defined, the Chaninae contains five of the seven chanid genera, as in Poyato-Ariza (1996a). This clade is supported by two synapomorphies, both of them unambiguous characters: presence of a maxillary process for articulation with autopalatine (convergent with the Kneriini) and presence of a ridge on the anteroventral limb of the preopercular bone. These diagnostic characters are quite different from that diagnosis by Poyato-Ariza (1996c) because of removal of some maxillary and mandibular characters that were eliminated from the present analysis since they could not be accurately compared with the gonorynchoids (see Withdrawn Characters above). A future, chanid-oriented revision of their relationships might result in additional characters with which to diagnose the subfamily.

Within the Chaninae, there is a strongly supported sister-group relationship between *Chanos* and †*Tharrhias* that appears in the strict consensus tree and even in both trees of the Bremer support test (Figs. 7.9–7.14, node D). This relationship appears consistently in all analyses since Poyato-Ariza (1996a, c). In the present one, it is supported by five synapomorphies, four of which are unambiguous, including two autapomorphies. The tribe Chanini thus defined is supported by: exoccipitals that present a posterior concave-convex border, forming a projection above basioccipital (A); mesethmoid large, with broad posterolateral wing-like expansions (A); symplectic and quadrate separated (through cartilage at least in the Recent *Chanos*); preopercular expansion present in posteroventral corner and part of the posterodorsal limb of the bone; and four infraorbitals (a loss interpreted convergent with gonorynchoids, but unique among chanids where the number of infraorbitals is accessible).

In turn, the subfamily †Rubiesichthyinae (Figs. 7.9–7.14, node E) is formed by the sister genera †*Rubiesichthys* and †*Gordichthys*, as in Poyato-Ariza (1996a) and in Grande and Poyato-Ariza (1999). In the present analysis, it is confirmed as a relatively strong clade that holds the Bremer support test as explained above (Figs. 7.13, 7.14). This node is supported by four unambiguous characters, one of which is an autapomorphy: nasal bone small but flat; dorsal and ventral borders of the maxillary articular process very curved, almost describing an angle; angle formed by preopercular limbs acute; and posterior process on the posterior border of first supraneural present (A) (Fig. 7.7).

Comments on Some Chanid Genera

†*Aethalionopsis* appears in an indeterminate position within the Chanidae only in the strict consensus tree including the problematic forms commented on above (Fig. 7.9), but it appears well determined in the rest of the consensus trees (Figs. 7.10–7.12), and even in the 50% majority rule consensus tree of the Bremer support test (Fig. 7.14).

The traditional view on †*Aethalionopsis* emphasized its affinities with chanids (e.g., Taverne 1981), although it was later considered stem-group gonorynchiform. Both Blum (1991a: 283) and Fink and Fink (1996: 212, fig. 1), following Patterson's (1984) hypothesis that all supposed fossil chanids were not actually chanids, considered this genus to be more primitive than all the other gonorynchiforms mostly on the basis of two characters: larger parietals and three epurals instead of two. The interpretation of character 15, parietal reduction, in †*Aethalionopsis* is admittedly problematic. It was coded as exhibiting the primitive condition by Grande and Poyato-Ariza (1999) following previous interpretations (Taverne 1981). In fact, †*Aethalionopsis* material has revealed that the precise limits of the parietals are very difficult to establish with certainty, and new definition of this character has resulted in its being coded as first derived state; it is true that parietals of this genus are not as reduced as those of other gonorynchiforms, but they present reduction when compared with the primitive condition (see discussion of character 15). However, this genus does present the primitive condition in the number of uroneurals (see discussion of character 97 above), not only of epurals (Taverne 1981: fig. 8; pers. obs.). The caudal skeleton of †*Aethalionopsis* presents characters that are both primitive (presence of fringing fulcra, three uroneurals) and derived (uroneural two separated from second ural centrum). However, it still appears as a chanid in the present analysis, as it presents all the diagnostic characters of the group. About the exceptional presence in a genus of structures normally absent in a group, let us note that Fink and Fink (1981: 322), commenting on a few exceptions to the absence of supramaxillae in ostariophysans, stated that "if absence (…) is due to suppression of a developmental pathway, presence of a small, separate element may be due to re-expression of that pathway". In addition, the characters in the caudal skeleton are often convergent in different groups, so the alternative explanation of independent loss of elements within the Chanidae is also possible.

The phylogenetic position of †*Dastilbe* and †*Parachanos* with respect to each other and to the Chanini is not solved by the present analysis. The different alternatives (i.e., sister group to each other or either one sister group to the Chanini) are supported by very few, ambiguous characters, so we do not prefer any of them. However, according to this analysis, each of these genus can be diagnosed by apomorphic characters (see Appendix 3) and thus cannot be considered synonyms. In addition, they exhibit differences in the bones of the oral region, such as the relative length of the maxillary articular process and the morphology of the anteroventral border of the dentary (Fig. 7.6). These characters were not appropriate in an analysis with gonorynchoids, but may be important in a future analysis of chanids alone.

†Dastilbe presents all the chanid synapomorphies, although Dietze (2007: 15) claimed that it lacked three of them: (1) Position of the mandibular articulation, which is claimed to be in the anterior border of the orbit, and not anterior to it; however, in her illustration of the skull (Dietze 2007: fig. 3), the articulation lies anterior to the orbital rim as restored from the limits of the infraorbital bones, and, in addition, the anatomic relevance of the character is the forward displacement of the quadrate and the mandible (see character 35 above). (2) The absence of an anterior mandibular notch, which is an observational artefact (see character 31 above). (3) The lack of suprapreopercular bone, but this is a delicate bone, very rarely observable, and, in any case, its absence, if confirmed, may be an autapomorphic reversion of the genus. The most parsimonious position of †Dastilbe is within the Chanidae and, furthermore, within the Chaninae. The real problem of this genus at present, as repeatedly seen in the discussion of characters above, is the uncertainty of its specific composition and the variation of its characters, some of which had to be coded as polymorphic for the present analysis. For instance, the variability in the crucial characters of the caudal endoskeleton (Poyato-Ariza 1996a, Davis and Martill 1999) has never been tested to be either individual or specific (their own observations on this variability by the latter are inconsistent with their hypothesis of a single species for this genus). The genus †Dastilbe still remains, at present, in need of revision, including comparison of the type material of all nominal species previous to any new specific rearrangement.

†Prochanos was not included in the present analysis because of its morphological incompleteness and poor preservation, and because it is in need of revision. It most likely belongs to the Chanidae, because it presents a forwarded quadrato-mandibular articulation and an expansion of the opercular bone, two of the autapomorphies of the family (see also Poyato-Ariza 1996a: 41–42). It does not have the autapomorphies of the †Rubiesichthyinae and its size is well beyond the range of that small-sized group. Its preopercular looks morphologically closer to that of the Chaninae, so it seems suitable to consider it a Chaninae *incertae sedis* at present, pending further revision of the genus.

Suborder Gonorynchoidei

In a recent phylogeny by Diogo (2007: fig. 3), the Gonorynchiformes appeared as a monophyletic group, with *Chanos* and *Gonorynchus* as sister groups, and *Phractolaemus* as sister group to the other Kneriidae, whose interrelationships agree with those resulted in the present analysis. The only difference in that phylogeny is the fact that *Gonorynchus* appears more closely related to *Chanos* than to other Gonorynchoids. We think that this is probably the result of bias sampling, as fossil gonorynchiforms are not included, and the main aim of that paper is the origin of higher clades, so the study is very large-scaled. Or it may be "long branches attraction" bias, as recently observed by Cavin and Suteethorn (2006) in other actinopterygians. In any case, the Gonorynchoidei is a strong clade that appears well supported in all our consensus trees, including those of the Bremer support test (Figs. 7.9–7.14, node F). The same goes for its two internal clades, Gonorynchidae and Kneriidae.

The clade Gonorynchoidei is supported by 14 characters, 11 of which are unambiguous and eight autapomorphic. They include: reduction and separation of pterosphenoids (A); absence of mandibular sensory canal (A); reduction of ectopterygoids that results in a loosely articulated suspensorium (A); articular head of hyomandibular bone double, with the anterior articular surface forming a separate head from the posterior articular surface; unossified first basibranchial (A); unossified first pharyngobranchial (A); partial to near complete skin connection of opercular borders to the side of the head (A); four or fewer infraorbitals (including lacrimal and excluding dermosphenotic); infraorbital bones not including lacrimal reduced to small, tubular ossifications (A); anterior neural arches abutting contact laterally with adjoining arches; first two anterior parapophyses fused to their centra; absence of postcleithra (convergent with *Chanos*); and adductor mandibulae Aω absent (A).

Let us note that some of the characters above are not accessible in fossils, and a few must be checked in adult specimens (e.g., lack of ossification in branchial bones; see discussion of characters 49 and 51 above). These diagnostic characters essentially coincide with those in Grande and Poyato-Ariza (1999), although the character involving fusions in the caudal endoskeleton does not appear, because of reinterpretation of the caudal endoskeleton of †*Notogoneus* (Grande and Grande 2008).

Family Gonorynchidae

It is a well-supported clade present in all consensus trees and standing the Bremer support test as explained above (Figs. 7.9–7.14, node G). It is supported by eight characters, with six unambiguous and five autapomorphic: brush-like cranial intermuscular bones (*sensu* Patterson and Johnson 1995) present (A); frontals elongate and narrow except in postorbital region (A); notch between the dentary and the angulo-articular bones present; patch of about 20 conical teeth on endopterygoids and basibranchial 2 present (A); triangular opercular bone (A); and modified ctenoid scales (A).

Some of these characters, such as the cranial intermuscular bones or the type of scales, are difficult to assess in fossils. On the other hand, a number of them are relatively easy to assess, and account for what we familiarly consider the "gonorynchid look", mainly the narrow frontals, endopterygoid teeth, and triangular opercular bone. The diagnostic characters are the same as in Grande and Poyato-Ariza (1999), except for the lack of caudal endoskeleton characters due to their reinterpretation in †Notogoneus (see Gonorynchoidei above).

Gonorynchus and †Notogoneus

These two genera were previously considered sister groups (Grande and Poyato-Ariza 1999: 231). However, the present analysis provides a different result, with †Notogoneus as the basal genus and Gonorynchus as the sister group to all other gonorynchids. This is due to the effect of the reinterpretation of the caudal endoskeleton of †Notogoneus (Grande and Grande 2008) as a structure essentially without fusions: the node separating †Notogoneus from all other gonorynchids (Figs. 7.10, 7.12, 7.14, node H) is nearly completely supported by non-ambiguous characters from the caudal endoskeleton (see more details in Appendix 3). Only one of the characters formerly uniting the two genera is not used in the present analysis (shape of supraoccipital bone in dorsal view; see comments in Withdrawn Characters section above); the distribution of the other characters (e.g., V-shaped dentary) is reinterpreted in the present analysis according to the new position of these two genera (see Appendix 3 for details). In any case, this is the most parsimonious position of these two genera according to the information available at present. Future work, notably further revision of the closely related Middle East forms (see next paragraph), may confirm this new hypothesis of phylogenetic relationships.

Comments on the Fossil Gonorynchids from the Middle East

The clade formed by †Charitopsis, †Charitosomus, †Hakeliosomus, †Judeichthys, and †Ramallichthys does not appear in strict consensus trees, but it consistently does in the 50% majority rule consensus trees, including the Bremer support one (Figs. 7.10, 7.12, 7.14, node I). These five nominal genera present all the apomorphic characters of the Gonorynchidae; currently, there is no sense in granting any of them a familial status.

This five-genera clade appears as the sister group to *Gonorynchus* and is supported by five characters, all of them unambiguous, including one autapomorphy: maxillary process for articulation with autopalatine present (convergent with the Chaninae); presence of at least one enlarged retroarticular process in the mandible (the inferior one) (A); absence of spine on the neural arch of first vertebra; and anterior neural arches in overlapping contact with adjoining arches laterally. This clade is new, since the relationships of these forms with the other gonorynchids remained unsolved in Grande and Poyato-Ariza (1999: fig. 2).

However, the interrelationships of these genera are presently impossible to resolve in the strict consensus trees of the present analyses (Figs. 7.9, 7.11, 7.13), even after removal of †*Lecceichthys* from the data matrix (see comments on this genus above). It is worth mentioning the ways in which †*Lecceichthys* affects the interrelationships of the other five. A clade formed by the unresolved trichotomy [†*Hakeliosomus*, †*Judeichthys*, †*Ramallichthys*] distinctly appears in the analysis with †*Lecceichthys* (Figs. 7.9, 7.10, node Ibis), but, when this genus is removed, the result in the strict consensus tree is a polytomy of all five genera (Figs. 7.11, 7.13). The 50% majority rule consensus tree (Fig. 7.12) provides a little more resolution but is still unsatisfactory (only 53% of trees). The trichotomy [†*Hakeliosomus*, †*Judeichthys*, †*Ramallichthys*] is the sister group to [†*Charitopsis* + †*Charitosomus*].

The trichotomy is supported by a single character, vomer extending anteriorly beyond the level of anterior margin of mesethmoid, a character that is difficult to interpret (see Gonorynchidae in Appendix 3). In turn, †*Charitopsis* is the sister group to †*Charitosomus* on the basis of two autapomorphic characters: presence of spine on posterior border of subopercular bone; and inferior and superior mandibular retroarticular processes present (only with ACCTRAN). These are the only two clades within the present analysis of gonorynchiform fishes that do not stand the test of the Bremer support (Fig. 7.14).

These interrelationships, although not completely satisfactory, agree with those proposed by Grande and Grande (2008) in that *Gonorynchus* forms the sister group to the Middle Eastern clade, which consists of †*Ramallichthys* as the sister group to (†*Charitosomus* + †*Charitopsis*). In that paper, †*Ramallichthys*, †*Hakeliosomus*, and †*Judeichthys* are considered synonyms. Our results in the present analysis support that synonymy. Further nomenclatural problems involving these genera need to be solved at species level. For instance, "†*Hakeliosomus*" *hakelensis* was considered as a species of a distinct genus by Gayet (1993b) but as a species of †*Charitosomus* by le Danois (1966), Patterson (1970), Grande (1996) and Grande and Poyato-Ariza (1999). The results of the present analysis and of Grande and Grande (2008) support a synonymy with †*Ramallichthys*.

Family Kneriidae

There are currently some disagreements concerning the composition of this family. Grande and Poyato-Ariza (1999) proposed a monophyletic Kneriidae including *Cromeria*, *Grasseichthys*, *Kneria*, *Parakneria*, and *Phractolaemus*. Nonetheless, *Phractolaemus* was traditionally assessed to a family of its own (e.g., Thys van den Audenaerde 1961), occasionally even far from gonorynchids. Even after the phylogenetic relationships of this genus are disclosed, *Phractolaemus* is sometimes placed in its own monotypic family within the Gonorynchiformes in recent works as well ("This order contains four families": Moyle and Cech 2004: 301), or, with the same framework of phylogenetic interrelationships, it is proposed as a distinct monotypic family within a suborder Kneroidei (Nelson 2006: 135, 137).

The results of the present analysis confirm the strength of the clade formed by these five Recent genera. It is more strongly supported than the node that excludes *Phractolaemus* (see subfamily Kneriinae below). We acknowledge the morphological uniqueness of *Phractolaemus* (see Appendix 3), and that its evolutionary history must have been long and independent with respect to the rest of Kneriidae, but we think that this is acknowledged by its subfamily status within the Kneriidae. In addition, we prefer to follow the principle of subordination by granting a familiar rank to the sister group of the family Gonorynchidae (see Fig. 7.12 for further details).

The Kneriidae including *Cromeria*, *Grasseichthys*, *Kneria*, *Parakneria*, and *Phractolaemus* is a strong clade supported by 11 characters, 10 of which are unambiguous, including six autapomorphies: parietals reduced to tubular ossifications or absent as independent ossifications; foramen magnum enlarged and dorsally bounded by supraoccipital (A); presence of wings (extensions) on lateral ethmoids (A); presence of a metapterygoid process of hyomandibular bone placed in ventral position (A); presence of posterodorsal ascending process in the interopercular bone (A); reduction in the number of supraneurals to two or fewer (A); only one uroneural (convergent with *Gonorynchus*); haemal arch and preural centrum 2 fused (convergent with *Gonorynchus*); and presence of a peculiar adductor mandibulae A1-OST-M (A). These characters basically differ from those in Grande and Poyato-Ariza (1999) in those revised (e.g., contact of neural arches) or not used (e.g., mandibular height) in the present analysis. In turn, myological characters help support this node as well. Lavoué *et al.* (2005) comment on the problems posed by reductive characters in phylogeny, particularly in the study of the Kneriidae. While we acknowledge that reductive characters can pose difficulties in establishing homologies in some cases, this diagnosis of the Kneriidae is largely based on non-reductive apomorphies.

Subfamily Kneriinae

This taxonomic status replaced the sense of the family Kneriidae previous to Grande and Poyato-Ariza (1999). The Kneriidae excluding *Phractolaemus* is a strong clade, supported by seven characters, with four unambiguous and autapomorphic ones: absence of parietals as independent ossifications; mesethmoid long and slender, with anterior elongate lateral extensions (A); ossification of fifth basibranchial in adult specimens (A); contact of neural arch of first vertebra and supraoccipital (A); and great reduction in size of supraneurals, if present at all. These results are very similar to those by Grande and Poyato-Ariza (1999) and therefore confirm the taxonomic rank of this group.

As mentioned earlier, the distribution of characters within the Kneriinae was agressively criticised by Britz and Moritz (2007). However, these authors did not provide any alternative distribution of characters or an alternative hypothesis of phylogenetic relationships. Therefore, we will simply emphasize here that, even after revision of characters with correction of errors and polymorphic character states (see discussion of characters above), the phylogenetic interrelationships of the Kneriinae remain the same, although, effectively, some characters have changed their distribution in the present analysis (e.g., inclination of the vomer, lack of ossification of first basibranchial). We will present the distribution of characters provided by the analysis, which is never a simple "assumption" (e.g., Britz and Moritz 2007: 45) but the most parsimonious distribution according to the cladistic analysis with the information of the data matrix. Needless to say, phylogenetic relationships are not absolute truths, so future research will most likely provide additional evidence that may, or may not, result in dissimilar hypotheses. However, alternative hypotheses of character distribution can only be discussed in the frame of an alternative hypothesis of phylogenetic relationships or, at least, in the frame of a revised hypothesis of character distribution in the most parsimonious trees.

In the present analysis, both the tribe Cromeriini (*Cromeria* + *Grasseichthys*) and the tribe Kneriini (*Kneria* + *Parakneria*) are strong clades that appear in all consensus trees (Figs. 7.9–7.12), and they also pass the Bremer support test as explained above (Figs. 7.13, 7.14). The Cromeriini are supported by six unambiguous characters, four of which are autapomorphies: absence of nasal bone as independent ossification; presence of interfrontal fontanelle; absence of ectopterygoid bone as independent ossification (A); absence of ossified interhyal (A); reduction of interopercular bone to a long thin spine positioned lateroventral to preopercular bone (A); and absence of body scales (A). The presence of a peculiar tendon of adductor mandibulae A2 running perpendicular to the main body of this section and connecting it to the anteroventral surface of the quadrate is interpreted as an additional autapomorphy by the ACCTRAN option. Curiously, the Kneriini are also supported by six unambiguous characters, two of which are autapomorphic: nasal bone small but flat (reverted): presence of a maxillary process for the articulation with the autopalatine (convergent with the Chaninae); anterior portion of vomer anteroventrally inclined, nearly vertical (A); suprapreopercular bone(s) present and reduced to tubular ossification(s); six hypurals (reversion); and hyohyoideus abductor with significant part of its fibers attaching on pectoral girdle (A).

Additional Discussion

Supplementary Analyses: Myological Characters

In the results of the analyses just presented, the new myological characters added to the osteological data matrix (numbers 112–130) reinforce the diagnosis of certain taxa, such as the Gonorynchoidei, the Kneriidae, and the Kneriini. The situation is less clear concerning other taxa; for instance, the Gonorynchidae can be supported by three additional myological autapomorphic characters. Because myological characters are observable in Recent forms only, they are coded for *Gonorynchus* only within the family, so they can be alternatively interpreted as autapomorphic of this genus. Nonetheless, this study for the first time examines two independent data bases to better understand gonorynchiform interrelationships. Results from this study show that these data sets are largely congruent, providing strong support for the systematic hierarchy presented in this chapter.

Analysis Without Myological Characters

The first additional analysis consisted of another heuristic search with all taxa (except the problematic ones) without the myological characters (numbers 112–130). The result was 90 most parsimonious trees (the same number) with 215 steps (logically shorter). Both the strict consensus and the 50% majority rule consensus trees were identical to those with the myological characters (Figs. 7.11, 7.12). This allows us to conclude that myological characters do not change the hypothesis of phylogenetic interrelationships as based on osteological characters only.

Analysis with Myological Characters Alone

The second additional analysis consisted of another heuristic search, this time with myological characters only, by deleting all osteological characters (numbers 1–111). Fossil taxa were also deleted; since all of their myological characters are unknown, the result would necessarily be a complete indetermination and a great source of noise. There is no sense in processing fossil forms with myological characters alone. The result was rather interesting. There were six most parsimonious trees (a much lower number) of 20 steps (very short because there are very few characters). The strict consensus tree (Fig. 7.15A) consists of *Chanos* in an indetermination with the outgroup genera plus the monophyletic Gonorynchoidei; within these, only the clade (*Kneria* + *Parakneria*) escapes the general indetermination. The 50% majority rule consensus tree (Fig. 7.15B) improves by placing *Gonorynchus* as the sister group to the monophyletic Kneriidae (in 67% of the trees). Therefore, we can conclude that myological characters alone are not enough to provide a good resolution of the interrelationships of Recent forms, although the resolution provided follows the general relationships established by the complete data matrix, and is totally congruent with them.

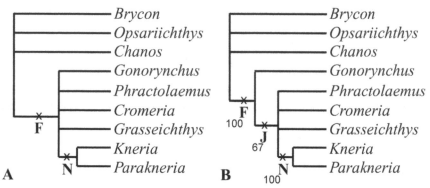

Fig. 7.15 Consensus trees obtained when processing myological characters and Recent genera alone. **A**, strict consensus tree; **B**, 50% majority rule consensus tree. Nodes: **F**, Gonorynchoidei; **J**, Kneriidae; **N**, Kneriini. See further explanations in main text and character distribution in Appendix 3.

Molecular Phylogenies

Results of molecular phylogenies are, at present, quite inconclusive (see Introduction). The conflicting molecular phylogenetic hypotheses currently available probably reflect the inevitable bias of the absence of fossil forms, one of the many problems in comparing morphological and molecular phylogenies. We do think that the evolutionary history of any given group cannot be accurately known without the fossil forms, so we feel more comfortable with a hypothesis of phylogenetic relationships between *Chanos* and *Gonorynchus* where fossil forms are included. On the other hand, we also believe molecular phylogenies are of interest, especially for groups with some morphological problems (paedomorphic characters) and without fossil record: in our case, the Kneriidae. Some unpublished work by T. Grande may substantiate the hypothesis that *Cromeria* is more closely related to the Kneriini than it is to *Grasseichthys*.

Molecular evidence mostly corroborates the monophyly of the Gonorynchiformes and its sister-group relationship with the Otophysi (with exceptions such as Saitoh *et al.* 2003; see in-depth review in Lecointre present volume), but the molecular interrelationships of gonorynchiforms are still uncertain. We hope to contribute to future comparisons with less conflicting molecular hypothesis with the updated morphological analyses herein provided, which support the traditional hypothesis of *Chanos* as the most basal Recent gonorynchiform, with *Gonorynchus* as sister group to kneriids. As stated by Lecointre (present volume), a clade cannot be considered reliable unless it has been positively tested through several independent sources of data; in the case that there is no agreement with the molecular studies about a topology different from that expected from morphological topologies, the classical morphological hypothesis should be the preferred one; that is, one molecular tree obtained from one molecular data set cannot falsify a tree based on morphological data, and only several independent congruent molecular phylogenies can do that, in a long-term process of scientific research. We agree completely with this point of view, and we do think that this is currently the case with Gonorynchiform interrelationships.

Historical Biogeography

The Gonorynchiformes no doubt have a long and complicated history. As discussed in Grande (1999), present-day distribution patterns most likely resulted from a combination of dispersal and vicariant events. We will follow the relationships illustrated in Fig. 7.10 for this discussion.

The chanid *Chanos* exhibits a geographically widespread distribution and, as an adult, inhabits marine waters throughout the Indo-Pacific. Interestingly, it breeds in inshore waters, produces pelagic eggs, and, when the larvae reach about 10 mm (Morioka *et al.* 1993), they enter brackish creeks that have minimum contact with the open water. As adults they return to the sea (Patterson 1975). Unlike *Chanos*, fossil chanid species are known from discrete localities in Brazil, Europe, and western Africa; most chanid localities date back to the Early Cretaceous. Acknowledging that there is some debate concerning the salinity of these ancient waters, it seems clear that these localities may have been influenced by marine systems, with the exception of Las Hoyas, the locality of †*Gordichthys* and one of the localities of †*Rubiesichthys* (Poyato-Ariza *et al.* 1998), and Bernissart, the locality of †*Aethalionopsis* (e.g., Taverne 1981). Cavin (2008) tried to distinguish an initial dispersal of basal chanoids from Europe and a subsequent vicariance between South American- and African-derived chanoids.

As illustrated (Fig. 7.10), the Chanidae (node B) forms the sister group to the Gonorynchidae (node G) + Kneriidae (node J). The Kneriidae consists of extant taxa found in freshwater streams and rivers of western and central Africa. The Gonorynchidae, on the other hand, is a predominately marine group consisting of one extant genus, *Gonorynchus*, and several fossil forms dating back to the Late Cretaceous.

The phylogenetic results of the Gonorynchidae described here differ from those of Grande and Poyato-Ariza (1999) in that †*Notogoneus* forms the sister group to the rest of Gonorynchidae, and *Gonorynchus* forms the sister group to the Middle Eastern forms (Fig. 7.10, node I). Whether this phylogenetic change has historical biogeographic implications still needs to be determined. †*Notogoneus* (eight nominal species) is known from Late Cretaceous-Oligocene freshwater deposits of North America, Europe, Asia, and Australia, and some possible brackish and marine deposits in France and Germany. The monophyletic Middle Eastern forms (e.g., †*Ramallichthys*, †*Charitopsis*, and †*Charitosomus*) are exclusively marine and all date to the Late Cretaceous. This distribution pattern has led researchers to speculate that the Gonorynchidae has a Tethys origin (Jerzmańska 1977, Gaudant 1993).

Now, does the inclusion of fossil taxa such as †*Apulichthys*, †*Halecopsis*, †*Sorbininardus*, and †*Lecceichthys* contribute to a better understanding of the historical biogeography of Gonorynchiformes? The answer is yes and no. †*Halecopsis* is found in Eocene marine deposits from the London Clay (England), Argiles de Flandres (Belgium), and Nord (France) (e.g., Poyato-Ariza 1996a). †*Apulichthys*, †*Sorbininardus*, and †*Lecceichthys* are from Late Cretaceous marine deposits of Nardò, in southern Italy (Taverne 1997, 1998, 1999). If †*Sorbininardus* and †*Lecceichthys* are gonorynchids, they might push back the time line for gonorynchids in Europe a bit. †*Notogoneus* is known from Oligocene-Eocene deposits in North America, Germany, and France, while †*Charitosomus formosus*, known from Westphalian deposits, is slightly younger than its conspecifics in the Middle East. These presumptive gonorynchids, along with †*Halecopsis*, increase the presence of gonorynchids in Europe and suggest that the group was more diverse than previously thought. Until the taxonomic identity and phylogenetic relationships of †*Apulichthys* can be clearly determined, this taxon adds nothing to the historical biogeography of the group.

In summary, data indicate that the Gonorynchiformes evolved before the Cretaceous, since its two sister-suborders must have separated before the Early Cretaceous. The Chanidae forms the gonorynchiform subgroup with the oldest known record, dating back to the earliest Early Cretaceous (†*Rubiesichthys*; see Fara *et al.* present volume for an updated account of the gonorynchiform fossil record). Acknowledging that the paleoecologies of several chanoid localities are debatable, many researchers agree that like *Chanos*, chanoid taxa were probably capable of tolerating wide salinity changes, making them more euryhaline than exclusively freshwater (with the exception of †*Aethalionopsis* and †*Gordichthys*). Gonorynchiformes thus contain a combination of freshwater, marine, and brackish water taxa, with many of them exhibiting wide salinity ranges. Patterson (1975) hypothesized that the primitive condition for gonorynchiform or ostariophysan fishes is a euryhaline life style with an inland phase in fresh or brackish water. This suggests that the marine *Gonorynchus* may have eliminated the "fish pond" state as Patterson (1975) describes it, while kneriids would have had to mature in fresh water by eliminating the marine stage. Patterson (1975) suggested that, if these fishes were primitively euryhaline (with *Chanos* exhibiting the primitive life history condition), the hypothetical ancestor of Ostariophysi was an inshore fish of warm seas, whose life-cycle involved an inland phase in fresh or brackish water. Patterson (1975) also argued for a Pangean origin for the group (Patterson 1975), since the order must have evolved before the Cretaceous. This hypothesis does not discount a Tethys influence, as suggested above. Many gonorynchiform taxa have been collected from European and Middle Eastern localities (e.g., †*Gordichthys*, †*Rubiesichthys*, †*Charitosomus*, †*Aethalionopsis*, †*Apulichthys*, †*Halecopsis*, †*Sorbininardus*, †*Lecceichthys* = Europe; †*Ramallichthys*, †*Judeichthys*, †*Charitosomus*, †*Charitopsis* = Middle East), suggesting that a Tethys origin for some taxa or a dispersal route for others is possible. This model, in combination with Patterson's (1975) Pangean hypothesis, may well be plausible historical biogeographic explanations.

Conclusions: The Evolutionary History of the Gonorynchiformes

The Gonorynchiformes, as an ensemble, has a widespread geographic distribution extending to Europe, North and South America, Africa, and Asia; and a fossil record that dates back to the Early Cretaceous. This fossil record indicates that the two main evolutionary lineages of the Gonorynchiformes (Chanoidei and Gonorynchoidei) have been separated at least since the earliest Early Cretaceous. There has therefore been a long time for complex and diverse anatomical variations to appear and mask their common bauplan. This has been precisely the main problem in recognizing their shared kinship.

The gonorynchiform structures have undergone remarkable morphological (and consequently functional) modifications within the different evolutionary lineages of the order. For example, some of them (Chanidae) have a very small, terminal mouth gap; *Phractolaemus* has evolved morphological specializations resulting on the capability of breathing atmospheric air and a uniquely protractile mouth, which has been acquired independently from other teleostean groups, and these characters have probably evolved independently of each other.

Diverse types of fusions in the caudal endoskeleton of gonorynchiforms have been acquired independently from other teleostean groups, and at least twice independently within the order. In the Recent *Chanos chanos*, the two ural centra, preural centrum 1, and first uroneural are fused, whereas these and all other caudal endoskeletal elements are totally autogenous in fossil chanids (e.g., Fig. 7.8). Most gonorynchoids also show fusions in their caudal skeleton, clearly acquired independently from those of *Chanos*, as supported by the present analysis. However, the lack of kneriid fossil record prevents us from knowing whether fusions appeared independently in this family or in the common ancestor shared with gonorynchids. Recent revision of †*Notogoneus* (Grande and Grande 2008) indicates that the caudal skeletal elements are not fused in this gonorynchid genus, which would imply that fusions in the caudal skeleton were independently acquired in gonorynchids and in kneriids (or alternatively, but less likely, acquired once in gonorynchoids and reverted in †*Notogoneus*). Thus interpreted, this would account for three independent occurrences of fusions in the caudal skeleton within Gonorynchiformes: once in chanids, once in gonorynchids, and once in kneriids, all independent from Otophysi and other teleostean lineages.

The derived characters of the Chanidae appear consistently in the family since their first known fossil record, in the Early Cretaceous, clearly indicating that the Chanidae are a homogeneous, very conservative evolutionary lineage. The two chanid subfamilies are also separated at least since the Early Cretaceous, but their differences are very slight compared with the common pattern shared by all chanids (see Poyato-Ariza 1996a for further details). In turn, the evolutionary histories of the Gonorynchidae and the Kneriidae are very different. The former has a large fossil record; the latter lacks it. The fossil record of the Gonorynchidae (e.g., Fara *et al.* present volume) indicates that the two families are separated at least since the earliest Late Cretaceous. The bias in the fossil record of the Kneriidae can be due to many factors: less potential for preservation by reason of their lower degree of ossification; environments (freshwater currents) that, in general, do not favor sedimentation processes; and, most of all, the general lack of knowledge on the fish fossil record from Central Africa basically due to shortage of specimens. The Kneriidae have many reductive characters, mostly paedomorphic (e.g., Grande 1994), so that they are a very interesting group for the study of evolutionary processes and their mechanisms.

The evolutionary history of the gonorynchiforms has, for a long time, been understood in neontological terms only and, as a consequence, often misinterpreted, even today. The evolutionary connotations of the Gonorynchiformes are still often presented in neontological terms only, especially in general books of ichthyology, thus leading to potential confusion. For instance: "the kneriids are of special interest, because they provide some idea of what the early ancestors of the otophysan fishes may have been like" (Berra 2001 *fide* Moyle and Cech 2004: 301). By including fossils in the discussion of the evolutionary history of the Gonorynchiformes, we realize that all Recent gonorynchiforms are fairly, even exceptionally, derived within their own distinct evolutionary lineages. Therefore, their morphology is misleading when trying to interpret the "generalized" or primitive morphology of the Ostariophysi. As in other groups, the best insight into their evolutionary history is provided by their fossil record, which will unveil morphologies closer to those of the common ancestors (e.g., Arratia 1997, basal ostariophysan from the Late Jurassic of Bavaria).

The phylogenetic relationships and the evolutionary history of the Gonorynchiformes have been a matter of intense debate for over 150 years. These fishes are rather different and varied in external aspect, behavior, distribution, fossil record, and even general anatomy. As a consequence, the group looks superficially very heterogeneous. The Gonorynchiformes were, for a long time, a systematic enigma; their placement within the Teleostei was largely debated. Their monophyly was questioned or simply not even addressed as a possibility for a very long time. Such uncertainty was most likely due to the impressive amount of morphological and behavioral variation within the group.

There are still many issues to resolve before we fully understand the interrelationships and evolutionary history of the Gonorynchiformes. It is to be expected that future, separate approaches to the Chanoidei and the Gonorynchoidei, or even to the Chanidae, Gonorynchidae, and Kneriidae, will result in a higher resolution of their corresponding interrelationships. Further revision of the fossil record is still necessary for the Chanidae and the Gonorynchidae, while ontogenetic studies and molecular phylogenies may help solve the current problems of the Kneriidae, which lack a fossil record. However, it is not especially strange that we face these many problems. The morphological, ecological, and behavioral disparity of the Gonorynchiformes is truly remarkable for so small a group, as is their rich evolutionary history. High disparity, low diversity: what a motivating combination.

Systematics of the Gonorynchiformes

Superorder Ostariophysi *sensu* Rosen and Greenwood, 1970

Series Anotophysi Rosen and Greenwood, 1970

Order Gonorynchiformes *sensu* Rosen and Greenwood, 1970

Diagnosis
Ostariophysan fishes with: no Weberian apparatus; orbitosphenoid lost; pterosphenoids reduced and separated; cephalic ribs present; parietals reduced and separated by supraoccipital or absent; teeth on fifth ceratobranchial, premaxilla, maxilla, and dentary absent; premaxillary ascending process lost; neural arch of first vertebra contacting exoccipitals; rib on third vertebral centrum wider and shorter than others; and paired intermuscular bones consisting of three series, namely: epipleurals, epicentrals, and epineurals.

Taxa included
Suborder Chanoidei and suborder Gonorynchoidei.

Type family
Gonorynchidae.

Suborder Chanoidei *sensu* Poyato-Ariza, 1996a

Diagnosis
As for family (monotypic suborder).

Taxa included
Family Chanidae (type and only family).

Family Chanidae *sensu* Poyato-Ariza, 1996a

Diagnosis
Gonorynchiform fishes with: large, very broad, concave-convex premaxilla with long oral process; enlarged maxilla, posteriorly swollen to a bulbous outline, with a curved posterior border; notch in the anterodorsal border of the dentary present; quadrate-mandibular articulation anteriorly displaced, anterior to the orbit (except in †*Aethalionopsis*), by displacement of the quadrate bone and correlative elongation of the symplectic; metapterygoid process of hyomandibular bone present, placed on anterior border of the bone; opercular bone expanded, at least one third of the head length; angle formed by preopercular limbs straight to acute; preopercular expansion present (except in †*Dastilbe*); suprapreopercular bone present as a relatively large, flat bone; and unmodified neural arches anterior to dorsal fin autogenous in adult individuals, at least laterally.

Taxa included
Subfamily Chaninae (type subfamily) and subfamily †Rubiesichthyinae.

Type genus
Chanos Lacepède, 1803; originally described as a species of the genus *Mugil* by Forsskål (1775). The type and only Recent species is *Chanos chanos* (Forsskål, 1775). There are several nominal fossil species, all in need of revision.

Subfamily Chaninae Poyato-Ariza, 1996a

Diagnosis
Chanid fishes with: presence of maxillary process for articulation with autopalatine; and presence of a ridge on the anteroventral limb of the preopercular bone.

Taxa included
Tribe Chanini, †*Aethalionopsis* Gaudant, 1966, †*Dastilbe* Jordan, 1910, †*Parachanos* Arambourg and Schneegans, 1935, and †*Prochanos* Bassani, 1879 (*incertae sedis*).

Tribe Chanini Grande and Poyato-Ariza, 1999

Diagnosis
Chanin fishes with: exoccipitals with a posterior concave-convex border that forms a projection above the basioccipital; mesethmoid large, with broad posterolateral wing-like expansions; separation of symplectic and quadrate (filled with cartilage at least in the Recent *Chanos*); and preopercular expansion present, occupying the posteroventral corner and a good part of the posterodorsal limb of preopercular bone.

Taxa included
Chanos and †*Tharrhias* Jordan and Branner, 1908.

Subfamily †Rubiesichthyinae Poyato-Ariza, 1996a

Diagnosis
Chanid fishes with: nasal bone small but flat, not reduced to a tubular ossification; maxillary articular process very curved, almost describing an angle; preopercular limbs forming an acute angle; and presence of a posterior process on the posterior border of first, enlarged supraneural.

Genera included
†*Gordichthys* Poyato-Ariza, 1994 and †*Rubiesichthys*.

Type genus
†*Rubiesichthys* Wenz, 1984. Type and only species: †*Rubiesichthys gregalis* Wenz, 1984.

Suborder Gonorynchoidei *sensu* Rosen and Greenwood, 1970

Diagnosis

Gonorynchiform fishes with: pterosphenoids greatly reduced and broadly separated both anteroventrally and anterodorsally; absence of mandibular sensory canal; ectopterygoids reduced, articulating with the ventral surface of the autopalatine by at most 10% through cartilage, resulting in a loosely articulated suspensorium; articular head of hyomandibular bone double, with anterior articular surface forming a separate head from posterior articular surface; first basibranchial and first pharyngobranchial unossified in adult specimens, at least in Recent forms; opercular borders partly or almost completely connected to side of head with skin at least in Recent forms; four or fewer infraorbitals (including lacrimal and excluding dermosphenotic); infraorbital bones other than lacrimal reduced to small, tubular ossifications; anterior neural arches abutting contact laterally with adjoining arches; first two anterior parapophyses fused to the corresponding centra; postcleithra absent; and absence of muscle adductor mandibulae Aω.

Taxa included

Family Gonorynchidae and family Kneriidae.

Family Gonorynchidae *sensu* Grande and Poyato-Ariza, 1999

Diagnosis

Gonorynchoid fishes with: brush-like cranial intermuscular bones; frontals elongate and narrow except in their postorbital region; notch between the dentary and the angulo-articular bones present; patches of about 20 conical teeth on endopterygoids and basibranchial 2 present; opercular bone of triangular shape; and modified ctenoid scales.

Genera included

?†*Charitopsis* Gayet, 1993c (if valid, *incertae sedis*); †*Charitosomus* von der Marck, 1885; *Gonorynchus*; ?†*Hakeliosomus* Gayet, 1993b (if valid, *incertae sedis*); *Halecopsis* Delvaux and Ortlieb, 1887 (*incertae sedis*); ?†*Judeichthys* Gayet, 1985 (if valid, *incertae sedis*); †*Lecceichthys* Taverne, 1998 (*incertae sedis*); †*Notogoneus* Cope, 1885; and ?†*Ramallichthys* Gayet, 1982a.

Type genus

Gonorynchus Scopoli, 1777. Five nominal species, all Recent. The type species is *Gonorynchus gonorynchus* (Linnaeus, 1766).

Family Kneriidae *sensu* Grande and Poyato-Ariza, 1999

Diagnosis
Gonorynchoid fishes with: parietals highly reduced or lost; foramen magnum enlarged, dorsally bounded by supraoccipital; wings (extensions) on lateral ethmoids; metapterygoid process of hyomandibular bone present and placed in ventral position; posterodorsal ascending process in interopercular bone present; number of supraneurals reduced to two or fewer; number of uroneurals reduced to one; fusion of haemal arch and preural centrum 2; and presence of a peculiar adductor mandibulae A1-OST-M.

Taxa included
Subfamilies Phractolaeminae and Kneriinae.

Type genus
Kneria Steindachner, 1866. Up to 13 nominal species, all Recent. The type species is *Kneria angolensis* Stcindachner, 1866.

Subfamily Phractolaeminae Grande and Poyato-Ariza, 1999

Diagnosis
As for genus (monotypic family).

Type genus
Phractolaemus Boulenger, 1901. The type and only species is the Recent *Phractolaemus ansorgei* Boulenger, 1901.

Subfamily Kneriinae Grande and Poyato-Ariza, 1999

Diagnosis
Kneriid fishes with: parietals absent as distinct ossifications; mescthmoid long and slender, with anterior elongate lateral extensions; fifth basibranchial ossified in adult specimens; neural arch of first vertebra and supraoccipital in contact; and supraneurals greatly reduced in size or absent.

Taxa included
Tribe Cromeriini and tribe Kneriini.

Tribe Cromeriini Grande and Poyato-Ariza, 1999

Diagnosis
Kneriin fishes with: nasal bone absent as independent ossification; interfrontal fontanelle present; ectopterygoid absent as independent ossification; ossified interhyal absent as an independent ossification; interopercular bone reduced to a long thin spine and positioned lateroventral to preopercular bone; and body scales absent (naked body).

Genera included
Cromeria Boulenger, 1901 and *Grasseichthys* Géry, 1964.

Tribe Kneriini Grande and Poyato-Ariza, 1999

Diagnosis
Kneriin fishes with: nasal bone small but flat; maxillary process for articulation with autopalatine present; anterior portion of vomer anteroventrally inclined, nearly vertical; one or two suprapreopercular bones present as tubular ossifications; six hypurals; and hyohyoideus abductor muscle with significant part of its fibers attaching on pectoral girdle.

Genera included
Kneria and *Parakneria* Poll, 1965.

Acknowledgements

F.J.P.A. thanks: E. Bultynk and D. Nolf for their kind assistance during his research visit to the Institut Royal des Sciences Naturelles de Belgique in 2004 for additional study of the †*Aethalionopsis* specimens in that collection; L. Cavin for information on the new genus, and for commenting on some conflicting characters; H. J. Martín-Abad for assistance with typing the different data matrices; and L. Taverne for his comments on the fossil genera from Nardò. Many thanks to the reviewers of this manuscript, whose comments greatly improved the final paper. This project was partly funded by the National Science Foundation through grants to T. Grande (DEB 0128794 and EF 7032589) and by the Spanish Ministerio de Educación y Ciencia to F. J. Poyato-Ariza (project number CGL2005-01121).

References

Arambourg, C. 1935. Observations sur quelques poissons fossiles de l'ordre des halécostomes et sur l'origine des clupeidés. C. R. Acad. Sci. Paris 200(25): 2110–2112.

Arambourg, C. and D. Schneegans. 1935. Poissons fossiles du Bassin Sédimentaire du Gabon. Ann. Paléontol. 24: 139–160.

Arratia, G. 1992. Development and variation of the supensorium of primitive Catfishes (Teleostei: Ostariophysi) and their phylogenetic relationships. Bonner zool. Monogr. 32: 1–149.

Arratia, G. 1997. Basal teleosts and teleostean phylogeny. Palaeo Ichthyol. 7: 1–168.

Bassani, F. 1879. Vorläufige Mittheinlungen über die Fischfauna der Insel Lesina. Verhandlungen der K. K. Geologischen Reichsanstalt 1879(8): 162–170.

Berg, L.S. 1940. [Russian original; German translation, 1958]. System der rezenten und fossilien Fischartigen und Fische: XI + 311 pp. Deutscher Verlag, Berlin, Germany.

Bertin, L. and C. Arambourg. 1958. Super-Ordre des Téléostéens, pp. 2204–2500. In: P.P. Grassé [ed.]. Traité de Zoologie, 13(3). Masson et Cie., Paris.

Blum, S. 1991a. *Dastilbe* Jordan, 1910, pp. 274–285. *In*: J.G. Maisey [ed.]. Santana Fossils: An Illustrated Atlas. T.F.H. Publications, Neptune City, New Jersey.

Blum, S. 1991b. *Tharrhias* Jordan and Branner, 1908, pp. 286–298. *In*: J.G. Maisey [ed.]. Santana Fossils: An Illustrated Atlas. T.F.H. Publications, Neptune City, New Jersey.

Boulenger, G.A. 1901. Diagnoses of new fishes discovered by Mr. W.L.S. Loat in the Nile. Ann. Mag. Nat. Hist. ser. 7, 8: 444–446.

Brito, P. and C.R.L. Amaral. 2008. An overview of the specific problems of *Dastilbe* Jordan, 1910 (Gonorynchiformes: Chanidae) from the Lower Cretaceous of western Gondwana, pp. 279–294. *In*: G. Arratia, H.-P. Schultze and M.W.H. Wilson [eds.]. Mesozoic Fishes 4-Homology and Phylogeny. Verlag Dr. Friedrich Pfeil, Munich.

Britz, R. and T. Moritz. 2007. Reinvestigation of the osteology of the miniature African freshwater fishes *Cromeria* and *Grasseichthys* (Teleostei, Gonorynchiformes, Kneriidae), with comments on kneriid relationships. Mitt. Mus. Naturk. Berlin, Zool. Reihe 83(1): 3–42.

Cavin, L. 2008. Palaeobiogeography of Cretaceous bony fishes (Actinistia, Dipnoi and Actinopterygii). Geological Society of London, Special Publications 295: 165–183.

Cavin, L, and V. Suteethorn. 2006. A new semionotiform (Actinopterygii, Neopterygii) from Upper Jurassic–Lower Cretaceous deposits of North-East Thailand, with comments on the relationships of semionotiforms. Palaeontology 49: 339–353.

Chang, M.-M. and J. Maisey. 2003. Redescription of †*Ellimma branneri* and †*Diplomystus shegliensis*, and relationships of some basal clupeomorphs. Am. Mus. Novitates 3404: 1–35.

Cope, E.D. 1885. Eocene paddle-fish and Gonorhynchidae. Am. Natur. 19: 1090–1091.

Cuvier, Le B. and A. Valenciennes. 1846. Histoire Naturelle des Poissons, 19: XXII + 544 pp. P. Bertrand, Paris.

da Silva-Santos, R. 1947. Uma redescricão de *Dastilbe elongatus*, com algumas considerações sobre o gênero *Dastilbe*. Notas preliminares e estudos, Ministerio da Agricultura, Divisâo de Geologia e Mineralogia 42: 1–7.

da Silva-Santos, R. 1979. Actinopterygii do Aptiano do Estado de Minas Gerais. Tese de Concurso, Departamento de Biologia Animal e Vegetal do Instituto de Biologia da Universidade do Estado do Rio de Janeiro, Rio de Janeiro.

Davis, S.P. and D.M. Martill. 1999. The gonorynchiform fish *Dastilbe*, from the Lower Cretaceous of Brazil. Palaeontology 42(4): 715–740.

Delvaux, E. and J. Ortlieb. 1887. Les poissons fossils de l'argile yprésienne de Belgique. Ann. Soc. géol. Nord 15: 50–66.

de Oliveira, A.F. 1978. O gênero *Tharrhias* no Cretáceo da Chapada do Araripe. Anais da Academia Brasileira de Ciencias 50(4): 537–552.

Dietze, K. 2007. Redescription of *Dastilbe crandalli* (Chanidae, Euteleostei) from the Early Cretaceous Crato Formation of North-Eastern Brazil. J. Vert. Paleontol. 27(1): 8–16.

Diogo, R. 2007. The origin of higher clades. Osteology, myology, phylogeny and evolution of bony fishes and the rise of tetrapods: XVIII + 367 pp. Science Publishers, Enfield, New Hampshire.

Eastman, C.R. 1905. Les types de Poissons fossiles du Monte Bolca au Muséum d'Histoire Naturelle de Paris. Mém. Soc. Géol. Fr., Paléontol. 34: 1–33.

Fink, S.V. and W.L. Fink. 1981. Interrelationships of the Ostariophysan Fishes (Teleostei). Zool. J. Linn. Soc. London 72(4): 297–353.

Fink, S.V. and W.L. Fink. 1996. Interrelationships of Ostariophysan Fishes (Teleostei), pp. 209–249. *In*: M.L.J. Stiassny, L.R. Parenti and G.D. Johnson [eds.]. Interrelationships of Fishes. Academic Press Inc., San Diego.

Forsskål, P. 1775. Descriptiones animalium avium, amphibiorum, piscium, insectorum, vermium; quae in itinere orientali observait: 20 + XXXIV + 164 pp. Post mortem auctoris edidit Carsten Niebuhr, Hauniae (Copenhagen).

Gaudant, J. 1966. Sur la nécessité d'une subdivision du genre *Anaethalion* White (Poisson Téléostéen). C.R. Somm. Soc. Géol. Fr. 8: 308–310.

Gaudant, J. 1993. The Eocene freshwater fish-fauna of Europe: from paleobiogeography to paleoclimatology. Kaupis 3: 231–244.

Gayet, M. 1980. Hypothèses sur l'origine des ostariophysaires. C. R. Acad. Sci. Paris série D 290: 1197–1199.

Gayet, M. 1982a. Cypriniformes ou Gonorhynchiformes? *Ramallichthys* nouveau genre du Cénomanien inférieur de Ramallah (Monts de Judée). C. R. Acad. Sci. Paris 295(2): 405–407.

Gayet, M. 1982b. Considération sur la phylogénie et la paléobiogéographie des ostariophysaires. Geobios, mémoire spécial 6: 39–52.

Gayet, M. 1985. Gonorhynchiformes nouveau du Cénomanien inférieur marin de Ramallah (Monts de Judée): *Judeichthys haasi* nov. gen. nov. sp. (Teleostei, Ostariophysi, Judeichthyidae nov. fam.). Bull. Mus. natl. Hist. nat. Paris sér. 4, 7C(1): 65–85.

Gayet, M. 1986a. *Ramallichthys* Gayet du Cénomanien inférieur de Ramallah (Judée), une introduction aux relations phylogénétiques des Ostariophysi. Mém. Mus. natl. Hist. nat. Paris n.s. C, Sciences de la Terre 51: 1–81.

Gayet, M. 1986b. About ostariophysan fishes: a reply to S.V. Fink, P.H. Greenwood and W.L. Fink's criticisms. Bull. Mus. natl. Hist. nat. Paris sér. 4, 8C(3): 393–409.

Gayet, M. 1989. Note préliminaire sur le matériel paléoichthyologique éocretacé du Río Benito (sud de Bata, Guinée Équatoriale). Bull. Mus. natl. Hist. nat. Paris sér.4, 11(1): 21–31.

Gayet, M. 1993a. Relations phylogénétiques des Gonorynchiformes (Ostariophysi). Belg. J. Zool. 123(2): 165–192.

Gayet, M. 1993b. Gonorynchoidei du Crétacé supérieur marin du Liban et relations phylogénétiques des Charitosomidae nov. fam. Documents des laboratoires de Géologie, 126: 131 pp., Lyon.

Gayet, M. 1993c. Nouveau genre de Gonorynchidae du Cénomanien inférieur marin de Hakel (Liban). Implications phylogénétiques. C. R. Acad. Sci. Paris, série II, 432: 57–163.

Géry, J. 1964. Une nouvelle famille de poissons dulçaquicoles africains: les Grasseichthyidae. C. R. Acad. Sci. Paris 259(25): 4805–4807.

Gill, M.D. 1872. Arrangement of the families of fishes, or classes Pisces, Marsipobranchii, and Leptocardii. Smithsonian Miscellaneous Collections 11(247): XLVI + 49 pp.

Gill, M.D. 1893. Families and subfamilies of fishes. Mem. Nat. Acad. Sci. 6(6): 125–138.

Goodrich, E.S. 1909. Vertebrata Craniata (First fascicle: Cyclostomes and Fishes), pp. I–XVI + 1–518. *In*: R. Lankester [ed.] 1909. A treatise on zoology, 7. Adam and Charles Black, London, England.

Gosline, W.A. 1960. Contribution toward a classification of modern Isospondylous Fishes. Bull. Brit. Mus. (Nat. Hist.), Zool. 6(6): 325–365.

Gosline, W.A. 1971. Functional Morphology and Classification of Teleostean Fishes. The University Press of Hawaii, Honolulu.

Grande, L. 1982. A revision of the fossil genus †*Diplomystus*, with comments on the interrelationships of clupeomoph fishes. Am. Mus. Novitates 2728: 1–34.

Grande, L. and W.E. Bemis. 1998. A comprehensive phylogenetic study of amiid fishes (Amiidae) based on comparative skeletal anatomy. An empirical search for interconnected patterns of natural history. J. Vert. Paleontol. Mem. 4, suppl. to 18(1): X + 690 pp.

Grande L. and T. Grande. 2008. Redescription of the type species for the genus †*Notogoneus* (Teleostei) based on much, well-preserved material. J. Paleontol. 82(70): 1–31.

Grande, T. 1992. Higher interrelationships of Recent and fossil gonorynchiform fishes. Unpublished doctoral thesis, University of Illinois at Chicago, Chicago.

Grande, T. 1994. Phylogeny and paedomorphosis in an African family of freshwater fishes (Gonorynchiformes: Kneriidae). Fieldiana, Zool. N.S. 78: I–III + 1–20.

Grande, T. 1996. The interrelationships of fossil and Recent gonorynchid fishes with comments on two Cretaceous taxa from Israel, pp. 299–318. *In*: G. Arratia and G. Viohl [eds.]. Mesozoic Fishes—Systematics and Paleoecology. Verlag Dr. Friedrich Pfeil, Munich.

Grande, T. and F.J. Poyato-Ariza. 1999. Phylogenetic relationships of fossil and Recent gonorynchiform fishes (Teleostei: Ostariophysi). Zool. J. Linn. Soc. 125: 197–238.

Grande, T. and B. Young. 1997. Morphological development of the opercular apparatus in *Kneria wittei* (Ostariophysi: Gonorynchiformes) with coments on its possible function. Acta Zool. 78(2): 145–162.

Greenwood, P.H., D.E. Rosen, S.H. Weitzman and G.S. Myers. 1966. Phyletic studies of teleostean fishes, with a provisional classification of living forms. Bull. Am. Mus. Nat. Hist. 131(4): 339–456.

Gregory, W.K. 1933. Fish skulls. A study of the evolution of natural mechanisms. Transactions of the American Philosophical Society, 23(2): 75–481. Facsimile re-issue, 1959, Eric Lundberg, Laurel, Florida.

Günther, A.C.L.G. 1868. Catalogue of the Fishes in the British Museum, vol. 7: XX + 512 pp. Trustees of the British Museum, London.

Günther, A.C.L.G. 1880. An Introduction to the Study of Fishes. 720 pp. Adam and Charles Black, Edinburgh.

Howes, G.J. 1985. Cranial muscles of gonorynchiform fishes, with comments on generic relationships. Bull. Brit. Mus. nat. Hist. (Zool.) 49(2): 273–303.

Jerzmańska, A. 1977. Süiswasserfische des älteren Tertiärs von Europa. *In*: H.W. Matthes and B. Thaler [eds.]. Eozäne Wirbeltiere des Geiseltales: 67–76. Martin-Luther Universität Halle-Wittenberg, Wittenberg.

Jordan, D.S. 1910. Description of a collection of fossil fishes from the bituminous shales at Riacho Doce, state of Alagôas, Brazil. Ann. Carnegie Mus. 7: 23–34.

Jordan, D.S. 1923. A classification of fishes, including families and genera as far as known. Standford Univ. Publ., Univ. Ser., Biol. Sci. 3(2): 77–243.

Jordan, D.S. and J.C. Branner. 1908. The Cretaceous fishes of Ceará, Brazil. Smithsonian Misc. Coll. 52: 1–29.

Lacepède, B.G.E. (1803). Histoire Naturelle des Poissons, 5: XLVIII + 803 pp. Furne et Cie., Paris.

Lauder, G.V. and K.F. Liem. 1983. The evolution and interrelationships of the actinopterygian fishes. Bull. Mus. Comp. Zool., Harvard Univ. 150(3): 95–197.

Lavoué, S., M. Miya, J.G. Inoue, K. Saitoh, N.B. Ishiguro and M. Nishida. 2005. Molecular systematics of the gonorynchiform fishes (Teleostei) based on whole mitogenome sequences: Implications for higher-level relationships within the Otocephala. Mol. Phylogenet. Evol. 37(1): 165–177.

Le Danois, Y. 1966. Remarques anatomiques sur la region céphalique de *Gonorynchus gonorynchus* (Linné, 1766). Bull. Inst. Fr. Afr. Noire, sér. A 28(1): 283–342.

Lehman, J.-P. 1966. Actinopterygii, pp. 1–242. *In*: J. Piveteau [ed.]. Traité de Paléontologie, 4(3). Masson et Cie., Paris.

Lenglet, G. 1974. Contribution à l'étude ostéologique des Kneriidae. Ann. Soc. Roy. Zool. Belg. 104 : 51–103.

Linnaeus, C. 1766. Systema naturae sive regna tria naturæ, secundum classes, ordines, genera, species, cum characteribus, differentiis, synonymis, locis : XII + 532 pp., Laurentii Salvii, Holmiae/Stockholm.

Monod, T. 1963. Sur quelques points de l'anatomie de *Gonorhynchus gonorhynchus* (Linné 1766). Mel. Ichthyol. Mém. inst. Fr. Afr. Noire 66: 255–313.

Monod, T. 1968. Le complexe urophore des poissons téléostéens. Mém. inst. fond. Afr. noire, Dakar 81: VI + 705 pp.

Morioka, S., A. Ohno, H. Khono and Y. Taki. 1993. Recruitment and survival of milkfish *Chanos chanos* larvae in the surf zone. Japan J. Ichthyol. 40(2): 247–260.

Moyle, P.B. and J.J. Cech. 2004. Fishes: an Introduction to Ichthyology (5th ed.). XVI + 726 pp. Prentice-Hall Inc., Upper Saddle River, New Jersey.

Nelson, J.S. 2006. Fishes of the World (4th ed.): XIX + 601 pp., John Wiley and Sons, Inc., Hoboken, New Jersey.

Patterson, C. 1970. Two Upper Cretaceous salmoniform fishes from the Lebanon. Bull. Brit. Mus. (Nat. Hist.), Geol. 19(5): 208–296.

Patterson, C. 1975. The distribution of Mesozoic freshwater fishes. Mém. Mus. Natl. Hist. Nat. Paris, sér. A 88: 156–174.

Patterson, C. 1984a. Family Chanidae and other teleostean fishes as living fossils, pp. 132–139. In: N. Eldredge and S.M. Stanley [eds.]. Living Fossils, Springer Verlag, Berlin.

Patterson, C. 1984b. Chanoides, a marine Eocene otophysan fish (Teleostei: Ostariophysi). J. Vert. Paleontol. 4(3): 430–456.

Patterson, C. and G.D. Johnson. 1995. The intermuscular bones and ligaments of teleostean fishes. Smithsonian Contrib. Zool. 559: IV + 83 pp.

Peng, Z., S. He, J. Wang, W. Wang and R. Diogo. 2006. Mitochondrial molecular clocks and the origin of the major Otocephalan clades (Pisces: Teleostei): A new insight. Gene 370: 113–124.

Perkins, P.L. 1970. Equitability and trophic levels in an Eocene fish population. Lethaia 3: 301–310.

Perrier, E. 1903. Traité de Zoologie, fascicule 6, Poissons, pp. 2357–2727. Masson et Cie., Paris.

Poll, M. 1965. Contribution à l'étude des Kneriidae et description d'un nouveau genre, le genre Parakneria (Pisces, Kneriidae). Mém. Acad. roy. Belg. 36: 1–28.

Poyato-Ariza, F.J. 1991. Teleósteos primitivos del Cretácico inferior español: órdenes Elopiformes y Gonorynchiformes. Unpublished doctoral thesis, Departamento de Biología, Facultad de Ciencias, Universidad Autónoma de Madrid (2 vols.).

Poyato-Ariza, F.J. 1994. A new Early Cretaceous Gonorynchiform Fish (Teleostei: Ostariophysi) from Las Hoyas (Cuenca, Spain). Occ. Pap. Mus. Nat. Hist., Univ. Kansas 164: 1–37.

Poyato-Ariza, F.J. 1996a. A revision of the ostariophysan fish family Chanidae, with special reference to the Mesozoic forms. Palaeo Ichthyol. 6: 1–52.

Poyato-Ariza, F.J. 1996b. A revision of Rubiesichthys gregalis Wenz 1984 (Ostariophysi, Chanidae), from the Early Cretaceous of Spain, pp. 319–328. In: G. Arratia and G. Viohl [eds.]. Mesozoic Fishes—Systematics and Paleoecology. Verlag Dr. Friedrich Pfeil, Munich.

Poyato-Ariza, F.J. 1996c. The phylogenetic relationships of Rubiesichthys gregalis and Gordichthys conquensis (Ostariophysi, Chanidae), from the Early Cretaceous of Spain, pp. 329–348. In: G. Arratia and G. Viohl [eds.]. Mesozoic Fishes—Systematics and Paleoecology. Verlag Dr. Friedrich Pfeil, Munich.

Poyato-Ariza, F.J. 2005. Controversies on the evolutionary history of pycnodont fishes— A response to Kriwet. Trans. Roy. Soc. Edinburgh: Earth Sci. 95: 543–546.

Poyato-Ariza, F.J., M.A. López-Horgue and F. García-Garmilla. 2000. A new early Cretaceous clupeomorph fish from the Arratia Valley, Basque Country, Spain. Cretaceous Res. 21: 571–585.

Poyato-Ariza, F.J., M.R. Talbot, M.A. Fregenal-Martínez, N. Meléndez and S. Wenz. 1998. First isotopic and multidisciplinary evidence for nonmarine coelacanths and pycnodontiform fishes: palaeoenvironmental implications. Palaeogeogr., Palaeoclimatol., Palaeoecol. 144(1–2): 65–84.

Rabor, D.R. 1938. Studies on the anatomy of the bangos, Chanos chanos (Forsskål), I. The skeletal system. Philippine J. Sci. 67: 351–377.

Ridewood, W.G. 1904a. On the cranial osteology of the fishes of the familes Elopidae and Albulidae, with remarks on the morphology of the skull in the lower teleostean fishes generally. Proc. Zool. Soc. London 2: 35–81.

Ridewood, W.G. 1904b. On the cranial osteology of the clupeoid fishes. Proc. Zool. Soc. London 2: 448–493.

Ridewood, W.G. 1905. On the skull of *Gonorhynchus Greyi*. Ann. Mag. Nat. Hist. 7(15): 361–372.

Roberts, T.R. 1973. Interrelationships of ostariophysans, pp. 373–395. *In*: P.H. Greenwood, R.S. Miles and C. Patterson [eds.]. Interrelationships of Fishes. Zool. J. Linn. Soc. London 53(suppl. 1).

Rojo, A.L. 1988. Diccionario enciclopédico de anatomía de peces. Monografías del Instituto Español de Oceanografía, 3: 566 pp. (English translation: 1991. Dictionary of Evolutionary Fish Osteology. C.R.C. Press, Boca Raton, Florida.)

Rosen, D.E. and P.H. Greenwood. 1970. Origin of the Weberian apparatus and the relationships of the ostariophysan fishes. Am. Mus. Novitates 2428: 1–25.

Saitoh, K., M. Miya, J.G. Inoue, N.B. Ishiguro and M. Nishida. 2003. Mitochondrial genomics of ostariophysan fishes: perspectives on phylogeny and biogeography. J. Mol. Evol. 56: 464–472.

Schuster, W.H. 1960. Synopsis of biological data on milkfish *Chanos chanos* (Forsskål), 1775. F.A.O. Fisheries Biology Synopsis 4: VI + 52 pp.

Scopoli, J.A. 1777. Introdvctio ad historiam natvralem sistems genera lapidvm, plantarvm, et animalivm hactenvs detecta, caracteribvs essentialibvs donata, in tribvs divisa, svbinde ad leges natvrae: IX + 506 pp., Gerle, Prague.

Steindachner, F. 1866. Ichthyologische Mittheilungen (IX). Verhandlungen der K.K. Bot. Zool. Reichsanstalt 1866: 761–796.

Taverne, L. 1974a. Sur l'origine des Téléostéens Gonorhynchiformes. Bull. Soc. Belg. Géol. 83(1): 55–60.

Taverne, L. 1974b. *Parachanos* Arambourg et Scheneegans (Pisces Gonorhynchiformes) du Crétacé inférieur du Gabon et de Guinée Equatoriale et l'origine des Téléostéens Ostariophysi. Rev. Zool. afr. 88(3): 683–688.

Taverne, L. 1981. Ostéologie et position systématique d'*Aethalionopsis robustus* (Pisces, Teleostei) du Cretacé inférieur de Bernissart (Belgique) et considérations sur les affinités des Gonorhynchiformes. Bull. Acad. Roy. Belg., Classe Sci., sér. 5, 67(12): 958–982.

Taverne, L. 1984. À propos de *Chanopsis lombardi* du Crétacé inférieur du Zaïre (Teleostei, Osteoglossiformes). Rev. Zool. afr. 98(3): 578–590.

Taverne, L. 1997. Les poissons crétacés de Nardo. 4°: †*Apulichthys gayeti* gen. nov., sp. nov. (Teleostei, Ostariophysi, Gonorhynchiformes). Boll. Mus. Civ. Stor. nat. Verona 21: 401–436.

Taverne, L. 1998. Les poissons crétacés de Nardo. 7°: †*Lecceichthys wautyi* gen. nov., sp. nov. (Teleostei, Ostariophysi, Gonorhynchiformes) et considérations sur la phylogénie des Gonorhynchidae. Boll. Mus. Civ. Stor. nat. Verona 22: 291–316.

Taverne, L. 1999. Les poissons crétacés de Nardò. 8°: †*Sorbininardus apulensis* gen. nov., sp. nov. (Teleostei, Ostariophysi, Anotophysi, Sorbininardiformes, nov. ord.). Studi e Ricerche sui Giacimenti Terziari di Bolca, Memor. Vol. Lorenzo Sorbini, Museo Civico di Storia Naturale, Verona 8: 77–103.

Taverne, L. and M. Gayet. 2006. Nouvelle description d'†*Halecopsis insignis* de l'Éocène marin de l'Europe et les relations de ce taxon avec les Gonorynchiformes (Teleostei, Ostariophysi). Cybium 30(2): 109–114.

Thys van den Audenaerde, D.F.E. 1961. L'anatomie de *Phractolæmus ansorgei* Blgr. et la position systématique des Phractolæmidae. Ann. Mus. Roy. Afr. Centr., Sci. Zool. sér. 8, 103: 101–167.

Traquair, R.H. 1911. Les poissons wealdiens de Bernissart. Mém. Mus. Roy. Hist. Nat. Belg. 5(1908), IV + 65 pp. Brussels.

von der Marck, W. 1885. Fische der Oberen Kreide Westfalen. Palaeontographica 31: 233–267.

Weitzman, S.H. 1962. The osteology of *Brycon meeki*, a generalized characid fish, with an osteological definition of the family. Standford Ichthyol. J. 8(1): 1–77.

Wenz, S. 1984. *Rubiesichthys gregalis* n.g., n.sp., Pisces Gonorynchiformes, du Crétacé inférieur du Montsech (Province de Lérida, Espagne). Bull. Mus. Natl. Hist. Nat. Paris, sér. 4, 6C(3): 275–285.

Woodward, A.S. 1896. On some extinct fishes of the teleostean family Gonorhynchidae. Proc. Zool. Soc. London 1896: 500–504.

Woodward, A.S. 1901. Catalogue of the Fossil Fishes in the British Museum (Natural History), vol. 4: XXXIV + 636 pp. Trustees of the British Museum (Natural History), London.

Zittel, K.A. 1893. Traité de Paléontologie. Vol. 3. Paléozoologie. Vertebrata (Pisces, Amphibia, Reptilia, Aves). Octave Doin, Paris.

APPENDIX 1

Abbreviated List of Characters

1. Orbitosphenoid: present [0]; absent [1].
2. Basisphenoid: present [0]; absent [1].
3. Pterosphenoids: well developed and articulating with each other [0]; slightly reduced, not articulating anteroventrally but approaching each other anterodorsally [1]; greatly reduced and broadly separated both anteroventrally and anterodorsally [2].
4. Posterolateral expansion of exoccipitals: absent [0]; present [1].
5. Exoccipitals: posteriorly smooth with no projection above the basioccipital [0]; with a posterior concave-convex border, and a projection above basioccipital [1].
6. Cephalic ribs: absent [0]; present and all articulating with the exoccipitals [1]; present and articulating with both the exoccipitals and basioccipital [2].
7. Brush-like cranial intermuscular bones (*sensu* Patterson and Johnson 1995): absent [0]; present [1].
8. Nasal bone: small but flat [0]; just a tubular ossification around the canal [1]; absent as independent ossification [2].
9. Frontals: wide through most of their length, narrowing anteriorly to form a triangular anterior border [0]; elongate and narrow except in postorbital region [1]; wide, anteriorly shortened, anterior border roughly straight [2]; roughly rectangular in outline, narrow throughout their length [3].
10. Interfrontal fontanelle: absent [0]; present [1].
11. Frontal bones: paired in adult [0]; co-ossified, with no median suture [1].
12. Foramen for olfactory nerve in frontal bones: absent [0]; present [1].
13. Relative position of the parietals: medioparietal (in full contact with each other along their midline) [0]; mesoparietal [1]; lateroparietal (completely separated from each other by the supraoccipital) [2].
14. Parietal portion of the supraorbital canal: absent [0]; present [1].
15. Parietals: large [0]; reduced but flat and blade-like in shape [1]; reduced to canal-bearing bones [2]; absent as independent ossifications [3].

16. Supraoccipital crest: small, short in lateral view [0]; long and enlarged, projecting above occipital region and first vertebrae, forming a vertical, posteriorly deeply pectinated blade [1].
17. Foramen magnum: dorsally bounded by exoccipitals [0]; enlarged and dorsally bounded by supraoccipital [1].
18. Mesethmoid: wide and short [0]; long and slender, with anterior elongate lateral extensions [1]; large, with broad posterolateral wing-like expansions [2]; elongated and thin [3].
19. Wings (extensions) on lateral ethmoids: absent [0]; present [1].
20. Teeth in premaxilla, maxilla, and dentary: present [0]; absent [1].
21. Premaxilla: consisting of one solid portion [0]; premaxilla consisting of two distinct portions, with a shorter, non-osseous element lying ventral to a much longer osseous portion, which in turn articulates with the maxilla [1].
22. Premaxillary "gingival teeth": absent [0]; present [1].
23. Premaxilla: small, flat and roughly triangular [0]; large, very broad, concave-convex, with long oral process [1]; narrow and elongated, its length more than one half the length of the maxilla [2].
24. Premaxillary ascending process: present [0]; absent [1].
25. Dorsal and ventral borders of the maxillary articular process: straight or slightly curved [0]; very curved, almost describing an angle [1].
26. Maxillary process for articulation with autopalatine: absent [0]; present [1].
27. Posterior region of the maxilla: slightly and progressively expanded to form a thin blade, with roughly straight posterior border [0]; very enlarged, swollen to a bulbous outline, with curved posterior border [1].
28. Supramaxilla(e): present [0]; absent [1].
29. Notch between the dentary and the angulo-articular bones: absent [0]; present [1].
30. Articulation between dentary and angulo-articular: strong, dentary not V-shaped posteriorly [0]; loose, with a posteriorly V-shaped dentary [1].
31. Notch in the anterodorsal border of the dentary: absent [0]; present [1].
32. Mandibular sensory canal: present [0]; absent [1].
33. Inferior and superior enlarged retroarticular processes of mandible: both absent [0]; inferior retroarticular process present, superior retroarticular process absent [1]; both inferior and superior retroarticular processes present [2].
34. Quadrate with: posterior margin smooth [0]; elongated forked posterior process [1].

35. Quadrate-mandibular articulation: below or posterior to orbit, no elongation or displacement of quadrate [0]; anterior to orbit, quadrate displaced but not elongate [1]; anterior to orbit, with elongation of the body of quadrate instead of displacement [2].
36. Symplectic: elongated in shape but relatively short [0]; very long, about twice the length of the ingroup [1]; absent as an independent ossification [2].
37. Symplectic and quadrate: articulating directly with each other [0]; separated through cartilage [1]; no contact due to absence of symplectic [2].
38. Metapterygoid: large, broad and in contact with quadrate and symplectic through cartilage [0]; reduced to a thin rod [1].
39. Dermopalatine: present [0]; absent [1].
40. A patch of about 20 conical teeth on endopterygoids and basibranchial 2: absent [0]; present [1].
41. Ectopterygoids: well developed, ectopterygoid overlapping with the ventral surface of the autopalatine by at least 50% [0]; well developed, with three branches in lateral view, reduced but direct contact with autopalatine [1]; reduced, articulating with the ventral surface of the autopalatine by at most 10% through cartilage, resulting in a loosely articulated suspensorium [2]; absent as distinct ossifications [3].
42. Teeth on vomer and parasphenoid: absent [0]; present [1].
43. Anterior portion of vomer: horizontal [0]; anteroventrally inclined, nearly vertical [1]; dorsally curved [2].
44. Spatial relationship between vomer and mesethmoid anteriorly: vomer and mesethmoid ending at about the same anterior level [0]; mesethmoid extending anteriorly beyond the level of anterior margin of vomer [1]; vomer extending anteriorly beyond the level of anterior margin of mesethmoid [2].
45. Articular head of hyomandibular bone: double, with both articular surfaces placed on the dorsal border of the main body of the bone [0]; double, with the anterior articular surface forming a separate head from the posterior articular surface [1].
46. Metapterygoid process of hyomandibular bone: absent [0]; present, anterior [1]; present, ventral [2].
47. Ossified interhyal: present [0]; absent as an independent ossification [1].
48. Teeth on fifth ceratobranchial: present [0]; absent [1].
49. First basibranchial in adult specimens: ossified [0]; unossified [1].
50. Fifth basibranchial in adult specimens: cartilaginous [0]; ossified [1].
51. First pharyngobranchial in adult specimens: ossified [0]; unossified [1].

52. Size of opercular bone: normal, about one quarter the head length [0]; expanded, at least one third the head length [1].
53. Shape of opercular bone in lateral view: rounded/oval [0]; triangular [1].
54. Opercular spines: absent [0]; present [1].
55. Opercular apparatus on external surface of operculum: absent [0]; present [1].
56. Opercular borders: free from side of head [0]; partly or almost completely connected to side of head with skin [1].
57. Angle formed by preopercular limbs: obtuse [0]; approximately straight [1]; acute [2].
58. Posterodorsal limb of preopercular bone: well developed [0]; reduced, correlated with expansion of anteroventral limb that meets its fellow along the ventral midline [1].
59. Ridge on anteroventral limb of preopercular bone: absent [0]; present [1].
60. Preopercular expansion: absent, preopercular not enlarged [0]; present, restricted to the posteroventral corner [1]; present in posteroventral corner and part of the posterodorsal limb [2]; present in anteroventral limb only [3].
61. Suprapreopercular bone: absent [0]; present as a relatively large, flat bone [1]; present as tubular ossification(s) [2].
62. Spine on posterior border of subopercular bone: absent [0]; present [1].
63. Major axis of subopercular bone in lateral view: inclined [0]; subhorizontal [1]; subvertical [2].
64. Subopercular clefts: absent [0]; present [1].
65. Interopercular bone: relatively broad and positioned medioventral to preopercular bone [0]; reduced to a long thin spine and positioned mediodorsal to preopercular bone [1]; reduced to a long thin spine and positioned lateroventral to preopercular bone [2].
66. Spine on posterior border of interopercular bone: absent [0]; present [1].
67. Posterodorsal ascending process of interopercular bone: absent [0]; present [1].
68. Number of infraorbitals: five or more [0]; four [1]; three or fewer [2].
69. Infraorbital bones not including lacrimal: well developed [0]; reduced to small, tubular ossifications [1]; hypertrophied [2].
70. Lacrimal: flat and comparable in length to subsequent infraorbitals [0]; tube-like and extremely long, without keel [1]; flat, long and large, with keel near lower edge [2]; long and large, with spines and crests [3].
71. Supraorbital: present [0]; absent [1].

72. Two anteriormost vertebrae: as long as posterior ones [0]; shorter than posterior ones [1].
73. Second abdominal centrum: as long as first [0]; shorter than first [1].
74. Autogenous neural arch anterior to arch of first vertebra: present [0]; absent [1].
75. Neural arch of first vertebra and exoccipitals: separate [0]; in contact [1].
76. Neural arch of first vertebra and supraoccipital: separated [0]; in contact [1].
77. Spine on the neural arch of first vertebra: present, well developed [0]; present but reduced [1]; absent [2].
78. Anterior neural arches: no contact with adjoining arches [0]; abutting contact laterally with adjoining arches, no overlapping [1]; overlapping contact with adjoining arches laterally [2].
79. Unmodified neural arches anterior to dorsal fin in adults: fused to centra [0]; autogenous, at least laterally [1].
80. Neural arches of vertebrae posterior to the dorsal fin in adults: fused to centrum [0]; autogenous, at least laterally [1].
81. First two anterior parapophyses: autogenous [0]; fused to centra [1].
82. Rib on third vertebral centrum: similar to posterior ones [0]; widened and shortened [1]; modified into Weberian apparatus [2].
83. Paired intermuscular bones consisting of three series: epipleurals, epicentrals, and epineurals: absent [0]; present [1].
84. Anterior (first six) epicentral bones: unmodified, no differences in size from others [0]; highly modified, much larger than posterior ones [1]; epicentrals in anterior vertebrae absent [2].
85. Size and arrangement of anterior supraneurals (whatever the number present): large, separate from each other if more than one supraneural present [0]; larger, in contact with neighbors if more than one supraneural present [1]; supraneurals greatly reduced in size [2].
86. Posterior process on the posterior border of first supraneural: absent [0]; present [1].
87. Number of supraneurals: several supraneurals in a long series [0]; two or fewer supraneurals [1].
88. Postcleithra: present [0]; absent [1].
89. Lateral line and supracleithrum: supracleithrum pierced through dorsal region [0]; supracleithrum pierced all through its length [1]; lateral line does not pierce supracleithrum [2].
90. Fleshy lobe of paired fins: absent [0]; present [1].
91. Caudal fin morphology: elongated, posteriorly forked [0]; higher than long, slightly incurved posteriorly [1]; crescent-shaped [2].
92. Fringing fulcra in dorsal lobe of caudal fin: present [0]; absent [1].

93. Caudal scutes: absent [0]; present [1].
94. Ural centra, preural centrum one, and uroneural one: autogenous [0]; fused [1]; fused except for ural centrum two, which is autogenous [2].
95. Neural arch and spine of preural centrum one: both well developed, spine about half as long as preceding ones [0]; arch complete and closed, spine rudimentary [1]; arch open, no spine [2].
96. Uroneurals: arranged in a linear series [0]; arranged in a double series [1].
97. Number of uroneurals: three [0]; two or one [1]; none [2].
98. Anterior extent of first uroneural: to anterior end of first preural [0]; to anterior end of second preural [1]; to anterior end of third preural [2]; uroneural fused to caudal fin complex [3].
99. Uroneural two and second ural centrum: in contact [0]; separated [1]; uroneural two absent [2].
100. Parahypural and preural centrum 1: independent in adults [0]; fused only in large adults [1]; fused since early ontogenetic stages [2].
101. Reduction in the number of hypurals: six [0]; fewer than six [1].
102. Hypurals 1 and 2: independent [0]; partly fused to each other [1]; totally fused to each other [2].
103. Hypural 1 and terminal centrum: articulating [0]; separated by a hiatus [1]; fused [2].
104. Hypural 2 and terminal centrum: fused [0]; articulating [1].
105. Hypural 5: of comparable size to preceding ones [0]; considerably larger [1].
106. Hypural 5 (plus 6 if present) and second ural centrum: separate [0]; articulating [1].
107. Haemal arch and preural centrum 2: fused [0]; independent [1].
108. Posterolateral process of caudal endoskeleton: absent [0]; present [1].
109. Scales on body: present [0]; absent [1].
110. Type of scales: cycloid [0]; modified ctenoid [1].
111. Lateral line: not extending to posterior margin of hypurals [0]; extending to posterior margin of hypurals [1].
112. Intermandibularis: mainly attaching on the dentary [0]; exclusively attaching to angulo-articular [1].
113. Protractor hyoidei: not inserting on coronoid process [0]; inserting on coronoid process [1].
114. Hyohyoideus inferioris of both sides: mostly overlapping each other [0]; mostly mixing mesially with each other [1].
115. Hyohyoideus abductor: not attaching on pectoral girdle [0]; with significant part of its fibers attaching on pectoral girdle [1].

116. Adductor profundus: not subdivided into different sections [0]; subdivided into different sections [1].
117. Attachment of adductor profundus: on first pectoral ray only [0]; on first and second pectoral rays [1].
118. Most lateral bundles of adductor mandibulae: inserting on mandible and/or primordial ligament [0]; attaching also, or even exclusively, on other bones such as the maxilla or the premaxilla [1].
119. Position of adductor mandibulae A1-OST: mostly horizontal [0]; with a peculiar anterior portion almost perpendicular to its posterior portion [1].
120. Section A2 of adductor mandibulae: present [0]; absent [1].
121. Several small tendons branching off from adductor mandibulae A2: absent [0]; present [1].
122. Peculiar adductor mandibulae A1-OST-M: absent [0]; present [1].
123. Direct insertion of adductor mandibulae A2 far anteriorly on the anteromesial surface of dentary: absent [0]; present [1].
124. Dilatator operculi: mainly mesial and/or dorsal to adductor mandibulae A2 [0]; markedly lateral to A2 [1].
125. Peculiar tendon of adductor mandibulae A2 running perpendicular to the main body of this section and connecting it to the anteroventral surface of the quadrate: absent [0]; present [1].
126. Distinct section A3 of adductor mandibulae: absent [0]; present [1].
127. Adductor mandibulae Aω: present [0]; absent [1].
128. Adductor arcus palatini: not inserting on preopercle [0]; inserting also on preopercle [1].
129. Levator arcus palatini: not divided [0]; divided into two well-differentiated bundles [1].
130. Origin of dilatator operculi: on ventrolateral surface of neurocranium [0]; on dorsal margin of cranial roof [1].

APPENDIX 2

Data Matrix

This data matrix shows the character states coded for all the taxa included in the analysis. Each first column shows first the three genera of the ougroup, then the genera of the ingroup, in alphabetical order. For an abbreviated list of characters, see Appendix 1. For discussion of character states, see main text. Numbers in the data matrix correspond to the numbers of the characters (first line, italics) and to the character states (remaining lines, not italics); for character states, cells with more than one numeral indicate simultaneous codification of two states (see main text for discussion).

	1	2	3	4	5	6	7	8	9	10	11	12	13	14	15	16	17	18	19	20	21
†Diplomystus	0	0	0	0	?	0	0	0	0	0	0	0	0	0	0	0	0	0	0	0	0
Brycon	0	1	0	0	0	0	0	1	0	1	0	0	2	0	0	0	0	0	0	0	0
Opsariichthys	0	1	0	0	0	0	0	1	0	0	0	0	0	0	0	0	0	0	0	1	0
†Aethalionopsis	?	?	?	?	?	?	?	?	0	0	0	?	2	1	1	0	?	0	?	1	0
†Apulichthys	1	0	2	?	?	12	1	1	1	0	0	0	2	0	2	0	?	3	0	1	?
Chanos	1	1	1	1	1	2	0	1	0	0	0	0	2	0	2	1	0	2	0	1	0
†Charitopsis	1	1	?	?	?	?	?	?	1	0	0	?	2	?	1	?	?	0	0	1	0
†Charitosomus	1	1	2	0	?	1	1	?	1	0	0	0	2	0	1	0	0	0	0	1	0
Cromeria	1	1	2	0	0	1	0	2	0	1	0	0	2	0	3	0	1	1	1	1	0
†Dastilbe	1	1	1	0	0	?	?	?	0	0	0	0	2	0	1	1	0	0	?	1	0
Gonorynchus	1	1	2	0	0	1	1	1	1	0	1	1	2	0	1	0	0	0	0	1	0
†Gordichthys	1	1	?	0	?	?	?	0	0	0	0	0	1	0	1	0	?	0	?	1	0
Grasseichthys	1	1	2	0	0	0	0	2	0	1	0	0	2	0	3	0	1	1	1	1	0
†Hakeliosomus	1	?	?	0	0	?	?	?	1	0	0	0	2	0	1	0	?	0	0	1	?
†Halecopsis	1	?	2	?	?	?	?	?	0	0	0	0	2	0	1	0	?	3	0	1	?
†Judeichthys	1	1	?	0	?	1	1	?	1	0	0	0	2	0	1	0	?	0	0	1	0
Kneria	1	1	2	0	0	1	0	0	0	0	0	0	2	0	3	0	1	1	1	1	0
†Lecceichthys	1	1	?	0	?	?	?	?	1	?	?	0	2	?	1	0	?	0	0	?	?
†New genus	?	?	?	?	?	?	?	0	3	0	0	?	?	?	?	0	?	3	0	1	?
†Notogoneus	1	?	2	0	0	1	1	?	1	0	0	0	2	0	1	0	0	0	0	1	?
†Parachanos	?	?	?	0	?	?	?	1	0	0	0	0	2	0	1	?	?	0	?	1	0
Parakneria	1	1	2	0	0	1	0	0	0	0	0	0	2	0	3	0	1	1	1	1	0
Phractolaemus	1	1	2	0	0	1	0	1	2	0	0	0	2	0	2	0	1	0	1	1	0
†Ramallichthys	1	1	?	0	0	1	1	?	1	0	0	0	2	0	1	0	?	0	0	1	0
†Rubiesichthys	1	1	?	0	?	?	?	0	0	0	0	0	2	0	1	0	?	0	?	1	0
†Sorbininardus	0	1	0	?	?	?	?	?	1	0	0	?	2	0	1	?	?	0	0	1	?
†Tharrhias	1	1	1	0	1	?	0	1	0	0	0	0	2	0	1	1	0	2	?	1	0

	22	23	24	25	26	27	28	29	30	31	32	33	34	35	36	37	38	39
†Diplomystus	0	0	0	0	0	0	0	0	0	0	0	0	0	0	0	0	0	0
Brycon	0	0	0	0	0	0	1	0	0	0	0	0	0	0	0	0	0	1
Opsariichthys	0	0	0	0	0	0	1	0	0	0	0	0	0	0	0	0	0	1
†Aethalionopsis	?	1	?	0	1	1	1	0	0	?	0	0	0	1	1	0	0	?
†Apulichthys	?	2	0	0	1	0	1	0	0	?	0	0	0	1	0	0	0	1
Chanos	0	1	1	0	1	1	1	0	0	1	0	0	0	1	1	1	0	1
†Charitopsis	?	0	1	0	1	0	1	1	0	0	?	?	0	0	?	?	0	?
†Charitosomus	?	0	1	0	1	0	1	1	0	0	?	2	0	0	0	0	0	1
Cromeria	0	0	1	0	0	0	1	1	0	0	1	0	1	0	2	2	0	1
†Dastilbe	?	1	1	0	1	1	1	0	0	1	0	0	0	1	1	0	0	1
Gonorynchus	1	0	1	0	0	0	1	1	1	0	1	0	0	0	0	0	1	1
†Gordichthys	?	1	1	1	0	1	1	0	0	1	0	0	0	1	1	0	0	1
Grasseichthys	0	0	1	0	0	0	1	1	0	0	1	0	0	0	1	0	N	1
†Hakeliosomus	?	0	1	0	1	0	1	0	0	0	?	1	0	0	0	0	0	1
†Halecopsis	?	?	?	?	?	?	?	?	?	?	1	0	0	2	?	?	?	?
†Judeichthys	?	0	1	0	1	0	1	1	0	0	?	1	0	0	?	0	0	1
Kneria	0	0	1	0	1	0	1	1	0	0	1	0	0	0	0	0	0	1
†Lecceichthys	?	?	?	?	?	?	?	?	0	?	?	0	?	1	0	0	0	1
†New genus	?	?	?	0	0	0	0	?	?	0	0	0	0	0	0	0	?	?
†Notogoneus	?	0	1	0	?	0	1	1	1	0	?	0	0	0	?	0	?	?
†Parachanos	?	1	1	0	1	1	1	0	0	?	0	0	0	1	1	0	0	?
Parakneria	0	0	1	0	1	0	1	1	0	0	1	0	0	0	0	0	0	1
Phractolaemus	0	0	1	0	0	0	1	0	0	0	1	0	0	2	2	2	0	1
†Ramallichthys	?	0	1	0	1	0	1	1	0	0	?	1	0	0	?	0	0	1
†Rubiesichthys	?	1	1	1	0	1	1	0	0	1	0	0	0	1	1	0	0	1
†Sorbininardus	?	2	0	0	0	0	1	0	0	?	0	0	0	0	0	0	0	?
†Tharrhias	?	1	1	0	1	1	1	0	0	1	0	0	0	1	1	1	0	1

	40	41	42	43	44	45	46	47	48	49	50	51	52	53	54	55	56	57
†Diplomystus	0	0	0	0	0	0	0	?	?	?	?	?	0	0	0	0	0	0
Brycon	0	0	0	0	0	0	0	0	0	0	0	0	0	0	0	0	0	0
Opsariichthys	0	0	0	0	0	0	0	0	0	0	0	0	0	0	0	0	0	0
†Aethalionopsis	0	0	?	?	?	0	?	?	?	?	?	?	1	0	0	0	?	1
†Apulichthys	?	1	0	0	2	0	1	?	?	?	?	?	0	0	0	0	?	0
Chanos	0	0	0	0	2	0	1	0	1	0	0	0	1	0	0	0	0	1
†Charitopsis	1	2	0	0	1	?	0	?	?	?	?	?	0	1	1	0	?	0
†Charitosomus	1	2	0	0	1	1	0	?	?	?	?	?	0	1	0	0	?	0
Cromeria	0	3	0	01	02	1	2	1	1	1	1	1	0	0	0	0	1	0
†Dastilbe	0	0	?	?	?	0	1	?	?	?	?	?	1	0	0	0	?	1
Gonorynchus	1	2	0	0	1	1	0	1	1	0	1	0	1	0	0	0	1	0
†Gordichthys	0	0	1	?	?	0	1	?	?	?	?	?	1	0	0	0	?	2
Grasseichthys	0	3	0	0	2	1	2	1	1	1	1	1	0	0	0	0	1	0
†Hakeliosomus	1	2	0	0	2	1	0	0	?	?	?	?	0	1	0	0	?	0
†Halecopsis	?	?	?	?	?	1	0	?	?	?	?	?	0	1	0	0	?	1
†Judeichthys	1	2	0	0	2	1	?	?	?	?	?	?	0	1	0	0	?	0
Kneria	0	2	0	1	2	1	2	0	1	1	1	1	0	0	0	1	1	0
†Lecceichthys	1	0	0	?	?	0	0	?	?	?	?	?	0	1	0	0	?	0
†New genus	1	?	0	1	?	?	?	?	?	?	?	?	0	0	0	0	?	1
†Notogoneus	1	2	0	0	1	1	0	?	?	?	?	?	0	1	0	0	?	0
†Parachanos	0	0	?	?	?	0	1	?	?	?	?	?	1	0	0	0	?	1
Parakneria	0	2	0	1	2	1	2	0	1	1	1	1	0	0	0	0	1	1
Phractolaemus	0	2	0	2	2	0	2	0	1	1	0	1	0	0	0	0	1	0
†Ramallichthys	1	2	0	0	2	1	?	?	?	?	?	?	0	1	0	0	?	0
†Rubiesichthys	0	0	?	?	?	0	1	?	?	?	?	?	1	0	0	0	?	2
†Sorbininardus	?	2	0	1	1	?	2	?	?	?	?	?	0	1	1	0	?	0
†Tharrhias	0	0	0	0	2	0	1	?	1	?	?	?	1	0	0	0	?	1

	58	59	60	61	62	63	64	65	66	67	68	69	70	71	72	73	74	75
†Diplomystus	0	0	0	0	0	0	0	0	0	0	0	0	0	?	0	0	0	0
Brycon	0	0	0	0	0	0	0	0	0	0	0	0	0	0	1	0	1	0
Opsariichthys	0	0	0	0	0	0	0	0	0	0	0	0	0	0	1	0	1	0
†Aethalionopsis	0	1	1	1	0	0	0	0	0	0	0	0	0	0	?	?	?	?
†Apulichthys	0	0	0	2	0	0	0	0	0	0	1	0	0	0	1	1	1	?
Chanos	0	0	2	1	0	1	0	0	0	0	1	0	0	0	1	0	1	1
†Charitopsis	0	0	0	0	1	0	0	0	0	0	1	1	?	1	1	0	?	?
†Charitosomus	0	0	0	0	1	0	0	0	0	0	1	1	0	0	1	0	1	1
Cromeria	0	0	0	0	0	0	0	2	0	1	12	1	0	0	1	0	1	1
†Dastilbe	0	1	0	1	0	1	0	0	0	0	?	0	0	0	1	0	1	?
Gonorynchus	0	0	0	2	0	0	0	0	0	0	2	1	2	1	1	0	1	1
†Gordichthys	0	0	1	1	0	0	0	0	0	0	?	0	0	0	1	0	?	?
Grasseichthys	0	0	0	0	0	0	0	2	0	1	12	1	0	0	1	0	1	1
†Hakeliosomus	0	0	0	0	0	0	0	0	0	0	1	1	0	0	1	0	1	?
†Halecopsis	0	0	2	0	?	0	0	0	0	0	1	0	0	?	?	?	?	?
†Judeichthys	0	0	0	0	0	0	0	0	0	0	1	1	0	0	1	0	1	?
Kneria	0	0	0	2	0	0	0	0	0	1	1	1	0	0	1	0	1	1
†Lecceichthys	0	0	0	0	0	0	0	0	1	0	01	01	0	0	?	?	?	?
†New genus	0	1	?	0	0	1	0	0	0	?	1	0	0	0	1	?	0	?
†Notogoneus	0	0	0	0	0	0	1	0	0	0	1	1	?	1	1	?	1	1
†Parachanos	0	1	1	1	0	1	0	0	0	0	?	0	0	0	?	?	1	?
Parakneria	0	0	0	2	0	0	0	0	0	1	1	1	1	0	1	0	1	1
Phractolaemus	1	0	3	1	0	0	0	1	0	1	1	2	0	0	1	1	1	1
†Ramallichthys	0	0	0	0	0	0	0	0	0	0	1	1	0	0	1	0	1	?
†Rubiesichthys	0	0	1	1	0	0	0	0	0	0	?	0	0	0	1	0	1	?
†Sorbininardus	0	0	2	0	1	2	0	0	0	0	?	0	3	1	?	?	?	?
†Tharrhias	0	1	2	1	0	1	0	0	0	1	0	1	0	0	1	0	1	1

	76	77	78	79	80	81	82	83	84	85	86	87	88	89	90	91	92	93
†Diplomystus	0	0	0	0	0	0	0	0	0	0	0	0	0	0	0	0	0	1
Brycon	0	1	0	0	0	1	2	0	0	1	0	0	0	0	0	0	1	0
Opsariichthys	0	1	0	0	0	1	2	0	0	1	0	0	0	0	0	0	1	0
†Aethalionopsis	?	?	?	1	1	0	?	?	0	?	?	0	?	?	0	0	0	1
†Apulichthys	?	?	0	0	0	?	0	0	2	0	0	0	0	2	0	0	1	0
Chanos	0	1	0	1	0	0	1	1	0	0	0	0	1	0	0	0	1	1
†Charitopsis	?	2	2	?	0	?	1	1	0	?	?	?	1	?	0	0	1	0
†Charitosomus	?	2	2	?	0	?	1	1	0	1	0	0	1	0	0	0	1	0
Cromeria	1	1	1	0	0	1	1	1	0	N	0	1	1	0	0	0	1	0
†Dastilbe	0	1	0	1	0	0	1	1	0	0	0	0	0	0	0	0	1	0
Gonorynchus	0	1	1	0	0	1	1	1	0	1	0	0	1	0	1	0	1	0
†Gordichthys	?	1	0	1	0	?	1	?	0	0	1	0	0	1	0	1	1	0
Grasseichthys	1	1	1	0	0	1	1	0	0	0	0	1	1	0	0	0	1	0
†Hakeliosomus	?	2	2	?	0	?	1	1	0	1	0	0	1	?	0	0	1	0
†Halecopsis	?	?	?	?	?	?	?	?	?	?	?	?	1	?	?	?	?	?
†Judeichthys	?	2	2	?	0	?	1	1	0	1	0	0	1	0	0	0	1	0
Kneria	1	1	1	0	0	1	1	1	1	0	0	1	1	0	0	0	1	0
†Lecceichthys	?	?	?	0	0	?	1	1	2	1	0	0	0	?	0	0	1	0
†New genus	0	?	?	1	?	0	?	?	?	?	?	?	?	0	?	?	?	?
†Notogoneus	?	0	0	0	0	1	1	1	0	0	0	0	1	0	0	0	1	0
†Parachanos	?	?	?	1	0	?	1	?	0	?	?	0	0	0	0	0	1	?
Parakneria	1	1	1	0	0	1	1	1	0	0	0	1	1	0	0	0	1	0
Phractolaemus	0	1	1	0	0	1	1	1	0	1	0	1	1	0	0	2	1	0
†Ramallichthys	?	2	2	?	0	?	1	1	0	1	0	0	1	0	0	0	1	0
†Rubiesichthys	?	1	0	1	0	?	1	?	0	0	1	0	0	?	0	0	1	0
†Sorbininardus	?	?	?	0	0	?	0	1	2	1	?	0	1	?	0	0	1	0
†Tharrhias	0	1	0	1	0	0	1	1	0	0	0	0	0	0	0	0	1	0

	94	95	96	97	98	99	100	101	102	103	104	105	106	107	108	109
†Diplomystus	0	0	0	0	0	0	?	0	0	0	0	0	0	0	0	0
Brycon	1	1	0	1	3	1	2	0	0	1	0	0	0	0	0	0
Opsariichthys	1	1	0	1	3	1	0	0	0	0	0	0	0	1	0	0
†Aethalionopsis	0	1	0	0	?	1	0	0	0	0	1	0	0	1	0	0
†Apulichthys	2	0	0	1	3	0	0	0	2	0	1	0	0	0	0	0
Chanos	1	2	0	1	3	1	1	0	0	1	1	0	0	1	1	0
†Charitopsis	?	?	?	?	?	1	?	?	?	?	?	?	?	?	0	0
†Charitosomus	1	1	0	1	3	1	2	1	1	2	0	0	0	1	0	0
Cromeria	1	2	0	2	3	2	0	1	0	1	1	0	0	0	0	1
†Dastilbe	0	1	0	1	1	0	0	0	0	0	1	0	0	1	0	0
Gonorynchus	1	1	0	2	3	2	2	1	1	2	0	0	0	0	0	0
†Gordichthys	0	1	0	1	1	0	0	0	0	0	1	0	1	1	0	0
Grasseichthys	1	1	0	2	3	2	0	1	0	2	1	0	0	0	0	1
†Hakeliosomus	1	1	0	1	3	1	2	1	0	2	0	0	0	0	0	0
†Halecopsis	?	?	?	?	?	?	?	?	?	?	?	?	?	?	?	0
†Judeichthys	1	1	0	1	3	?	2	1	?	2	0	0	0	1	0	0
Kneria	1	1	0	2	3	2	0	0	0	1	1	0	0	0	0	0
†Lecceichthys	1	?	?	?	3	0	2	?	1	2	?	0	0	?	0	0
†New genus	?	?	?	?	?	?	?	?	?	?	?	?	?	?	?	0
†Notogoneus	1	1	0	1	3	0	0	1	0	0	1	0	0	1	0	0
†Parachanos	0	1	0	1	2	0	0	0	0	0	1	0	0	1	0	0
Parakneria	1	2	0	2	3	2	0	0	0	1	1	0	0	0	0	0
Phractolaemus	1	2	0	2	3	2	0	1	0	1	0	1	0	0	0	0
†Ramallichthys	1	1	0	1	3	1	2	1	1	2	0	0	0	1	0	0
†Rubiesichthys	0	1	0	1	1	0	0	0	0	0	1	0	0	1	0	0
†Sorbininardus	1	1	0	1	3	1	2	0	0	2	0	0	0	1	0	0
†Tharrhias	0	0	1	0	1	0	0	0	0	0	1	0	0	1	0	0

	110	111	112	113	114	115	116	117	118	119	120	121	122	123	124
†Diplomystus	0	0	?	?	?	?	?	?	?	?	?	?	?	?	?
Brycon	0	0	0	0	0	0	1	1	0	0	0	0	0	0	1
Opsariichthys	0	0	0	0	1	0	0	0	1	0	0	0	0	0	0
†Aethalionopsis	0	?	?	?	?	?	?	?	?	?	?	?	?	?	?
†Apulichthys	0	?	?	?	?	?	?	?	?	?	?	?	?	?	?
Chanos	0	0	0	0	1	0	1	1	1	0	0	0	0	0	0
†Charitopsis	?	0	?	?	?	?	?	?	?	?	?	?	?	?	?
†Charitosomus	?	0	?	?	?	?	?	?	?	?	?	?	?	?	?
Cromeria	N	0	0	0	1	0	1	1	1	0	0	0	1	0	0
†Dastilbe	0	0	?	?	?	?	?	?	?	?	?	?	?	?	?
Gonorynchus	1	1	0	1	1	0	1	1	1	0	0	0	0	1	0
†Gordichthys	0	0	?	?	?	?	?	?	?	?	?	?	?	?	?
Grasseichthys	N	0	0	0	1	0	1	1	1	0	1	N	?	N	N
†Hakeliosomus	?	0	?	?	?	?	?	?	?	?	?	?	?	?	?
†Halecopsis	0	?	?	?	?	?	?	?	?	?	?	?	?	?	?
†Judeichthys	?	0	?	?	?	?	?	?	?	?	?	?	?	?	?
Kneria	0	0	0	0	1	1	1	1	1	0	0	0	1	0	0
†Lecceichthys	0	?	?	?	?	?	?	?	?	?	?	?	?	?	?
†New genus	0	?	?	?	?	?	?	?	?	?	?	?	?	?	?
†Notogoneus	1	0	?	?	?	?	?	?	?	?	?	?	?	?	?
†Parachanos	?	?	?	?	?	?	?	?	?	?	?	?	?	?	?
Parakneria	0	0	0	0	1	1	1	1	1	0	0	0	1	0	0
Phractolaemus	0	0	1	0	1	0	1	1	1	1	0	1	1	0	0
†Ramallichthys	?	0	?	?	?	?	?	?	?	?	?	?	?	?	?
†Rubiesichthys	0	0	?	?	?	?	?	?	?	?	?	?	?	?	?
†Sorbininardus	0	?	?	?	?	?	?	?	?	?	?	?	?	?	?
†Tharrhias	0	0	?	?	?	?	?	?	?	?	?	?	?	?	?

	125	126	127	128	129	130
†*Diplomystus*	?	?	?	?	?	?
Brycon	0	1	0	0	0	1
Opsariichthys	0	0	0	0	1	0
†*Aethalionopsis*	?	?	?	?	?	?
†*Apulichthys*	?	?	?	?	?	?
Chanos	0	0	0	0	0	0
†*Charitopsis*	?	?	?	?	?	?
†*Charitosomus*	?	?	?	?	?	?
Cromeria	1	0	1	0	0	0
†*Dastilbe*	?	?	?	?	?	?
Gonorynchus	0	0	1	1	0	0
†*Gordichthys*	?	?	?	?	?	?
Grasseichthys	N	0	1	0	0	0
†*Hakeliosomus*	?	?	?	?	?	?
†*Halecopsis*	?	?	?	?	?	?
†*Judeichthys*	?	?	?	?	?	?
Kneria	0	0	1	0	0	1
†*Lecceichthys*	?	?	?	?	?	?
New genus	?	?	?	?	?	?
†*Notogoneus*	?	?	?	?	?	?
†*Parachanos*	?	?	?	?	?	?
Parakneria	0	0	1	0	0	0
Phractolaemus	0	0	1	0	0	0
†*Ramallichthys*	?	?	?	?	?	?
†*Rubiesichthys*	?	?	?	?	?	?
†*Sorbininardus*	?	?	?	?	?	?
†*Tharrhias*	?	?	?	?	?	?

APPENDIX 3

Character Distribution (Nodes Support)

The present appendix shows the character distribution according to the individual most parsimonious trees whose consensus trees are shown by Figs. 7.11 and 7.12. As commented on in the text, the support of the most relevant nodes is essentially the same with and without the problematic forms. The differences in character distribution caused by these problematic forms correspond to the unlettered nodes in the figures and are not listed here, but they are commented on in the main text. Each node is supported below according to the character distribution provided by the DELTRAN optimization option; the particular differences with the ACCTRAN optimization option are commented on for each node. A indicates autapomorphy, R indicates reversion, and the numbers in brackets reflect derived character states other than 1.

Node A = Order Gonorynchiformes: 1(A), 3(A), 13(2), 15, 20, 24(A), 44(2), 48(A), 75(A), 82, and 83(A). With ACCTRAN it also presents characters 6, 68, 104, and 107.

Node B = Suborder Chanoidei = Family Chanidae: 23(A), 27(A), 31(A), 35(A), 36, 46(A), 52(A), 57, 60, 61, 79(A), 98(1,R), 104, and 107. With ACCTRAN it also presents character 6(2), and lacks characters 104 and 107.

Node C = Subfamily Chaninae: 26 and 59. With ACCTRAN it presents the same characters.

Node C bis = Subfamily Chaninae minus †Aethalionopsis: 16(A) and 63(A). With ACCTRAN it has the same characters.

Node D = Tribe Chanini: 5(A), 18(2,A), 37, 60(2), and 68. With ACCTRAN it does not present additional characters and lacks character 68.

Node E = Subfamily †Rubiesichthyinae: 8(R), 25, 57(2), 86(A), and 99(R). With ACCTRAN it also presents characters 42(A) and 89(A), and lacks character 99.

Node F = Superorder Gonorynchoidei: 3(2,A), 6, 32(A), 41(2,A), 49(A), 51(A), 56(A), 68, 69(A), 81, 88, 94, 101, and 127(A). With ACCTRAN it also presents characters 29, 45, 78, and 99, and lacks characters 6 and 68.

Node G = Family Gonorynchidae: 7(A), 9(A), 29, 40(A), 44(R,1), 45, 53(A), and 110(A). With ACCTRAN it also presents characters 22, 30, 71, 113(A), 123(A), and 128(A), and lacks characters 29 and 45.

Node H = Gonorynchidae minus †*Notogoneus*: 78, 85, 99, 100(2), 102, and 103(2). With ACCTRAN it also presents character 104(R) and lacks characters 78 and 99.

Node I = Middle Eastern genera: 26, 33(A), 77(2), and 78(2). With ACCTRAN it also presents characters 30(R) and 71(R).

Node I bis = [†*Hakeliosomus*, †*Judeichthys*, †*Ramallichthys*]: 44(2), same with ACCTRAN.

[†*Charitopsis* + †*Charitosomus*]: 62(A), plus 33(2,A) with ACCTRAN.

Node J = Family Kneriidae: 15(2), 17(A), 19(A), 46(2,A), 67(A), 78, 87(A), 97(2), 99(2), 103, and 122(A). With ACCTRAN it also presents characters 36(2), 95(2), and 107(R), and it lacks character 78.

K = Subfamily Phractolaeminae: see terminal autapomorphies for *Phractolaemus* below.

Node L = Subfamily Kneriinae: 15(3), 18(A), 29, 45, 50(A), 76(A), and 104. With ACCTRAN it lacks characters 29, 45, and 104.

Node M =Tribu Cromeriini: 8(2), 10, 41(3,A), 47(A), 65(2,A), and 109(A). With ACCTRAN it also has character 125(A).

Node N =Tribu Kneriini: 8(R), 26, 43(A), 61(2), 101(R), and 115(A). With ACCTRAN it also has character 36(R).

List of generic apomorphies (= characters for terminal branches): genera of the ingroup are listed below in alphabetical order:

†*Aethalionopsis*: 14(A), 80(A), 92(R), 93, 97(R), and 99. With ACCTRAN it also shows character 68(R).

†*Apulichthys*: variable according to its position (see comments in text). When stem-group gonorynchiform, it presents: 2(R), 7, 9, 15(2), 18(3), 23(2), 26, 41, 61(2), 73, 82(R), 84(2), 89(2,A), 94(2), 95(R), and 102(2). When stem-group Chanoidei, it does not present character 26, and it also presents characters 24(R) and 83(R).

Chanos: 4(A), 6(2), 15(2), 59(R), 88, 93, 94, 95(2), 98(3), 99, 100, 103, and 108(A). With ACCTRAN it lacks character 6.

†*Charitopsis*: 54(A) and 71, same with ACCTRAN.

†*Charitosomus*: 33(2, A) and 107. No characters in this terminal branch with ACCTRAN.

Cromeria: 34(A), 36(2), 37(2), 95(2), and 125(A). With ACCTRAN it lacks characters 36, 95, and 125.

†*Dastilbe*: 60(R), same with ACCTRAN.

Gonorynchus: 11(A), 12(A), 21(A), 22(A), 30, 38(A), 61(2), 68(2), 70(2,A), 71, 90(A), 97(2), 99(2), 111(A), 113(A), 123(A), and 128(A). With ACCTRAN it also has character 107(R), and lacks characters 22, 30, 71, 113, 123, and 128.

†*Gordichthys*: 13(1,R), 42(A), 89(A), 91(A), and 106(A). With ACCTRAN it lacks characters 42 and 89.

Grasseichthys: 6(R), 36, 83(R), 103(2), and 120 (A). With ACCTRAN it also has character 95(1,R).

†*Hakeliosomus*: 29(R) and 102(R); with ACCTRAN it also has character 107(R).

†*Halecopsis*: 9(R), 18(3), 35(2), and 51.

†*Judeichthys*: no apomorphic characters in the present analysis.

Kneria: 55(A), 84(A), and 130. With ACCTRAN it also has character 95(1,R).

†*Lecceichthys*: 35, 41(R), 45(R), 66(A), 84(2), 88(R), and 99(R).

†New genus: see Pittet and Cavin present volume.

†*Notogoneus*: 30, 64(A), 71, 77(R), 104, and 107. With ACCTRAN it also has characters 78(R) and 99(R), and it lacks characters 30, 71, 104, and 107.

†*Parachanos*: 98(2), same with ACCTRAN.

Parakneria: 57, 70(A), and 95(2). With ACCTRAN it lacks character 95.

Phractolaemus: 9(2,A), 35(2,A), 36(2), 37(2), 43(2,A), 58(A), 60(3), 61, 65(A), 69(2,A), 73(A), 85, 91(2,A), 95(2), 105(A), 112(A), 119(A), and 121(A). With ACCTRAN it also has characters 29(R), 45(R), and 104(R), and it lacks characters 36 and 95.

†*Ramallichthys*: no apomorphic characters in the present analysis.

†*Rubiesichthys*: no apomorphic characters in the present analysis (for diagnostic differences with its sister group †*Gordichthys*, see Poyato-Ariza 1994, 1996a, b, c).

†*Sorbininardus*: 1(R), 3(R), 32(R), 46(2), 54, 63(2), and 70(3,A).

†*Tharrhias*: 95(R), 96(A), and 97(R). With ACCTRAN it has the same ones.

8

A New Teleostean Fish from the Early Late Cretaceous (Cenomanian) of SE Morocco, with a Discussion of Its Relationships with Ostariophysans

Frédéric Pittet[1], Lionel Cavin[2,*] and
Francisco José Poyato-Ariza[3]

Abstract

†*Erfoudichthys rosae* gen. and sp. nov. is described on the basis of a single isolated head found in an unknown locality of the early Cenomanian Kem Kem beds, southeast of Morocco. The new species shows a combination of plesiomorphic and apomorphic characters among ostariophysans, such as a thin and elongated mesethmoid, a nasal formed by a cylindrical unit for the sensory canal extending laterally as a bony lamina, a large lacrimal, a small second infraorbital triangular in shape and wedged between the lacrimal and the third infraorbital, clusters of large conical teeth on the endopterygoids and basibranchial 2, a mandible with a ventral symphysal process and a deep coronoid process. A phylogenetic analysis of †*E. rosae* among Gonorynchiformes provides two very different patterns, with the new taxon located either as a stem Chanidae (= Chanoidei) or outside the

[1] Quai de la Thièle 7, 1400 Yverdon-les-Bains, Switzerland.
[2] Dept. de géologie et paléontologie, Muséum de la Ville de Genève, CP 6434, 1211 Genève 6, Switzerland. e-mail: Lionel.cavin@ville-ge.ch.
[3] Unidad de Paleontología, Departamento de Biología, C/Darwin s/n, 2, Universidad Autónoma de Madrid, Cantoblanco, 28049-Madrid, Spain.
* author for correspondence.

Ostariophysi. This problematic phylogenetic resolution rests on the fragmentary condition of the available material and the lack of non-ostariophysan teleosts for comparison, and prevents any analysis of the palaeobiogeographical signal that this taxon can provide.

Introduction

Vertebrate assemblage of the Moroccan "continental intercalaire" Formation, or Kem Kem beds, are known since the mid-20th century (Lavocat 1954), and most of the discoveries of vertebrate remains have been made in the past few decades. This is due to the increasing assistance provided by local people in collecting fossils for commercial purpose. Unfortunately, most of these discoveries have been made with no data about their geographic and stratigraphic contexts.

The Kem Kem beds form a series of continental and deltaic deposits that outcrop along a cliff around the Tafilalt basin, extending to the S-SW towards the Kem Kem area. Field trips conducted in this area for more than 10 years have shown that vertebrate remains occur along more than 200 km. Lateral variations of facies and of faunal contents occur, but their precise description can not be made so far.

The Kem Kem beds are divided into two units. The lowest one consists of detritic sand and sandstones showing cross-bedding structures. It has yielded a rich vertebrate assemblage with tetrapods and a variety of fishes such as sharks (Dutheil 1999a), lungfishes (Tabaste 1963, Martin 1984a, 1984b), coelacanths (Wenz 1981, Cavin and Forey 2004), gars (Cavin and Brito 2001), amiids (Forey and Grande 1998), notopterids (Forey 1997, Taverne and Maisey 1999, Taverne 2000, 2004, Cavin and Forey 2001), semionotid, ichthyodectiform (Forey and Cavin 2007) and tselfatiiform (Cavin and Forey, 2008). The upper marly unit has yielded fewer vertebrate remains, although there have appeared isolated †Onchopristis teeth and other vertebrate fragments, together with an assemblage of articulated small fish (Dutheil 1999a) comprising cladistia (Dutheil 1999b), actinopterygian with uncertain affinities (Filleul and Dutheil 2004) and basal acanthomorph (Filleul and Dutheil 2001), among others.

On the basis of the shark assemblage (Sereno et al. 1996) and comparisons with other Late Cretaceous vertebrate assemblages of North Africa (Wellnhofer and Buffetaut 1999), the lower unit is regarded as early Cenomanian in age.

Here we describe a new teleostean fish based on a single isolated head. The specimen was discovered by local researchers and acquired together with a lot of other fish remains. The precise location of the specimen is unknown but the matrix attached to the fossil and the mode of preservation clearly show that it has been found in the lower sandy unit of the Kem Kem beds. The specimen is housed in the Musée des Dinosaures of Espéraza, France, under the catalogue number MDE F43.

Systematic Palaeontology

Subdivision Teleostei Müller
incertae sedis

†*Erfoudichthys rosae* gen. and sp. nov.

Locality: Southeast Morocco, unknown spot.
Horizon: Kem Kem beds
Age: early Cenomanian, based on a shark assemblage (Sereno *et al.* 1996) and comparisons with other Late Cretaceous vertebrate assemblages of North Africa (Wellnhofer and Buffetaut 1999).
Etymology: from the city of Erfoud, close to the Kem Kem beds, and Greek, *ichthys*, fish and *rosae*, in memory of the senior author's mother.
Holotype and only known specimen: MDE F43.

Diagnosis

Ostariophysan fish with mesethmoid thin and elongated, with no lateral processes; nasal formed by a cylindrical unit for the sensory canal extending laterally as a bony lamina; lacrimal large; second infraorbital small, triangular in shape and wedged between the lacrimal and the third infraorbital, but not reaching the ventral margin of the circumorbital ring; clusters of large teeth on the endopterygoids and basibranchial 2; mandible with a ventral symphysal process and deep coronoid process; first two pairs of ribs enlarged.

Since the phylogenetic position of this taxon is unresolved so far (see "phylogenetic relationships" below), we are unable to determine which of the diagnostic characters listed above are apomorphic for the new taxon.

Description

The material comprises a single head preserved in three dimensions with the anteriormost eight vertebrae still preserved. The head with the opercular series is ca. 9 cm long and ca. 5.5 cm deep.

The Braincase

Anteriorly the mesethmoid (*Mes*: Figs. 8.1, 8.2, 8.3 and 8.5) fuses with both frontals through an interdigitated suture. The bone is triangular in shape and extends forwards as a thin rostrum. It does not bear prominent lateral processes and it is proportionally much narrower than in any other

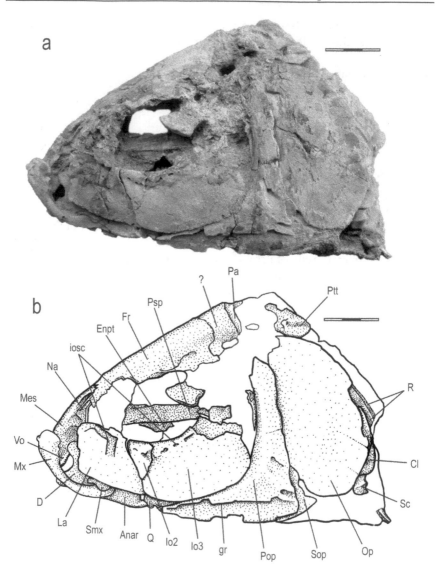

Fig. 8.1 †*Erfoudichthys rosae* gen. and sp. nov., holotype, MDE F43. Photograph (a) and semi-interpretative line drawing (b) of the head in left lateral view. Anar, anguloarticular; Cl, cleithrum; D, dentary; Enpt, endopterygoid; Fr, frontal; gr, groove; Io, infraorbital (numbered); iosc, infraorbital sensory canal; La, lacrimal; Mes, mesethmoid; Mx, maxilla; Na, nasal; Op, opercle; Pa, parietal; Pop, preopercle; Psp, parasphenoid; Ptt, post-temporal; Q, quadrate; R, rib; Sc, scapula; Smx, supramaxilla; Sop, subopercle; Vo, vomer. Scale bars: 15 mm.

gonorynchiforms. In *Gonorhynchus* the snout anterior to the lateral ethmoid is also elongated, but this part of the skull roof is formed dorsally by the frontals only. In †*Hakeliosomus hakelensis* (Davis, 1887) and †*Charistosomus lineolatus* (Pictet and Humbert, 1866), the mesethmoid is elongated but broader than in †*E. rosae*.

Both lateral ethmoids (*Le*, Figs. 8.2 and 8.5) are incompletely preserved, but the right one shows that no lateral extension is present as in kneriids and *Phractolaemus*.

The anterior extremity of the vomer is visible (*Vo*, Fig. 8.1). It shows paired anterolaterally oriented circular facets and an unpaired anteriorly oriented one. The anterolateral facets articulate with the maxilla and the anterior one is possibly for the attachment of ligaments binding a lost rostral cartilage, or hypoethmoid. In lateral view the anterior extremity of the vomer is ventrally inclined and forms an angle of 30 degrees with the ventral margin of the parasphenoid (Fig. 8.1). The ventral inclination of the vomer is similar in *Kneria*, *Parakneria*, *Cromeria* and *Chanos*. In the latter, moreover, the anterior extremity shows a similar tripartite arrangement of the circular areas. In *Phractolaemus* the vomer is dorsally bent. We cannot see in †*E. rosae* whether the vomer bears teeth.

The parasphenoid is barely visible (*Psp*, Figs. 8.1–8.5). It is apparently rather wide in the orbital region, but we cannot discern whether teeth are present.

Although incompletely preserved, the nasal (*Na*, Figs. 8.1, 8.2, 8.3 and 8.5) shows a cylindrical tube that accommodated the sensory canal and extends laterally as a wing-like lamina of bone. This structure differs from the nasal of most gonorynchiforms, in which it is tubular in shape, but is similar to the kneriids that show a lateral wing.

The frontal (*Fr*, Figs. 8.1, 8.2 and 8.5) is elongated and rectangular in shape. It is much narrower than in *Chanos* and †*Tharrhias*, and especially *Phractolaemus* and kneriids. In *Gonorhynchus*, the frontal is narrow at the orbital level, but the bone widens posteriorly in the otic region, unlike in †*E. rosae*. The posterior part of the skull roof is badly preserved and we cannot observe the shape and the extension of the parietals. In particular, we cannot determine whether a transversal limit at a level just posterior to the orbit corresponds to the suture between the frontal and parietal, or is a fracture (Fig. 8.1). Only the posterior part of the parietal, forming a transversal ridge against which rested the extrascapulars, is visible on the left side of the skull roof (*Pa*, Fig. 8.1).

The pterotic (*Pto*, Fig. 8.2) is barely visible. It appears to have a small extension on the skull roof and to end posteriorly as a blunt spine.

The supraoccipital shows a small area on the dorsal face of the skull roof behind the large paired bones anteriorly (frontals or parietals?). The supraoccipital spine is broken but the posterior extremity is still preserved,

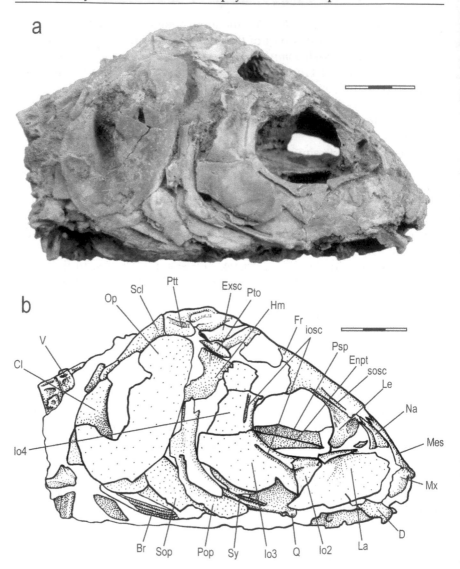

Fig. 8.2 †*Erfoudichthys rosae* gen. and sp. nov., holotype, MDE F43. Photograph (a) and semi-interpretative line drawing (b) of the head in right lateral view. Br, branchiostegal ray; Cl, cleithrum; D, dentary; Enpt, endopterygoid; Exsc, extrascapula; Fr, frontal; Hm, hyomandibula; Io, infraorbital (numbered); iosc, infraorbital sensory canal; La, lacrimal; Le, lateral ethmoid; Mes, mesethmoid; Mx, maxilla; Na, nasal; Pop, preopercle; Psp, parasphenoid; Pto, pterotic; Q, quadrate; Sop, subopercle; sosc, supraorbital sensory canal; Sy, symplectic; V, vertebra. Scale bars: 15 mm.

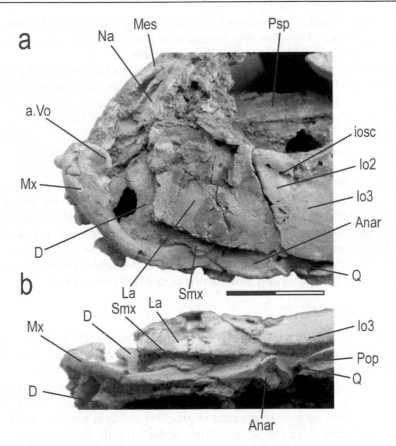

Fig. 8.3 †*Erfoudichthys rosae* gen. and sp. nov., holotype, MDE F43. Detail of the snout region in lateral (a) and ventral (b) views. Anar, anguloarticular; a.Vo, articular head for the vomer on the maxilla; D, dentary; Io, infraorbital (numbered); iosc, infraorbital sensory canal; La, lacrimal; Mes, mesethmoid; Mx, maxilla; Na, nasal; Psp, parasphenoid; Q, quadrate; Smx, supramaxilla. Scale bars: 10 mm.

although slightly shifted. The complete spine was posteriorly elongated and rather deep in lateral view. An elongated supraoccipital spine occurs in *Chanos* and †*Tharrhias*, but not in other gonorynchiforms.

The Circumorbital Series

Specimen MDE F43 shows four, rather thick ossifications preserved along the posterior and ventral margins of the orbit. An unpreserved dermosphenotic was probably originally present, as well as a supraorbital as indicated by the excavated lateral margin of the frontal. We regard here the anteriormost large infraorbital as a lacrimal. However, we cannot definitively rule out the possibility that this bone is actually the second infraorbital, which was preceded by a small, now lost, tubular lacrimal as in *Phractolaemus* (Thys Van den Audenaerde 1961). This makes the original number of circumorbitals between five and seven.

The anteriormost preserved bone regarded as a lacrimal (*La*, Figs. 8.1, 8.2, 8.3 and 8.5) is a large ossification, trapezoidal in shape, with the anterior and posterior borders inclined backward. Lacrimals of most gonorynchiforms are proportionally smaller, except in *Chanos*. The second infraorbital (*Io2*, Figs. 8.1, 8.2, 8.3 and 8.5) is a small, triangular ossification wedged between the lacrimal and the third infraorbital. Its ventral border does not reach the ventral margin of the circumorbital ring. This pattern appears to be unique among gonorynchiforms, and among teleosts as a whole. The third infraorbital (*Io3*, Figs. 8.1, 8.2 and 8.5) is a long and deep ossification that completely fills the gap between the orbit and the opercular series. As usual in teleosts, this bone forms the ventroposterior corner of the orbit. An incomplete fourth infraorbital is visible on the right side of the specimen (*Io4*, Figs. 8.2 and 8.5). It is deep and probably completely filled the space between the orbit and the preopercle in the living fish.

Suspensorium

Only the dorsalmost part of the hyomandibula is visible (*Hm*, Figs. 8.2 and 8.5), but it shows no salient feature. Its articular head with the braincase appears to be simple. Most of the suspensorium is hidden by the circumorbital series, except for a few elements in the ventralmost part. The quadrate (*Q*, Figs. 8.1, 8.2, 8.3 and 8.5) is anteroposteriorly developed. It bears a large anteriorly orientated articular condyle for the mandible, and it is located at the level of the anterior half of the orbit. The ventroposterior margin of the quadrate is elongated and extends laterally as a narrow lamina that rests on the dorsal margin of the preopercle (the so-called "quadratojugal process"). The symplectic (*Sy*, Fig. 8.2), only partly visible, appears to be thin and elongate. *Gonorynchus* have a reduced symplectic, while it is very elongate in chanids (*Chanos*, †*Tharrhias*), but in the latter fishes there is no direct contact between the symplectic and the quadrate but a cartilaginous connection. The situation in MDE F43 is similar to the situation in basal teleosts, and probably plesiomophic for gonorynchiformes. The endopterygoid (*Enpt*, Figs. 8.1, 8.2, 8.3 and 8.5) is partly visible through the orbit. A cluster of proportionally large and closely spaced conical teeth are present on each endopterygoid (*Enpt.t*, Figs. 8.4 and 8.5). They face ca. 11 teeth, similar in shape, borne by the basibranchial 2. The arrangement and shape of these teeth show similarities with those observed in *Gonorynchus* and all extinct gonorhynchids, although in these forms the teeth are tightly packed, small and rounded (conical), and not pointed (Terry Grande, pers. comm. 2008).

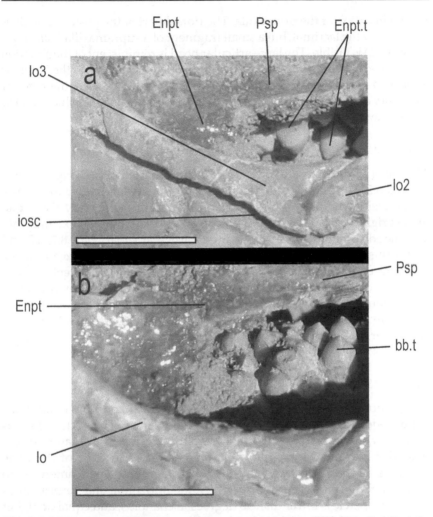

Fig. 8.4 †*Erfoudichthys rosae* gen. and sp. nov., holotype, MDE F43. Detail of the endopterygoid teeth in oblique (lateroventral) view (a) and of the basibranchial teeth in oblique (laterodorsal view (b). bb.t, basibranchial teeth; Enpt, endopterygoid; Enpt.t, endopterygoid teeth; Io, infraorbital (numbered); iosc, infraorbital sensory canal; Psp, parasphenoid. Scale bars: 5 mm.

Upper Jaw

The premaxilla is missing. The edentulous maxilla (*Mx*, Figs. 8.1, 8.2, 8.3 and 8.5) has a peculiar shape. The anterior part forms a curved cylindrical rod of bone, with a large articular head at its anterior extremity, and the posterior part forms a broad, thin blade of bone. The anterior articular head bears medially a large globular articular facet, probably for the articulation with the vomer (*a.Vo*, Fig. 8.3), and anterodorsally a shallow groove probably for the insertion of the premaxilla. The dorsal part of the posterior blade is overlaid by the lacrimal, but a small fragment of a supramaxilla (*Smx*, Figs. 8.1 and 8.3) is visible. The large articular head is reminiscent of the situation in *Kneria* and *Parakneria* and, to a lesser degree, *Chanos*. But in the latter, as well as in the fossil gonorynchids described by Gayet (1993), the anterior extremity of the maxilla bears a ventromedial process against which rest the blade-like premaxilla.

Lower Jaw

The lower jaw is short and deep. The edentulous dentary (*D*, Figs. 8.1, 8.2, 8.3 and 8.5) has a shallow symphyseal extremity extending ventrally as a well-developed blunt process. The oral margin of the dentary rises at an almost right angle at the level of its first anterior quarter and expands into a rounded coronoid process. The anguloarticular (*Anar*, Figs. 8.1, 8.3 and 8.5) forms the posterior third of the mandible. It is thin, except at the posterior end, where it widens to form the articular facet. In ventral view, the facet forms a Y-shaped articular head, with a thin lateral branch and a more robust medial branch (Fig. 8.3b). The mandible is reminiscent of the mandible of *Chanos*, but in this Recent taxon the symphyseal region is shallower, the anguloarticular is proportionally smaller, the coronoid process is shallower and it bears a notch at its anterior edge.

Opercular Series

The preopercle (*Pop*, Figs. 8.1 and 8.5) has vertical and horizontal limbs almost equal in size and arranged at ca. 90 degrees. The vertical limb is narrow with almost parallel margins, although the preservation of the posterior margin is incomplete. The posteroventral edge of the bone develops a rounded process and the horizontal limb tapers anteriorly to end as a blunt spine. The angle between the horizontal and vertical limbs of the preopercle is lower than 90 degrees in chanoids, and equal or higher than 90 degrees in gonorhynchoids and kneriids. The opercle is ovoid in shape (*Op*, Figs. 8.1, 8.2 and 8.5), with a dorsal border slightly narrower than the ventral border. The posteroventral margin of the bone is regularly rounded. The posteroventral area of the bone is ornamented with very faint ridges. The articular facet is hardly visible in lateral view and bears no process as in *Chanos* and some fossil gonorynchids (Gayet 1985, 1986). The subopercle (*Sop*, Figs. 8.1, 8.2 and 8.5) is mainly hidden by the opercle, and the interopercle is visible as a strip of bone underneath the horizontal limb of the preopercle.

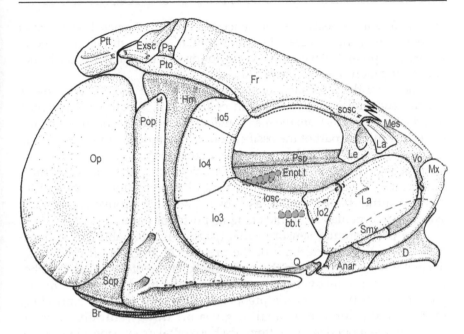

Fig. 8.5 †*Erfoudichthys rosae* gen. and sp. nov., holotype, MDE F43. Reconstruction of the head in right lateral view. Endopterygoid and basibranchial teeth, as well as the outline of the mandible, are visible by transparency. Anar, anguloarticular; bb.t, branchiostegal ray; D, dentary; Enpt.t, endopterygoid teeth; Exsc, extrascapula; Fr, frontal; Hm, hyomandibula; Io, infraorbital (numbered); iosc, infraorbital sensory canal; La, lacrimal; Le, lateral ethmoid; Mes, mesethmoid; Mx, maxilla; Na, nasal; Pop, preopercle; popsc, preopercular sensory canal; Psp, parasphenoid; Pto, pterotic; Ptt, post-temporal; Q, quadrate; Smx, supramaxilla; Sop, subopercle; sosc, supraorbital sensory canal; Vo, vomer.

Pectoral Girdle

The extrascapula (*Exsc*, Figs. 8.2 and 8.5) is badly preserved. It is formed by a well-developed transversal limb, carrying the extrascapular commissure, that rests against a transversal ridge running on the posterior part of the skull roof, and by a short posterior process carrying the lateral sensory canal. The post-temporal (*Ptt*, Figs. 8.2 and 8.5) shows a well-developed dorsal branch and a posterior plate bearing the sensory canal, while the ventral branch is not visible. The supracleithrum (*Scl*, Fig. 8.2) is large, but its precise shape cannot be determinate. The cleithrum (*Cl*, Figs. 8.1 and 8.2) is well developed but shows no salient feature. The scapula is visible on the left side (*Sc*, Fig. 8.1).

Cephalic Sensory Canal System

The supraorbital sensory canal (*sosc*, Fig. 8.2) runs in a canal along the medial part of the frontal, then in a groove on the anteriormost part of the bone, before it connects to a bone-enclosed canal in the nasal. The path of the otic canal is not visible. The extrascapular commissure is enclosed in the extrascapular, but we cannot verify for preservational reasons whether both ossifications meet in the midline.

The infraorbital sensory canal (*iosc*, Figs. 8.1, 8.2 and 8.5) is visible as a canal running along the margin of the orbit. A ventrally oriented diverticulum diverts from the main canal in the lacrimal. The diverticulum reachs the middle part of the ossification, where it opens through a broad opening. The infraorbital sensory canal crosses the second infraorbital, along a ventroposterior line as in *Phractolaemus*, and joins the third infraorbital. It possibly gives off a ventral diverticulum at the border between infraorbitals 2 and 3, as shown by a faint groove on the right side. No diverticula are visible along its posterior path.

The preopercular sensory canal (*popsc*, Figs. 8.1 and 8.5) is bone-enclosed in the vertical limb of the preopercle. It gives off a ventrally orientated large pore above the right-angled curve of the canal. Another pore opens posteroventrally and extends into a groove in the posteroventral process of the preopercle. The sensory canal runs along the horizontal limb in a canal. It gives off three ventrally oriented diverticula that open in the bottom of a horizontal groove marking a ridge at the external side of the preopercle. The pattern formed by the horizontal ridge and the large posteroventral opening recalls the patterns present in notopterids (Taverne 1978).

Vertebral Column

The first eight anterior-most vertebral centra are visible (*V*, Fig. 8.2). A neural arch is present anterior to centrum one, but it is probably associated with the occipital condyle. The anteriormost centrum is mainly covered with matrix. The second one is slightly shorter than the subsequent ones. The centra are laterally scored with deep grooves and a network of interconnected small ridges. Four neural arches are preserved associated with the first four centra. They articulate in large pockets excavated in the dorsal face of each centrum, but they are not fused to it. Each neural arch articulates with the following one through a large pre- and post-zygapohysis complex. The articulations between zygapohyses are made via well-developed articular facets. The neural spines are not preserved, but the broken bases show that they were probably well developed. The neural arch associated with the occipital condyle is difficult to observe, but it looks like the subsequent neural arch. There is no trace of a posteriorly extended roof formed by the exoccipitals above the neural canal as described in *Chanos* (Patterson 1984). We cannot see whether a parapophysis is associated with the first free centrum. The second and third centra bear parapophyses with large excavated articular facets. The more posterior parapophyses bear much smaller articular facets for the associated ribs. On the left side of the specimen lie two pleural ribs slightly shifted from their anatomical position. The anteriormost one is very broad and the second one narrower, although broader than a common pleural rib. These ribs originally articulated in the large articular facets on the parapophyses associated with the second and third centra. *Chanos* and †*Tharrhias* also have a first pair of ribs that are proportionally very large.

Phylogenetic Relationships

The new genus described herein was included in the new phylogenetic analysis by Poyato-Ariza *et al.* (present volume). In this analysis, the outgroup taxa are the clupeomorph †*Diplomystus*, and the otophysi *Brycon* and *Opsariichthys*. In the present discussion we do not retain all the terminal ingroup taxa included in Poyato-Ariza *et al.*'s analysis (present volume), but only those relevant for understanding the interrelationships of †*Erfoudichthys*, i.e., the Chanidae, the Gonorynchidae and †*Apulichthys* Taverne, 1997 from the Late Cretaceous of Italy. For greater consistency, we specify the coding of the characters for †*Erfoudichthys rosae* in Appendix 1 of this chapter. The phylogenetic resolutions are not described in detail here; only the main results are briefly presented. The numbers in brackets mentioned below correspond to the list of characters in Poyato-Ariza *et al.* (present volume) and in Appendix 1 of this chapter.

The results of the analysis concerning the new taxon are complex, rather unexpected, yet quite interesting. The position of the new genus appears totally unresolved in the strict consensus tree, in a multiple polytomy with the outgroup (†*Diplomystus*, *Brycon*, and *Opsariichthys*), †*Apulichthys*, the Chanidae, and the Gonorynchidae (Fig. 8.6 left). This obviously lessens the resolution of this tree. In the 50% consensus tree (Fig. 8.6 right) it appears in a polycotomy together with part of the outgroup (†*Diplomystus*) and the Ostariophysi.

Based on this analysis, there are only two possible, very different, positions for the new genus. First, †*Erfoudichthys rosae* appears as the stem-group Chanidae (= Chanoidei), in 35 percent of the trees. This happens only in situations in which †*Apulichthys* is not the stem-group; that is, one genus excludes the other from this position. In other trees neither taxon forms the chanid stem-group. This is probably the result of the sampling effect, especially because almost no postcranial characters are known in †*Erfoudichthys*. In the discussion below, only characters visible on MDE F43 are mentioned. In this position, the new genus is joined to the Chanidae by only two characters with DELTRAN optimization: angle formed by preopercular limbs approximately straight (57, 1) and unmodified neural

Fig. 8.6 Simplified phylogenetic interrelationships of †*Erfoudichthys* with the Gonorynchiformes. Modified from Poyato-Ariza *et al.* (present volume). Left: strict consensus tree, right: 50% consensus tree.

arches anterior to dorsal fin autogenous in adults, at least laterally (79, 1). In turn, with the ACCTRAN optimization option, additional visible characters are present: nasal bone small but flat, not reduced (8, reversal); presence of ridge on anteroventral limb of preopercular bone (59, 1); and first two autogenous anterior parapophyses (81, reversal). Other characters are actually unknown in the new taxon, and therefore, their distribution cannot be known, but, if one chooses the ACCTRAN optimization, this affects the definition of the Chanidae. In turn, †*Erfoudichthys* is separated from the Chanidae by the following synapomorphies of that family: a greatly enlarged maxilla, posteriorly swollen to a bulbous outline, with a clearly curved posterior border (27, 1, autapomorphic); presence of a notch in the anterodorsal border of the dentary (31, autapomorphic); a quadrate-mandibular articulation anteriorly displaced, due to displacement of the quadrate bone, which is not elongated and maintains the usual triangular shape of its main body, and anterior to the orbit (35); symplectic correlatively elongated, at least twice the normal length (36, 1); opercular bone expanded, at least one third of the head length (52, autapomorphic); and supraopercular bone present as a relatively large, flat bone (61, 1).

The apomorphic characters of the new genus as terminal branch, in this position, are highly homoplastic. With DELTRAN optimization option, they are: nasal bone small but flat, not reduced to a tubular ossification (8, reversion); frontals roughly rectangular in outline, narrow throughout their length (character 9, state 3); mesethmoid elongated and thin (character 18, state 3; convergent with †*Apulichthys* and with †*Halecopsis*); presence of supramaxillae (28, reversion); a patch of about 20 conical teeth on endopterygoids and basibranchial 2 present (40, 1; convergent with the Gonorynchidae); anterior portion of vomer anteroventrally inclined, nearly vertical (43, 1; convergent with the Knerinii); ridge on anteroventral limb of preopercular bone present (59, 1; convergent with the Chaninae); major axis of subopercular bone subhorizontal in lateral view (63, 1; convergent with the Chaninae minus †*Aethalionopsis*); and presence of autogenous neural arch anterior to arch of first vertebra (74, reversion).

In any case, this possible position is quite unexpected for the new genus, since it does show gonorynchid autapomorphic characters, which necessarily appear as a convergent character, e.g., teeth on endopterygoids and basibranchial 2 (see above).

The other possible position of the new taxon is less unexpected and has been mentioned before; it may appear outside the Ostariophysi (in 65 percent of the trees), a position that is congruent with its lack of some ostariophysian autapomorphic characters. †*Erfoudichthys rosae* differs from ostariophysans because it lacks several autapomorphies of this clade: nasal bone reduced to a tubular ossification around the canal (8, 1); absence of supramaxillae (28, 1; autapomorphic for the Ostariophysi); and absence of autogenous neural arch anterior to arch of first vertebra (74, 1; autapomorphic for the Ostariophysi). With ACCTRAN, it also lacks the first two anterior parapophyses fused to their respective centra (81, 1).

It is interesting to have a look at the visible apomorphic characters of the new genus in this position, as terminal branch, because they are also highly convergent. With DELTRAN optimization option, they are: frontals roughly rectangular in outline, narrow throughout their length (character 9, state 3); mesethmoid elongated and thin (character 18, state 3; convergent with †*Apulichthys* and with †*Halecopsis*); a patch of about 20 conical teeth on endopterygoids and basibranchial 2 present (40, 1; convergent with the Gonorynchidae); anterior portion of vomer anteroventrally inclined, nearly vertical (43, 1; convergent with the Knerinii); angle formed by preopercular limbs approximately straight (57, 1; convergent with the Chanidae); ridge on anteroventral limb of preopercular bone present (59, 1; convergent with the Chaninae); major axis of subopercular bone subhorizontal in lateral view (63, 1; convergent with the Chaninae minus †*Aethalionopsis*); and unmodified neural arches anterior to dorsal fin autogenous in adults, at least laterally (79, 1; convergent with the Chanidae). With ACCTRAN, it also presents character 68 (four infraorbitals; convergent with the Gonorynchiformes according to this optimization). As we see, all characters are convergent, confirming the problematic nature of this phylogenetic position.

Conclusions

The problematic phylogenetic relationships of the new genus are the result of two unfavourable factors. The first is the incompleteness of the remains, which makes most characters of the post-cranial skeleton unknown in the new genus; therefore, their distribution is impossible to know, which consequently can affect the definition of the Ostariophysi or, alternatively, of the Chanidae, especially if one chooses the ACCTRAN optimization option (see discussion of characters above). This could only be solved by finding additional, more complete specimens.

The second factor involves the characters that are actually known in the new genus: it is the unexpected combination of primitive and derived character states. The new genus does lack some relevant ostariophysan autapomorphies: for instance, it has supramaxillae and an autogenous neural arch anterior to arch of first vertebra, which are always absent in the Ostariophysi (see discussion of characters above). These characters contrast with some mixed apomorphic characters, such as the shape and crest of the preopercular bone (typical of chanids) or the presence of teeth on the endopterygoid and basibranchial 2 (typical of gonorynchids). If the new genus is not a gonorynchiform, then these characters must obviously be interpreted as convergent, but, even if it were a gonorynchiform, then part of them are convergent with other gonorynchiforms. This mosaic of primitive and derived, or, to put it otherwise, of non-ostariophysan plus chanid and gonorynchid characters, causes major problems when trying to interpret the character distribution in a gonorynchiform-only phylogenetic analysis. Therefore, it seems that the best option to make substantial conclusions about the phylogentic relationships of the new genus is to attempt additional analyses by including a number of varied non-ostariophysan teleosts. This would be the only way to really test whether the new genus is a gonorynchiform fish, or not an ostariophysan fish at all. At present, we are unable to answer this question with the generic sample used as ingroup in the present phylogenetic analysis.

Acknowledgements

Funding was provided by a Marie Curie Individual Fellowship funded by the Swiss Federal Office for Education and Science (LC, grant no. 02.0335). We thank Terry Grande and Lance Grande (Chicago) for their comments on the manuscript. F.P. and L.C. thank Redouan Bshary (Neuchâtel) and Peter L. Forey (London) for their support during this work.

References

Cavin, L. and P.M. Brito. 2001. A new Lepisosteidae (Actinopterygii: ginglymodi) from the Cretaceous of the Kem Kem beds, Southern Morocco. Bull. Soc. géol. Fr. 172: 141–150.

Cavin, L. and D.B. Dutheil. 1999. A new Cenomanian ichthyofauna from southeastern Morocco and its relationships with other early Late Cretaceous Moroccan faunas. Geologie en Mijnbouw 78: 261–266.

Cavin, L. and P.L. Forey. 2001. Osteology and systematic affinities of *Palaeonotopterus greenwoodi* Forey, 1997. Zool. J. Linn. Soc. 149: 1–28.

Cavin, L. and P.L. Forey. 2004. New mawsoniid coelacanth (Sarcopterygii: Actinistia) remains from the Cretaceous of the Kem Kem beds, SE Morocco, pp. 493–506. *In*: A. Tintori and G. Arratia [eds.]. Mesozoic Fishes 3—Systematics, Paleoenvironments and Biodiversity. Dr. F. Pfeil Verlag, Munich.

Cavin, L. and P.L. Forey. 2008. A new tselfatiiform teleost from the late Cretaceous (Cenomanian) of the Kem Kem beds, Southern Morocco, pp. 199–216. In: G. Arratia, H.-P. Schultze and M.V.H. Wilson [eds.]. Mesozoic Fishes 4—Homology and Phylogeny. Dr. F. Pfeil Verlag, Munich.

Dutheil, D. B. 1999a. An overview of the freshwater fish fauna from the Kem Kem beds (Late Cretaceous: Cenomanian) of southeastern Morocco, pp. 553–563. In: G. Arratia and H.-P. Schultze [eds.]. Mesozoic Fishes 2—Systematics and Fossil Record. Dr. F. Pfeil Verlag, Munich.

Dutheil, D.B. 1999b. The first articulated fossil cladistian: Serenoichthys kemkemensis, gen. et sp. nov., from the Cretaceous of Morocco. J. Vert. Paleontol. 19: 243–246.

Filleul, A. and D.B. Dutheil. 2001. Spinocaudichthys oumtkoutensis a freshwater acanthomorph from the Cenomanian Morocco. J. Vert. Paleontol. 21: 774–780.

Filleul, A. and D.B. Dutheil. 2004. A very peculiar actinopterygian fish from Cretaceous of Morocco. J. Vert. Paleontol. 24: 290–298.

Forey, P.L. 1997. A Cretaceous notopterid (Pisces: Osteoglossomorpha) from Morocco. S. Afr. J. Sci. 93: 564–569.

Forey, P.L. and L. Cavin. 2007. A New Species of Cladocyclus (Teleostei: Ichthyodectiformes) from the Cenomanian of Morocco. Palaeontol. Electronica 10: 1–10.

Forey, P.L. and L. Grande. 1998. An African twin to the Brazilian Calamopleurus (Actinopterygii: Amiidae). Zoological Journal of the Linnean Society 123: 179–195.

Lavocat, R. 1954. Reconnaissance géologique dans les Hammadas des confins algéro-marocains du sud. Notes Mém. Serv. Géol. Maroc, Rabat 116: 1–147.

Martin, M. 1984a. Deux Lepdosirenidae (Dipnoi) crétacés du Sahara, Protopterus humei (Priem) et Protopterus protopteroides (Tabaste). Paläontol. Zeitschr. 58: 265–277.

Martin, M. 1984b. Révision des Arganodontidés et des Néoceratodontidés (Dipnoi, Ceratodontiformes) du Crétacé africain. Neu. Jahrb. Geol. Paläontol., Abhandl 169: 225–260.

Patterson, C. 1984. Chanoides, a marine Eocene Otophysan fish (Teleostei: Ostariophysi). J. Vert. Paleontol. 4: 430–456.

Sereno, P.C., D.B. Dutheil, M. Iarochène, H.C.E. Larsson, G.H. Lyon, P.M. Magwene, C.A. Sidor, D.J. Varricchio and J.A. Wilson. 1996. Predatory dinosaurs from the Sahara and Late Cretaceous faunal differentiation. Science 272: 986–991.

Tabaste, N. 1963. Etude de restes de poissons du Crétacé saharien. Mém. Inst. Fr. Afr. Nord, Mél. Ichthyol. 68: 437–485.

Taverne, L. 1997. Les poissons crétacés de Nardo. 4°: †Apulichthys gaycti gen. nov., sp. nov. (Teleostei, Ostariophysi, Gonorhynchiformes). Bolletino del Museo civico di Storia naturale di Verona 21: 401–436.

Taverne, L. 2000. Nouvelles données ostéologiques et phylogénétiques sur Palaeonotopterus greenwoodi, notoptéridé (Teleostei, Osteoglossomorpha) du Cénomanien inférieur continental (Crétacé) du Maroc. Stuttgarter Beitr. Naturk. Ser. B (Geol. Paläontol.) 293: 1–24.

Taverne, L. 2004. On a complete hyomandibular of the Cretaceous Moroccan notopterid Palaeonotopterus greenwoodi (Teleostei, Osteoglossomorpha). Stuttgarter Beitr. Naturk. Ser. B (Geol. Paläontol.) 348: 1–7.

Taverne, L. and J.G. Maisey. 1999. A Notopterid skull (Teleostei, Osteoglossomorpha) from the Continental Early Cretaceous of Southern Morocco. Am. Mus. Novitates 3260: 1–12.

Thys Van den Audenaerde, D.F.E. 1961. L'anatomie de Phractolaemus ansorgei Blgr. et la position systématique des Phractolaemidae. Ann. Mus. Roy. Af. Centr. (Zool.) 103: 99–167.

Wellnhofer, P. and E. Buffetaut. 1999. Pterosaur remains from the Cretaceous of Morocco. Paläontol. Zeitschr. 73: 133–142.

Wenz, S. 1981. Un coelacanthe géant, Mawsonia lavocati Tabaste, de l'Albien-base du Cénomanien du sud marocain. Ann. Paléontol. (Vert.) 67: 1–20.

APPENDIX 1

Characters used by Poyato-Ariza *et al.* (present volume) for analysing Gonorynchiform interrelationships. Character states for †*Erfoudichthys rosae* gen. and sp. nov. are in bold.

1. Orbitosphenoid: present [0]; absent [1]; **[?]**.
2. Basisphenoid: present [0]; absent [1]; **[?]**.
3. Pterosphenoids: well developed and articulating with each other [0]; slightly reduced; not articulating anteroventrally but approaching each other anterodorsally [1]; greatly reduced and broadly separated both anteroventrally and anterodorsally [2]; **[?]**.
4. Posterolateral expansion of exoccipitals: absent [0]; present [1]; **[?]**.
5. Exoccipitals: posteriorly smooth with no projection above the basioccipital [0]; with a posterior concave-convex border; and a projection above basioccipital [1]; **[?]**.
6. Cephalic ribs: absent [0]; present and all articulating with the exoccipitals [1]; present and articulating with both the exoccipitals and basioccipital [2]; **[?]**.
7. Brush-like cranial intermuscular bones (*sensu* Patterson and Johnson 1995): absent [0]; present [1]; **[?]**.
8. Nasal bone: small but flat **[0]**; just a tubular ossification around the canal [1]; absent as independent ossification [2].
9. Frontals: wide through most of their length; narrowing anteriorly to form a triangular anterior border [0]; elongate and narrow except in postorbital region [1]; wide; anteriorly shortened; anterior border roughly straight [2]; roughly rectangular in outline; narrow throughout their length **[3]**.
10. Interfrontal fontanelle: absent **[0]**; present [1].
11. Frontal bones: paired in adult **[0]**; co-ossified; with no median suture [1]; **[?]**.
12. Foramen for olfactory nerve in frontal bones: absent [0]; present [1]; **[?]**.
13. Relative position of the parietals: medioparietal (in full contact with each other along their midline) [0]; mesoparietal [1]; lateroparietal (completely separated from each other by the supraoccipital) [2]; **[?]**.
14. Parietal portion of the supraorbital canal: absent [0]; present [1]; **[?]**.

15. Parietals: large [0]; reduced but flat and blade-like in shape [1]; reduced to canal-bearing bones [2]; absent as independent ossifications [3]; [?].

16. Supraoccipital crest: small; short in lateral view [0]; long and enlarged; projecting above occipital region and first vertebrae; forming a vertical, posteriorly deeply pectinated blade [1].

17. Foramen magnum: dorsally bounded by exoccipitals [0]; enlarged and dorsally bounded by supraoccipital [1]; [?].

18. Mesethmoid: wide and short [0]; long and slender; with anterior elongate lateral extensions [1]; large; with broad posterolateral wing-like expansions [2]; elongated and thin [3].

19. Wings (extensions) on lateral ethmoids: absent [0]; present [1].

20. Teeth in premaxilla; maxilla; and dentary: present [0]; absent [1].

21. Premaxilla: consisting of one solid portion [0]; premaxilla consisting of two distinct portions; with a shorter, non-osseous element lying ventral to a much longer osseous portion, which in turn articulates with the maxilla [1]; [?].

22. Premaxillary "gingival teeth": absent [0]; present [1]; [?].

23. Premaxilla: small; flat and roughly triangular [0]; large; very broad; concave-convex; with long oral process [1]; narrow and elongated; its length more than half the length of the maxilla [2]; [?].

24. Premaxillary ascending process: present [0]; absent [1]; [?].

25. Dorsal and ventral borders of the maxillary articular process: straight or slightly curved [0]; very curved; almost describing an angle [1].

26. Maxillary process for articulation with autopalatine: absent [0]; present [1].

27. Posterior region of the maxilla: slightly and progressively expanded to form a thin blade; with roughly straight posterior border [0]; very enlarged; swollen to a bulbous outline; with curved posterior border [1].

28. Supramaxilla(e): present [0]; absent [1].

29. Notch between the dentary and the angulo-articular bones: absent [0]; present [1]; [?].

30(28). Articulation between dentary and angulo-articular: strong; dentary not V-shaped posteriorly [0]; loose; with a posteriorly V- shaped dentary [1]; [?].

31. Notch in the anterodorsal border of the dentary: absent [0]; present [1].

32. Mandibular sensory canal: present [0]; absent [1].

33. Inferior and superior enlarged retroarticular processes of mandible: both absent [0]; inferior retroarticular process present; superior

retroarticular process absent [1]; both inferior and superior retroarticular processes present [2].

34. Quadrate with: posterior margin smooth [0]; elongated forked posterior process [1].

35. Quadrate-mandibular articulation: below or posterior to orbit; no elongation or displacement of quadrate [0]; anterior to orbit; quadrate displaced but not elongate [1]; anterior to orbit; with elongation of the body of quadrate instead of displacement [2].

36. Symplectic: elongated in shape but relatively short [0]; very long; about twice the length of the ingroup [1]; absent as an independent ossification [2].

37. Symplectic and quadrate: articulating directly with each other [0]; separated through cartilage [1]; no contact due to absence of symplectic [2].

38. Metapterygoid: large; broad and in contact with quadrate and symplectic through cartilage [0]; reduced to a thin rod [1]; [?].

39. Dermopalatine: present [0]; absent [1]; [?].

40. A patch of about 20 conical teeth on endopterygoids and basibranchial 2: absent [0]; present [1].

41. Ectopterygoids: well developed; ectopterygoid overlapping with the ventral surface of the autopalatine by at least 50% [0]; well developed; with three branches in lateral view; reduced but direct contact with autopalatine [1]; reduced; articulating with the ventral surface of the autopalatine by at most 10% through cartilage; resulting in a loosely articulated suspensorium [2]; absent as distinct ossifications [3]; [?].

42. Teeth on vomer and parasphenoid: absent [0]; present [1].

43. Anterior portion of vomer: horizontal [0]; anteroventrally inclined; nearly vertical [1]; dorsally curved [2].

44. Spatial relationship between vomer and mesethmoid anteriorly: vomer and mesethmoid ending at about the same anterior level [0]; mesethmoid extending anteriorly beyond the level of anterior margin of vomer [1]; vomer extending anteriorly beyond the level of anterior margin of mesethmoid [2]; [?].

45. Articular head of hyomandibular bone: double; with both articular surfaces placed on the dorsal border of the main body of the bone [0]; double; with the anterior articular surface forming a separate head from the posterior articular surface [1]; [?].

46. Metapterygoid process of hyomandibular bone: absent [0]; present; anterior [1]; present; ventral [2]; [?].

47. Ossified interhyal: present [0]; absent as an independent ossification [1]; [?].

48. Teeth on fifth ceratobranchial: present [0]; absent [1]; [?].
49. First basibranchial in adult specimens: ossified [0]; unossified [1]; [?].
50. Fifth basibranchial in adult specimens: cartilaginous [0]; ossified [1]; [?].
51. First pharyngobranchial in adult specimens: ossified [0]; unossified [1]; [?].
52. Size of opercular bone: normal; about one quater of the head length [0]; expanded; at least one third of the head length [1].
53. Shape of opercular bone in lateral view: rounded/oval [0]; triangular [1].
54. Opercular spines: absent [0]; present [1].
55. Opercular apparatus on external surface of operculum: absent [0]; present [1].
56. Opercular borders: free from side of head [0]; partly or almost completely connected to side of head with skin [1]; [?].
57. Angle formed by preopercular limbs: obtuse [0]; approximately straight [1]; acute [2].
58. Posterodorsal limb of preopercular bone: well developed [0]; reduced; correlated with expansion of anteroventral limb that meets its fellow along the ventral midline [1].
59. Ridge on anteroventral limb of preopercular bone: absent [0]; present [1].
60. Preopercular expansion: absent; preopercular not enlarged [0]; present; restricted to the posteroventral corner [1]; present in posteroventral corner and part of the posterodorsal limb [2]; present in anteroventral limb only [3]; [?].
61. Supraopercular bone: absent [0]; present as a relatively large, flat bone [1]; present as tubular ossification(s) [2].
62. Spine on posterior border of subopercular bone: absent [0]; present [1].
63. Major axis of subopercular bone in lateral view: inclined [0]; subhorizontal [1]; subvertical [2].
64. Subopercular clefts: absent [0]; present [1].
65. Interopercular bone: relatively broad and positioned medioventral to preopercular bone [0]; reduced to a long thin spine and positioned mediodorsal to preopercular bone [1]; reduced to a long thin spine and positioned lateroventral to preopercular bone [2].
66. Spine on posterior border of interopercular bone: absent [0]; present [1].
67. Posterodorsal ascending process of interopercular bone: absent [0]; present [1]; [?].
68. Number of infraorbitals: five or more [0]; four [1]; three or fewer [2].

69. Infraorbital bones not including lacrimal: well developed [0]; reduced to small; tubular ossifications [1]; hypertrophied [2].
70. Lacrimal: flat and comparable in length to subsequent infraorbitals [0]; tube-like and extremely long; without keel [1]; flat; long and large; with keel near lower edge [2]; long and large; with spines and crests [3].
71. Supraorbital: present [0]; absent [1].
72. Two anteriormost vertebrae: as long as posterior ones [0]; shorter than posterior ones [1].
73. Second abdominal centrum: as long as first [0]; shorter than first [1]; [?].
74. Autogenous neural arch anterior to arch of first vertebra: present [0]; absent [1].
75. Neural arch of first vertebra and exoccipitals: separate [0]; in contact [1]; [?].
76. Neural arch of first vertebra and supraoccipital: separated [0]; in contact [1].
77. Spine on the neural arch of first vertebra: present; well developed [0]; present but reduced [1]; absent [2]; [?].
78. Anterior neural arches: no contact with adjoining arches [0]; abutting contact laterally with adjoining arches; no overlapping [1]; overlapping contact with adjoining arches laterally [2]; [?].
79. Unmodified neural arches anterior to dorsal fin in adults: fused to centra [0]; autogenous; at least laterally [1].
80. Neural arches to vertebrae posterior to the dorsal fin in adults: fused to centrum [0]; autogenous; at least laterally [1]; [?].
81. First two anterior parapophyses: autogenous [0]; fused to centra [1].
82. Rib on third vertebral centrum: similar to posterior ones [0]; widened and shortened [1]; modified into Weberian apparatus [2]; [?].
83. Paired intermuscular bones consisting of three series: epipleurals; epicentrals; and epineurals: absent [0]; present [1]; [?].
84. Anterior (first six) epicentral bones: unmodified; no differences in size from others [0]; highly modified; much larger than posterior ones [1]; epicentrals in anterior vertebrae absent [2]; [?].
85. Size and arrangement of anterior supraneurals (whatever the number present): large; separate from each other if more than one supraneural present [0]; larger; in contact with neighbours if more than one supraneural present [1]; supraneurals greatly reduced in size [2]; [?].
86. Posterior process on the posterior border of first supraneural: absent [0]; present [1]; [?].

87. Number of supraneurals: several supraneurals in a long series [0]; two or fewer supraneurals [1]; [?].
88. Postcleithra: present [0]; absent [1]; [?].
89. Lateral line and supracleithrum: supracleithrum pierced through dorsal region [0]; supracleithrum pierced all through its length [1]; lateral line does not pierce supracleithrum [2].
90. Fleshy lobe of paired fins: absent [0]; present [1]; [?].
91. Caudal fin morphology: elongated; posteriorly forked [0]; higher than long; slightly incurved posteriorly [1]; crescent-shaped [2]; [?].
92. Fringing fulcra in dorsal lobe of caudal fin: present [0]; absent [1]; [?].
93. Caudal scutes: absent [0]; present [1]; [?].
94. Ural centra; preural centrum one; and uroneural one: autogenous [0]; fused [1]; fused except for ural centrum two; which is autogenous [2]; [?].
95. Neural arch and spine of preural centrum one: both well developed; spine about half as long as preceding ones [0]; arch complete and closed; spine rudimentary [1]; arch open; no spine [2]; [?].
96. Uroneurals: arranged in a linear series [0]; arranged in a double series [1].
97. Number of uroneurals: three [0]; two [1]; one [2]; [?].
98. Anterior extent of first uroneural: to anterior end of first preural [0]; to anterior end of second preural [1]; to anterior end of third preural [2]; uroneural fused to caudal fin complex [3]; [?].
99. Uroneural two and second ural centrum: in contact [0]; separated [1]; uroneural two absent [2]; [?].
100. Parahypural and preural centrum 1: independent in adults [0]; fused only in large adults [1]; fused since early ontogenetic stages [2]; [?].
101. Reduction in the number of hypurals: six [0]; less than six [1]; [?].
102. Hypurals 1 and 2: independent [0]; partly fused to each other [1]; totally fused to each other [2]; [?].
103. Hypural 1 and terminal centrum: articulating [0]; separated by a hiatus [1]; fused [2]; [?].
104. Hypural 2 and terminal centrum: fused [0]; articulating [1]; [?].
105. Hypural 5: of comparable size to preceding ones [0]; considerably larger [1]; [?].
106. Hypural 5 (plus 6 if present) and second ural centrum: separate [0]; articulating [1]; [?].
107. Haemal arch and preural centrum 2: fused [0]; independent [1]; [?].
108. Posterolateral process of caudal endoskeleton: absent [0]; present [1]; [?].
109. Scales on body: present [0]; absent [1].

110. Type of scales: cycloid **[0]**; modified ctenoid [1].
111. Lateral line: not extending to posterior margin of hypurals [0]; extending to posterior margin of hypurals [1]; [?].
112. Intermandibularis: mainly attaching on the dentary [0]; exclusively attaching to angulo-articular [1]; [?].
113. Protractor hyoidei: not inserting on coronoid process [0]; inserting on coronoid process [1]; [?].
114. Hyohyoideus inferioris of both sides: mostly overlapping each other [0]; mostly mixing mesially with each other [1]; [?].
115. Hyohyoideus abductor: not attaching on pectoral girdle [0]; with significant part of its fibers attaching on pectoral girdle [1]; [?].
116. Adductor profundus: not subdivided into different sections [0]; subdivided into different sections [1]; [?].
117. Attachment of adductor profundus: on first pectoral ray only [0]; on first and second pectoral rays [1]; [?].
118. Most lateral bundles of adductor mandibulae: inserting on mandible and/or primordial ligament [0]; attaching also, or even exclusively, on other bones such as the maxilla or the premaxilla [1]; [?].
119. Position of adductor mandibulae A1-OST: mostly horizontal [0]; with a peculiar anterior portion almost perpendicular to its posterior portion [1]; [?].
120. Section A2 of adductor mandibulae: present [0]; absent [1]; [?].
121. Several small tendons branching off from adductor mandibulae A2: absent [0]; present [1]; [?].
122. Peculiar adductor mandibulae A1-OST-M: absent [0]; present [1]; [?].
123. Direct insertion of adductor mandibulae A2 far anteriorly on the anteromesial surface of dentary: absent [0]; present [1]; [?].
124. Dilatator operculi: mainly mesial and/or dorsal to adductor mandibulae A2 [0]; markedly lateral to A2 [1]; [?].
125. Peculiar tendon of adductor mandibulae A2 running perpendicular to the main body of this section and connecting it to the anteroventral surface of the quadrate: absent [0]; present [1]; [?].
126. Distinct section A3 of adductor mandibulae: absent [0]; present [1]; [?].
127. Adductor mandibulae Aω: present [0]; absent [1]; [?].
128. Adductor arcus palatini: not inserting on preopercle [0]; inserting also on preopercle [1]; [?]. 129.- Levator arcus palatini: not divided [0]; divided into two well-differentiated bundles [1]; [?].
130. Origin of dilatator operculi: on ventrolateral surface of neurocranium [0]; on dorsal margin of cranial roof [1]; [?].

9

Gonorynchiformes in the Teleostean Phylogeny: Molecules and Morphology Used to Investigate Interrelationships of the Ostariophysi

Guillaume Lecointre[1]

Abstract

The phylogenetic position of the Gonorynchiformes is linked to a number of more general phylogenetic problems. From the beginning of phylogenetic systematics in ichthyology, the monophyly of ostariophysans was part of the discussion of the gonorynchiform interrelationships and must be considered first, then the position of the Ostariophysi among teleosts. Results from molecular, as well as morphological, systematics of teleosteans of the last fifteen years showed that the sister group of ostariophysans was the Clupeomorpha. It then became necessary to redefine euteleosts, which, in turn, required an assessment of the position of esocoids among teleosts. Finally, some discussions among ichthyologists about the monophyly of otophysans involved hypotheses about the position of gonorynchiformes and therefore led to a discussion of the phylogeny within ostariophysans. All these problems are reviewed by considering that a molecular phylogeny obtained through a single data set is not enough to falsify a morphological

[1] Equipe "Phylogénie", UMR 7138 CNRS-UPMC-MNHN-IRD "Systématique, Adaptation, Evolution", département "Systématique et Evolution", Muséum National d'Histoire Naturelle, CP26, 57 rue Cuvier, 75005 Paris.

one. A clade is considered reliable when it has been corroborated through several independent sources of data. In other words, morphological synapomorphies are falsified and have to be revised only when they have been contradicted by several independent molecular phylogenies. Gonorynchiformes is the sister group of otophysans, forming the monophyletic Ostariophysi. Ostariophysans and clupeomorphs are sister groups and form the Otocephala. The sister group of the Otocephala is the Euteleostei, among which esocoids are sister group of salmonoids. Finally, something new recently appeared in that general picture. The Alepocephaloidea, which is generally considered the sister group of the Argentinoidea, could be either a member of the Otocephala or the closest sister group of it. However, those two possibilities cannot be considered reliable at present and must be taken as working hypotheses. More independent data are needed to test them.

Reliability of Phylogenies

The Fifty Years that Changed Our Understanding of Teleosts

In their treatise of zoology edited by Pierre-Paul Grassé, Bertin and Arambourg (1958) prophesized:

"Le groupe des téléostéens, en particulier, est tellement polymorphe et polyphylétique qu'il ne sera peut-être jamais possible d'en réaliser une classification naturelle." (The teleostean group, in particular, is so polymorphic and polyphyletic that it probably will never be possible to achieve its natural classification.) Note that here "polyphyletic" means that there are many lineages.

This prediction was soon contradicted. In the late 1960s, systematic ichthyology entered a new era, both because of new techniques of bone, cartilage and nerve staining (Taylor 1967, Dingerkus and Uhler 1977, Filipski and Wilson 1984, Song and Parenti 1995), and because the phylogenetic systematics of Hennig (1950, 1966) was transferred from entomology to ichthyology in the years 1967–1973. Studies were carried out, in the beginning, without matrices and computerized searches of the most parsimonious trees, but the concept of synapomorphy and the subsequent rebuttal of paraphyletic groups were enough to change the way we would subsequently understand teleostean interrelationships. The organisms themselves were rather favourable to these advances, because samples were easily accessible, the fossil record was rich (in comparison with, for instance, those of bats or birds), and new characters were revealed, leading to a reassessment of the group. Moreover, there is another unique feature for this group, its "taxonomic density" (Dupuis 1992), that is, there are many subgroups each composed of numerous closely related species; these groups are diverse and individually well studied by specialists.

Classical landmarks in the literature of the hennigian era concerning large-scale teleostean phylogenetics are Greenwood *et al.* (1966, though still with paraphyletic groups), Nelson (1969), Greenwood, Miles and Patterson (1973), Patterson and Rosen (1977), Moser *et al.* (1984), Johnson and Patterson (1993), Stiassny, Parenti and Johnson (1996), and Arratia (1997, 1999). Voluminous reviews are available from Lauder and Liem (1983), Lecointre (1994), Janvier (1996), and Arratia (2000). Interestingly, they did not really influence books devoted to general zoology (Nelson 1976, 1984, 1994, Kardong 1998, Fishbase.org) or anatomy (e.g., Monod 1968, Fujita 1990) written by non-phylogenetists. Only recently do we find non-paleontological textbooks that fully take into account results from phylogenetic classifications (Tudge 2000, Liem *et al.* 2001, Lecointre and Le Guyader 2001, 2006a, b).

It is remarkable that the increase of phylogenetic knowledge almost followed the teleostean tree from the bottom to the top. Roughly described, the picture of the interrelationships of basal actinopterygians and basal teleosts arose during the 1970s, the relationships among euteleosts during the 1980s, and those among acanthomorphs during the 1990s. We find a sign of this trend in Nelson (1989), writing about teleosts: "Thus, recent work has resolved the bush at the bottom, but the bush at the top persists." The 1990s and the early years of the 21st century are the years of massive contributions of molecular data to systematics. Huge taxonomic samplings and large sequence data sets are changing our views of the tree of life. Completely new, unexpected and reliable clades are emerging in groups of animals (e.g., in placental mammals, squamates or higher teleosts) or plants (e.g., in angiosperms) that were thought to be already well known and well studied. Ostariophysans are among those cases. They were classically considered members of the Euteleostei. However, they became the sister group of clupeomorphs, a non-euteleost clade, and the euteleosts had to be redefined without ostariophysans. This change is one of the topics of the present review. But why prefer this picture over another? This question leads to a discussion of the extent to which phylogenetic science is reliable.

Evolving Concepts of Reliability

If all the comparative data set produced by systematists unambiguously supported a single phylogeny, there would be no need to question the reliability of our phylogenies. There are actually conflicts among data sets, and even conflicts among characters within each data set. It is useful to note with Hillis (1987) that "conflicts among morphological or among molecular studies are probably as common as real conflicts between morphological and molecular studies." We should not always focus on incongruence between molecular data and morphological data. In other words, the classical focus on discrepancy between morphological trees and molecular trees is more sociological than epistemological.

Reviewing the phylogeny of any group implies making choices. One is forced to present the relationships that are thought to be the most reliable. But what is reliability? The reliability of phylogenetic inferences is measured differently depending on what epistemological principles are assigned to systematics. Two main schools can be identified, with their corresponding approaches to data phylogenetic analysis. The "total evidence" approach, under its most extreme form (Kluge 1989, Kluge and Wolff 1993, Siddall 1997), does not allow the formal acceptance of any background knowledge, leading to the refusal of any data partition as natural. Consequently, third and second positions of the codon are not recognized; the reasons why we have selected a particular gene for carrying out a phylogeny are not formally accepted; the distinction between "morphological" and "molecular" data sets is no longer considered valid. So only the *maximization of character congruence* by simultaneous analysis (Nixon and Carpenter 1996) of all the data is considered valid. As all the available data has already been integrated into the analysed matrix, this approach can only rely on branch support or clade robustness (measured with Bremer index or jacknife/bootstrap proportions according to authors) to assess the reliability of our inferences (Fig. 9.1). This reliance exclusively on *coherence* among all the available characters is associated with coherentism by Rieppel (2004a, b, 2005a, b; see also Kearney and Rieppel 2006).

A different approach is to accept background knowledge (Mahner and Bunge 1997: 132), as long as it is explicit and justified. This foundationalist point of view (Rieppel 2003a, b, 2004a, b, Kearney and Rieppel 2006) recognizes arguments for naturalness of data partitions and for the use of models in phylogenetic reconstruction. The accumulated knowledge about the evolution of the nucleotide sequences justifies the assumption of the naturalness of partitions. Various genes evolve in different ways and many authors recommend the use of several genes for performing a phylogenetic inference (e.g., Bielawski and Gold 1996, Baker *et al.* 2001, Seo *et al.* 2005). If the selective pressures characterizing the mutational space at each position are homogeneous within genes and heterogeneous among genes, the observation that a clade is recurrently inferred from independent genes in spite of a possible directionality in the homoplastic changes within each of the individual "process partitions" (Doyle 1992, 1997) can be considered an indication of its reliability (Chen *et al.* 2003, Dettaï and Lecointre 2005, Lecointre and Deleporte 2005). This is also based on the fact that recovering a given clade several times from independent data just by chance is highly

Characters

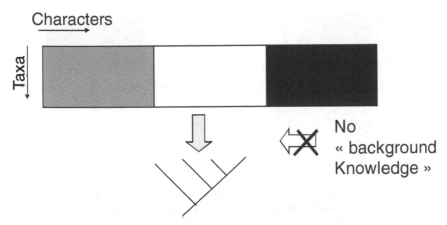

« Total Evidence » (sensu Kluge, 1989) maximizes congruence
among characters («coherentism » of O. Rieppel)

**Reliability of clades = statistical robustness =
sensitivity of clades to data perturbation**

Fig. 9.1 The procedure of simultaneous analysis consists of gathering all the available data
into a single data matrix, following the principle of "total evidence" (Kluge 1989). This principle
does not recognize background knowledge justifying data partitioning and considers that
interrelationships are only known through maximizing congruence among characters. In such
a framework, reliability of a clade is its degree of support (e.g., Bremer Index, branch length)
or its statistical robustness (e.g., bootstrap or jackknife proportion).

improbable (Page and Holmes 1998: 214). The probability of obtaining
exactly the same tree reconstruction artefact from independent genes is also
low. If these facts are recognized, repetition of a clade from different separate
analyses is a better indicator of reliability (Fig. 9.1) than bootstrap values
alone, even when based on the whole available data. Indeed, a tree
reconstruction artefact such as long branch attraction can result in clades
supported by high bootstrap values (Philippe and Douzery 1994) no matter
what the number of characters analysed. Furthermore, comparison of
separate analyses of independent partitions is a good way to detect cases in
which one marker imposes its biases (single positively misleading signal;
Grande 1994, Chen et al. 2000, 2001, 2003) on a part of the tree inferred from
the combined data, resulting in the presence of artefactual branches on that
tree. Consequently, with such an approach genes are recognized as
independent partitions, as well as any morphological or karyological data
set. However, the value of separate analyses for reliability assessment, as
well as for detection of artefactual groupings, is directly dependent on the
number of markers used and on their adequacy to both the group and the
studied divergence times (Chen et al. 2003, Dettaï and Lecointre 2004, 2005).

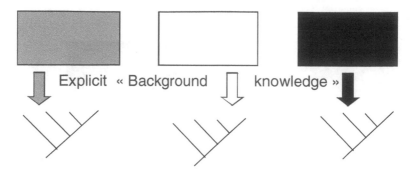

Taxonomic congruence: metalanguage that maximizes coherence among statements about relationships

Two successive coherence-driven processes : among characters, then among statements

Reliability of a clade = repeatability

Fig. 9.2 The procedure of separate analyses fully recognizes background knowledge (Mahner and Bunge 1997) needed for data sets delineations and partitioning. Phylogenetic trees are constructed separately. There are two successive coherence-driven processes. Each tree maximizes coherence among characters of each matrix. Then taxonomic congruence is a summarizing process that maximizes coherence among statements about relationships. Technically this process can use consensus trees, supertrees, or just summary trees that exhibit those clades considered reliable. In such a framework, reliability of a clade is its repeatability across trees obtained from independent data.

Note that this dependence is also valid for any simultaneous analysis. More specifically, separate analyses are more prone to stochastic errors if the data partitions are small. That is why simultaneous analysis also presents advantages: allowing the emergence of phylogenetic signal hidden in separate analyses by marker-specific biases (for an example of such biases, see Chang and Campbell 2000), or stochastic error (review of pitfalls and advantages of both approaches in Miyamoto and Fitch 1995 and Lecointre and Deleporte 2005).

Among results of ostariophysan molecular phylogenetics, the following review will select the clades that are repeated across independent genes and teams (Fig. 9.2) (i.e., reliable clades). If there is no agreement among molecular studies on a topology that was not initially expected from our past morphological knowledge, the classical morphological hypothesis will be preferred. In other words, a single molecular phylogenetic tree obtained from a single molecular data set (however large it is, Steinke *et al.* 2006) cannot falsify a tree based on morphological data. Only several independent congruent molecular phylogenies obtained at same or different times of the long-term process of scientific discovery can do that. In addition, only studies of the era of phylogenetic systematics will be considered (i.e., since Greenwood *et al.* 1966).

The Monophyly of Ostariophysi *sensu* Rosen and Greenwood (1970)

In Greenwood *et al.* (1966), the Ostariophysi was separated from gonorynchiformes, the latter being included in the Protacanthopterygii. The content of the Ostariophysi is that of today's otophysans and the origin of the group is unknown. The taxon is justified by 13 features, most of them being known today as highly homoplastic. Among them are "the otophysic connections involving the intercalation of bony elements" and the "swim bladder primitively subdivided, reduced in many species". Greenwood *et al.* (1966) further mentioned the lack of an ossified basisphenoid and presence of an orbitosphenoid. Rosen and Greenwood (1970) proposed a new taxonomic content for the Ostariophysi that has been corroborated by molecular studies and widely accepted since then. The Gonorynchiformes were included within the new, broader, Ostariophysi. The taxa of the previous Ostariophysi then formed a clade named the Otophysi, after Garstang (1931). Features retained to support the new Ostariophysi were the caudal skeleton anatomy (discussed by Roberts 1973), presence of fright cells and fright substance, chambers of the swim bladder, and histologically disctinctive nuptial tubercles. It may seem surprising not to find the otophysic connection among these features. After having discussed the anatomy of these connections in otophysans and gonorynchiformes, the authors concluded that the common ancestor must have had a number of cephalic ribs differentiated from the epineural and epicentral series, an enlargement of the first pleural rib; however, the ways to achieve the otophysic connection were considered different in both groups.

This definition of ostariophysans was "considered with reservations" by Roberts (1973), but supported by new arguments from Fink and Fink (1981), with 15 characters, among which six are homologies in the otophysic connection (i.e., linked to the structure of the gas bladder, anteriormost pleural ribs and otic region). From this starting point, the gonorhynchiform condition, in which the expanded first pleural rib articulates with the third vertebra and supports a thickened peritoneum that partly invests the anterior chamber of the swim bladder, appeared to many authors to have occurred along the evolution of the Weberian apparatus of the otophysans (Lauder and Liem 1983). Gayet (1982, 1986a, b) recognized that the gonorynchiformes were more closely related to otophysans than to any other teleostean group, but argued for another point of view (i.e., gonorynchiforms are the sister group of cypriniforms *sensu stricto*). She recognized five (over 15) of Fink and Fink's ostariophysan synapomorphies, one of them linked to the otophysic connection (Gayet 1993: 170, Gayet 1986a: 73): the fact that the peritoneal tunic of the anterior chamber of the gasbladder is attached to the anteriormost two pleural ribs. From that feature and anatomical analysis of fossil material, Gayet (1982: 50, 1986a: 74) considered that the otophysic connection, and even the Weberian apparatus, could have occurred separately in different ostariophysan groups. Fink and Fink (1996) revised their own data, including the synapomorphies defining the Ostariophysi. They maintained the monophyly of the Ostariophysi with the following synapomorphies:

— the loss of basisphenoid;
— the position of sacculi and lagenae more posterior and nearer to midline (but see Grande and de Pinna 2004: this could be an otocephalan synapomorphy);
— absence of dermopalatine;
— gasbladder with anterior and posterior chambers, anterior chamber partly or completely covered by a shiny peritoneal tunic (but see also Grande and de Pinna 2004);
— peritoneal tunic of anterior chamber of gasbladder attached to anterior two pleural ribs (but see also Grande and de Pinna 2004);
— dorsal mesentery suspending the gasbladder heavily thickened anterodorsally near its attachment to the vertebral column;
— absence of supraneural anterior to the first neural arch, roof over neural canal formed by anterior neural arch;
— haemal spines of preural centrum 3 and anterior vertebrae fused with their respective centra (though this character has also been proposed as a possible otocephalan synapomorphy, see below);
— alarm substance;
— keratinized nuptial tubercles;
— keratinous unicellular projections of the epidermis (uniculi).

From Arratia (1999), two additional characters were coded "another condition":

— Supramaxillae not dorsal to the dorsal margin of maxilla;
— First uroneural fused with a compound centrum apparently formed by preural centrum 1 and ural centrum (or centra). This feature appears as an ostariophysan synapomorphy as long as the pleurostyle is no longer recognized as an otocephalan synapomorphy (see below).

Grande and de Pinna (2004) proposed/revised all synapomorphies for the clade Ostariophysi linked to the Weberian apparatus, but each of them was homoplastic. The gasbladder with a silvery peritoneal tunic covering at least the anterior chamber was also found in the Clupeoidei. The anterior 1–2 vertebral centra shorter than posterior ones were also found in the Pristigasteroidea. The first pleural rib attached to the peritoneal tonic of the gasbladder was also found in the Pristigasteroidea and the Coiliidae. The second pleural rib attached to the peritoneal tunic of gasbladder was also found in the Pristigasteroidea. The constriction of the gasbladder into two chambers was also found in the Pristigasteroidea and the Engrauloidea. Interestingly, these homoplastic features were shared by clupeomorph components, showing parallelisms in otophysic connections of the two groups (see below): different elements of the Weberian apparatus may have evolved in different groups, but only became functional all together as a "Weberian apparatus" in the Otophysi.

The monophyly of the ostariophysans *including Gonorynchiformes* has never been seriously challenged since then. Inoue *et al.* (2004) performed partitioned bayesian analyses of DNA sequences of 12 mitochondrial protein-coding genes, 22 tRNAs genes and 2 rDNA of 30 teleosts. From these "mitogenomic" data they recovered, among otocephalans, two gonorynchiformes more closely related to clupeomorphs than to the two otophysans sampled. However, the support for that hypothesis of paraphyletic ostariophysans was low. The "mitogenomic" data from Saitoh *et al.* (2003) presented a better sampling of otophysans: all otophysan orders and gonorynchiformes were present. The same kind of analysis led to the same results. The node challenging the ostariophysans was, however, poorly supported and cannot really be a serious alternative hypothesis. Unfortunately, Peng *et al.* (2006) used that topology to estimate divergence times among otocephalans from nearly the same mitochondrial DNA sequence data. They noted, though without fully recognizing all of the consequences, that a Kishino-Hasegawa test performed on these mitogenomic data did not significantly distinguish the two phylogenetic hypotheses (i.e., gonorynchiforms as the sister group of clupeomorphs *versus* gonorynchiforms as the sister group of otophysans). Using mitogenomic data and a taxonomic sampling focusing on the definition of the Protacanthopterygii (Ishiguro *et al.* 2003, Fig. 9.3l) found paraphyletic ostariophysans with *Gonorynchus* more closely related to the two clupeomorphs *Engraulis* and *Sardinops* than to the three otophysans *Crossostoma*, *Cyprinus* and *Carassius*. Moreover, the Alepocephaloidei (*Alepocephalus* and *Platytroctes*) were inserted between otophysans and the two other groups (surprisingly alepocephaloids were deleted from the same data used by Peng *et al.* 2006). However, these relationships were not highly supported. A far better ostariophysan molecular sampling was reached by Lavoué *et al.* (2005, Fig. 9.3m), which included 13 otophysans and seven gonorynchiformes sequences of the mitogenome and

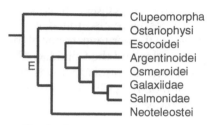

Fig. 3a. Rosen, 1973, 1974.

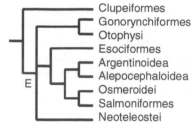

Fig. 3b. Rosen, 1982.

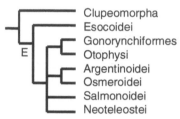

Fig. 3c. Fink and Weitzman, 1982.

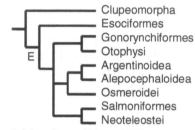

Fig. 3d. Lauder and Liem, 1983.

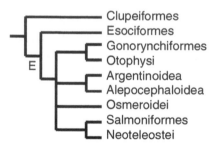

Fig. 3e. Fink, 1984.

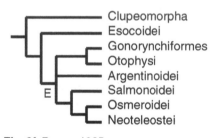

Fig. 3f. Rosen, 1985.

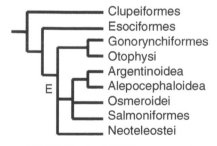

Fig. 3g. Sanford, 1990.

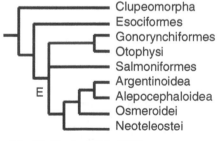

Fig. 3h. Begle, 1991, 1992.

Fig. 9.3 contd....

Fig. 9.3 contd....

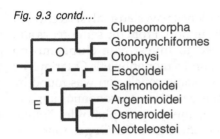

Fig. 3i. Lecointre, 1995 (after Lê *et al.* 1993), Lecointre and Nelson, 1996.

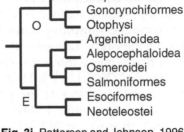

Fig. 3j. Patterson and Johnson, 1996.

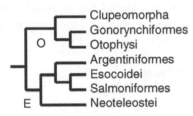

Fig. 3k. Zaragüeta-Bagils *et al.* 2002.

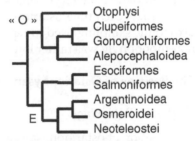

Fig. 3l. Ishiguro *et al.* 2003.

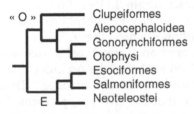

Fig. 3m. Lavoué *et al.* 2005

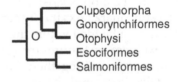

Fig. 3n. Arratia, 1997, 1999.

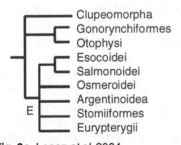

Fig. 3o. Lopez *et al.* 2004.

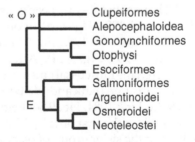

Fig. 3p. Lavoué *et al.* 2008

Fig. 9.3 Cladograms showing the various hypotheses given for clupeocephalan interrelationships. E: Euteleostei. O: Otocephala. "O": Otocephala *sensu lato* (i.e., admitting an additional member).

mitochondrial ribosomal DNA. The partitioned bayesian analysis recovered a monophyletic Ostariophysi with high support. From the same genes and a taxonomic sampling more designed to investigate clupeomorph intrarelationships, Lavoué *et al.* (2007) found a monophyletic Ostariophysi *sensu* Rosen and Greenwood (1970). The same conclusion was found by Lavoué *et al.* (2008, Fig. 9.3p) from the same genes and a taxonomic sampling designed to investigate the position of the Alepocephaloidei among clupeocephalans. From 1,444 base pairs of the nuclear RAG1 gene and 12S-16S mitochondrial rDNA sequenced for an appropriate sampling of 50 teleosteans, Lopez *et al.* (2004, Fig. 9.3o) also found a strong support for monophyletic ostariophysans *sensu* Rosen and Greenwood (1970). To summarize, each time ostariophysans were not found monophyletic, the taxonomic sampling used to represent them was poor and the support for non-monophyly was rather low.

The Position of Ostariophysans among Teleosts

Confusing Beginnings

The position of ostariophysans among teleosts has always been related to the definition of euteleosts. In Greenwood *et al.* (1966), the "Ostariophysi" (actually the otophysans) have an unknown origin. In their famous fig. 1, a question mark links the "ostariophysans" to the gonorynchiform lineage, which is embedded within the protacanthopterygian cluster. However, in the text there are features recognized as common to gonorynchiformes and "ostariophysans" together: "the caudal fin skeleton, the occipitocervical specializations, and the trend towards a divided swim bladder." These features were interpreted as suggesting that the two groups formed a "derivation from a common stem, the dichotomy being very near the base." The authors concluded: "The stem itself was probably derived from some ancestral salmoniform." The Ostariophysi *sensu* Rosen and Greenwood (1970) have subsequently been considered the most basal euteleostean group (Rosen 1973, 1974, Fig. 9.3a), or one of the most basal (esocoids being within euteleosts in Rosen 1982, Fig. 9.3b, esocoids being excluded from euteleosts in Rosen 1985, Fig. 9.3f). Alternatively, the esocoids have also been considered the most basal euteleosts, ostariophysans being embedded within a multifurcation (Fink and Weitzman 1982, Fig. 9.3c; Lauder and Liem 1983, Fig. 9.3d; Fink 1984, Fig. 9.3e; Begle 1992, Fig. 9.3h). A challenging picture proposed for the first time by Lê *et al.* (1993) and Lecointre (1993, 1995, Fig. 9.3i), that the sister group of ostariophysans was a non-euteleost clade, the Clupeomorpha, was later confirmed (see below).

This new picture contradicted the clade Euteleostei as defined by Rosen (1985), but it was not a serious conflict. Indeed, the euteleostean clade has generally been considered poorly defined (see comments in Patterson and Rosen 1977, Lauder and Liem 1983, Fink 1984). Patterson and Rosen (1977) retained the stegural, the presence of nuptial tubercles on the head and body, and an adipose fin posterior to the dorsal fin. Rosen (1985) came back to the euteleostean concept as defined by Patterson and Rosen (1977) and retained only one synapomorphy—the adipose dorsal fin—therefore excluding esocoids from his new Euteleostei because they lack the adipose fin. This synapomorphy could be considered doubtful because esocoids have the dorsal fin situated very posteriorly and could have simply lost the adipose fin. Moreover, presence of an adipose fin in euteleosts is homoplastic, lacking in gonorynchiformes, cypriniformes, and extant gymnotoids and sparsely spread among neoteleosts.

The problem of the sister group of the Ostariophysi was related not only to the definition of the Euteleostei, but also to the relationships of esocoids, argentinoids, and clupeomorphs. Before 1993, the relationships of ostariophysans among teleosteans remained imprecise. The well-established points were (1) ostariophysans were clupeocephalans (Patterson and Rosen 1977) and (2) they were not neognaths (Rosen 1973). Their branching point also depended on the imprecise relationships of esocoids and argentinoids. Fink and Weitzman (1982, Fig. 9.3c) proposed to branch ostariophysans within a multifurcation containing three other clades: (1) salmonids, (2) a new clade grouping argentinoids and osmeroids, and (3) neoteleosts. Esocoids were the sister group of the whole, but the authors admitted that they could be included within this multifurcation. Lauder and Liem (1983, Fig. 9.3d) reduced this multifurcation to (1) ostariophysans plus two other clades: (2) (argentinoids + osmeroids) and (3) (salmonids + neoteleosts = neognaths). Esocoids were the sister group of the whole. Fink (1984, Fig. 9.3e) proposed another multifurcation of four clades: (1) ostariophysans, (2) argentinoids, (3) osmeroids, and (4) neognaths; esocoids being the sister group of the whole. Rosen (1985, Fig. 9.3f) proposed a trifurcation within his euteleosteans: (1) ostariophysans, (2) argentinoids, (3) neognaths (comprising salmonids, osmeroids and neoteleosts), esocoids being excluded from euteleosts. Sanford (1990, Fig. 9.3g) proposed a multifurcation of three clades: (1) ostariophysans, (2) neoteleosts, and (3) a clade made of argentinoids, alepocephaloids, osmeroids and salmoniforms; esocoids being the sister group of the whole. Finally, Begle (1991, 1992, Fig. 9.3h) proposed synapomorphies supporting the monophyly of the Argentinoidei, the monophyly of the Osmeroidei, and the monophyly of the Osmerae, a clade uniting argentinoids and osmeroids. This author considers the Osmerae the sister group of the Neoteleostei, the subsequent clade being comprised within a multifurcation also including salmonids and ostariophysans. Begle's matrices have been revisited by Johnson and Patterson (1996, Fig.

9.3j). These authors found the Osmeroidei the sister group of the Salmonoidei, and the Argentinoidei the sister group of both.

Relationships of esocoids were as poorly known as those of ostariophysans. Rosen (1973) placed esocoids "provisionally" close to the salmonoids, as previously accepted (Greenwood *et al.* 1966). Rosen (1974) placed esocoids within his "salmoniformes" (the "protacanthopterygians" of Greenwood *et al.* (1966) but reduced to argentinoids, galaxiids, esocoids, salmonids and osmeroids), as the sister group of his Salmonae (Fig. 3a). Then, the Salmoniformes of Rosen (1974) were split up by Fink and Weitzman (1982: 86, Fig. 3c), who notably excluded the esocoids ("We find no evidence to consider esocoids closely related to the other members of the group") without finding their relationships: "Esocoids seem to share no unique specializations with the other included taxa (Ostariophysi, Argentinoidei, Osmeroidei (including galaxiids), Salmonidae, Stomiiformes); we could list esocoids as sedis mutabilis at the euteleostean level or as the sister group of all other teleosts, depending on placement of the ostariophysans." Because Rosen (1985, Fig. 9.3f) defined his euteleosteans with the dorsal adipose fin, esocoids were excluded and were collapsed in a clupeocephalan three-taxa multifurcation: (1) clupeomorphs (2) esocoids, and (3) euteleosteans. Later, Begle (1991, 1992, Fig. 9.3h) provided synapomorphies linking Argentinoidei and Osmeroidei, a grouping already suggested by Fink and Weitzman (1982), and in his scheme placed esocoids as the sister group of euteleosts. A completely different result was obtained by Johnson and Patterson (1996, Fig. 9.3j), where argentinoids were the sister group of alepocephaloids, salmonoids the sister group of osmeroids and esocoids the sister group of neoteleosts.

Following the conclusions of Begle (1991, 1992) and rejecting the presence of a dorsal adipose fin as a reliable synapomorphy of euteleosteans (the character "6" of Rosen 1985) (see Lecointre and Nelson 1996), this leads to the poorly resolved situation of the early 1990s where clupeocephalan interrelationships were a multifurcation involving four clades: (1) clupeomorphs, (2) esocoids, (3) ostariophysans, and (4) neognaths (Salmonoidei, Osmerae and Neoteleostei). At that time, the first molecular phylogenies provided new hypotheses.

Ostariophysi with Clupeomorphs: Molecular Evidence

The first molecular results showing clear relationships of ostariophysans among teleosts were those of Lê *et al.* (1993). In a parsimony-bootstrap tree containing 38 species they found a strong grouping of the clupeines with otophysans (bootstrap proportion of 92 percent) from 190 informative sites from 28S rRNA sequences. This grouping included two clupeomorph species (*Sardina, Clupea*) and three otophysan species (*Tinca, Gobio, Ictalurus*). The corresponding sequences have been found to evolve slightly faster than others. This raised doubts about the interpretation of that grouping, because long-branch attraction artefact was suspected by the authors. At that time the authors had a single molecular data set at hand and had to be prudent in their conclusions. These results were nevertheless fruitful because they pointed out the very poor definition of the Euteleostei *sensu* Rosen (1985) and Lauder and Liem (1983). Lê *et al.* (1993: 47) concluded that if the same grouping was confirmed from a wider taxonomic sampling of each of the two sister groups, then "the possibility of such relationships should be seriously considered by morphologists". The confirmation came from the work of Patterson *et al.* (1994, communication at the fish phylogeny meeting of the 13th Willi Hennig Society meeting), who sequenced the 18S rRNA of a collection of teleosts. Their unpublished results showed an extraordinarily high number of molecular synapomorphies shared by *Chanos* and *Clupea*, in a 31–species-Hennig 86–tree. The number of taxa sampled in each of the two sister groups was very small, but the only ostariophysan species represents the earliest branch of the group, the gonorynchiforms. The two studies (Lê *et al.* 1993 sampling otophysans, Patterson *et al.* 1994 sampling gonorynchiforms) reanalysed by Lecointre (1995) and Lecointre and Nelson (1996, Fig. 9.3i) converged to the same conclusion: the sister-group relationship between clupeomorphs and ostariophysans.

Very soon, other molecular data were collected. Müller-Schmid *et al.* (1993) found "no indication for the existence of a Euteleost infradivision" from about 200 sites of the amino-acid sequences of the ependymins of six teleost species. Four families were sampled: Clupeidae, Cyprinidae, Esocidae, and Salmonidae. They found a different sister group of the clupeids depending on the tree-construction method used, each contradicting the euteleost clade. The UPGMA-bootstrap tree apparently showed a clupeomorph-ostariophysan sister-group relationship under the form ((Clupeidae, Cyprinidae) (Esocidae, Salmonidae)). However, these results are due to bias. UPGMA arbitrarily roots at the midpoint of the tree. The authors indicated a higher rate of evolution in salmonid and esocid ependymin sequences (*sic*) while using UPGMA. It is thus obvious that the arbitrary midpoint rooting was automatically set on the internal long branch and led to the apparent grouping of salmonids and esocids on the one side, clupeids and cyprinids on the other side. It is actually impossible to test the monophyly of euteleosts with these data using *Neighbor Joining* and/or parsimony procedures because one has to choose a root. In phylogenetics roots are justified according to previous knowledge. If one considers that the tree must be rooted on the Clupeidae, one would automatically obtain monophyletic euteleosts. The authors found the apparent clupeid-cyprinid sister-group relationship because they left the software root at the midpoint and did not justify it. Worse yet, by evoking evolution rate inequalities, which is precisely the case when midpoint rooting should be avoided, they annihilate their own claims. *Neighbor Joining* and parsimony trees are lacking in the paper, although the authors indicated that these two procedures were used and led to an irresolution (*sic*), without giving any information about the outgroup chosen in those cases. Because of the phenetic procedure used the cyprinid-clupeid grouping is likely to be due to either symplesiomorphies or undetected otocephalan synapomorphies; this paper is clearly inconclusive. Curiously, the same journal published in 2006 a paper seemingly based on a poor taxonomic sampling. The study concluded that euteleosts were not supported by the "phylogenomic analysis" of 10 fish models (Steinke *et al.* 2006). Both euteleosts *sensu* Rosen (1985) and euteleosts as redefined below appeared contradicted because salmonids were the sister group of ostariophysans in a 10–taxon tree rooted on *Homo*. This "phylogenomic analysis of a set of 42 orthologous genes from 10 available fish model systems (...)" is clearly a data-driven study. The taxonomic sampling was not designed to address an interesting systematic question; it was rather an exploitation of the expressed sequence tags (ESTs) made available in sequence data banks. The aim of the authors was "to test the power of multilocus approaches to reveal phylogenies". However, when the approach did not give the expected tree, they did not conclude that the approach was inappropriate but instead challenged current classifications. Whatever the logical and technical problems with this study, the main weakness is the taxonomic sampling. The best explanation of the results is that the too distant outgroup attracted atherinomorphs and tetraodontiforms to the base, artificially making salmonids the sister group of ostariophysans in the absence of other teleosteans. Indeed, several atherinomorphs and tetraodontiforms generally exhibit long branches from many nuclear genes. Large amounts of sequence data (sequence length) do not prevent long-branch attractions if the taxonomic sampling remains poor. This artefact is even more powerful when the number of terminals is low (Lecointre *et al.* 1993, Philippe and Douzery 1994).

Zaragüeta-Bagils *et al.* (2002, Fig. 9.3k) found repeated ostariophysan-clupeomorph sister-group relationships from five different genes analysed separately (28S rDNA, 18S rDNA, MLL amino-acid sequences, RAG1 amino-acid sequences, Rhodopsin gene DNA sequences), the sixth (12S–16S mitochondrial rDNA) providing no resolution. From 1,444 base pairs of the RAG1 gene, and 12S–16S mitochondrial rDNA, Lopez *et al.* (2004, Fig. 9.3o) found no resolution at the corresponding depth of the teleostean tree. Since then, the clupeomorph-ostariophysan relationship was confirmed by mitochondrial DNA sequences of 12 protein-coding genes, 22 tRNAs genes and 2 rDNA in various analytical contexts and taxonomic samplings, focusing on actinopterygians in Inoue *et al.* (2003), ostariophysans in Saitoh *et al.* (2003), or elopomorphs in Inoue *et al.* (2004). These "mitogenomic data" brought some novel, unexpected results. Focusing on the concept of "protacanthopterygii", Ishiguro *et al.* (2003, Fig. 9.3l) found the Alepocephaloidei within the clade Clupeomorpha + Ostariophysi. However, as the corresponding branch support was low, it seemed not to deserve much attention. Later, from the same type of data and better taxonomic sampling, Lavoué *et al.* (2005, Fig. 9.3m; 2007) confirmed the monophyly of clupeomorphs and the monophyly of ostariophysans, but the Alepocephaloidei was again among the Otocephala as the sister group of ostariophysans, and the Clupeomorpha was the sister group of the clade Alepocephaloidei + Ostariophysi. In other words, clupeomorphs and ostariophysans were still closely related, but the Alepocephaloidei was inserted in between. From the same kind of data with a better taxonomic sample of the Alepocephaloidei, Lavoué *et al.* (2008) found the same results, but the clade Alepocephaloidei + Ostariophysi was not so highly supported as in Lavoué *et al.* (2007). It is still possible that the Otocephala is monophyletic with the Alepocephaloidei as its sister group (Fig. 9.3p). Such new results were obtained from a single type of data, the "mitogenomic" data, and the exact position of the Alepocephaloidei seems sensitive to taxonomic sampling effects. For the moment it is more prudent to consider it a working hypothesis (dashes in Fig. 9.5) that needs to be confirmed (as suggested by Lavoué *et al.* 2005), and preferably by nuclear genes.

Ostariophysans with Clupeomorphs: Anatomical Evidence

Morphological evidence contrasts with molecular evidence in the sense that it is far more ambiguous with regard to the question of ostariophysan interrelationships. After the first molecular results (Lê *et al.* 1993), a sister-group relationship of clupeomorphs and ostariophysans was proposed from anatomical data by Lecointre (1993, 1995, Fig. 9.3i), Lecointre and Nelson (1996, Fig. 9.3i), Johnson and Patterson (1996, Fig. 9.3j), Arratia (1997, 1999, Fig. 9.3n), Zaragüeta-Bagils *et al.* (2002, Fig. 9.3k), and Grande and de Pinna (2004).

These findings imply a criticism of the synapomorphies given for the "classical" euteleosteans (Rosen, 1985, Fig. 9.3f) because although ostariophysans had always been considered euteleosts, clupeomorphs have not. They also imply a new definition of the Euteleostei and a review of the possible synapomorphies for the new clade Clupeomorpha + Ostariophysi.

Criticism of the Usual Definition of Euteleosteans

Nuptial Breeding Tubercles on the Head and Body. This character (Patterson and Rosen 1977: 130; Lauder and Liem 1983) is too variable to define euteleosts in the sense of Rosen (1985). Salmonids and osmeroids have nuptial tubercles, while argentinoids do not (Patterson and Rosen 1977, Fink 1984). This character is not shared by all ostariophysans: it is present in some gonorynchiforms (e.g., *Phractolaemus*), a large number of cypriniforms, and a few characiforms and siluroids (Fink and Fink 1981). This character is rejected by Rosen (1985) as a synapomorphy of euteleosteans. Fink (1984) suggests that it could be a synapomorphy of a clade uniting ostariophysans, salmonids and osmeroids, but he also considered that too many other characters contradicted this grouping.

Stegural. In the caudal skeleton, an outgrowth of laminar bone (or membranous lamina) is defined extending anteriorily from the anterodorsal margin of the first uroneural. The exact definition of this character as a euteleostean synapomorphy is "anterior membranous outgrowth of first uroneural (stegural) not joining opposite member in midline" (Patterson and Rosen 1977). When the stegural is impaired and median (as in clupeomorphs), it is considered a synapomorphy of the clupeocephalan clade (Patterson et Rosen 1977, Rosen 1985; but criticized by Arratia 1991). When there are "paired stegural outgrowths of the first uroneural" (Rosen 1985), it is a synapomorphy of neognaths, a clade uniting salmonids, osmeroids, and neoteleosts. The paired stegural cannot be taken as a definition of euteleosts in the classical sense (i.e., including esocoids and ostariophysans: Patterson and Rosen 1977). Indeed, the anterodorsal outgrowth of the first uroneural of esocoids may not be homologous to the outgrowth of other euteleosts (Rosen 1985). Moreover, ostariophysans do not have the stegural (Monod 1968, Rosen and Greenwood 1970). In argentinoids, a supraneural lamina identified by Monod (1968, p. 304) as a stegural may not be homologous to the stegural of salmonids (Greenwood and Rosen 1971). Indeed, this lamina in argentinoids is linked to the underlying vertebral centrum through one or two rudiments of neural arches and is rarely fused with the anterodorsal margin of the first uroneural (Greenwood et Rosen 1971: 14, 20, 21, fig. 12). Consequently, and according to Rosen's (1985) cladogram (Fig. 9.3f), the paired stegural from the first uroneural is found exclusively in neognaths.

Adipose Fin Posterior to the Dorsal Fin. This character was considered by Rosen (1985) to be the only one that could define euteleosts (excluding esocoids because they lack an adipose fin). However, it remains too variable a character at the taxonomic level of interest. The adipose fin is not regularily distributed among ostariophysans: it is absent in gonorynchiforms, cypriniforms and extant gymnotoids, it is present in characiforms, siluroids and fossil gymnotoids. The adipose fin is present in argentinoids, salmonids, osmeroids, but lost in most of the neoteleostean clades.

The ostariophysans-clupeomorphs sister-group relationship contradicts another synapomorphy that "ostatiophysans share with all non-esocoid euteleosts" (Fink and Weitzman 1982): the loss of the dentigerous toothplate over the fourth basibranchial. Clupeomorphs (Nelson 1970: 2, 7), and esocoids (Rosen 1974: 274) retain this plate, while ostariophysans and neognath groups have lost it (Fink and Weitzman 1982). Most of the neognath fishes have also lost the whole fourth basibranchial (see for instance the basibranchial elements in Salmonids in Rosen 1974: 275), which makes the interpretation of this character difficult.

Possible Morphological Synapomorphies for the Clade Clupeomorpha + Ostariophysi

Presence of a Pleurostyle in the Caudal Skeleton. The pleurostyles in these organisms are paired lateral processes that extend dorsally and posteriorily from preural centrum 1 and are not homologous with the pseudurostyles of acanthopterygians (Monod 1967). They were found by Monod (1967, 1968) to be exclusively present in "clupes" and ostariophysans (including gonorynchiforms). Fujita (1990) also recorded the presence of pleurostyles exclusively in clupeomorphs and ostariophysans, in every taxa from the families he studied (Clupeidae, Engraulididae, Chirocentridae, Chanidae, Gonorynchidae, Cyprinidae, Cobitidae, Gasteropelecidae, Characidae, Ictaluridae, Bagridae, Siluridae), except Chacidae and Plotosidae. Neither Monod (1968: 128, 275, 597, 599) nor Grande (1985: 259) found pleurostyles in *Denticeps*, while Fujita (1990) did not include *Denticeps* in his taxonomic sample. Monod (1968) insisted that the caudal skeleton of *Denticeps* was in many traits (pp. 275, 599) closer to that of salmonids ("type I") than of "clupes" ("type II"). His synthetic (basically phenetic) fig. 856 (*Denticeps*, p. 597) is more similar to fig. 859 (*Salmo*, p. 601) than to fig. 853 (*Clupea*, p. 597) and opens the question of the relationship of *Denticeps* (p. 275). Grande (1985: 258) considered that "the general shape of the *Denticeps* uroneural, and the slight branched appearance of its posterior end suggest that the three uroneurals have fused in Denticipitoids", while "all other Clupeomorphs observed have three uroneurals". Grande considers the pleurostyle as the first uroneural (see for instance Grande 1982b: 7, 13, 19). Uroneural 1 in hypothetical ancestors of *Denticeps* did not give rise to a pleurostyle, but to another single structure with the uroneurals 2 and 3. Grande (1985: 258) answered the above question of the relationships of *Denticeps* by concluding: "Although *Denticeps* has many peculiar skeletal features, it is clearly a clupeiform Clupeomorph because of the presence of pterotic and prootic bullae, abdominal scutes, and recessus lateralis. The many peculiarities of the skull and caudal skeleton are independently derived features of Denticipitoids." Lavoué *et al.* (2005, 2007) from "mitogenomic" data, Di Dario and de Pinna (2006) from the pattern of ramification of cephalic sensory canals and Diogo and Doadrio (2008) from cephalic and pectoral girdle muscles all fully confirmed that *Denticeps* is a clupeomorph. Pleurostyles are also absent in the fossil clupeomorph †*Diplomystus* (Patterson and Rosen 1977: 138; Grande 1982a) but present in other fossil clupeomorphs, for instance in †*Knightia* and †*Gosiutichthys* (Grande 1982b). Monod (1968: 99) noticed that in †*Diplomystus brevissimus* urodermals 1 and 2 contact preural centrum 1 (urodermals are Patterson's and Fujita's uroneurals, Monod 1994, pers. comm.). From this observation, he considered that the sole fusion of the second pair of urodermals (UN2) with preural centrum 1 could form a pleurostyle. Only that fusion would virtually make of †*Diplomystus* caudal skeleton a typical clupeomorph one. This interpretation of the uroneurals of †*Diplomystus* is an interesting conjecture, but leads to two problems. First, the UN1 and UN2 in †*Diplomystus longicostatus* do not reach preural centrum 1 proximally (Patterson and Rosen 1977), while UN2 does in †*D. dentatus* and †*D. birdi* (Grande 1982a). Second, Monod (1968) contradicted his own nomenclature, because while supposing that the origin of the pleurostyle is the fusion of uroneural 2 ("urodermal 2") to preural centrum 1, independent bones called "urodermals 1 and 2" are identified in all clupeomorph caudal skeletons of his study, with the presence of a well-developed pleurostyle. It is therefore more appropriate to follow Grande and Patterson's nomenclature for the same bones (UN2 and UN3), supposing (as Grande did implicitly) that the pleurostyle comes from the fusion of the first pair of uroneurals (and not the second) with the preural centrum 1. Arratia (1999) kept the fusion of UN1 to the centra as a synapomorphy of ostariophysans (her character 138 optimized onto her fig. 19) while keeping a separate character for the "pleurostyle" (her character 141). These authors generally accept the uroneural origin of the pleurostyles, but an overview of their studies shows two difficulties. Primary homologies among the different pleurostyles found in modern clupeomorphs and modern ostariophysans depend on bone nomenclature (also discussed in Arratia 1999: 307); and pleurostyles as a synapomorphy for the group Clupeomorpha + Ostariophysi is difficult to maintain as long as *Denticeps* and a number of fossil clupeomorphs (†*Diplomystus*, †*Paraclupea*, †*Santanaclupea*) and gonorynchiforms (†*Dastilbe*, †*Tharrias*) lack a pleurostyle (Arratia 1997, 1999). For these reasons Zaragüeta-Bagils *et al.* (2002), although mistaken about the absence of a pleurostyle in gonorynchiformes, considered this character convergent in ostariophysans and clupeioid clupeomorphs. Neither Johnson and Patterson (1996) nor Arratia (1997,

1999) retained this character as a synapomorphy for the clade called "otocephala" by the former and "ostarioclupeomorpha" by the latter.

Fusion of Hypural 2 with Ural Centrum 1. The caudal skeletons of a great variety of teleosts presented in the studies of Monod (1968) and Fujita (1990) show that unlike other extant teleosts, clupeomorphs (even *Denticeps*) and otophysans all have the second hypural fused to ural centrum 1. The survey of Fujita (1990: 824–845) precisely shows that except in cases of global fusion of the caudal skeleton in some teleosts, fusion of these two bones is restricted to clupeomorphs and otophysans. However, this fusion is not present in all anotophysans. It is present in *Gonorynchus* (Monod 1968: 199), but not in *Chanos* (Monod 1968: 70; Fink and Fink 1981: 340A; Fujita 1990: 211), *Phractolaemus* (Monod 1968: 71), or *Kneria* (Gayet pers. comm.). This fusion is found in some fossil clupeomorphs (†*Diplomystus*, Patterson and Rosen 1977, Grande 1982a; †*Ellimmichthys*, †*Knightia*, Grande 1982a, b; but not in †*Sorbinichthys elusivo*, Bannikov and Bacchia 2000) and in some fossil ostariophysans (†*Ramallichthys*, †*Lusitanichthys*, Gayet 1986; †*Chanoides*, Patterson 1984; but not in †*Prochanos*, †*Rubiesichthys*, †*Tharrhias*, Gayet pers. comm.). This fusion is also found in some Jurassic non-clupeomorph teleosts (Arratia 1991). Arratia (1997: 155) retained the feature "primitively one hypural fuses to the first ural centrum" as a synapomorphy for the "Ostarioclupeomorpha" (= "Otocephala"), but problems of coding this character were pointed out by Patterson (1998) and Zaragüeta-Bagils *et al.* (2002). This lack of stability among teleosts led Johnson and Patterson (1996), Arratia (1999: 326) and Zaragüeta-Bagils *et al.* (2002) to reject this feature as a synapomorphy of the Otocephala.

Fusion of (Mesial?) Extrascapulars with Parietals. Basically, the supratemporal commissural sensory canal extends transversally from one pterotic to the other passing through the extrascapulars (for instance in *Lepisosteus, Amia,* the anabantid acanthomorph *Ctenopoma* (Daget 1964: 272). When this canal passes through the parietals (sometimes received laterally from the supratemporals), at least one of the extrascapular bones is supposed to have fused with parietals (Patterson 1970). As illustrated by Gayet (1986a: 16–17), this canal passes through the parietals in extant gonorynchiforms like *Chanos,* and *Phractolaemus* (not observed in *Kneria* because of the loss of parietals), the fossil cypriniform *Beaufortia,* the fossil gonorynchiform †*Ramallichthys* (p. 13). This is also clearly shown in the fossil otophysan †*Chanoides* (Patterson 1984: 434–435), and the extant characiforms *Brycon* (Daget 1964: 272), *Acestrorhynchus, Ctenolucius, Hoplias, Hepsetus* (Roberts 1969). The "supratemporal commissural sensory canal primitively passing through parietals and supraoccipitals" is considered by Grande (1985: 326) as a synapomorphy of the clupeomorpha. This opinion is based on a large survey of fossil and extant clupeomorphs and is illustrated in Grande (1985) with *Dorosoma* and *Odaxothrissa*, otherwise in Patterson's study of the fossil Clupeomorph *Spratticeps* (Patterson 1970) or the study of *Denticeps* by Di Dario and de Pinna (2006). Arratia and Gayet (1995) also found this feature in various ostariophysans. As a consequence, Arratia (1999: 326) retained this character as a synapomorphy of the "Ostarioclupeomorpha", which are "Clupeocephalans in which primitively there is an ankylosis or fusion between the mesial extrascapular and parietal alone or parietal and supraoccipital" (her character 32, state 1). This was based on analyses including fossil and extant teleosts as well as analyses including extant teleosts only. As noticed by Zaragüeta-Bagils *et al.* (2002), it is difficult to consider this character as independent from her character 31 (state 1), which is "supratemporal commissure (primitively) passing through parietals or through parietals and supraoccipital" (Patterson and Rosen, 1977, Grande, 1985). Arratia and Gayet (1995) discussed the difficulties of homologizing lateral and mesial extrascapulars among various ostariophysans. These difficulties may explain some contradiction among authors concerning which extrascapular—lateral or mesial or both—was fused with parietals, and even among different writings of Arratia, between what is observed and what is coded. This point was not really discussed by Patterson and Johnson (1996), who retained this character under the form "Parietals carrying the occipital commissural sensory canal" (synapomorphy n°3 in their fig. 23, with one convergence in argentinoids). Zaragüeta-Bagils *et al.* (2002) considered "the supratemporal canal passing through the parietals" as "the most reliable otocephalan synapomorphy", but they noticed that this feature must have appeared at least twice among teleosts, since it is also found in most osteoglossomorphs. Without further ontological investigation of these fusions, especially in clupeomorphs, this character should provisionally be cited as did Johnson and Patterson (1996) and Zaragüeta-Bagils *et al.* (2002): "Parietals carrying the occipital commissural sensory canal".

Haemal Spines Anterior to that of the Second Preural Centrum are Fused to the Centra. This synapomorphy was suggested by Fink and Fink (1981: 339–341), although they (too rapidly) concluded that this character was an ostariophysan synapomorphy: "In ostariophysans, all haemal spines anterior to that of the second preural centrum are fused to the centra from a young juvenile stage (...). In the primitive members of most other primitive teleostean lineages, including Scleropages, Elops, Esox and Diplophos, four or more haemal spines are autogenous in much larger juveniles. An adult Esox specimen has five autogenous haemal spines, and an adult Salmo two (dry skeletal material). Clupeomorphs have all haemal spines fused to the centra, suggesting relationships between the clupeomorpha and the ostariophysi. However, clupeomorphs lack the adipose fin and breeding tubercles which link ostariophysans with other members of the euteleostei (Patterson and Rosen 1977)." The two latter characters that Fink and Fink (1981) held as true at the time are now rejected (see above). These fusions are therefore interpreted as a potential clupeomorpha-ostariophysi synapomorphy by Lecointre and Nelson (1996), not really discussed by Johnson and Patterson (1996). The distribution observed by Fink and Fink (1981) can be confirmed with the work of Fujita (1990: 824–845), showing that fusions of haemal spines anterior to PU2 do occur in all clupeomorphs and ostariophysans. However, the distribution suffers from some occurrences in other teleostean groups. Fujita (1990: 253) did not observe four or more autogenous haemal spines in *Diplophos*: only the HPU2 was autogenous. Without taking into account fishes in which the caudal skeleton is highly derived, specialized in multiple fusions of various bones, the fusion of the third haemal spine to its centrum (and that of each haemal spine more anteriorly positioned) occurs sparsely in the following non-acanthomorph fishes: *Umbra* (but not in other esocoids), *Bathylagus*, Gonostomatidae, Sternoptychidae, two genera of the Photichthyidae, *Bathypterois*, *Bathysaurus*, *Omosudis*, *Alepisaurus*, *Centrobranchus*, *Lampanyctus*. These cases, however, have to be considered in terms of their frequency of occurrence and distribution among the large taxonomic sample of Fujita (1990). Of the 39 families and 72 genera of non-acanthomorph, non-clupeomorph and non-ostariophysan fishes sampled by Fujita (1990), they were a minority of taxa showing these fusions. Moreover, all of them were isolated cases within their family or their order (perhaps except in Gonostomatidae and Sternoptychidae) (i.e., these fusions are likely to be due to isolated convergences and cannot constitute a synapomorphy of an entire non-acanthomorph order or super-order). By contrast, these fusions were constantly found through all the 39 genera and 14 families of clupeomorphs and ostariophysans that Fujita sampled. These multiple, taxonomically isolated occurrences show that these fusions of the third and further haemal spines to their centra seem to have evolved independently several times in different clades. Arratia (1997, 1999) did not maintain this feature as a synapomorphy of the "ostarioclupeomorpha". Zaragüeta-Bagils *et al.* (2002) temporarily accepted this feature but objected that it was also found in most Jurassic basal teleosts (Arratia 1996, 1997, 1999) and some osteoglossomorphs (Taverne 1977, 1978, 1979).

Presence of Intermuscular Epicentral Bones. Clupeomorphs have intermuscular epicentral bones. Gonorynchiforms have these bones, but they are lacking in most of the otophysans (Gayet *et al.* 1994, Johnson and Patterson 1996). They are present in gymnotoids (Meunier and Gayet 1991: fig. 2, p. 225; Gayet *et al.* 1994). There is another teleostean clade in which these bones occur: the Elopomorpha, more specifically in *Anguilla*, and maybe more generally in anguilliforms (Blot 1978, his fig. 2, p. 18). This character must therefore be considered doubtful and was retained neither by Arratia (1997, 1999) nor by Zaragüeta-Bagils *et al.* (2002). However, "ossified epicentrals" is one of the synapomorphies (n°20) given for the Otocephala in the tree fig. 23 of Johnson and Patterson (1996). This synapomorphy probably results from the optimization of the question mark coded for this feature for ostariophysans as a character state "1".

Otophysic Connection between the Swimbladder and the Inner Ear. In many teleosts, the swimbladder is related to the inner ear, directly or indirectly. The direct connection can be described as a diverticulum extending from the swimbladder to the labyrinth in some osteoglossomorphs (Notopteroidea), in all clupeomorphs, and in some holocentrids. The indirect connection is achieved in ostariophysans through a paired chain of bones, the Weberian apparatus, and perilymphatic cavities, positioned between the swimbladder and the labyrinth. One must admit that the anatomy of the otophysic connection is very different in clupeomorphs and ostariophysans, and they are generally not taken as possibly homologous (Fink and Fink 1981, p. 343). However, if one wants to understand the evolution of the Weberian apparatus, one must suppose the pre-existence of a soft direct otophysic connection before the development of the indirect otophysic connection, as suggested by Gayet (1986a). Can we imagine the soft otophysic connection of the last hypothetical common ancestor to clupeomorphs and ostariophysans as similar to that of clupeomorphs? Only developmental studies can help. The development of the "hard part" of the connection, the Weberian ossicles, has been described (Vandewalle *et al.* 1990, Grande 1996, Grande and Poyato-Ariza 1999, Grande and Shardo 2002, Grande and Young 2004, Grande and de Pinna 2004). Grande and de Pinna (2004), who conducted a remarkable scientific strategy specifically devoted to solving these questions of homology between the two otophysic connections, concluded that although the adult functioning Weberian apparatus seems homologous within Otophysi, the development of some Weberian ossicles is not the same in all otophysan subgroups. So it is clear that the Weberian apparatus did not evolve all at once. If all the otophysic connections cannot be taken as such as an otocephalan synapomorphy, the patchy evolution within the otophysic connections in the Clupeomorpha and the Ostariophysi provides several cases of parallelisms and two synapomorphies of the Otocephala. The gasbladder with a silvery peritoneal tunic covering at least the anterior chamber of the Ostariophysi parallels that of the Clupeoidei. The anterior 1–2 vertebral centra shorter than posterior ones were found in the Ostariophysi and the Pristigasteroidea. The first pleural rib attached to the peritoneal tunic of the gasbladder, classically described as an ostariophysan feature, was also found in the Pristigasteroidea and the Coiliidae. The second pleural rib attached to the peritoneal tunic of gasbladder was also found in the Pristigasteroidea. The constriction of the gasbladder into two chambers is found in the Ostariophysi, Pristigasteroidea and Engrauloidea. A number of these features could appear as parallelisms because of unobserved character states in some taxa covered by the study of Grande and de Pinna (2004). Once the anatomical gaps filled, some of these features could perhaps become otocephalan synapomorphies. Grande and de Pinna (2004) proposed two features as synapomorphies of the Otocephala. The saccular and lagenar otoliths are in a posterior and median position (previously proposed as an ostariophysan synapomorphy by Fink and Fink 1996). The labyrinth of the inner ear is posteriorly elongated.

Autopalatines Ossify Early in Ontogeny (Arratia 1997, 1999: 326 and her fig. 19: clade J in the tree including fossil and extant teleosts). Zaragüeta-Bagils *et al.* (2002) objected that this character is difficult to test in fossils.

The Bases of Hypurals 1 and 2 Are Not Joined by Cartilage in Any Growth Stage (Arratia 1999: 326, her figs. 19 and 24: clade J in the tree including fossil and extant teleosts, clade E in the tree including extant taxa only). Zaragüeta-Bagils *et al.* (2002) objected that this character is difficult to assess in fossil material, which limits the reliability of the character but cannot be a logical argument for rejection.

The Neural Arches of Most Abdominal Vertebrae with Fused Halves Forming a Medial Neural Spine, and Not with Separate Halves (Arratia 1999: clade E of her fig. 24, seventh analysis including extant taxa only). Zaragüeta-Bagils *et al.* (2002) noticed that this feature is found in most basal clupeomorphs but also in most non-clupeocephalans and must therefore be considered as a teleostean symplesiomorphy. This is a strong argument for rejecting this character as a synapomorphy of the Otocephala. Arratia (1999: 326) did not retain this character in her synthesis.

Anteriormost Uroneurals Present as One Long Uroneural (Loss or Fusion of Three) (Arratia 1999: clade E of her fig. 24, seventh analysis including extant taxa only). This character appears as a synapomorphy because the matrix contains extant taxa only: fossils do not have this character state. Basal fossil clupeomorphs and anotothysans show two or three long uroneurals. Arratia (1999: 326) did not retain this character in her synthesis.

First Uroneural Reaches Preural Centrum One (Arratia 1999: clade E of her fig. 24, seventh analysis including extant taxa only). Same as above. Arratia (1999: 326) did not retain this character in her synthesis.

Neural Spine of Preural Centrum 2 (NPU2) Leaflike, as Long as NPU3 (Johnson and Patterson 1996). This is synapomorphy n°25 given for the otocephalans in their tree figure 23B. In that tree this character suffers from a certain level of homoplasy: it is also present in esocoids and argentinoids.

Urodermal Absent (Johnson and Patterson 1996). This is synapomorphy n°34 given for the otocephalans in their fig. 23B. In that tree the character suffers from a certain level of homoplasy: absence of urodermal (*sensu* Johnson and Patterson 1996) is also found in esocoids and alepocephaloids. Note that the absence of urodermal in alepocephaloids is congruent with recent molecular results in which alepocephaloids are found among otocephalans.

The Problem of the Alepocephaloidei

Greenwood *et al.* (1966) expressed the difficulties they had in assigning the Alepocephaloidei to any teleostean division. Indeed, alepocephaloids seem to be generalized fishes in which special features seem to be restricted to adaptive responses to the deep-sea environment. They considered alepocephaloids "primitive salmoniformes" mainly from the anatomy of the caudal skeleton. The group was placed by earlier authors either close to stomiatoids, among "Clupeiformes", or among "Salmoniformes". However, at those times those groups had wider meanings than they have now. Gosline (1969) proposed a relationship of alepocephaloids with osmeroids. Greenwood and Rosen (1971) remarked that Gosline's survey did not involve argentinoids and "filled the gap". They concluded in placing the Alepocephaloidea as the sister group of the Argentinoidea within the Argentinoidei, a position followed by Rosen (1982, Fig. 9.3b), Lauder and Liem (1983, Fig. 9.3d), Fink (1984, Fig. 9.3e), Sanford (1990, Fig. 9.3g), Patterson (1994), and Nelson with slight differences in ranking (1984, 1994, 2006). After having corrected Begle's (1991, 1992, Fig. 9.3h) matrices, Johnson and Patterson (1996, Fig. 9.3j) investigated the position of the Alepocephaloidea among clupeocephalans using a morphological matrix of 42 characters and nine taxa and confirmed its sister-group relationship with the Argentinoidea, forming the Argentinoidei. This was based on three synapomorphies: accessory cartilage between ceratobranchial five and epibranchial five, distal parts of first two to four epineurals descended, and caudal median cartilages together supporting the lowermost ray of the upper caudal fin lobe (though some alepocephalids have lost these cartilages). Diogo (2008) found the same relationships with a different ranking, i.e., the Alepocephaloidei sister group of the Argentinoidei, both making the Argentiniformes. He recorded six synapomorphies supporting the group. One is the peculiar configuration of the hypoaxialis, the second is the presence of autopterotic and dermopterotic bones as independent, distinct ossifications, another one is the first one cited above (accessory cartilage of Johnson and Patterson

1996), and three others are homoplastic characters. The result is obtained from "more than 70 extant and fossil teleostean terminal taxa and more than 270 morphological characters", but the matrix and character descriptions are not published in that article and are mentioned as being "in press". In their "mitogenomic" approach to molecular phylogeny, Ishiguro *et al.* (2003, Fig. 9.3l) placed the Alepocephaloidei (actually *Alepocephalus* and *Platytroctes*) within the Otocephala. They were sister group of a clade containing gonorhynchiforms (*Gonorynchus*) and clupeomorphs with low support. Using a far better sampling of clupeomorphs and gonorynchiforms for the same mitogenomic approach, Lavoué *et al.* (2005, 2007, Fig. 9.3m) found alepocephaloids as the sister group of monophyletic ostariophysans. This is in conflict with the position assigned to alepocephaloids by Johnson and Patterson (1996) and by Diogo (2008). The molecular position of Lavoué *et al.* (2005, 2007) raises the problem of finding otocephalan synapomorphies in alepocephaloids. However, Lavoué *et al.* (2008) obtained from the same genes a better alepocephaloid taxonomic sample and found an unresolved interrelationship among Clupeomorpha, Alepocephaloidei and Ostariophysi (Fig. 9.3p), leaving open the possibility of monophyletic Otocephala. Whatever the molecular sister group of the Alepocephaloidei (i.e., Ostariophysi or Otocephala), this challenges the interpretation of the characters shared by argentinoids and alepocephaloids.

From their matrix of nine clupeocephalan taxa and 42 morphological characters, Johnson and Patterson (1996) obtained a single most parsimonious tree of 89 steps (Fig. 9.3j) where the Alepocephaloidea was the sister group of the Argentinoidea. Transferring the alepocephaloids to the otophysans contradicts four of the six internal nodes of that tree (their fig. 23) and requires six extra steps (95 steps). However, that contradiction is not the only one produced by molecular results, and focusing solely on alepocephaloids would not exhaust outcomes of that matrix. Johnson and Patterson (1996) found the esocoids as the sister group of neoteleosts. The sister-group relationship between esocoids and salmonoids, a more traditional grouping, repeatedly found by various molecular phylogenies (see below) based on different genes, contradicts three of the six internal nodes and requires 10 extra steps. This is not a justification for transferring alepocephaloids to the otocephalans but stresses that hypotheses must not be judged from what is expected from our "common sense" but from real levels of contradictions contained in data matrices. Here an unexpected result costs fewer extra steps than the more expected one. The same exercise could not be done on the matrix of Diogo (2008) because it was still unpublished at the time.

Are the putative otocephalan synapomorphies listed above found in alepocephaloids? The Alepocephaloidei (Alepocephaloidea according to some authors, but this is a matter of ranking, not a matter of content) lack a pleurostyle, and fusion between hypural two and ural centrum one (Greenwood and Rosen 1971). In alepocephaloids the first uroneural reaches preural centrum two, not preural centrum one. The neural spine of preural centrum 2 (NPU2) is not leaflike (Greenwood and Rosen 1971, Johnson and Patterson 1996). In the ostarioclupeomorphs of Arratia, the anteriormost uroneurals are "present as one long uroneural (loss or fusion of three)". In most alepocephaloids there is a long free first uroneural followed by one of medium length and one short (Greenwood and Rosen 1971), which does not correspond to the "loss or fusion of three uroneurals" as stated by Arratia. Presence of fully ossified intermuscular epicentral bones is an otocephalan synapomorphy in the scheme of Johnson and Patterson (1996), though this depends on the choice of character optimization (the character $n°20$ is coded "1" for clupeomorphs and "?" for ostariophysans because epicentrals are lacking in most ostariophysans). They did not code alepocephaloids in the same way: alepocephaloids, argentinoids, salmonoids and osmeroids were coded "2", i.e. "epicentrals with cartilage rods distally". There is no occipital commissural sensory canal passing through the parietals in alepocephaloids. They exhibit neither otophysic connection nor remnants of such a connection. If the position of the Alepocephaloidei as recovered by the first "mitogenomic" approach (Ishiguro et al. 2003, Fig. 9.3l) is confirmed by nuclear genes, this would seriously challenge putative homologies of the Weberian apparatuses of Gonorynchiformes and Ostariophysi (Rosen and Greenwood 1970). However, Rosen and Greenwood (1970) trusted the sister-group relationship between these two groups on the basis of the caudal skeleton and considered the otophysic connection of poor phylogenetic importance. If the position of the Alepocephaloidei as the sister group of ostariophysans, as recovered by the second "mitogenomic" approach (Lavoué et al. 2005, 2007), is confirmed by nuclear genes, nothing will change with regard to homologies in otophysic connections among otocephalans: gonorynchiforms remain the sister group of otophysans and the patchy evolution of otophysic connections in clupeomorphs and ostariophysans (Grande and de Pinna 2004) will tend to advocate for parallelisms rather than for secondary homologies. For the moment, there is no ontogenetic data establishing whether the bases of hypurals 1 and 2 are joined by cartilage in any growth stage, or whether autopalatines ossify early in ontogeny.

Is there any otocephalan character found in alepocephaloids? There are only two of them. The absence of a urodermal is noted in all alepocephaloids described to date (Greenwood and Rosen 1971, Johnson and Patterson 1996). If this appears as a synapomorphy of otocephalans (now in a wider meaning: including alepocephaloids), one must admit a convergence in esocoids (Johnson and Patterson 1996). Haemal spines anterior to that of the second preural centrum are fused to the centra in all caudal skeletons of species having fully ossified adult stages as observed by Greenwood and Rosen (1971): *Talismania, Binghamichthys, Xenodermichthys, Rouleina*. This character was considered doubtful above because it is possibly too variable, but it is not rejected. Finally, one must keep in mind that the good taxonomic sample of the mitogenomic approach of Lavoué *et al.* (2008) leaves open the question of the otocephalan monophyly: the unresolved relationships among the Alepocephaloidei, the Clupeomorpha and the Ostariophysi include the solution of the Alepocephaloidei as the sister group of the Otocephala. If that solution were to be confirmed by phylogenies based on nuclear genes, the above question would become unnecessary.

The Position of Esocoids

The position of ostariophysans among teleosts is linked to the definition of euteleosts because they became the sister group of a non-euteleostean clade, the Clupeomorpha. In the same way the definition of the Euteleostei cannot be treated without considering the problem of the position of esocoids because esocoids have been considered inside or outside euteleosts according to authors and periods.

Some molecular studies have included esocoid sequences, but no ostariophysan and/or clupeomorph ones. Lê *et al.* (1989) sequenced portions of the 28S rRNA for a collection of 12 fishes, among which were seven teleosts: *Clupea, Esox* and five neoteleosteans. The tree based on a distance-matrix method showed, within the monophyletic teleosts, the (*clupea* (*Esox*—neoteleosteans)) groupings, thus resolving the clupeocephalan basal trifurcation of Rosen (1985). However, some problems remained: no robustness indicators were given, the *Esox* 28S rRNA sequence suffered from a slightly higher number of undetermined nucleotides, and no justification was given for discarding five species (among them *Clupea*) from the data set used for the parsimony analysis. Bernardi *et al.* (1993) analysed a data set of 27 growth-hormone amino-acid sequences (96 informative sites), among them 25 teleosts: *Anguilla*, five otophysans, *Esox*, four salmonids, and 14 acanthopterygians. In the parsimony-bootstrap tree, the intra-teleostean relationships are (*Anguilla* (otophysans ((*Esox* salmonids) acanthopterygians))). Among these nodes of interest, each is well supported except the monophylies of salmonids (62 percent), and the salmonids + *Esox* grouping (the salmoniforms of Rosen 1974) (69 percent). Unfortunately, there is no clupeomorph sequence: the sister group of *Anguilla* is well supported (100 percent). This sister group could be the clupeocephalan clade as well as the euteleostean clade (including *Esox*, as in Rosen 1973, 1974). This study, therefore, does not answer the question of the monophyly of euteleosts.

The following studies included esocoids, ostariophysans and clupeomorphs. The study of Müller-Schmid *et al.* (1993) included an ependymin sequence of *Esox* and found an *Esox*-salmonids sister-group relationship, but these results are inconclusive because of a rooting problem (see above). Other studies have used ependymin sequence data (Orti and Meyer 1996) but did not investigate large-scale teleostean interrelationships. They rather investigated interrelationships within otophysans. Nevertheless, Orti and Meyer (1996) used ependymin sequences of *Salmo* and *Esox* and, while rooting their tree on *Clupea*, found a *Salmo-Esox* sister-group relationship. In the abstract they claimed that "ependymin DNA sequences have established the first molecular evidence for the monophyly of a group containing salmonids and esociforms". This result must actually be interpreted with caution because they had only three non-otophysans in the sample (*Clupea*, *Esox*, and *Salmo*) and no crown teleosts: if *Esox* really were the sister group of neoteleosts (as in Johnson and Patterson 1996, for example), in the absence of any neoteleost the topology would have been exactly the same. The claim appears today to be based on a false rooting option. This work is inconclusive concerning otocephalans and esocoid relationships. In the unpublished Hennig-86 tree based on 18S rRNA sequences of Patterson *et al.* (1994: communication at the fish phylogeny meeting of the 13th Willi Hennig Society meeting), the intra-clupeocephalan topology was ((*Clupea-Chanos*) (*Esox* (*Salmo*, Neoteleosteans))). This is in agreement with Lê *et al.* (1989), but not with Bernardi *et al.* (1993). The esocoids and salmonoids were sister groups in several trees published by Zaragüeta-Bagils *et al.* (2002, Fig. 9.3k) from independent data sets (28SrDNA, Rhodopsin, RAG1, and the MLL gene, which sequence of *Plecoglossus altivelis* of Venkatesh *et al.* 1999 actually is a contamination). Lopez *et al.* (2004, Fig. 9.3o) clearly confirmed these results using 12S–16S data and RAG1 sequences with a rich taxonomic sampling including argentinoids, osmeroids, galaxioids, stomiiforms and basal and crown teleosts. This sister-group relationship was also confirmed independently by the "mitogenomic approach" of Ishiguro *et al.* (2003) and Lavoué *et al.* (2005) from large data sets of mitochondrial sequences including the genera *Esox*, *Dallia*, *Coregonus*, *Oncorhynchus*, and *Salmo*.

The Esocoidei as the sister group of the Salmonoidei is therefore a highly reliable hypothesis, as recovered from several independent molecular markers and studies (Bernardi *et al.* 1993, Zaragüeta-Bagils *et al.* 2002, Ishiguro *et al.* 2003, Lopez *et al.* 2004).

From a morphological point of view, the picture is not so clear. Traditionally the Esocoidei is positioned within Salmoniformes (for instance in Greenwood *et al.* 1966, Rosen 1974, Nelson 1976). However, Salmoniformes and Esociformes were separated during the 1980s and 1990s (Fink and Weitzman 1982, Rosen 1982, 1985, Lauder and Liem 1983,

Fink 1984, Nelson 1984, Sanford 1990, Patterson 1994, Lecointre 1994, Johnson and Patterson 1996, Figs. 9.3b–j). From their matrix of 42 characters coded for eight virtual clupeocephalan taxa and one virtual outgroup, Johnson and Patterson (1996, Fig. 9.3j) found the Esocoidei to be the sister group of the Neoteleostei, and the Salmonoidei the sister group of the Osmeroidei. Esocoids shared three derived, non-homoplastic states with neoteleosts: tooth attachment type IV, uroneural three absent, and cheek and operculum scaled (though coded "?" in neoteleosts). Salmonoids and osmeroids shared nine derived, non-homoplastic character states. The sister-group relationship between esocoids and salmonoids, as found in several molecular phylogenies, contradicts three of the six internal nodes of their tree and requires 10 extra steps (99 steps). Their anatomical matrix is therefore in strong conflict with several independent molecular phylogenies. From a matrix of 135 anatomical characters coded for 32 extant and fossil taxa, Arratia (1997: 143) found the Salmonoidei and the Esocoidei linked by 10 synapomorphies. The result is the same in Arratia (1999: 314). From a matrix of 196 anatomical characters coded for 48 fossil and extant taxa, the two groups are linked by nine synapomorphies. However, there are no neoteleosts sampled in the two later studies (Fig. 9.3n) and Johnson and Patterson's hypothesis (esocoids sister group of neoteleosts) was not really tested by these studies.

What are Euteleosts?

Euteleosts can be defined once we have a clear picture of positions of the ostariophysans and esocoids. Today there is a significant amount of data supporting the following hypotheses (Fig. 9.5):

— Gonorynchiformes are members of the Ostatiophysi.
— Ostariophysans and clupeomorphs are sister groups (the question of the position of the Alepocephaloidea within or as a sister group of the Otocephala has to be studied further).
— Esocoids and salmonoids are sister groups.

The euteleosts as defined by Patterson and Rosen (1977, separating clupeomorphs from the ostariophysans), or by Rosen (1985, Fig. 9.3f, separating clupeomorphs from ostariophysans and esocoids from salmonoids) can no longer be considered valid. Given this, a new definition of the Euteleostei including both the Esocoidei (esocoids being sister group of salmonoids) and the Clupeomorpha (clupeomorphs being sister group of ostariophysans) would be required. Johnson and Patterson (1996) have chosen the following option (also followed by Arratia 1999: 325, Zaragüeta-Bagils *et al.* 2002, Ishiguro *et al.* 2003, Lopez *et al.* 2004, Lavoué *et al.* 2005): ostariophysans are put outside euteleosts and the Otocephala (or

Ostarioclupeomorpha) is the sister group of the Euteleostei. *Euteleosts are therefore all clupeocephalans except otocephalans* (Figs. 9.3i, 9.3j, 9.5). Characters retained are the following:

— The "pattern 2" of development of the supraneurals as described by Johnson and Patterson (1996) (interestingly alepocephaloids exhibit several patterns and could be with the otocephalans with regard to that character). The first supraneural (anterior to the neural spine of the first vertebrae) develops independently, and the remainder differentiate in rostral and caudal gradients from a focus roughly midway between the occiput and dorsal fin origin. In adults resulting from this mode of development, the first supraneural is usually differentiated from the second and they are separated by two (or more) neural spines.

— Membranous anterodorsal outgrowth of uroneural 1. This does not directly correspond to "the stegural" of Monod (1968). The membranous outgrowth considered here is when the first uroneural is autogenous. While Rosen (1985) did not consider the anterodorsal outgrowth of the first uroneural of esocoids homologous to the outgrowth of other euteleosts, Johnson and Patterson (1996) did. The structure is absent in ostariophysans (Monod 1968, Rosen and Greenwood 1970, Johnson and Patterson 1996: 287), so it fits the new Euteleostei. However, there are two homoplastic features to retain. First, the outgrowth is considered absent and secondarily lost in argentinoids (Johnson and Patterson 1996: 287). Second, alepocephaloids do have the membranous anterodorsal ourgrowth of autogenous uroneural 1. If the Alepocephaloidea are among the Otocephala (Lavoué *et al.* 2005), this character must be considered independently acquired.

— Presence of caudal median cartilages (secondarily lost in at least six euteleostean lineages, Johnson and Patterson 1996: 288). The reliability of this synapomorphy is weakened by these multiple losses and, above all, by the fact that the gains and losses of these cartilages on to the tree of Johnson and Patterson (1996) depend on character state optimization. If the gain of these cartilages is a synapomorphy of euteleosts we must consider it secondary loss in esocoids (reversions favoured, Johnson and Patterson 1996, their fig. 23B). However, this feature can also be seen as a synapomorphy of the Protacanthopterygii by the authors (i.e., salmonoids, osmeroids, alepocephaloids, argentinoids) with a convergence in neoteleosts (convergences favoured, Johnson and Patterson, 1996, their fig. 23A). The interpretation of caudal median cartilages as a euteleostean synapomorphy is therefore doubtful.

Phylogeny within Ostariophysans

Interrelationships within ostariophysans have been mostly debated with regard to three points:

— First, the position of gonorynchiforms, as sister group of otophysans (Regan 1911, Greenwood and Rosen 1970, Roberts 1973, Novacek and Marshal 1976, Fink and Fink 1981, 1996, Grande and de Pinna 2004, Figs. 9.4ab, 9.5), or sister group of clupeomorphs (Ishiguro *et al.*, 2003, Saitoh *et al.*, 2003, Fig. 9.3l), or sister group of cypriniforms (Gayet 1982, 1986a, b, Fig. 9.4d).

— Second, the position of cypriniforms, either as sister group of the clade grouping gymnotiforms and characiforms (Regan 1911, Greenwood and Rosen 1970, Roberts 1973, Novacek and Marshall 1976, Fig. 9.4a), or the most basal otophysans (Fink and Fink 1981, 1996, Orti and Meyer 1996, Dimmick and Larson 1996, Saitoh *et al.* 2003, Lavoué *et al.* 2005, Fig. 9.4b).

— Third, gymnotiforms, more closely related to siluriforms (Fink and Fink 1981, 1996, Fig. 9.4b), to characiforms (Regan 1911, Greenwood and Rosen 1970, Roberts 1973, Novacek and Marshall 1976, Fig. 9.4a, Dimmick and Larson 1996 (molecular data only); Saitoh *et al.* 2003, Fig. 9.4c), or to a clade grouping characiformes and siluriformes (Lavoué *et al.* 2005).

From the morphological point of view, since Regan (1911a, b), most ichthyologists recognized the siluroids as the sister group of all other otophysans and the cypriniforms as the sister group of a group containing the characiforms and the gymnotoids. Greenwood and Rosen (1970) changed the content of the Ostariophysi by introducing the gonorynchiforms as the sister group of the rest, called the Otophysi (see above). Roberts (1973) accepted the characiforms as more closely related to the gymnotoids (Fig. 4a). A significant breakthrough was made by Fink and Fink (1981), who analysed 127 anatomical characters and found the Cypriniformes to be the sister group of all other otophysans, and grouped gymnotoids and siluroids as sister groups in their "siluriformes" (Fig. 9.4b). They recognized the Gonorynchiformes as the sister group of otophysans. From fossil material, Gayet (1982, 1986a, b) discussed Fink and Fink's synapomorphies and promoted a conflicting hypothesis in which gonorynchiforms were a sister group of cypriniforms, and gymnotoids more related to characiforms than to siluriforms (Fig. 9.4d). The opposition was more methodological (Fink *et al.* 1984, Gayet 1986b) than osteological, but Fink and Fink (1996) revised their data incorporating data from fossils and maintained their scheme. The hypothesis of Fink and Fink (1996) concerning the position of the Cypriniformes was soon corroborated by all molecular results, but not the

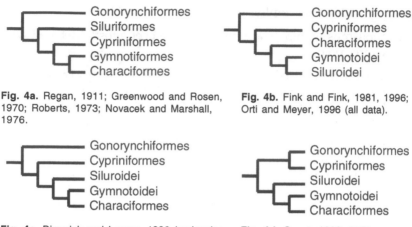

Fig. 4a. Regan, 1911; Greenwood and Rosen, 1970; Roberts, 1973; Novacek and Marshall, 1976.

Fig. 4b. Fink and Fink, 1981, 1996; Orti and Meyer, 1996 (all data).

Fig. 4c. Dimmick and Larson, 1996 (molecules only); Orti and Meyer, 1996 (sorted data); Saitoh *et al.* 2003.

Fig. 4d. Gayet, 1982, 1986.

Fig. 9.4 Cladograms showing the various hypotheses given for ostariophysan intrarelationships.

hypothesis of the sister-group relationship between siluroids and gymnotoids, for which molecular results are ambiguous.

Orti and Meyer (1996) analysed 359 phylogenetically informative positions in 25 sequences of the ependymin gene, a single-copy gene coding for a highly expressed glycoprotein in the brain matrix of teleosts (193 informative positions when third codon positions of this highly variable gene are removed). The taxonomic sampling included three cyprinids, two gymnotoids, four siluroids and 13 characiforms (plus three outgroups). Cypriniforms appeared as the most basal otophysans with high support. Relationships among characiforms, siluroids and gymnotoids were poorly supported. Gymnotoids were the sister group of siluroids (as in Fink and Fink 1981, Fig. 9.4b), but gymnotoids were the sister group of characiforms when transitions at third codon positions were removed (Fig. 9.4c). Dimmick and Larson (1996) analysed 160 phylogenetically informative sites in nuclear ribosomal DNA and 208 phylogenetically informative sites in the mitochondrial 12S and 16S ribosomal DNA and valine transfer RNA gene for nine otophysans and one gonorynchiform outgroup. There were two characiformes, two cypriniformes, two gymnotoids, three siluroids in the sample. The complete matrix of molecules yielded a parsimony tree where cypriniformes were basal and gymnotoids the sister group of characiforms (Fig. 9.4c). The addition of 85 morphological characters from Fink and Fink (1981) to the molecular matrix yielded the same tree except that gymnotoids became the sister group of siluroids (Fig. 9.4b). In their "mitogenomic" approach, Saitoh *et al.* (2003)

analysed 8,196 positions of the mitochondrial genome (7,286 from 13 mitochondrial protein coding genes (third codon positions removed) and 910 from 22 tRNA genes) for 24 taxa (two gonorynchiforms, seven cyprinids, two gymnotoids, two characiforms, and two siluroids). Their maximum likelihood analysis yielded gonorynchiformes as the sister group of clupeomorphs (two species sampled) with poor support (as in Fig. 9.3l), cypriniformes as the most basal otophysans with high support, and gymnotoids as a sister group of characiforms with poor support (Fig. 9.4c). From the same type of data, Peng et al. (2006) analysed 6,918 mitochondrial positions for 30 species (among them two gonorynchiforms, eight cyprinids, two gymnotoids, two characiforms, and five siluroids). The maximum likelihood analysis yielded the same results, though interpretations made by the authors must be treated with caution. More convincing is the mitogenomic approach of Lavoué et al. (2005) with a better taxonomic sampling (three clupeiforms including Denticeps, seven gonorynchiforms, seven cyprinids, two gymnotoids, two siluroids, and two characiforms) and 14,025 mitochondrial positions analysed (12 protein-coding genes, 22 tRNA genes, and 12S–16S rDNA) for 40 teleosts. When basal ostariophysans are better sampled, good support is obtained for monophyletic ostariophysans, gonorynchiforms being the sister group of otophysans with high support. Cypriniformes are the most basal otophysans with high support (Fig. 9.3m). For the first time, Characiformes and Siluroidei are sister groups; gymnotoidei is the sister group of both, but the result is not discussed.

The Position of the Gonorynchiformes

Both from the molecular and the morphological point of view, the best character sampling strategies converge to the same results: gonorynchiformes are monophyletic and they are the sister group of otophysans (Fig. 9.5). This result was reached by the richest morphological data of Fink and Fink (1996) taking both fossil and extant material into account, and Lavoué et al. (2005), which is the best molecular gonorynchiform sampling to date. Both ostariophysans and otophysans are monophyletic and the synapomorphies to take into account are those of Fink and Fink (1996).

The Position of the Cypriniformes

Cladistic morphological approaches to the problem of the position of cypriniformes among ostariophysans have yielded results corroborated by all molecular approaches. Three independent sources of molecular data (ependymin gene, nuclear and mitochondrial ribosomal DNA, mitochondrial coding genes coupled with tRNA genes) analysed by three

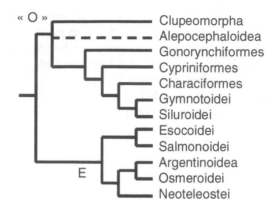

Fig. 9.5 Cladogram showing a synthesis of clupeocephalan interrelationships. Only clades that have been corroborated through multiple sources of independent data are shown, except for the position of the Alepocephaloidea, which is shown here as a working hypothesis. Otherwise, the Alepocephaloidea should be the sister group of the Argentinoidea.

independent teams led to the same conclusion. Cypriniformes are the sister group of the rest of the otophysans (as in Fink and Fink 1996, Lavoué *et al.* 2005, Fig. 5).

The Position of the Gymnotoidei

From the morphological point of view, cladistic approaches to the problem proposed a sister-group relationship between gymnotoids and siluroids (Fink and Fink 1981, 1996) and have not been challenged by new cladistic analyses of morphological characters. Molecular approaches to that problem yielded ambiguous results. Only the full ependymin data of Orti and Meyer (1996) corroborated this result, though with poor support. Other molecular approaches have shown poor support for a gymnotoid-characiform (Dimmick and Larson 1996: molecular data only, Saitoh *et al.* 2003) or a siluroid-characiform (Lavoué *et al.* 2005) sister-group relationship with rather poor taxonomic samplings of each of these groups. We must consider that there is no serious challenge to the morphological hypothesis of Fink and Fink (1996); nevertheless, the question remains open.

Conclusion: Systematists, Conservatism and Scientific Discovery

Gonorynchiforms are ostariophysans and they are the sister group of otophysans (Rosen and Greenwood 1970, see synapomorphies in Fink and Fink 1996, Grande and de Pinna 2004, Fig. 5). That hypothesis has been independently verified by various molecular studies. This view could appear conservative because it was proposed long ago. This peculiar conservatism in scientists comes from ongoing acceptance of a given hypothesis because it has resisted many challenges. Each molecular matrix with relevant taxonomic sample can be seen as a challenge. This conservatism in scientists comes from their tendency to challenge new findings, not because they refuse them on principle, but because new findings must be robust to be accepted as objective knowledge.

Ostariophysans are the sister group of clupeomorphs, forming a clade called Otocephala or Ostarioclupeomorpha, depending on the authors (Fig. 9.5). Here again, several independent molecular analyses found these relationships. *Seven anatomical synapomorphies* can be retained for the group: (1) parietals carrying the occipital commissural sensory canal, (2) haemal spines anterior to that of the second preural centrum fused to the centra, (3) autopalatines ossifying early in ontogeny, (4) bases of hypurals 1 and 2 not joined by cartilage in any growth stage, (5) absence of urodermal, (6) saccular and lagenar otoliths in a posterior and median position, and (7) labyrinth of the inner ear posteriorly elongated.

To date the "mitogenomic" approaches to molecular phylogeny were the only studies of molecular systematics including alepocephaloids, and they found alepocephaloids either among otocephalans or a sister group of it (Lavoué *et al.* 2007, 2008). It may be too early to choose one of the two possibilities, but it should guide our research efforts. The first option, if confirmed, needs to check the seven synapomorphies cited above in alepocephaloids. Two of them are observed (haemal spines anterior to that of the second preural centrum fused to the centra, urodermal absent), four of them remain to be investigated (autopalatines ossifying early in ontogeny, bases of hypurals 1 and 2 not joined by cartilage in any growth stage, saccular and lagenar otoliths in a posterior and median position, labyrinth of the inner ear posteriorly elongated), and one of them is contradictory and requires one more step (alepocephalan parietals do not carry the occipital commissural sensory canal). We therefore must consider the hypothesis with an open mind and push the investigations forward: we need ontogenetic data on alepocephaloids and must look at the problem using nuclear molecular markers, with taxonomic samplings including argentinoids.

That position of alepocephaloids is also challenging the new definition of euteleosts. The new Euteleostei (excluding ostariophysans and including esocoids) are supported by two non-ambiguous synapomorphies: (1) a certain pattern of development of the supraneurals (type "2" of Johnson and Patterson 1996), and (2) membranous anterodorsal outgrowth of autogenous uroneural 1. One of them is conflicting because it requires one convergence in alepocephaloids if alepocephaloids are within the Otocephala (membranous anterodorsal outgrowth of autogenous uroneural 1). The other is not really conflicting because there are several patterns of supraneural development within alepocephaloids, including type "2". Here again, the new position of alepocephaloids has to be envisaged with an open mind.

Finally, the sister-group relationship between the Argentinoidea and the Alepocephaloidea of Johnson and Patterson (1996) was based on three synapomorphies: (1) accessory cartilage between ceratobranchial five and epibranchial five, (2) distal parts of first two to four epineurals descended, and (3) caudal median cartilages together supporting the lowermost ray of the upper caudal fin lobe. If alepocephaloids are separated from argentinoids, the first two characters should then be considered as convergences in the two groups (as the configuration of the hypoaxialis (Diogo 2008) and the autopterotic and dermopterotic bones present as distinct ossifications (Diogo 2008). The third synapomorphy is not really strongly contradictory because it is a rather homoplastic feature as some alepocephalids have lost caudal median cartilages. This is the same situation for three other synapomorphies proposed by Diogo (2008), which are homoplastic. The degree of conflict can only be assessed if the corresponding matrix is published.

Another conservatism exists, which consists of rejecting an unexpected hypothesis because it does not conform to one's habits or pre-conceived views, and one has not the technical background to be able to judge the quality of the data and the rigour of the methods used for proposing it. It is more often encountered than we think. In the absence of the ability to evaluate the rigour of the methods used, we are all naturally inclined to accept results that conform to our own views and reject unexpected ones. Only methodological competence can protect us from these tendencies. That kind of competence tells us that molecular phylogenies of poor quality are as common as morphological studies of poor quality and shows that real conflicts between "molecules" and "morphology" are actually rare.

References

Arratia, G. 1991. The caudal skeleton of jurassic teleosts; a phylogenetic analysis, pp. 249–340. In: M.M. Chang, Y.H Liu and G.R. Zhang [eds.]. Early vertebrates and related problems of evolutionary biology. Science Press, Beijing.

Arratia, G. 1996. Reassessment of the phylogenetic relationships of Jurassic teleosts and their implications in teleostean phylogeny, pp. 219–242. In: G. Arratia and G. Viohl. [eds.]. Mesozoic Fishes. Systematics and Paleoecology. Verlag Dr. Pfeil, Munich.

Arratia, G. 1997. Basal teleosts and teleostean phylogeny. Palaeoichthylogica 7: 5–168.

Arratia, G. 1999. The monophyly of Teleostei and stem-group teleosts. Consensus and disagreements, pp. 265–334. In: G. Arratia and H.-P Schultze [eds.]. Mesozoic Fishes 2. Systematics and Fossil Record. Verlag Dr. Friedrich Pfeil, Munich.

Arratia, G. 2000. Phylogenetic relationships of Teleostei. Past and present. Estud. Oceanol. 19: 19–51.

Arratia, G. and M. Gayet, 1995. Sensory canals and related bones of Tertiary siluriform crania from Bolivia and North America and comparison with recent forms. J. Vert. Paleontol. 15: 482–505.

Baker, R.H., G.S. Wilkinson and R. DeSalle. 2001. Phylogenetic utility of different types of molecular data used to infer evolutionary relationships among stalk-eyed flies (Diopsidae). Syst. Biol. 50: 87–105.

Bannikov, F. and F. Bacchia. 2000. A remarkable clupeomorph fish (Pisces, Teleostei) from a new Upper Cretaceous marine locality in Lebanon. Senckenbergiana Lethaea 80: 3–11.

Begle, D.P. 1991. Relationships of osmeroid fishes and the use of reductive characters in phylogenetic analysis. Syst. Zool. 40: 33–53.

Begle, D.P. 1992. Monophyly and relationships of the argentinoid fishes. Copeia 1992: 350–366.

Bernardi, G., G. D'Onofrio, S. Caccio and G. Bernardi. 1993. Molecular phylogeny of bony fishes, based on the amino acid sequence of the growth hormone. J. Mol. Evol. 37: 644–649.

Bertin, L. and C. Arambourg. 1958. Systématique des Poissons. In P.P. Grassé [ed.]. Traité de Zoologie VIII (3). Masson, Paris.

Bielawski, J.P. and J.R. Gold. 1996. Unequal synonymous substitution rates within and between two protein-coding mitochondrial genes. Mol. Biol. Evol. 13(6): 889–892.

Blot, J. 1978. Les Apodes fossiles du Monte Bolca. Studi e ricerche sui giacimenti terziari di Bolca. 3(1). Museo civico di storia naturale. Verona, Italy.

Chang, B.S. and D.L. Campbell. 2000. Bias in phylogenetic reconstruction of vertebrate rhodopsin sequences. Mol. Biol. Evol. 17(8): 1220–1231.

Chen, W.J. 2001. La répétitivité des clades comme critère de fiabilité: application à la phylogénie des Acanthomorpha (Teleostei) et des Notothenioidei (acanthomorphes antarctiques). Ph.D. thesis (Thèse de doctorat), Univ. Paris VI, Paris.

Chen, W.J., C. Bonillo and G. Lecointre. 2000. Taxonomic congruence as a tool to discover new clades in the acanthomorph (Teleostei) radiation, p. 369. In: The Proceedings of the 80th ASIH meeting, La Paz, Mexico.

Chen, W.J., C. Bonillo and G. Lecointre. 2003. Repeatability as a criterion of reliability of new clades in the acanthomorph (Teleostei) radiation. Mol. Phylogenet. Evol. 26: 262–288.

Daget, J. 1964. Le crâne des Téléostéens. Mém. Mus. Natl. Hist. Nat. n.s. A 31(2): 163–342.

Dettaï, A. and G. Lecointre. 2004. In search for notothenioid relatives. Antarctic Sci. 16(1): 71–85.

Dettaï, A. and G. Lecointre. 2005. Further support for the clades obtained by multiple molecular phylogenies in the acanthomorph bush. C.R. Biol. 328: 674–689.

Di Dario, F. and M.C.C. de Pinna. 2006. The supratemporal system and the pattern of ramification of cephalic sensory canals in Denticeps clupeoides (Denticipitoidei, Teleostei): additional evidence for monophyly of Clupeiformes and Clupeoidei. Papéis Avulsos de Zoologia. Mus. Zool. Univ. Sao Paulo 46(10): 107–123.

Diogo, R. 2008. On the cephalic and pectoral girdle muscles of the deep sea fish Alepocephalus rostratus, with comments on the functional morphology and phylogenetic relationships of the Alepocephaloidei (Teleostei). Anim. Biol. 58: 23–39.

Diogo, R. and I. Doadrio. 2008. Cephalic and pectoral girdle muscles of the clupeiform Denticeps clupeoides, with comments on the homologies and plesiomorphic states of these muscles within the Otocephala (Teleostei). Anim. Biol. 58: 41–66.

Dimmick, W.W. and A. Larson. 1996. A molecular perspective on the phylogenetic relationships of the otophysan fishes. Mol. Phylogenet. Evol. 6: 120–133.

Dingerkus, G. and L.D. Uhler. 1977. Enzyme clearing of alcian blue stained whole small vertebrates for demonstration of cartilage. Stain Technol. 52: 229–232.

Doyle, J.J. 1992. Gene trees and species trees: Molecular systematics as one-character taxomony. Syst. Bot. 17: 144–163.

Doyle, J.J. 1997. Trees within trees: genes and species, molecules and morphology. Syst. Biol. 46: 537–553.

Dupuis, C. 1992. Regards épistémologiques sur la taxinomie cladiste. Addresse à la XIème session de la Willi Hennig Society (Paris, 1992). Cahiers des Naturalistes. Bull. N. P. n.s. 48(2): 29–56.

Fink, S.V. and W.L. Fink. 1981. Interrelationships of the ostariophysan fishes (Teleostei). Zool. J. Linn. Soc. 72(4): 297–353.

Fink, S.V. and W.L. Fink. 1996. Interrelationships of ostariophysan fishes (Teleostei), pp. 209–249. In: M.L.J. Stiassny, L.R. Parenti and G.D. Johnson [eds.]. Interrelationships of Fishes II. Academic Press, New York.

Fink, S.V., P.H. Greenwood and W. Fink. 1984. A critique of recent work on fossil ostariophysan fishes. Copeia 1984(4): 1033–1041.

Fink, W.L. 1984. Basal euteleosts: relationships, pp. 202–206. In: H.G. Moser, W.J. Richards, D.M. Cohen, M.P. Fahay, A.W. Kendall Jr. and S.L. Richardson [eds.]. Ontogeny and systematics of fishes based on an international symposium dedicated to the memory of Elbert Halvor Ahlstrom. Am. Soc. Ichthyol. Herpetol. Spec. Pub. 1.

Fink, W.L. and S.H. Weitzman. 1982. Relationships of the stomiiform fishes (Teleostei) with a description of Diplophos. Bull. Mus. Comp. Zool. 150: 31–93.

Fujita, K. 1990. The Caudal Skeleton of Teleostean Fishes. Tokai Univ. Press, Tokyo.

Filipski, G.T. and M.V.H. Wilson. 1984. Sudan Black B as a nerve stain for whole cleared fishes. Copeia 1984: 204–208.

Garstang, W. 1931. The phyletic classification of Teleostei. Proc. Leeds Philos. Soc. Lit. Soc. (Scientific section), 2 (1929–1934): 240–260.

Gayet, M. 1982. Considération sur la phylogénie et la paléobiologie des ostariophysaires. Geobios spec. vol. 6: 39–52.

Gayet, M. 1986a. Ramallichthys Gayet du Cénomanien inférieur marin de Ramallah (Monts de Judée), une introduction aux relations phylogénétiques des Ostariophysi. Mém. Mus. Ser. C 51: 1–82.

Gayet, M. 1986b. About ostariophysan fishes: a reply to S.V. Fink, P.H. Greenwood and W.L. Fink's criticisms. Bull. Mus. Hist. Nat., Paris, ser. 4, 8(C, 3): 393–409.

Gayet, M. 1993. Relations phylogénétiques des Gonorhynchiformes (Ostariophysi). Belg. J. Zool. 123: 165–192.

Gayet, M., F. Meunier and F. Kirschbaum. 1994. Ellisella kirschbaumi Gayet and Meunier, 1991, gymnotiforme fossile de Bolivie et ses relations phylogénétiques au sein des formes actuelles. Cybium 18: 273–306.

Gosline, W.A. 1969. The morphology and systematic position of the alepocephaloid fishes. Bull. Br. Mus. (Nat. Hist.) Zool. 18(6): 185–218.

Grande, L. 1982a. Revision of the fossil genus †Diplomystius, with comments on the interrelationships of Clupeomorph fishes. Am. Mus. Novitates 2728: 1–34.

Grande, L. 1982b. A revision of the fossil Genus Knightia, with a description of a new genus from the Green River formation (Teleostei, Clupeidae). Am. Mus. Novitates 2731: 1–22.

Grande, L. 1985. Recent and fossil clupeomorph fishes with materials for revision of the subgroups of clupeoids. Bull. Am. Mus. Nat. Hist. 181(2): 231–372.

Grande, L. 1994. Repeating patterns in nature, predictability, and 'impact' in science, pp. 61–84. In: L.A. Grande and O. Rieppel [eds.]. Interpreting the Hierarchy of Nature. Academic Press, New York.

Grande, T. 1996. The interrelationships of fossil and recent gonorynchid fishes with comments on two Cretaceous taxa from Israel, pp. 299–318. *In*: G. Arratia and G. ViohlG. [eds.]. Mesozoic Fishes. Systematics and Paleoecology. Verlag Dr. Pfeil, Munich.

Grande, T. and B. Young. 2004. The ontogeny and homology of the Weberian apparatus in the zebrafish *Danio rerio* (Ostariophysi: Cypriniformes). Zool. J. Linn. Soc. 140: 241–254.

Grande, T. and F. Poyato-Ariza. 1999. Phylogenetic relationships of fossil and recent gonorynchiform fishes (Teleostei: Ostariophysi). Zool. J. Linn. Soc. 125: 197–238.

Grande, T. and J.D. Shardo. 2002. Morphology and development of the postcranial skeleton in the channel catfish *Ictalurus punctatus* (Ostariophysi: Siluriformes). Fieldiana 1518 (99): 1–30.

Grande, T. and M.C.C. de Pinna. 2004. The evolution of the Weberian apparatus: a phylogenetic perspective, pp. 429–448. *In*: G. Arratia and A. Tintori [eds.]. Mesozoic Fishes 3. Systematics, Paleoenvironments and Biodiversity. Verlag Dr. Friedrich Pfeil, Munich.

Greenwood, P.H. 1973. Interrelationships of osteoglossomorphs, pp. 307–332. *In*: P.H. Greenwood, R.S. Miles and C. Patterson [eds.]. Interrelationships of Fishes, Academic Press, London.

Greenwood, P.H. and D.E. Rosen. 1971. Notes on the structure and relationships of the alepocephaloid fishes. Am. Mus. Novitates 2473: 1–41.

Greenwood, P.H., D.E. Rosen, S.H. Weitzman and G.S. Myers. 1966. Phyletic studies of teleostean fishes, with a provisional classification of living forms. Bull. Am. Mus. Nat. Hist. 131(4): 339–456.

Greenwood, P.H., R.S. Miles and C. Patterson. 1973. Interrelationships of Fishes. Academic Press, London.

Hennig, W. 1950. Grundzüge einer theorie der phylogenetischen systematik. Deutscher Zentralverlag, Berlin.

Hennig, W. 1966. Phylogenetic Systematics. Urbana Univ., Illinois.

Hillis, D.M. 1987. Molecular versus morphological approaches to systematics. Ann. Rev. Ecol. Syst. 18: 23–42.

Inoue J.G., M. Miya, K. Tsukamoto and M. Nishida. 2003. Basal actinopterygian relationships: a mitogenomic perspective on the phylogeny of the "ancient fish". Mol. Phylogenet. Evol. 26: 110–120.

Inoue J.G., M. Miya, K. Tsukamoto and M. Nishida. 2004. Mitogenomic evidence for the monophyly of elopomorph fishes (Teleostei) and the evolutionary origin of the leptocephalus larva. Mol. Phylogenet. Evol. 32: 274–286.

Ishiguro, N.B., M. Miya and M. Nishida. 2003. Basal euteleostean relationships: a mitogenomic perspective on the phylogenetic reality of the "Protacanthopterygii". Mol. Phylogenet. Evol. 27: 476–488.

Janvier, P. 1996. Early Vertebrates. Clarendon Press, Oxford.

Johnson, G.D. and C. Patterson. 1993. Percomorph phylogeny: a survey of acanthomorphs and a new proposal. Bull. Mar. Sci. 52(1): 554–626.

Johnson, G.D. and C. Patterson. 1996. Relationships of lower euteleostean fishes, pp. 251–332. *In*: M.L.J. Stiassny, L.R. Parenti and G.D. Johnson [eds.]. Interrelationships of Fishes II. Academic Press, New York.

Kardong, K. 1998. Vertebrates. Comparative Anatomy, Function, Evolution, 2nd ed. McGraw Hill, Boston.

Kearney, M. and O. Rieppel. 2006. Rejecting "the given" in systematics. Cladistics 22: 369–377.

Kluge, A.G. 1989. A concern for evidence and a phylogenetic hypothesis of relationships among *Epicrates* (Boidae, Serpentes). Syst. Zool. 38: 7–25.

Kluge, A.G. and A.J. Wolf. 1993. Cladistics: what's in a word? Cladistics 9: 183–199.
Lauder, G.V. and K.F. Liem. 1983. The evolution and interrelationships of the actinopterygian fishes. Bull. Mus. Comp. Zool. Cambridge (Mass.) 150(3): 95–197.
Lavoué, S., M. Miya, J.G. Inoue, K. Saitoh, N.B. Ishiguro and M. Nishida. 2005. Molecular systematics of the gonorynchiform fishes (Teleostei) based on whole mitogenome sequences: implications for higher-level relationships within the Otocephala. Mol. Phylogenet. Evol. 37: 165–177.
Lavoué, S., M. Miya, K. Saitoh, N.B. Ishiguro and M. Nishida. 2007. Phylogenetic relationships among anchovies, sardines, herrings and their relatives (Clupeiformes), inferred from whole mitogenome sequences. Mol. Phylogenet. Evol. 43: 1096–1105.
Lavoué, S., M. Miya, J.Y. Poulsen, P.R. Moller and M. Nishida. 2008. Monophyly, phylogenetic position and inter-familial relationships of the Alepocephaliformes (Teleostei) based on whole mitogenome sequences. Mol. Phylogenet. Evol. In Press.
Lê, H.L.V., G. Lecointre and R. Perasso. 1993. A 28S rRNA-based phylogeny of the gnathostomes: first steps in the analysis of conflict and congruence with morphologically based cladograms. Mol. Phylogenet. Evol. 2(1): 31–51.
Lê, H.L.V., R. Perasso and R. Billard. 1989. Phylogénie moléculaire préliminaire des "poissons" basée sur l'analyse de séquences d'ARN ribosomique 28S. C.R. Acad. Sci. 309: 493–498.
Lecointre, G. 1993. Etude de l'impact de l'échantillonnage des espèces et la longueur des séquences sur la robustesse des phylogénies moléculaires. Implications sur la phylogénie des téléostéens. Ph.D. thesis, Univ. Paris VII, Paris.
Lecointre, G., H. Philippe, H.L.V. Lê and H. le Guyader. 1993. Species sampling has a major impact on phylogenetic inference. Mol. Phylogenet. Evol. 2: 205–224.
Lecointre, G. 1994. Aspects historiques et heuristiques de l'ichtyologie systématique. Cybium 18(4): 339–430.
Lecointre, G. 1995. Molecular and morphological evidence for a Clupeomorpha-Ostariophysi sister-group relationship (Teleostei). Geobios spec. pub. 19: 205–210.
Lecointre, G. and G. Nelson. 1996. Clupeomorpha, sister-group of Ostariophysi. Pp. 193–207. In: M.L.J. Stiassny, L.R. Parenti and G.D. Johnson [eds.]. Interrelationships of Fishes II. Academic Press, New York.
Lecointre, G. and H. Le Guyader. 2001. Classification Phylogénétique du Vivant, 1st ed. Belin, Paris.
Lecointre, G. and H. Le Guyader. 2006a. Classification Phylogénétique du Vivant. 3rd ed. Belin, Paris.
Lecointre, G. and H. Le Guyader. 2006b. The Tree of Life. A Phylogenetic Classification. The Belknap Press of Harvard University Press. Cambridge, Massachusetts, London.
Lecointre, G. and P. Deleporte. 2000. Le principe du "total evidence" requiert l'exclusion de données trompeuses, pp. 129–151. In: V. Barriel and T. Bourgoin [eds.]. Biosystema 18: Caractères, Société Française de Systématique, Paris.
Lecointre, G. and P. Deleporte. 2005. Total evidence requires exclusion of phylogenetically misleading data. Zool. Scr. 34(1):101–117.
Liem, K.L., W.E. Bemis, W.F. Walker and L.A. Grande. 2001. Functional Anatomy of the Vertebrates, 3rd ed. Brooks/Cole – Thomson Learning, Belmont, California.
Lopez, J.A., W.J. Chen and G. Orti. 2004. Esociform phylogeny. Copeia 2004(3): 449–464.
Mahner, M. and M. Bunge. 1997. Foundations of Biophilosophy. Springer Verlag, Berlin, Heidelberg.
Meunier, F.J. and M. Gayet. 1991. Premier cas de morphogénèse réparatrice de l'endosquelette caudal d'un poisson gymnotiforme du miocène supérieur bolivien. Geobios, ms. 13: 223–230.
Miyamoto M. and W.M. Fitch. 1995. Testing species phylogenies and phylogenetic methods with congruence. Syst. Biol. 44: 64–76.

Monod, T. 1967. Le complexe urophore des Téléostéens: typologie et évolution (note préliminaire), pp. 111–131. In: Problèmes Actuels de Paléontologie (Evolution des Vertébrés). Colloques CNRS n°163.

Monod, T., 1968. Le complexe urophore des poissons Téléostéens. Mém. Inst. Fond. Afr. Noire 81.

Moser, H.G., W.J. Richards, D.M. Cohen, M.P. Fahay, A.W. Kendall Jr. and S.L. Richardson. 1984. Ontogeny and systematics of fishes based on an international symposium dedicated to the memory of Elbert Halvor Ahlstrom. Am. Soc. Ichthyol. Herpetol. Spec. Pub. 1: 1–760.

Müller-Schmid, A., B. Ganss, T. Gorr and W. Hoffmann. 1993. Molecular analysis of ependymins from the cerebrospinal fluid of the orders Clupeiformes and Salmoniformes: no indication for the existence of the euteleost infradivision. J. Mol. Evol. 36: 578–585.

Nelson, G.J. 1969. Origin and diversification of teleostean fishes. Ann. New York Acad. Sci. 167: 18–30.

Nelson, G.J. 1970. The hypobranchial apparatus of teleostean fishes of the families Engraulidae and Chirocentridae. Am. Mus. Novitates 2410: 1–30.

Nelson, G.J. 1989. Phylogeny of major fish groups, pp. 325–336. In: B. Fernholm, K. Bremer and H. Jörnvall [eds.]. The Hierarchy of Life. Molecules and morphology in phylogenetic analysis. Proc. Nobel Symp. 70, International congress series 824. Excerpta Medica, Amsterdam, New York, Oxford.

Nelson, J.S. 1976. Fishes of the World. 1st ed. John Wiley and Sons, Chichester, New York.

Nelson, J.S. 1984. Fishes of the World. 2nd ed. John Wiley and Sons, Chichester, New York.

Nelson, J.S. 1994. Fishes of the World. 3rd ed. John Wiley and Sons, New York.

Nelson, J.S. 2006. Fishes of the world. 4th ed. John Wiley and Sons, Inc., New York.

Nixon, K. and J.M. Carpenter. 1996. On simultaneous analysis. Cladistics 12: 221–241.

Novacek, M.J. and L.G. Marshall. 1976. Early biogeographic history of ostariophysan fishes. Copeia 1976: 1–12.

Orti, G. and A. Meyer. 1996. Molecular evolution of ependymin and the phylogenetic resolution of early divergences among euteleost fishes. Mol. Biol. Evol. 13(4): 556–573.

Page, R.D.M. and E.C. Holmes. 1998. Molecular evolution: a phylogenetic approach. Blackwell Science, Abingdon, UK.

Patterson, C. 1970. A clupeomorph fish from the Gault (Lower Cretaceous). Zool. J. Linn. Soc. 49(3): 161–182.

Patterson, C. 1984. Chanoides, a marine eocene otophysan fish (Teleostei: Ostariophysi). J. Vert. Paleontol. 4(3): 430–456.

Patterson, C. 1998. Comments on basal teleosts and teleostean phylogeny, by Gloria Arratia. Copeia 1998 (1): 1107–1113.

Patterson, C. and D.E. Rosen. 1977. Review of ichthyodectiform and other Mesozoic teleost fishes and the theory and practice of classifying fossils. Bull. Am. Mus. Nat. Hist. 158(2): 83–172.

Peng, Z., S. He, J. Wang, W. Wang and R. Diogo. 2006. Mitochondrial molecular clocks and the origin of the major otocephalan clades (Pisces: Teleostei): A new insight. Gene 370: 113–124.

Philippe, H. and E. Douzery. 1994. The pitfalls of molecular phylogeny based on four species, as illustrated by the Cetacea/Artiodactyla relationships. J. Mammal. Evol. 2: 133–152.

Regan, C.T. 1911a. The classification of teleostean fishes of the order Ostariophysi—1. Cyprinoidea. Ann. Mag. Nat. Hist. 8: 13–32.

Regan, C.T. 1911b. The classification of teleostean fishes of the order Ostariophysi—2. Siluroidea. Ann. Mag. Nat. Hist. 8: 553–557.

Rieppel, O. 2003a. Popper and Systematics. Syst. Biol. 52: 259–271.

Rieppel, O. 2003b. Semaphoronts, cladograms and the roots of total evidence. Biol. J. Linn. Soc. 80: 167–186.

Rieppel, O. 2004a. The language of systematics, and the philosophy of 'total evidence'. Syst. Biodivers. 2: 9–19.

Rieppel, O. 2004b. What happens when the language of science threatens to break down in systematics: a popperian perspective, pp. 57–100. In: D. Williams and P. Forey [eds.]. Milestones in Systematics. CRC Press, London.

Rieppel, O. 2005a. The philosophy of total evidence and its relevance for phylogenetic inference. Papéis Avulsos de Zoologia. Mus. Zool. Univ. Sao Paulo 45(8): 77–89.

Rieppel, O. 2005b. Le cohérentisme en systématique, pp. 115–126. In: P. Deleporte and G. Lecointre [eds.]. Philosophie de la Systématique, Biosystema 24. Société Française de Systématique, Paris.

Roberts, T. 1969. Osteology and relationships of characoid fishes, particularly the genera Hepsetus, Salminus, Hoplias, Ctenolucius, and Acestrorhynchus. Proc. Calif. Acad. Sci., 4th ser. 36(15): 391–500.

Roberts, T. 1973. Interrelationships of ostariophysans, pp. 373–395. In: P.H. Greenwood, R.S. Miles and C. Patterson [eds.]. Interrelationships of Fishes. Academic Press, London.

Rosen, D.E. 1973. Interrelationships of higher euteleostean fishes, pp. 397–513. In: P.H. Greenwood, R.S. Miles and C. Patterson [eds.]. Interrelationships of Fishes II. Academic Press, London.

Rosen, D.E. 1974. Phylogeny and zoogeography of salmoniform fishes and relationships of Lepidogalaxias salamandroides. Bull. Am. Mus. Nat. Hist. 153: 265–326.

Rosen, D.E. 1982. Teleostean interrelationships, morphological function and evolutionary inference. Am. Zool. 22: 261–273.

Rosen, D.E. 1985. An essay on euteleostean classification. Am. Mus. Novitates 2827: 1–57.

Rosen, D.E. and P.H. Greenwood. 1970. Origin of the Weberian apparatus and the relationships of the ostariophysan and gonorynchiform fishes. Am. Mus. Novitates 2428: 1–25.

Saitoh, K., M. Miya, J.G. Inoue, N.B. Ishiguro and M. Nishida. 2003. Mitochondrial genomics of ostariophysan fishes: perspectives on phylogeny and biogeography. J. Mol. Evol. 56: 464–472.

Sanford, P.J. 1990. The phylogenetic relationships of the salmonoid fishes. Bull. Br. Mus. (Nat. Hist.), Zool. 56: 145–153.

Seo, T.K., H. Kishino and J.L. Thorne. 2005. Incorporating gene-specific variation when inferring and evaluating optimal evolutionary tree topologies from multilocus sequence data. Proc. Natl. Acad. Sci.,102(12): 4436–4441.

Siddall, M.E. 1997. Prior agreement: arbitration or arbitrary? Syst. Biol. 46: 765–769.

Song, J. and L.R. Parenti. 1995. Clearing and staining whole fish specimens for simultaneous demonstration of bone, cartilage and nerves. Copeia 1995: 114–118.

Steinke, D., W. Salzburger and A. Meyer. 2006. Novel relationships among ten fish model species revealed based on a phylogenomic analysis using ESTs. J. Mol. Evol. 62: 772–784.

Stiassny, M.L.J, L.R. Parenti and G.D. Johnson. 1996. Interrelationships of Fishes II. Academic Press, New York.

Taverne, L. 1977. Ostéologie, phylogenèse et systématique des téléostéens fossiles et actuels du super-ordre des Ostéoglossomorphes. Première partie. Ostéologie des genres Hiodon, Eohiodon, Lycoptera, Osteoglossum, Scleropages, Heterotis et Arapaima. Mém. Class. Sci. Acad. Roy. Belg. 42: 1–235.

Taverne, L. 1978. Ostéologie, phylogenèse et systématique des téléostéens fossiles et actuels du super-ordre des Ostéoglossomorphes. Deuxième partie. Ostéologie des genres Phareodus, Phareoides, Brychaetus, Musperia, Pantodon, Singida, Notopterus, Xenomystus et Papyrocranus. Mém. Class. Sci. Acad. Roy. Belg. 42: 1–213.

Taverne, L. 1979. Ostéologie, phylogenèse et systématique des téléostéens fossiles et actuels du super-ordre des Ostéoglossomorphes. Troisième partie. Evolution des structures ostéologiques et conclusions générales relatives à la phylogénèse et à la systématique du Super-Ordre. Addendum, Mém. Class. Sci. Acad. Roy. Belg. 43(2): 1–168.

Taylor, W.R. 1967. An enzyme method of clearing and staining small vertebrates. Proc. U.S. Nat. Mus. 122: 1–17.

Tudge, C. 2000. The Variety of Life. Oxford University Press, Oxford.

Vandewalle, P., R. Radermaker, C. Surlemont and M. Chardon. 1990. Apparition of the Weberian characters in *Barbus barbus* (Linne 1758) (Pisces Cyprinidae). Zool. Anz. 225: 362–376.

Zaragüeta-Bagils, R., S. Lavoué, A. Tillier, C. Bonillo and G. Lecointre. 2002. Assessment of otocephalan and protacanthopterygian concepts in the light of multiple molecular phylogenies. C.R. Biol. 325: 1–17.

10

Systematics and Phylogenetic Relationships of Cypriniformes

Andrew M. Simons[1] and Nicholas J. Gidmark[2]

Abstract

The Cypriniformes is a large group of fishes containing eight families, 280 genera, and over 3,500 species. These fishes are diverse, occupying a wide range of freshwater aquatic habitats in North America, Europe, Asia, India, and Africa. The taxonomy and classification of Cypriniformes is complex and phylogenetic relationships of the included groups are largely unresolved. This lack of resolution stems from the large geographic distribution of taxa, the large number of species, and the difficulty of obtaining specimens of critical taxa. In this chapter we review previous phylogenetic studies and classifications within and among cypriniform families, discuss problematic taxa, and present recent changes to the taxonomy. Recent morphological and molecular studies have provided hypotheses of relationships among included families and forced a reconsideration of the previous five-family classification. Current phylogenetic hypotheses indicate that Cypriniformes can be classified into two major monophyletic groups, Cyprinoidea, containing the Cyprinidae and Cobitoidea, containing Catostomidae, Gyrinocheilidae, Botiidae, Vailantellidae, Cobitidae, Balitoridae, and Nemacheilidae. The Cyprinidae contains a large number of species divided into several subfamilies. There

[1] Bell Museum of Natural History & Department of Fisheries, Wildlife, and Conservation Biology, University of Minnesota, Saint Paul, MN 55108, U.S.A.
[2] Department of Ecology and Evolutionary Biology, Brown University, Providence, RI 02912, U.S.A.

is little agreement on the content or relationships among these taxa. Cobitoidea contains fewer species but is also the subject of taxonomic debate. Issues include the relationships of catostomids and gyrinocheilids with respect to the remaining families; species relationships within catostomids; balitorids, cobitids, and nemacheilids; and the phylogenetic position of the cobitids.

Introduction

The Cypriniformes is a very large clade of fishes containing over 3,500 species that are almost entirely restricted to fresh water. Cypriniformes are native to Africa, Asia, Europe, India, and North America but absent from Antarctica, Australia, and South America. These fishes are widely distributed on continents where they occur and occupy nearly every available freshwater habitat. Cypriniformes have been found in streams high in the Himalayas and several taxa have invaded cave environments. They occur in hot springs and proglacial lakes, small creeks and springs, as well as very large rivers. Although considered primarily freshwater fishes, several species have been reported from brackish water including *Mylocheilus caurinus* (Richardson), *Tribolodon hakonensis* (Günther), and *Pogonichthys macrolepidotus* (Ayres) (Carl *et al.* 1976, Meng and Moyle 1995, Okada 1960). Cypriniform fishes exhibit a wide range of sizes. At 7.9 mm, *Paedocypris progenetica* Kottelat, Britz, Hui and Witte is the smallest known vertebrate (Kottelat *et al.* 2006), whereas *Catlocarpio siamensis* Boulenger reaches at least 2.5 m and *Tor putitora* (Hamilton) reaches 2.7 m in length (Nelson 1994). Cypriniformes have the reputation of being taxonomically intractable and many ichthyology students have expressed despair when faced with a tray or net full of silvery minnows. The taxonomic difficulties posed by this group are reflected in the small number of higher taxonomic categories in cypriniform classifications compared to other similar-sized clades.

Cypriniform fishes are of great economic importance. Several species are used as food fishes. Many are important as forage for game fishes, and others are raised for use as bait in sports fisheries. In Asia, carps have been raised for food in multispecies aquaculture systems for thousands of years, and domesticated carp, *Cyprinus carpio* Linnaeus, and goldfish, *Carassius auratus* (Linnaeus), are widely used in the pet trade. In addition, many other cypriniform fishes, particularly those from Southeast Asia, are widely available in the pet trade. The ease of culture, and the variation in life history, ecology, and trophic variation of cypriniform fishes make them of great interest to biologists. The zebra danio, *Danio rerio* (Hamilton), for example, is a model organism for developmental biologists. However, placing this information into an evolutionary context is hampered by the lack of a comprehensive phylogeny for Cypriniformes. In this chapter, we review evidence supporting monophyly of the Cypriniformes, the diversity within the group, previous phylogenetic hypotheses, and other issues in cypriniform systematics.

There are few papers that deal with higher order relationships within the Cypriniformes. Hensel (1970) provided a thorough review of the taxonomic history of the included taxa. Fink and Fink (1981) presented a hypothesis of relationships of ostariophysan fishes and provided character evidence for the monophyly of the Cypriniformes. They identified seven unreversed unique synapomorphies and eight homoplasic characters supporting monophyly of the Cypriniformes (Fink and Fink 1981). The unique and unreversed characters that support monophyly of Cypriniformes are: kinethmoid bone; preethmoid tightly articulated between the vomer and mesethmoid; dorso-medial palatine process that abuts mesethmoid; palatine articulates posteriorly with mesopterygoid via a concave facet; ectopterygoid does not overlap palatine anteriorly; premaxilla extends furthest dorsally adjacent to midline of fish; fifth ceratobranchial enlarged, extending much further dorsally than other ceratobranchials (not true of gyrinocheilids [Ramaswami 1952a]); teeth on fifth ceratobranchial ankylosed to the bone (not true of gyrinocheilids [Ramaswami 1952a, Siebert 1987]); elongate lateral process on second centrum. The non-unique (homoplasious) characters include maxillary barbels; absence of teeth in the jaw; loss of two posterior pharyngobranchial toothplates; no teeth on 2nd or 3rd pharyngobranchial toothplates; no toothplate associated with basibranchials 1–3 (gyrinocheilids do have presumed dermal ossifications associated with copula 3 although these lack teeth [Siebert 1987]); postcleithra reduced to one; two or fewer epurals; no adipose fin. Note that Fink and Fink (1981) focused on two genera of cypriniformes, *Ospariichthys* and *Zacco*, both cyprinids. This sampling may have biased their results, as it assumes these taxa adequately represent the plesiomorphic condition of Cyprinidae and that Cyprinidae represents the plesiomorphic condition of Cypriniformes. Chen *et al.* (1984) considered *Opsariichthys* and *Zacco* to be members of their Danionine clade (= Rasborinae), which validates Fink and Fink's (1981) use of these taxa. However, Howes (1980, 1983) included *Opsariichthys* and *Zacco* in a bariliine clade and Cavender and Coburn (1992) considered these genera to be members of the Xenocyprinae plus Cultrinae clade, nested well within Cyprinidae. Cavender and Coburn's (1992) hypothesis was corroborated by Satioh *et al.*'s (2006) analysis, which placed *Opsariichthys* plus *Zacco* sister to *Xenocypris* and *Ischikauia* plus *Culter*. Fink and Fink (1981) mentioned two characteristics plesiomorphic for Cypriniformes not observed in *Opsariichthys* or *Zacco*; first pharyngobranchial is present (Gyrinocheilidae, Catostomidae) and a posterior cranial fontanelle is present (Gyrinocheilidae, some catostomids, some cobitids, some balitorids, and some cyprinids). In this review, we recognize eight families in the Cypriniformes: Catostomidae, Gyrinocheilidae, Cobitidae, Botiidae, Balitoridae, Nemacheilidae, Vaillantellidae and Cyprinidae (Šlechtová et al. 2007).

Catostomidae

The family Catostomidae contains the suckers, a group composed of approximately 76 species (Mayden et al. 1992, Jenkins and Burkhead 1993). These are classified in approximately 14 genera, although some are paraphyletic (Harris and Mayden 2001). The number of recognized genera is bound to change as we obtain a clearer picture of the relationships within the family. Catostomids are widely distributed in North and Central America with one North American species, Catostomus catostomus (Forster), also found in eastern Siberia, a presumed dispersal across the Bering land bridge during the Pleistocene (Lindsey and McPhail 1986). A single catostomid species, Myxocyprinus asiaticus (Bleeker), inhabits the Yangtze River basin in China. This trans-Pacific distribution is reflected in the fossil record; Amyzon, a fossil catostomid, is known from the Eocene and Paleocene of western North America and China (Grande et al. 1982, Cope 1872, Wilson 1977, Chang et al. 2001). Catostomids are tetraploid with a diploid chromosome count of 100 (Uyeno and Smith 1972, Suzuki 1992). Most have fleshy, papillose or plicate lips and are benthic feeders although Ictiobus cyprinellus (Valenciennes) is a planktivore. Eastman (1977) recognized three trophic categories: microphagous filter feeding, generalized benthic feeding, and specialization on large mollusks. In addition, the mountain suckers are specialized algae scrapers (Smith 1966). Suckers have a single row of teeth on the fifth ceratobranchial; these act against a basioccipital process. The basioccipital process extends anteriorly and supports a large palatal organ. The palatal organ, together with the epibranchials, form an epibranchial organ (Weisel 1960, Eastman 1977) believed to be important in sorting and retaining food particles. Sibbing (1988) demonstrated this function in Cyprinus carpio, a cyprinid. The anatomy of the palatal organ is similar in Cyprinus and catostomids, although the function in catostomids has not been determined experimentally. In catostomids the basioccipital process is a delicate network of fenestrated bone rather than a solid bony plate as observed in many cyprinids. It may or may not surround the dorsal aorta depending on the species (Lo and Wu 1979, Wu et al. 1981). The process also lacks a horny pad on its base, with the exception of Moxostoma carinatum (Cope), M. hubbsi Legendre, and the undescribed sicklefin sucker, Moxostoma sp., which presumably feed on mollusks and exhibit trophic morphologies similar to common carp (Cyprinus carpio). In these species, the pharyngeal teeth are large and molariform, acting against a horny pad supported by a robust basioccipital process (Eastman 1977).

Howes (1991), Berrebi *et al.* (1996), and Smith (1992) all raised the possibility that the Catostomidae may be nested within the Cyprinidae. Howes (1991) suggested that catostomids may the North American representatives of cyprinines or gobionines. Smith (1992) listed several similarities between cyprinines and catostomids, including lobed epibranchials, presence of paired basipterygoid processes articulating with the first branchial arches, and fusion of the first supraneurals with the Weberian apparatus. Berrebi *et al.* (1996) cited the issue of tetraploidy in cyprinines—particularly *Barbus*—and catostomids. None of these hypotheses have been corroborated by phylogenetic analyses and there is no evidence that Catostomidae are nested within Cyprinidae.

Smith (1992) recognized three subfamilies within the Catostomidae: Ictiobinae, containing the fossil genus *Amyzon* and the extant genera *Ictiobus* and *Carpiodes*; Cycleptinae, containing *Myxocyprinus* and *Cycleptus*; and Catostominae, containing *Catostomus, Chasmistes, Deltistes, Xyrauchen, Erimyzon, Minytrema, Moxostoma, Thoburnia, Scartomyzon,* and *Hypentelium. Thoburnia, Scartomyzon* and *Hypentelium* are considered subgenera within *Moxostoma* by some authors. Systematic work on catostomids has been based on morphological, biochemical, and molecular data. In particular, a great deal of work has been done on the phylogenetic pattern of gene silencing or functional diploidization (Buth 1979, 1980, Clements *et al.* 2004, Ferris 1984, Ferris and Whitt 1977, 1979, 1980).

Smith (1992) presented the most thorough phylogenetic analysis of catostomid relationships (Fig. 10.1). He included 62 extant taxa and one extinct genus in his matrix and coded 157 characters, including osteological, biochemical, soft tissue, morphometric and ontogenetic data. These were analyzed using parsimony as the optimality criterion. Smith's analysis found two equally parsimonious trees that differed in the relationships among *Scartomyzon ariommum, Hypentelium,* and *Thoburnia.* Note that *Moxostoma* was paraphyletic with respect to *Scartomyzon, Thoburnia,* and *Hypentelium.* Harris and Mayden (2001) examined phylogenetic relationships of catostomids using DNA sequences of mitochondrial genes (Fig. 10.2). The results from these analyses illustrate difficult issues in catostomid systematics including (1) phylogenetic position of *Myxocyprinus,* (2) relationships of *Catostomus, Chasmistes, Deltistes,* and *Xyrauchen,* and (3) *Moxostoma* and its relatives.

Phylogenetic Position of the Asian Sucker *Myxocyprinus*

This taxon has long been considered to be the sister taxon to all other catostomids. This perspective appears to have been based more on

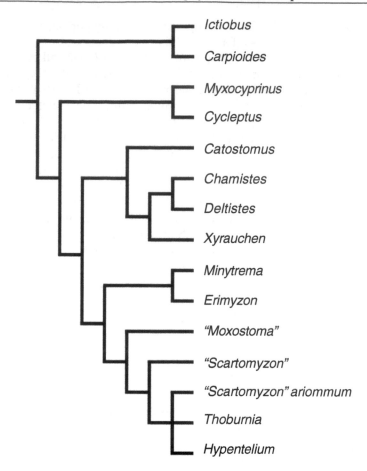

Fig. 10.1 Phylogenetic relationships of extant catostomid genera based on osteological biochemical, soft tissue, morphometric and ontogenetic characters, modified from Smith (1992). *Moxostoma* is paraphyletic, *Scartomyzon* is rendered paraphyletic by the relationship of *S. ariommum* to *Thoburnia* and *Hypentelium*.

biogeographic considerations than on rigorous phylogenetic analyses. Smith (1992) considered *Myxocyprinus* sister to *Cycleptus*. Harris and Mayden (2001) found *Myxocyprinus* was sister to all remaining catostomids based on analysis of both large and small mitochondrial ribosomal RNAs; however, analysis of only the small ribosomal RNA sequences recovered Smith's *Myxocyprinus* plus *Cycleptus* hypothesis. In an analysis of complete mitochondrial genomes of 53 cypriniform fishes, Saitoh *et al.* (2006) found *Myxocyprinus* sister to *Cycleptus* plus *Carpiodes* (they did not include *Ictiobus*). These conflicting hypotheses raise questions of the validity and membership of the subfamilies Cycleptinae and Ictiobinae.

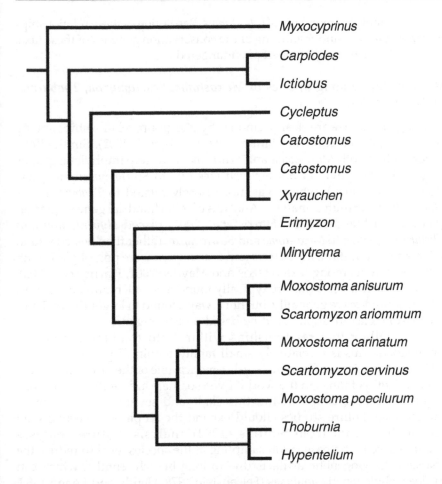

Fig. 10.2 Phylogenetic relationships of catostomid genera based on mitochondrial DNA sequences, modified from Harris and Mayden (2001).

Monophyly and Relationships of the Genera *Catostomus*, *Chasmistes*, *Deltistes*, and *Xyrauchen*

Smith (1992) recovered a monophyletic *Catostomus* sister to a clade containing *Chasmistes*, *Deltistes*, and *Xyrauchen*, all of which are morphologically distinctive suckers from western North America. Harris and Mayden (2001) and Harris *et al.* (2002) did not recover a monophlyetic *Catostomus* with respect to *Deltistes* and *Xyrauchen*; *Chasmistes* was not included in their study. Restricted taxon sampling of *Catostomus* and the possibilites of hybridization may have misled the molecular analyses (Harris *et al.* 2002). However, it is possible that *Chasmistes*, *Deltistes*, and *Xyrauchen* are simply highly apomorphic members of *Catostomus*. Resolution of these relationships is important to an understanding of the conservation genetics of these taxa, several of which are threatened or endangered.

Relationships among Species of *Moxostoma, Scartomyzon, Thoburnia,* and *Hypentelium*

Although these taxa are clearly a monophyletic group, relationships among them are not well resolved (Smith 1992, Harris *et al.* 2002). Smith (1992) reported that both *Moxostoma* and *Scartomyzon* were paraphyletic groups (Fig. 10.1). Paraphyly of *Scartomyzon* was due to *Scartomyzon ariommum* (Robins and Raney), which was more closely related to *Thoburnia* plus *Hypentelium* in Smith's analysis. Analysis of mitochondrial gene sequences (Harris and Mayden 2001, Harris *et al.* 2002) placed *Hypentelium* and *Thoburnia* as sister to *Moxostoma* and *Scartomyzon* rather than nested within this clade. In addition, *Moxostoma* and *Scartomyzon* were polyphyletic with little resolution among taxa (Harris and Mayden 2001, Harris *et al.* 2002). *Scartomyzon* and *Thoburnia* are typically found in small, highland streams, while *Moxostoma* are generally, but not always, found in larger rivers. Thus, the existing classification could be based on convergence.

It is telling that relationships within Catostomidae are poorly understood. This is a relatively small family within Cypriniformes and all but one species are found in North America, one of the most intensively studied ichthyofaunas in the world. Even so, there are multiple questions regarding relationships within this clade and several species remain undescribed. Future studies should expand the scope of molecular data and include data from Smith's (1992) analysis. Future analyses should strive to increase taxon sampling at the species level to reduce the chance of phylogenetic artifacts due to long branch lengths, which can mislead phylogenetic analyses (Felsenstein 1978, Hendy and Penny 1989, Graybeal 1998).

Gyrinocheilidae

The family Gyrinocheilidae was erected by Hora (1923) and contains the algae eaters. These are unique among cypriniform fishes as they lack teeth and the fifth ceratobranchial is unspecialized and rodlike (Ramaswami 1952a, Siebert 1987). Gyrinocheilids are native to Southeast Asia and are found in fast-flowing waters, usually in or near rapids (Roberts and Kottelat 1993). They exhibit several specializations for this environment. The mouth is surrounded by an oral sucker equipped with keratinized rasps (Benjamin 1986). The rasps enable gyrinocheilids to remove algae from the surface of rocks and the sucker also enables the fish to hold on to rocks in a fast current. The sucker chamber can be isolated from the buccal cavity and is capable of creating a true vacuum (Benjamin 1986), unlike in most fishes with suctorial mouths. The opercular slit of gyrinocheilids is divided into a dorsal and ventral opening. The dorsal opening allows the intake of water for gas exchange (Hora 1923, Smith 1945) (The dorsal opening is sometimes referred to as a spiracle but it is not homologous to the true gnathostome spiracle.) Ramaswami (1952a) described cranial osteology and Benjamin (1986) presented a detailed description of the anatomy of the oral sucker. Siebert (1987) does not provide synapomorphies for Gyrinocheilidae but the absence of pharyngeal teeth, the unique jaw morphology, and the specialized opercular opening all support the monophyly of this family. Three species are recognized in this group: *Gyrinocheilus aymonieri* (Tirant), *G. pennocki* (Fowler), and *G. pustulosus* Vaillant (Roberts and Kottelat 1993).

Cobitidae

The Cobitidae contains many popular aquarium species. There are approximately 150 species widespread across Eurasia and extending into northern Africa. Cobitids have an unusual, mobile, spine-like lateral ethmoid that is presumed to deter predators. Cobitids are generally elongate, round-bodied fishes with a single pair of rostral barbels (Nelson 1994). The family includes some cave-dwelling species that are eyeless and lack pigment (Roberts 1989, Yang *et al.* 1994). Sawada (1982) lists seven unique and unreversed synapomorphies for Cobitidae: absence of coronomeckelian; orbitosphenoid separated from pterosphenoid; rod-shaped entopterygoid; reduced cleithrum; first hypural fused with first preural centrum; second preural centrum fused with its haemal spine.

Botiidae

Botiids, in contrast, are stout, laterally compressed fishes and have two or more pairs of rostral barbels (Kottelat and Chu 1987, Nelson 1994). They are benthic fishes usually found in slow to moderately swift streams over rocky substrates (Kottelat and Chu 1987). The family contains approximately 55 described species and like the cobitids have a moveable spine-like lateral ethmoid and a socket-like articulation between the supraethmoid-complex and the frontals (Sawada 1982). In a phylogenetic analysis of Botiids, Šlechtová *et al.* (2006) divided botiids into two lineages. The first contains the diploid species in the genera *Parabotia* and *Leptobotia*. The second clade contains the tetraploid species in the genera *Botia*, *Chromobotia*, *Yasuhikotakia*, *Syncrossus*, and *Sinibotia*. *Yasuhikotakia* is paraphyletic with two species, *Y. nigrolineata* and *Y. sidthimunki* forming the sister group to *Sinibotia*. Sawada (1982) identified two synapomorphies for the Botiidae, a sesamoid bone dorsal to the second preethmoid (Ramaswami 1953) and the separation of the parietal from the sphenotic.

Balitoridae

The Balitoridae contains the flat loaches. These fishes are known from fast-flowing streams in India through Southeast Asia. Balitorids generally have flattened bodies, subterminal or inferior mouths, and enlarged pectoral and pelvic fins. The anteriormost rays of the pectoral and pelvic fins have large friction pads (Roberts 1989) that are used to hold the fishes in place on the stream bottom. In some species, the pectoral and pelvic fins are greatly enlarged and modified to form a sucking disk. Traditionally, the balitorids have been divided into two groups, Balitorini and Gastromyzontini, based on the number of unbranched rays in the pectoral and pelvic fins. Balitorins have two or more unbranched anterior rays in both pectoral and pelvic fins, whereas gastromyzontins have a single unbranched anterior ray in the pectoral and pelvic fins (Nelson 1994). However, according to Kottelat and Chu (1988a), *Balitora* has a single unbranched pelvic ray. A recurring issue in balitorine systematics is the relationship of the Asian and Bornean forms. Ramaswami (1952b) cites Fang (1934), who considered the gastromyzontins to be polyphyletic containing two groups: a *Crossostoma* group derived from nemachiline loaches and a gastromyzontin group. Hora (1932, 1952a, 1952b) considered the balitorins to be derived from a cyprinid ancestor and the gastromyzontins to be derived from a cobitid ancestor. He furthered this by arguing for independent origins of the Bornean and mainland gastromyzontids from a cobitid ancestor (Hora 1952b). Silas (1953) supported Hora's (1932) hypothesis and argued further that the gastromyzontins had evolved independently in Asia and Borneo. Roberts (1989) pointed out that in Bornean balitorins, the enlarged pelvic rib is borne on the 10th vertebra, while in the gastromyzontins it is borne on the 12th vertebra.

Siebert (1987) reported that Gastromyzontini differs from *Ellopostoma* and Balitorini in that it lacks a subtemporal fossa. Most systematic work is focused on alpha taxonomy and little is known of the phylogenetic relationships among balitorids. It is clear that the group would benefit from a comprehensive examination of morphology across all taxa. The current distinction between the Balitorini and Gastromyzontini appears artificial and requires a more inclusive examination of these taxa from their entire range.

Nemacheilidae

Nemacheilids are small benthic fishes distributed in Europe, Asia, and northeast Africa. They occupy a wide range of habitats from hot springs in Tibet, 5,200 m above sea level (Zhu 1981), to subterranean creeks, 400 m below the surface (Chu and Chen 1979). There are approximately 300 valid species within the group (Kottelat and Chu 1988b, Kottelat 1990). Kottelat (1990) presents an excellent overview of the classification and taxonomic history of nemacheilid loaches. Little is known of the phylogenetic relationships among taxa and most systematic work has been focused on alpha taxonomy (Kottelat 1984, 1990, Banarescu and Nalbant 1995).

Vaillantellidae

Vaillantella contains three species: *V. cinnamomea* Kottelat, *V. euepiptera* (Vaillant), and *V. maassi* Weber and de Beaufort. These unusual fishes have an extremely long dorsal fin, unlike other loaches. Little is known of their biology. Roberts (1989) reports that *V. eupiptera* is found in lowland streams, while *V. maasii* occurs in highland streams; according to Kottelat (1994), vaillantellids can be found in accumulations of leaf litter in forest streams.

Cyprinidae

The Cyprinidae is the most species-rich family in the Cypriniformes. There are over 2,400 described species in approximately 210 genera, a number that, no doubt, will rise dramatically with further study. Cyprinids are classified into several subfamilies although the number of recognized subfamilies varies with the investigator and there is little evidence for monophyly of several of these taxa. The number of subfamilies ranges from as few as two (Cavender and Coburn 1992) to 12 (Arai 1982). Subfamilial names include Acheilognathinae, Alburninae, Barbinae, Cultrinae, Cyprininae, Gobioninae, Labeoninae, Leucisinae, Psiloryhnchinae, Rasborinae (=Danioninae [Gosline 1978]), Schizothoracinae, Tincinae, and Xenocyprinae. Our understanding of higher-level relationships of cyprinids is based largely on morphological studies including Chen *et al.* (1984), Cavender and Coburn (1992), and several works by Howes (summarized in Howes 1991).

Characters supporting monophyly of the Cyprinidae include: absence of an uncinate process on epibranchials I and II (Siebert 1987); absence of infrapharyngobranchial I (Siebert 1987); overlap of infrapharyngobranchial II by infrapharyngobranchial III (Siebert 1987); deep, well-developed subtemporal fossae (Siebert 1987); interorbital septum formed by both orbitosphenoid and parasphenoid (Cavender and

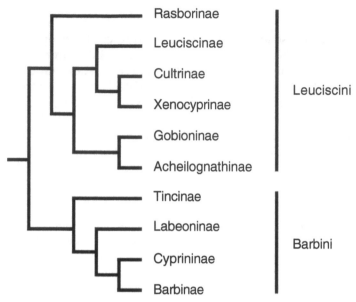

Fig. 10.3 Phylogenetic relationships of cyprinid subfamilies, modified from Chen *et al.* (1984).

Coburn 1992); fused second and third weberian vertebrae, although this character is homoplastic and the derived condition is observed in cobitoids, derived catostomids, and varies within cyprinids (Cavender and Coburn 1992); anterior opening of the trigeminal-facial chamber positioned between the prootic and pterosphenoid (Cavender and Coburn 1992); loss of contact between infraorbital five and the supraorbital (Cavender and Coburn 1992); ossified preethmoids (Cavender and Coburn 1992); and presence of an opercular canal (Cavender and Coburn 1992). The opercular canal does not occur in *Tinca*, acheilognathins, or phoxinins (a group contained within Leuciscinae) but Cavender and Coburn (1992) argue that this is a secondary loss, a hypothesis corroborated by phylogenies published by Cunha *et al.* (2002) and Saitoh *et al.* (2006).

Chen *et al.* (1984) examined phylogenetics of Cyprinidae based on 25 osteological characters for 92 species. They recognized two major clades, Leuciscini and Barbini (Fig. 10.3). Tincinae was included in the Barbini, sister to Labeoninae and Barbinae plus Cyprininae. Cavender and Coburn (1992) expanded Chen *et al.*'s (1984) data to include 47 characters. They recovered a very similar set of subfamilial relationships to those recovered by Chen *et al.* (1984); the major difference between these analyses was the position of Tincinae. Cavender and Coburn (1992) considered tincins to be the sister to the Leuciscini rather than the Barbini (Fig. 10.4). Like Chen *et al.* (1984), Cavender and Coburn (1992) recognized only two subfamilies

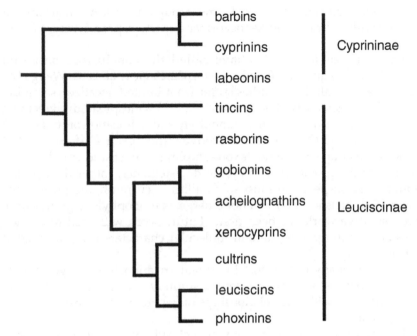

Fig. 10.4 Phylogenetic relationships of cyprinid subfamilies, modified from Cavender and Coburn (1992).

within Cyprinidae, Leuciscinae and Cyprininae. Saitoh *et al.* (2006) also found strong support for these two lineages within Cyprinidae.

Cunha *et al.* (2002) examined phylogenetic relationships of Eurasian and North American cyprinids using mitochondrial cytochrome *b* sequences. Their analysis included representatives from most major groups of Cyprinidae. They found little support for monophyly of cyprinid subfamilies as presently defined. While North American and European leuciscins (including European alburnins) formed a monophyletic group in their analysis, Asian leuciscins were more closely related to other Asian taxa than to European leuciscins, which supported the analysis of Cavender and Coburn (1992). Cunha *et al.* (2002) remarked that Asian species were generally more closely related to each other than to European or North American species from the same genera or subfamilies. Similar results, albeit with lower taxon sampling, were reported by Wang *et al.* (2002) and S. He *et al.* (2004). S. He *et al.* (2004) also used cytochrome *b* to examine relationships of 54 species of cyprinids. They found support for the monophyly of Cultrinae, Xenocyprinae, and Gobioninae, unlike Cuhna *et al.* (2002); however, they had much lower taxon sampling, weakening the test of monophyly for included taxa. S. He *et al.* (2004) also remarked that the leuciscins contained two distinct groups, an East Asian group that clustered with Cultrinae and Xenocyprinae and a European, Siberian, North American group.

These molecular studies have called the conclusions based on morphology into question (Cunha *et al.* 2002, S. He *et al.* 2004, Wang *et al.* 2002). However, all these studies suffer from limited, localized sampling and small character sets. It is clear that there is widespread disagreement among all these studies as to the monophyly and delineation of subfamilial groupings in the Cyprinidae. Moreover, the absence of substantial taxonomic overlap and limited taxon sampling of most studies only serves to illustrate the problems with cyprinid systematics, rather than posing solutions. We suspect that most subfamilies as currently recognized (see Howes 1991, Nelson 1994) will not emerge as monophyletic groups once more extensive work has been done. Future work will need to include assessment of morphological and molecular characters using a range of outgroup taxa.

There are many papers that investigate relationships of cyprinids at finer taxonomic scales. We present a summary of these, emphasizing papers that deal with interrelationships of genera or relationships within large genera.

The Acheilognathins are a phylogenetically enigmatic group. There are 40 described species, of which all but one (*Rhodeus sericeus*) are restricted to Asia. They are well recognized as monophyletic and engage in an interesting spawning symbiosis with freshwater mussels whereby the female lays eggs on the gills of the mussel through the excurrent siphon (Smith *et al.* 2004). Miyazaki *et al.* (1998) considered them basal cyprinids based on liver protein gel electrophoresis; however, their analysis only contained representatives from the Acheilognathinae, Leuciscinae, Gobioninae, and Cyprininae. Gilles *et al.* (1998) used cytochrome *b* and 16S sequences and placed Acheilognathins sister to a Leuciscinae plus Alburninae clade, though this placement is not well supported. Gilles *et al.* (2001) added mitochondrial control region sequence data to the Gilles *et al.* (1998) data set and found a sister relationship of Acheilognathinae to a clade containing Gobioninae, Leuciscinae and Alburninae. Liu and Chen (2003) examined the mitochondrial control region and found a weakly supported sister relationship with Tincinae in some analyses. S. He *et al.* (2004) were similarly unable to place this subfamily using the cytochrome *b* gene. Saitoh *et al.* (2006) recovered Acheilognathinae as sister to a large clade containing Phoxininae, Leuciscinae, Tincinae, Gobioninae, Cultrinae, Xenocyprinae, and *Aphocypris*, *Opsariichthys*, and *Zacco*. Within Acheilognathinae, Okazaki *et al.* (2001) found two major clades using 12S ribosomal DNA. The genus *Acheilognathus* was sister to a *Tanakia* plus *Rhodeus* clade, though the latter was not well resolved. Cunha *et al.* (2002)

did not recover a monophyletic Acheilognathinae in their analysis, although this may be a function of restricted taxon sampling.

Howes (1991) divided Cyprininae into six lineages: Barbinae, Labeoninae, Squaliobarbinae, Schizothoracinae, *Cyprinion/Onychostoma*, and *Barbus sensu lato*. Chen (1989) examined morphology and phylogeny of *Onychostoma*, *Scaphiodonichthys*, *Semiplotus*, and *Cyprinion*. Howes (1987) described morphology and discussed phylogenetic relationships of *Aulopyge* with respect to *Barbus* and reviewed characteristics of the Cyprininae and Leuciscinae.

Barbus contains over 340 species and there is tremendous diversity in size and morphology among fishes in this genus, leading to questions regarding its validity. There are also marked differences in ploidy among *Barbus* species including diploids, tetraploids, and hexaploids, further suggesting that current classifications may be incorrect. Classification of *Barbus* is based largely on presence or absence of fin spines, barbel number, and ploidy. These characters have not produced a stable classification of the genus. Berrebi and Tsigenopoulous (2002) considered this classification artificial. The African species do not appear to be related to *Barbus sensu stricto* and Berrebi *et al.* (1996) point out that the African *"Barbus"* present a substantial taxonomic problem that will require focused study to resolve. Tsigenopoulous *et al.* (2002) examined the evolution of polyploidy in European and African barbines. They presented evidence that tetraploidy has evolved at least twice. Their phylogenetic analysis recovered three monophyletic lineages, South African tetraploids plus small African diploids; African large hexaploids; and Mediterranean tetraploids. However, there is no evidence that these three lineages together form a monophyletic group. Berrebi and Tsigenopoulous (2002) restrict *Barbus* to peri-Mediterranean tetraploid species.

Schizothoracinae, commonly referred to as snow-trouts, are found in fast-flowing streams and rivers of East Asia. Wu (1984) examined morphology and presented a phylogenetic analysis of Chinese schizothoracines. Chen and Chen (2001) presented detailed morphology of the specialized Schizothoracine fishes, a group containing three genera, *Diptychus*, *Ptychobarbus*, and *Gymnodiptychus*. D. He *et al.* (2004) presented a molecular phylogeny of the schizothoracines based on cytochrome *b* sequences. They demonstrated that the specialized schizothoracines were paraphyletic with respect to the highly specialized schizothoracines and that some of the genera within the Schizothoracinae are not monophyletic. Liu and Chen (2003) argued that Schizothoracinae should be included within Barbinae.

Labeonins are widespread in Asia, Europe and Africa. Li *et al.* (2005) used 16S rRNA sequence data to examine phylogenetic relationships of 18 Labeoninae species in 13 genera. They found two major clades: one containing the genera *Cirrhinus, Crossocheilus,* and *Garra* and the other containing the genera *Labeo, Sinlabeo, Osteocheilus, Pseudocrossocheilus, Parasinilabeo, Ptychidio, Semilabeo, Pseudogyrinocheilus, Rectori,* and *Discogobio.* Liu and Chen (2003) described this group as basal within the cyprinins.

Gobionins are widespread throughout Europe and Asia. These have been placed as sister to the Cyprinines (Miyazaki *et al.* 1998, Gilles *et al.* 1998). However, Gilles *et al.* (2001) and Liu and Chen (2003) argued that gobionins were sister to the European Leuciscinae (including the alburnins). Saitoh *et al.* (2006) placed the gobionins as sister to the tincins and phoxinins plus leuciscins.

The Leuciscinae is a large, widespread group, and there is little evidence for its monophyly. Morphology and systematics of several taxa in this group are described and discussed by Howes (1978, 1984a, 1985) and reviewed by Howes (1991). This group includes the North American cyprinids, all of which, except *Notemigonus crysoleucas,* were placed in the phoxinin lineage by Cavender and Coburn (1992). Mayden (1989) examined morphology of primarily eastern North American phoxinins and presented a phylogenetic hypothesis for an open posterior myodome clade. Simons and Mayden (1999) presented molecular data supporting the homology of the open posterior myodome character. Coburn and Cavender (1992) presented the first phylogenetic hypothesis based on morphology that included all North American phoxinins and some Asian phoxinins. Using molecular data, Simons *et al.* (2003) examined phylogenetic relationships of nearly all North American cyprinid genera, and Smith *et al.* (2002) presented a hypothesis of western North American cyprinids. Much work remains to be done in North American cyprinid systematics, particularly the resolution of the shiner clade, which contains *Notropis* and its relatives (Simons *et al.* 2003). The North American genus *Notemigonus* was placed in the leuciscins and is more closely related to European Leucisinae than the rest of the North American fauna (Briolay *et al.* 1998). Gilles (1998) and Briolay *et al.* (1998) examined relationships of European leucisins and demonstrated that alburnins are nested within Leuciscins (see also Cunha *et al.* 2002).

The xenocyprins have been included in the abramin lineage within the Leuciscinae (Howes 1991). Cunha *et al.* (2002) presented evidence that xenocyprins are not part of the Leucisinae and are more closely related to other Asian taxa. Xenocyprins are restricted to East Asia with most diversity in China (Xiao *et al.* 2001). The group contains approximately 12 species classified in four genera: *Pseudobrama, Xenocypris, Distoechodon,* and *Xenocyprinoides.* Xiao *et al.* (2001) estimated species-level relationships in this group based on mitochondrial gene sequences. The molecular phylogeny is inconsistent with the generic classification. The monophyly of this group could not be corroborated because *Myxocyprinus* was the only outgroup included in the analysis. Liu and Chen (2003) suggested that xenocyprins were sister to cultrins. S. He *et al.* (2004) considered cultrins to be monophyletic but nested in a clade with East Asian leuciscins and xenocyprins. Saitoh *et al.* (2006) recovered xenocyprinis sister to the cultrins (*Ischikauia* plus *Culter*).

The Rasborinae is a large, probably polyphyletic group that contains familiar aquarium fishes including the genera *Rasbora*, *Esomus*, *Brachydanio*, and *Danio*. Howes (1991) included the bariliine and neobolines in the Rasborinae. The anatomy and relationships of African neobolines were described by Howes (1984b) and the anatomy and relationships of the bariliines were reviewed by Howes (1980, 1983). Saitoh *et al.* (2006) included five rasborines in their analysis: *Danio*, *Esomus*, *Aphyocypris*, *Opsariichthys*, and *Zacco*. These did not form a monophyletic group; *Aphyocypris*, *Opsariichthys*, and *Zacco* form a monophyletic group with xenocyprins and cultrins, while *Danio* plus *Esomus* formed the sister to a large clade containing acheilognathins, *Aphyocypris*, *Ischikauia*, *Zacco*, xenocyprins, cultrins, gobionins, tincins, leuciscins and phoxinins. Rasborinae as currently defined is clearly not monophyletic.

Although these papers contribute to our understanding of cyprinid relationships, their sampling is not sufficient to illuminate relationships of major groups of cyprinids and thus serve largely to illustrate the confusing aspects of cyprinid systematics and identify glaring taxonomic and systematic issues.

Problematic Taxa

There are a number of taxa whose taxonomic position has been controversial and still not clearly resolved. These include *Psilorhynchus* and its relatives, *Tinca*, and four enigmatic loach genera: *Ellopostoma*, *Vaillantella*, *Barbucca*, and *Serpenticobitis*.

The group containing *Psilorhynchus* and its relatives includes five or six species found in India, Nepal, China, and Myanmar (Rainboth 1983). Two species (*P. homaloptera* Hora and Misra and *P. pseudechenis* Menon and Datta) were assigned to the genus *Psilorhynchoides* based on body shape and small scale size (Yazdani *et al.* 1989 and 1990). Little is known about these fishes, but Rainboth (1983) describes some details of habitat preferences. *Psilorhynchus* and the closely related *Psilorhynchoides* exhibit similarities to both cyprinids and loaches. This has resulted in early confusion over their phylogenetic position and they were placed in several families—Cyprinidae, Balitoridae, and Cobitidae—by early researchers (see review in Hora 1925). Hora (1920) considered *Psilorhynchus* to be a member of Cyprinidae based on the structure of the gas bladder, the pharyngeal arches, and the presence of three rows of pharyngeal teeth. Hora (1925) later changed his position and reported a single row of four pharyngeal teeth. This and the presence of "a number of simple rays in the paired fins" led Hora to conclude that *Psilorhynchus* did not belong in Cyprinidae and erected the family Psilorhynchidae to contain it. Ramaswami (1952c) examined cranial osteology of *P. sucatio* (Hamilton) and compiled a list of eight characteristics unique to *Psilorhynchus* justifying its position in a separate family.

Chen (1981) examined cranial osteology of *Psilorhynchus homaloptera* and concluded that this species was a highly apomorphic cyprinid. Chen considered Psilorhynchinae a subfamily of Cyprinidae. Howes (1991) noted that maxillary innervation is similar to that observed in some cyprinids (cyprinins) and saw no reason to justify a separate family. Conway and Mayden (2007) provided an extensive review of the systematic history of *Psilorhynchus* and described the gill arches of two species. In a reanalysis of a subset of Siebert's (1987) data, modified to include *Psilorhynchus*, Conway and Mayden (2007) found no support for a sister-group relationship between *Psilorhynchus* and Cyprinidae and suggest that *Psilorhynchus* should be classified in a separate family within Cypriniformes, included in the Cobitoidea, together with Catostomidae, Gyrinocheilidae, Cobitidae, and Balitoridae.

Šlechtová *et al.* (2007) included *Psilorhynchus* in a phylogenetic analysis of Cypriniformes based on RAG-1 sequences. *Psilorhynchus* was sister to the four species of cyprinid included in this analysis, suggesting *Psilorhynchus* is a cyprinid or sister to the Cyprinidae. However, the low numbers of cyprinids included in their study prevent a thorough assessment of the phylogenetic position of *Psilorhynchus*.

Tinca could justifiably be considered the most problematic taxon within the Cyprinidae. The genus contains a single species, *Tinca tinca*, that is widely distributed in Europe. The phylogenetic position of *Tinca* varies depending on the analysis and the included data. Neither morphological nor molecular data provide a satisfactory resolution to the problem. Chen *et al.* (1984) included *Tinca* in the Barbini, one of the two major clades of Cyprinidae, sister to Labeoninae and Barbinae plus Cyprininae (Fig. 10.4). Cavender and Coburn (1992) placed *Tinca* in the other major clade of cyprinids, sister to the Leuciscini (Fig. 10.4). Cunha *et al.* (2002) found *Tinca* to be sister to gobionins and phoxinins plus leuciscins while Saitoh *et al.* (2006) reported *Tinca* as sister to phoxinins plus leuciscins. The difficulty of identifying the phylogenetic relationships of this species is reflected by its classification in a monotypic subfamily, the Tincinae.

Ellopostoma contains two species, *E. megalomycter* (Vaillant) and *E. mystax* Tam and Lim. Siebert (1987) included *Ellopostoma megalomycter* in the Balitoridae although Roberts (1989) placed this species in Cobitidae.

The phylogenetic position of *Ellopostoma* within Balitoridae was unclear in Siebert's (1987) analyses. *Ellopostoma* is known from the Kaupas Basin in western Borneo and the Tapi basin of Thailand (Roberts 1989). *Ellopostoma* is rarely collected and there are very few available specimens in museum collections.

Nalbant and Banarescu (1977) placed *Vaillantella* in a separate subfamily, Vaillantellinae, and identified similarities with both Botiinae and Nemacheilinae within the Cobitidae. Similarities with Botiinae include the incomplete ossification of the gas-bladder capsule and both pairs of rostral barbels originating from a common base. Nalbant and Banarescu (1977) noted that this condition is also observed in "some primitive Nemacheilinae." *Vaillantella* is similar to nemacheilines in that it lacks a moveable suborbital spine (Nalbant and Banarescu 1977). Sawada (1982) placed *Vaillantella* in the Nemacheilinae within the family Balitoridae. He based his comparisons on six characters but only compared *Vaillantella* to botiids and nemacheilids. The characters included (1) sesamoid bone dorsal to preethmoid (present in botiids), (2) mobile lateral ethmoid (present in botiids and cobitids), (3) socket-like articulation between the supraethmoid/ethmoid complex and frontals (present in botiids and cobitids, (4) second descending process (produced from lateral part of second centrum) forms part of the gas bladder capsule (observed in botiids and cobitids), (5) second horizontal process (also produced from the lateral part of the second centrum) forms part of the gas bladder capsule (observed in botiids and cobitids), and (6) gas bladder capsule subdivided (observed in vaillantellids, nemacheilids, and balitorids). Sawada argued that the incomplete ossification of the gas bladder was plesiomorphic and that vaillantellids did not exhibit any similarities with botiids. However, Sawada only had radiographs available and did not examine an actual specimen. Siebert (1987) examined a stained-and-cleared individual and found no evidence to justify classifying *Vaillantella* as a nemacheiline. He presented five shared similarities between *Vaillantella* and botiines and cobitines. Siebert (1987) placed *Vaillantella* in Cobitidae. Interestingly, although both Siebert and Sawada comment on the phylogenetic position of *Vaillantella*, neither of them included this species on their phylogenies or in their data matrices. Saitoh *et al.* (2006) and Šlechtová *et al.* (2007) included *Vaillantella* in their analyses based on whole mitochondrial genomes and RAG-1 sequences respectively, and recovered it as sister to Cobitidae plus Balitoridae and Nemacheilidae.

Barbucca and *Serpenticobitis* also have an unusual suite of morphologies. *Serpenticobitis* resembles nemacheilids but has a mobile, spine-like lateral ethmoid similar to botiids and cobitids. *Barbucca* also resembles nemacheilids but this may be a function of miniaturization obscuring its relationships. Šlechtová *et al.* (2007) present evidence that both taxa are nested within Balitoridae, an unusual result given previous hypotheses based on morphology.

Relationships among Cypriniform Families

Wu *et al.* (1981) provided the first explicitly phylogenetic classification of Cypriniform fishes based on cladistic argumentation of eight characters. They recognized five families: Cyprinidae, Catostomidae, Gyrinocheilidae, Cobitidae, and Balitoridae. (Wu *et al.*'s [1981] conception of Cobitidae included taxa now classified in Cobitidae, Botiidae, and Nemacheilidae.) Within Cypriniformes they recognized two major clades. The first contained Cobitidae, sister to Catostomidae plus Gyrinocheilidae, and the second contained the sister taxa Cyprinidae plus Balitoridae (Fig. 10.5). The relationship between Cyprinidae and Balitoridae was based on a similarity in the shape of the fifth ceratobranchial, presumed similarity in the basioccipital process of cyprinds and some balitorids, and the development of the subtemporal fossa. Wu *et al.* (1981) used a limited number of characters and did not include outgroups in their analysis.

Sawada (1982) examined phylogenetic relationships among loaches (Botiidae, Cobitidae, Balitoridae, Nemacheilidae, and Vaillantellidae) using a range of morphological characters. Character states were polarized using catostomids as the outgroup. Sawada used strict Hennigian (Hennig 1966) argumentation to arrive at most parsimonious explanation of character distribution among these taxa. Sawada (1982) divided the loaches into two families, including botiids and cobitids into the family Cobitidae and removing nemacheilids from Cobitidae and placing them, together with balitorids, in family Balitoridae.

Siebert (1987) examined the phylogenetic relationships of Cypriniformes and included nine OTUs: Cyprinidae; Gyrinocheilidae; Catostomidae; Botiidae; Cobitidae; Nemacheilidae; *Ellopostoma*; and Gastromyzontini and Balitorini in the family Balitoridae. (Given issues with classification in the Balitoridae, specifically the recognition of Gastromyzontini and Balitorini as mutually exclusive taxa, we refer Gastromyzontini and Balitorini to Balitorinae.) Siebert used an explicitly phylogenetic approach based on parsimony analysis of 41 characters, primarily cranial osteology. His results determined conceptions of cypriniform relationships for the following 19 years (Fig. 10.6). Siebert recognized two suborders within Cypriniformes, Cyprinoidea containing Cyprinidae and Cobitoidea containing Catostomidae, Gyrinocheilidae, Botiidae, Cobitidae Vaillantellidae (botiids, cobitids, and vaillantellids were included in Cobitidae), Nemacheilidae, and Balitoridae (nemacheilids and balitorids were included in Balitoridae). Unfortunately, his analysis suffers from a number of problems including character independence,

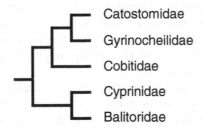

Fig. 10.5 Phylogenetic relationships of cypriniform families, modified from Wu *et al.* (1981). Note that Cobitidae included taxa now classified in Botiidae, Cobitidae, and Nemacheilidae.

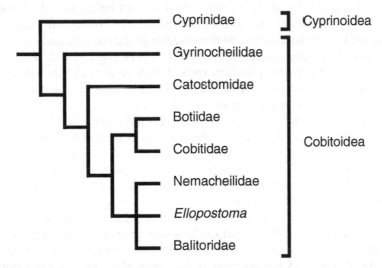

Fig. 10.6 Phylogenetic relationships of major cypriniform groups, modified from Siebert (1987).

unclear character descriptions, strictly binary coding, and the use of a hypothetical outgroup.

Character independence is an assumption of phylogenetic systematics and not all Siebert's (1987) characters are independent. For example, character 4 (p. 93), consolidation of the infrapharyngobranchial series, is described as an "end-to-end relationship between infrapharyngo-branchials II and III with a consequent modification of the epibranchial I and II articulations with infrapharyngobranchials and lack a separate infrapharyngobranchial IV." The derived condition was coded for Nemacheilidae and Balitoridae. Siebert's character 5 is the absence of infrapharyngobranchial IV, which is clearly part of the description for character 4. The derived condition was coded for Nemacheilidae and Balitoridae. Infrapharyngobranchial IV is absent in several cyprinids but this was considered secondarily derived (Siebert 1987). The plesiomorphic condition, infrapharyngobranchial IV present, was observed in gyrinocheilids, catostomids, cobitids, botiids, and *Ellopostoma*. Cyprinids were also considered to exhibit the pleisomorphic condition. This element is absent in *Vaillantella*, and the infrapharyngobranchial IV may not develop in some cobitids.

A number of character descriptions are unclear, particularly those taken from previous studies. Siebert's character 38, the supraethmoid (= mesethmoid) was taken from Sawada (1982). The derived condition was reported in Botiidae and Cobitidae. Sawada described the supraethmoid-ethmoid complex as a bone that forms the anteriormost part of the median dorsal surface of the skull. In nemacheilids and balitorids, the bone is firmly joined to the anterior part of the frontals. In botiids and cobitids it articulates in a socket formed by the anterior of the frontals and can move with respect to the frontals. Siebert's (1987) coding indicates that *Ellopostoma* exhibits the plesiomorphic fused condition but this is not figured or described in the text. Siebert did not describe or discuss the condition in other Cypriniformes.

Siebert's character 39 is described as "shoulder girdle" and taken from Sawada (1982). The derived condition was reported for Balitoridae. Sawada presented an 11 page description and discussion of the pectoral girdle of loaches but it is unclear which aspect of pectoral girdle morphology Siebert refers to and the condition in other Cypriniformes and the outgroups is not clear.

All of Siebert's characters are coded as binary, which resulted in oversimplification of the situation in some instances. This problem is amplified by the use of a hypothetical ancestor, rather than coding the situation observed in the outgroups. For example, Siebert's character 18 refers to the ossification of basibranchial IV. The derived condition is the presence of an ossified basibranchial IV. This condition is observed in loaches and *Psilorynchus*. It is cartilaginous (plesiomorphic) in catostomids. This element is missing from Cyprinidae and *Gyrinocheilus*. For the latter taxa it was coded as plesiomorphic, yet coding as missing or as a third state would be more informative.

One of the presumed synapomorphies for Cypriniformes is the presence of teeth ankylosed to the fifth ceratobranchial. The ceratobranchial is modified with broad surfaces for muscle attachment. The teeth are present in a single row in catostomids and loaches. In cyprinids, teeth may be present in one, two, or three rows. In *Gyrinocheilus*, the pharyngeal teeth are absent and the fifth ceratobranchial is unmodified. Siebert (character 23) coded a single row of ceratobranchial V teeth as the derived condition (state one) for catostomids and loaches. He coded both *Gyrinocheilus* and Cyprinidae as plesiomorphic in spite of the obvious differences between these two. It is unclear what this coding is supposed to represent. If cyprinids are coded as plesiomorphic this assumes that multiple rows are the plesiomorphic condition for the family. This may be the case; however, there is not a robust phylogeny of the Cyprinidae that supports this assumption. Regardless of the plesiomorphic condition for Cyprinidae, it is different from the condition observed in Gyrinocheilidae. Adding to the complexity of this character is the true outgroup condition. The outgroups do not have ceratobranchial teeth ankylosed to the ceratobranchials and therefore this character cannot be polarized. Although Siebert's thesis represents a major step forward in our understanding of interrelationships of Cypriniformes, the included characters and codings need review across a wider array of taxa.

Molecular Systematics of Cypriniformes

Clements *et al.* (2004) used growth hormone sequences to construct a phylogeny of some teleosts. Included Cypriniformes were cyprinids, catostomids, and cobitids. Their analysis found cobitids (*Misgurnus*) sister to cyprinids plus catostomids. They state that this result is consistent with Smith (1992) and Uyeno and Smith (1972), but these papers did not test relationships of Cypriniformes. Smith (1992) rooted his trees with *Cyprinus* and *Carrasius* and included a cobitid loach in his analysis. Uyeno and Smith speculated that catostomids arose from a cyprinid-like ancestor but did not present evidence pertinent to cypriniform relationships.

Harris and Mayden (2001) commented on relationships within the Cypriniformes in their examination of catostomid relationships using sequences of the mitochondrial ribosomal 12S and 16S genes. In addition to catostomids, their study included *Cyprinus* and *Carrassius* (Cyprinidae), *Gyrinocheilus* (Gyrinocheilidae), *Crossostoma* (Balitoridae), and *Botia* (Botiidae). All analyses used weighted and unweighted parsimony as the optimality criteria and were rooted using *Cyprinus* and *Carrassius*. Harris and Mayden (2001) reported that they consistently recovered a monophyletic Cobitoidea (*sensu* Siebert 1987) containing Balitoridae, Botiidae, Catostomidae, and Gyrinocheilidae. However, Siebert (1987) found catostomids sister to balitorids plus cobitids and botiids, whereas Harris and Mayden (2001) found catostomids sister to gyrinocheilids (Fig. 10.7). We do not believe that this represents a valid test of Siebert's hypothesis or cypriniform relationships. No outgroup outside of the Cypriniformes was included in this work, and thus relationships among cypriniform families could not be assessed.

Our understanding of Cypriniform relationships was dramatically changed by three papers published in 2006 and 2007 (Tang *et al.* 2006, Saitoh *et al.* 2006, Šlechtová *et al.* 2007). These papers all included much greater taxon sampling than previous molecular studies, particularly of

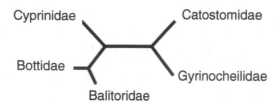

Fig. 10.7 Unrooted network of cypriniform families, modified from Harris and Mayden (2001).

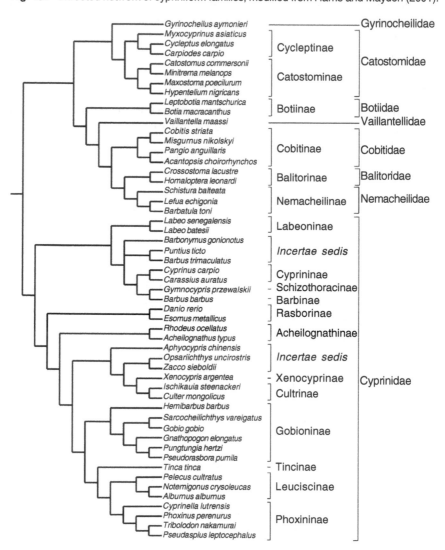

Fig. 10.8 Phylogenetic relationships of Cypriniformes, modified from Saitoh *et al.* (2006).

loaches, and have resulted in substantial changes in taxonomy. Tang *et al.* (2006) presented data suggesting that neither Cobitidae nor Balitoridae as previously construed were monophyletic groups. Their analysis, based on mitochondrial cytochrome *b* sequences, suggested that Botiidae is sister to Balitoridae and Cobitidae plus Nemacheilidae (*Vaillantella* was not included in their analysis). This is in stark contrast to previous work that suggested cobitids and botiids were sister taxa (formerly contained in family Cobitidae) and Nemacheilidae and Balitoridae were sister taxa (formerly contained in family Balitoridae).

Saitoh *et al.* (2006) published a phylogenetic analysis of 53 cypriniform mitochondrial genomes. They included a broad sampling of cypriniform taxa and produced a robust hypothesis of relationships among cypriniform families and subfamilies (Fig. 10.8). Saitoh *et al.* (2006) recovered a monophyletic Cobitoidea sister to the Cyprinoidea, containing only the Cyprinidae. The Gyrinocheilidae was sister to the Catostomidae, these formed the sister group to the loaches. This reflects Harris and Mayden's (2001) hypothesis but conflicts with Siebert's hypothesis of cobitoidean relationships. Within the loaches, the following relationships were obtained: (Botiidae (Vailantellidae (Cobitidae (Balitoridae plus Nemacheilidae)))). The sister-group relationship between balitorids and nemacheilids is contrary to Tang *et al.* (2006) but corroborates Sawada's (1992) hypothesis. Botiidae is sister to all remaining loaches corroborating Tang *et al.* (2006). The analysis of Šlechtová *et al.* (2007) was based on nuclear RAG-1 sequences. Their results reflected Saitoh *et al.* (2006) with the exception of the phylogenetic position of Catostomidae and Gyrinocheilidae. Whereas Saitoh *et al.* (2006) recovered Gyrinocheilidae and Catostomidae as a monophyletic group, Šlechtová *et al.* (2007) placed Catostomidae as sister to Gyrinocheilidae plus the loaches.

Conclusion

It is clear that resolution of Cypriniform relationships will require a substantial effort at a variety of taxonomic levels. The large number of taxa, the complexity of the existing taxonomy, and the wide geographic distribution of the clade have hampered previous work. The Cypriniformes Tree of Life project is a large multi-collaborator initiative bringing together morphologists, paleontologists, developmental biologists and molecular systematists to review existing data, develop new phylogenetically informative data, and synthesize this information to arrive at a comprehensive phylogenetic hypothesis for this large group of fishes. The family Cyprinidae will continue to be a focus of systematists at many levels, from species delineation and diversity to identification of higher-level monophyletic groups within the family. Molecular analyses indicate that several of the currently recognized subfamilies are not monophyletic; particularly problematic taxa include Rasborinae and Barbinae. Dense taxon sampling and additional character systems will be necessary to clarify these issues.

The monophyly of Cobitoidea (*sensu* Siebert 1987), containing Catostomidae, Gyrinocheilidae, Botiidae, Vaillantellidae, Balitoridae, Cobitidae, and Nemacheilidae, has been corroborated by a number of studies based on molecular data. Relationships among the included families await resolution. Are catostomids and gyrinocheilids a monophyletic group? If not, which family is the sister to the remaining cobitoids? Within the loaches, botiids appear to be the sister of all remaining taxa but the phylogenetic position of Cobitidae has not been settled.

A well-corroborated phylogenetic hypothesis of the families will provide a framework for examination of more difficult issues in Cypriniformes, particularly the phylogenetic relationships within the Cyprinidae. A robust phylogenetic hypothesis will also provide a framework for addressing a number of interesting evolutionary issues within the Cypriniformes. These include the loss of teeth on the fifth ceratobranchial of gyrinocheilids. These teeth are present in all other cypriniform fishes, their absence in gyrinocheilids may be due to their extremely specialized mode of feeding (Benjamin, 1986), but this does not explain why teeth would be retained in all other cypriniform fishes including other algae scraping or detrivorous taxa. Another interesting trophic morphological characteristic is the development of a pronounced basioccipital process in both cyprinids and catostomids. Did this process evolve independently in the common ancestors of both cyprinids and catostomids or is the presence of the basioccipital process plesiomorphic for Cypriniformes with subsequent losses in the ancestors of Gyrinocheilids and loaches?

One of the morphological characters used to justify a monophyletic group containing Cobitidae and Botiidae is the presence of a lateral ethmoid modified to form a mobile spine (Sawada 1982, Siebert 1987). In most fishes, the lateral ethmoid forms the anterior wall of the orbit. A mobile spine-like lateral ethmoid is present in all botiids, cobitids, and *Serpenticobitis* yet molecular data indicated that *Serpenticobitis* is nested within family Balitoridae (Šlechtová *et al.* 2007) and Botiidae and Cobitidae do not form a monophyletic group. These results, together with the phylogenetic position of *Vaillantella*, which lacks a mobile lateral ethmoid (Siebert 1987), imply multiple gains or losses of this condition.

Given the many unique characteristics of these fishes, their economic and scientific importance, and increasing concerns with the conservation status of some taxa, it is imperative that we work towards a clear, consistent phylogeny and classification for this group.

Acknowledgements

We thank Terry Grande for the invitation to participate in this volume. We are grateful to Elizabeth M. Johnson and Brett C. Nagle for help with library work. Kevin W. Conway, Miles M. Coburn, and an anonymous reviewer provided helpful comments. Partial support was provided by the National Science Foundation (EF 0431132 to AMS).

References

Arai, R. 1982. A chromosome study on two cyprinid fishes *Acrossocheilus labiatus* and *Pseudorasbora pumila pumila*, with notes on Eurasian cyprinids and their karyotypes. Bull. Nat. Sci. Mus., Tokyo (A) 8: 131–152.

Banarescu, P.M. and T.T. Nalbant. 1995. A generical classification of Nemacheilinae with description of two new genera (Teleostei: Cypriniformes: Cobitidae). Trav. Mus. Hist. nat. "Grigore Antipa" 35: 429–496.

Benjamin, M. 1986. The oral sucker of *Gyrinocheilus aymonieri* (Teleostei: Cyprinformes). J. Zool. Soc. Lond. Series B, 1:211–254.

Berrebi, P. and C.S. Tsigenopoulos. 2002. Contribution des séquences mitochondriales à l'éclatement phylogénétique du genre polyploïde *Barbus* (Téléostéens Cyprinidés) implications biogéographiques. Biosystema 20. Systémat. biogéogr. 2002: 49–56.

Berrebi, P., M. Kottelat, P. Skelton and P. Ráb. 1996. Systematics of *Barbus*: State of the art and heuristic comments. Folia Zool. 45 (Suppl. 1): 5–12.

Briolay, J., N. Galtier, R.M. Brito and Y. Bouvet. 1998. Molecular phylogeny of Cyprinidae inferred from cytochrome b DNA sequences. Mol. Phylogenet. Evol. 9: 100–108.

Buth, D.G. 1979. Duplicate gene expression in tetraploid fishes of the tribe Moxostomatini (Cypriniformes: Catostomidae). Comp. Biochem. Physiol. Series B, 63: 7–12.

Buth, D.G. 1980. Evolutionary genetics and systematic relationships in the catostomid genus *Hypentelium*. Copeia 1980: 280–290.

Carl, G.C., W.A. Clemens and C.C. Lindsey. 1967. The freshwater fishes of British Columbia. British Columbia Provincial Museum, Victoria.

Cavender, T.M. and M.M. Coburn. 1992. Phylogenetic relationships of North American Cyprinidae, pp. 293–327. In: R.L. Mayden [ed.]. Systematics, Historical Ecology, and North American Freshwater Fishes. Stanford University Press, Stanford.

Chang, M-M, D. Miao, Y. Chen, J. Zhou and P. Chen. 2001. Suckers (Fish, Catostomidae) from the Eocene of China account for the family's current disjunct distributions. Science in China (Series D) 44: 577–586.

Chen, X.L., P.Q. Yue and R.D. Lin. 1984. Major groups within the family Cyprinidae and their phylogenetic relationships. Acta Zootaxon. Sinica 9: 424–440.

Chen, Y. 1981. Investigation on the systematic position of *Psilorhynchus* (Cyprinoidei, Pisces). Acta Hydrobiol. Sinica 7: 371–376.

Chen, Y. 1989. Anatomy and phylogeny of the cyprinid fish genus *Onychostoma* Günther, 1896. Bull. Br. Mus. Nat. Hist. (Zool.) 55: 109–121.

Chen, Z.M. and Y.F. Chen 2001. Phylogeny of the specialized schizothoracine fishes (Teleostei: Cypriniformes: Cyprinidae). Zool. Stud. 40: 147–157.

Chu, X.L. and Y.-R. Chen. 1979. A new blind cobitid fish (Pisces, Cypriniformes) from subterranean waters in Yunnan, China. Acta Zool. Sinica 25: 285–287.

Clements, M.D., H.L.J. Bart and D.L. Hurley. 2004. Isolation and characterization of two distinct growth hormone DNAs from the tetraploid smallmouth buffalofish (*Ictiobus bubalus*). Gen. Comp. Endocrinol. 136: 411–418.

Coburn, M.M. and T.M. Cavender. 1992. Interrelationships of North American cyprinid fishes, pp. 328–373. In: R.L. Mayden [ed.]. Systematics, Historical Ecology, and North American Freshwater Fishes. Stanford University Press, Stanford.

Conway, K.W. and R.L. Mayden. 2007. The gill arches of *Psilorhynchus* (Ostariophysi: Psilorhynchidae). Copeia 2007: 267–280.

Cope, E.D. 1872. On the Tertiary coal and fossils of Osino, Nevada. Proc. Amer. Phil. Soc. 12: 478–481.

Cunha, C., N. Mesquita, T.E. Dowling, A. Gilles and M.M. Coelho. 2002. Phylogenetic relationships of Eurasian and American cyprinids using cytochrome *b* sequences. J. Fish Biol. 61: 929–944.

Eastman, J.T. 1977. The pharyngeal bones and teeth of catostomid fishes. Am. Midland Naturalist 97: 68–88.

Fang, P.W. 1934. Study on the crossostomoid fishes of China. Sinensia 6: 44–97.

Felsenstein, J. 1978. Cases in which parsimony or compatibility methods will be positively misleading. Syst. Zool. 27: 401–410.

Ferris, S.D. 1984. Tetraploidy and the evolution of the catostomid fishes, pp. 55–93. *In*: B.J. Turner [ed.]. Evolutionary Genetics of Fishes. Plenum, New York.

Ferris, S.D. and G.S. Whitt. 1977. The evolution of duplicate gene expression after polyploidization. Nature 265: 258–260.

Ferris, S.D. and G.S. Whitt. 1979. Evolution of the differential regulation of duplicate genes after polyploidization. J. Mol. Evol. 12: 267–317.

Ferris, S.D. and G.S. Whitt. 1980. Phylogeny of tetraploid catostomid fishes based on the loss of duplicate gene expression. Syst. Zool. 27: 189–206.

Fink, S.V. and W.L. Fink. 1981. Interrelationships of the ostariophysan fishes (Teleostei). Zool. J. Linn. Soc. 72: 297–353.

Gilles, A., B. Lecointre, E. Faure, R. Chappaz and G. Brun. 1998. Mitochondrial phylogeny of the European cyprinids: Implications for their systematics, reticulate evolution, and colonization time. Mol. Phylogenet. Evol. 10: 132–143.

Gilles, A., G. Lecointre, A. Miquelis, M. Loerstcher, R. Chappaz and G. Brun. 2001. Partial combination applied to phylogeny of European cyprinids using mitochondrial control region. Mol. Phylogenet. Evol. 19: 22–33.

Gosline, W.A. 1978. Unbranched dorsal-fin rays and subfamily classification of the fish family Cyprinidae. Occ. Pap. Mus. Zool Univ. Michigan 684: 1–21.

Grande, L., J.T. Eastman and T.M. Cavender. 1982. †*Amyzon gosiutensis*, a new catostomid fish from the Green River formation. Copeia 1982: 523–532.

Graybeal, A. 1998. Is it better to add taxa or characters to a difficult phylogenetic problem? Syst. Biol. 47: 9–17.

Harris, P.M. and R.L. Mayden. 2001. Phylogenetic relationships of major clades of Catostomidae (Teleostei: Cyprinformes) as inferred from mitochondrial SSU and LSU rDNA sequences. Mol. Phylogenet. Evol. 20: 225–237.

Harris, P.M., R.L. Mayden, H.S. Espinosa Pérez and F. García de Leon. 2002. Phylogenetic relationships of *Moxostoma* and *Scartomyzon* (Catostomidae) based on mitochondrial cytochrome *b* sequence data. J. Fish Biol. 61: 1433–1452.

He, D., Y. Chen, Y. Chen and Z. Chen. 2004. Molecular phylogeny of the specialized schizothoracine fishes (Teleostei: Cyprinidae), with their implications for the uplift of the Qinghai-Tibetan Plateau. Chin. Sci. Bull. 49: 39–48.

He, S., H. Liu, Y. Chen, K. Masayuki, N. Tsuneu and Y. Zhong. 2004. Molecular phylogenetic relationships of Eastern Asian Cyprinidae (Pisces: Cypriniformes) inferred from cytochrome *b* sequences. Science in China Series C Life Sciences 47(2): 130–138.

Hendy, M.D. and D. Penny. 1989. A framework for the study of evolutionary trees. Syst. Zool. 38: 297–309.

Hennig, W. 1966. Phylogenetic Systematics. Univ. Illinois Press, Urbana.

Hensel, K. 1970. Review of the classification and of the opinions on the evolution of Cyprinoidei (Eventognathi) with an anotated list of genera and subgenera described since 1921. Annot. Zool. Bot. 57: 1–45.

Hora, S.L. 1920. Revision of the Indian Homalopteridae and of the genus *Psilorhynchus* (Cyprinidae). Rec. Ind. Mus. 19: 195–215.

Hora, S.L. 1923. On a collection of fish from Siam. J. Nat. Hist. Soc. Siam 6: 143–184, plates 10–12.

Hora, S.L. 1925. Notes on fishes in the Indian Museum. XII. The systematic position of the cyprnoid genus Psilorhynchus McClelland. Rec. Ind. Mus. 19: 457–460.

Hora, S.L. 1932. Classification, bionomics, and evolution of homalopterid fishes. Mem. Ind. Mus. 12: 247–248.

Hora, S.L. 1952a. Parallel evolution in the crossostomid fishes on the mainland of Asia and in Borneo. Proc. Nat. Inst. Sci. India 18: 417–421.

Hora, S.L. 1952b. Parallel evolution of the gastromyzonid fishes on the mainland of Asia and in the island of Borneo. Proc. Nat. Inst. Sci. India 18: 407–416.

Howes, G.J. 1978. The anatomy and relationships of the cyprinid fish Luciobrama macrocephalus (Lacepède). Bull. Br. Mus. Nat. Hist. (Zool.) 34: 1–64.

Howes, G.J. 1980. The anatomy, phylogeny and classification of bariliine cyprinid fishes. Bull Br. Mus. Nat. Hist. (Zool.) 37: 129–198.

Howes, G.J. 1983. Additional notes on bariliine cyprinid fishes. Bull Br. Mus. Nat. Hist. (Zool.) 45: 95–101.

Howes, G.J. 1984a. Phyletics and biogeography of the aspinine cyprinid fishes. Bull Br. Mus. Nat. Hist. (Zool.) 47: 283–303.

Howes, G.J. 1984b. A review of the anatomy, taxonomy, phylogeny and biogeography of the African neoboline cyprinid fishes. Bull. Br. Mus. Nat. Hist. (Zool.) 47: 151–185.

Howes, G.J. 1985. A revised synonymy of the minnow genus Phoxinus Rafinesque, 1820 (Teleostei: Cyprinidae) with comments on its relationships and distribution. Bull Br. Mus. Nat. Hist. (Zool.) 48: 57–74.

Howes, G.J. 1987. The phylogenetic position of the Yugoslavian cyprinid fish genus Aulopyge Heckel, 1841, with an appraisal of the genus Barbus Cuvier and Cloquet, 1816 and the subfamily Cyprininae. Bull. Br. Mus. Nat. Hist. (Zool.) 52: 165–196

Howes, G.J. 1991. Systematics and biogeography: an overview, pp. 1–33. In: I.J. Winfield and J.S. Nelson [eds.]. Cyprinid Fishes: Systematics, Biology and Exploitation. Chapman and Hall, London.

Jenkins, R.E. and N.M. Burkhead. 1993. Freshwater Fishes of Virginia. American Fisheries Society, Bethesda, Maryland.

Kottelat, M. 1984. Revision of the Indonesian and Malaysian loaches of the subfamily Nemacheilinae. Japan. J. Ichthyol. 31: 225–260.

Kottelat, M. 1990. Indochinese nemacheilines: A revision of nemacheiline loaches (Pisces: Cypriniformes) of Thailand, Burma, Laos, Cambodia and southern Viet Nam. Verlag Dr. Friedrich Pfeil, Munich.

Kottelat, M. 1994. Vaillantella cinnamomea, a new species of balitorid loach from eastern Borneo. Japan. J. Ichthyol. 40: 427–431.

Kottelat, M., R. Britz, T.H. Hui and K.-E. Witte. 2006. Paedocypris, a new genus of Southeast Asian cyprinid fish with a remarkable sexual dimorphism, comprises the world's smallest vertebrate. Proc. Roy. Soc. Lond., Ser. B, published online.

Kottelat, M. and X.-L. Chu. 1987. The botiine loaches (Osteichthys: Cobitidae) of the Lancangjiang (Upper Mekong) with description of a new species. Zool. Res. 8: 393–400.

Kottelat, M. and X.-L. Chu. 1988a. A synopsis of Chinese balitorine loaches (Osteichthyes: Homalopteridae) with comments on their phylogeny and description of a new genus. Rev. Suisse Zool. 95: 181–201.

Kottelat, M. and X.-L. Chu. 1988b. Revision of Yunnanilus with descriptions of a miniature species flock and six new species from China (Cypriniformes: Homalopteridae). Env. Biol. Fish. 23: 65–93.

Li, J., X. Wang, S. He and Y. Chen. 2005. Phylogenetic studies of Chinese Labeoninae fishes (Teleostei: Cyprinidae) based on mitochondrial 16S rRNA gene. Prog. Nat. Sci. 15(3): 213–219.

Lindsey, C.C. and J.D. McPhail. 1986. Zoogeography of fishes of the Yukon and Mackenzie Basins, pp. 639–674. In: C.H. Hocutt and E.O. Wiley [eds.]. The Zoogeography of North American Freshwater Fishes. John Wiley and Sons, New York.

Liu, H. and Y. Chen. 2003. Phylogeny of the East Asian cyprinids inferred from sequences of the mitochondrial DNA control region. Can. J. Zool. 81: 1938–1946.

Lo, Y.L. and H.W. Wu. 1979. Anatomical features of *Myxocyprinus asiaticus* and its systematic position. Acta Zootaxon. Sinica 4: 195–203.

Mayden, R.L. 1989. Phylogenetic studies of North American minnows, with emphasis on the genus *Cyprinella* (Teleostei: Cypriniformes). Misc. Publ. Univ. Kansas Mus. Nat. Hist. 80: 1–189.

Mayden, R.L., B.M. Burr, L.M. Page and R.R. Miller. 1992. The native fishes of North America, pp. 827–863. In: R.L. Mayden [ed.]. Systematics, Historical Ecology, and North American Freshwater Fishes. Stanford University Press, Stanford.

Meng, L. and P.B. Moyle. 1995. Status of splittail in the Sacramento-San Joaquin estuary. Trans. Am. Fish. Soc. 124: 538–549.

Miyazaki, J-I., T. Hirabayashi, K. Hosoya and T. Iwami. 1998. A study of the systematics of cyprinid fishes by two-dimensional gel electrophoresis. Env. Biol. Fish. 52: 173–179.

Nalbant, T.T. and P.M. Banarescu. 1977. Vaillantellinae, a new subfamily of Cobitidae (Pisces, Cypriniformes). Zoologische Mededelingen, Rijksmuseum van Natuurlijke Historie te Leiden 52: 99–105.

Nelson, J.S. 1994. Fishes of the World. John Wiley and Sons, Inc., New York.

Okada, Y. 1960. Studies on the freshwater fishes of Japan II. Special Part. J. Fac. Fish. Pref. Univ. Mie-Tsu 4: 267–588.

Okazaki, M., K. Naruse, A. Shima and R. Arai. 2001. Phylogenetic relationships of bitterlings based on mitochondrial 12S ribosomal DNA sequences. J. Fish Biol. 58: 89–106.

Rainboth, W.J. 1983. *Psilorhynchus gracilis*, a new cyprinoid fish from the Gangetic lowlands. Proc. Cal. Acad. Sci. 43: 67–76.

Ramaswami, L.S. 1952a. Skeleton of cyprinoid fishes in relation to phylogenetic studies. The systematic postition of the genus *Gyrinocheilus* Vaillant. Proc. Nat. Inst. Sci. India 18.

Ramaswami, L.S. 1952b. Skeleton of cyprinoid fishes in relation to phylogenetic studies. IV. The skull and other skeletal structures of gastromyzonid fishes. Proc. Nat. Inst. Sci. India 18: 519–538.

Ramaswami, L.S. 1952c. Skeleton of cyprinoid fishes in relation to phylogenetic studies. II. The systematic position of *Psilorhynchus* McClelland. Proc. Nat. Inst. Sci. India 18: 141–150.

Ramaswami, L.S. 1953. Skeleton of cyprinoid fishes in relation to phylogenetic studies. V. The skull and the gasbladder capsule of the Cobitidae. Proc. Nat. Inst. Sci. India 19: 323–347.

Roberts, T.R. 1989. The freshwater fishes of western Borneo (Kalimantan Barat, Indonesia). Mem. Cal. Acad. Sci. 14: 1–210.

Roberts, T.R. and M. Kottelat. 1993. Revision of the southeastern Asian freshwater fish family Gyrinocheilidae. Ichthyol. Explor. Freshwaters 4: 375–383.

Saitoh, K., T. Sado, R.L. Mayden, N. Hanzawa, K. Nakamura, M. Nishida and M. Miya. 2006. Mitogenomic evolution and interrelationships of the Cypriniformes (Actinopterygii: Ostariophysi): The first evidence toward resolution of higher-level relationships of the world's largest freshwater fish clade based on 59 whole mitogenome sequences. J. Mol. Evol. 63: 826–841.

Sawada, Y. 1982. Phylogeny and zoogeography of the superfamily Cobitoidea (Cyprinoidei, Cypriniformes). Mem. Fac. Fish., Hokkaido University 28: 65–223.

Sibbing, F.A. 1988. Specializations and limitations in the utilization of food resources by the carp, *Cyprinus carpio*, a study of oral food processing. Env. Biol. Fish. 22: 161–178.

Siebert, D.J. 1987. Interrelationships among families of the order Cypriniformes (Teleostei). Ph.D. thesis, University of New York, New York.

Silas, E.G. 1953. Classification, zoogeography and evolution of the fishes of the cyprinoid families Homalopteridae and Gastromyzonidae. Rec. Ind. Mus. 50: 173–264.

Simons, A.M. and R.L. Mayden. 1999. Phylogenetic relationships of North American cyprinds and assessment of homology of the open posterior myodome. Copeia 1999: 13–21.

Simons, A.M., P.B. Berendzen and R.L. Mayden. 2003. Molecular systematics of North American phoxinin genera (Actinopterygii: Cyprinidae) inferred from mitochondrial 12S and 16S ribosomal RNA sequences. Zool. J. Linn. Soc. 139: 63–80.

Šlechtová, V., J. Bohlen, J. Freyhof and Petr Ráb. 2006. Molecular phylogeny of the Southeast Asian freshwater fish family Botiidae (Teleostei: Cobitoidea) and the origin of polyploidy in their evolution. Mol. Phylogenet. Evol. 39: 529–541.

Šlechtová, V., J. Bohlen and H.H. Tan. 2007. Families of Cobitoidea (Teleostei; Cypriniformes) as revealed from nuclear genetic data and the position of the mysterious genera from *Barbucca, Psilorhynchus, Serpenticobitis* and *Vaillantella*. Mol. Phylogenet. Evol. 44: 1358–1365.

Smith, C., M. Reichard, P. Jurjada and M. Przybylski. 2004. Review: the reproductive ecology of the European bitterling (*Rhodeus sericeus*). J. Zool. (Lond.) 262: 107–124.

Smith, G.R. 1966. Distribution and evolution of the North American catostomid fishes of the subgenus *Pantosteus*, genus *Catostomus*. Misc. Publ. Mus. Zool., Univ. Michigan 129: 1–132.

Smith, G.R. 1992. Phylogeny and biogeography of the Catostomidae, freshwater fishes of North America and Asia, pp. 778–826. *In*: R.L. Mayden [ed.]. Systematics, Historical Ecology, and North American Freshwater Fishes. Stanford University Press, Stanford.

Smith, G.R., T.E. Dowling, K.W. Gobalet, T. Lugaski, D.K. Shiozawa and R.P. Evans. 2002. Biogeography and timing of evolutionary events among Great Basin fishes, pp. 175–234. *In*: R. Hershler, D.B. Madsen and D.R. Currey [eds.]. Great Basin Aquatic Systems History. Smithsonian Institution Press, Washington.

Smith, H.M. 1945. The fresh-water fishes of Siam or Thailand. Bull. US Natl. Mus. 188: 1–622.

Suzuki, A. 1992. Karyotype and DNA contents of the chinese catostomid fish, *Myxocyprinus*. Chromosome Info. Serv. 53.

Tang, Q., H. Liu, R.L. Mayden, B. Xiong. 2006. Comparison of evolutionary rates in the mitochondrial DNA cytochrome b gene and control region and their implications for phylogeny of the Cobitoidea (Teleostei: Cypriniformes). Mol. Phylogenet. Evol. 39: 347–357.

Tsigenopoulos, C.S., P. Ráb, D. Naran and P. Berrebi. 2002. Multiple origins of polyploidy in the phylogeny of southern African barbs (Cyprinidae) as inferred from mtDNA markers. Heredity 88: 466–473.

Uyeno, T. and G.R. Smith. 1972. Tetraploid origin of the karyotype of catostomid fishes. Science 175: 644–646.

Wang, X., S. He. and Y. Chen. 2002. Sequence variations of the S7 ribosomal protein gene in primitive cyprinid fishes: Implications on phylogenetic analysis. Chin. Sci. Bull. 47: 1638–1643.

Weisel, G. F. 1960. The osteocranium of the catostomid fish, *Catostomus macrocheilus*: A study in adaptation and natural relationship. J. Morph. 106: 109–130.

Wilson, M.V.H. 1977. Middle Eocene freshwater fishes from British Columbia. Life Sci. Contr. Roy. Ontario Mus. 113: 1–61.

Wu, Y. 1984. Systematic studies on the cyprinid fishes of the subfamily Schizothoracinae from China. Acta Biol. Plateau Sinica 3: 119–140.

Wu, X., Y. Chen, X. Chen and J. Chen. 1981. A taxonomical system and phylogenetic relationship of the families of the suborder Cyprinoidei (Pisces). Scientia Sinica 24: 563–572.

Xiao, W., Y. Zhang and H. Liu. (2001). Molecular systematics of Xenocyprinae (Teleostei: Cyprinidae): Taxonomy, biogeography, and coevolution of a special group restricted in east Asia. Mol. Phylogenet. Evol. 18: 163–173.

Yang, J.X., Y.R. Chen and J.H. Lan. 1994. *Protocobitis typhlops*, a new genus and species of cave loach from China (Cypriniformes: Cobitidae). Ichthyol. Explor. Freshwaters 5: 91–96.

Yazdani, G.M., D.F. Singh and M.B. Rao. 1989 and 1990. *Psilorhynchoides*, a new genus for the cyprinid fish, *Psilorhynchus homaloptera* Hora and Mukherji and *P. pseudecheneis* Menon and Dutta, with a definition of the subfamily Psilorhynchinae (Cyprinidae). Matsya 15 and 16: 14–20.

Zhu, S.Q. 1981. Notes on the scaleless loaches (Nemachilinae, Cobitidae) from Qinghai-Xizang Plateau and adjacent territories in China. Geol. Ecol. Stud. Qinghai-Xizang Plateau 2: 1061–1070.

11

Review of the Phylogenetic Relationships and Fossil Record of Characiformes

Wasila M. Dahdul[1]

Abstract

The Order Characiformes is a diverse lineage of ostariophysan fishes presently distributed in the freshwater lakes and rivers of the Neotropics and Africa. Fossil characiforms are also known from Europe. The order contains 19 families and close to 1,700 species, the majority endemic to South American drainages. Aside from a few well-supported subgroups of characiforms, relationships among families are not settled, and the taxonomy of some families, such as the complex Characidae, have undergone considerable recent revision. Most studies place the Citharinoidei (families Citharinidae + Distichodontidae) as the sister group to Characoidei, i.e., all other characiforms. Within the Characoidei, the Anostomoidea (families Anostomidae, Curimatidae, Prochilodontidae and Chilodontidae) is a well-supported lineage. The trans-Atlantic group consisting of the Neotropical families Ctenoluciidae, Erythrinidae, Lebiasinidae and African Hepsetidae is also well supported. Despite lack of resolution, three trans-Atlantic sister clades involving the African Citharinoidei, Alestidae, and Hepsetidae are consistently recovered among higher level studies, although with differing Neotropical sister groups among the published studies. Characiform fossils are known from the

[1] Department of Biology, University of South Dakota, Vermillion SD 57069 and National Evolutionary Synthesis Center, Durham, NC. e-mail: wasila.dahdul@usd.edu

Late Cretaceous and much of the Cenozoic throughout their modern range, pointing to the deep history of the group. Controversial hypotheses on the early history of characiforms are difficult to test presently.

Introduction

Fishes of the order Characiformes (characins, tetras, tigerfishes, piranhas) are a diverse lineage of fishes endemic to the freshwaters of the Americas and Africa. In the New World, characiforms range from southern North America to Central and South America, and in the Old World they extend throughout the African continent and are also known from the European fossil record. The order is most diverse in Central and South America, with over 1,400 species compared to approximately 260 species in Africa (Reis *et al.* 2003, Nelson 2006). Characiforms are found in a variety of freshwater habitats, from pelagic waters of large rivers to small streams, calm river margins, lakes, and caves. Body sizes range from over 1 m length in the African tigerfish to less than 2 cm adult length in miniature species (Weitzman and Vari 1988). Characiforms are impressively diverse in feeding ecology and associated ecomorphology. Beyond the notorious piranhas, which possess triangular blade-like teeth suited to biting prey, are equally remarkable detritivorous prochilodontids with fleshy lips used to suck up food (Castro and Vari 2004); herbivorous pacus with complex molar-like teeth used to crush seeds and fruit (Goulding 1980); lepidophagous characins and serrasalmids with specialized dentitions for dislodging scales (Sazima 1983, Nico and Taphorn 1988); carnivores with large canines such as the African tigerfish; and anostomids with a range of jaw orientations from sub- to supra-terminal (Sidlauskas and Vari 2008). Economically, characiforms are popular aquarium fishes (e.g., tetras, *Metynnis* silverdollars) and important food fishes in aquaculture, and contribute to a significant portion of the fish catch in South America.

Monophyly of the Characiformes is well accepted, whereas outside of a few well-corroborated clades, intrarelationships among characiform subgroups are largely unresolved. Fink and Fink (1981, 1996) presented the following seven synapomorphies as evidence for the monophyly of characiforms: presence of auditory foramen in prootic; posttemporal fossa with mediodorsal opening; lagenar capsule well developed; replacement tooth trenches present on dentary and premaxilla; teeth multicuspidate; transverse process present on third neural arch; and hypural 1 separated from compound centrum. Higher-level studies of characiform relationships that include nearly complete sampling of representatives of most families include the morphology-based work of Buckup (1998), and the unpublished dissertations by Lucena (1993) and Uj (1990). Molecular sequence–based studies of characiform relationships to date include Ortí and Meyer's (1997) work based on 12S and 16S mitochondrial ribosomal DNA sequences and Calcagnotto *et al.*'s (2005) analysis based on a set of four nuclear and two mitochondrial genes. In addition, Hubert *et al.* (2005), in a study not specifically focused on reconstructing characiform relationships, reanalyzed Ortí and Meyer's (1997) mitochondrial dataset with additional taxon sampling of African characiforms to test a method that accounts for molecular saturation. Other higher-level studies analyze relationships among characiform subgroups consisting of several families, e.g., distichodontids and citharinids (Vari 1979); curimatids, prochilodontids, anostomids, and chilodontids (Vari 1983); and ctenoluciids, erythrinids, hepsetids, and lebiasinids (Vari 1995). In total, these studies have corroborated three African-South American sister clades, though the recovered Neotropical relatives of two of the sister clades differ among published studies. Fossil characiforms are known from South American, African, and European deposits of minimally the Upper Cretaceous through the Cenozoic, mostly as isolated teeth or fragmentary jaw elements (Arratia and Cione 1996, Gayet and Meunier 1998, Lundberg 1998). The historical biogeography of characiforms, given the freshwater affinities of Recent characiforms and their present distribution in the Neotropics and Africa, has traditionally been interpreted against the backdrop of drift-vicariance scenarios related to the separation of Africa and South America (Lundberg 1993, Buckup 1998). Recently, the possibility of marine dispersal to explain trans-Atlantic relationships has been raised, based on fossils from brackishwater deposits identified as belonging to early characiforms or otophysans and new phylogenetic results pertaining to characiform relationships (e.g., Gayet *et al.* 2003, Calcagnotto *et al.* 2005).

In addition to being a model group for biogeographic studies, characiforms have also been the focus of recent investigation into the factors involved in unequal morphological diversity among sister groups of fishes (Sidlauskas 2007, 2008), and the published systematic characters for characiforms, as part of an effort focused on the Ostariophysi, are being coded into a computable format for a phenotype database linked to zebrafish mutant data (Mabee *et al.* 2007). A well-supported framework of characiform relationships is essential to investigations of characiform biogeography and evolutionary history. Vari (1998) and Weitzman and Malabarba (1998) reviewed in detail the historical concepts of classification and evidence for relationships in Characiformes and Characidae, respectively. In the 10 years since their publication, 332 new characiform species have been described, 68% of which belong to the Characidae (Eschmeyer and Fong 2008), the taxonomically complex and likely non-monophyletic family. Here I review the evidence for the monophyly of and relationships among characiform families and summarize the fossil evidence related to characiforms. Emphasis is on published work on higher-level characiform relationships (e.g., Ortí and Meyer 1997, Fig. 11.1; Buckup 1998, Fig. 11.2; Calcagnotto *et al.* 2005, Fig. 11.3). Well-supported monophyletic subgroups of characiforms are discussed first, followed by families.

Suborder Citharinoidei

The African Citharinidae contains three genera (*Citharidium, Citharinops,* and *Citharinus*) and eight species of primarily detritivores. The African Distichodontidae consists of 18 genera and 90 species with diverse feeding habits, including herbivores and carnivores. These two families form a monophyletic clade (Vari 1979) that is the sister group to all other characiforms (Fink and Fink 1981). The basal position of this African clade has been widely corroborated by morphological evidence (Buckup 1998) and molecular evidence (Ortí and Meyer 1997, Calcagnotto *et al.* 2005). †*Eocitharinus macrognathus* was described from a partially preserved specimen from the Eocene of Tanzania and identified as a citharinoid (Distichodontidae + Citharinidae), although presence of a Weberian apparatus could not be confirmed due to poor preservation in the anterior region of the vertebral column (Murray 2003a).

Suborder Characoidei

Anostomoidea

The Neotropical Anostomoidea is a large clade comprising four families: Anostomidae, Chilodontidae, Curimatidae, and Prochilodontidae. Monophyly of anostomoids is supported by morphological evidence (e.g., Vari 1983, Buckup 1998) and molecular evidence (Ortí and Meyer 1997, Calcagnotto *et al.* 2005). Vari (1983) presented morphological synapomorphies that support Anostomidae + Chilodontidae as sister to Curimatidae + Prochilodontidae. These relationships were also recovered by Buckup's (1998) morphology-based work. In contrast, the molecular sequence–based work by Calcagnotto *et al.* (2005) recovered ((Prochilodontidae, Chilodontidae), Anostomidae)), although Curimatidae was not sampled in their study. Anostomoid monophyly was not recovered by Ortí and Meyer's (1997) analysis based on mitochondrial sequences.

The four families of Anostomoidea are each well supported as monophyletic. The Anostomidae contains 15 genera and 140 nominal species distributed in the trans-Andean waters of Colombia (Magdalena and Atrato rivers) and Venezuela (Lake Maracaibo) and cis-Andean drainages of the Orinoco, Amazon, Guianas, São Francisco, Río de La Plata, and coastal rivers of Brazil (Sidlauskas and Vari, 2008). Anostomids feed on a variety of food items and some are primarily detritivores that feed in a vertical orientation, hence the common name "headstanders." Monophyly of the Anostomidae is well supported (Winterbottom 1980, Vari 1983) and relationships among anostomid species were most recently studied by Sidlauskas and Vari (2008). Anostomids are known from isolated fossil teeth of the Miocene La Venta fauna (Lundberg 1997) and Miocene of the Cuenca Basin in Ecuador (Roberts 1975).

Chilodontidae contains two genera (*Caenotropus* and *Chilodus*) and eight species with a cis-Andean distribution in the Orinoco, Amazon, and Guiana drainages, and Paraíba River in northeastern Brazil (Vari and Raredon 2003). Monophyly of Chilodontidae is well supported by morphological synapomorphies (Vari 1983, Vari *et al.* 1995, Vari and Raredon 2003), and chilodontid relationships were analyzed in Vari *et al.* (1995). The vertical body orientation of these fishes in swimming also inspired the common name "headstanders."

Curimatidae are toothless characiforms with 99 nominal species contained in eight genera (*Curimata, Curimatella, Curimatopsis, Cyphocharax, Potamorhina, Psectrogaster, Pseudocurimata,* and *Steindachnerina*). Most species are detritivores. Curimatids are distributed on both sides of the Andes, from Costa Rica, Peru, Colombia and Venezuela (Lake Maracaibo basin) to eastern slope drainages including Orinoco, Amazon, Guianas, São Francisco drainages, and coastal Brazilian rivers. Curimatid monophyly and intrarelationships were analyzed in Vari (1983, 1988). The fossil curimatid †*Cyphocharax mosesi* is known from the Oligocene of the Tremembé Formation of southeastern Brazil (Malabarba 1996), as is †*Plesiocurimata alvarengai* (Figueiredo and da Costa-Carvalho 1999).

Prochilodontidae contains three genera (*Ichthyoelephas, Prochilodus,* and *Semaprochilodus*) and 21 species distributed on both slopes of the Andes, in trans-Andean Lake Maracaibo basin in Venezuela and rivers of Colombia and Ecuador, and in cis-Andean drainages of the Orinoco, Amazon, Río de La Plata river system, Guianas, São Francisco, and Brazilian coastal rivers. Morphological synapomorphies that support the monophyly of Prochilodontidae are discussed in Castro and Vari (2004). Prochilodontids are predominately detritivores that possess distinctively fleshy lips with loosely attached teeth (Vari 1983, Castro and Vari 2004).

Families Ctenoluciidae, Erythrinidae, Lebiasinidae, and Hepsetidae

Consistent among morphology-based studies with the most complete taxon sampling is the clade formed by the Neotropical Ctenoluciidae, Erythrinidae, and Lebiasinidae, and African Hepsetidae. The African Hepsetidae + Ctenoluciidae were recovered as sister to Lebiasinidae + Erythrinidae by Lucena (1993) and Buckup (1998). Vari (1995) recovered the Hepsetidae as sister to Ctenoluciidae + Erythrinidae, with Lebiasinidae as sister to these three families. The monophyly of the four families was not corroborated by either of the large-scale molecular studies on characiforms, although some families are hypothesized to share close relationships. Ortí and Meyer (1997) recovered Hepsetidae as sister to Erythrinidae. Calcagnotto *et al.* (2005) recovered Hepsetidae as sister to the clade formed by Ctenoluciidae and Lebiasinidae, with the African Alestidae in turn sister to this clade, whereas Erythrinidae was recovered as sister to Crenuchidae.

Ctenoluciidae contains two genera (*Ctenolucius* and *Boulengerella*) and seven species distributed from the trans-Andean drainages of Panama, Colombia, and Lake Maracaibo basin in Venezuela to eastern slope drainages including Orinoco, Amazon, and Guianas. Ctenoluciids are fish predators that possess conical teeth and are pike-like in appearance. Vari (1995) presented morphological evidence for the ctenoluciid monophyly and relationships.

Erythrinidae (trahiras) contain three genera (*Erythrinus*, *Hoplerythrinus*, and *Hoplias*) and 12 species with trans-Andean distribution in Ecuador, Costa Rica, Panama, Colombia (Río Magdalena) and Venezuela (Lake Maracaibo), and cis-Andean distribution in Orinoco, Amazon, Guianas, São Francisco, and Paraná-Paraguay drainages. Erythrinids are also found on the island of Trinidad. These fish predators inhabit a variety of riverine, lagoon, and lake habitats. Erythrinid relationships were analyzed by Weitzman (1964) and characters supporting the monophyly of the family were recognized by Vari (1995). Fossil teeth identified as *Hoplias* are reported from the Colombian La Venta Fauna (Lundberg 1997) and †*Paleohoplias assisbrasiliensis* is described from fossil premaxilla and dentary bones from the Solimões Formation, Acre Brazil (Gayet *et al.* 2003). Among the oldest characiform fossils are fragmentary jaw and teeth described as erythrinid from the Late Cretaceous to Paleocene of Bolivia (Gayet 1991, Gayet and Meunier 1998).

Lebiasinids, commonly known as pencilfishes, are contained in seven genera (*Copeina*, *Copella*, *Derhamia*, *Lebiasina*, *Nannostomus*, *Piabucina*, and *Pyrrhulina*) with 62 species. Lebiasinid relationships were analyzed by Weitzman (1964) and monophyly of the group discussed in Vari (1995). Lebiasinids are distributed throughout Central and South America, from Costa Rica to Argentina.

The African family Hepsetidae contains one species, *Hepsetus odoe* (African pike). *Hepsetus* is a fish predator found in calm waters of lagoons and river backwaters (Winemiller and Winemiller 1994). Fossil teeth referable to *Hepsetus* are reported by Otero and Gayet (2001) from the Lower Oligocene and Miocene of the Arabian Peninsula.

Family Acestrorhynchidae

Acestrorhynchidae contains one genus (*Acestrorhynchus*) with 14 species inhabiting the major cis-Andean drainages of South America, including the Amazon, Orinoco, Guianas, São Francisco, Paraná-Paraguay, and Uruguay basins (Menezes 2003). The predatory acestrorhynchids are piscivores, with strong canines on the premaxilla and dentary, that inhabit the still waters of lakes, lagoons, and calm river margins. Monophyly of Acestrorhynchidae is well supported (Menezes 1969, Menezes and Géry 1983, Lucena and Menezes 1998) and phylogenetic relationships among acestrorhynchid species was most recently analyzed by Toledo-Piza (2007) based on osteological and other morphological characters.

Family Alestidae

The African family Alestidae contains 19 genera and 105 species (Zanata and Vari, 2005), including the tigerfish *Hydrocynus forskahlii*, a predator found in open waters, and the dwarf alestids (e.g., *Virilia* species). Géry (1977) placed the African components of Greenwood *et al.*'s (1966) Characidae in the separate family Alestidae, and monophyly of the Alestidae is well supported by morphological synapomorphies (Murray and Stewart 2002, Zanata and Vari 2005). Zanata and Vari (2005) provided the most comprehensive analysis of this family to date and did not find support for the monophyly of the subfamily Petersiini. In their analysis, the New World *Chalceus* was recovered as the sister group to the African members of Alestidae. This conflicts with the placement of *Chalceus* as sister to Neotropical Characidae by Calcagnotto *et al.* (2005) based on molecular evidence. Both Zanata and Vari (2005) and Calcagnotto *et al.* (2005) agreed in the placement of *Arnoldichthys* as sister to all other African alestids but otherwise differed in the relationships recovered.

Fossil alestids are known from numerous European and African localities. Zanata and Vari (2005) recently reviewed a subset of these fossils. Isolated teeth from the Lower and Middle Eocene of France originally described as cf. *Alestes* by Cappetta *et al.* (1972) and identified as †*Alestoides eocaenicus* in Monod and Gaudant (1998) were reevaluated as close to *Alestes* by Zanata and Vari (2005). Among fossils Zanata and Vari (2005) identified as closely related to the clade formed by *Alestes*, *Brycinus*, and *Bryconaethiops* are numerous isolated multicuspidate fossil teeth from the Lower Eocene of Spain that were assigned to aff. *Alestes* by De La Peña Zarzuelo (1996), and alestid fossils from the Lower Oligocene and Miocene of the Arabian Plate (Otero and Gayet 2001). †*Sindacharax*, with six species known from the Miocene-Pliocene of numerous African localities (Greenwood 1972, 1976, Greenwood and Howes 1975, Stewart 1994, 1997,

2003, Stewart and Murray 2008), was originally suggested as having a close relationship to the South American Serrasalminae (Greenwood and Howes 1975), but reevaluation of †*Sindacharax* by Zanata and Vari (2005) indicates a close relationship with the clade formed by *Alestes, Brycinus,* and *Bryconaethiops*. Revaluation of the fossil †*Mahengecharax carrolli* from the Eocene of Tanzania, originally identified as a fossil alestid (Murray 2003b), did not find support for its placement in the Alestidae; further, the lack of evidence of a Weberian apparatus, among other diagnostic features, due to fossil preservation calls the identity of †*Mahengecharax carrolli* as an otophysan into question (Zanata and Vari 2005).

Other reported alestid fossils include those attributable to the genus *Hydrocynus* known from Miocene to Pliocene deposits of Lothagam, Kenya (Stewart 1994, 2003). Fossils presently identified only to the level of Alestidae are reported from Upper Eocene and Lower Oligocene of Egypt (Murray 2004) based on isolated multicuspidate teeth. †*Arabocharax baidensis* from Oligocene of the Arabian Peninsula (Micklich and Roscher 1990) is identified as an alestid but its relationships within the family are uncertain. Isolated teeth from the Maastrichtian of southern France were described as possibly alestid (Otero *et al.* 2008) and, if confirmed, would represent the oldest fossils reported for the family.

Family Characidae

The Characidae is an uncertainly monophyletic and taxonomically complex assemblage that contains many well-supported monophyletic, subfamily-level (and below) clades of characiforms. Characids are typically small, colorful fishes such as neon tetras, but also include larger fish such as the predatory dourado (*Salminus*), which is popular in sport fishing. Several subfamilies formerly contained in Characidae have recently been removed and raised to the level of family based on their closer relationships to other characiform families (e.g., Crenuchidae, African Alestidae, Serrasalmidae). A large number of taxa have also recently been placed as *incertae sedis* within Characidae as a result of taxonomic revision, e.g., Cheirodontinae (Malabarba 1998a), or lack of evidence for their monophyly, e.g., Tetragonopterinae, which presently contains only the genus *Tetragonopterus* (Reis *et al.* 2003). Among the subfamilies of Characidae recognized in Reis *et al.* (2003) are: Glandulocaudinae with monophyly established by Weitzman and Menezes (1998), Cheirodontinae with monophyly established by Malabarba (1998a), Rhoadsiinae, Tetragonopterinae, Stethaprioninae with monophyly established by Reis (1989), Characinae, Aphyocharacinae, Bryconinae, Iguanodectinae with monophyly established by Vari (1977), Clupeacharacinae, and Agoniatinae.

Numerous fossil tricuspidate teeth from throughout the Neogene of South America (Lundberg *et al.* in press) are identified as tetragonopterine. Fossil characids from the Oligocene of the Tremembé Formation of southeastern Brazil (Malabarba 1998b) include †*Brycon avus*, the cheirodontin †*Megacheirodon unicus*, and †*Lignobrycon ligniticus* (and the modern *L. myersi*), which are sister to *Triportheus*.

Family Crenuchidae

Crenuchids (South American darters) are broadly distributed on both sides of the Andes, from Panama to the major drainages of the Orinoco, Amazon, Guianas, and Río de La Plata basins, typically in swiftly flowing streams (Buckup 2003). Crenuchidae contains 12 genera and 81 species in two subfamilies, Crenuchinae (*Crenuchus* and *Poecilocharax*) and Characidiinae (*Ammocryptocharax, Characidium, Elachocharax, Geryichthys, Klausewitzia, Leptocharacidium, Melanocharacidium, Microcharacidium, Odontocharacidium, Skiotocharax*). Buckup (1993a, b, 1998) presented evidence on the monophyly and interrelationships of crenuchids.

Family Cynodontidae

Cynodontids are fish predators distributed in the trans-Andean waters of Colombia (Río Atrato and Río Magdalena) and Venezuela (Lake Maracaibo), and in cis-Andean drainages including the Amazon, Orinoco, Guianas, and Paraná-Paraguay basins. The 14 species of cynodontids are placed in two subfamilies: Cynodontinae (*Cynodon* and *Rhaphiodon*) and Roestinae (*Gilbertolus, Hydrolycus, Roestes*) (Lucena and Menezes 1998, Menezes 1998). Fossil cynodontids are known from the Miocene La Venta fauna in Colombia (Lundberg 1997), and Miocene of Argentina (Cione and Casciotta 1995, 1997). Intrarelationships of the subfamily Cynodontinae were examined by Toledo-Piza (2000).

Family Gasteropelecidae

Gasteropelecidae (freshwater hatchetfish) contains three genera (*Carnegiella, Gasteropelecus, Thoracocharax*) and nine species distributed widely in Central and South America from Panama to Argentina. Weitzman (1954, 1960) presented morphological characters that support the monophyly of this family. Gasteropelecids are found in the open waters of rivers and lakes and have a highly modified, keel-like pectoral girdle and elongate pectoral fins that help them jump out of the water.

Family Hemiodontidae

Hemiodontidae contains five genera (*Anodus, Argonectes, Bivibranchia, Hemiodus,* and *Micromischodus*) and 29 species with cis-Andean distribution in the Orinoco, Amazon, Guianas, and Paraná-Paraguay drainages. Hemiodontids are typically pelagic and feed on a variety of food items including plant material, insects, and detritus. Relationships within the family have been the focus of studies by Roberts (1974) and Langeani (1996, 1998).

Family Parodontidae

Parodontidae contains three genera (*Apareiodon, Parodon,* and *Saccodon*) and 20 species. Parodontids were previously grouped as a subfamily of the Hemiodontidae (Géry 1977), but this relationship was not supported by morphological evidence (Roberts 1974, Buckup 1998). Fossil parodontid teeth, similar to those found in the genus *Parodon*, are known from the Miocene of Ecuador (Roberts 1975). Parodontids are widely distributed in Central and South America, from Panama to Argentina.

Family Serrasalmidae

Well known for the carnivorous biting habits of *Serrasalmus* and *Pygocentrus* piranhas, the 16 genera and approximately 80 species (Jégu 2003) of Serrasalmidae also include fin- and scale-eaters, rock-scrapers, and herbivorous pacus that feed on fruits, seeds, and nuts (Goulding 1980, Nico and Taphorn 1988). Serrasalmids are distributed in the trans-Andean Lake Maracaibo basin of Venezuela and cis-Andean drainages of the Orinoco, Guianas, Amazon, São Francisco, and Paraná-Paraguay. The group was previously considered a subfamily of Characidae (e.g., Buckup 1998), but most higher-level studies that included serrasalmids have suggested relationships with other characiforms. The well-supported monophyletic lineage (Machado-Allison 1983, Ortí *et al.* 1996) is here considered a family.

Serrasalmid intrarelationships have been the focus of numerous studies based on morphological and molecular evidence. Recent work based on osteological characters and nuclear gene sequences found high support for three monophyletic subgroups of serrasalmids: (pacus ("Myleus" clade, piranhas)) (Dahdul 2007), which corroborates work based on mitochondrial gene and D-loop sequences (Ortí *et al.* 1996, 2008). This placement of the pacu lineage (*Colossoma, Mylossoma, Piaractus*) as sister to all other serrasalmids conflicts with the placement of the pacu lineage in a derived position in Machado-Allison's (1982, 1983) morphology-based work. Jégu (2004), based on morphological evidence, also recovered a monophyletic "*Myleus*" and piranha clade, but not a monophyletic pacu clade. Hubert *et al.* (2007) analyzed the relationships and phylogeography of *Serrasalmus* and *Pygocentrus* piranhas using mitochondrial ribosomal 16S gene and control region sequences. They found support for the monophyly of these two genera, though noting that inclusion of a species of *Pristobrycon* in the genus *Serrasalmus* rendered *Pristobrycon* polyphyletic.

Fossil serrasalmids are known from the Upper Cretaceous (Gayet and Meunier 1998) to Pliocene of South America and their fossil record was reviewed recently in Dahdul (2004) and Lundberg *et al.* (in press). Among the oldest serrasalmid fossils are the isolated pacu-like teeth from the Upper Cretaceous to Paleocene of Bolivia (Gayet 1991, Gayet and Meunier 1998). Other serrasalmid taxa represented by isolated fossil teeth and fragmentary jaw elements are *Colossoma macropomum*, *Piaractus*, *Mylossoma*, and fossil teeth indistinguishable from modern piranhas. An extinct genus, proposed as closely related to modern piranhas, has recently been described by Cione *et al.* (in press) from the Upper Miocene of Argentina. Some of these records indicate that serrasalmids were distributed in drainages from which they are absent today, such as the Magdalena drainage of Colombia as evidenced by fossil *Colossoma macropomum* and piranhas of the La Venta fauna (Lundberg *et al.* 1986, Lundberg 1997) and fossil *Colossoma* from the Paraná drainage basin in Argentina (Cione 1986, Cione *et al.* 2000). As discussed above, a relationship between the African †*Sindacharax* and Neotropical serrasalmids is not supported based on comparisons of tooth and jaw morphologies.

Characiform Fossil Record

The oldest records of characiforms date from the Late Cretaceous (e.g., erythrinids and serrasalmids). The fossil record of characiforms in South America, including records of anostomids, erythrinids, serrasalmids, and tetragonopterines indicate that components of these groups were strikingly modern by the mid-Miocene (Lundberg *et al.* in press, Lundberg 1998). Studies based on phylogenetic evidence and geographic distributions of subgroups of characiforms also agree with the deep history of characiform clades (Weitzman and Weitzman 1982, Vari 1988). In addition, the record of other freshwater fish taxa, including catfishes, lungfishes and cichlids, reveals the modernization of these lineages by the Miocene (Lundberg 1998). In Africa and Europe, fossil characiforms from the Oligocene to Miocene age also reveal the deep history of African characiforms, including alestids and hepsetids. For some of these groups, fossils can be assigned to subgroup lineages within families, e.g., Alestidae (Zanata and Vari 2005) or included in a phylogenetic analysis, e.g., Serrasalmidae (Cione *et al.* in press), to provide minimum ages of origin for lineages within their families.

In South America, the fossil record does not indicate that distinct lineages of characiforms became extinct; instead, the record indicates that characiforms were extirpated from peripheral habitats (e.g., Magdalena river drainage; Lundberg *et al.* 1986, 1997). This pattern is also observed in Africa and Europe, where alestids were formerly distributed in Europe and the Arabian Peninsula.

The majority of fossil characiforms are known from isolated teeth or fragmentary jaw elements. Tooth shape, position, and jaw structure of characiforms are known to be useful in fossil identification (e.g., Lundberg 1997, Dahdul 2004, Zanata and Vari 2005). Numerous fossils attributable to Characiformes may be more finely identified and phylogenetically placed pending comparison to modern African and Neotropical characiforms. Among the fossils that may be more finely identified are small, isolated characiform teeth from the Maastrichtian of southern France (Otero *et al.* 2008) and a fossil multicuspidate characiform tooth from Upper Miocene of Portugal (Antunes *et al.* 1995). Fossils that might be placed in a more detailed phylogenetic context upon reevaluation include the alestid †*Arabocharax baidensis* from Oligocene of the Arabian Peninsula (Micklich and Roscher 1990), †*Bunocharax* from the Mio-Pliocene Albertine Rift Valley in Uganda (Van Neer 1994), and †*Tiupampichthys intermedius*, known from fossil premaxillary and dentary bones from Upper Cretaceous and Paleocene of Bolivia and hypothesized to share a close relationship to Acestrorhynchidae (Gayet *et al.* 2003).

Older fossil characiforms remain to be evaluated for their confident placement in a phylogenetic context and, given their age and importance for characiform biogeography, need to be compared not only to characiforms but also to other otophysans. These include †*Santanichthys* from the Lower Cretaceous Santana Formation of Brazil recently described as an early characiform (Filleul and Maisey 2004), †*Salminops* from the Cenomanian of Portugal described as a characiform (Gayet 1985), †*Sorbinicharax* from the Upper Cretaceous of Italy (Taverne 2003), and †*Lusitanichthys* originally described as a characiform (Gayet 1981) but more recently discussed as *incertae sedis* in Otophysi (Fink and Fink 1996). As some of these fossils were deposited in brackish water or marine sediments, their reevaluation is necessary for deciphering the biogeographic history of characiforms.

Phylogenetic Relationships among Characiform Families

Among the published higher-level studies of characiform relationships that employ explicit phylogenetic criteria, several well-supported groupings among characiforms are generally accepted (Fig. 11.4): (1) The African Distichodontidae and Citharinidae are monophyletic (Vari 1979) and are

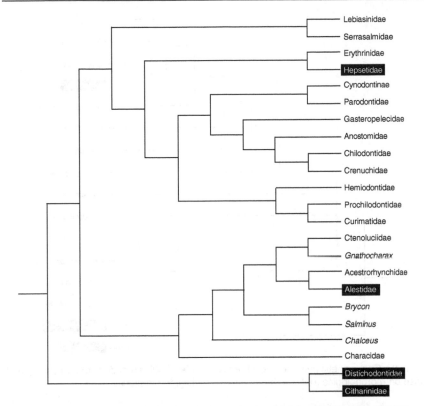

Fig. 11.1 Phylogenetic hypothesis of characiform relationships modified from Ortí and Meyer (1997) based on mitochondrial DNA sequences, with African taxa highlighted in black.

the sister group to all other characiforms (Fink and Fink 1981). This relationship was also supported in the studies of Ortí and Meyer (1997), Buckup (1998) and Calcagnotto *et al.* (2005) (Figs. 11.1–11.3). (2) The monophyletic Anostomidae and Chilodontidae form the sister group to the Curimatidae and Prochilodontidae (Vari 1983). This clade was also recovered by Buckup (1998; Fig. 11.2), but not supported by Ortí and Meyer (1997; Fig. 11.1) or Calcagnotto *et al.* (2005; Fig. 11.3), as discussed above. (3) The Neotropical Ctenoluciidae, Erythrinidae, Lebiasinidae and African Hepsetidae form a monophyletic assemblage, though with differing intrarelationships among morphological studies, e.g., Lucena (1993), Buckup (1998; Fig. 11.2) and Vari (1995). Neither of the large-scale molecular studies of characiform relationships (Ortí and Meyer 1997, Calcagnotto *et al.* 2005; Figs. 11.1, 11.3) recovered the monophyly of these four families, as discussed previously.

Other characiform relationships receiving some support across multiple studies include the sister-group relationship of the

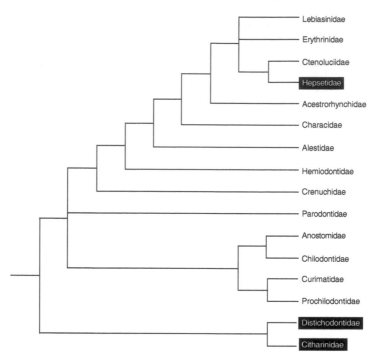

Fig. 11.2 Phylogenetic hypothesis of characiform relationships modified from Buckup (1998) based on morphological data, with African taxa highlighted in black.

Acestrorhynchidae and Cynododontidae, which was recovered by Lucena and Menezes (1998) and Lucena (1993). However, Ortí and Meyer (1997) recovered Acestrorhychidae as sister to the African Alestidae (Fig. 11.1), whereas Buckup (1998; Fig. 11.2) recovered Acestrorhynchidae as sister to a clade containing Ctenoluciidae, Hepsetidae, Lebiasinidae, and Erythrinidae. Calcagnotto *et al.* (2005; Fig. 11.3) recovered *Acestrorhynchus* as nestled within their Characidae clade.

In addition to the foregoing, several other outstanding questions in characiform phylogenetics remain to be resolved. The position of the Neotropical *Chalceus* is in conflict among the published studies. Based on molecular sequence data, Calcagnotto *et al.* (2005) recovered *Chalceus* as sister to the Neotropical Characidae with high node support values. Zanata and Vari (2005) proposed *Chalceus* as sister to the African Alestidae and presented a number of morphological synapomorphies that support this relationship. Its inclusion in this African family would make Alestidae the only characiform family with a trans-Atlantic distribution (Zanata and Vari, 2005). The phylogenetic position of the Serrasalmidae is also unresolved. On the basis of morphology, Lucena (1993) recovered the placement of serrasalmids within the Neotropical Characidae, and

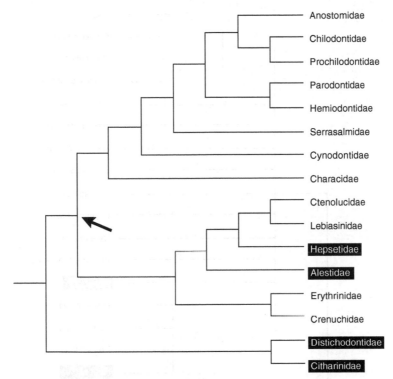

Fig. 11.3 Phylogenetic hypothesis of characiform relationships modified from Calcagnotto *et al.* (2005) based on nuclear and mitochondrial DNA sequences, with African taxa highlighted in black.

serrasalmid position was not resolved in the molecular sequence–based study of Ortí and Meyer (1997). Calcagnotto *et al.* (2005) recovered serrasalmids as the sister group to their Anostomoidea (Anostomidae, Chilodontidae, Prochilodontidae, Hemiodontidae, and Parodontidae). Serrasalmids represent one of the oldest known characiforms in the fossil record, dating from the Late Cretaceous (Gayet and Meunier 1998). Their placement in a well-supported phylogeny of characiform fishes will lend an important calibration point to reconstructing characiform evolutionary history.

Trans-Atlantic Relationships

As previously noted, characiforms in Africa and South America do not form reciprocally monophyletic groups as might be expected if the separation of the southern continents occurred early in the evolution of the group. Rather, the higher-level studies of characiform relationships recover three trans-Atlantic sister clades (Buckup 1991, Ortí and Meyer 1997, Calcagnotto *et al.*

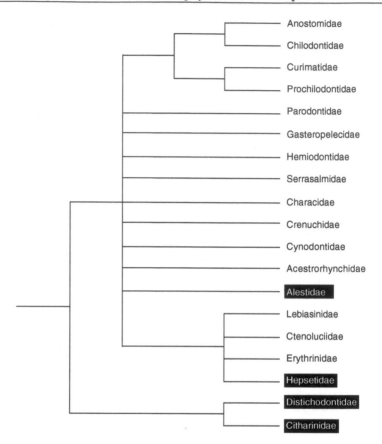

Fig. 11.4 Summary tree of published phylogenetic hypotheses for characiform relationships, with African taxa highlighted in black.

2005). Common to all studies is agreement on the sister-group relationship between the African Citharinoidei (Citharinidae + Distichodontidae) and all other characiforms. The second trans-Atlantic sister-group pair involves the African Hepsetidae and a sister-group relationship with either: (1) the Ctenoluciidae within a clade that also contains Lebiasinidae and Erythrinidae (Buckup 1998); (2) the Erythrinidae within a large clade of Neotropical characiforms (Ortí and Meyer 1997); or (3) the Ctenoluciidae + Lebiasinidae, which is in turn sister to the African Alestidae (Calcagnotto *et al.* 2005). The third trans-Atlantic sister-group pair consists of the African Alestidae as the sister group to either: (1) a large Neotropical clade that also contains the African Hepsetidae (Buckup 1998); (2) the clade consisting of ((Ctenolucidae+Lebiasinidae), Hepsetidae)) (Calcagnotto *et al.* 2005); or (3) *Acestrorhynchus* within an entirely Neotropical clade (Ortí and Meyer 1997).

These preceding hypotheses recovered Alestidae as a strictly African lineage, whereas Zanata and Vari (2005) found support for the inclusion of the Neotropical *Chalceus* in Alestidae.

Given the strictly freshwater affinities of modern members of the order and their geographic distribution, it is generally accepted that characiforms originated prior to the break-up of Africa and South America (110 to 90 million years ago) (Fink and Fink 1981, Lundberg 1993, Buckup 1998), and the biogeographic history of characiforms has been interpreted in the context of Gondwanan drift-vicariance scenarios. In discussing the results of Buckup (1991), Lundberg (1993) hypothesized that if the divergence of the African Hepsetidae + Ctenoluciidae clade resulted from the separation of South America and Africa, then all subtending lineages to this terminally positioned clade had originated prior to continental drift and, subsequent to the vicariance event, there was a greater extinction of African characiforms compared to their South American counterparts.

Recently, a few authors have questioned the vicariance model to explain the modern distribution of strictly freshwater characiforms (Gayet *et al.* 2003, Calcagnotto *et al.* 2005). Gayet *et al.* (2003) cited evidence from the European fossil record, including *Salminops* of the Cenomanian of Portugal (Gayet 1985) and *Eurocharax tourainei* from the Oligocene of France (Gaudant 1979, 1980), as evidence that characiforms historically were tolerant of brackish or marine waters, and therefore capable of dispersal in the early history of the group. As noted above, many of these fossils require reevaluation (e.g., Fink and Fink 1996) for their confident identification as characiforms and phylogenetic placement. Calcagnotto *et al.* (2005), citing early otophysan fossils as evidence of the potential for marine dispersal in the early history of characiforms, also point out that the topology of relationships obtained in their study require only two marine dispersal events and one extinction event to account for the distribution of modern characiforms. According to this scenario, if the separation of Africa and South America corresponded to the divergence indicated by an arrow in Fig. 11.3, then the extinction of Citharinoids in South America and the dispersal of Crenuchidae + Erythrinidae and Ctenoluciidae + Lebiasinidae from Africa to South America would follow.

As the foregoing demonstrates, relationships among characiforms are not yet firmly resolved to permit detailed analysis of and confident conclusions about characiform biogeographic history. The fact of saltwater intolerance among Recent characiforms and the phylogenetic results to date, although conflicting, imply that some of the major lineages of characiforms originated prior to continental separation.

Conclusions

Higher-level studies of characiform relationships to date have supported the relationships of subgroups of characiforms but a well-supported hypothesis for the group as a whole is lacking. Morphological and molecular data have both proven useful in corroborating the relationships among subunits of characiforms, and efforts using both types of data will contribute to resolving outstanding conflicts. Additional morphological characters from a wide variety of character systems will be helpful in further investigating characiform relationships (Weitzman and Malabarba, 1998). Future molecular sequence–based work will undoubtedly involve the search for new genetic markers, along with exploration and use of markers that have been demonstrated as useful for resolving deep relationships in other taxa, e.g., usefulness of the nuclear *rag* genes in siluriform phylogenetics (Sullivan *et al.* 2006).

Thorough evaluation of scenarios regarding characiform evolution and biogeographic history will require well-supported and corroborated phylogenetic evidence for their relationships and confidently identified fossils. Comparative study of tooth morphology among modern characiforms is necessary for the finer identification of the hundreds of teeth that make up a great majority of the fossilized remains of characiforms. Molecular dating with a well-supported phylogeny and confidently identified and placed fossils will also be an important evaluation of the biogeographic history of characiforms and a test of the hypotheses of their evolutionary history.

Acknowledgments

I appreciate Terry Grande's invitation to contribute to this volume. I am grateful to John Lundberg, Luiz Malabarba, Maria Claudia Malabarba, and an anonymous reviewer for their helpful comments and suggestions on the chapter.

References

Antunes, M.T., A. Balbino and J. Gaudant. 1995. Découverte du plus récent poisson Characiforme européen dans le Miocène terminal due Portugal. Comunicacões del Instituto de Geológico e Mineiro 81: 79–84.

Arratia, G. and A. Cione. 1996. The record of fossil fish of southern South America. Münchner Geowissenschaftliche Abhandlungen A, Geologische Paläontologische 30: 9–72.

Buckup, P.A. 1991. The Characidiinae: A phylogenetic study of the South American darters and their relationships with other characiform fishes. Ph.D. Thesis, The University of Michigan.

Buckup, P.A. 1993a. Phylogenetic interrelationships and reductive evolution in Neotropical characidiin fishes (Characiformes, Ostariophysi). Cladistics—Int. J. Willi Hennig Soc. 9: 305–341.

Buckup, P.A. 1993b. The monophyly of the Characidiinae, a neotropical group of characiform fishes (Teleostei, Ostariophysi). Zool. J. Linn. Soc. 108: 225–245.

Buckup, P.A. 1998. Relationships of the Characidiinae and phylogeny of characiform fishes (Teleostei: Ostariophysi). pp. 123–144. In: L.R. Malabarba, R.E. Reis, R.P. Vari, Z.M.S. Lucena and C.A.S. Lucena [eds.]. Phylogeny and Classification of Neotropical Fishes. EDIPUCRS, Porto Alegre, Brazil.

Buckup P.A. 2003. Family Crenuchidae. p. 742. In: R.E. Reis, S.O. Kullander and C.J. Ferraris Jr. [eds.]. Checklist of the Freshwater Fishes of South and Central America. EDIPUCRS, Porto Alegre, Brazil.

Calcagnotto, D., S.A. Schaefer and R. DeSalle. 2005. Relationships among characiform fishes inferred from analysis of nuclear and mitochondrial gene sequences. Mol. Phylogenet. Evol. 36: 135–153.

Cappetta, H., D.E. Russel and J. Braillon. 1972. Sur la découverte de Characidae dans l'Eocène inférieur français (Pisces, Cypriniformes). Bull. Mus. nat. Hist. nat. Paris, sér. 3, 51, Sciences de la Terre 9: 37–51.

Castro, R.M.C. and R.P. Vari. 2004. Detritivores of the South American fish family Prochilodontidae (Ostariophysi: Characiformes): a phylogenetic and revisionary study. Smithsonian Contr. Zool. 622: 1–190.

Cione, A. 1986. Los peces continentales del Cenozoico de Argentina: su significación paleoambiental y paleobiogeográfica. pp. 101–106. Evolución de los Vertebrados Cenozoicos de América del Sur [simposio], Actas IV Congreso Argentino de Paleontología y Bioestratigrafía, 2.

Cione, A., M.M. Azpelicueta, M. Bond, A.A. Carlini, J.R. Casciotta, M.A. Cozzuol, M. de la Fuente, Z. Gasparini, F.J. Goin, J. Noriega, G.J. Scillato-Yané, L. Soibelzon, E.P. Tonni, D. Verzi and M.G. Vucetich. 2000. The Miocene vertebrates from Paraná, eastern Argentina. In: F.G. Aceñolaza and R. Herbst [eds.]. El Neógeno de Argentina. Serie Correlación Geológica 14: 191–237.

Cione, A. and J. Casciotta. 1995. Freshwater teleostean fishes from the Miocene of the Quebrada de la Yesera, Salta, Northwestern Argentina. Neues Jahrbuch für Geologie und Paläontologie. Stuttgart. Abhandlungen 196: 377–394.

Cione, A. and J. Casciotta. 1997. Miocene cynodontids (Osteichthyes: Characiformes) from Parana, central eastern Argentina. J. Vert. Paleontol. 17: 616–619.

Cione, A., W.M. Dahdul, J.G. Lundberg and A. Machado-Allison. in press. Megapiranha paranensis, a new genus and species of Serrasalminae (Characidae, Teleostei) from the upper Miocene of Argentina. J. Vert. Paleontol.

Dahdul, W.M. 2004. Fossil serrasalmine fishes (Teleostei : Characiformes) from the Lower Miocene of north-western Venezuela. pp. 23–28. Fossils of the Miocene Castillo Formation, Venezuela: Contributions on Neotropical Palaeontology.

Dahdul, W.M. 2007. Phylogenetics and Diversification of the Neotropical Serrasalminae (Ostariophysi: Characiformes). Ph.D. thesis, University of Pennsylvania.

De La Peña Zarzuelo, A. 1996. Characid teeth from the Lower Eocene of the Ager Basin (Lérida, Spain): paleobiogeographical comments. Copeia 746–750.

Eschmeyer, W.N. and J.D. Fong. 2008. Species of Fishes by family/subfamily. On-line version dated 29 August 2008. http://research.calacademy.org/research/ichthyology/catalog/SpeciesByFamily.html.

Figueiredo, F.J. d. and B.C.M. da Costa-Carvalho. 1999. Plesiocurimata alvarengai gen. et sp. nov. (Teleostei: Ostariophysi: Curimatidae) from the Tertiary of Taubaté Basin, São Paulo State, Brazil. Anais de la Academia Brasileira de Ciencias 71: 885–894.

Filleul, A. and J.G. Maisey. 2004. Redescription of Santanichthys diasii (Otophysi, Characiformes) from the Albian of the Santana Formation and comments on its implications for otophysan relationships. American Museum Novitates 3455.

Fink, S.V. and W.L. Fink. 1981. Interrelationships of the Ostariophysan fishes (Teleostei). Zool. J. Linn. Soc. 72: 297–353.

Fink, S.V. and W.L. Fink. 1996. Interrelationships of Ostariophysan fishes (Teleostei). pp. 209–249. *In*: M.L.J. Stiassny, L.R. Parenti and G.D. Johnson [eds.]. Interrelationships of Fishes. Academic Press, San Diego.

Gaudant, J. 1979. Sur la présence de dents de Characidae Poissons Téléostéens, Osteriophysi) dans les "Calcaires à Bythinies" et les "Sables bleuntés du Var. Geobios 12: 451–457.

Gaudant, J. 1980. *Eurocharax tourainei* nov. gen. nov. sp. (Poissons Teleosteen, Ostariophysi): Nouveau Characidae fossile des "Calcaires à Bythinies" du Var. Geobios 13: 686–703.

Gayet, M. 1981. Contribution à l'étude anatomique et systématique de l'ichthyofaune cénomanienne du Portugal. II: Complément à l'étude des Ostariophysaires. Comunicacões del Servicio Geologico de Portugal 67: 173–190.

Gayet, M. 1985. Contribution à l'étude anatomique et systématique de l'ichthyofaune cénomanienne du Portugal. 3e partie: Complément à l'étude des Ostariophysaires. Comunicacões del Servicio Geologico de Portugal, 71: 91–118.

Gayet, M. 1991. "Holostean" and teleostean fishes of Bolivia. Revista Técnica de Yacimientos Petroliferos Fiscales Bolivianos 12: 453–494.

Gayet, M., M. Jégu, J. Bocquentin and F.R. Negri. 2003. New characoids from the late Cretaceous and Paleocene of Bolivia and the Mio-Pliocene of Brazil: phylogenetic position and paleogeographic implications. J. Vert. Paleontol. 23.

Gayet, M. and F.J. Meunier. 1998. Maastrichtian to Early Late Paleocene freshwater Osteichthyes of Bolivia: additions and comments. pp. 85–110. *In*: L.R. Malabarba, R.E. Reis, R.P. Vari, Z.M.S. Lucena and C.A.S. Lucena [eds.]. Phylogeny and Classification of Neotropical Fishes. EDIPUCRS, Porto Alegre.

Géry, J. 1977. Characoids of the World. Tropical Fish Hobbyist Publications, Neptune City, New Jersey.

Goulding, M. 1980. The Fishes and the Forest: Explorations in Amazonian Natural History. University of California Press, Berkeley, California.

Greenwood, P.H. 1972. New fish fossils from the Pliocene of Wadi Natrun, Egypt. J. Zool. 168: 503–519.

Greenwood, P.H. 1976. Notes on *Sindacharax* Greenwood and Howes, 1975, a genus of fossil African characid fishes. Revue de Zoologie Africaine 90: 1–13.

Greenwood, P.H. and G.J. Howes. 1975. Neogene fossi fishes from the Lake Albert-Lake Edward Rift (Zaire). Bull. Brit. Mus. (Nat. Hist.), Geol. 26: 69–126.

Greenwood, P.H., D.E. Rosen, S.H. Weitzman and G.S. Myers. 1966. Phyletic studies of teleostean fishes with a provisional classification of living forms. Bull. Am. Mus. Nat. Hist. 131: 339–456.

Hubert, N., C. Bonillo and D. Paugy. 2005. Does elison account for molecular saturation: Case study based on mitochondrial ribosomal DNA among characiform fishes (Teleostei: Ostariophysi). Mol. Phylogenet. Evol. 35: 300–308.

Hubert, N., F. Duponchelle, J. Nunez, C. Garcia-Davila, D. Paugy and J.F. Renno. 2007. Phylogeography of the piranha genera *Serrasalmus* and *Pygocentrus*: implications for the diversification of the Neotropical ichthyofauna. Mol. Ecol. 16: 2115–2136.

Jégu, M. 2003. Subfamily Serrasalminae (pacus and piranhas). pp. 182–196. *In*: R.E. Reis, C.J.J. Ferraris and S.O. Kulander [eds.]. CLOFFSCA, Check List of Freshwater Fishes from South and Central America.

Jégu, M. 2004. Taxinomie des Serrasalminae phytophages et phylogenie des Serrasalminae (Teleostei: Characiformes: Characidae). Ph.D., Museum National d'Histoire Naturelle.

Langeani, F. 1996. Estudo filogenético e revisão taxonômica da família Hemiodontidae Boulenger, 1904 (sensu Roberts, 1974) (Ostariophysi, Characiformes). Ph.D., Universidade de São Paulo.

Langeani, F. 1998. Phylogenetic study of the Hemiodontidae (Ostariophysi: Characiformes). pp. 145–160. *In*: L.R. Malabarba, R.E. Reis, R.P. Vari, Z.M.S. Lucena and C.A.

Lucena [eds.]. Phylogeny and Classification of Neotropical Fishes. EDIPUCRS, Porto Alegre.

Lucena, C.A.S. 1993. Estudo filogenético da família Characidae com uma discussão dos grupos naturais propostos (Teleostei, Ostariophysi, Characiformes). PhD, Universidade de São Paulo.

Lucena, C.A.S. and N.A. Menezes. 1998. A phylogenetic analysis of *Roestes* Günther and *Gilbertolus* Eigenmann with a hypothesis on the relationships of the Cynodontidae and Acestrorhynchidae (Teleostei, Ostariophysi, Characiformes). *In*: L.R. Malabarba, R.E. Reis, R.P. Vari, Z.M.S. Lucena and C.A.S. Lucena [eds.]. Phylogeny and Classification of Neotropical Fishes. EDIPUCRS, Porto Alegre, Brazil.

Lundberg, J. 1993. African-South American freshwater fish clades and continental drift: problems with a paradigm. pp. 156–199. *In*: P. Goldblatt [ed.]. Biological Relationships Between Africa and South America. Yale University Press, New Haven.

Lundberg, J.G. 1997. Fishes of the Miocene La Venta fauna: additional taxa and their biotic and paleoenvironmental implications. pp. 67–91. *In*: R.F. Kay, R.H. Madden, R.L. Cifelli and J.J. Flynn [eds.]. Vertebrate Paleontology in the Neotropics: The Miocene Fauna of La Venta, Colombia. Smithsonian Institution Press, Washington, DC.

Lundberg, J.G. 1998. The temporal context for diversification of Neotropical fishes. pp. 49–68. *In*: L.R. Malabarba, R.E. Reis, R.P. Vari, Z.M.S. Lucena and C.A.S. Lucena [eds.]. Phylogeny and Classification of Neotropical Fishes. EDIPUCRS, Porto Alegre, Brazil.

Lundberg, J.G., A. Machado-Allison and R.F. Kay. 1986. Miocene characid fishes from Colombia: evolutionary stasis and extirpation. Science 234: 208–209.

Lundberg, J.G., M. Sabaj-Pérez, W.M. Dahdul and O.A. Aguilera S. in press. The Amazonian Neogene Fish Fauna. *In*: C. Hoorn and F. Wesselingh [eds.]. Amazonia: Landscape and Species Evolution, Wiley-Blackwell.

Mabee, P.M., G. Arratia, M. Coburn, M. Haendel, E.J. Hilton, J.G. Lundberg, R.L. Mayden, N. Rios and M. Westerfield. 2007. Connecting evolutionary morphology to genomics using ontologies: A case study from cypriniformes including zebrafish. J. Exp. Zool. Part B-Mol. Dev. Evol. 308B: 655–668.

Machado-Allison, A. 1982. Studies on the systematics of the subfamily Serrasalminae (Pisces-Characidae). PhD, George Washington University.

Machado-Allison, A. 1983. Estudios sobre la sistemática de la subfamilia Serrasalminae (Teleostei, Characidae). Parte II. Discusión sobre la condición monofilética de la subfamilia. Acta Biol. Venez. 11: 145–195.

Malabarba, L.R. 1998a. Monophyly of the Cheirodontinae, Characters and Major Clades (Ostariophysi: Characidae). *In*: L.R. Malabarba, R.E. Reis, R.P. Vari, Z.M.S. Lucena and C.A.S. Lucena [eds.]. Phylogeny and Classification of Neotropical Fishes. EDIPUCRS, Porto Alegre, Brazil.

Malabarba, M. 1996. *Cyphocharax mosesi* (Travassos & Santos), a fossil curimatidae from Tertiary of São Paulo, Brazil. Anais-Academia Brasileira de Ciencias 68: 294–294.

Malabarba, M.C.S.L. 1998b. Phylogeny of fossil Characiformes and paleobiogeography of the Tremembé Formation, São Paulo. pp. 69–84. *In*: L.R. Malabarba, R.E. Reis, R.P. Vari, Z.M.S. Lucena and C.A. Lucena [eds.]. Phylogeny and Classification of Neotropical Fishes. EDIPUCRS, Porto Alegre.

Menezes, N.A. 1969. Systematics and evolution of the tribe Acestrorhynchini (Pisces, Characidae). Arquivos de Zoologia (São Paulo) 18: 1–150.

Menezes, N.A. 1998. Revision of the subfamily Roestinae (Ostariophysi: Characiformes: Cynodontidae). Ichthyol. Explor. Freshwaters 9: 279–291.

Menezes, N.A. 2003. Acestrorhynchidae. pp. 231–233. *In*: R.E. Reis, S.O. Kullander and C.J. Ferraris Jr. [eds.]. Check List of the Freshwater Fishes of South and Central America. EDIPUCRS, Porto Alegre.

462 Gonorynchiformes and Ostariophysan Relationships

Menezes, N.A. and J. Géry. 1983. Seven new Acestrorhynchin characid species (Osteichthyes, Ostariophysi, Characiformes) with comments on the systematics of the group. Rev. Suisse Zool. 90: 563–592.

Micklich, N. and B. Roscher. 1990. Neue Fischfunde aus der Baid-Formation (Oligozän; Tihamat Asir, SW Saudi-Arabien). Neues Jahrbuch für Geologie und Paläontologie. Stuttgart. Abhandlungen 180: 139–175.

Monod, T. and J. Gaudant. 1998. Un nom pour les poissons Characiformes de l'Éocene inférieur et moyen du Bassin de Paris et du Sud de la France: Alestoides eocaenicus nov. gen., nov. sp. Cybium 22: 15–20.

Murray, A.M. 2003a. A new Eocene citharinoid fish (Ostariophysi: Characiformes) from Tanzania. J. Vert. Paleontol. 23: 501–507.

Murray, A.M. 2003b. A new characiform fish (Teleostei: Ostariophysi) from the Eocene of Tanzania. Canad. J. Earth Sci. 40: 473–481.

Murray, A.M. 2004. Late Eocene and early Oligocene teleost and associated ichthyofauna of the Jebel Qatrani Formation, Fayum, Egypt. Palaeontology 47: 711–724.

Murray, A.M. and K.M. Stewart. 2002. Phylogenetic relationships of the African genera Alestes and Brycinus (Teleostei, Characiformes, Alestidae). Canad. J. Earth Sci. 80: 1887–1899.

Nelson, J.S. 2006. Fishes of the World, 4th ed. John Wiley & Sons, Hoboken, NJ.

Nico, L.G. and D.C. Taphorn. 1988. Food habits of piranhas in the low llanos of Venezuela. Biotropica 20: 311–321.

Ortí, G. and A. Meyer. 1997. The radiation of characiform fishes and the limits of resolution of mitochondrial ribosomal DNA sequences. Syst. Biol. 46: 75–100.

Ortí, G., P. Petry, J.I.R. Porto, M. Jegu and A. Meyer. 1996. Patterns of nucleotide change in mitochondrial ribosomal RNA genes and the phylogeny of piranhas. J. Mol. Evol. 42: 169–182.

Ortí, G., A. Sivasundar and M. Jégu. 2008. Phylogeny of the Serrasalmidae (Characiformes) based on mitochondrial DNA sequences. Genet. Mol. Biol. 31: 343–351.

Otero, O. and M. Gayet. 2001. Palaeoichthyofaunas from the Lower Oligocene and Miocene of the Arabian Plate: palaeoecological and palaeobiogeographical implications. Palaeogeogr. Palaeoclimatol. Palaeoecol. 165: 141–169.

Otero, O., X. Valentin and G. Garcia. 2008. Cretaceous characiform fishes (Teleostei: Ostariophysi) from Northern Tethys: description of new material from the Maastrichtian of Provence (Southern France) and palaeobiogeographical implications. Geol. Soc. London Spec. Pub. 295: 155.

Reis, R.E. 1989. Systematic revision of the neotropical characid subfamily Stethaprioninae (Pisces, Characiformes). Commun. Mus. Ciênc. Tecnol. PUCRS, Sér. Zool., Porto Alegre 2: 3–86.

Reis, R.E., S.O. Kullander and C.J. Ferraris Jr. [eds.] 2003. Checklist of the freshwater fishes of South and Central America. EDIPUCRS, Porto Alegre, Brazil.

Roberts, T.R. 1974. Osteology and classification of the Neotropical characoid fishes of the families Hemiodontidae (Including Anodontinae) and Parodontidae. Bull. Mus. Comp. Zool. 146: 411–472.

Roberts, T.R. 1975. Characoid fish teeth from the Miocene deposits in the Cuenca Basin, Ecuador. J. Zool. 175: 265–271.

Sazima, I. 1983. Scale-eating in characoids and other fishes. Environ. Biol. Fishes 9: 87–101.

Sidlauskas, B. 2008. Continuous and arrested morphological diversification in sister clades of characiform fish: a phylomorphospace approach. Evolution 62: 3135–3156.

Sidlauskas, B.L. 2007. Testing for unequal rates of morphological diversification in the absence of a detailed phylogeny: a case study from characiform fishes. Evolution 61: 299–316.

Sidlauskas, B.L. and R.P. Vari. 2008. Phylogenetic relationships within the South American fish family Anostomidae (Teleostei, Ostariophysi, Characiformes). Zool. J. Linn. Soc. 154: 70–210.

Stewart, K.M. 1994. A Late Miocene fish fauna from Lothagam, Kenya. J. Vert. Paleontol. 14: 592–594.

Stewart, K.M. 1997. A new *Sindacharax* from Lothagam, Kenya, and some implications for the genus. J. Vert. Paleontol. 17: 34–38.

Stewart, K.M. 2003. Fossil fish remains from Mio-Pliocene deposits at Lothagam, Kenya. pp. 75–111. *In*: M.G. Leakey and J.M. Harris [eds.]. Lothagam: the dawn of humanity in Eastern Africa. Columbia University Press, New York.

Stewart, K.M. and A.M. Murray. 2008. Fish remains from the Plio-Pleistocene Shungura Formation, Omo River basin, Ethiopia. Geobios 41: 283–295.

Sullivan, J.P., J.G. Lundberg and M. Hardman. 2006. A phylogenetic analysis of the major groups of catfishes (Teleostei : Siluriformes) using rag1 and rag2 nuclear gene sequences. Mol. Phylogenet. Evol. 41: 636–662.

Taverne, L. 2003. Les poissons Crétacés de Nardò. 160. *Sorbinicharax verraesi* gen. et sp. nov. (Teleostei, Ostariophysi, Otophysi, Characiformes). Bolletino del Museo Civicio di Storia Naturale di Verona 27: 29–45.

Toledo-Piza, M. 2000. The Neotropical Fish Subfamily Cynodontinae (Teleostei: Ostariophysi: Characiformes): A Phylogenetic Study and a Revision of *Cynodon* and *Rhaphiodon*. American Museum Novitates 3286: 1–88.

Toledo-Piza, M. 2007. Phylogenetic relationships among *Acestrorhynchus* species (Ostariophysi: Characiformes: Acestrorhynchidae). Zool. J. Linn. Soc. 151: 691–757.

Uj, A. 1990. Etude comparative de l'osteologie cranienne des poissons de la famille des Characidae et son importance phylogenetique. PhD, Université de Geneve.

Van Neer, W. 1994. Cenozoic fish fossils from the Albertine Rift Valley in Uganda. Geol. Palaeobiol. Albertine Rift Valley, Uganda–Zaire 2: 89–127.

Vari, R.P. 1977. Notes on the characoid subfamily Iguanodectinae, with a description of a new species. American Museum Novitates 2612: 1–6.

Vari, R.P. 1979. Anatomy, relationships and classification of the families Citharinidae and Distichodontidae (Pisces, Characoidea). Bull. Brit. Mus. (Nat. Hist.) Zool. 36: 261–344.

Vari, R.P. 1983. Phylogenetic relationships of the families Curimatidae, Prochilodontidae, Anostomidae, and Chilodontidae (Pisces: Characiformes). Smithsonian Contr. Zool. 378: 60.

Vari, R.P. 1988. The Curimatidae: A lowland Neotropical fish family (Pisces: Characiformes); distribution, endemism, and phylogenetic biogeography. pp. 343–377. *In*: W.R. Heyer and P.E. Vanzolini [eds.]. Proceedings of a workshop on Neotropical distribution patterns. Academia Brasileira de Ciencias, Rio de Janeiro.

Vari, R.P. 1995. The Neotropical fish family Ctenoluciidae (Teleostei: Ostariophysi: Characiformes): Supra and Intrafamilial phylogenetic relationships, with a revisionary study. Smithsonian Contr. Zool. 1–97.

Vari, R.P. 1998. Higher level phylogenetic concepts within characiforms (Ostariophysi), a historical review. pp. 111–122. *In*: L.R. Malabarba, R.E. Reis, R.P. Vari, Z.M.S. Lucena and C.A.S. Lucena [eds.]. Phylogeny and Classification of Neotropical Fishes. EDIPUCRS, Porto Alegre, Brazil.

Vari, R.P., R.M.C. Castro and S.J. Raredon. 1995. The Neotropical fish family Chilodontidae (Teleostei: Characiformes): a phylogenetic study and a revision of *Caenotropus* Günther. Smithsonian Contr. Zool. 577: 32.

Vari, R.P. and S.J. Raredon. 2003. Chilodontidae (headstanders). pp. 85–86. *In*: R.E. Reis, S.O. Kullander and C.J. Ferraris Jr. [eds.]. Check List of the Freshwater Fishes of South and Central America. EDIPUCRS, Porto Alegre.

Weitzman, S.H. 1954. The osteology and the relationships of the South American characid fishes of the subfamily Gasteropelecinae. Stanford Ichthyol. Bull. 4: 213–263.

Weitzman, S.H. 1960. Further notes on the relationships and classification of the South American characid fishes of the subfamily Gasteropelecinae. Stanford Ichthyol. Bull. 7: 114–123.

Weitzman, S.H. 1964. Osteology and relationships of the South American characid fishes of the subfamilies Lebiasinidae and Erythrinidae, with special reference to subtribe Nannostomina. Proc. US Natl Mus. 116: 127–169.

Weitzman, S.H. and L.R. Malabarba. 1998. Perspectives about the Phylogeny and Classification of the Characidae (Teleostei: Characiformes). pp. 161–170. In: L.R. Malabarba, R.E. Reis, R.P. Vari, Z.M.S. Lucena and C.A.S. Lucena [eds.]. Phylogeny and Classification of Neotropical Fishes. EDIPUCRS, Porto Alegre, Brazil.

Weitzman, S.H. and R.P. Vari. 1988. Miniaturization in South American freshwater fishes: an overview and discussion. Proc. Biol. Soc. Washington 101: 444–465.

Weitzman, S.H. and M.J. Weitzman. 1982. Biogeography and evolutionary diversification in Neotropical freshwater fishes, with comments on the refuge theory. pp. 403–422. In: G.T. Prance [ed.]. Biological Diversification in the Tropics. Columbia University Press, New York.

Weitzman, S.H.W. and N.A. Menezes. 1998. Relationships of the tribes and genera of the Glandulocaudiinae (Ostariophysi: Characiformes: Characidae), with a description of a new genus. pp. 171–192. In: L.R. Malabarba, R.E. Reis, R.P. Vari, Z.M.S. Lucena and C.A.S. Lucena [eds.]. Phylogeny and Classification of Neotropical Fishes. EDIPUCRS, Porto Alegre.

Winemiller, K.O. and L.C. Winemiller. 1994. Comparative ecology of the African pike, Hepsetus odoe, and tigerfish, Hydrocynus forskahlii, in the Zambezi River floodplain. J. Fish Biol. 45: 211–225.

Winterbottom, R. 1980. Systematics, osteology and phylogenetic relationships of fishes of the ostariophysan subfamily Anostominae (Characoidei, Anostomidae). R. Ontario Mus. Life Sci. Contr. 123: 112.

Zanata, A.M. and R.P. Vari. 2005. The family Alestidae (Ostariophysi, Characiformes): a phylogenetic analysis of a trans-Atlantic clade. Zool. J. Linn. Soc. 145: 1–144.

12

State of the Art of Siluriform Higher-level Phylogeny

Rui Diogo[1] and Zuogang Peng[2]

Abstract

The Siluriformes, or catfishes, with more than 430 genera and 2750 species, represent about one third of all freshwater fishes. The relationships between the various siluriform families have long been studied. However, the number of works focused on this subject increased considerably in the past few decades because of the renewed impetus provided by the advent of cladistics in the second half of the 20th century. In this chapter, we provide an overview of those cladistic studies on the higher-level phylogeny of the Siluriformes, as a foundation for an overall discussion on the state of the art of catfish phylogeny. As will be shown, considerable progress has recently been achieved in this field, and we are arriving at some consensus concerning certain aspects of catfish higher-level phylogeny: e.g., the close relationship between ictalurids and cranoglanidids, between anchariids and ariids, and between the pimelodids, pseudopimelodids and heptapterids; the basal position within the order of the Diplomystidae, the Loricarioidei, the Cetopsidae and the Hypsidoridae; the monophyly of a clade including *Heteropneustes* and clariids, of a clade including auchenipterids and doradids, of a clade

---

[1] Rui Diogo (corresponding author): Department of Anthropology, The George Washington University, 2110 G St. NW, Washington, DC 20052, USA. email: Rui_Diogo@hotmail.com
[2] Zuogang PENG: Laboratory of Fish Phylogenetics and Biogeography, Institute of Hydrobiology, Chinese Academy of Sciences, Wuhan, Hubei 430072, P. R. China. email: catfish@ihb.ac.cn

including erethistids, sisorids, akysids, amblycipitids and perhaps aspredinids; and the sister-group relationship between the clade Nematogenyidae + Trichomycteridae and the clade Callichthyidae + (Scoloplacidae + (Astroblepidae + Loricariidae)). However, despite the recent progress in catfish phylogeny, much remains to be done. For example, we are far from reaching a consensus about the relationships, and in some cases even the monophyly (e.g., Claroteidae, Bagridae, Schilbidae), of families such as the Siluridae, Austroglanididae, Schilbidae, Claroteidae, Malapteruridae, Bagridae, Mochokidae, Lacantuniidae, Chacidae, Plotosidae, and Amphiliidae.

Introduction

The Siluriformes, or catfishes, with more than 430 genera and 2750 species, represent about one third of all freshwater fishes (Teugels 2003). Diogo (2003b) recognized 37 catfish families. However, it should be noted that, as it will be seen below, there is much controversy concerning the validity of some of these families (e.g., according to De Pinna 1998 and Sullivan *et al.* 2006, the Claroteidae are not monophyletic; for Diogo 2004, the Pimelodidae, Heptapteridae and Pseudopimelodidae should be grouped in a single family, the genus *Ancharius* of the Anchariidae should be included in the Ariidae, and the genus *Heteropneustes* of the monogeneric family Heteropneustidae should be placed in the family Clariidae: see below). Also, it should be stressed that Rodiles-Hernández *et al.* (2005: 1) described a new catfish species, *Lacantunia enigmatica*, that, in the opinion of these authors, "cannot be placed within or as a basal sister lineage to any known catfish family or multifamily clade except Siluroidei". Consequently, they opted to place this species in its own family, the Lacantuniidae (see Table 12.1), which according to them is diagnosed by five "autapomorphic and anatomically complex structures": "the fifth (last) infraorbital bone is relatively large, anteriorly convex and remote from a prominent sphenotic process"; "the lateral margin of the frontal, lateral ethmoid and sphenotic bones are thick at the origins of much enlarged adductor mandibulae and levator arcus palatini muscles"; "one pair of cone-shaped 'pseudo-pharyngobranchial' bones is present at the anterior tips of enlarged cartilages medial to the first epibranchial"; "a hypertrophied, axe-shaped uncinate process emerges dorsally from the third epibranchial"; the gas bladder has paired spherical, unencapsulated diverticulae protruding from its anterodorsal wall" (Rodiles-Hernandéz *et al.* 2005). The inclusion of *Lacantunia enigmatica* in a new family Lacantuniidae is followed in the present work; this family is thus listed, together with the 37 catfish families recognized by Diogo (2003a), in Table 12.1.

The relationships between the various families of siluriforms have been studied for a long time. The number of works focused on this subject has increased considerably in the last three decades because of the renewed impetus provided by the advent of cladistics in the second half of the 20th century (De Pinna 1998). The principal aim of this chapter is to provide an updated overview of those cladistic studies published in the last few decades dealing with the higher-level phylogeny of the Siluriformes, as a foundation for an overall discussion on the state of the art of catfish phylogeny. It should be noted that this chapter is partially based on the work of Diogo (2003a). Since that work was written, numerous studies concerning catfish phylogeny have been published (e.g., Brito 2003, Diogo 2003b, 2004, 2005, Diogo and Chardon 2003, 2006a, b, Diogo et al. 2003a, b, c, 2004a, b, c, d, Hardman and Page 2003, Ng 2003, Schaefer 2003, Shibatta 2003, Teugels and Adriaens 2003, Armbruster 2004, Hardman 2004, Kailola 2004, Moyer et al. 2004, Peng et al. 2004, 2006, Shimabukuro-Dias et al. 2004, Wilcox et al. 2004, Agnese and Teugels 2005, Diogo and Bills 2006, Ng and Sparks 2005, Hardman 2005, Zawadzki et al. 2005, Zhou and Zhou 2005, Day and Wilkinson 2006, Jansen et al. 2006, Hardman and Lundberg 2006, Sullivan et al. 2006, Ku et al. 2007). Most of these studies deal with the intra-relationships and/or monophyly of individual catfish families, which is the topic of the first section of this chapter (see Table 12.1: studies published after 2003 are given in italic). Three of these studies (Diogo 2004, Hardman 2005, Sullivan et al. 2006) provide, however, cladistic analyses on the higher-level phylogeny of siluriforms (see below). We will thus take into account all these phylogenetic studies published after Diogo (2003a) in this updated overview.

Intra-relationships and Monophyly of the Different Catfish Families

As explained above, the main subject of this chapter is a discussion of the phylogenetic relationships between the different catfish families. However, we consider that it is convenient to present in this section a brief overview of the principal published cladistic studies providing relevant information on the autapomorphies of these families, as well as on the relationships among their genera (Table 12.1). It should be noted that in Table 12.1, as well as throughout the text, when we refer to phylogenetic studies dealing either with the relationships among the genera of each catfish family or with the relationships between these families, we refer only to published cladistic analyses. Among the several reasons to follow this procedure, one of the most important is to reduce the confusion associated with such a puzzling issue, as is the phylogeny of the Siluriformes. In fact, many of the "pre-cladistic" (see De Pinna 1998) studies dealing with catfish phylogeny are highly confusing, grouping certain taxa according to the presence of both

Table 12.1 List of the principal cladistic studies published to date providing relevant information concerning the phylogenetic relationships among the genera and/or the autapomorphies of the various catfish families.

Family	Relationships among the various genera of the family	Autapomorphies to support monophyly of the family
Akysidae	De Pinna 1996	Mo 1991, De Pinna 1996
Amblycipitidae	Chen and Lundberg 1994	Mo 1991, Chen and Lundberg 1994, De Pinna 1996, *Diogo et al. 2003a*
Amphiliidae	He *et al.* 1999, *Diogo 2003b*	*Diogo 2003b*
Anchariidae	*Ng and Sparks 2005*	*Diogo 2004, Ng and Sparks 2005* (but see comments in text)
Andinichthyidae	Family with a single genus	Gayet 1988
Ariidae	*Kailola 2004*	Mo 1991, Oliveira *et al.* 2002, *Kailola 2004*
Aspredinidae	De Pinna 1998	De Pinna 1996, Diogo *et al.* 2001
Astroblepidae	Family with a single genus	Schaefer and Lauder 1986, Schaefer 1990, Howes 1983, De Pinna 1998, *Schaefer 2003*
Auchenipteridae	Curran 1989, De Pinna 1998, Soares-Porto 1998 (p)	Curran 1989, De Pinna 1998, *Diogo et al. 2003b*
Austroglanididae	Family with a single genus	Mo 1991, *Diogo and Bills 2006*
Bagridae	Mo 1991, Peng *et al.* 2002 (p), Ng 2003, Hardman 2005 (p), Ku et al. 2007 (p)	Mo 1991, Diogo *et al.* 1999
Callichthyidae	Reis 1998, *Brito 2003, Shimabukuro-Dias et al. 2004*	Schaefer 1990, Reis 1998
Cetopsidae	*Not available*	De Pinna and Vari 1995, *Diogo and Chardon 2006a*
Chacidae	Family with a single genus	Brown and Ferraris 1988, *Diogo et al. 2004a*
Clariidae	Agnese and Teugels 2001a (p), 2001b (p), *2005 (p), Diogo and Chardon 2003 (p), Teugels and Adriaens 2003; Jansen et al. 2006 (p)*	*Not available*
Claroteidae	Mo 1991	Mo 1991
Cranoglanididae	Family with a single genus	Diogo *et al.* 2002a
Diplomystidae	Arratia 1987, 1992	Arratia 1987, 1992
Doradidae	De Pinna 1998, *Moyer et al. 2004*	De Pinna 1998, *Diogo et al. 2004b*
Erethistidae	De Pinna 1996	De Pinna 1996, *Diogo et al. 2003c*
Heptapteridae	Ferraris 1988 (p), Lundberg *et al.* 1991a, Bockmann 1994 (p)	Lundberg and McDade 1986, Ferraris 1988, Lundberg *et al.* 1988, 1991a, De Pinna 1998 (but see comments in text)
Heteropneustidae	Family with a single genus (but see comments in text)	*Diogo and Chardon 2003* (but see comments in text)
Hypsidoridae	Family with a single genus	Grande 1987, Arratia 1992 (but see Mo 1991: 195, Grande and De Pinna 1998: 471)

Family	Relationships among the various genera of the family	Autapomorphies to support monophyly of the family
Ictaluridae	Lundberg 1975, 1982, 1992, *Hardman and Page 2003 (p)*, *Hardman 2004 (p)*, *Wilcox et al. 2004 (p)*	Grande and Lundberg 1988, Lundberg 1992
Lacantuniidae	Family with a single genus	*Rodiles-Hernández et al. 2005*
Loricariidae	Howes 1983 (p), Schaefer 1987, 1991 (p), 1998 (p), Armbruster 1998 (p), Montoya-Burgos et al. 1997 (p), 1998, 2002 (p), *Schaefer 2003 (p)*,*Armbruster 2004(p)*, *Zawadzki et al. 2005(p)*	Howes 1983, Schaefer and Lauder 1986, 1996, Schaefer 1987, 1990, De Pinna 1998, Aquino and Schaefer 2002, *Schaefer 2003*
Malapteruridae	Family with a single genus	Howes 1985
Mochokidae	*Day and Wilkinson 2006 (p)*	Mo 1991
Nematogenyidae	Family with a single genus	Arratia 1992, De Pinna 1998, *Diogo and Chardon 2006b*
Pangasiidae	Pouyaud et al. 2000 (p)	*Not available*
Pimelodidae	Lundberg et al. 1991b, De Pinna 1998, Perdices et al. 2002 (p), *Hardman and Lundberg 2006 (p)*	Lundberg et al. 1988, 1991b, De Pinna 1998, *Diogo 2005* (but see comments in text)
Plotosidae	*Not available*	Oliveira et al. 2001
Pseudopimelodidae	*Shibatta 2003*	Lundberg et al. 1991a, De Pinna 1998, *Diogo et al. 2004c* (but see comments in text)
Schilbidae	*Not available*	*Diogo et al. 2004d*
Scoloplacidae	Family with a single genus	Schaefer et al. 1989, Schaefer 1990
Siluridae	Bornbusch and Lundberg 1989 (p), Bornbusch 1991a (p), 1995, Howes and Fumihito 1991 (p)	Bornbusch 1991b, Howes and Fumihito 1991
Sisoridae	De Pinna 1996 (p), He 1996 (p), *Peng et al. 2004 (p), 2006 (p)*, *Guo et al. 2005 (p)*, *Zhou and Zhou 2006 (p)*	De Pinna 1996, *Diogo et al. 2002b*
Trichomycteridae	De Pinna 1988 (p), 1989ab (p), 1992, 1998, De Pinna and Starnes 1990 (p), Costa 1994 (p), Costa and Bockmann 1994 (p)	De Pinna 1989b, 1992, 1998

The references given in Italic were published after the writing of Diogo's 2003a paper; a (p) after a reference indicates that it provides information about the relationships among only part of the family (see text for more details).

plesiomorphic and highly homoplasic characters, or simply (several times) without giving a clear explanation. Moreover, it would be somewhat complicated, and also highly confusing, to compare, in a clear and objective way, the conclusions of those studies using such "pre-cladistic" methodologies with the phylogenetic results of those studies following a cladistic methodology. It should, however, be noted that this does not mean that all those "pre-cladistic" studies should be ignored (see Diogo 2004). On the contrary, it simply means that, in the specific case of this chapter, the authors consider it more appropriate to present a more detailed overview of all the cladistic studies published to date on the relationships among the genera of each of the catfish families (Table 12.1) and/or among these families (see next section in this chapter). For a detailed overview of the most important "pre-cladistic" works on the taxonomy and classification of siluriforms, see De Pinna (1998).

As can be seen in Table 12.1, there has been significant progress concerning the intra-relationships/autapomorphies of the various catfish families in the past few years (see, e.g., the numerous papers cited in italic, which were published after the writing of Diogo's 2003a paper). For instance, Kailola (2004) and Shibatta (2003) wrote the first published cladistic analysis on the intra-relationships of the Ariidae and Pseudopimelodidae, respectively. Diogo (2003b) and Diogo *et al.* (2004d) provided some autapomorphies to support the monophyly of the Amphiliidae and Schilbidae, two groups that have been the subject of much controversy in the past (see below). Another example of the progress made recently is that there are now various published studies on the intra-relationships of the Clariidae. Also relevant, Ng and Sparks (2005) described a new genus of the previously monogeneric family Anchariidae, which now includes the genera *Ancharius* and *Gogo* and five species, provided some features to diagnose the anchariids, and investigated the relationships among these fishes. And it should also be emphasized here that, as explained above, various autapomorphies were listed in the recent study of Rodiles-Hernández *et al.* (2005) to diagnose a new catfish family, the Lacantuniidae. However, from the analysis of Table 12.1 it is also evident that, in general, there is a still an imbalance between the number of phylogenetic studies dedicated to the intra-relationships of the New World families (although a phylogenetic analysis of the intra-relationships of the Cetopsidae is still lacking) and the Old World taxa (e.g., the intra-relationships within the family Schilbidae have never been the subject of a cladistic analysis; the intra-relationships of the Mochokidae have been the subject of a single cladistic analysis, and this analysis is focused in a very restricted subgroup of that family). With respect to the autapomorphic characters supporting the monophyly of the various catfish families, there is also a pronounced imbalance between the studies dedicated to the New versus Old World families. In fact, the two multigeneric families for which there are no available, well-defined autapomorphies published to date are from the Old World: Clariidae and Pangasiidae (Table 12.1).

Higher-level Phylogeny of the Siluriformes: A Historical Account

Among all cladistic works published to date on catfish phylogeny, the vast majority were dedicated to the study of the intra-relationships of either a part or the entirety of a particular catfish family (Table 12.1; see above). The only published cladistic studies presenting original, explicit cladograms on the interfamilial relationships of either a significant part or the whole of the order Siluriformes are those of Howes (1983), Grande (1987), Schaefer (1990), Mo (1991), Arratia (1992), De Pinna (1992, 1996, 1998), Lundberg (1993), He *et al.* (1999), Diogo (2004), Hardman (2005), and Sullivan *et al.* (2006). A brief description of each of these studies follows.

Howes 1983 (Fig. 12.1)

In his fig. 22, Howes presented a hypothesis on loricarioid relationships (Fig. 12.1), which was based both on his own observations and on an unpublished thesis of Baskin dated 1972 (Howes 1983: 341–342). According to this hypothesis, the Astroblepidae and Loricariidae form a clade that is the sister-group of Scoloplacidae, with these three families being, in turn, the sister-group of Callichthyidae. Also according to this hypothesis, the clade formed by these four families is the sister-group of Trichomycteridae, the Nematogenyidae being the next sister-group, and, consequently, the most basal of the six loricarioid families. In order to support the phylogenetic hypothesis illustrated in Fig. 12.1, Howes (1983: 342) listed several derived morphological characters to support its different nodes (involving mainly the osteology, myology and external morphology of the cephalic region,

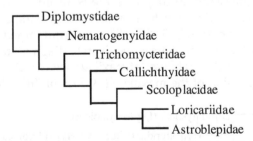

Fig. 12.1 Hypothetical relationships among the loricarioid families according to Howes (1983).

Weberian apparatus, and the caudal fin). The clade including all six loricarioid families was diagnosed by "swimbladder encapsuled, divided into separate vesicles; some part of the cranium contributing to encapsulation". The clade including all loricarioid families excluding the Nematogenyidae was defined by "claustrum and intercalarium lacking from the Weberian vertebral series". The characters uniting the Callichthyidae, Scoloplacidae, Loricariidae, and Astroblepidae were: "posttemporal contributing to distal portion of swimbladder capsule; derived hypural fusion pattern; low number of principal caudal fin rays". Finally, the Scoloplacidae, Loricariidae and Astroblepidae were united by the presence of a "connecting bone between the first rib and the second pterygophore", the two latter families being grouped into a monophyletic clade by the presence of a retractor premaxillae muscle, the medial division of the protractor hyoideus, as well as to the "reverted lip; lateropterygium; 6 fused anterior centra".

However, Howes called attention to some derived characters that conflicted with this phylogenetic hypothesis. The single derived character defining the clade including all non-nematogenyid loricarioids, for example, conflicted with the "lateral insertion of the dilatator operculi muscle", a derived character shared by both Nematogenyidae and Trichomycteridae (Howes 1983: 341–342).

Grande 1987 (Fig. 12.2)

In 1987, Grande, based on his own osteological observations of a fossil catfish from the Eocene Green River Formation of Wyoming originally described by Lundberg and Case (1970), *Hypsidoris farsonensis*, as well as on his own comparisons between this fossil and other catfishes, proposed an original hypothesis (Fig. 12.2). According to this hypothesis, the fossil catfish family Hypsidoridae is the sister-group of all other non-diplomystid catfish families (= Siluroidea), the clade formed by the latter plus Hypsidoridae (= Siluroidei) being the sister-group of Diplomystidae (Fig. 12.2). Grande (1987: 48) listed three characters to diagnose the Siluroidei: (1) "17 or fewer principal caudal rays"; (2) "an extension of lamellar bone over the ventral surface of the fifth centrum"; (3) "the fifth centrum joined closely to the complex centrum". Five characters were given to diagnose the superfamily Siluroidea (Grande 1987: 48): (1) loss "of maxillary teeth"; (2) loss or reduction of "the distal expansion of the maxilla"; (3) loss of "the elongate mesial

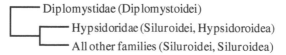

Fig. 12.2 Hypothetical relationships among the Diplomystidae, Hypsidoridae and the other catfish families according to Grande (1987).

process of the maxilla"; (4) reduction of "the palatine either to an extremely small bone, or to a rod-shaped bone"; (5) "long, interdigitating sutural contacts between the ceratohyal and epihyal". The hypothesis of Grande (1987) contradicted the phylogenetic hypothesis formulated in the original description of *Hypsidoris farsonensis*, in which this fossil species was described as a member of the family Ictaluridae (Lundberg and Case 1970: 451–456).

Schaefer 1990 (Fig. 12.3)

Schaefer (1990) undertook a phylogenetic analysis to infer the relationships between the loricarioid families (Fig. 12.3) and also between the different scoloplacid species. That study was based on osteological, myological and arthrological structures of the cephalic region, as well as on osteological structures of the axial and caudal fin. The numerical analysis (using PAUP) of a data matrix of 72 characters × nine terminal taxa, which did not include autapomorphic characters, resulted in a single, most parsimonious cladogram with 77 steps and a CI = 0.842. Concerning the interrelationships between the Loricariidae, Astroblepidae, Scoloplacidae and Callichthyidae, this cladogram (Fig. 12.3) is similar to that of Howes (1983). The synapomorphies listed by Schaefer (1990: 204) to diagnose the clade including these four families were: (1) "loss of the mesethmoid lateral cornua"; (2) "fusion of the pterotic and supracleithrum"; (3) "loss of the canal in the lachrimal-antorbital"; (4) "loss of the tight attachment of the premaxillae with the neurocranium"; (5) "four or fewer branchiostegal rays"; (6) "dorsal hypurals fused with the compound caudal centrum"; (7) "presence of the mesethmoid-maxillary ligament"; (8) "presence of a mesethmoid-premaxillary ligament"; (9) "presence of the retractor tentaculi muscle". Synapomorphies listed by Schaefer (1990: 204) to define the clade formed by astroblepids, loricariids and scoloplacids included: (1) "loss of the open cranial fontanels"; (2) "loss of the cranial aperture which receives the cleithral dorsal process"; (3) "presence of bifid jaw teeth"; (4) "loss of the

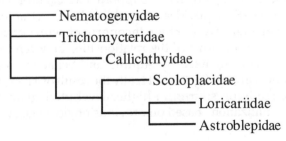

Fig. 12.3 Hypothetical relationships among the loricarioid families according to Schaefer (1990).

interopercle"; (5) "ventrolateral shift in the articulation of the rib on the sixth centrum"; (6) "presence of a lateral bone"; (7) "loss of the pterygoethmoid ligament"; (8) "loss of cranial attachment"; (9) "loss of the interoperculomandibular ligament"; (10) "shift in origin of the retractor tentaculi muscle"; (11) "bifurcation of the hyohyoideus muscles". Lastly, the synapomorphies listed by Schaefer (1990: 204) to support the sister-group relationship between loricariids and astroblepids included: (1) "loss of the contact of the mesethmoid posterior process with the frontals"; (2) "presence of a hyomandibula-metapterygoid suture"; (3) "ventromedial rotation of the mandibles"; (4) "ankylosis or suture between the sixth centrum and Weberian complex centra"; (5) "loss of the vertebral parapophyses"; (6) "presence of expanded transverse shelf on the first anal fin"; (7) "geniohyoideus bilaterally subdivided"; (8) "presence of an expanded oral disk"; (9) "right and left sides of the lower jaws not tightly associated at the midline"; (10) "presence of the intermandibular cartilage"; (11) "presence of juxtaposed nostrils".

The most significant difference between the studies of Schaefer (1990) and Howes (1983) is that Schaefer (1990) did not place the Trichomycteridae as the sister-group of the clade formed by these four families, but instead in an unresolved trichotomy including that clade, the Trichomycteridae and the Nematogenyidae (Fig. 12.3). However, as stressed by Schaefer (1990: 174), such an unresolved trichotomy was the consequence not of his own phylogenetic results, but, instead, of considering the relationships between nematogenyids, trichomycterids and the remaining loricarioids "unresolved *a priori*". It should be noted that Schaefer's (1990) phylogenetic analysis, together with the work of De Pinna's (1998) (see below), served as a basis for the cladogram of loricarioid relationships shown in fig. 8 of the recent paper by Aquino and Schaefer (2002). In fact, Aquino and Schaefer's (2002) cladogram resulted from a cladistic analysis including 57 characters from Schaefer (1990), 21 characters from De Pinna (1998), and eight new characters, a total of 86 characters. Of these eight new characters (which concerned the temporal region of the cranium), only one (compound pterotic-supracleithrum) did, in fact, supported the clade of Schaefer (1990), the Callichthyidae + Scoloplacidae + Astroblepidae + Loricariidae. The other seven new characters were either incongruent with Schaefer's (1990) cladogram or uninformative of the relationships between the loricarioid families. Thus, as recognized by Aquino and Schaefer (2002: 239), the cladogram shown in their fig. 8 is not really the result of a new, independent phylogenetic study on siluriform higher-level phylogeny, but rather a "cladogram optimization" based on previous phylogenetic works.

Mo 1991 (Fig. 12.4)

One year after the publication of Schaefer's 1990 paper, Mo (1991) published a study dealing mainly with the phylogenetic relationships of the Bagridae, including a generic-level revision and phylogeny of the family. In addition, Mo (1991) included a somewhat brief analysis (46 pages in a total of 216 pages plus 63 unnumbered figures) of the higher-level phylogeny of siluriforms. This analysis, based on osteological characters of the cephalic region, Weberian apparatus, pectoral girdle, and the various fins, but also on a few soft and/or myological characters, resulted in two considerably different cladograms. One (Fig. 12.4A) was based on an "unweighted" numerical analysis (using Hennig 1986) of a data matrix of 126

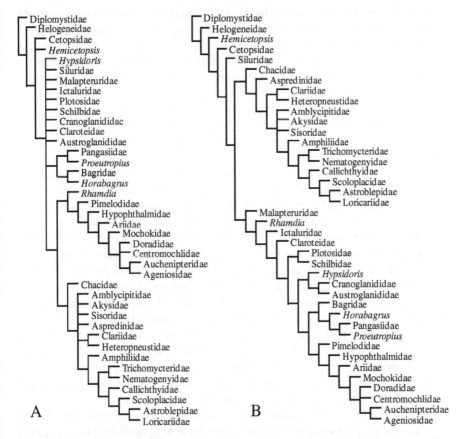

Fig. 12.4 Hypothetical relationships among the major groups of the Siluriformes according to Mo (1991). **A:** Cladogram produced from the numerical analysis of 126 unweighted characters. **B:** Cladogram produced from the numerical analysis of 126 characters with a weighting (4) on one of them, namely the "number of vertebrae united to the complex vertebra" (Mo 1991: 193).

characters and 40 terminal taxa (which did not include autapomorphic characters). It had 602 steps, CI = 0.31, RI = 0.64 and was only poorly resolved (Fig. 12.4A). The other (Fig. 4B) was based on a "weighted" numerical analysis (also using Hennig 1986) of the same data matrix in which one of the 126 characters ("number of vertebrae united to the complex vertebra") was given a phylogenetic weight four times that of the other characters. It had 549 steps, CI = 0.36, RI = 0.72, and, apart from two trichotomies, was completely resolved (Fig. 12.4B). According to Mo (1991: 193), this "weighting" was due to the "morphologically stability and consistent distribution of this character in comparisons with those conflicting features".

One of the main conclusions of Mo (1991) was the separation of the "Bagridae" into three monophyletic units, namely the Bagridae (*sensu* Mo 1991), Claroteidae and Austroglanididae, which, according to Mo, are more closely related with other catfish families than with each other (Fig. 12.4).

Another important conclusion of Mo's study was the suggestion that the Cetopsidae occupies a markedly basal position in the order Siluriformes (although the "Helogeneidae", "Cetopsidae" and "*Hemicetopsis*" of Mo were not, as they are commonly accepted today, grouped in a single monophyletic taxon, they were placed in a rather basal position among Siluriformes in both Mo's cladograms). In Mo's cladogram I (which clearly seems to be preferred by Mo) the cetopsids and diplomystids were separated from all the other siluriforms by the presence, in these latter, of two characters: (1) "interdigital union of the two coracoids" and (2) "ramus mandibularis nerve runs inside hyomandibular for a distance" (Mo 1991: 204). Of these two characters, only the latter is listed in Mo's cladogram II to diagnose the clade composed by all non-diplomystid, non-cetopsid catfishes.

Another important, but confusing, conclusion of Mo (1991) is the phylogenetic position of the fossil catfish *Hypsidoris*. In Mo's cladogram II (Fig. 12.4B), *Hypsidoris* was placed in a far more derived position than in Grande's 1987 cladogram (Fig. 12.2). However, in Mo's cladogram I (Fig. 12.4A), *Hypsidoris* was placed in an unresolved polichotomy, the phylogenetic position of this genus being, thus, uncertain.

With respect to the other catfish groups, their phylogenetic position is, with only a few exceptions, also quite uncertain, not only as a consequence of the poor resolution of Mo's cladogram I (Fig. 12.4A), but primarily because of the significant differences between this cladogram and Mo's cladogram II (Fig. 12.4B). These few exceptions are discussed below.

One of these exceptions concerns the relationships among the loricarioid families, which are essentially similar to those proposed by Howes (1983) and Schaefer (1990). The only difference is that in Mo's cladograms the Trichomycteridae and Nematogenyidae were considered sister-groups (Fig. 12.4). This sister-group relationship was supported, according to Mo (1991:

204, 208), by the fact that trichomycterids and nematogenyids, contrarily to other loricarioids, have the "nasal barbels situated at anterior nostrils".

In both cladograms of Mo (1991), the loricarioids and amphiliids were grouped in a clade that is closely related to the Sisoridae, Akysidae, Amblycipitidae, Clariidae, Heteropneustidae, Aspredinidae and Chacidae (Fig. 12.4). In Mo's cladogram I, the clade including the Amphiliidae and Loricarioidei was diagnosed by the "posterior portion of the palatine reduced into a bony lamina or short spinelike process without distal cartilage" (Mo 1991: 204). In Mo's cladogram II, no character defined this clade. In Mo's cladogram I, the clade including loricarioids, amphiliids, sisorids, akysids, amblycipitids, clariids, heteropneustids, aspredinids and chacids was justified by a "computer generated node" (Mo 1991: 204). In Mo's cladogram II, the clade including all these groups was diagnosed by the "absence of extrascapular" (Mo 1991: 207).

Both cladograms of Mo suggested a monophyletic clade consisting of the Auchenipteridae, Doradidae, Mochokidae, Ariidae, Hypophthalmidae, and Pimelodidae (although the "Auchenipteridae", "Ageneiosidae", and "Centromochlidae" of Mo were not, as is commonly accepted, grouped in the family Auchenipteridae, all these three groups were placed in this clade in both of Mo's cladograms). In Mo's cladogram I, this clade was defined by a single character, namely the "anteriorly thickened and rounded or convex mesethmoid" (Mo 1991: 204). In Mo's cladogram II, the clade was defined by this character, but also by two other characters: the presence of "four infraorbitals" and the "enclosed aortic canal in the complex vertebra" (Mo 1991: 208).

De Pinna 1992 (Fig. 12.5)

One year after the publication of Mo's work, De Pinna (1992) described a new subfamily of the Neotropical catfish family Trichomycteridae, the Copionodontinae. In the same work, De Pinna provided a phylogenetic analysis of the interrelationships among trichomycterids, as well as among those catfishes and other loricarioids. The 27 characters included in that analysis consisted mainly of osteological characters of the cephalic region, Weberian apparatus, dorsal fin, pelvic fin and pectoral girdle, but also on a few soft and/or myological characters. The hand-made comparison of these characters resulted in a fully resolved cladogram with a CI = 0.78 (autapomorphic characters not included). As both cladograms of Mo (1991), this cladogram suggested a sister-group relationship between the Trichomycteridae and Nematogenyidae (Fig. 12.5). However, it should be noted that De Pinna's 1992 phylogenetic analysis was completely independent from that of Mo (1991) (when De Pinna was writing his paper, he was unaware of Mo's results). In fact, the main reason that led De Pinna

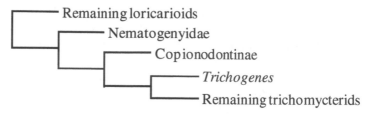

Fig. 12.5 Hypothetical relationships among the trichomycterids, as well as among these fishes and other loricarioids according to De Pinna (1992).

to propose a sister-group relationship between the Trichomycteridae and the Nematogenyidae was "to a major extent induced by the inclusion of copionodontines and *Trichogenes* in the analysis of lower loricarioid relationships" (De Pinna 1992: 175). According to De Pinna (1992: fig. 23), this inclusion indicated that, of the four derived characters traditionally used to support a sister-group relationship between trichomycterids and the other non-nematogenyid loricarioids, only one ("transformator process of tripus absent") indeed represented the plesiomorphic situation for trichomycterids. The other three ("intercalarium absent", "ductus pneumaticus absent" and "superficial ossification covering ventral surface of articulation between complex vertebrae and basioccipital") represented, instead, an apomorphic configuration exclusively present in a restricted group of derived trichomycterids (Fig. 12.5: "Remaining trichomycterids"). Consequently, the grouping of all non-nematogenyid loricarioid families that could no longer be supported than by a single derived character was parsimoniously discarded by De Pinna (1992) in favor of a sister-group relationship between the Trichomycteridae and the Nematogenyidae, supported by three derived characters. These three characters are: (1) "mesial juncture between scapulo-coracoids without interdigitations"; (2) "first dorsal-fin pterygophore inserted posterior to neural spine of ninth free vertebra"; and (3) the "absence of dorsal-fin spine and locking mechanism" (De Pinna 1992: fig. 23).

Arratia 1992 (Fig. 12.6)

Arratia's 1992 work is a detailed, extensive study dedicated to the development, morphological variation and homologies of the suspensorium of certain siluriform and non-siluriform ostariophysans. In addition, it provided an analysis, based on the suspensorial features examined, as well as on some other characters described previously in other studies (e.g., Fink and Fink 1981, Arratia 1987, Grande 1987), of the phylogenetic relationships among the different ostariophysan orders and among certain catfish groups. With respect to the relationships among the catfish groups, Arratia's (1992) analysis resulted in four practically identical cladograms, the only difference

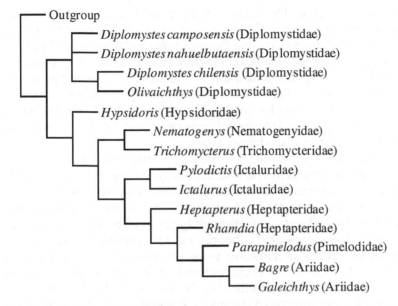

Fig. 12.6 Hypothetical relationships among certain catfish taxa according to Arratia (1992).

concerning the relationships among diplomystids. As this section is essentially dedicated to the interfamilial relationships of the Siluriformes, we will only refer to the cladogram illustrated in Arratia's (1992) fig. 46A (for a detailed explanation on the methodology followed to produce the other three cladograms, as well as for a discussion on the differences between them, see Arratia 1992: 126–129). This cladogram (Fig. 12.6) was based on a numerical analysis (using PAUP) of a data matrix of 75 characters × 15 terminal taxa, in which 69 characters were ordered and six unordered. It corresponds to the consensus of two most parsimonious trees (CI = 0.672) with 137 steps. It supports Grande's (1987) hypothesis, according to which the fossil catfish *Hypsidoris* occupies a rather basal position within the Siluriformes. Arratia (1992: fig. 46) listed eight uniquely derived characters and two homoplasic ones to diagnose the clade constituted by all non-diplomystid, non-hypsidorid catfishes examined in her study. These are: (1) maxilla without the long anterior process; (2) maxilla rudimentary; (3) articulation between autopalatine and maxilla double, lateroventrally oriented; (4) absence of the autopalatine extension dorsal to the dermal entopterygoid; (5) absence of a dermal ectopterygoid; (6) absence of a dermal entopterygoid; (7) presence of a link between the "entopterygoid" and the vomer (homoplasic); (8) loss of the notch separating the processus basalis and the posterodorsal part of the metapterygoid (homoplasic); (9) presence of three or four pairs of barbels; and (10) absence of a supraneural bone above the Weberian apparatus in adults.

Arratia's cladogram (Fig. 12.6) also supported Mo's (1991) phylogenetic results, according to which the ariids are somewhat closely related to pimelodids. The characters uniting *Parapimelodus* and the two ariid genera, *Bagre* and *Galeichthys*, in Arratia's cladogram (Fig. 12.6) are: (1) presence of a sesamoid ectopterygoid joining the autopalatine and "entopterygoid"; (2) presence of ectopterygoid process of metapterygoid (homoplasic); (3) blood vessels running in a tube-like lamellar formation ventral to the Weberian apparatus (homoplasic); (4) fusion of hypurals 1 and 2 (homoplasic); and (5) branched sensory canals (homoplasic).

However, Arratia's cladogram (Fig. 12.6) attributed a rather basal position to the Nematogenyidae and Trichomycteridae, two families that occupy a rather derived position in Mo's (1991) cladograms (Fig. 12.4). Arratia (1992: fig. 46) listed nine derived characters to separate the diplomystids, hypsidorids, nematogenyids and trichomycterids from all the other catfishes represented in her cladogram: (1) presence of a rod-like autopalatine; (2) no articulation between autopalatine and vomer (homoplasic); (3) presence of a ligament and/or connective tissue between "entopterygoid" and lateral ethmoid (homoplasic); (4) presence of a metapterygoid-"entopterygoid" ligament (homoplasic); (5) hyomandibula articulating with autosphenotic; (6) absence of prootic in the hyomandibular fossa; (7) presence of bony extension over the ventral surface of the fifth centrum (homoplasic); (8) presence of suture between pterosphenoid and parasphenoid (homoplasic); and (9) blood vessels in a groove partially surrounded by lamellar walls in the ventral part of the Weberian apparatus (homoplasic).

Another significant aspect of Arratia's cladogram is the fact that the heptapterid genus *Rhamdia* appeared more closely related to the clade formed by *Parapimelodus* (Pimelodidae), *Bagre* (Ariidae) and *Galeichthys* (Ariidae) than to the heptapterid genus *Heptapterus*. The characters listed by Arratia (1992: fig. 46) to unite the genera *Galeichthys*, *Bagre*, *Parapimelodus* and *Rhamdia*, and, thus, to separate these genera from *Heptapterus*, are (1) the fusion of abdominal centra two-six or more and (2) the presence of a small, elongate pharyngobranchial attached to the epibranchial and the medial aspect of the hyomandibula (homoplasic).

Lundberg 1993 (Fig. 12.7)

In an overview of certain clades formed by African and South-American freshwater fishes and their respective implications on the continental drift theory, Lundberg (1993) provided a phylogenetic hypothesis (Fig. 12.7) on the relationships among some catfish taxa. This hypothesis was based on a hand-made analysis of 12 osteological and myological characters of the cephalic region, dorsal fin and Weberian apparatus, which had been

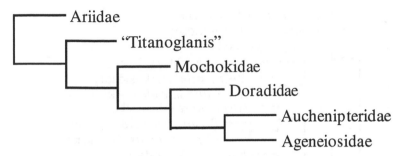

Fig. 12.7 Hypothetical relationships among certain catfish taxa according to Lundberg (1993).

previously described by other authors and/or personally observed by Lundberg (Lundberg 1993: 180). Lundberg's hypothesis (Fig. 12.7) is practically identical to that of Mo (Fig. 12.4), with the addition of the Eocene fossil catfish "*Titanoglanis*" as the sister-group of the clade constituted by the Mochokidae, Auchenipteridae and Doradidae (the "Auchenipteridae" and the "Ageneiosidae" of Lundberg correspond to the Auchenipteridae of this work). It should be noted, however, that Lundberg (1993) was seemingly unaware of Mo's study, since this latter paper was not cited by Lundberg. The two characters listed by Lundberg (1993: 180) to support the sister-group relationship between "*Titanoglanis*" and the clade including Mochokidae, Auchenipteridae and Doradidae were (1) "posterior edge of supraoccipital truncated, not draw out to form a process" and (2) "middle nuchal plate with anterolateral processes contacting posttemporal-epioccipital region of skull".

De Pinna 1996 (Fig. 12.8)

De Pinna's (1996) study was based on a phylogenetic comparison of numerous characters of the cephalic region, Weberian apparatus, pectoral fins and girdle, vertebrae, dorsal fin, pelvic fins and girdle and caudal fin. It provided a hypothesis on the relationships among the Asiatic Amblycipitidae, Akysidae, Sisoridae and Erethistidae and the South-American Aspredinidae, as well as among certain genera of these families. De Pinna's numerical analysis (using Hennig 1986) of a data matrix with 112 characters × 21 terminal taxa resulted in a single, completely resolved, most parsimonious cladogram with 167 steps, CI = 0.70 and RI = 0.79 (autapomorphic characters included) (Fig. 12.8).

One of the most significant conclusions of De Pinna's work was that the Sisoridae of previous authors was a paraphyletic assemblage, a submit of it (subsequently named Erethistidae by De Pinna) being more closely related to the Neotropical Aspredinidae than to the remaining taxa previously assigned to the Sisoridae (Fig. 12.8). Five synapomorphies were

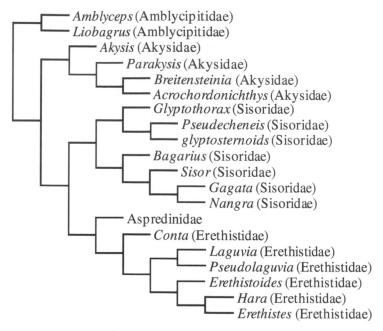

Fig. 12.8 Hypothetical relationships among the Sisoroidea according to De Pinna (1996).

listed by De Pinna (1996: 64) to diagnose the clade constituted by the Erethistidae and Aspredinidae, of which only the last is non-homoplasic: (1) "mandibular laterosensory canal absent"; (2) "second hypobranchial unossified"; (3) "anterior margin of pectoral spine with serrations"; (4) "internal support for pectoral fin rays small in size"; (5) "anterior portion of lateral line running closely in parallel to lateral margin of Weberian lamina". In turn, ten synapomorphies were listed by De Pinna (1993: 61) to diagnose the clade formed by these two families plus the Sisoridae *sensu stricto*, eight of which are homoplasic: (1) "posterior portion of supracleithrum ankylosed to margin of Weberian lamina" (homoplasic); (2) "parapophysis of fifth vertebra strongly flattened and expanded" (homoplasic); (3) "parapophysis of fifth vertebra long, almost or quite reaching lateral surface of body wall"; (4) "humeral process or region around it connected to anterior portion of vertebral column by well-defined ligament-state 3" (homoplasic); (5) "posterior part of Weberian lamina extensively contacting parapophysis of fifth vertebra"; (6) "(reversal of) anterior half of segments of pectoral-fin spine elongate, almost parallel to axis of spine" (homoplasic); (7) coracoid with ventral anterior process" (homoplasic); (8) "(reversal of) second dorsal-fin spine with medial ridge along its anterior surface, forming bilateral longitudinal pouches" (homoplasic); (9) "ventral arms of first dorsal-fin spine with posterior subprocesses attached dorsal to their tip" (homoplasic); (10) "basipterygium with ventral longitudinal keel, anteriorly extending alongside internal arm" (homoplasic).

In addition, De Pinna (1996) suggested the existence of a monophyletic clade formed by the Sisoridae, Erethistidae, Aspredinidae and Akysidae, which, in turn, together with the Amblycipitidae formed the superfamily Sisoroidea (Fig. 12.8). Three synapomorphies were listed to define the clade including the Sisoridae, Erethistidae, Aspredinidae and Akysidae: (1) "supratemporal fossae present" (homoplasic); (2) "supracleithrum strongly attached to skull"; 3) "posterior nuchal plate with anterior process forming facet for articulation with anterior nuchal plate" (De Pinna 1996: 60). With respect to the superfamily Sisoroidea, De Pinna (1996: 59–60) listed seven synapomorphies: (1) "posterior center of ossification of palatine compressed and expanded vertically" (homoplasic); (2) "articular region of lateral ethmoid elongated as a process, with articular facet for palatine at tip"; (3) "parapophysis of fifth vertebra strong and attached to ventral side of centrum, directed directly transversely to centrum"; (4) "humeral process or soft tissue around it connected to anterior portion of vertebral column by well-defined ligament"; (5) "segments of pectoral fin spine very oblique, almost parallel to axis of spine, not evident" (homoplasic); (6) the "dorsal spine with medial ridges along its anterior surface, forming bilateral longitudinal pouches" (homoplasic); (7) "ventral tip of first dorsal fin pterygophore and corresponding neural spines with contacting facets".

De Pinna 1998 (Fig. 12.9)

Two years after the publication of De Pinna's (1996) work, this author published an overview on the phylogenetic relationships of the Neotropical Siluriformes, which included a not completely resolved cladogram expressing the relationships among the major groups of the whole order (Fig. 12.9). As explained by De Pinna (1998: 289–290), this cladogram was mainly derived from De Pinna's (1993) unpublished thesis, "with some resolution added on the sisoroid-aspredinidid part of the tree based on the results of De Pinna (1996)", the "position of the Ariidae from Lundberg (1993)" and the position of the Hypsidoridae left unresolved.

There are some interesting aspects in which De Pinna's 1998 cladogram (Fig. 12.9) is similar to Mo's 1991 results (Fig. 12.4): (1) the rather basal position of Cetopsidae within the siluriforms; (2) the relationships among the various loricarioid families and the close relationship between these families and the Sisoridae (*sensu lato*), Akysidae, Aspredinidae and Amblycipitidae; and (3) the relationships among the Mochokidae, Auchenipteridae, Doradidae and Ariidae (it should be noted, however, that the position of Ariidae in De Pinna's 1998 cladogram is based on Lundberg's 1993 paper: see below).

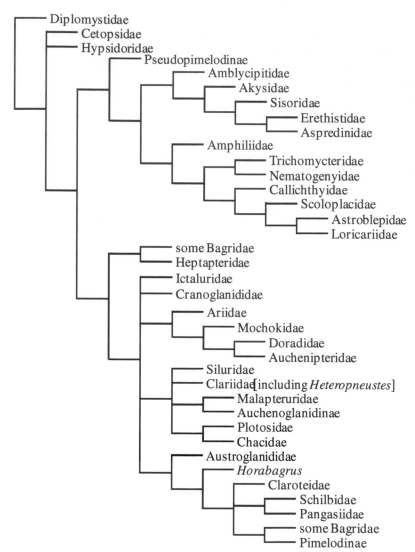

Fig. 12.9 Hypothetical relationships among the major groups of the Siluriformes according to De Pinna (1998).

But there are also some significant differences between the cladogram of De Pinna (1998) and the phylogenetic results of Mo (1991), of which one of the most notable is De Pinna's suggestion that both the Bagridae and Claroteidae *sensu* Mo (1991) are, in fact, polyphyletic groups (Fig. 12.9).

Another interesting aspect of De Pinna's 1998 cladogram (Fig. 12.9) is that it constituted the first published cladogram providing an explicit hypothesis on the phylogenetic position of the three "Pimelodidae" groups, i.e., Pseudopimelodidae (De Pinna's Pseudopimelodinae), Pimelodidae (De Pinna's Pimelodidae) and Heptapteridae (De Pinna's Heptapterinae) (see Introduction, above). In De Pinna's 1998 cladogram, the pseudopimelodids formed, together with loricarioids and sisoroids, a monophyletic unit that is the sister-group of a clade with the heptapterids and some bagrids as its more basal taxa (Fig. 12.9). With respect to the pimelodids, De Pinna suggested a sister-group relationship between these catfishes and some bagrids. The clade formed by these two groups was included, together with claroteines (see above), schilbids, pangasiids, *Horabagrus* and austroglanidids, in a clade included in a large, unresolved pentatomy (Fig. 12.9). Unfortunately, De Pinna's 1998 cladogram was mainly based on De Pinna's 1993 unpublished results and, excepting the interrelationships among the loricarioid families, as well as some other specific points, De Pinna's 1998 paper did not directly provide the phylogenetic characters that support the interfamilial relationships illustrated in that cladogram (these characters were provided in De Pinna's unpublished thesis of 1993). Consequently, neither the characters concerning the polyphyly of the Bagridae and Claroteidae *sensu* Mo (1991) nor the characters concerning the phylogenetic position of the Pimelodidae, Pseudopimelodidae and Heptapteridae within the Siluriformes were published by De Pinna.

He et al. 1999 (Fig. 12.10)

He *et al.* (1999) was a study mainly dedicated to the phylogeny of the African family Amphiliidae, but that also included an analysis of the relationships between this family and some other taxa, namely the Diplomystidae, Amblycipitidae, Hypsidoridae, Bagridae, Sisoridae and *Leptoglanis* (the phylogenetic position of this genus, which was transferred from the Bagridae to the Amphiliidae in Bailey and Stewart's 1984 paper, was considered uncertain *a priori* by He *et al.*). The characters used in this study concerned osteological features of the cephalic region, Weberian apparatus, vertebrae, pectoral girdle, dorsal fin, caudal skeleton and the pelvic girdle. The phylogenetic study of He *et al.* (1999), based on a numerical analysis (using PAUP) of a data matrix of 73 characters × 14 terminal taxa, resulted in a single most parsimonious cladogram with 190 steps and CI = 0.616 (0.603 excluding autapomorphic characters).

  According to this cladogram (Fig. 12.10), neither the Amphiliidae nor the Sisoridae are monophyletic groups, the doumein amphiliids being more closely related to *Leptoglanis* and to the sisorid *Glyptothorax* than to the sisorid *Euchiloglanis* and the amphiliine amphiliids (Fig. 12.10). The characters listed by He *et al.* (indirectly given in their Table I) to support the clade composed by the doumein amphiliids, *Leptoglanis* and the sisorid *Glyptothorax* were: (1) no posterior fontanel (homoplasic); (2) posterodorsal

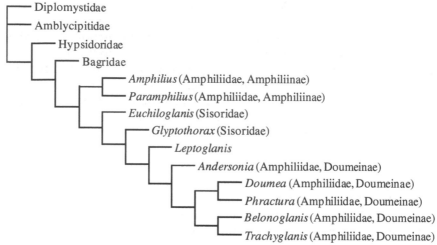

Fig. 12.10 Hypothetical relationships among the amphiliid genera, as well as among these genera and some non-amphiliid taxa, according to He *et al.* (1999).

process of supraoccipital short, slightly forked at its posterior end (homoplasic); (3) short maxillary without enlarged fan-like or forked posterior part; (4) fourth and fifth parapophyses of Weberian apparatus partly fused, thin and long (homoplasic); (5) proximal one and two of dorsal fin with independent nuchal plates; and (6) all the units of second dorsal spine fused. Another interesting aspect of He *et al.* (1999) was the placement of the Amblycipitidae in an unresolved trichotomy leading to this family, the Diplomystidae, and a clade constituted by the remaining catfishes examined by these authors, including the fossil catfish family Hypsidoridae (Fig. 12.10).

Diogo 2004 (Fig. 12.11)

Diogo's (2004) phylogenetic analysis was based on an extensive phylogenetic comparison of characters concerning myological, osteological and soft structures of the cephalic region, Weberian apparatus, and the pectoral fins and girdle. It provided a hypothesis on the relationships among the extant catfish families, as well as among some genera of these families. Diogo's numerical analysis (using Hennig 1986 and Nona and Winclada 2002) of a data matrix of 440 characters × 21 terminal taxa (genera) resulted in 12 most parsimonious trees with a length of 898 steps, CI=0.52 and RI=0.78 (autapomorphic characters included). The strict consensus of these 12 most parsimonious trees resulted in an almost completely resolved cladogram with a length of 902 steps, CI=0.52, and RI=0.78 (autapomorphic characters included). This resulting cladogram

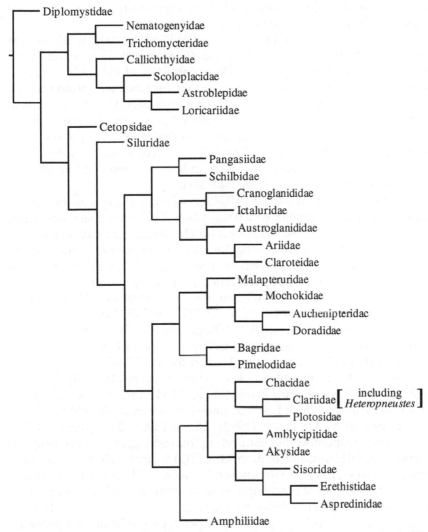

Fig. 12.11 Hypothetical relationships among the major siluriform groups according to Diogo (2004).

presented only three trichotomies. Two of them referred to intrafamilial relationships (one inside the Schilbidae, leading to *Laides*, to *Ailia*, and to *Siluranodon*, and the other inside the Sisoridae, leading to *Gagata*, to *Bagarius*, and to a clade *Glyptosternon* + *Glyptothorax*). Thus, only one trichotomy concerned directly the interfamilial relationships within the Siluriformes: that leading to the Akysidae, to the Amblycipitidae, and to the clade including the Sisoridae, Erethistidae and Aspredinidae (Fig.

12.11). Figure 12.11 shows the hypothetical relationships among the extant catfish families according to Diogo's (2004) results.

One of the most interesting outcomes of Diogo's (2004) work was the placement of the Loricarioidea as the sister-group of all extant non-diplomystid catfishes (Fig. 12.11). The monophyly of the clade including non-diplomystid and non-loricarioid extant catfishes was supported by four characters, of which only the last is homoplasic: (1) "differentiation of the protractor hyoidei in a well-differentiated pars ventralis, pars lateralis and pars dorsalis"; (2) "articulatory surface of autopalatine for neurocranium essentially directed mesially"; (3) "coronomeckelian reduced in size"; and (4) "transition from no mandibular barbels to two pairs of mandibular barbels" (Diogo 2004: 240–241).

Another important aspect of Diogo's (2004) study concerned the monophyly of the Schilbidae, which was supported by five characters. Three of these five characters were homoplasic: (1) "levator operculi originating on both the neurocranium and the dorsolateral surface of hyomandibulo-metapterygoid"; (2) "maxilla reduced in size"; and (3) "anterior cartilage of autopalatine markedly elongated anteroposteriorly". Two were unique, autapomorphic characters within the Siluriformes: (1) "adductor mandibulae A2 essentially lateral to A1-OST" and (2) "Meckel's cartilage markedly extended posteriorly" (Diogo 2004: 255). As noted above, Schilbidae monophyly was questioned by authors such as Mo (1991), but was supported in De Pinna's unpublished thesis of 1993.

Also worthy of mention is the sister-group relationship between *Ancharius* and the remaining ariids (in Fig. 12.11, *Ancharius* is included in the family Ariidae), which was questioned by, for example, Mo (1991) but was supported by De Pinna (1993). Diogo (2004: 261) provided five homoplasic synapomorphies supporting this sister-group relationship: (1) "marked lateral bifurcation of premaxilla"; (2) "complete foramen between dorsal surfaces of lateral ethmoid and frontal"; (3) "presence of suture between mesial limb of posttemporo-supracleithrum and neurocranium"; (4) "mesocoracoid arch reduced to thin structure fused with main body of scapulo-coracoid"; and (5) "presence of protractor of Müllerian process". The grouping of the family Ariidae, including *Ancharius*, and the families Schilbidae, Pangasiidae, Ictaluridae, Cranoglanididae, Austroglanididae, and Claroteidae in a well-supported clade, also constituted an original outcome of Diogo (2004) (Fig. 12.11). Four characters supported this clade, the first two being non-homoplasic: (1) "presence of a well-developed, deep fossa between the dorsomesial limb of the posttemporo-supracleithrum, extrascapular and pterotic"; (2) "massive, somewhat cartilaginous or cord-like tissue connecting the coronoid process and the maxilla"; (3) "presence of a particularly conspicuous posterior process of the parieto-supraoccipital"; and (4) "presence of a well-developed posterior laminar projection of the mesial limb of the posttemporo-supracleithrum" (Diogo 2004: 252). Within this well-supported clade, it is interesting to note the position of ariids, which are often associated with mochokid, doradid and auchenipterid catfishes, but appear close to the claroteids (Fig. 12.11). The two characters uniting the claroteids and ariids in Diogo (2004: 260), both homoplasic, were: (1) "presence of a muscle depressor of the internal mandibular barbels" and (2) "insertion of a significant part of the fibers of the extensor tentaculi on the mesial and/or dorsal surface of sesamoid bone one of the suspensorium". With respect to the phylogenetic position of *Malapterurus*, it is interesting to verify that in Diogo (2004) this genus essentially comes back to the position in which it was assigned by Günther in 1864, that is, close to the clade formed by the African Mochokidae and the Neotropical Doradidae and Auchenipteridae (Fig. 12.11). However, as stressed by Diogo (2004), the evidence supporting this hypothesis was not strong, as the three characters diagnosing the clade Malapteruridae + Mochokidae + Doradidae + Auchenipteridae are homoplasic: (1) hypertrophiation of muscle hyohyoideus inferior; (2) posterior truncation of parieto-supraoccipital; and (3) presence of a muscle protractor of Müllerian process.

Like the Bagridae, the Claroteidae *sensu* Mo (1991) appeared monophyletic in Diogo's (2004) work (contra De Pinna 1998). The four homoplasic characters supporting claroteid monophyly were (Diogo 2004: 263): (1) "abductor superficialis 1 does not reach anteroventral surface of cleithrum"; (2) "absence of well-defined dorsal concavity between lateral ethmoid and frontal"; (3) "presence of well-developed cartilage between dorsal processes of cleithrum"; and (4) "presence of prominent anteroventrolateral projection of autopalatine".

Apart from the points referred to above, one of the most interesting outcomes of Diogo's (2004) work was the grouping of the Pimelodidae, Heptapteridae and Pseudopimelodidae in a monophyletic unit, which thus corresponds to the "Pimelodidae" *sensu lato* (see above). This monophyletic unit was strongly supported by five characters, the first two being homoplasy-free with Siluriformes (Diogo 2004: 270–271): (1) "presence of cartilaginous plates carrying the mandibular barbels"; (2) "presence of a muscle 4 of the mandibular barbels"; (3) "presence of markedly developed anterolateral laminar projection of cleithrum"; (4) "presence of well-developed adductor hyomandibularis"; and (5) "levator operculi originating on both the neurocranium and the dorsolateral surface of the hyomandibulo-metapterygoid". Another interesting aspect of Diogo (2004) is that it corroborated De Pinna's (1993) unpublished thesis in that *Heteropneustes*, usually included in its own family Heteropneustidae, lies in the very core of the family Clariidae (Fig. 12.11). In Diogo (2004: 279–280), clariids such as *Clarias* and *Heterobranchus* appeared, in fact, more closely related to

Heteropneustes than to the clariid genus *Uegitglanis*. The characters supporting such a close relationship between the two former genera and *Heteropneustes* are: (1) "posttemporo-supracleithrum has a prominent posterodorsal projection that is firmly associated with the anterolateral margin of the parapophysis of fourth vertebra" (homoplasic) and (2) "both infraorbital four and supraopercle are highly developed, markedly enlarged bones firmly ankylosed to the neurocranium" (non-homoplasic).

Hardman 2005 (Fig. 12.12)

Hardman (2005) provided the first comprehensive investigation of siluriform higher-level phylogeny based on molecular data. He analyzed complete sequences (1170 contiguous nucleotides) of mitochondrial gene cytochrome *b* and partial sequences of the downstream threonine tRNA for 170 species representing most extant catfish families (see Fig. 12.12). Not all terminal taxa had complete sequences and the mean±standard deviation sequence length was 1125.3±87.6 nucleotides. His parsimony analysis (using PAUP 2001) of a data matrix of 650 informative characters × 170 terminal taxa (species) resulted in three most parsimonious trees with a length of 17,595 steps, a RI of 0.387 and a rather low CI of 0.085. As recognized by Hardman, the main aim of his cladistic analysis was to investigate which are the closest relatives of ictalurids. In order to do that, he made a praiseworthy effort to include representatives of several non-ictalurid taxa in that analysis. But, as stressed by him, the results obtained regarding the relationships among those taxa should be taken with some caution. For instance, although there is some congruence between the results of his parsimony analysis and those of his Bayesian likelihood analysis, they differ in many aspects. In fact, it is evident from Harman's paper that he considered the results of his parsimony analysis, which are shown in Fig. 12.12, as more "warrantable" than those of the Bayesian analysis. As stated by Hardman (2005: 713), "several nodes in the Bayesian consensus suggest rate heterogeneity to have misled the analysis and yielded erroneous resolution; these nodes include the inferred non-monophyly of Callichthyidae, Siluridae, Sisoridae, Akysidae, and Loricarioidea". Apart from these taxa, families such as the Bagridae, Loricariidae, Amblycipitidae, Cetopsidae, and Schilbidae also appeared as non-monophyletic in his Bayesian analysis. It should be noted that many of these taxa also appeared as non-monophyletic in his parsimony analysis, e.g., the Cetopsidae, Loricariidae, Loricarioidea, Schilbidae, Amblycipitidae, Sisoridae, and Bagridae (Fig. 12.12).

But there are some interesting points in which the results of the molecular cladistic analysis of Hardman (2005) are in agreement with those of morphological phylogenetic analyses discussed above. For example,

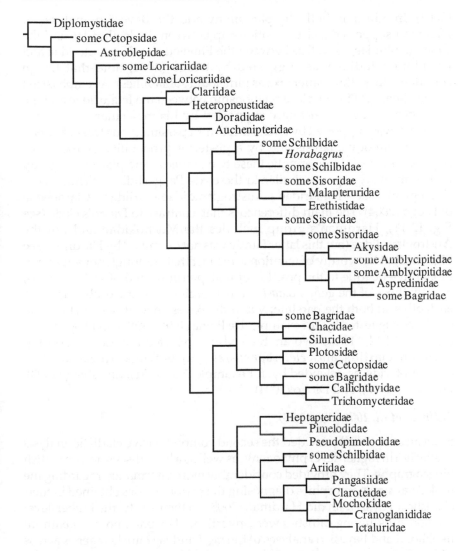

Fig. 12.12 Hypothetical relationships among the major siluriform groups according to the parsimony analysis of Hardman (2005).

although the cetopsids do not appear as monophyletic, in the parsimony analysis of Hardman some of them (namely those of the genus *Helogenes*) are placed in a rather basal position within the Siluriformes, as suggested by, for example, Mo (1991), De Pinna (1998), and Diogo (2004) (Fig. 12.12). And, although the loricarioids also did not appear as monophyletic, some of these fishes (namely the loricariids and astroblepids) were placed in a rather plesiomorphic position within the order, as suggested by Diogo

(2004). In addition, both the parsimony and the Bayesian analyses of Hardman supported a close relationship between the Ictaluridae and the Cranoglanididae, as well as between the Pimelodidae, Heptapteridae and Pseudopimelodidae, as suggested by Diogo (2004; note that Diogo considered these three latter taxa as pimelodid subfamilies). As emphasized by Hardman (2005), the close relationship between the Ictaluridae and the Cranoglanididae was strongly supported by his molecular study. This because it was supported by all the three most parsimonious trees obtained in his parsimony analysis and by a posterior probability of 100 in his Bayesian analysis. Interestingly, these two families were placed in a group that is, in many aspects, similar to the clade Pangasiidae + Schilbidae + Cranoglanididae + Ictaluridae + Austroglanididae + Ariidae + Claroteidae of Diogo (2004). The main difference is that, contrary to Diogo's clade (see Fig. 12.11), Hardman's group includes the Mochokidae and not the Austroglanididae (but this latter family was not examined by Hardman: see Fig. 12.12). Also worthy of mention is the fact that, although not appearing monophyletic (due to the position of malapterurids and of some bagrids, namely those of the genus *Batasio*), the sisoroids appear closely related to each other in both the parsimony and the Bayesian analyses of Hardman (2005), as suggested by, for example, De Pinna (1996, 1998) and Diogo (2004) (see Fig. 12.12). These two analyses also supported a close relationship between clariids and *Heteropneustes* and between doradids and auchenipterids, as defended by, for example, Mo (1991), Lundberg (1993), De Pinna (1998), and Diogo (2004).

Sullivan et al. 2006 (Fig. 12.13)

Sullivan *et al.* (2006) provided the second comprehensive cladistic analysis of siluriform higher-level phylogeny, as well as a brief discussion on catfish biogeography. They provided considerably more information regarding the molecular synapomorphies diagnosing the various clades obtained in their cladistic analysis than did Hardman (2005). In their study, the higher-level relationships among catfishes were investigated by parsimony, maximum likelihood and Bayesian analyses of the *rag 1* and *rag2* nuclear genes across 110 species representing all extant siluriform families except the Austroglanididae and Lacantuniidae (see Fig. 12.13). The parsimony analysis (using PAUP 2001) of 3,660 aligned bases of *rag1* and *rag2* for the 110 species resulted in eight most parsimonious trees with a length of 15,608 steps, RI=0.574 and CI=0.274. As noted by Sullivan *et al.* (2006: 642–643), there is a "broad congruence" among the results of this parsimony analysis and those of the maximum likelihood bootstrap and Bayesian analyses, "with no conflict among those nodes receiving strong support in each (maximum parsimony or maximum likelihood bootstrap support between 75 and 100%, Bayesian posterior probabilities between 0.9 and 1.0)

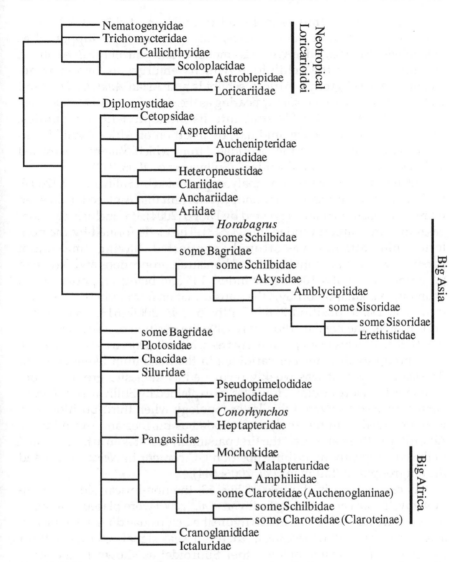

Fig. 12.13 Hypothetical relationships among the major siluriform groups according to the trees shown in figs. 1 and 2 of Sullivan *et al.* (2006).

and few cases of conflict among nodes receiving moderate support (bootstrap support between 50 and 75%, posterior probabilities between 0.5 and 0.9)". Figure 12.13 is based on figs. 1 and 2 of Sullivan *et al.* (2006); these latter figures showed those clades obtained in their parsimony analysis that were supported at or above the 50% bootstrap proportion.

Of the catfish families included in Sullivan *et al.*'s (2006) study represented by two or more species, only four were not supported as monophyletic: the Bagridae, which is "monophyletic with only *Rita* excluded and unresolved"; the Schilbidae, which "emerges as two separate monophyletic subgroups, African schilbids and Asian *Ailia+Laides*" (and with the genus *Pseudeutropius* appearing as the sister-group of *Horabagrus*); the Claroteidae, which "is split into its monophyletic subfamilies, Claroteinae and Auchenoglanidinae, by insertion of African schilbids as sister to claroteines"; and the Sisoridae, "from which *Nangra* is extracted and placed as sister to Erethistidae" (Sullivan *et al.* 2006: 243).

Interestingly, in all the three analyses promoted by Sullivan *et al.* (2006), the Loricarioidei appears as a monophyletic unit that is placed in an even more basal position than suggested by Diogo (2004): it is inclusively more basal than the Diplomystidae (Fig. 12.13). The clade formed by the non-loricarioid catfishes was strongly supported, having "maximum parsimony and maximum likelihood bootstrap proportions and Bayesian posterior probability=100, decay index=18" and being supported by an unambiguous autapomorphy, "the transformation from serine to threonine in *rag1* at *Danio* position 278" (Sullivan *et al.* 2006: 643). Within the Loricarioidei, the *rag* data did not resolve the position of *Nematogenys*. In the strict consensus of all parsimony trees, *Nematogenys* was placed as the sister-group to all other Loricarioidei; in the maximum likelihood and Bayesian analyses it was weakly recovered as the sister-group of non-trichomycterid loricarioids. However, as explained by Sullivan *et al.* (2006), parsimony also recovered this latter topology when third-position sites were excluded from the analysis, or coded as purines and pyrimidines. According to these authors, "the first parsimony result is an artifact related to the convergently high third-position GC content in *Nematogenys* and the outgroup taxa (Sullivan *et al.* 2006: 644).

As can be seen in the tree of Fig. 13, the non-loricarioid and non-diplomystid taxa analyzed by Sullivan *et al.* (2006) were placed in a rather unresolved clade. Although in this figure the Cetopsidae does not appear in a basal position within these taxa, "there is a weak signal in the *rag* data for placement of cetopsids below other Siluroidei as shown in the strict consensus of the maximum parsimony trees and in the maximum likelihood tree; on these topologies non-cetopsid siluroids are united by few characters" but "MacClade identifies three unambiguous amino acid substitutions that are synapomorphies of non-cetopsid siluroids: one, the substitution of serine for glycine at *Danio* position 726, is unique and invariant" (Sullivan *et al.* 2006: 644). The Bayesian consensus analysis did not resolve the position of cetopsids within the non-loricarioid and non-diplomystid taxa. All the three analyses promoted by Sullivan *et al.* (2006) supported a clade Aspredinidae

+ (Doradidae + Auchenipteridae). According to these authors, the *rag* support for this clade was high. The maximum parsimony analysis constraining the two latter South-American families to be the sister-group of the African Mochokidae (see above) produced shortest trees 48 steps longer than the shortest trees obtained in the unconstrained analyses. The maximum likelihood analysis with a constraint placing the Aspredinidae with Asian Sisoroidea (see above) produced shortest trees 87 steps longer than the shortest trees obtained in the unconstrained analyses. Both of these alternative hypotheses were rejected by parsimony-based Templeton and likelihood-based SH tests (Sullivan *et al.* 2006: 645).

As suggested by, for example, Mo (1991), De Pinna (1998), Diogo (2004), and Hardman (2005), the parsimony, maximum likelihood and Bayesian analyses of Sullivan *et al.* (2006: 645) indicated a close relationship between clariids and *Heteropneustes*, which was supported by "a unique and invariant synapomorphy in the *rag2* gene at *Danio* position 183 where phenylalanine replaces the plesiomorphic tyrosine". These three analyses also corroborated the close relationship between ariids and anchariids, as suggested by, for example, De Pinna (1993), Diogo (2004), and Ng and Sparks (2005); these catfishes are characterized by "an unique and invariant synapomorphic amino acid substitution at *rag1 Danio* position 304 where serine has replaced the plesiomorphic phenylalanine" (Sullivan *et al.* 2006: 645). They also supported a close relationship between the Cranoglanididae and the Ictaluridae and between the Pimelodidae, Pseudopimelodidae and Heptapteridae, as defended by Diogo (2004) and Hardman (2005) (no molecular synapomorphies were, however, provided by Sullivan *et al.* 2006 to support these clades). Interestingly, in all these analyses *Conorhynchos conirostris*, a species formerly classified within Pimelodidae but then removed from that family, was placed as the sister-group of the Heptapteridae (Fig. 12.13; see Sullivan *et al.* 2006).

One of the most interesting results of Sullivan *et al.* (2006) was the strong support, in all their analyses, for a large "Big Asia" clade including the akysids, the amblycipitids, the non-monophyletic sisorids, the erethistids, *Horabagrus*, and some of the non-monophyletic bagrids and schilbids, and for a large "Big Africa" clade comprising the remaining schilbids, the non-monophyletic claroteids, the amphiliids, the malapterurids and the mochokids. These analyses also strongly supported the monophyly of the Asian sisoroids (i.e., Sisoroidea without Aspredinidae) as well as of the Erethistidae (see above), this latter taxon appearing as the sister-group of the sisorid *Nangra*. According to Sullivan *et al.* (2006: 645), the Asian sisoroids are characterized by the "synapomorphic deletion of one codon (5_ *rag1* dataset position 203)" and the erethistids by the "large deletion of six codons (*rag1* codons at dataset positions 205–210)".

Higher-level Phylogeny of the Siluriformes: Consensus, Contradictions and Perspectives

As in Diogo's 2003a overview (see Introduction, above), after presenting the phylogenetic hypotheses of those papers dealing with the interfamilial relationships of siluriforms, we provide here a brief discussion on the consensus and contradictions between these phylogenetic hypotheses. Of the five most taxonomically complete surveys on siluriform higher-level phylogeny now available (Mo 1991, De Pinna 1998, Diogo 2004, Hardman 2005, Sullivan *et al.* 2006), only two were published when Diogo's 2003a paper was written (Mo 1991 and De Pinna 1998). So, some points that were tentatively listed as "general consensus" in that paper may well not be considered consensual in view of the results of the new studies now available. For example, the close relationship between amphiliids and loricarioids and between the Ariidae and the clade Mochokidae + Auchenipteridae + Doradidae, listed as "general consensus" by Diogo (2003a), have been contradicted by Diogo (2004), Hardman (2005), and Sullivan *et al.* (2006). However, it should be stressed that these latter works have also supported phylogenetic hypotheses that were commonly accepted in 2003, as shown below.

Sister-group Relationship between the Diplomystidae and the Other Siluriformes

The diplomystids appear as the most basal catfishes in all studies listed in the section above dealing with the phylogenetic position of these fishes (Grande 1987, Mo 1991, Arratia 1992, De Pinna 1998, He *et al.* 1999, Diogo 2004, Hardman 2005), except Sullivan *et al.* (2006). Thus, despite this latter work, the sister-group relationship between diplomystids and all other siluriforms can still be considered a "general consensus" *sensu* Diogo (2003a). However, it should be stressed that in some of the studies mentioned above, such as Diogo (2004), the diplomystids were considered *a priori* the sister-group of other catfishes, that is, they were used as outgroup of all other siluriforms. But it should also be noted that in a new cladistic analysis of Teleostei phylogeny promoted by Diogo (in press) that did not considered *a priori* the diplomystids as the most basal siluriforms and that included several representatives of all the other ostariophysan orders as well as numerous non-ostariophysan teleosts, the diplomystids did appear as the sister-group of all other catfishes (the relationships among the other catfishes included in that analysis are similar to those obtained by Diogo 2004). Despite the "general consensus" regarding the sister-group relationship between diplomystids and other catfishes, there are some contradictions concerning the morphological characters supporting the grouping of all non-diplomystid catfishes. For example, one of the characters listed by

Grande (1987) and Arratia (1992) as a synapomorphy of non-diplomystid catfishes (Siluroidei) is the ankylosis between the fifth centrum and the complex centrum of the Weberian apparatus. However, according to Mo (1991: 207), this character does not constitute a synapomorphy of the Siluroidei, but, instead, of all non-helogenine siluroids. In turn, some characters listed by Mo to diagnose the Siluroidei, such as the "T-shaped vomer" (Mo 1991: 203), were contradicted by, for example, Arratia (1992: 122) and De Pinna (1998: 291). Actually, the only morphological character that is commonly accepted as an unambiguous synapomorphy of the clade formed by non-diplomystid siluriforms is the "17 or fewer principal caudal rays (vs. 18 or more in *Diplomystes* and other primitive teleosts)" (Grande, 1987: 48). Hardman did not list any molecular synapomorphy to support this clade.

Basal Position of the Cetopsidae and Hypsidoridae

The analyses of Mo (1991), De Pinna (1998), Diogo (2004), and Hardman (2005) place the family Cetopsidae, or at least part of it, in a markedly basal position within the Siluriformes. As explained above, the Bayesian consensus analysis of Sullivan *et al.* (2006) did not resolve the position of cetopsids, but in their maximum parsimony trees and in the maximum likelihood tree these fishes appeared as the most basal non-loricarioid and non-diplomystid siluriforms. Three unambiguous amino acid substitutions unite the extant catfishes excepting diplomystids, loricarioids and cetopsids in these trees, one of them, the substitution of serine for glycine at *Danio* position 726, being "unique and invariant" (see above). De Pinna (1998: 292) stated that cetopsids "lack some synapomorphies of all other catfishes except for diplomystids and in some instances also hypsidorids", but did not specify which are these synapomorphies. Mo (1991: 204) listed two derived characters not present in the cetopsid and diplomystid catfishes to support the rather basal position of the Cetopsidae: (1) "the ramus mandibularis nerve runs inside hyomandibular for a distance" (according to both cladograms of Mo); (2) "interdigital union of the two coracoids" (according to Mo's cladogram I). Diogo (2004: 249) provided two characters to unite the extant catfishes excluding diplomystids, loricarioids and cetopsids, and, thus, supporting a rather plesiomorphic position of the Cetopsidae within siluriforms: (1) "horizontal lamina of scapulo-coracoid present and similar to that of cleithrum"; (2) "posterodistal margin of maxilla not markedly concave". With respect to the fossil hypsidorids, there is a general, but not strict, consensus concerning their phylogenetic position. In fact, of the six cladograms presenting a hypothesis on the phylogenetic relationships of these fossil catfishes, two (Mo's 1991 cladogram II and He *et al.*'s 1999 cladogram) place them in a somewhat derived position, one

(Mo's 1991 cladogram I) leaves their phylogenetic position unresolved, and only three (Grande 1987, Arratia 1992, and De Pinna 1998) place them in a markedly basal position within siluriforms. However, it should be noted that the phylogenetic hypotheses represented on both the cladogram II of Mo (1991) and the cladogram of He *et al.* (1999) are strongly contested by some authors (e.g., De Pinna and Ferraris 1992, De Pinna 1993, 1998, Diogo 2004). Thus, the phylogenetic position of the Hypsidoridae on these cladograms should be regarded with some caution. With respect to the three studies suggesting a markedly basal position of hypsidorids, that is, Grande (1987), Arratia (1992), and De Pinna (1998), only the first two listed characters to support such a basal position; Grande (1987) listed five characters and Arratia (1992) added various others to that list (see above). However, it should be noted that one of the characters listed by Grande (1987) as a synapomorphy of the non-hypsidorid siluroids, the absence of maxillary teeth, was questioned by Arratia (1992: 108–109). This was because, according to Arratia (1992), the absence of maxillary teeth constitutes a catfish plesiomorphy.

Close Relationship between Anchariids and Ariids

The genus *Ancharius* from Madagascar was traditionally included in the family Ariidae, but some authors suggested that it should be included in its own family, Anchariidae (e.g., Glaw and Vences 1994), or in the African family Mochokidae (e.g., Mo 1991). As explained by Diogo (2004), in the unpublished thesis by De Pinna (1993) *Ancharius* was placed as the sister-group of the ariids (Diogo 2004 did not, however, mention which characters were provided in that unpublished thesis to support that hypothesis). This sister-group relationship was then supported by five unambiguous synapomorphies in Diogo's (2004) work: (1) marked lateral bifurcation of premaxilla; (2) complete foramen between dorsal surfaces of lateral ethmoid and frontal; (3) presence of suture between mesial limb of posttemporo-supracleithrum and neurocranium; (4) mesocoracoid arch reduced to thin structure fused with main body of scapulo-coracoid; and (5) presence of protractor of Müllerian process. In 2005, Ng and Sparks published a study (not including a cladistic analysis) in which they described a new anchariid genus, *Gogo*, and listed the five synapomorphies proposed by Diogo (2004) as evidence of a sister-group relationship between *Ancharius* + *Gogo* and ariids. Of the two recent molecular studies promoted by Hardman (2005) and Sullivan *et al.* (2006), only the latter included anchariids (namely a species of the genus *Gogo*). As explained above, the three analyses of Sullivan *et al.* (2006) supported a close relationship between ariids and anchariids. According to these authors, these catfishes are characterized by "a unique and invariant synapomorphic amino acid substitution at *rag1 Danio* position 304 where serine has replaced the plesiomorphic phenylalanine" (Sullivan *et al.* 2006: 645). Thus, it can be said that phylogenetically there is now a general consensus regarding the close relationship of anchariids and ariids, although taxonomically there is still some controversy on whether the anchariids should effectively be placed in their own family Anchariidae (e.g., Glaw and Vences 1994, Ng and Sparks 2005), or instead in the family Ariidae (e.g., De Pinna 1993, Diogo 2004).

Close Relationship between Pimelodids, Pseudopimelodids and Heptapterids

The heptapterids, pimelodids, and pseudopimelodids *sensu* the present work were traditionally placed in a single family, the "Pimelodidae". However, the unpublished thesis by De Pinna (1993) suggested that these three groups do not form a monophyletic unit, and in recent studies these groups have commonly been placed in the families Heptapteridae, Pimelodidae, and Pseudopimelodidae, respectively (see Table 12.1). But the morphological cladistic analysis of Diogo (2004) and the molecular cladistic analyses of Hardman (2005) and Sullivan *et al.* (2006) suggested that these three groups do form a monophyletic clade (which according to Sullivan *et al.*'s analysis includes the species *Conorhynchos conirostris*, also formerly classified within the "Pimelodidae": see above). Diogo (2004) listed five unambiguous synapomorphies to support this clade (the first two being homoplasy-free within the Siluriformes): (1) presence of cartilaginous plates carrying the mandibular barbels; (2) presence of a muscle four of the mandibular barbels; (3) presence of markedly developed anterolateral laminar projection of cleithrum; (4) presence of well-developed adductor hyomandibularis; and (5) levator operculi originating on both the neurocranium and the dorsolateral surface of the hyomandibulo-metapterygoid. Hardman (2005) and Sullivan *et al.* (2006) did not provide a list of molecular synapomorphies supporting this clade, but according to Sullivan *et al.* (2006: 650) their study did "provide compelling evidence" to support it. These latter authors also referred to an unpublished work by Alves-Gomes and Lundberg using 12S and 16S mt rDNA and seemingly also corroborating this clade. In summary, the recent studies do support a close relationship between pseudopimelodids, pimelodids, and heptapterids. However, as in the case of anchariids/ariids, there is still some controversy, among those studies, on whether these three groups should be given a family status, as commonly done nowadays (see Table 12.1), or instead be considered subfamilies of the family Pimelodidae, as suggested by, for example, Diogo (2004).

Close Relationship between *Heteropneustes* and the Clariids

The five studies dealing with the interrelationships of *Heteropneustes* (Mo 1991, De Pinna 1998, Diogo 2004, Hardman 2005, Sullivan *et al.* 2006) strongly support a sister-group relationship between this genus and clariids. De Pinna (1998) and Diogo (2004) go as far as including *Heteropneustes* in the very core of the family Clariidae (see above). Of these five studies, only three (Mo 1991, Diogo 2004, Sullivan *et al.* 2006) provided a list of derived characters supporting the close relationship between *Heteropneustes* and clariids. The characters listed by Mo (1991) were: (1) "pterotic sutured with the frontal on the cranial roof"; (2) "accessory respiratory organ"; (3) "enlarged laminar last infraorbital"; and (4) "rib on the 6th vertebra vestigial". The characters listed by Diogo (2004) to support the view that clariids such as *Clarias* and *Heterobranchus* are more closely related to *Heteropneustes* than to clariids such as *Uegitglanis* were given in the section above. Sullivan *et al.* (2006: 645) provided a "unique and invariant synapomorphy" supporting the close relationship between clariids and *Heteropneustes*: the replacement of tyrosine by phenylalanine at *Danio* position 183 of the *rag2* gene.

Close Relationship between Erethistids, Sisorids, Akysids, Amblycipitids and Perhaps Aspredinids

Five of the six cladistic studies dealing with the interrelationships of the Aspredinidae (Mo 1991, De Pinna 1996, 1998, Diogo 2004, Hardman 2005) suggest a close relationship between that Neotropical family and the Asiatic families Erethistidae, Sisoridae, Akysidae, and Amblycipitidae (the one not supporting this hypothesis being that of Sullivan *et al.* 2006). Mo (1991) suggested a somewhat close, but highly unresolved (Fig. 12.4), relationship between these families (see above). These five families were later the subject of a detailed phylogenetic analysis by De Pinna (1996), who provided seven synapomorphies to support their inclusion in a monophyletic clade (Sisoroidea) (see above), the relationships within this clade being (Amblycipitidae + (Akysidae + (Sisoridae + (Aspredinidae + Erethistidae)))). This phylogenetic hypothesis was subsequently strengthened by De Pinna (1998: 319), who listed one additional character (the "development of the second ural centrum") corroborating the close relationship between aspredinids and Asian sisoroids. This close relationship was also supported by Diogo (2004). It should be noted that in this latter work the Amblycipitidae, the Akysidae, and the clade Sisoridae + Aspredinidae + Erethistidae appeared in an unresolved trichotomy (Fig. 12.11). Four additional characters were listed by Diogo (2004: 282) to support the Sisoroidea monophyly: (1) "presence of well-developed anteromesial process of cleithrum"; (2) "adductor mandibulae A2 directly inserted on mesial surface of mandible"; (3) "entoectopterygoid remarkably reduced in size, with medial surface completely surrounded by lateral surface of sesamoid bone one of suspensorium"; and (4) "coronoid process of mandible essentially constituted by posterodorsal surface of dentary". Hardman (2005) did not clarify which molecular synapomorphies supported, in his study, the close relationship between aspredinids and the Asian sisoroids. It should be noted that in a recent anatomical study on the configuration of the second ural centrum in siluriforms (which did not include a cladistic analysis), De Pinna and Ng (2004: 1) pointed out a new potential synapomorphy to support the monophyly of the Sisoroidea: the second ural centrum "well formed" and constituting "a complete intervertebral joint anteriorly with the compound caudal centrum".

Close Relationship between Auchenipterids and Doradids

As explained in the section above, the cladograms of Mo (1991), Lundberg (2003) and De Pinna (1998) indicated a close relationship between the African Mochokidae and the clade formed by the Neotropical Auchenipteridae and Doradidae. The main characters proposed to support this close relationship were: (1) distal end of elastic spring expanded to form a disc; (2) depressor dorsalis muscle of second dorsal-fin lepidotrich, or spine, inserts on base of first lepidotrich, or spinelet; (3) first dorsal-fin lepidotrich with greatly elongated and recurved limbs; (4) first and second dorsal-fin basals tightly bound or sutured to fourth, fifth, and sometimes sixth neural spines; (5) tripus with a recurved transformator process that enters the peritoneal tunic of swim bladder; (6) compound of Weberian complex strongly sutured to exoccipitals; and (7) epioccipital with a large superficial dermal component in the skull roof (see Mo 1991, Lundberg 1993, De Pinna 1998). The last of these seven characters was contested by Diogo (2004), who argued that it constitutes, in fact, a synapomorphy of Doradidae + Auchenipteridae and not of these two families plus the Mochokidae. In the same work, however, Diogo (2004: 266) pointed out five additional synapomorphies supporting the monophyly of the clade constituted by Doradidae + Auchenipteridae + Mochokidae: (1) "hyohyoideus abductor hypertrophied, with median aponeurosis firmly attached to pectoral girdle"; (2) "abductor superficialis 1 not reaching anteroventral surface of cleithrum"; (3) "presence of markedly developed anterolateral laminar projection of cleithrum"; (4) "adductor mandibulae A2 directly inserted on mesial surface of mandible" (char. 210); and (5) "adductor operculi or part differentiated from it not contacting hyomandibulo-metapterygoid". Thus, all the four morphological published cladistic analyses that have focused on the relationships between the Doradidae, Auchenipteridae, and Mochokidae support the close relationship between these three groups. However, the recent molecular analyses of Hardman and Sullivan *et al.* (2006) contradict this close relationship. In the parsimony analysis of Hardman (2005), the Mochokidae and the clade Auchenipteridae + Doradidae appear separated by several catfish taxa (see, e.g., Fig. 12.12). Hardman did not clarify which molecular characters support, in that analysis, a much closer relationship between the Mochokidae and various other siluriform families than between the Mochokidae and the Auchenipteridae + Doradidae. Interestingly, in Hardman's (2005) Bayesian analysis the Auchenipteridae + Doradidae appeared closely related to the Aspredinidae, as suggested in the parsimony, maximum likelihood and Bayesian analyses of Sullivan *et al.* (2006). However, neither Hardman (2005) nor Sullivan *et al.* (2006) clarified which molecular features supported such a close relationship in those analyses. Therefore, it can be said that there is an ample consensus concerning the clade doradids + auchenipterids, but there is still much controversy on whether these fishes are more closely related to mochokids or to aspredinids (or eventually to other catfish taxa).

Phylogenetic Relationships among the Loricarioid Families

There is a strict consensus in studies by Howes (1983), Schaefer (1990), Mo (1991), De Pinna (1992, 1998), Diogo (2004), and Sullivan *et al.* (2006) regarding the relationships between the Astroblepidae, Loricariidae, Scoloplacidae, and Callichthyidae. All these studies suggest that the Scoloplacidae is the sister-group of a monophyletic unit including the Loricariidae and Astroblepidae, the clade formed by these three families being, in turn, the sister-group of the Callichthyidae. With respect to the relationships of Nematogenyidae and Trichomycteridae, Howes (1983) suggested that the former were probably the sister-group of all other loricarioids. However, he clearly stressed that this hypothesis was weakly supported, and even contradicted, by some derived features present in both Trichomycteridae and Nematogenyidae (see above). Schaefer (1990) was fully aware of this problem and, thus, preferred to consider the relationships of the Trichomycteridae and Nematogenyidae as unresolved a *priori* (see above). Mo (1991) analyzed this problem and pointed out that the nematogenyids and trichomycterids were probably sister-groups, the clade formed by these two groups being the sister-group of the other loricarioids. This hypothesis was subsequently strongly supported by De Pinna (1992), who included in his analysis a new, undescribed group of trichomycterids, the Copionodontinae. This group was proposed to be the most plesiomorphic taxa within the Trichomycteridae. De Pinna's (1992) phylogenetic analysis pointed out that some of the characters commonly used to place the trichomycterids as the sister-group of the remaining non-nematogenyid loricarioids were, in reality, plesiomorphically absent in the Trichomycteridae (see above). The sister-group relationship between Nematogenyidae and Trichomycteridae was further supported by De Pinna (1998), who provided some new data to corroborate that hypothesis (De Pinna 1998: 296–297). The hypothesis was also corroborated by Diogo (2004: 135, 162, 164, 204, 207, 208, 220, 244), who provided seven additional characters supporting it: (1) "presence of well-developed anterodorsolateral salience of sphenotic"; (2) "presence of prominent dorsomesial process of first pectoral ray"; (3) "dilatator operculi markedly lateral to adductor mandibulae A2"; (4) "presence of markedly developed, elongated posterodorsal projection of hyomandibulo-metapterygoid firmly attaching to neurocranium by massive, strong connective tissue"; (5) "sharply-pointed, dorsally oriented projection of opercle"; (6) "anterodorsal surface of interopercle connected by well-defined, long, strong ligament to posterodorsal surface of preopercle"; and (7) "dorsal tip of coronoid process markedly curved mesially". As explained above, Hardman's 2005 molecular study did not include *Nematogenys* and Sullivan *et al.*'s 2006 molecular work did not allow clarification of the position of this genus inside the Loricarioidea. In summary, it can be said that although there is no strict consensus on this subject, most of the studies listed in the Section above suggest that the Trichomycteridae and Nematogenyidae are more closely related to each other than to the other loricarioids. Therefore, there is a general (but not strict) consensus regarding the relationships among the loricarioid families, these being ((Nematogenyidae + Trichomycteridae) + (Callichthyidae + (Scoloplacidae + (Loricariidae + Astroblepidae)))). An extensive compilation of the numerous characters supporting the major clades within the Loricarioidea is given by De Pinna (1998; see Fig. 12.6) and by Diogo (2004: 243–248).

Close Relationship between Cranoglanidids and Ictalurids

The phylogenetic relationships between cranoglanidids and other catfishes have been one of the most puzzling questions regarding catfish phylogeny. One of the first authors dealing with this issue was Jayaram (1956). He suggested that the Cranoglanididae could be closely related to the Pangasiidae, Bagridae, and/or Ictaluridae. However, as noted by, for example, Chardon (1968), the arguments given by Jayaram (1956) to support his hypotheses were rather fragile. Based on characters of the Weberian apparatus, Chardon (1968) suggested that the Cranoglanididae are probably closely related to the Bagridae, although he did also not provide convincing arguments supporting this hypothesis (i.e., he did not provide a list of derived features shared by both the cranoglanidids and the bagrids). Mo (1991) presented the first cladistic analysis dealing with the relationships of cranoglanidids. In his cladogram I the relationships of the Cranoglanididae were unresolved, while in his cladogram II the Cranoglanididae were grouped with the Austroglanididae (see Fig. 12.4). But, again, Mo did not give a convincing argument for this latter hypothesis (the grouping of cranoglanidids and austroglanidids in his cladogram II is based on a synapomorphy [synapomorphy number 49], which is not described subsequently by the author). The unpublished thesis of de Pinna (1993) also did not clarify the relationships of cranoglanidids: these fishes were placed in a "large pentatomy" including several catfish taxa (see Fig. 12.9). Diogo (2004) published the first cladistic analysis on catfish higher-level phylogeny supporting a close relationship between the Cranoglanididae and the Ictaluridae, thus corroborating one of the three alternative hypotheses proposed by Jayaram (1956) (see above). Two unambiguous characters were listed by Diogo to support such a close relationship: adductor mandibulae A3" presenting a large anterior tendon inserting on mandible and on posterior portion of the primordial ligament; levator operculi originating on both the neurocranium and the dorsolateral surface of the hyomandibulo-metapterygoid. The close relationship between ictalurids and cranoglanidids was subsequently corroborated by Hardman's (2005) and Sullivan *et al.*'s (2006) molecular studies. According to their authors, both these molecular studies provided strong evidence for this close relationship, although they did not clarify which molecular synapomorphies supported it. It should be noted here that the close relationship between cranoglanidids and ictalurids is also corroborated by a new cladistic analysis on catfish higher-level phylogeny based on mitochondrial DNA, currently conducted by one of us (Peng, in progress).

It can thus be said that there is some general consensus concerning the phylogenetic position and/or interrelationships of certain siluriform groups among those cladistic studies dealing with catfish higher-level phylogeny (see Fig. 12.14). In fact, apart from the consensual points discussed just above, one should call attention to certain interesting similar results obtained in some of the studies listed in the preceding section. For instance, the results of De Pinna (1998: based on his unpublished thesis of 1993) and Diogo (2004) suggest a close relationship between pangasiids and schilbids, as proposed in various non-cladistic works (see Diogo 2004 for more details on this subject), as well as between chacids and plotosids (see Figs. 12.9, 12.11). And, as explained above, there is some agreement in the molecular cladistic analysis of Hardman (2005) and in the morphological cladistic analysis of Diogo (2004) regarding a possible close relationship between austroglanidids, ariids, anchariids, claroteids, at least some schilbids, pangasiids, cranoglanidids and ictalurids (see Figs. 12.11, 12.12, 12.14).

However, despite the recent progress in catfish higher-level phylogeny, much remains to be done in this field, as can be seen in Fig. 12.14. For

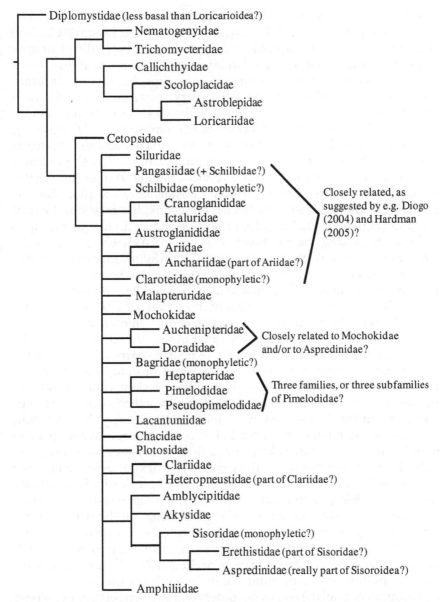

Fig. 12.14 Tree summarizing the most consensual points, in view of the discussion provided in this chapter, as well as some of the major controversies, regarding the present knowledge of the higher-level phylogeny of extant catfishes (for more details, see text).

instance, catfish specialists are far from reaching a consensus about the relationships, and in some cases even the monophyly (e.g., Claroteidae, Bagridae, Schilbidae), of families such as the Siluridae, Austroglanididae,

Schilbidae, Claroteidae, Malapteruridae, Bagridae, Mochokidae, Lacantuniidae, Chacidae, Plotosidae, and Amphiliidae. Diogo (2003a: 379) stated that "one of the most problematic issues concerning catfish phylogeny is surely the fact that a major fundamental question, with important implications not only for the choice of outgroups, but also for decisions regarding the polarity and/or evolutionary transformations of certain characters, remains unanswered: which is the most basal taxon within the suborder Siluroidei, that is, within the non-diplomystid taxa?" Fortunately, the morphological cladistic analysis of Diogo (2004) and the molecular cladistic analyses of Hardman (2005) and Sullivan *et al.* (2006), published after the writing of Diogo's (2003a) overview, have shed some light on this issue: it now begins to be somewhat consensually accepted that the loricarioids are, together with the diplomystids, the most basal extant catfishes (Figs. 12.11, 12.12, 12.13, 12.14) (see above). In fact, according to Sullivan *et al.* (2006), the question now is not so much whether or not the loricarioids occupy a rather basal position within Siluriformes, but whether they are phylogenetically more derived (as suggested by most studies) or more basal (as suggested by their study) than the diplomystids (see Fig. 12.14). This stresses a major problem of those cladistic analyses on catfish higher-level phylogeny listed in the section above: almost none of them has actually included terminal taxa from other ostariophysan orders, in order to appropriately investigate which are, in reality, the most basal catfishes. As explained above, there is a large amount of data in the literature suggesting that the diplomystids are the most plesiomorphic siluriforms. But, as stressed by Sullivan *et al.* (2006), this hypothesis does need to be tested in future cladistic analyses including not only diplomystids, loricarioids and other catfish taxa, but also various representatives of all the other otophysan orders. In a new cladistic analysis of Diogo (in press) including representatives of all these orders and of various non-otophysan taxa, as well as catfishes such as diplomystids, loricarioids, cetopsids, silurids, claroteids, bagrids, and pimelodids, the diplomystids effectively appeared as the sister-group of a clade including, on the one hand, the loricarioids and, on the other, the remaining catfishes (as mentioned above, with respect to the relationships among these remaining catfishes, the results of this new analysis are similar to those of Diogo 2004). It is thus interesting to see whether new molecular cladistic analyses including various representatives of all the otophysan orders will also support this hypothesis or will, instead, support that loricarioids are the most basal extant catfishes, as suggested by Sullivan *et al.*'s (2006) molecular study (which, it should be noted, besides the siluriforms included only one characiform species and two gymnotiform species).

As pointed out by Diogo (2003a), the more and more efficient computer techniques available, the increasing number of new researchers working on

Siluriformes, and particularly the increasing cooperation between them (the "All Catfishes Species" project is a very good example of that) seem to augur a promising future for this area of research. Regarding future works, one of the simplest ways to try to clarify the uncertainties involving catfish phylogeny is to simply include a greater number of terminal taxa. Apart from including more non-siluriform terminal taxa (see above), it would, for example, be interesting to include particularly "problematic" catfish taxa such as *Lacantunia* (none of the cladistic analyses published so far included this taxon), or key fossils such as †*Hypsidoris* or †*Andinichthys* (see, e.g., De Pinna 1998, Diogo 2004). Also, in order to test the monophyly of certain families, it is necessary to include, in explicit cladistic analyses, a greater number of representatives of those families. For instance, Sullivan *et al.* (2006: 648) stated that "morphological and molecular data provide good evidence for monophyly" of the families Heteropneustidae and Clariidae and for their sister-group relationship. However, only two explicit cladistic analyses have in fact included *Horaglanis* and/or *Uegitglanis*, i.e., the two clariid genera that according to some authors are less closely related to other clariids than is the genus *Heteropneustes* (see above). One is that provided in the unpublished thesis of De Pinna (1993); the other is that presented by Diogo (2004). And the results of these two analyses supported that clariids such as *Clarias* and *Heterobranchus* are effectively more closely related to *Heteropneustes* than to *Horaglanis* and/or *Uegitglanis*. That is, according to these results the family Clariidae is not monophyletic, unless *Heteropneustes* is included in it. Therefore, in order to appropriately test the results of De Pinna (1993) and Diogo (2004), future morphological and/or molecular cladistic analyses will necessarily need to include either *Horaglanis* or *Uegitglanis*, or both. An elegant example showing how an increase of our knowledge in catfish phylogeny could arise from the inclusion of more taxa in phylogenetic analyses was provided by De Pinna (1992) (see above). Another simple way of trying to clarify the uncertainties regarding catfish phylogeny is, of course, to include a greater number of characters, of different types, in the cladistic analyses. Examples of this are the recent works of Diogo (2004), which included numerous new, undescribed morphological characters concerning structures such as the cephalic and pectoral girdle muscles, and of Hardman (2005) and Sullivan *et al.* (2006), which provided the first comprehensive cladistic analyses on catfish higher-level phylogeny based on molecular data. However, many morphological features that can eventually reveal new and useful information to clarify siluriform relationships remain to be examined in a phylogenetic sense (e.g., the musculature of the branchial apparatus and/or of the caudal skeleton). And molecular studies based on sequences other than those examined by Hardman (2005) and Sullivan *et al.* (2006) are also needed to test the results of these authors and/or of the morphological cladistic analyses available so far. This is, for instance, precisely what is being done in a new study by Zardoya and colleagues (in progress). The combination of morphological and molecular data will, it is hoped, help to clarify catfish higher-level phylogeny.

Acknowledgements

We would like to acknowledge the helpful advice, discussions, and assistance of P. Vandewalle, M. Chardon, R.P. Vari, S. Weitzman, T. Abreu, A. Zanata, B.G. Kapoor, F. Meunier, S. He, D. Adriaens, F. Wagemans, C. Oliveira, E. Parmentier, M.M. De Pinna, P. Skelton, M.J.L. Stiassny, F.J. Poyato-Ariza, G. Arratia, T. Grande, M.G.F. Santini, J.C. Briggs, J.G. Maisey, S. Hughes, M. Gayet, J. Alves-Gomes, G. Lecointre, M. Brito, J.F. Agnese, A.E. Aquino, M.R. Brito, C.J. Ferraris, L.M. Soares-Porto, C. Oliveira, C. Borden, L. Taverne, M. Hardman and specially G.G. Teugels. We would also like to thank J. Lundberg and an anonymous reviewer for providing useful remarks on a previous version of this paper.

References

Agnese, J.F. and G.G. Teugels. 2001a. Genetic evidence for monophyly of the genus *Heterobranchus* and paraphyly of the genus *Clarias* (Siluriformes, Clariidae). Copeia 2001: 548–552.

Agnese, J.F. and G.G. Teugels. 2001b. The *Bathyclarias—Clarias* species flock. A new model to understand rapid speciation in African Great lakes. C.R. Acad. Sci. Paris 324: 683–688.

Agnese, J.F. and G.G. Teugels. 2005. Insight into the phylogeny of African Clariidae (Teleostei, Siluriformes): implications for their body shape, evolution, biogeography, and taxonomy. Mol. Phylogenet. Evol. 36: 546–553.

Aquino, A.E. and S.A. Schaefer. 2002. The temporal region of the cranium of loricarioid catfishes (Teleostei: Siluriformes): morphological diversity and phylogenetic significance. Zool. Anz. 241: 223–244.

Armbruster, J.W. 1998. Phylogenetic relationships of the suckermouth armoured catfishes of the *Rhinelepis* group (Loricariidae: Hypostominae). Copeia 1998: 620–636.

Armbruster, J.W. 2004. Phylogenetic relationships of the suckermouth armoured catfishes (Loricariidae) with emphasis on the Hypostominae and the Ancistrinae. Zool. J. Linn. Soc. 141: 1–80.

Arratia, G. 1987. Description of the primitive family Diplomystidae (Siluriformes, Teleostei, Pisces): morphology, taxonomy and phylogenetic implications. Bonn. Zool. Monogr. 24: 1–120.

Arratia, G. 1992. Development and variation of the suspensorium of primitive catfishes (Teleostei: Ostariophysi) and their phylogenetic relationships. Bonn. Zool. Monogr. 32: 1–148.

Bailey, R.M. and D.J. Stewart. 1984. Bagrid catfishes from Lake Tanganyika, with a key and descriptions of new taxa. Misc. Publ. Mus. Zool. Zool. Univ. Michigan 167: 1–41.

Baskin, J.N. 1972. Structure and relationships of the Trichomycteridae. Ph.D. thesis, City Univ. New York, New York.

Bockmann, F.A. 1994. Description of *Mastiglanis asopos*, a new pimelodid catfish from northern Brazil, with comments on phylogenetic relationships inside the subfamily Rhamdiinae (Siluriformes: Pimelodidae). Proc. Biol. Soc. Wash. 107: 760–777.

Bornbusch, A.H. 1991a. Redescription and reclassification of the silurid catfish *Apodoglanis furnessi* Fowler (Siluriformes: Siluridae), with diagnoses of three intrafamilial silurid subgroups. Copeia 1991: 1070–1084.

Bornbusch, A.H. 1991b. Monophyly of the catfish family Siluridae (Teleostei: Siluriformes), with a critique of previous hypotheses of the family's relationships. Zool. J. Linn. Soc. 101: 105–120.

Bornbusch, A.H. 1995. Phylogenetic relationships within the Eurasian catfish family Siluridae (Pisces: Siluriformes), with comments on generic validities and biogeography. Zool. J. Linn. Soc. 115: 1–46.

Bornbusch, A.H. and J.G. Lundberg. 1989. A new species of *Hemisilurus* (Siluriformes, Siluridae) from the Mekong river, with comments on its relationships and historical biogeography. Copeia 1989: 434–444.

Brito, M.R. 2003. Phylogeny of the subfamily Cordoradinae Hoedeman, 1952 (Siluriformes: Callichthyidae), with a definition of its genera. Proc. Acad. Nat. Sci. Philadelphia 153: 119–154.

Brown, B.A. and C.J. Ferraris. 1988. Comparative osteology of the Asian catfish family Chacidae, with the description of a new species from Burma. Am. Mus. Novitates 2907: 1–16.

Chardon M. 1968. Anatomie comparée de l'appareil de Weber et des structures connexes chez les Siluriformes. Ann. Mus. Roy. Afr. Centr. 169: 1–273.

Chen, X. and J.G. Lundberg. 1994. *Xiurenbagrus*, a new genus of amblycipitid catfishes (Teleostei: Siluriformes), and phylogenetic relationships among the genera of Amblycipitidae. Copeia 1994: 780–800.

Costa, W.J.E.M. 1994. A new genus and species of Sarcoglanidinae (Siluriformes: Trichomycteridae) from the Araguaia basin, central Brazil, with notes on subfamilial phylogeny. Ichthyol. Explor. Freshwaters 5: 207–216.

Costa, W.J.E.M. and F.A. Bockmann. 1994. A new genus and species of Sarcoglanidinae (Siluriformes: Trichomycteridae) from southeastern Brazil, with a re-examination of subfamilial phylogeny. J. Nat. Hist. 28: 715–730.

Curran, D.J. 1989. Phylogentic relationships among the catfish genera of the family Auchenipteridae (Teleostei: Siluroidea). Copeia 1989: 408–419.

Day, J.J. and M. Wilkinson. 2006. On the origin of the *Synodontis* catfish species flock from Lake Tanganyika. Biol. Lett. 2: 548–552.

De Pinna, M.C.C. 1988. A new genus of trichomycterid catfish (Siluroidei, Glanapterygidae), with comments on its phylogentic relationships. Rev. Suisse Zool. 95: 113–128.

De Pinna, M.C.C. 1989a. A new sarcoglanidine catfish, phylogeny of its subfamily, and an appraisal of the phyletic status of the Trichomycterinae (Teleostei, Trichomycteridae). Am. Mus. Novitates 2950: 1–39.

De Pinna, M.C.C. 1989b. Redescription of *Glanapteryx anguilla*, with notes on the phylogeny of Glanapteryginae (Siluriformes, Trichomycteridae). Proc. Acad. Nat. Sci. Philadelphia 141: 361–374.

De Pinna, M.C.C. 1992. A new subfamily of Trichomycteridae (Teleostei: Siluriformes), lower loricaroid relationships and a discussion on the impact of additional taxa for phylogentic analysis. Zool. J. Linn. Soc. 106: 175–229.

De Pinna, M.C.C. 1993. Higher-level phylogeny of Siluriformes, with a new classification of the order (Teleostei, Ostariophysi). Ph.D. thesis, City Univ. New York, New York.

De Pinna, M.C.C. 1996. A phylogenetic analysis of the Asian catfish families Sisoridae, Akysidae and Amblycipitidae, with a hypothesis on the relationships of the Neotropical Asprenidae (Teleostei, Ostariophysi). Fieldiana, Zool. 84: 1–82.

De Pinna, M.C.C. 1998. Phylogenetic relationships of Neotropical Siluriformes: History, overview and synthesis of hypotheses, pp. 279–330. *In*: L.R. Malabarba, R.E. Reis, R.P. Vari, Z.M. Lucena and C.A.S. Lucena [eds.]. Phylogeny and Classification of Neotropical Fishes. Edipucrs, Porto Alegre, Brazil.

De Pinna, M.C.C. and C.J. Ferraris. 1992. [Review of] Anatomy, relationships and systematics of the Bagridae (Teleostei: Siluroidei) with a hypothesis of Siluroid phylogeny, by T. Mo. Copeia 1992: 1132–1134.

De Pinna, M.C.C. and H.H. Ng. 2004. The second ural centrum in Siluriformes and its implication for the monophyly of superfamily Sisoroidea (Teleostei, Ostariophysi). Am. Mus. Novitates 3437: 1–23.

De Pinna, M.C.C. and W.C. Starnes. 1990. A new genus and species of Sarcoglanidinae from the Rio Mamoré, Amazon Basin, with comments on subfamilial phylogeny (Teleostei, Trichomycteridae). J. Zool. (Lond.) 222: 75–88.

De Pinna, M.C.C. and R.P. Vari. 1995. Monophyly and phylogenetic diagnosis of the family Cetopsidae, with synonymization of the Helogenidae (Teleostei: Siluriformes). Smithsonian Contr. Zool. 571: 1–26.

Diogo R. 2003a. Higher-level phylogeny of Siluriformes: an overview, pp. 353–384. *In*: B.G. Kapoor, G. Arratia, M. Chardon, and R. Diogo [eds.]. Catfishes. Science Publishers, Enfield, USA.

Diogo R. 2003b. Anatomy, phylogeny and taxonomy of Amphiliidae, pp. 401–438. *In*: B.G. Kapoor, G. Arratia, M. Chardon, and R. Diogo [eds.]. Catfishes. Science Publishers, Enfield, USA.

Diogo, R. 2004. Morphological Evolution, Aptations, Homoplasies, Constraints, and Evolutionary Trends: Catfishes as a Case Study on General Phylogeny and Macroevolution. Science Publishers, Enfield, USA.

Diogo, R. 2005. Osteology and myology of the cephalic region and pectoral girdle of *Pimelodus blochii*, comparison with other pimelodines, and comments on the synapomorphies and phylogenetic relationships of the Pimelodinae (Ostariophysi: Siluriformes). Eur. J. Morphol. 42: 115–126.

Diogo, R. In press. On the Origin and Evolution of Higher-Clades: Osteology, Myology, Phylogeny and Macroevolution of Bony Fishes and the Rise of Tetrapods. Science Publishers, Enfield, USA.

Diogo, R. and R. Bills. 2006. Osteology and myology of the cephalic region and pectoral girdle of the South African catfish *Austroglanis gilli*, with comments on the autapomorphies and phylogenetic relationships of the Austroglanididae (Teleostei: Siluriformes). Anim. Biol. 56: 39–62

Diogo, R. and M. Chardon. 2003. Osteology and myology of the cephalic region and pectoral girdle of *Heteropneustes fossilis* (Teleostei: Siluriformes), with comments on the phylogenetic relationships between *Heteropneustes* and the clariid catfishes. Anim. Biol. 53: 379–396.

Diogo, R. and M. Chardon. 2006a. Osteology and myology of the cephalic region and pectoral girdle of *Cetopsis coecutiens*, comparison with other cetopsids, and discussion on the synapomorphies and phylogenetic position of the Cetopsidae (Teleostei: Siluriformes). Belg. J. Zool. 136: 3–13.

Diogo, R. and M. Chardon. 2006b. On the osteology and myology of the cephalic region and pectoral girdle of *Nematogenys inermis* (Ghichenot, 1848), with comments on the autapomorphies and phylogenetic relationships of the Nematogenyidae (Teleostei: Siluriformes). Belg. J. Zool. 136: 15–24.

Diogo, R., P. Vandewalle and M. Chardon. 1999. Morphological description of the cephalic region of *Bagrus docmak*, with a reflection on Bagridae (Teleostei: Siluriformes) autapomorphies. Neth. J. Zool. 49: 207–232.

Diogo, R., M. Chardon and P. Vandewalle. 2001. Osteology and myology of the cephalic region and pectoral girdle of *Bunocephalus knerii*, and a discussion on the phylogenetic relationships of the Aspredinidae (Teleostei: Siluriformes). Neth. J. Zool. 51: 457–481.

Diogo, R., M. Chardon and P. Vandewalle. 2002a. Osteology and myology of the cephalic region and pectoral girdle of the Chinese catfish *Cranoglanis bouderius*, with a discussion on the autapomorphies and phylogenetic relationships of the Cranoglanididae (Teleostei: Siluriformes). J. Morphol. 253: 229–242.

Diogo, R., M. Chardon and P. Vandewalle. 2002b. Osteology and myology of the cephalic region and pectoral girdle of *Glyptothorax fukiensis* (Rendahl, 1925), comparison with other sisorids, and comments on the synapomorphies of the Sisoridae (Teleostei: Siluriformes). Belg. J. Zool. 132: 93–101.

Diogo, R., M. Chardon and P. Vandewalle. 2003a. Osteology and myology of the cephalic region and pectoral girdle of *Liobagrus reini* Hilgendorf 1878, with a discussion on the phylogenetic relationships of the Amblycipitidae (Teleostei: Siluriformes). Belg. J. Zool. 133: 77–84.

Diogo, R., M. Chardon and P. Vandewalle. 2003b. Osteology and myology of the cephalic region and pectoral girdle of *Centromochlus heckelii*, comparison with other auchenipterids, and comments on the synapomorphies and phylogenetic relationships of the Auchenipteridae (Teleostei: Siluriformes). Anim. Biol. 53: 397–416.

Diogo, R., M. Chardon and P. Vandewalle. 2003c. Osteology and myology of the cephalic region and pectoral girdle of *Erethistes pusillus*, comparison with other erethistids, and discussion on the autapomorphies and phylogenetic relationships of the Erethistidae (Teleostei: Siluriformes). J. Fish Biol. 63: 1160–1176.

Diogo, R., M. Chardon and P. Vandewalle. 2004a. On the osteology and myology of the cephalic region and pectoral girdle of *Chaca bankanensis* Bleeker 1852, with comments on the autapomorphies and phylogenetic relationships of the Chacidae (Teleostei: Siluriformes). Anim. Biol. 54: 159–174.

Diogo, R., M. Chardon and P. Vandewalle. 2004b. On the osteology and myology of the cephalic region and pectoral girdle of *Franciscodoras marmoratus* (Lütken 1874), comparison with other doradids, and comments on the synapomorphies and phylogenetic relationships of the Doradidae (Teleostei: Siluriformes). Anim. Biol. 54: 175–193.

Diogo, R., M. Chardon and P. Vandewalle. 2004c. Osteology and myology of the cephalic region and pectoral girdle of *Batrachoglanis raninus*, with a discussion on the synapomorphies and phylogenetic relationships of the Pseudopimelodinae and the Pimelodidae (Teleostei: Siluriformes). Anim. Biol. 54: 261–280.

Diogo, R., M. Chardon and P. Vandewalle. 2004d. Osteology and myology of the cephalic region and pectoral girdle of *Schilbe mystus*, comparison with other schilbids, and discussion on the monophyly and phylogenetic relationships of the Schilbidae (Teleostei: Siluriformes). Anim. Biol. 54: 91–110.

Ferraris, C.J. 1988. Relationships of the Neotropical catfish genus *Nemuroglanis*, with a description of a new species (Osteichthyes, Siluriformes, Pimelodidae). Proc. Biol. Soc. Wash. 101: 509–516.

Fink, S.V. and W.L. Fink. 1981. Interrelationships of ostariophysan fishes (Teleostei). Zool. J. Linn. Soc. 72: 297–353.

Gayet, M. 1988. Le plus ancien crâne de siluriforme: *Andinichthys bolivianensis* nov. gen., nov. sp. (Andinichthyidae nov. fam.) du Maastrichtien de Tiupampa (Bolivie). C.R. Acad. Sci. Paris 307: 833–836.

Glaw, F. and M. Vences. 1994. A fieldguide to the amphibians and reptiles of Madagascar, including mammals and freshwater fish (2nd ed.). Köln: Privately published.

Grande, L. 1987. Redescription of *Hypsidoris farsonensis* (Teleostei: Siluriformes), with a reassessment of its phylogenetic relationships. J. Vert. Paleontol. 7: 24–54.

Grande, L. and J.G. Lundberg. 1988. Revision and redescription of the genus *Astephus* (Siluriformes: Ictaluridae) with a discussion of its phylogenetic relationships. J. Vert. Paleontol. 8: 139–171.

Grande, L. and M.C.C. De Pinna. 1998. Description of a second species of the catfish †*Hypsidoris* and a revaluation of the genus and the family †Hypsidoridae. J. Vert. Paleontol. 18: 451–474.

Guo, X., S. He and Y. Zhang. 2005. Phylogeny and biogeography of Chinese sisorid catfishes re-examined using mitochondrial cytochrome b and 16S rRNA gene sequences. Mol. Phylogenet. Evol. 35: 344–362.

Hardman, M. 2004. The phylogenetic relationships among *Noturus* catfishes (Siluriformes: Ictaluridae) as inferred from mitochondrial gene cytochrome b and nuclear recombination activating gene 2. Mol. Phylogenet. Evol. 30: 395–408.

Hardman, M. 2005. The phylogenetic relationships among non-diplomystid catfishes as inferred from mitochondrial cytochrome b sequences; the search for the ictalurid sister taxon (Otophysi: Siluriformes). Mol. Phylogenet. Evol. 37: 700–720.

Hardman, M. and J.G. Lundberg. 2006. Molecular phylogeny and a chronology of diversification for "phractocephaline" catfishes (Siluriformes: Pimelodidae) based on mitochondrial DNA and nuclear recombination activating gene 2 sequences. Mol. Phylogenet. Evol. 40: 410–418.

Hardman, M. and L.M. Page. 2003. Phylogenetic relationships among bullhead catfishes of the genus *Ameiurus* (Siluriformes: Ictaluridae). Copeia 2003: 20–33.

He, S. 1996. The phylogeny of the glyptosternoid fishes (Teleostei: Siluriformes, Sisoridae). Cybium 20: 115–159.

He, S., M. Gayet and F.J. Meunier. 1999. Phylogeny of the Amphiliidae (Teleostei: Siluriformes). Ann. Sci. Nat. 20: 117–146.

Howes, G.J. 1983. The cranial muscles of the loricarioid catfishes, their homologies and value as taxonomic characters. Bull. Br. Mus. Nat. Hist. (Zool.) 45: 309–345.

Howes, G.J. 1985. The phylogenetic relationships of the electric family Malapteruridae (Teleostei: Siluroidei). J. Nat. Hist. 19: 37–67.

Howes, G.J. and A. Fumihito. 1991. Cranial anatomy and phylogeny of the South-East Asian catfish genus *Belodontichthys*. Bull. Br. Mus. Nat. Hist. (Zool.) 57: 133–160.

Jansen, G., S. Devaere, P.H.H. Weekers and D. Adriaens. 2006. Phylogenetic relationships and divergence time estimate of African anguilliform catfish (Siluriformes: Clariidae) inferred from ribosomal gene and spacer sequences. Mol. Phylogenet. Evol. 38: 65–78.

Jayaram, K.C. 1956. Taxonomic status of the Chinese catfish family Cranoglanididae Myers, 1931. Proc. Natl. Inst. Sci. India 21: 256–263.

Kailola, P. J. 2004. A phylogenetic exploration of the catfish family Ariidae (Otophysi: Siluriformes). Beagle Rec. Mus. and Art Galleries Northern Territory 20: 87–166.

Ku, X., Z. Peng, R. Diogo, and S. He. 2007. MtDNA phylogeny provides evidence of generic polyphyleticism for East Asian bagrid catfishes. *Hydrobiologia* 579: 147–159.

Lundberg, J.G. 1975. The fossil catfishes of North America. Univ. Mich. Mus. Paleont. Pap. Paleontol. 11: 1–51.

Lundberg, J.G. 1982. The comparative anatomy of the toothless blindcat, *Trogloglanis pattersoni* Eigenmann, with a phylogentic analysis of the ictalurid catfishes. Misc. Publ. Mus. Zool. Univ. Mich. 163: 1–85.

Lundberg, J.G. 1992. The phylogeny of ictalurid catfishes: a synthesis of recent work., pp. 392–420. *In:* R.L. Mayden [ed.]. Systematics, Historical Ecology and North American Freshwater Fishes. Stanford University Press, Stanford, California.

Lundberg, J.G. 1993. African-South American freshwater fish clades and continental drift: problems with a paradigm, pp. 156–199. *In:* P. Goldblatt [ed.]. Biological Relationships between Africa and South America. Yale University Press, New Haven, Connecticut.

Lundberg, J.G. and G.R. Case. 1970. A new catfish from the Eocene Green River formation, Wyoming. J. Paleontol. 44: 451–457.

Lundberg, J.G. and L. McDade. 1986. A redescription of the rare Venezuelan catfish *Brachyrhamdia imitator* Myers (Siluriformes, Pimelodidae), with phylogenetic evidence for a large intrafamilial lineage. Notulae Naturae 463: 1–24.

Lundberg, J.G., O. Linares and P. Nass. 1988. *Phractocephalus hemiliopterus* (Pimelodidae, Siluriformes) from the upper Miocene Urumaco formation, Venezuela: a further case of evolutionary stasis and local extinction among South American fishes. J. Vert. Paleontol. 8: 131–138.

Lundberg, J.G., A.H. Bornbusch and F. Mago-Leccia. 1991a. *Gladioglanis conquistador* n. sp. from Ecuador, with diagnoses of the subfamilies Rhamdiinae Bleeker and Pseudopimelodinae n. subf. (Siluriformes: Pimelodidae). Copeia 1991: 190–209.

Lundberg, J.G., F. Mago-Leccia and P. Nass. 1991b. *Exallodontus aguanai*, a new genus and species of Pimelodidae (Pisces: Siluriformes) from deep river channels of South America, and delimitation of the subfamily Pimelodinae. Proc. Biol. Soc. Wash. 104: 840–869.

Mo, T. 1991. Anatomy, relationships and systematics of the Bagridae (Teleostei: Siluroidei) with a hypothesis of siluroid phylogeny. Theses Zool. 17: 1–216.

Montoya-Burgos, J.-I., S. Muller, C. Weber and J. Pawlowski. 1997. Phylogenetic relationships between Hypostominae and Ancistrinae (Siluroidei: Loricariidae): first results from mitochondrial 12S and 16S rRNA gene sequences. Rev. Suisse Zool. 104: 185–198.

Montoya-Burgos, J.-I., S. Muller, C. Weber and J. Pawlowski. 1998. Phylogenetic relationships of the Loricariidae (Siluriformes) based on mitochondrial rRNA gene sequences, pp. 363–374. *In*: L.R. Malabarba, R.E. Reis, R.P. Vari, Z.M. Lucena and C.A.S. Lucena [eds.]. Phylogeny and Classification of Neotropical Fishes. Edipucrs, Porto Alegre, Brazil.

Montoya-Burgos, J.-I., C. Weber and P-Y. Le Bail. 2002. Phylogenetic relationships within *Hypostomus* (Siluriformes: Loricariidae) and related genera based on mitochondrial D-loop sequences. Rev. Suisse Zool. 109: 368–382.

Moyer, G.R., B.M. Burr and C. Krajewski. 2004. Phylogenetic relationships of thorny catfishes (Siluriformes: Doradidae) inferred from molecular and morphological data. Zool. J. Linn. Soc. 140: 551–575.

Ng, H.H. 2003. Phylogeny and systematics of Bagridae, pp. 439–463. *In*: B.G. Kapoor, G. Arratia, M. Chardon, and R. Diogo [eds.]. Catfishes. Science Publishers, Enfield, USA.

Ng, H.H. and J.S. Sparks. 2005. Revision of the endemic Malagasy catfish family Anchariidae (Teleostei: Siluriformes), with descriptions of a new genus and three new species. Ichthyol. Explor. Freshwaters 16: 303–323.

Oliveira, C., R. Diogo, P. Vandewalle and M. Chardon. 2001. Osteology and myology of the cephalic region and pectoral girdle of *Plotosus lineatus*, with comments on Plotosidae (Teleostei: Siluriformes) autapomorphies. J. Fish. Biol. 59: 243–266.

Oliveira, C., R. Diogo, P. Vandewalle and M. Chardon. 2002. On the myology of the cephalic region and pectoral girdle of three ariid species, *Arius heudeloti*, *Genidens genidens* and *Bagre marinus*, with a comparison with other catfishes (Teleostei: Siluriformes). Belg. J. Zool. 59: 243–266.

Peng, Z., S. He and Y. Zhang. 2002. Mitochondrial cytochrome b sequence variations and phylogeny of East Asian bagrid catfishes. Progress Nat. Sci. 12: 421–425.

Peng, Z., S. He and Y. Zhang. 2004. Phylogenetic relationships of glyptosternoid fishes (Siluriformes: Sisoridae) inferred from mitochondrial cytochrome b gene sequences. Mol. Phylogenet. Evol. 31: 979–987.

Peng, Z., S.Y.W. Ho, Y. Zhang and S. He. 2006. Uplift of the Tibetan plateau: Evidence from divergence times of glyptosternoid catfishes. Mol. Phylogenet. Evol. 39: 568–572.

Perdices, A., E. Bermingham, A. Montilla and I. Doadrio. 2002. Evolutionary history of the genus *Rhamdia* (Teleostei: Pimelodidae) in Central America. Mol. Phylogenet. Evol. 25: 172–189.

Pouyaud, L., G.G. Teugels, R. Gustiano and M. Legendre. 2000. Contribution to the phylogeny of pangasiid catfishes based on allozymes and mitochondrial DNA. J. Fish. Biol. 56: 1509–1538.

Reis, R.E. 1998. Anatomy and phylogenetic analysis of the Neotropical callichthyid catfishes (Ostariophysi, Siluriformes). Zool. J. Linn. Soc. 124: 105–168.

Rodiles-Hernández, R., D.A. Hendrickson, J.G. Lundberg and J.M. Humphries. 2005. *Lacantunia enigmatica* (Teleostei: Siluriformes) a new and phylogenetically puzzling freshwater fish from Mesoamerica. Zootaxa 1000: 1–24.

Schaefer, S.A. 1987. Osteology of *Hypostomus plecostomus* (Linnaeus), with a phylogenetic analysis of the loricariid subfamilies (Pisces, Siluroidei). Contrib. Sci. 394: 1–31.

Schaefer, S.A. 1990. Anatomy and relationships of the scoloplacid catfishes. Proc. Acad. Nat. Sci. Philadelphia 142: 167–210.

Schaefer, S.A. 1991. Phylogenetic analysis of the loricariid subfamily Hypoptopomatinae (Pisces: Siluroidei: Loricariidae), with comments on generic diagnoses and geographic distribution. Zool. J. Linn. Soc. 102: 1–41.

Schaefer, S.A. 1998. Conflict and resolution: impact of new taxa on phylogenetic studies of the Neotropical cascudinhos (Siluroidei: Loricariidae), pp. 375–400. In: L.R. Malabarba, R.E. Reis, R.P. Vari, Z.M. Lucena and C.A.S. Lucena [eds.]. Phylogeny and Classification of Neotropical Fishes. Edipucrs, Porto Alegre, Brazil.

Schaefer, S.A. 2003 Relationships of *Lithogenes villosus* Eigenmann, 1909 (Siluriformes, Loricariidae): Evidence from High-Resolution Computed Microtomography. Am. Mus. Novitates 3401: 1–55.

Schaefer, S.A. and G.V. Lauder. 1986. Historical transformation of functional design: evolutionary morphology of feeding mechanisms in loricarioid catfishes. Syst. Zool. 35: 489–508.

Schaefer, S.A. and G.V. Lauder. 1996. Testing historical hypotheses of morphological change: biomechanical decoupling in loricarioid catfishes. Evolution 50: 1661–1675.

Schaefer, S.A., H.S. Weitzman and H.A. Britski. 1989. Review of the Neotropical catfish genus *Scoloplax* (Pisces: Loricarioidea: Scoloplacidae) with comments on reductive characters in phylogenetic analysis. Proc. Acad. Nat. Sci. Philadelphia 141: 181–211.

Shibatta, O.A. 2003. Phylogeny and classification of 'Pimelodidae', pp. 385–400. In: B.G. Kapoor, G. Arratia, M. Chardon, and R. Diogo [eds.]. Catfishes. Science Publishers, Enfield, USA.

Shimabukuro-Dias, C.K., C. Oliveira, R.E. Reis and F. Foresti. 2004. Molecular phylogeny of the armored catfish family Callicthyidae (Ostariophysi, Siluriformes). Mol. Phylogenet. Evol. 32: 152–163.

Soares-Porto, L.M. 1998. Monophyly and interrelationships of the Centromochlinae (Siluriformes: Auchenipteridae), pp. 331–350. In: L.R. Malabarba, R.E. Reis, R.P. Vari, Z.M. Lucena and C.A.S. Lucena [eds.]. Phylogeny and Classification of Neotropical Fishes. Edipucrs, Porto Alegre, Brazil.

Sullivan, J.P., J.G. Lundberg and M. Hardman. 2006. A phylogenetic analysis of the major groups of catfishes (Teleostei: Siluriformes) using rag1 and rag2 nuclear gene sequences. Mol. Phylogenet. Evol. 41: 636–662.

Teugels, G.G. 2003. State of the art of recent siluriform systematics, pp. 317–352. In: B.G. Kapoor, G. Arratia, M. Chardon and R. Diogo [eds.]. Catfishes. Science Publishers, Enfield, USA.

Teugels, G.G. and D. Adriaens. 2003. Taxonomy and phylogeny of Clariidae—an overview, pp. 465–487. *In*: B.G. Kapoor, G. Arratia, M. Chardon, and R. Diogo [eds.]. Catfishes. Science Publishers, Enfield, USA.

Wilcox, T.P., F.J. Garcia de Leon, D.A. Hendrickson and D.M. Hillis. 2004. Convergence among cave catfishes: long-branch attraction and a Bayesian relative rates test. Mol. Phylogenet. Evol. 31: 1101–1113.

Zawadzki, C.H., E. Renesto, R.E. Reis, M.O. Moura and R.P. Mateus. 2005. Allozyme relationships in hypostomines (Teleostei: Loricariidae) from the Itaipu Reservoir, Upper Rio Parana' basin, Brazil. Genetica 123: 271–283.

Zhou, W. and Y.W. Zhou. 2005. Phylogeny of the genus *Pseudecheneis* (Sisoridae) with an explanation of its distribution pattern. Zool. Stud. 44: 417–433.

13

The Mitochondrial Phylogeny of the South American Electric Fish (Gymnotiformes) and an Alternative Hypothesis for the Otophysan Historical Biogeography

José A. Alves-Gomes[1]

Abstract

In order to address the mitochondrial phylogeny of the order Gymnotiformes within the Otophysi, two gonorynchiform genera were used to root the tree, and 15 genera from each remaining otophysan order (Cypriniformes, Characiformes, Siluriformes and Gymnotiformes) were sampled. The final data matrix consisted of two fragments (701 aligned nucleotides) of the 12S and 16S mitochondrial ribosomal RNA genes. Both molecules appear to be under functional/structural constraints, that can limit their efficiency as molecular marker for speciation events within a short time window. Nevertheless, 12S and 16S rRNA appear to retain enough phylogenetic information to determine branching order within the Otophysi, which is congruent with morphological data. Cypriniformes is the sister group of (Characiformes + Siluriformes + Gymnotiformes), and Characiformes is the sister group of the Siluriphysi (Siluriformes + Gymnotiformes). Adopting this hypothesis as the correct

[1] Laboratório de Fisiologia Comportamental e Evolução (LFCE), Instituto Nacional de Pesquisas da Amazônia. e-mail: puraque@inpa.gov.br

alternative, the fossil records of each order were reviewed in conjunction with their present distribution and current theories about the fragmentation of Pangea. The following scenario has been proposed: the ancestral Otophysi differentiated in freshwater habitats in Eurasia, from a lineage that invaded continental waters during the Early Cretaceous. Cypriniformes differentiated in Asia, colonized Africa but never reached South America. Cypriniform representatives arrived in North America by the Beringian land bridge and/or using a less obvious path through the Europe–Greenland–Labrador connection. The final separation of South America and Africa, between 80 and 125 million years ago, represented a vicariant event for already differentiated characiform clades, and sister-group relationships are found between African and South American genera. Characiforms reached North America only after the uplift of the Panamanian Isthmus, and not using the same route as the cypriniforms, probably for reasons associated with the current absence of characiforms in high latitudes. Siluriformes and Gymnotiformes differentiated within the isolating South America. It is proposed that catfish were able to disperse through shallow brackish/marine water by the Late Cretaceous and attained a cosmopolitan distribution by the Early Tertiary. Gymnotiforms never left South America, which may be related to functional constraints placed upon the efficiency of the Electrogenic and Electrosensory Systems (EES) by brackish (low resistivity) water.

Introduction

Among the many fascinating challenges associated with the evolutionary biology of the South American electric fishes (order Gymnotiformes) is to elucidate the time and the place of their differentiation. This is particularly relevant as one considers that the current accepted hypothesis about their phylogeny (Fink and Fink 1981, 1996) places them as the sister group of the catfish (order Siluriformes) and, therefore, as the same age. If this is correct, then we should explain why or how catfish are found in almost every continent of the world, whereas gymnotiforms are restricted to the Neotropics. Or is it possible that gymnotiforms differentiated from a South American catfish lineage, after the separation of Africa and South America? This hypothesis has never been tested and would imply Siluriformes to be paraphyletic and that gymnotiforms would be more closely related to some South American catfish lineage than this particular Neotropical catfish lineage to other catfish in other Continents. If this is not the case, did catfish and electric fish differentiate prior or after the fragmentation of Pangea? Or maybe Fink and Fink's (1981, 1996) hypothesis is not correct, and electric fish are more closely related to characiforms instead, as several previous and more recent studies suggest (Regan 1911, Greenwood et al. 1966, Rosen and Greenwood 1970, Dimmick and Larson 1966, Orti 1997, Saitoh *et al.* 2003, Peng *et al.* 2006a,b).

This chapter deals, in essence, with three basic questions related to the gymnotiform evolution: who is their sister group, when did they differentiate, and where did this cladogenetic event take place? Using the most extensive molecular data set for the Otophysi up to date, and assuming that Gonorynchiformes is their sister group (Rosen and Greenwood 1970, Fink and Fink 1981, 1986, Grande and Poyato-Ariza 1995), I addressed the branching order among the otophysan orders by means of mitochondrial DNA sequence data. The results obtained were evaluated together with morphological studies and additional molecular information that has been published more recently. The resultant phylogenetic hypothesis was further analyzed under the auspices of our current knowledge about the present and past distributions of the clades involved, as well as the existing views of plate tectonics and the fragmentation of Pangea. By combining phylogeny, current and past distribution and plate tectonics, a hypothesis about otophysan historical biogeography is proposed. As consequence, the time and place of the cladogenetic event that gave origin to the Gymnotiformes' lineage and its sister group is hypothesized. Subsequent arguments and evidence justifying their present distribution are also presented.

General Considerations about Ostariophysan Natural History

The superorder Ostariophysi, as it is currently conceived (Rosen and Greenwood 1970, Fink and Fink 1981, 1996), forms a monophyletic assemblage (but see Inoue *et al.* 2004) of teleost fish constituted by two sister clades: the series Anotophysi, containing the single order Gonorynchiformes (milkfish and relatives), and the series Otophysi, embracing the orders Cypriniformes (carps, minnows and relatives), Characiformes (piranhas, tetras and relatives), Siluriformes (catfishes), and Gymnotiformes (South American electric fishes). Together, these five orders represent about three-fourths of all extant teleost species currently living in fresh waters, with approximately 7,900 recognized species and 68 families distributed over all major land masses except Antarctica, Greenland, and New Zealand (Rosen and Greenwood 1970, Fink and Fink 1981, 1996, Nelson 2006). There are about 130 species ($\approx$ 1.6% of the total), which either are marine or tolerate salty waters. Saltwater tolerance evolved more than once within the group, as it can be found in distantly related clades such as the basal Gonorynchiformes (all species of Gonorynchidae and in the monotypic Chanidae), one cypriniform genus (*Tribolodon*), one characiform genus (*Astyanax*), and a few representatives of five catfish families (Novacek and Marshall 1976, Briggs 1979, de Pinna 1998, Nelson 2006). Among the extant gonorynchiforms, seven out of the about 35 species actually live in oceanic waters (Nelson 1994). *Chanos* occasionally invade fresh water, but it spawns in the ocean. In relation to catfishes, half of the Plotosidae and the majority of the Ariidae spend most of their life cycle in the ocean, and some members of Pangasidae, Auchenipteridae and Aspredinidae tolerate estuarine/coastal waters (Novacek and Marshall 1976, Briggs 1979, Nelson 2006, de Pinna 1998). Despite the exceptions mentioned, ostariophysans are considered primarily freshwater fishes, and salt water has been considered a dispersal barrier for them since Myers' work (1949).

Despite the obvious biological and economical importance of ostariophysans around the world and the intensive studies carried out on them, critical issues still remain to be clarified about their evolutionary history. Among those are the puzzling aspects of their historical biogeography, i.e., how did these clades attain their current geographical distribution, as one takes simultaneously into consideration their phylogenetic history and the currently accepted ideas about continental drift? In fact, a consistent and consensual biogeographical scenario aggregating our current understanding about these topics into a single hypothesis is still lacking despite the fact that the reconstitution of the ostariophysan historical biogeography is a relatively well addressed topic in fish evolution. None of the available hypotheses can be currently considered fully satisfactory because they have either relied upon a phylogenetic hypothesis (branching order) which is not fully accepted for the ostariophysans (Nichols 1930, Gosline 1944, Patterson 1975, Novacek and Marshall 1976, Briggs 1979, 2005, Brooks and McLennan 1991, Saitoh *et al.* 2003), or have only addressed intra-ordinal aspects of ostariophysan biogeography (Lundberg 1993, Ortí and Meyer 1997, Lundberg *et al.* 2007). Up to the moment, no biogeographic hypothesis for ostariophysans or otophysans reconciles Fink and Fink's phylogenetic hypothesis with continental drift.

Ostariophysan fossil records have been identified in sedimentary beds dating from the Late Jurassic, 144–163 million years ago (Ma) (Arratia 1997). Considering that the fragmentation of Pangea started between 200 and 220 Ma (Pitmann *et al.* 1993) and the final separation of Africa and South America is estimated to have occurred between 80 and 125 Ma, the past and present distribution of the group is intimately related to the distribution of freshwater habitats in the moving earth's crust since the Mesozoic.

A consistent hypothesis about the ostariophysan historical biogeography requires information and support from at least three main lines of substantiation. First, a phylogenetic hypothesis for the five orders, which will determine the branching order (sister group relationships) among them, needs to be defined; second, the present and past distribution of each lineage, retrieved from surveys and/or fossil records, can provide information about their historical geographical range and allow inference about extinctions or even possible dispersal paths; and third, the information obtained previously needs to be combined with the most accurate data available about the timing of the geological (vicariant) events shaping the earth's crust since each clade's existence.

Current Views about the Branching Order within the Ostariophysi

The ostariophysan classification and the implicit inter-ordinal relationships have gone through significant changes since the clade was first defined by Sagemehl (1885 in Gosline 1944) more than a century ago (Fig. 13.1). It is not the purpose of the present chapter to thoroughly review the studies about ostariophysan relationships since then. Besides the chapters in the present volume, the reader may refer to the papers by Regan (1911), Jordan (1923), Greenwood *et al.* (1966), Rosen and Greenwood (1970), Roberts (1973), and Fink and Fink (1981, 1996) for a comprehensive review of the subject, as well as to the historical development of the argument culminating in the currently accepted ideas. Here, I shall briefly address only the most recent studies about ostariophysan phylogeny, particularly those that have employed a cladistic and/or molecular approach.

The prevalent hypothesis about ostariophysan phylogeny is that originally proposed by Fink and Fink (1981). The authors analyzed about 120 morphological characters and their resulting topology is the one depicted in Fig. 13.1C. In 1996, the authors published a study about ostariophysan relationships reinforcing their previous findings (Fink and Fink 1996). The hypothesis established by Fink and Fink represented a turning point in the study of ostariophysan evolution for two particularly important reasons: their study was the first that explicitly employed a cladistic rationale and encompassed the whole group, and second their phylogenetic hypothesis departed radically from preceding ideas. In the "pre-cladistic" view preceding Fink and Fink's 1981 study (Figs. 13.1A and 13.1B), gymnotiforms and characiforms were considered closely related taxa and, in several classifications (Regan 1911, Greenwood *et al.* 1966, Rosen and Greenwood 1970), the catfish (Siluriformes) were given the same rank as the group formed by cypriniforms, characiforms, and gymnotiforms. Roberts (1973) objected to such classification arguing that there was no evidence "... indicating that characoids and cyprinoids are more closely related to each other than either is to siluroids", and assigned the same rank (Suborder) for characoids, cyprinoids, and catfish. Roberts considered the electric fish (gymnotiforms) as part of the characoids, i.e., a sister group of characiforms, but he also observed, in advance of what Fink and Fink were to establish formally in 1981, that the general assumption considering the resemblances between gymnotiforms and siluriforms as parallel adaptation still remained untested. The notion that the South American electric fishes represented "specialized" characiforms had been historically endorsed as the correct hypothesis by generations of ichthyologists prior to Fink and Fink's study,

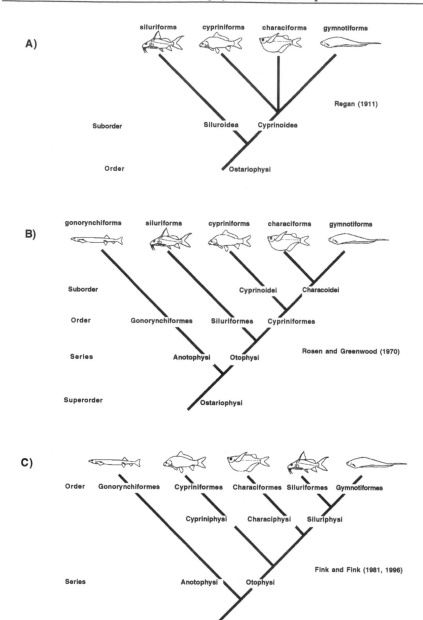

Fig. 13.1 The phylogenetic hypothesis for the Ostariophysi, based on the relationships conceived by several authors since the early 20th century. Fink and Fink's hypothesis meant a major departure from previous studies by proposing a sister-group relationship between the the Siluriformes and Gymnotiformes. For decades, despite the lack of cladistic evidence, Gymnotiformes were assumed to be closely related to Characiformes.

although no morphological synapomorphy supporting such grouping was ever proposed in the literature.

The phylogeny proposed by Fink and Fink also demanded a critical reevaluation of the existing ideas about the past and present biogeography of the Ostariophysi. Before 1981, all the authors formulated hypotheses in which a gymnotiform-characiform common ancestor differentiated from a common ancestor shared with the cypriniforms somewhere either in Africa or South America (Greenwood *et al.* 1966, Novacek and Marshall 1976, Briggs 1979, Brooks and McLennan 1991). If Fink and Fink's (1981) hypothesis is correct, we have to account for a particular time and place in which the siluriform-gymnotiform common ancestor lived, or even more intriguing, to resolve how siluriforms can be found in every continent in the world, whereas their sister group is confined to South America. Few authors after Fink and Fink's studies have proposed alternative biogeographic scenarios for the Otophysi (or Ostariophysi), but still considering gymnotiforms and sister group of the characiforms (Brooks and McLennan 1991, Saitoh *et al.* 2003, Briggs 2005).

Since it was first published, Fink and Fink's hypothesis has been tested by studies based upon DNA sequences, and the results were not completely congruent. Alves-Gomes (1995), in an unpublished thesis chapter, used a data matrix containing the most conserved regions of the 12S and 16S rRNA gene fragments from 10 taxa of each Characiphysi + Siluriphysi order plus three cypriniform genera as outgroups. With that data set and several weighting schemes for transitions and transversions, most of the resultant topologies pointed to a sister group relationship between Siluriformes and Gymnotiformes. However, *Electrophorus* and *Gymnotus* for the Gymnotiformes, *Diplomystes* and *Loricaria* for Siluriformes and the two African citharinids *Nannaethiops* and *Neolebias* for the Characiformes, were consistently placed in an unorthodox position in the phylogenetic trees, changing their positions according to the relative weight of transitions and transversions. By analyzing the molecular results in conjunction with other sources of evidences, the author proposed, as final hypothesis, the same topology as Fink and Fink (1981) for (Characiformes (Siluriformes + Gymnotiformes)). Dimmick and Larson (1996) used 1260 nucleotide sites corresponding to several discontinued segments of the mitochondrial genome, including portions of the 12S, 16S and Valine tRNA genes plus 1216 sites from segments of the nuclear ribosomal subunits 18S and 28S. The molecular data was analyzed separately and in conjunction with the same morphological characters used by Fink and Fink (1981). When the entire data matrix (molecular + morphological) is used to estimate phylogenies, Fink and Fink's hypothesis of a monophyletic Gymnotiformes + Siluriformes is corroborated. However, when only the molecular data is used, the most parsimonious topology depicts characiforms and gymnotiforms as sister groups. Ortí and Meyer (1996, 1997) and Ortí (1997) report results obtained with the phylogenetic analysis of two mitochondrial genes (12S and 16S rRNA) and the nuclear gene ependymin. The most parsimonious topology obtained with the rRNA genes corroborates Fink and Fink's hypothesis, but the two characiform genera *Citharinus* and *Distichodus*, regarded by several authors (Vari 1979, Fink and Fink 1981, Buckup 1991) as the most plesiomorphic lineage among characiforms, are located awkwardly in the tree, as sister group of the catfishes. When the nuclear gene ependymin was analyzed under several types of weighting schemes and data partitioning, Ortí and Meyer (1996) and Ortí (1997) give preference to a topology contrasting with Fink and Fink's (1981) and in agreement with the older literature, where the Gymnotiformes is the sister group of the Characiformes. However, when all epedymin nucleotide sites are used, the most parsimonious and the tree with best likelihood place Gymnotiformes as the sister group of the Siluriformes (table 4 in Ortí and Meyer 1996). More recently, Saitoh *et al.* (2003) published a paper about ostariophysan mitochondrial phylogeny, using a total of 8,196 base pairs of the mitochondrial genome for 15 ostariophysan taxa. From the complete mitochondrial genome, the authors discarded third site positions and sites with indels of 13 protein-coding genes (total of 7,286 sites) and also used 910 sites of stem regions of 22 tRNA genes. The methodological procedure of alignment and tree search adopted by the authors is unclear, but their final results present two alternative hypothesis for the branching order within the otophysans: for maximum likelihood, the resulting topology depicts (Cypriniformes (Siluriformes (Characiformes + Gymnotiformes))), whereas for parsimony the topology shows (Cypriniformes (Gymnotiformes (Characiformes + Siluriformes))). Also using the mitochondrial genome, Peng *et al.* 2006a composed a DNA data matrix of 11,100 base pairs from all 13 protein-coding mitochondrial genes (excluded the D-loop, the 22 tRNAs and the 12S and 16S rRNAs) for 13 otophysan fish plus the gonorynchiform *Chanos chanos*, used as outgroup. The phylogenies were inferred by a Bayesian approach from a concatenated mtDNA protein-coding gene sequences and the resultant topology is (Gonorynchiformes (Cypriniformes (Siluriformes (Characiformes + Gymnotiformes)))). Lavoué *et al.* (2005) addressed the systematics of gonorynchiform fishes based on a final matrix with 14,025 base pairs of the mitochondrial genome (ND6, stop codons and regions of difficult alignment, i.e. with posterior probabilities ≤ 50%, were excluded from the data set) for 40 terminal taxa. A phylogenetic hypothesis was generated by Bayesian analysis of two character matrices and alternative tree topologies were compared with the resultant Bayesian tree with the aid of the likelihood based SH test. Among the taxa used in the study, there were 20 ostariophysan genera. The paper was directed to gonorynchiform phylogeny, but the results depict the following topology for ostariophysans:

(Gonorynchiformes (Cypriniformes (Gymnotiformes (Siluriformes + Characiformes). Finally, Peng *et al.* 2006b examined 6,918 base pairs of the mitochondrial genome of 19 ostariophysans (+ 11 other taxa from 8 additional fish orders). The alignment of the data set was done with Clustal X with default parameters and subsequently inspected by eye for obvious misalignments. From the complete mitochondrial genome, the authors used only protein-coding genes and excluded ND3, ND4L, and ATP8 due to their small size, and the ND6 due to its unreliable alignment. From the remaining matrix, 3rd codon positions were also excluded. Phylogenies were inferred by maximum likelihood and Bayesian methods. The final topology presented by the authors depicts three interesting results associated to ostariophysans: first, the Gonorynchiformes appear as sister groups of the Clupeiformes (Ostariophysi non monophyletic); second, Characiformes are also non-monophyletic, with *Chalceus macrolepidotus* inserted between the two other gymnotiform genera included and third, the (Gymnotiformes + Characiformes) clade is the sister group of Siluriformes. The final topology is ((Gonorynchiformes + Clupeiformes) (Cypriniformes (Siluriformes (Characiformes + Gymnotiformes)))). It is worth to note that the results of all molecular studies mentioned above agree that Cypriniformes is the sister group of (Characiformes + Siluriformes + Gymnotiformes), i.e., in conformity with morphological characters, the Characiphysi + Siluriphysi *sensu* Fink and Fink (1981, 1996), was always recovered as a monophyletic assemblage by DNA datasets.

The results of the molecular phylogenies mentioned above recovered all the three possible alternatives for the branching order among the Characiphysi + Siluriphysi, i.e.: (Characiformes (Gymnotiformes + Siluriformes)), (Siluriformes (Characiformes + Gymnotiformes)), and (Gymnotiformes (Characiformes + Siluriformes)). This may, at first glance, shed undeserved doubt over the molecular systematic methods used in the cited studies or over the evolutionary information content of the DNA matrices. However, it is argued here that the contradictory results are, greatly, products of methodological constraints related to sampling bias. The possible sources of errors associated with phylogenies estimated from molecular data have been addressed extensively in the literature (see Felsenstein 2004), but one particular methodological problem that could be promoting these conflicting results is the fact that in all studies above, only a small number of taxa were used to genetically characterize each one of the orders used. The only relatively well-represented clade in those studies was the order Characiformes in Ortí and Meyer's (1996) work. Those authors used 3 cypriniform, 13 characiform, 4 siluriform, and 2 gymnotiform genera in their study with ependymin, whereas Ortí and Meyer (1997) used 2, 11, 4, and 3 genera respectively with the 12S and 16S rRNA plus 2 gonorynchiform genera as outgroups. Dimmick and Larson (1996) used 1 gonorynchiform, 2 cypriniform, 2 characiform, 3 siluriform, and 1 gymnotiform genera (2 species of *Apteronotus*) in their estimations. Saitoh *et al.* (2003) used 2 gonorynchiforms, 7 cypriniforms, 2 characiforms, 2 gymnotiforms and 2 siluriforms. Peng *et al.* (2006a) included 1 gonorynchiform, 5 cypriniforms, 2 characiforms, 2 gymnotiforms and 4 siluriforms, while Peng *et al.* (2006b) used 2, 8, 2, 2 and 5 taxa from each order, respectively. Finally, Lavoue *et al.* (2005) included 7 gonorynchiforms, 7 cypriniforms, 2 characiforms, 2 gymnotiforms and 2 siluriforms. Phylogenetic hypotheses based on few genera representing clades as large and diverse as the case of the otophysan orders may produce awkward topologies due to possible misrepresentation of the genetic diversity within each clade and to long branch attraction (Felsenstein 2004). When a larger number of representatives of each clade are used, the genetic variability sampled will be larger, the chance of breaking up long branches will be greater, and the genetic parameters found are more likely to represent the "average" values for a given clade, as result of common evolutionary history. This is particularly important as one uses evolutionary models based on several investigational assumptions to interpret the differences currently detected in DNA sequences.

In summary, despite the fact that there seems to be a tendency favoring Fink and Fink's topology, molecular information alone has not produced a consensual hypothesis. Therefore, it seems worthwhile to test Fink and Fink's phylogenetic hypothesis again with a larger molecular data matrix, including the largest variability possible found within each order. In order to reach this potential variability for this study, the following steps were carried out: first, the number of intra-ordinal clades was increased in the data matrix, including the most distantly related clades, according to existing morphological and molecular hypotheses; and second, for each order, representatives from the widest geographical range of distribution were also included in the sampling.

Methodological Approach

The taxa used in the present study and the sources of their mitochondrial DNA sequences are depicted in Table 13.1. Most of the sequences were obtained from published studies through GenBank, whereas a few others were specially generated for the present study. Fifteen genera representing each one of the otophysan orders were included in the data matrix, and two gonorynchiform genera were used as outgroups. In order to assess the potential molecular diversity and the geographical range within each of the ingroup orders, at least one representative was sampled from each continent that respective order currently inhabits (Table 13.1).

DNA was extracted, purified, amplified and sequenced according to protocols described elsewhere (Meyer *et al.* 1993, Alves-Gomes *et al.* 1995, Alves-Gomes and Hopkins 1997), following standardized procedures. Two segments of the mitochondrial genome were used, namely two fragments of the 12S and 16S rRNA genes. The 12S primers were modified from Kocher *et al.* (1989) and the sequences for the 16S primers were obtained from Palumbi *et al.* (1991). The sequences are: **12S:** L1091: 5'-AAACTGGGATTAGATACCCCACTAT-3' and H1478: 5'- GAGGGT GACGGGCGGTGTGT-3'; **16S:** 16Sa-L: 5'-CGCCTGTTTATCAAAAACAT-3', and 16Sb-H:5'-CCGGTCTGAACTCAGATCACGT-3'. The positions of the 3' end of each primer in the L strand of the human mitochondrial genome (Anderson *et al.* 1981) are, respectively, 1091, 1478, 2501 and 3059.

Sequences were aligned according to the secondary structure of their respective molecules, as described by Alves-Gomes *et al.* (1995). By comparing the individual sequences of both rRNA fragments to their respective secondary structure (see figs. 3A and 3B in Alves-Gomes *et al.* 1995), regions forming stems and loops were defined, and base pair formation was prioritized for the stems. Minor adjustments within the unpaired regions were done by eye, with the criteria that transitions (TS) are more costly than transversions (TV), and the latter more costly than gaps (GP). Within the molecules, short segments of generally unpaired sites show flagrant length variation, which is regularly associated with problematic alignment. Doubtful alignments compromise the establishment of homology among sites and become a source of noise for phylogenetic analysis. Therefore, in this study, all the sites at which alignment was ambiguous were excluded from all the analyses. The final data matrix consisted of 294 bases of the 12S ribosomal RNA (rRNA) and 407 bases of the 16S rRNA from 62 taxa representing all five Ostariophysi orders. The aligned data set can be obtained from the author upon request.

For all analyses reported subsequently, the 12S and 16S rRNA sequences were combined into a unique data matrix, since there is no evidence that these two genes are evolving under different constraints, and their base compositions show very little variation within Otophysi (Alves-Gomes *et al.* 1995).

Phylogenetic Analysis

For estimation of phylogenetic relationships, two methodological approaches were used: maximum likelihood (ML) and maximum parsimony (MP). Hypotheses based upon each methodological approach were generated with the program PAUP* version 4.0b10 (Swoford 2002). For maximum parsimony, several cost matrices with different weights for TS in relation to TV, either considering GP as a fifth character state or as missing information, were used. When gaps were treated as a missing character, the following weighting schemes were used: in 'TS1TV1', the same cost of one step was given to each TS and each TV, and under 'TS1TV3' the cost for TS was 3 times less than the cost for each TV. In the other approach, considering

Table 13.1 List of taxa used in the present study, their current geographical occurrence, and the source of their DNA sequences. S.A., South America, N.A., North America, C.A., Central America, Afr., Africa, S. Asia, South Asia, Col., Colombia, Aust., Australia.

Siluriformes	Family	Abbr.	Occurrence	DNA source
Dyplomystes	Dyplomistidae	Dyp	S.A.	This study[1]
Peckoltia	Loricariidae	Pec	S.A.	This study[1]
Helogenes	Helogeneidae	Hel	S.A.	GB: AF072136, AF072150
Hemicetopsis	Cetopsidae	Hem	S.A.	GB: U15265, U15241
Malapterurus	Malapteruridae	Mal	Africa	GB: U15261, U15237
Synodontis	Mochokidae	Syn	Africa	This study[2]
Liocassis	Bagridae	Lio	Africa	This study[1]
Schilbe	Schilbeidae	Sch	Afr./S. Asia	This study[3]
Pangasius	Pangasiidae	Pan	Asia	This study[2]
Silurus	Siluridae	Sil	Eurasia	This study[4]
Kryptopeterus	Siluridae	Kry	Eurasia	This study[5]
Ictalurus	Ictaluridae	Ict	N.A.	This study[6]
Tandanus	Plotosidae	Tan	Coastal Aust.	This study[7]
Plotosus	Plotosidae	Plo	Coastal Japan	This study[8]
Arius	Ariidae	Ari	Coastal Col.	This study[1]

Gymnotiformes	Family	Abbr.	Occurrence	DNA source
Sternopygus 1	Sternopygidae	Spy1	S.A.	GB: U15252, U15228
Sternopygus 2	Sternopygidae	Spy2	S.A.	GB: AF072139, AF072153
Eigenmannia	Eigenmanniidae	Eig	S.A.	GB: U15269, U15245
Archolaemus	Eigenmanniidae	Arc	S.A.	GB: AF072149, AF072163
Rhabdolichops	Eigenmanniidae	Rha	S.A.	GB: U15258, U15234
Apteronotus	Apteronotidae	Aal	S.A.	GB: U15275, U15226
Adontosternarchus	Apteronotidae	Ado	S.A.	GB: U15274, U15250
Sternarchogiton	Apteronotidae	Ste	S.A.	GB: U15255, U15231
Gymnorhamphichthys	Rhamphichthyidae	Gyr	S.A.	GB: U15267, U15243
Hypopygus	Rhamphichthyidae	Hyp	S.A.	GB: U15264, U15240
Rhamphichthys	Rhamphichthyidae	Rhp	S.A.	GB: U15257, U15233
Steatogenys	Rhamphichthyidae	Ste	S.A.	GB: U15253, U15229
Brachyhypopomus	Hypopomidae	Bra	S.A.	GB: U15262, U15238
Electrophorus	Electrophoridae	Ele	S.A.	GB: U15268, U15244
Gymnotus	Gymnotidae	Gym	S.A.	GB: U15266, U15242

Characiformes	Family	Abbr.	Occurrence	DNA source
Hoplias	Erythrinidae	Hop	S.A.	GB: U33976, U34013
Steindachnerina	Curimatidae	Ste	S.A.	GB: U33986, U34023
Chilodus	Chilodontidae	Chi	S.A.	GB: U33989, U34027
Characidium	Crenuchidae	Cha	S.A.	GB: U33828, U34030
Pyrrhulina	Lebiasinidae	Pyr	S.A.	GB: U33980, U34017
Serrasalmus	Characidae	Ser	S.A.	GB: U33560, U33592
Colossoma	Characidae	Col	S.A.	GB: U33582, U33617
Phenacogrammus	Characidae	Phe	S.A./C.A	GB: U33830, U33996
Leporinus	Anostomidae	Lep	S.A.	GB: U34031, U34026
Boulengerella	Ctenolucidae	Bou	S.A./C.A	GB: U33978, U34015
Carnegiella	Gasteropelecidae	Car	S.A./C.A	GB: U33983, U34020
Distichodus	Distichodontidae	Dis	Afr.	GB: U33827, U33994
Citharinus	Citharinidae	Cit	Afr.	GB: U33826, U33993
Alestes	Characidae	Ale	Afr.	GB: U33829, U33995
Hepsetus	Hepsetidae	Hep	Afr.	GB: U33825, U33992

Cypriniformes	Family	Abbr.	Occurrence	DNA source
Crossostoma	Balitoridae	Cro	Asia	GB: M91245
Cyprinus	Cyprinidae	Cyp	Asia/Afr.	GB: X61010
Tanichthys albonubes	Cyprinidae	Tan	Afr./Eurasia	GB: U21379, U21387
Danio malabaricus	Cyprinidae	Dan1	Afr./Eurasia	GB: U21376, U21384
Danio	Cyprinidae	Dan2	Afr./Eurasia	GB: U21375, U21370
Phoxinus 1	Cyprinidae	Pho1	N.A./Eurasia	GB: AF038490
Phoxinus 2	Cyprinidae	Pho2	N.A./Eurasia	GB: AF038493
Gila	Cyprinidae	Gil	N.A.	GB: AF038481
Notropis	Cyprinidae	Not	N.A.	GB: AF023195, AF038486
Relictus	Cyprinidae	Rel	N.A.	GB: AF038496
Abramis	Cyprinidae	Abr	N.A.	GB: AF038468
Luxilus	Cyprinidae	Lux	N.A.	GB: U09469
Leuciscus	Cyprinidae	Leu	N.A.	GB: AF038484
Platygobio	Cyprinidae	Pla	N.A.	GB: AF023197, AF038491
Snyderichthys	Cyprinidae	Sny	N.A.	GB: AF023201, AF038498
Gonorynchiformes	Family	Abbr.	Occurrence	DNA source
Kneria	Kneriidae	Kne	Africa	GB: U33990, U34028
Parakneria	Kneriidae	Par	Africa	GB: U33991, U34029

Note: [1]: Tissues provided by John Lundberg: *Diplomystes nahuelbutaensis*, voucher ANSP 180476; *Peckoltia cf. vittata*, CCF-96-341, ANSP uncat.; *Leiocassis sp.*, ANSP uncat.; *Arius sp*, ANSP uncat. [2]: Tissues obtained from aquarium trade fish. No voucher. [3]: Tissue provided by A.S. Golubtsov and Walter Dimmick, voucher A16/1415T Museum of Natural History, Divisions of Fishes, The University of Kansas, Lawrence, Kansas. [4]: Tissue obtained by Walter Heiligenberg, vouchered as a picture JAG#162, LFCE/INPA. [5]: *Kryptopterus sp* voucher JAG#136, LFCE/INPA. [6]: Tissue obtained from Glenn Northcutt, no voucher. [7]: Tissue obtained from Paul Manger, no voucher. [8]: Tissue provided by Shosaku Obara, no voucher. GenBank Accession Numbers (12S and 16S rRNA, respectively): *Diplomystes*: EU307870, EU307882; *Peckoltia*: EU307864, EU307876; *Synodontis*: EU307861, EU307873; *Leiocassis*: EU307866, EU307878; *Schilbe*: EU307859, EU307871; *Pangasius*: EU307869, EU307881; *Silurus*: EU307865, EU307877; *Kryptopeterus*: EU307860, EU307872; *Ictalurus*: EU307867, EU307879; *Tandanus*: EU307863, EU307875; *Plotosus*: EU307868, EU307880; *Arius*: EU307862, EU307874.

GP as a fifth character, two matrices were also used: 'TS1TV1GP1' and 'TS1TV3GP3'. In the former weighting scheme, gaps were counted as one step and in the second treatment GP were given the same weight as TV, i.e., 3 times the cost of TS. In all approaches, GP costs were independent of their length, i.e., insertion/deletion events (indels) in two adjacent sites were counted as two GP. Only one taxon (*Liocassis*) had gaps in adjacent sites when the variable (potentially informative) sites were considered. Such differential weighting schemes were used for the following reasons: first, to compensate the possible saturation effects of TS in the data set; second, to test the phylogenetic information contained in the three different types of mutational events (TS, TV and GP); and third, to verify which groupings were consistently recovered regardless of the weight matrix used. For each cost matrix in PAUP, 100 heuristic searches were performed with the following settings of the program: starting trees were constructed with random addition of taxa; the tree bisection-reconnection (TBR) algorithm was used as the swapping algorithm, and branch swapping was performed in all starting trees with the "steepest descent" option selected. The best tree was kept at each step, with the "collapse" option in effect. Uninformative and invariant characters were ignored in parsimony analyses. The remaining parameters were set to default values in PAUP*.

The first results obtained from parsimony analysis were used to orient further searches in PAUP*. For instance, as will be reported in detail below, MP revealed that some genera have a peculiar, erratic behavior in the resulting phylogenetic trees, depending upon the weighting scheme used. More specifically, as long as the characiform genera *Distichodus* and *Citharinus*, and the gymnotiform genera *Electrophorus* and *Gymnotus*, remain in the data set, doubtful topologies are generated. Although less dramatically, the siluriform genus *Peckoltia* also produces unexpected topologies. These lineages either attract or are attracted to other branches of the trees and disrupt otherwise well-supported clades. Although some phylogenetic uncertainty may still exist in relation to the two characiform genera Distichodus and Citharinus due to their morphological peculiarities (P. Buckup, pers. com.), few, if any, would disagree that *Electrophorus* and *Gymnotus* belong into Gymnotiformes, and that *Peckoltia* is a siluriform. A more elaborate interpretation about why the presence of these genera in the data matrix produces improbable results is presented in the next section. However, in order to test the effect of these fish in the resulting topologies under different weighting, a series of complementary analyses were performed, using the same weighting schemes previously defined, and progressively excluding the genera mentioned above from the data matrix (see results). In order to access the robustness of the phylogenetic signal in the data, 100 bootstrap replicates (Felsenstein 1985) were performed for the TS1TV1GP1 and TS1TV3GP3 cost matrices after the five genera mentioned before were removed from the data matrix. Other matrices were not bootstrapped due to computational time constraints. In the bootstrap replicates, 10 heuristic searches were performed with 10 random additions of taxa at each replicate.

The results obtained from parsimony analysis were also used to guide the likelihood searches in PAUP*, once the computational algorithms associated to ML analysis can be significantly more time consuming than MP. Because of computational time constraints, a relatively simple evolutionary model was used to estimate phylogenies. The substitution model selected was a variant of HKY85, with two types of substitution defined. The transition/transversion ratio was 2. The nucleotide frequencies were calculated from the data matrix, the distribution of rates at variable sites was set to equal, and there was no predefined proportion of invariable sites. Starting trees were generated with random stepwise addition of taxa.

For ML, GP were treated as missing characters. Ten searches were performed for the complete data set and another ten searchers were done for the reduced data matrix, i.e., excluding the same six genera as in MP. All other ML parameters were set to default in PAUP*.

Results

A total of 701 sites were aligned from the combined 12S and 16S rRNA sequences for the 62 taxa employed. Insertion/deletion events (gaps) had to be assigned a minimum of four times for one taxon (*Kneria*, Gonorynchiformes), and a maximum for 16 positions for another single taxon (*Luxilus*, Cypriniformes). The nucleotide composition showed little range among the taxa, in agreement with what was reported previously by Alves-Gomes *et al.* (1995) for gymnotiforms and siluriforms alone. The range of nucleotide frequency as the percentage of each nucleotide for all taxa was (mean % ± SD): A = 29.02 ± 0.83; G = 24.95 ± 0. 87; C = 24.49 ± 0.79; T = 21.51 ± 0.77. There were 260 informative sites when GP were considered as a missing character, and 6 sites became informative when GP were treated as fifth character in maximum parsimony analyses when all the 62 taxa were considered in the data set.

When no differential weights were assigned for TS, TV and GP, i.e., under the TS1TV1 and TS1TV1GP1 weighting schemes, the larger number and the less resolved trees were obtained (188 and 974 trees respectively). For these two matrices the strict consensus trees depict, in the ingroup, only Cypriniformes and Characiformes as monophyletic assemblages. In the 50% majority rule consensus tree, Siluriformes also become monophyletic for both weighting matrices, but the Gymnotiformes remain separated over the resulting trees. Further, Siluriformes and Characiformes are recovered as sister groups in the 50% majority rule consensus trees. In these consensus topologies, *Gymnotus* is "attracted" to the base of the tree, as the sister group of all remaining ostariophysans excluding the gonorynchiforms, whereas *Electrophorus* is placed as the sister group of Cypriniformes. The other gymnotiforms are distributed into three clades, between the Cypriniformes and a (Siluriformes + Characiformes) clade. Despite the fact that *Electrophorus* is a very peculiar fish, with several morphological, behavioral and molecular autapomorphies, few would disagree that this fish is a gymnotiform. Although less autapomorphic, the same is valid for *Gymnotus*. Probably because of constraints related to "*tempo* and *modo*" of the evolutionary changes in their 12S and 16S rRNA, these two fish caused an enormous fragmentation of the phylogenetic structure within the Gymnotiformes. Long branches associated with these gymnotiforms were already reported in the literature (Alves-Gomes 1995, Alves-Gomes *et al.* 1995). Such a situation provokes a significant amount of uncertainty over the other relationships established in the rest of the tree. In order to avoid a detailed and redundant discussion of each strict and majority rule consensus tree obtained for each weighting scheme, for the remainder of the chapter, unless explicitly stated, I only will discuss the topologies obtained by the 50% majority rule consensus trees.

As the weight of TV and GP were increased in relation to TS, for the TS1TV3 and TS1TV3GP3 matrices, the resulting trees decrease in number (38 and 36 respectively) and gain phylogenetic additional structure. Cypriniformes is depicted as the sister group of (Characiphysi + Siluriphysi), *Gymnotus* and *Electrophorus* are grouped as sister groups, as expected, but as sister group of all the remaining Characiphysi + Siluriphysi. Siluriformes and Characiformes remain monophyletic and as sister groups. The gymnotiforms remain distributed into three clades placed in between (*Electrophorus* + *Gymnotus*) and the (Characiformes + Siluriformes) clade.

Based upon these results, an alternative approach was taken for the phylogenetic analysis. A second data matrix was assembled, by excluding the genera *Electrophorus* and *Gymnotus* from the original data set. The same weight schemes were used under maximum parsimony. For TS1TV1 (841 tree found) and TS1TV1GP1 (32 trees found), Cypriniformes is the sister group of (Characiformes + Siluriformes + Gymnotiformes); Characiformes and Siluriformes remain, and Gymnotiformes becomes monophyletic. However, departing from the Fink and Fink hypothesis, Siluriformes and Characiformes are still depicted as sister clades. When the TV and GP are weighted three times the value given to TS (70 trees found for TS1TV3 and two trees found for TS1TV3GP3), the hypothesis proposed by Fink and Fink (1981, 1996) is recovered with one exception: the characiform genera *Distichodus* (Distichodontidae) and *Citharinus* (Citharinidae) are depicted as monophyletic clade, sister group of the Siluriphysi (Gymnotiformes + Siluriformes). Also, an unstable position for the South American catfish genus *Peckoltia*, within the Siluriformes, is noticed, as this genus *Peckoltia* "jumps" from being the most basal catfish lineage to an interior node, as sister group of *Silurus glanis*, as the weight of TV and GP goes from one to three.

If the two characiforms (*Distichodus* and *Citharinus*) are excluded from the matrix and the same searches are performed, there is a further approach of the resulting topologies from Fink and Fink's hypothesis. Only under TS1TV1(277 trees found) the Siluriformes continue to appear as sister group of Characiformes, in 70% of the most parsimonious trees. When gaps are included and/or the weight of transitions is lowered in relation to TV and GP, a (Siluriformes + Gymnotiformes) sister-group relationship is found in 80% of the most parsimonious trees under TS1TV1GP1 (100 trees found), in 86% for TS1TV3 (51 trees found), and in 100% for TS1TV3GP3 (four trees found). If *Peckoltia* is excluded from the data matrix as well, Siluriformes and Gymnotiformes are depicted as sister groups in 69% of the TS1TV1 (405 trees), in 97% of the TS1TV1GP1 (693 trees), and in 100% of the TS1TV3 (45 trees)and TS1TV3GP3 (88) most parsimonious trees.

In summary, under MP, as the relative weight of TV is increased in relation to TS, and GP are included in the analyses, there is an improvement in the resolution of the resultant trees. Furthermore, there is a clear disruption in the phylogenetic signal of the data matrix as far as the following genera are kept in the database: *Electrophorus, Gymnotus, Distichodus, Citharinus* and to a lesser extent *Peckoltia*. As these problematic, noise-generators taxa are progressively removed from the data matrix, then the resultant topologies converge to corroborate the hypothesis originally proposed by Fink and Fink in 1981: (Gonorynchiformes (Cypriniformes (Characiformes (Siluriformes + Gymnotiformes)))).

The ML results conform greatly with the topologies generated by parsimony. When the complete data set is analyzed, the resulting tree places *Gymnotus* as sister group of the Otophysi, *Electrophorus* as sister group of Cypriniformes, and *Sternopygus* as sister group of the (Siluriformes + Characiformes) clade. In this tree, Cypriniformes, Characiformes and Siluriformes are recovered as monophyletic groups. *Cyprinus* is the sister group of the remaining cypriniforms, the (*Distichodus* + *Citharinus*) clade is the sister group of the remaining characiforms, and *Peckoltia* and *Dyplomystes* are the two most basal siluriforms. The remaining gymnotiforms are grouped into three clades, positioned between the (Electrophorus + Cypriniformes) and the (Sternopygus (Siluriformes + Characiformes) clades. When the six genera mentioned previously are excluded from the matrix, the tree with best likelihood recovers Fink and Fink's hypothesis.

From the results above, it seems reasonable to assume that there is a not fully understood phenomenon taking place in the mitochondrial sequences of some representatives of the Characiformes and siluriphysan orders included in the original data matrix. The 12S and 16S rRNA mtDNA sequences of *Electrophorus, Gymnotus, Distichodus, Citharinus* and to a lesser extent *Peckoltia* and *Diplomystes* appear to have some kind of still uncategorized property that is generating noise in the phylogenetic signal. This assertion is different from concluding that these genes have not enough phylogenetic information to resolve the branching order of the Ostariophysi. The most obvious explanation for the observed results is the long branch attraction (Felsenstein 2004) possibly associated to these lineages. Long branches can result from either long periods of independent evolutionary history (old clades) or rapid evolutionary rates. Interestingly, and perhaps not coincidentally, all the genera listed in the previous paragraph, as will be discussed in greater detail below, are believed to represent ancient lineages within their respective orders (Mago-Leccia 1976, 1978, Vari 1979, Fink and Fink 1981, Lundberg 1993, Nelson 2006, Alves-Gomes *et al.* 1995, de Pinna 1998) and, therefore, may represent taxa that have accumulated homoplasic substitutions in their respective sequences, especially at those functionally unconstrained sites of the genes utilized.

The present chapter does not intend to discuss in detail the molecular systematic or the phylogenetic relationships obtained within each respective ostariophysan order. The selection of taxa for this study was focused on addressing the branching order among the Otophysi and, although 15 taxa are sufficient to include representatives of all gymnotiform families in the data set, it is still a reduced number of representatives for specious clades such as the other otophysan orders.

Figure 13.2 shows the 50% majority rule consensus trees from all weighting schemes used in MP, once the five genera previously mentioned in the text (*Electrophorus, Gymnotus, Distichodus, Citharinus* and *Peckoltia*) were excluded from the analyses. Their respective positions in the figure were defined according to complementary morphological (in the case of gymnotiforms and characiforms) and molecular information (for siluriforms). The topology shown in Fig. 13.2 is essentially the most conservative tree topology possible for the results obtained, as it is a consensus of 1,231 trees that includes all the trees found under the less efficient cost matrices (those excluding gaps and using the same weight for the different types of mutations), which generated less resolved topologies. Notwithstanding, Fig. 13.2 shows a reasonable degree of structure within each order, and despite the fact that discussion of inter-generic relationships within each order is not the main concern of the present chapter, some of the relationships found will be emphasized, as they stand as relevant aspects for the biogeographical argumentation developed subsequently. In particular, I want to point out aspects of the trees associated to gymnotiforms and siluriforms, due to their sister-group relationship.

The phylogeny of the order Gymnotiformes and its constituting groups has been addressed elsewhere (Mago-Leccia 1978, 1994, Alves-Gomes *et al.* 1995, Alves-Gomes 1998, 1999, Albert and Campos-da-Paz, 1998, Albert 2001, Campos-da-Paz and Albert 1998, Albert and Crampton 2005, Crampton and Albert 2006, Hulen *et al.* 2005, Triques 1993, 2005, de Santana 2007, Sullivan 1997), and from these several studies published, the enduring and most relevant disputes can be narrowed to the phylogenetic position of three clades: (*Electrophorus* + *Gymnotus*), *Sternopygus* and the tribe Steatogenini (genera *Hypopygus, Stegostenopus* and *Steatogenys*). The results obtained here revisit previous studies based upon molecular data (Alves-Gomes *et al.* 1995, Alves-Gomes 1998, 1999, Sullivan 1997, Schmitt 2005) and still disagree with morphology-based hypotheses for these three clades. In relation to these discrepancies, first, I would argue that there still is no sufficient evidence (molecular or morphological) to establish with confidence the phylogenetic position of (*Electrophorus* + *Gymnotus*) within the order (see Discussion). Likewise, despite the recurrent proposition based upon morphological characters, grouping *Sternopygus* with eigenmanniids (Albert 2001, Triques 1993, Albert and Campos-da-Paz 1998, Albert and Crampton 2005, Crampton and Albert 2006), I consider that morphological evidence and the related character-state interpretation and codification are controversial, lack phylogenetic density and are still not convincing, as better explicated in a couple of papers (Alves-Gomes *et al.* 1995, Alves-Gomes 1998,1999). Instead, I favor the hypothesis of an independent evolutionary lineage for *Sternopygus* within the order, as recovered from molecular data and other complementary evidences including morphology and physiology (Alves-Gomes *et al.* 1995, Alves-Gomes 1998,1999). None of the resultant topologies in the present study, obtained with any of the data sets and/or weighing scheme used, grouped *Sternopygus* with the eigenmanniids (Fig 13.2). Concerning the position of the Tribe Steatogenini within the order, authors interpreting morphological characters have positioned this clade within the Family Hypopomidae (Mago-Leccia 1976, 1978, 1994, Triques 1993, Albert 2001, Albert and Crampton 2005, Crampton and Albert 2006), whereas DNA data, including mitochondrial and nuclear sequences, places the clade within the Rhamphicthyidae (Alves-Gomes *et al.* 1995, Sullivan 1997, Schmitt 2005). The results of this study strongly corroborate a monophyletic (Hypopomidae + Rhamphicthyidae) but places *Brachyhypopomus* (an undeniable hypopomid) as sister taxa of *Gymnorhamphicthys* (an undeniable rhamphichthyid). This result is not a surprise, because several variable segments containing relevant phylogenetic information for more closely related taxa were removed from the data matrix due to alignment problems when the 62 terminal taxa were considered. If those regions were included in the analyzes, additional phylogenetic information would become available to sort out the relationship among these genera, as shown by Alves-Gomes *et al.* (1995) and Alves-Gomes (1998, 1999), using the same genes. Nevertheless, for the purpose of the present study, the exact branching order within the Gymnotiformes is less relevant than the branching order between the otophysan orders.

For catfishes, the results obtained in the present study are greatly concordant with the previous results obtained by Alves-Gomes (1995) and the recent and comprehensive work of Sullivan *et al.* (2006), using the nuclear rag1 and rag2 genes. It is worth to note the following basic agreements: the most basal catfishes lineages occur in South America (Diplomystidae and Loricariidae), and there is no sister group relationship between African and South American genera, i.e. the split of the Africa and South America apparently did not represent a vicariant event for catfishes. *Malapterurus* and *Synodontis*, two electrogenic African catfish, are shown to be closely related in the results of both molecular studies. In the present study *Pangasius*, a marine clade stands between the basal South American and the remaining catfish. Other marine catfishes are intermingled among the remaining clades. Sullivan *et al.* (2006) place these clades with marine or salt water tolerant representatives in a large politomy within the Siluroidei (the remaining catfish after Diplomystes + Loricarioidei).

In relation to the intra-ordinal phylogeny of the other otophysan orders, only some aspects of the topologies obtained from mitochondrial sequences will be emphasized alongside the discussion, as they stand as important information to subsidize the biogeographical hypotheses developed subsequently.

In summary, evaluating the molecular evidence provided by the 12S and 16S mitochondrial gene under the surveillance of complementary information coming from additional sources such as morphological studies and under the theoretical framework associated to the molecular evolution, it is possible to interpret and to propose alternative solutions for the otherwise unrealistic hypotheses initially obtained. I'll argue, therefore, that 12S and 16S mitochondrial phylogenies greatly corroborate Fink and Fink's 1981 hypothesis, and for the remaining sections of this chapter I assume that the correct branching order within Ostariophysi is: (Gonorynchiformes (Cypriniformes (Characiformes (Siluriformes + Gymnotiformes)))). The final topology taking into consideration all the issues raised up to now is depicted in Fig. 13.2.

Discussion

Otophysan Phylogeny According to the 12S and 16S rRNA Sequences: Is It Reliable?

It has been suggested in the literature that the 12S and 16S rRNA may have been saturated with homoplasic substitutions and therefore does not contain enough phylogenetic information to elucidate the branching order within the Ostariophysi (Ortí and Meyer 1997) or more ancient cladogenetic events

---

Fig. 13.2 Phylogenetic hypothesis based upon the 50% majority rule consensus trees from all weighting schemes used in MP, once the five genera mentioned in the text were excluded from the analyses. The phylogenetic position of the five taxa (*Gymnotus, Electrophorus, Peckoltia, Distichodus* and *Citharinus*) in the present tree were defined *a posteriori* (dashed lines), according to morphological and/or molecular data. The intra-ordinal relationships are not the main focus of the present study, but the following results should be noted: *Sternopygus* represents a separate lineage in relation to eigenmanniids; basal catfish occur in South America; *Malapterurus* and *Synodontys*, two electrogenic African catfish are depicted as sister groups; *Pangasius*, a marine catfish separates the South American basal genera from the remaining catfish clades; other representatives of clades that include marine or salt water tolerant catfish lineages (*Arius, Tandanus* and *Plotosus*) are distributed over the tree; the North American *Ictalurus* appears closely related to marine genera; no sister group relationship can be established between African and South American catfish lineages; well supported sister group relationships are established between African and South American characiforms (*Hoplias + Hepsetus* and *Phenacogrammus + Alestes*); Asian cypriniforms are basal. The two numbers over the branches represent, respectively, the bootstrap values obtained for the TS1TV1GP1 and TV1TV3GP3 matrices, after excluding the genera listed above. Missing values indicate that a particular branch was not recovered in more than 50% of the bootstrap results.

Fig. 13.2 contd...

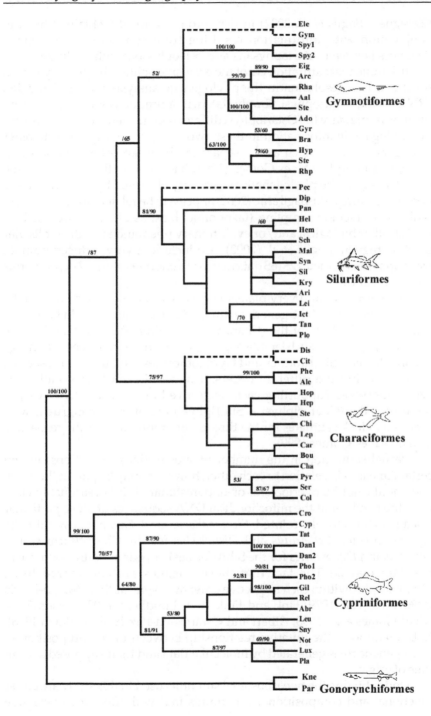

(Zaragüeta-Bagils *et al.* 2002) in fish. From the results obtained here, in conjunction with information resulting from comparisons of genetic distances between ostariophysans and osteoglossomorphs (Alves-Gomes 1999), I argue that saturation is not the exclusive neither the primary reason for the lack of resolution in the phylogenetic analysis with 12S and 16S rRNA for the Ostariophysi. Simply stated, if saturation was achieved within the time frame since the Ostariophysi differentiation, then we would observe no phylogenetic information in these two genes to clearly separate older cladogenetic events such as the separation between ostariophysans and osteoglossomorphs. Alves-Gomes (1999) has showed that this is not the case. Therefore, the phylogenetic uncertainty observed by Ortí and Meyer (1997) and Zaragueta-Bagil *et al.* (2002) appears to be related to other reasons, which may include sampling (taxonomic) bias, as already discussed, as well as other biological and/or evolutionary phenomena at the molecular level. Zaragueta-Bagils *et al.* (2002) also include a very reduced number ostariophysans in their 12S16S data matrix: 1 gonorynchiform, 2 cypriniforms and 1 siluriform.

An alternative and very plausible reason that may have provoked the awkward position of the genera mentioned in the results would be a rapid radiation in the base of the (Characiphysi + Siluriphysi) which, in turn, would imply short branch lengths separating not only each respective lineage (Characiformes, Siluriformes and Gymnotiformes) but also the most basal lineages within the different orders. This would imply few molecular synapomorphies supporting each respective branch in the tree. A short period for the (Characiphysi + Siluriphysi) radiation is congruent with molecular clock estimates for the time of differentiation for the respective orders (Alves-Gomes 1999).

Partial saturation in the number of substitutions (mutations) within certain lineages in relation to each other, however, may be part of the cause and should not be overlooked, or underestimated. It is reasonably clear from the results that the mitochondrial DNA sequences of those particular lineages (the closely related *Electrophorus* and *Gymnotus* within the Gymnotiformes, *Citharinus* and *Distichodus* for the Characiformes, and *Peckoltia* and *Diplomystes* for the Siluriformes) appear as source of noise in the data matrix. The fact that these lineages are considered to be plesiomorphic within each respective order by several authors (Mago-Leccia 1976, 1978, Vari 1979, Fink and Fink 1981, Lundberg 1993, Nelson 2006, Alves-Gomes *et al.* 1995, Albert and Campos-da-Paz 1998, de Pinna 1998). calls attention to the putative relationship between uncertain position in phylogenetic trees generated by molecular data and their supposed ancient time of origin.

Undoubtedly, the two ribosomal sub-units used in this study are under functional and compositional constraints that will affect the nucleotide positions that are free to undergo mutations without altering the function of the molecule. These two genes possess secondary structures that are conserved between fishes and tetrapods, yet some regions (the loops) contain length variation to such an extent that they are not unambiguously aligned among the otophysans (Alves-Gomes *et al.* 1995). Since each site has only five possible character states (A, G, C, T, and GP), it seems reasonable to assume that ancient lineages will tend to accumulate multiple mutations in the rapidly evolving sites that, in turn, might erase the original phylogenetic information. Given enough time, the taxa that have differentiated long ago will start to accumulate homoplasic substitutions in the pertinent sites. The same would also occur if a particular lineage had a higher mutational rate than others. Molecular evidence alone does not allow one to differentiate between these two (not exclusive) possibilities. Morphological and fossil evidence, therefore, is essential for addressing this problem and should be used as often as possible in conjunction with molecular data.

At the moment, there is no evidence suggesting that any particular clade within the Otophysi might have been evolving at a faster rate than the others. A possible exception might be the gymnotiform electric eel (*Electrophorus electricus*), which can discharge up to 600V during aggressive and defensive encounters. The putative role of electric potentials as a mutagenic agent for the eel's DNA is not known but should not be completely disregarded, especially because long branches associated to this particular taxon were detected (Alves-Gomes 1995, Alves Gomes *et al.* 1995) and there are not many alternative explanations for this phenomenon.

There is a large amount of complementary evidence indicating that the taxa mentioned previously could, in fact, represent lineages that differentiated very early in the history of their respective orders and, therefore, have long periods of independent evolution. In agreement with other studies based upon morphological data (Vari 1979, Fink and Fink 1981), *Distichodus* (Family Distichodontidae) and *Citharinus* (Family Citharinidae) are always depicted as a monophyletic clade in the topologies obtained with molecular data. However, they appear in several points of the tree as the weights for TS TV and GP change. Similarly, in Ortí and Meyer's study (1997), *Distichodus* (family Distichodontidae) was never depicted with the remaining characiforms. Fink and Fink (1981) have found that these two families (Citharinidae and Distichodontidae) possess several morphological features that are plesiomorphic for characiforms and for otophysans, and suggested that the assemblage Citharinidae + Distichodontidae form a monophyletic clade that contains the most ancient characiforms. The clade was therefore considered to be the sister group of all remaining characiform genera. *Distichodus* is one of the two characiform genera used by Dimmick and Larson (1996) to estimate the Ostariophysi phylogeny, which may explain why characiforms and gymnotiforms were depicted as sister groups in their study when only the molecular data was considered. The molecular results support the notion of an old separation of Distichodontidae and Citharinidae from the remaining characiforms. In the present chapter, I assume that both Citharinidae and Distichodontidae are basal characiforms (Fig. 13.2), but future studies may induce new interpretations in this regard.

Among Siluriformes, the South American *Dyplomystes* has been recognized as the living siluriform genus having the higher number of plesiomorphic characters (Lundberg and Baskin 1969, Fink and Fink 1981, Grande 1987, de Pinna 1993, 1998, Lundberg 1993). However, molecular data from the nuclear rag1 and rag2 genes (Sullivan *et al.* 2006) place the South American Loricarioidei as the most basal catfish clade and Diplomystidae as the next most basal taxa within the order. The results obtained in this study with molecular data reinforce this hypothesis. *Dyplomystes* is almost invariably placed in the base of the siluriform phylogeny, regardless of the weighting scheme adopted. Additionally, *Peckoltia*, a member of the South American family Loricariidae, also appears as a basal lineage within the Siluriformes in most of the results, but changes place in the trees as the relative weight between TS and TV changes. Alves-Gomes (1995) used a different loricariid genus (*Loricaria*) and obtained very similar results, i.e. *Loricaria* had an erratic behavior in the tree, but was depicted as the most basal siluriform, with *Diplomystes*, in most of the results, especially when the weight of TV was increased in relation to TS. Although the limited number of catfish species used does not allow major inferences about the catfish phylogeny, it is worth observing that mitochondrial phylogeny provides strong support for South America as the place of siluriform differentiation, as previously suggested by Lundberg (1993), since the basal taxa in the resultant molecular phylogenies are from the Neotropics. Nuclear markers also strongly corroborate this hypothesis (Sullivan *et al.* 2006), as already mentioned.

For Gymnotiformes, Mago-Leccia (1976, 1978, 1994) proposed that the suborder Gymnotoidei, formed by the monotypic families Electrophoridae and Gymnotidae, was the sister group of all remaining gymnotiforms, according to morphological characters. Later, Alves-Gomes *et al.* (1995) noticed exceptionally long branch lengths for *Electrophorus* (in *Gymnotus* the branches are not so long) in their study of mitochondrial phylogeny of the Gymnotiformes. Due to the erratic behavior of *Electrophorus* + *Gymnotus* in their trees, Alves-Gomes *et al.* (1995) suggested three alternative positions for the clade within the order, one of them as the sister group of all remaining gymnotiform families. Albert and Campos-da-Paz (1998) placed (*Gymnotus* + *Electrophorus*) as the sister group of the remaining gymnotiforms, and propose that the two genera should be combined into a single Family

Gymnotidae (suborder Gymnotoidei). The same topology was also adopted by Albert and Crampton (2005) and Crampton and Albert (2006). The results of the present study indicate that *Electrophorus* + *Gymnotus* are, indeed, sister groups and may represent a plesiomorphic lineage within gymnotiforms, but additional evidence is necessary to improve our degree of certainty about this specific assumption. The extreme long branch associated to *Electrophorus'* DNA sequences as well as the outstanding number of morphological, physiological and behavioral autapomorphies in this genus (Assunção and Schwassmann 1995, Farber and Rahn 1970, Mago-Leccia 1976, 1994) can mask their phylogenetic position and demand further investigation. For those main reasons (long branches, high number of autapomorphies and consequent uncertainty about their position in the gymnotiform tree), which are probably related, I'm still not convinced about the need or concrete advantage of combining Gymnotidae and Electrophoridae into a single Family at this point in time, as in the classification originally proposed by Albert and Campos-da-Paz (1998) and more recently adopted by Nelson (2006).

From the evidence analyzed, I would argue that 12S and 16S rRNA contain enough phylogenetic information to provide a reasonable sound hypothesis for the branching order within the Otophysi. In fact, 12S and 16S mtDNA phylogenies, when analyzed critically and under the scrutiny of additional morphological evidence, strongly support the following phylogenetic relationships for the Otophysi: Cypriniformes is the sister group of the Characiphysi + Siluriphysi; Characiformes is the sister group of the Siluriphysi (Siluriformes + Gymnotiformes), and they are all monophyletic, i.e. the South American electric fish differentiated from a Neotropical catfish lineage.

Phylogeny and Biogeography

When Fink and Fink (1981) first suggested their phylogenetic hypothesis for the Ostariophysi, they did not attempt to propose any new interpretation for the present ostariophysan distribution and stated that "the biogeographic history of the group [Ostariophysi] has been long and complex" (Fink and Fink 1981: 347). Until recently, the work of Lundberg (1993) stood as a unique attempt to summarize ostariophysi phylogeny and continental drift, although no biogeographic hypothesis was proposed. Saitoh *et al.* (2003) and Briggs (2005) also provided more recent interpretations, but both studies considered a branching order within the Otophysi that places Gymnotiformes as the sister group of Characiformes. Based upon the evidence interpreted in this study and explained previously, such hypothesis is assumed to be incorrect and, therefore, no further comments will be formulated.

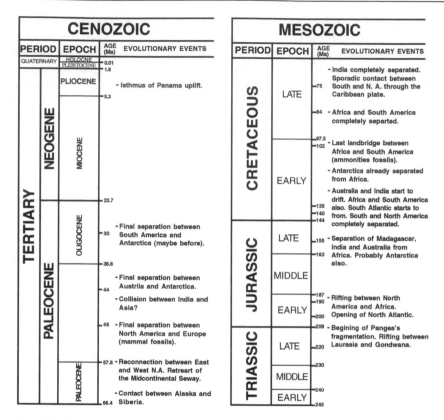

Fig. 13.3 Schematic review of principal events since the beginning of Pangea's fragmentation around the Triassic-Jurrassic boundary.

Table 13.2 summarizes our current knowledge of ostariophysan present and past distribution and Fig. 13.3 shows the main geological events since the beginning of the fragmentation of Pangean in the Mesozoic. In the following portion of the discussion, I will review and summarize the current evidence about the Pangea fragmentation, and what is known about fossil records for each order involved in this study. Finally I will combine these two different sources of information together with the phylogeny proposed previously and suggest a biogeographic hypothesis for the Otophysi and for the Gymnotiformes.

The Fragmentation of Pangea

In the last 210 Ma, the distribution of land masses on the earth changed from a continuous supercontinent (Pangea) to what is the present situation. Near the end of the Triassic (208 Ma), several rifting processes began at different points in Pangea, and the break-up and drift of several plates

Table 13.2 Summary of the fossil record and current distribution for each ostariophysan order.

	South/Central America		Africa		Europe-Asia		North America		Australia	
	Living	Fossil	Living	Fossil	Living	Fossil	Living	Fossil	Living	Fossil
Gonorynchiformes	No	Early Cretaceous[1]	No	Early Cretaceous[1]	No	Early Cretaceous[1,16]	No	Paleocene[10]	No	Not known
Cypriniformes	No	Not known[2,3,4]	Yes	Miocene[5]	Yes	Paleocene[5?] Basal Eocene[1]	Yes	Paleocene[10]	No	Not known
Characiformes	Yes	Early Cretaceous (Albian)[11]	Yes	Early Pleistocene[5]	No	Early Eocene[8]	Yes[1] one species	Not known	No	Not known
Siluriformes	Yes	Late Cretaceous[1,6,14,15]	Yes	Eocene[5,13]	Yes	Cretaceous[9,12]	Yes	Late Cretaceous[5]	Yes	Not known
Gymnotiformes	Yes	Late Miocene[7]	No	Not known	No	Not known	No	Not known	No	Not known

Note: ? indicates questionable identification according to the cited references. [1]Patterson (1975). [2]Gayet (1982b) has suggested a cypriniform fossil from South America, but other authors (Fink et al. 1984) disagree with the identification of the fossil as a cypriniform. [3]Fink et al. (1984). [4]Lundberg (1993). [5]Novacek and Marshall (1976). [6]Gayet (1988). [7]Gayet and Meunier (1991). [8]Cappetta et al. (1972). [9]De la Peña and Soler-Gijón (1996). [10]Cavender (1986). [11]Filleul and Masey (2004). [12]Cione and Prasad (2002). [13]Murray and Budney (2003). [14]Cione (1987). [15]Gayet and Brito (1989). [16]Poyato-Ariza (1994).

started to form the present geography of the earth. Since the fragmentation of Pangea, the current distribution of freshwater fishes results from two possibilities: either a group was present in a continuous land mass and was separated by a vicariant event, or it was able to disperse across salt water and colonize neighboring landmasses.

There are several details associated with the geology of Pangea's fragmentation that still are controversial, but a fairly accurate picture has emerged in the last few decades of paleontologic, stratigraphic, and seafloor-spreading research. The following paragraphs are an abbreviated review that summarizes the work of several authors, and I do not attempt to provide a full description of geological processes involved in rifting events and plate tectonics. Rather, I delineate the most important vicariant events and their respective geological times as a consensus between many critical papers and reviews (Le Pichon 1968, Martin 1969, Valencio and Vilas 1969, Dietz and Holden 1970, Dietz and Sproll 1970, McElhinny 1970, Smith and Hallam 1970, Vilas and Valencio 1970a, b, Jardine and McKenzie 1972, Cracraft 1973, Windley 1984, Briggs 1987, Pitman *et al.* 1993), which are also summarized in Fig. 13.3.

The fragmentation of Pangea began in the Upper Triassic-Lower Jurassic (220–200 Ma), with the formation of two major land masses: Laurasia (North America and Eurasia) and Gondwana (South America, Africa, Madagascar, India, Antarctica and Australia). The separation between Laurasia and Gondwana marks the initial phase of the opening of the North Atlantic, and the initial rift extended from the Tethys Sea westward, passing between Spain and Africa, then south between Africa and North America, and then westward again between North and South America. The next critical period of Pangea's fragmentation (between 160 and 130 Ma) is marked by several rifting processes. South America was completely separated from North America by the Early Cretaceous (144–135 Ma). About 155 Ma, in the southern portion of Gondwana, another rift system passed north of India and then turned south, separating Madagascar, India and Australia from Africa. There are various models about the relative position of Madagascar and India during the drift (review in Cracraft 1973), but it is believed that India separated from Australia about 135 Ma. India was completely isolated by the end of the Cretaceous (75 Ma) and probably joined Asia about 40–55 Ma (Windley 1984). Some authors suggest that on its way to Asia, India collided with the northwest tip of Africa in the Late Cretaceous (see Briggs 1987). About 140 Ma, the separation between Africa and South America starts at the southern portion of the two continents, initiating what became the South Atlantic. About 106 Ma, Africa began to rotate counterclockwise as a single plate and by 102 Ma it is believed that there still was a land bridge connecting Africa and South America, based upon differences in the ammonite fauna of the North and South Atlantic (Beurlen 1964 in Cracraft

1973). Pitman *et al.* (1993) suggest that Africa and South America were closed to migration about 95–100 Ma and completely separated by 84 Ma. The widening of the North Atlantic appears to have occurred at a slower pace than its southern counterpart. The North Atlantic opened along two main paths, one between Greenland and Labrador and the second between Greenland and Europe. Whereas there is evidence that there was some igneous activity associated with the rifting between Greenland and Labrador by the Middle to Late Cretaceous, the final separation between North America and Europe probably did not occur until the Early Eocene (47–49 Ma), according to mammal fossils on both sides (McKenna 1972). Only later (43–45 Ma) did Australia separate from Antarctica, and around 30 Ma Antarctica became isolated from South America.

Several other relevant events occurred during the fragmentation of Pangea: during the Late Cretaceous and extending to the Early Paleocene, sporadic contact between North and South America occurred with the migration of the Caribbean plate between the two continents (Pitman *et al.* 1993). Also, during the Cretaceous and Paleogene, a complex movement of small plates took place between Africa and Europe. It is believed that during the Cenozoic, Africa and Eurasia were in contact intermittently either in the East through the Iberian Peninsula or in the West between the Afro-Arab and Eurasian plates (Cooke 1972, Windley 1984). The Late Cretaceous also experienced a major marine transgression that created an epicontinental sea in North America, isolating the west side of it from the remainder of Pangea. The two parts of North America were reconnected during the Early Paleocene. Africa also was flooded by the Mid-Cretaceous and authors have suggested different degrees of isolation between the West Shield and the Afro-Arab plate during that period. In the Late Cretaceous to Early Paleocene, Alaska and Siberia made contact through the Beringian land bridge, which remained opened as a dispersal route until the Late Miocene (10–12 Ma), when a seaway was established (Hopkins 1967). The Bering land bridge became again available from 10 Ma until 4 Ma, when another marine transgression in the Pliocene reopened the contact between the Bering Sea and the Chukci Sea.

Current and Past Distribution of the Otophysan Clades

I now examine current views about the fossil records and present distribution for each otophysan order, focusing the approach mainly in the oldest known fossils of each order, which can provide minimal ages for the differentiation of the respective clades.

Living cypriniforms are currently found only in North America, Eurasia, and Africa. Fossils have been described from the Miocene of Africa (Novacek and Marshall 1976), from the basal Eocene of Eurasia (Patterson 1975), and from the Middle Eocene of North America (Grande *et al.* 1982, Grande 1984).

Gayet (1982b) examined a tooth-bearing bone from the Late Cretaceous sediments in Bolivia and has classified the bone fragment as a cypriniform, and suggested a new genus (*Molinichthys*). Other authors (Fink *et al*. 1984) suggested that *Molinichthys* might not even be a teleost. Except for this questionable evidence, it appears that the past distribution of cypriniforms is not different from their current distribution.

Living characiforms are currently restricted to Africa and South/Central America. One genus (*Astyanax*) reaches as far north as Texas. Late Cretaceous characiform fossils are known from South America (Novacek and Marshall 1976, Gayet 1982a). More recently, Filleul and Masey (2004) described the oldest characiform fossil known (*Santanichthys diasii*) from the Early Cretaceous (Albian) from South America. Lower Eocene fossils are known from Europe (Cappetta *et al*. 1972), implying that characiforms became extinct there. The oldest fossil from Africa is from the Pleistocene (Novacek and Marshall 1976). No characiform fossil is known from North America.

The order Siluriformes is the otophysan clade with the most extensive past and current geographic distribution. Catfishes constitute a cosmopolitan clade currently found in every continent, except Antarctica. Although living catfishes are absent from Antarctica, fossils of Eocene age have been described from Seymour Island (Grande and Eastman 1986). One crucial aspect stands up from the siluriform distribution: either catfish are old enough in order to have attained a worldwide distribution before the fragmentation of Pangea, or they were able to disperse through coastal, marine habitats. Several living catfish families with representatives inhabiting brackish or marine habitats (de Pinna 1998) and catfish fossils from marine deposits of Cretaceous confirm that this group has been able to explore marine ambient since the Mesozoic (Novacek and Marshall 1976, Lundberg 1993). Catfish fossils of several ages are known from South America, Africa, Eurasia, and North America (see Table 13.2). Among the oldest are Mesozoic (Campanian and Maastrichtian) fossils from Argentina, Bolivia and Brazil in South America (Cione 1987, Gayet 1988, 1990, Gayet and Brito 1989, Bertini *et al*. 1993). An otolith of Cretaceous age from marine deposits of South Dakota, United States, was identified as a siluriform fossil (*Vorhisia vulpes*) by Frizzel (1965) and Frizzel and Koenig (1973). The number and the age of catfish fossils from South America, in agreement with morphological and molecular evidence, suggests the Neotropic as the catfish's center of origin. The evidence also points to a worldwide distribution of catfish already by the Paleocene.

Very unlike its sister group, Gymnotiformes, living or extinct, are only found in South America. The only fossil known for the group comes from the Late Miocene deposits (8.3–12.4 Ma) from Bolivia (Gayet and Meunier 1991).

A Hypothesis for the Historical Biogeography of the Otophysi and the Time and Place for Gymnotiform Differentiation

Several authors have addressed in more or less detail the question of ostariophysan evolution and biogeography (Nichols 1930, Gosline 1944, 1973, Darlington 1957, Greenwood *et al.* 1966, Géry 1969, Patterson 1975, Novacek and Marshall 1976, Briggs 1979, 2005, Saitoh *et al.* 2003), but no consensus among the different hypotheses has been achieved (see also Lundberg 1993). Before Novacek and Marshall 1976 (but see also Géry 1969 and Briggs 1979), no author(s) explicitly considered continental drift as a main factor influencing the biogeographical patterns of the Ostariophysi distribution. Instead, they relied on other explanations such as intercontinental land bridges to justify faunistic similarities between the continents. Furthermore, all existing hypotheses regarding ostariophysan historical biogeography have been based on phylogenetic relationships that place Gymnotiformes as sister groups of Characiformes (Géry 1969, Novacek and Marshall 1976, Briggs 1979, 2005, Saitoh *et al.* 2003). The present work departs from the phylogenetic relationship (branching order) proposed by Fink and Fink (1981, 1994), which is also supported by the analysis of molecular evidence presented here.

The hypothesis presented below has several aspects that have been already proposed more or less explicitly by other authors (Gosline 1944, Darlington 1957, Greenwood *et al.* 1966, Géry 1969, Gosline 1973, Patterson 1975, Novacek and Marshall 1976, Briggs 1979, Lundberg 1993). The main difference from previous studies is that here I expressly consider the phylogenetic relationship of the Otophysi depicted in Fig. 13.2 in conjunction with continental drift, which implies major differences from previous hypotheses principally about the timing and place of divergence of catfishes and gymnotiforms.

Combining the phylogenetic hypothesis depicted in Fig. 13.2 with the data summarized in Fig. 13.3 and Table 13.2, I propose the following hypothetical scenario for the Otophysi historical biogeography, starting with its sister group:

(1) The Ostariophysi ancestor probably inhabited a marine or brackish environment on the Eurasian coast (Tethys Sea?), about the time of the separation of Antarctica, Australia, Madagascar and India from Africa (Late Jurassic). A coastal marine environment close to Laurasia is suggested for two main reasons. First, gonorhynchiform fossils are found in marine deposits of Cretaceous age in Italy (*Chanos leopoldi*), Yugoslavia (*Prochanos*), Lebanon and Germany (*Charitosomus*) (Patterson 1975), areas adjacent to the initial rift separating Gondwanaland and Laurasia. Second, among living gonorhynchiforms, the genus considered to be the sister group of the

remaining families (*Chanos*) occurs in salt water and has a life cycle similar to that of other basal teleosts (Elopomorphs) such as *Elops*, *Megalops*, *Tarpon*, and *Albula*. All these teleosts spawn in the ocean, and the post-larvae move to continental water bodies with variable salinity, before returning to the oceans to spawn (Patterson 1975, Fink and Fink 1981, Nelson 2006). It appears that the catadromy present in Elopomorphs preceded the full adaptation for fresh water currently found in the great majority of ostariophysans. As already suggested by Patterson (1975), if the ostariophysans did not originate in salt water, their ancestors were at least more tolerant to brackish environments in the past. The split between the Anotophysi and the Otophysi occurred at some point between the Late Jurassic and the Early Cretaceous in a location close to or within Laurasia, when the lineage that became the ancestor of the Otophysi became fully adapted to fresh water and started the differentiation within the Otophysi.

(2)   The first split within the Otophysi gave rise to what became the extant Cypriniformes and the ancestor of the Characiphysi + Siluriphysi. Cypriniforms may have differentiated in the central-eastern part of Eurasia, from where they radiated to other portions of the globe. By adopting the view that the abundance and diversity of a given clade reflects the time of existence of the clade in a given location, Asia is suggested as the locale of cypriniform differentiation based on the fact that there are near 600 cypriniform species in China alone (Wu *et al.* 1981 in Nelson 2006). Representatives of the families Catostomidae and Cyprinidae reached North America probably in the Tertiary using either of two possible dispersal routes: through Greenland and Labrador before the complete separation of North America and Europe (47–49 MA), or through the Beringian land bridge. The present occurrence of extant catostomids in Siberia (Nelson 2006) suggests that this family may also have used this alternative route in the past. Because older catostomid than cyprinid fossils are found in North America (Cavender 1986, Grande 1987), it is possible that cyprinids reached North America only at a later time, but also using either of the previously mentioned routes. Cypriniformes never became abundant in Africa, and I would suggest that they never reached South America. The fossil from the Late Cretaceous from Bolivia described by Gayet (1982b) as a cypriniform (*Molinichthys inopinatus*), as already suggested by other authors (Fink *et al.* 1984), must be a misinterpretation of the evidence. With the exception of one genus (*Tribolodon*), the Cypriniformes never developed tolerance to salt or brackish water, and their current continental distribution should be explained by connections between freshwater basins in the past. Because North America was separated

from South America before the differentiation of the Cypriniformes, the only possible path to South America was through Africa. Several possibilities can be evoked to justify why Cypriniformes never reached South America. It is possible to speculate about physiology, niche/ecological competition, lack of opportunity due to the absence of connection between river basins in the past, and other biotic factors such as predation, food availability and behavior. Maybe several of these factors contributed to some extent, but none can be currently tested or demonstrated.

(3) The common ancestor of all characiforms moved into Gondwana and Characiformes differentiated before the separation of Africa and South America. The last land bridge between the two continents existed at least until about 102 Ma (Pitman *et al.* 1993). The differentiation when still there was connection between the South American and African river basins allowed sister-group relationships between lineages currently living in these two continents, such as (*Hoplias* + *Hepsetus*) and (*Alestes* + *Phenacogrammus*), recovered with molecular data as in the present study (Fig. 13.2), and also proposed by other authors using molecular and morphological data (Roberts 1973, Vari 1979, Fink and Fink 1981, Lundberg 1993, Ortí 1997, Ortí and Meyer 1997). The final separation between Africa and South America represented a vicariant event for existing characiform sister clades. By the time of characiform differentiation, North America was isolated from South America and a single species, *Astyanax*, which is closely related to Central and South American clades, reached Texas (Nelson 2006) after the uplift of the Panamanian Isthmus. A few genera reached Europe in the lower Eocene probably using the connection through the Iberian Peninsula (which had been in intermittent contact with Africa during that period), but became extinct (Cappetta *et al.* 1972, Lundberg 1993). Characin fossils have not been described from North America or from Asia (Novacek and Marshall 1976, Lundberg 1993). Like the cypriniforms, characiforms were unable to develop saltwater tolerance and to disperse across marine environments. Furthermore, neither is extant characiform fish present nor have fossil records been described for high latitudes. The absence of both living and fossil characins from cold climates suggests some physiological constraint related to temperature tolerance. This conforms with the idea that, despite the availability of the Beringian land bridge and the contact between Europe and North America, characiforms did not colonize North America using any these routes going through high latitudes, as did the cypriniforms. However, the possibilities suggested previously to explain why cypriniforms never reached South America are also valid here. Many other biotic and

non-biotic factors besides low tolerance for cold may have contributed to the absence of characiforms from Asia and currently Europe.

The separation of South America and Africa represented a vicariant event to several characiform clades present on both continents. Several phylogenetic studies for the Characiformes, based on morphological as well as molecular characters, point to a close relationship between African and South American clades (Roberts 1973, Vari 1979, Fink and Fink 1981, Lundberg 1993, Ortí 1997, Ortí and Meyer 1997). The phylogeny proposed in Fig. 13.2 also corroborates several hypotheses based upon morphological data. The African family Hepsetidae is closely related to the South American family Erythrinidae, and the family Characidae, although probably not monophyletic, also has representatives in both continents. The gymnotiform/siluriform ancestor shared the South American rivers with characiforms probably before the Late Cretaceous.

(4) Sometime between the Early and Late Cretaceous (110–90 Ma), or about the time of final separation between Africa and South America, the differentiation between Siluriformes and Gymnotiformes took place within the recently isolated or isolating South America. As additional evidence supporting this hypothesis, it can be mentioned that the earliest fossils known for catfishes are from the Late Cretaceous deposits from South America (Lundberg 1993), and no fossil gymnotiform has ever been found outside the continent. Furthermore, loricarioids and *Diplomystes*, unanimously considered by several authors the morphologically most plesiomorphic catfish until recently are restricted to the central-southern portion of South America. Molecular evidence from the present study, from Hardman (2005), from Sullivan *et al.* (2006) and Lundberg *et al.* (2007) also place the South American clades (in particular the Suborder Loricarioidei, with more than 1,000 species, and Diplomystidae) as the basal catfish taxa. If siluriforms and gymnotiforms differentiated after the final separation of Africa and South America, then the only possible way to reach the other continents is by dispersal across coastal marine waters. Roberts (1973), addressing the worldwide distribution of siluriforms, already suggested that some clades could have reached distant land masses crossing salt water. I also argue in favor of dispersal through salt/brackish water as an explanation for the current catfish distribution. Among the oldest catfish fossils known, from Campanian (84–74.5 Ma) and Maastrichtian (74.5–66.4 Ma) deposits from South America, there are already marine representatives (Lundberg 1993). Marine catfish fossils are also found in Late Cretaceous deposits from North America and from the Eocene of southeastern Arkansas (Lundberg

1993). Five extant families of catfishes (Ariidae, Plotosidae, Auchenipteridae, Pangasiidae, and Aspredinidae) have representatives that either live in or tolerate coastal marine waters. The several fossils found in marine deposits have been putatively assigned to the current marine family Ariidae (Lundberg 1993). The present tolerance of many siluriform clades to salt water and the several fossils in marine deposits indicate a more malleable osmoregulatory physiology for this clade than the other otophysan orders already in the past. Siluriforms utilized marine habitats already in the Cretaceous and the present tolerance of several siluriform clades to salt waters may represent, at least for some taxa, the retention of an ancient condition.

If catfishes evolved before the separation of Africa and South America as proposed for the characiforms, current studies about catfish phylogeny should reveal sister-group relationships between African and South American clades. However, no monophyletic siluriform group having representatives in both Africa and South America is known (Lundberg 1993, Lundberg *et al.* 2007, Sullivan *et al.* 2006). Hardman (2005), estimating molecular phylogenies with cyt b sequences has provided strong evidence that North American ictalurids have a sister group relationship with Northern and Eastern Asia clades, more specifically with the genus *Cranoglanis*. This result was corroborated by Sullivan *et al.* (2006) with nuclear genes, reinforcing the hypothesis that ictalurids have reached North America probably using a land bridge connecting Northeastern Asia and Northwestern North America, by late Cretaceous. More recently, Lundberg *et al.* (2007) reported a surprising sister group relationship between the Central American new genus *Lacantunia* and the African Family Claroteidae, based upon the nuclear rag1 and rag2 genes. The authors suggest an ancient intercontinental passage to explain such relationship, proposing that the lacantuniid diverged from claroteids either by mid Eocene ($\approx$ 45 mya) or by late Cretaceous (75 – 94 mya), depending if the estimation is based upon fossil records or by the Bayesian relaxed clock (and penalized likelihood methods), respectively. Considering several possibilities to account for *Lacantunia* in Central America, the authors favor a route linking Africa to Central America by the North Atlantic land bridges (Africa – Europe – North America – Central America). These two studies are important to show that dispersal through salt water may have not been the only possible alternative for catfishes in their evolutionary history. It was, however, for the hypothesis formulated here, the necessary alternative to allow the early catfish lineage out of South America. If ictalurids and *Lacantunia* reached their current geographic distribution migrating from Africa and Asia to North and Central America, respectively, and the authors of both studies agree that South American is the center of origin of catfishes, then we still need to explain how catfishes reached Africa and/or Asia at the first place. In the case of *Lacantunia*, is also puzzling the fact that no fossil record or sister group taxon is found in its route from Africa to Asia to Central America (Lundberg *et al.* 2007). On the other hand, fig. 2 of Lundberg *et al.* (2007) shows pangasiids at the base of the African catfish + *Lacantunia* clade and some pangasiids are tolerant to marine waters. *Cranoglanis* + ictalurids in both studies also group within this African clade with pangasiids in the base. In the present study, *Pangasius* is placed between the basal South American lineages (*Diplomystes* + *Peckoltia*), suggesting that the phylogenetic connection between the basal catfish and the remaining genera may have gone through a marine link. On the other hand, similarly to Sullivan *et al.* (2006) this study also finds a close relationship between an African lineage (*Leiocassis*) with *Ictalurus* plus *Tandanus* and *Plotosus*, also two marine genera. Logically part of the results are probably due to the lack of additional taxa to fulfill the relevant phylogenetic positions in the tree (taxonomic sampling bias), but the relatively strong signal placing *Ictalurus* with these genera should not be neglected. In other words, I'm not claiming that either ictalurids or *Lacantunia* directly derived from marine ancestors, but I believe that our understanding about the phylogenetic position of marine or salt water tolerant catfish clades is still insufficient to discard them as main phylogenetic links between Continental siluriform faunas. Further, molecular phylogenetic trees cannot include fossil taxa and, therefore, it is possible that extant clades that are minimally tolerant to marine environments in the present, may have derived from extinct lineages better adapted to salt water or even fully marine. I would not find surprising if future studies encompassing the complete taxa currently under Ariidae, would reveal it as a non-monophyletic clade.

Understanding catfish biogeography would be much easier if there was a single unquestionable evidence that these fish existed before the separation between Africa and South America. However, up to the moment, there is no physical indication that catfishes evolved before the Afro-South American separation (but see Saitoh 2003 and Peng *et al.* 2006b). Here, combining molecular phylogenies, molecular clocks calibrated for ostariophysans (Alves-Gomes 1999) and evidence from fossil records, I propose that catfishes differentiated within South America, just after the existence of the last connection between Africa and South America. Antarctica was still attached to South America until the Oligocene, and Eurasia and

North America were not very far apart. If some freshwater or estuarine lineages were progressively becoming more tolerant to salt water by late Cretaceous, they could use the Atlantic as well as the Indian Ocean as dispersal corridors and reach essentially every other land mass. Further, if the life cycle of these salt water tolerant species incorporated an alternation of marine habits with spawning in fresh water—as currently found in most extant marine catfish species (M. de Pinna, personal communication)—some species could have invaded other continents instead of returning home to spawn. Some lineages then could have readapted to a fully freshwater condition, probably for having some kind of competitive advantage in the new habitats. The electrosensory system (see below) may have played a significant rule in conferring such advantage to catfish better explore new environments and potentially occupy free ecological niches both in costal environments as well as in new Continents. Siluriformes were then able to achieve a worldwide distribution using shallow marine, costal water to get out from South America. Migrations through existing land bridges should also be considered once they have reached other Continents and returned to a fully freshwater adaptation.

(5) Unlike their sister group, gymnotiforms were never able to leave South America. Even if electric fishes could have developed a similar osmoregulatory capability as actually found in catfishes, the main limiting factor for gymnotiforms entering the marine environment was related to the deficient operation of their electrogenic and electrosensory systems (EES) in water with very low electrical resistance, or high salt content. In order to develop the rationale behind such a hypothesis, I will assume that the common ancestor of siluriforms and gymnotiforms was electroreceptive (see Fink and Fink 1981, 1986), and second, that the common ancestor of all gymnotiforms had an electric organ (see Alves-Gomes 2001). This is consistent with the fact that every catfish examined so far possesses ampullary electroreceptors, and that electric organs as well as ampullary and tuberous electroreceptors are found in every gymnotiform (Bullock 1979, Fink and Fink 1981, Bullock et al. 1983, Alves-Gomes et al. 1995, Alves-Gomes 2001). I will argue that the ability to produce electric potentials above the threshold of their electroreceptors was among the first synapomorphies in the lineage leading to extant gymnotiforms. Very rapidly the EES became an essential motor-sensory mechanism during several types of behavior such as prey and object detection, social interactions, species recognition, hierarchical settlements, and courtship/spawning activities (Hopkins 1974a,b, Heiligenberg and Bastian 1981,

Hagedorn and Carr 1985, Hagedorn and Zelick 1989, Hopkins *et al.* 1990). This evolutionary novelty may have granted the gymnotiforms an adaptive advantage in the Neotropics but kept them away from exploring coastal waters.

Another important aspect related to the possibility of electric fish and catfish to enter brackish waters refers to the biophysical properties of electric organs and electroreceptors as a fish moves between environments with a high (low salt contents) or low water resistivity (high salt contents). An equivalent circuit for an idealized gymnotiform fish can be developed on the basis of experimental data. The total load of the electric organ can be interpreted as the sum of all the internal resistances associated with individual electrocytes, plus an external load represented by the body tissues, skin and surrounding water. Considering that the body and the skin have a constant composition, the water resistivity is the main variable determining the total load on the electric organ. Experiments have shown that for the broadest range of water resistivity found in their natural habitats, electric organs produce a constant current output, independently of the external load (Knudsen 1974, 1975, Bell *et al.* 1976). Knudsen (1975) showed that the electric organs of *Eigenmannia* and *Sternopygus* can be considered a current source in water with resistivity going up to 60,000 Ohms.cm (60 kΩ.cm), since the voltage measured external to the fish increases linearly as the water resistivity goes from about 2 kΩ.cm to 60 kΩ.cm. For *Apteronotus* the relationship is not as linear, perhaps because of the special properties of their electric organ, which is formed by neurons in the adults (Kirschbaum 1983). Under these circumstances, and applying Ohm's Law, as a gymnotiform fish enters an environment with high salt content water (low resistivity), the Voltage also drops, the electric field generated by their electric organs tend to collapse and the active portion of the EES looses its functionality. On top of that, in salt water, the electric organ becomes "short circuited" (the current lines tend to become flattened along the body, with current lines following the shortest distance between the two poles of the electric organ) and the fish can no longer use it for electrolocation or electrocommunication.

For electroreceptors alone, the problem is not so dramatic. Bell *et al.* (1976) considered that electroreceptors "respond to current flow" (Bell *et al.* 1976: 66), but other authors (Bennet 1971, Kalmijn 1974, Zakon 1987) have suggested that at least some types of electroreceptors are more properly described as voltage detectors. Here, it will suffice to consider that electroreceptors are driven by a voltage drop across their sensory cells. Several authors have addressed the theoretical, experimental, behavioral and evolutionary aspects of electroreception and electrocommunication since Lissmann (1951) first described this sensory modality for the mormyriform *Gymnarchus niloticus* (Heiligenberg 1973, 1975, 1977, 1990, Kalmijn 1974, 1987, Knudsen 1974, 1975, Bell *et al.* 1976, Bastian 1986, Zakon 1986, 1987, Alves-Gomes 2001).

Gymnotiforms have two types of electroreceptors: ampullary and tuberous. The physiology and structural features associated with each kind of receptor, including their sub-types, can be found in Szabo (1965), Bennet (1971b), Szabo *et al.* (1972), Kalmijn (1974), and Zakon (1986, 1987). Ampullary receptors are specialized to detect DC or very low-frequency AC fields and are found in several teleost fish living in marine as well fresh waters (Alves-Gomes 2001). However, physiological and structural adaptive differences have been described between ampullary receptors of fish living in these two environments as a way of maintaining their functionality in both environments (Bennet 1971, Szabo *et al.* 1972, Kalmijn 1974). In principle, the ampullary system of gymnotiform fish, if they ever happened to invade brackish water, would have to undergo an adaptation in the opposite direction to what occurred with the freshwater sting rays (Family Potamotrygonidae) that became adapted to fresh waters around the Early Miocene, between 15 and 23 Ma (Lovejoy *et al.* 1998).

In fresh water, the body tissues are more conductive than the water, and the current lines from any external source converge upon the fish. Because freshwater fish also have a skin with a resistivity much higher than their body tissues (also for osmoregulatory purposes), the most effective voltage drop occurs across the skin, and consequently the ampullary canals can be much shorter than in the marine species. In marine environment the internal tissues are less conductive than the water, and the skin resistance is not as high as in the freshwater species. Current lines tend to bend away from the fish and the voltage drop across skin is not as drastic as in the freshwater case. Consequently, marine species tend to have ampullary receptors with longer canals, as a way to have a voltage drop across the sensorial membrane of the ampulla that is sufficiently strong to drive the receptor cells (see Kalmijn 1974 for a more detailed description). In summary, the ampullary system would not be impaired if gymnotiforms invaded brackish water, once some structural modifications in the system were implemented.

Considering that Siluriformes also possess ampullary receptors, according to Fink and Fink (1981) a synapomorphy reinforcing the monophyletic nature of the Siluriphysi, it seems that some of the ancestral lineages were absolutely able to adapt their ampullary system to function in low-resistivity water. This would provide an enormous new possibility for those lineages that were able to take advantage of new ecological circumstances and open niches of shallow marine habitats. The shallow and coastal marine environments became great venues for catfish dispersion. However, as already mentioned, a more comprehensive analysis of the phylogenetic relationships among Siluriformes, with special emphasis on the relationships of the marine lineages, would provide further test for the present hypothesis. Morpho-physiological studies on marine catfish ampullary receptors would also help to clarify the adaptation of this group to marine environments.

Tuberous receptors are physiologically tuned to the fish's own electric signal (high frequency AC signals) and are essential during electrolocation and electric communication. The efficiency of these receptors is linked to their ability to discriminate distortions of the fish's electric field produced by any object or organism near the fish that has a different electrical conductivity from the water surrounding the fish (Heiligenberg 1977, Bastian 1986). If the fish's own electric field collapsed as the fish enters salty waters, the tuberous electroreceptors would loose their functionality as well. Electrolocation of objects and organisms as well as electrocommunication become impossible under these circumstances because the electric field around the fish will collapse and become too weak to drive the tuberous electroreceptors of conspecifics, even within a very short distance.

By having their EES deteriorated in water with very low resistivity, it seems predictable that gymnotiforms simply avoid such a habitat. Therefore, I predict that no fossil gymnotiform will be found outside South America and possibly Antarctica.

Summary

In the present study, the 12S and 16S rRNA sequences appeared to be under some type of constraint that limits the amount of phylogenetic information available to establish with confidence the phylogenetic position of the gymnotiforms *Electrophorus* and *Gymnotus*, the characiforms *Citharinus* and *Distichodus* and the siluriform *Peckoltia*. The uncertainty about their phylogenetic position could be related to a not yet fully understood phenomenon at molecular level in the two mitochondrial genes of these genera, to short branches (few synapomorphies) separating the basal otophysan clades and/or to long time of independent evolution of those lineages (long branches due to a long divergence time). Similar phenomenon may be affecting other mitochondrial genes and, therefore, mitochondrial (or nuclear) phylogenies of ostariophysan fishes should include as many taxa possible from each representative order, as an alternative to average the genetic parameters within each clade, as well as to break up possible long branches. Morphological as well as fossil data are critical and should be employed as an aid to discern between competing molecular hypotheses and to better evaluate, or even to reinterpret, improbable tree topologies.

Taking into account the particularities associated to the molecular phylogenies mentioned above and additional information from morphological studies, the DNA two segments analyzed in this study point to a phylogenetic relationship for the Otophysi that corroborate Fink and Fink's (1981, 1996) hypotheses. Cypriniformes is the sister group of the Characiphysi + Siluriphysi; Characiformes, Siluriformes and Gymnotiformes are monophyletic assemblages where Characiformes is the sister group of the Siluriformes + Gymnotiformes.

Cypriniformes evolved at some point in the Early Cretaceous and diversified in Eurasia. Representatives of two families reached North America, probably by the Beringian land bridge, but maybe using the Europe–Greenland–Labrador route. Several genera invaded Africa, but they never reached South America.

Characiformes evolved before the separation between Africa and South America. Unlike the Cypriniformes, live or extinct characins are not known from high latitudes. Some genera invaded Europe during the Tertiary but became extinct. The group apparently never used either the Beringian land bridge or the Europe–North America connection, and the presence of a single characiform genus in North America can be better explained by a later invasion from the South, after the closing of Central America in the Pliocene. The Afro–South American split represented a vicariant event for several Characiform clades.

Siluriformes evolved within South America about the time of the last connection between Africa and South America. Several marine fossils of catfishes indicate that the group was able to use coastal, shallow marine water as dispersal routes already by late Cretaceous. Some derived lineages returned to a fully freshwater condition and by the time of Paleocene and Eocene, catfishes already had a worldwide distribution. Because dispersal through salt water is hypothesized, phylogenetic relationships between certain clades, currently occurring in different continents, should be linked by marine clades either fossil or extant. Along their history, catfishes may also have used land bridges to reach particular geographical locations.

Gymnotiformes evolved within South America. The deterioration of the ecological/behavioral functionality of the EES in brackish or marine waters may be considered the main reason why gymnotiforms never left South America.

Despite being congruent with current data from various sources, the current hypothesis represents only an additional framework intended to stimulate further discussion. Additional data and further research are needed before we fully elucidate the complex evolutionary history of the gymnotiforms or the other otophysan orders. If the hypothesis provided here is correct, several predictions about the presence and absence of fossil records as well as phylogenetic relationships can be made which, in turn, make the hypothesis easily falsifiable. For instance, a single fossil of a gymnotiform outside South America, or a catfish fossil older than the Africa-South America separation, or even an unquestionable sister group relationship between an African and a South American catfish lineages, would promptly dismiss the present hypothesis. Until then, integrated approaches including fossil, mo lecular, physiological, morphological, behavioral and ecological information will always have a better likelihood of being closer of the true evolutionary history of these fish.

Acknowledgments

A significant portion of this chapter was conceived and executed while I was a graduate student at Scripps Institution of Oceanography, U.C. San Diego, where Margo Haygood kindly provided laboratory space for DNA sequencing. I thank the following people for tissue samples: John Lundberg (*Diplomystes, Peckoltia, Leiocassis* and *Arius preops*), Walter Dimmick and A. S. Golubtsov (*Schilbe intermedius*), Paul Manger (*Tandanus tandanus*) Shosaku Obara (*Plotosus sp*), Glenn Northcutt (*Ictalurus punctatus*) and Walter Heiligenberg (*Silurus glanis*). I thank the two anonymous reviewers for their valuable critiques and suggestions. This manuscript also benefited tremendously from discussions with John Lundberg, Guillermo Ortí, Margo Haygood, Calvin Wong, Jenz Astrup and Adrianus Kalmijn, and from critical comments from Richard Rosenblatt and Chris Braun. Linda Barlow and Carmen Pinuela provided helpful suggestions in the very early stages of the manuscript. Walter Heiligenberg and Ted Bullock will always be a source of inspiration. The UCSD academic Senate provided funds for a collecting field trip in the Brazilian Amazon. This research was also partially supported by a NSF grant INB 91-06705 to Walter Heiligenberg. My special thanks to Instituto Nacional de Pesquisas da Amazônia (INPA) and the Brazilian Research Council (CNPq) for having supported my graduated studies at Scripps Institution of Oceanography and a large portion of the present research.

References

Albert, J.S. 2000. Species diversity and phylogenetic systematics of American knifefishes (Gymnotiformes, Teleostei). Misc. Publ. Mus. Zool. Univ. Michigan 190: 1–127.

Albert, J.S. and R. Campos-da-Paz. 1998. Phylogenetic systematics of Gymnotiformes with diagnoses of 58 clades: a review of available data, pp. 419–446. In: L.R. Malabarba, R. Reis, R.P. Vari, Z.M. Lucena and C.A.S. Lucena [eds.]. Phylogeny and Classification of Neotropical Fishes. EDIPUCRS, Porto Alegre, Brazil.

Albert, J. S. and W.G.R. Crampton. 2005. Diversity and phylogeny of Neotropical electric fishes (Gymnotiformes). pp. 360–409. In: T.H. Bullock, C.D. Hopkins, A.N. Poper and R.R. Fay [eds.] 2005. Electroreception. Springer, New York, USA.

Alves-Gomes, J.A. 1995. The phylogeny and evolutionary history of the South American electric fishes (order Gymnotiformes). Ph.D. thesis, Univ. California, San Diego.

Alves-Gomes, J.A. 1998. The phylogenetic position of the South American electric fish genera Sternopygus and Archolaemus (Ostariophysi: Gymnotiformes) according to 12S and 16S mitochondrial DNA sequences, pp. 447–460. In: L.R. Malabarba, R. Reis, R.P. Vari, Z.M. Lucena and C.A.S. Lucena [eds.]. Phylogeny and Classification of Neotropical Fishes. EDIPUCRS, Porto Alegre, Brazil.

Alves-Gomes, J.A. 1999. Systematic biology of Gymnotiform and Mormyriform electric fishes: phylogenetic relationships, molecular clocks and rates of evolution in the mitochondrial rRNA genes. J. Exp. Biol. 202: 1167–1183.

Alves-Gomes, J.A. 2001. The evolution of electroreception and bioelectrogenesis in teleost fish: a phylogenetic perspective. J. Fish Biol. 58: 1489–1511.

Alves-Gomes, J.A. and C.D. Hopkins. 1997. Molecular insights into the phylogeny of mormyriform fishes and the evolution of their electric organs. Brain Behav. Evol. 49: 324–351.

Alves-Gomes, J.A., G. Ortí, M. Haygood, W. Heiligenberg and A. Meyer. 1995. Phylogenetic analysis of the South American electric fishes (order Gymnotiformes) and the evolution of their electrogenic system: a synthesis based on morphology, electrophysiology, and mitochondrial sequence data. Mol. Biol. Evol. 12(2): 298–318.

Anderson, S., A.T. Bankier, B.G. Barrell, M.H.L. de Bruijn, A.R. Coulson, J. Drouin, I.C. Eperon, D.P. Nierlich, B.A. Roe, F. Sanger, P.H. Schreier, A.J. Smith, R. Staden and I.G. Young. 1981. Sequence and organization of the human mitochondrial genome. Nature 290(5806): 457–465.

Arratia, G. 1996. The Jurassic and the early history of teleosts, pp. 243–259. In: G. Arratia and G. Viohl. [eds.]. Mesozoic Fishes—Systematics and Paleoecology. Verlag Dr. Pfeil, Munich.

Arratia, G. 1997. Basal teleosts and teleostean phylogeny. Palaeo Ichthyol. 7: 5–168.

Assunção, M.I.S. and H.O. Schwassmann. 1995. Reproduction and larval development of Electrophorus electricus on Marajo Island (Para, Brazil). Ichthyol. Explor. Freshwaters 6: 175–184.

Bastian, J. 1986. Electrolocation: behavior, anatomy and physiology, pp. 577–612. In: T.H. Bullock and W. Heiligenberg [eds.]. Electroreception. Wiley Series in Neurobiology, John Wiley and Sons, New York.

Bell, C.C., C.J. Bradbury and C.J. Russel. 1976. The electric organ of a mormyrid as a current and voltage source. J. Comp. Physiol. 110: 65–88.

Bennet, M.V.L. 1971. Electroreception, pp. 493–574. In: W.S. Hoar and D.S. Randall [eds.]. Fish Physiology. Academic Press, New York.

Bertini, R.J., L.G. Marshall, M. Gayet and P.M.M. Brito. 1993. Vertebrate faunas from the Adamantina and Marília (upper Bauru group, late Cretaceous, Brazil) in their stratigraphic and paleobiogeographic context. Neu. Jahrb. Geol. Paläontol. Mh. Stuttgart, Stuttgart 188(1):71–101.

Beurlen, K. 1964. Einige Bemerkungen zue erdeschichtlichen Entwicklung Nordost-Brasiliens. Geol. Palaont. 56(2): 82–105.

Briggs, J.C. 1979. Ostariophysan zoogeography: an alternative hypothesis. Copeia 1979(1): 111–118.

Briggs, J.C. 1987. Biogeography and plate tectonics. Vol. 10. Serie: Developments in Paleontology and Stratigraphy. Elsevier, Amsterdam.

Briggs, J.C. 2005. The biogeography of otophysan fishes (Ostariophysi: Otophysi): a new appraisal. J. Biogeogr. 32: 287–294.

Brooks, D.R. and D.A. McLennan. 1991. Phylogeny, Ecology, and Behavior: A Research Program in Comparative Biology. Univ. Chicago Press, Chicago.

Buckup, P.A. 1991. The Characidiinae: A phylogenetic study of the South American darters and their relationships with other characiform fishes. Ph.D. Thesis, Univ. Michigan, Ann Arbor, Michigan.

Bullock, T.H. 1979. Processing of ampullary input in the brain: comparison of sensitivity and evoked responses among elasmobranch and siluriform fishes. J. Physiol. (Paris) 75: 397–408.

Bullock, T.H., D.A. Bodznick and R.G. Northcutt. 1983. The phylogenetic distribution of electroreception: Evidence for convergent evolution of a primitive vertebrate sense modality. Brain Res. Rev. 6: 25–46.

Campos-da-Paz, R. and J.S. Albert. 1998. The gymnotiform 'eels' of tropical América: a history of classification and phylogeny of the South American knifefishes (Teleostei: Ostariophysi: Siluriphysi), pp. 401–418. In: L.R. Malabarba, R. Reis, R.P. Vari, Z.M. Lucena and C.A.S. Lucena [eds.]. Phylogeny and Classification of Neotropical Fishes. EDIPUCRS, Porto Alegre, Brazil.

Cappetta, H., D.E. Russel and J. Braillon. 1972. Sur la découverte de characidae dans l'Eocene inférieur Français (Pisces, Cypriniformes). Bull. Mus. Natl. Hist. Nat. Paris, 3rd sér.(51): 1–51.

Cavender, T. 1986. Review of the fossil history of North American freshwater fishes, pp. 699–724. In: C.H. Hocutt and E.O. Wiley [eds.]. The Zoogeography of North American Freshwater Fishes. John Wiley and Sons, New York.

Cione, A,L. 1987. The Late cretaceous fauna of Los Alamitos, Patagonia, Argentina, II. The Fishes. Rev. Mus. Argent. Cienc. Nat. "Bernerdino Rivadavia", Paleontol. 3: 111–120.

Cione, A.L. and V.R. Prasad. 2002. The oldest known catfish (Teleostei: Siluriformes) from Asia (India, Late Cretaceous). J. Paleontol. 76(1): 190–193.

Cooke, H.B.S. 1972. The fossil mammal fauna of Africa, pp. 89–139. In: A. Keast, F.C. Erk and B. Glass [eds.]. Evolution, Mammals, and Southern Continents. State Univ. New York Press, Albany, New York.

Cracraft, J. 1973. Continental drift, paleoclimatology, and the evolution and biogeography of birds. J. Zool. Lond. (169): 455–545.

Crampton, W.G.R. and J.S. Albert. 2006. Evolution of electric signal diversity in gymnotiform fishes. pp. 647–731. In: F. Ladich, S.P. Collin, P. Moller and B.G. Kapoor [eds.] 2006. Communication in Fishes. Volume 2. Science Publishers, New Hampshire, USA.

Darlington, P.J. 1957. Zoogeography: the Geographical Distribution of Animals. John Wiley and Sons, New York

De la Peña, A. and R. Soler-Gijón. 1996. The first siluriform fish from the Cretaceous—Tertiary boundary interval of Eurasia. Lethaia 29: 85–86.

de Pinna, M.C.C. 1993. Higher-level phylogeny of Siluriformes, with a new classification of the order (Teleostei, Ostariophysi). Ph.D. thesis, City Univ. New York, New York.

de Pinna, M.C.C. 1998. Phylogenetic Relationships of Neotropical Siluriformes (Teleostei: Ostariophysi): Historical Overview and Synthesis of Hypotheses, pp. 279– 330. In: L.R. Malabarba, R. Reis, R.P. Vari, Z.M. Lucena and C.A.S. Lucena [eds.]. Phylogeny and Classification of Neotropical Fishes. EDIPUCRS, Porto Alegre, Brazil.

de Santana, C.D. 2007. Sistemática e biogeografia da Família Apteronotidae Jordan 1900 (Otophysi: Gymnotiformes). Ph.D. thesis, INPA, Manaus, Brasil.

Dietz, R.S. and J.C. Holden 1970. Reconstruction of Pangea: breakup and dispersion of continents, Permian to present. J. Geophys. Res. 75(26): 4939–4956.

Dietz, R.S. and W.P. Sproll. 1970. Fit between Africa and Antarctica: a continental drift reconstruction. Science 167: 1612–1614.

Dimmick, W.W. and A. Larson. 1996. A molecular and morphological perspective on the phylogenetic relationships of the otophysan fishes. Mol. Phylogenet. Evol. 6(1): 120–133.

Farber, J. and H. Rahn. 1970. Gas exchange between air and water and the ventilation pattern in the electric eel. Respir. Physiol. 9: 151–161.

Felsenstein, J. 1985. Confidence limits on phylogenies: An approach using the bootstrap. Evolution 39: 783–791.

Felsenstein, J.F. 2004. Inferring Phylogenies. Sinauer Associates, Inc., Sunderland, Massachusetts.

Filleul, A. and J.G. Maisey. 2004. Redescription of Santanichthys diasii (Otophysi, Characiformes) from the Albian of the Santana Formation and comments on its implications for otophysan relationships. Am. Mus. Novitates 3455: 21.

Fink, S.V. and W.L. Fink. 1981. Interrelationships of the ostariophysan fishes (Pisces, Teleostei). Zool. J. Linn. Soc. 72: 297–353.

Fink, S.V. and W.L. Fink. 1996. Interrelationships of the ostariophysan fishes (Teleostei), pp. 209–249. In: M. Stiassny, L.R. Parenti and G.D. Johnson [eds.]. Interrelationships of Fishes. Academic Press, New York.

Fink, S.V., P.H. Greenwood and W.L. Fink. 1984. A critique of recent work on fossil ostariophysan fishes. Copeia 1984(4): 1033–1041.

Frizzel, D.L. 1965. Otoliths of a new fish (Vorhisia vulpes, n. gen., n. sp. Siluroidei?) from the upper Cretaceous of South Dakota. Copeia 1965(2): 178–181.

Frizzel, D.L. and J.W. Koenig. 1973. Upper Cretaceous ostariophysine (Vorhisia) redescribed from unique association of utricular and lagenar otoliths (lapillus and asteriscus). Copeia 1973(4): 692–698.

Gayet, M. 1982a. Découverte dans le Crétacé supérieur de Bolivie des plus anciens Characiformes connus. C.R. Acad. Sci. Paris 294(Série II): 1037–1040.

Gayet, M. 1982b. Cypriniformes Crétacés en Amérique du Sud. C.R. Acad. Sci. Paris 295(III): 1037–1040.

Gayet, M. 1988. Le le plus ancien crâne de siluriformes: Andinichthys boliviuanensis nov. gen., nov. sp. (Andinichthyidae nov. fam.) di Maatrichtien de Tiupampa (Bolivie). C.R. Acad. Sci. Paris 307(2): 833–836.

Gayet, M. 1990. Nouveaux Siluriformes du Maatrichtien de Tiupampa (Bolivie). C.R. Acad. Sci. Paris 310(2): 867–872.

Gayet, M. and P. Brito. 1989. Ichthyofaune nouvelle du Crétacé supérieur du Grup Baurú (Etats de São Paulo et Minas Gerais, Brésil). Geobios 22(6): 841–847.

Gayet, M. and F.J. Meunier. 1991. Première découverte de Gymnotiformes fossiles (Pisces, Ostariophysi) dans le Miocène supérieur de Bolivie. C.R. Acad. Sci. Paris 313(Ser. II): 471–476.

Géry, J. 1969. The fresh-water fishes of South America, pp. 828–848. In: E.J. Fittkau, J. Illes, H. Klinge, G.H. Schwabe and H. Sioli [eds.]. Biogeography and Ecology in South America. W. Junk Publ., The Hague.

Gosline, W.A. 1944. The problem of derivation of the South American and African fresh-water fish faunas. An. Acad. Bras. de Cien. XVI(8): 211–223.

Gosline, W.A. 1973. Considerations regarding the phylogeny of cypriniform fishes, with special reference to structures associated with feeding. Copeia 1973(4): 761–776.

Grande, L. 1984. Paleontology of the Green River formation with a review of the fish fauna. Bull. Geol. Surv. Wyoming. 2nd ed. 63: 1–333.

Grande, L. 1987. Redescription of *Hypsidoris farsonensis* (Teleostei: Siluriformes) with a reassessment of its phylogenetic position. J. Vert. Paleont. 7: 24–54.

Grande, L. and J.T. Eastman. 1986. A review of the Antarctic ichthyofaunas in light of new fossil discoveries. Paleontology 29(1): 113–117.

Grande, L., J.T. Eastman and T. Cavender. 1982. *Amyzon gosiutensis*, a new catostomid fish from the Green River Formation. Copeia 1982(3): 523–532.

Grande, T. and F. Poyato-Ariza. 1999. Phylogenetic relationships of fossil and Recent Gonorynchiform fishes (Teleostei: Ostariophysi). Zoological Journal of the Linnean Society. 125(2): 197–238.

Greenwood, P.H., D.E. Rosen, S.H. Weitzman and G.S. Myers. 1966. Phyletic studies of teleostean fishes, with a provisional classification of living forms. Bull. Am. Mus. Nat. Hist. 131(4): 339–455.

Hagedorn, M. and C. Carr. 1985. Single electrocytes produce a sexually dimorphic signal in South American electric fish, *Hypopomus occidentalis* (Gymnotiformes, Hypopomidae). J. Comp. Physiol. A 33: 254–265.

Hagedorn, M. and R. Zelick. 1989. Relative dominance among males is expressed in the electric organ discharge characteristics of a weakly electric fish. Anim. Behav. 38: 520–525.

Hagedorn, M., M. Womble and T.E. Finger. 1990. Synodontid catfish: a new group of weakly electric fish. Behavior and Anatomy. Brain Behav. Evol. 5: 268–277.

Hardman, M. 2005. The phylogenetic relationships among non-diplomystid catfishes as inferred from mitochondrial cytochrome b sequences; the search for the ictalurid sister taxon (Otophysi: Siluriformes). Mol. Phylogenetics and Evolution 37 (2005): 700–720.

Heiligenberg, W. 1973. Electrolocation of objects in the electric fish *Eigenmannia* (Rhamphichthyidae, Gymnotoidei). J. Comp. Physiol. 87: 137–164.

Heiligenberg, W. 1975. Theoretical and experimental approaches to spatial aspects of electrolocation. J. Comp. Physiol. 103: 247–272.

Heiligenberg, W. 1977. Principles of electrolocation and jamming avoidance in electric fish—a neuroethological approach. In: V. Braitenberg [ed.]. Serie: Studies of Brain Function. Springer-Verlag, Berlin.

Heiligenberg, W. 1990. Electrosensory systems in fish. Synapse 6(2): 196–206.

Heiligenberg, W. and J. Bastian. 1981. Especificidade das descargas do orgao eletrico em especies de gymnotiformes simpatricos do Rio Negro. Acta Amazonica 11(3): 429–437.

Hopkins, C. 1974a. Electric communication: function in the social behavior of *Eigenmannia virescens*. Behavior 50: 270–305.

Hopkins, C. 1974b. Electric communication in the reproductive behavior of *Sternopygus macrurus* (Gymnotiformes). Z. Tierpsychol. 35: 518–553.

Hopkins, C.D., N.C. Comfort, J. Bastian and A.H. Bass. 1990. Functional analysis of sexual dimorphism in an electric fish, *Hypopomus pinnicaudatus*, order Gymnotiformes. Brain Behav. Evol. 35(6): 350–67.

Hopkins, D.M. 1967. The Cenozoic history of Beringia—a synthesis, pp. 451–484. In: D.M. Hopkins [ed.].The Bering Land Bridge. Stanford University Press, Stanford, California.

Hulen, K.G., W.G.R. Crampton and J.S. Albert. 2005. Phylogenetic systematics and historical biogeography of the Neotropical electric fish *Sternopygus* (Teleostei: Gymnotiformes). Systematics and Biodiversity 3 (4): 407–432.

Inoue, J.G., M. Miya, K. Tsukamoto and M. Nishida. 2004. Mitogenic evidence for the monophyly of elopomorph fishes (Teleostei) and the evolutionary origin of the leptocephalus larva. Mol. Phylogenetics ad Evolution, 32 (2004): 274–286.

Jardine, N. and D. McKenzie. 1972. Continental drift and the dispersal and evolution of organisms. Nature 235: 20–24.

Jordan, D.S. 1923. Classification of fishes including families and genera as far as known. Stanford Univ. Publ., Univ. Ser. Biol. Sci. 3: 77–243.

Kalmijn, A. 1974. The detection of electric fields from inanimate and animate sources other than electric organs, pp. 147–200. In: A. Fessard [ed.]. Electroreceptors and other specialized receptors in lower vertebrates. Handbook of Sensory Physiology. Springer-Verlag, Berlin.

Kalmijn, A.J. 1987. Detection of weak electric fields, pp. 151–186. In: J. Atema, R.R. Fay, A.N. Popper and W.N. Tavolga [eds.]. Sensory Biology of Aquatic Animals. Springer-Verlag, Berlin.

Kirschbaum, F. 1983. Myogenic electric organ precedes the neurogenic organ in apteronotid fish. Naturwissenschaften 70: 205–207.

Knudsen, E.I. 1974. Behavioral thresholds to electric signals in high frequency electric fish. J. Comp. Physiol. 91: 333–353.

Knudsen, E.I. 1975. Spatial aspects of electric fields generated by weakly electric fish. J. Comp. Physiol. 99: 103–118.

Kocher, T.D., W.K. Thomas, A. Meyer, S.V. Edwards, S. Pääbo, F.X. Villablanca and A.C. Wilson. 1989. Dynamics of mitochondrial DNA evolution in animals. Proc. Natl. Acad. Sci. USA 86: 6196–6200.

Lavoué, S., M. Miya, J.G. Inoue, K. Saitoh, N.B. Ishiguro and M. Nishida. 2005. Molecular systematics of the gonorynchiform fishes (Teleostei) based on whole mitogenome sequences: Implications for higher-level relationships within the Otocephala. Mol. Phylogenetics and Evolution 37 (2005): 165–177.

Le Pichon, X. 1968. Sea-floor spreading and continental drift. J. Geophys. Res. 73: 3661–3697.

Lissmann, H. 1951. Continuous electrical signals from the tail of a fish Gymnarchus niloticus Cuv. Nature. 167(4240): 201–202.

Lovejoy, N.R., E. Bermingham and A. Martin. 1998. Marine incursion into South America. Nature 396: 421–422.

Lundberg, J. 1993. African-South American freshwater fish clades and continental drift: problems with a paradigm, pp. 157–199. In: P. Goldblatt [ed.]. Biological Relationships between Africa and South America. Yale University Press, New Haven, Connecticut.

Lundberg, J.G. and J.N. Baskin. 1969. The caudal skeleton of the catfishes, order Siluriformes. Am. Mus. Novitates 2399: 1–49.

Lundberg, J.G., J.P. Sullivan, R. Rodiles-Hernández and D. Hendrickson. 2007. Discovery of African roots for the Mesoamerican Chiapas catfish, Lacantunia enigmatica, requires an ancient intercontinental passage. Proc. Acad. Nat. Sci. Phil. 156: 39–53.

Mago-Leccia, F. 1976. Venezuelan gymnotiform fishes: a preliminary study for a revision of the group in South America. Ph.D., Universidad Central de Venezuela.

Mago-Leccia, F. 1978. Los peces de la familia Sternopygidae de Venezuela. Acta Cient. Venez. 29: 1–89.

Mago-Leccia, F. 1994. Electric fishes of the continental waters of America. Vol. XXIX. Biblioteca de la academia de ciencias fisicas, Caracas: Clemente editores, C. A.

Martin, H. 1969. A critical review of the evidence for a former direct connection of South America with Africa, pp. 25–53. In: E.J. Fittkau, J. Illes, H. Klinge, G.H. Schwabe and H. Sioli [eds.]. Biogeography and Ecology in South America. W. Junk Publ., The Hague.

McElhinny, M.W. 1970. Formation of the Indian Ocean. Nature 228(5275): 977–979.

McKenna, M.C. 1972. Was Europe connected directly to North America prior to Middle Eocene? Evol. Biol. 6: 179–189.

Meyer, A., C.H. Biermann and G. Ortí. 1993. The phylogenetic position of the zebrafish (Danio rerio), a model system in developmental biology: an invitation to the comparative method. Proc. Roy. Soc. Lond. B, Biol. Sci. 252(1335): 231–236.

Murray A.M. and L.A. Budney. 2003. A new species of catfish (Claroteidae, Chrysichthys) from an Eocene Crater lake in east Africa. Can. J. Earth Sci. 40(7): 983–993.

Myers , G.S. 1949. Salt-tolerance of fresh-water fish groups in relation to zoogeographical problems. Bijdr Dierk 28: 315–322.

Nelson, J.S. 2006. Fishes of the world. 4th Edition. John Wiley and Sons, Inc., Hoboken, New Jersey, USA.

Nichols, J.T. 1930. Speculation on the history of the Ostariophysi. Copeia 1930(4): 148–151.

Novacek, M.J. and L.G. Marshall. 1976. Early biogeographic history of ostariophysan fishes. Copeia 1976(1): 1–12.

Ortí, G. 1997. The radiation of Characiform fishes: evidence from mitochondrial and nuclear DNA sequences, pp. 215–239. In: T. Kocher and C. Stepien [eds.]. Molecular Systematics of Fishes. Academic Press, San Diego.

Ortí, G. and A. Meyer. 1996. Molecular evolution of ependymin and the phylogenetic resolution of early divergences among euteleost fishes. Mol. Biol. Evol. 13(4): 556–573.

Ortí, G. and A. Meyer. 1997. The radiation of Characiform fishes and the limits of resolution of mitochondrial ribosomal DNA sequences. Syst. Biol. 46(1): 75–100.

Palumbi, S., A. Martin, S. Romano, W.O. McMillan, L. Stice and G. Grabowski. 1991. The simple fool's guide to PCR. University of Hawaii, Honolulu.

Patterson, C. 1975. The distribution of Mesozoic freshwater fishes. Mem. Mus. Natl. Hist. Nat., Paris Serie A, Zool. 88: 155–174.

Peng, Z., J. Wang and S. He. 2006a. The complete mitochondrial genome of the helmet catfish Cranoglanis bouderius (Siluriformes: Cranoglanididae) and the phylogeny of otophysan fishes. Gene 376 (2006): 290–297.

Peng, Z., S. He, J. Wang, W. Wang and R. Diogo. 2006b. Motochondrial molecular clocks and the origin of the major Otocephalan clades (Pisces: Teleostei): A new insight. Gene 370 (2006): 113–124.

Pitmann, W.C., III, S. Cande, J. LaBrecque and J. Pindell. 1993. Fragmentation of Gondwana: the separation of Africa from South America, pp. 15–34. In: P. Goldblatt [ed.]. Biological Relationships between Africa and South America. Yale University Press, New Haven, Connecticut.

Poyato-Ariza, F.J. 1994. A new Early Cretaceous gonorynchiform fish (Teleostei: Ostariophysi) from Las Hoyas (Cuenca, Spain). Occ. Pap. Mus. Nat. Hist. n.164, 37 pp. University of Kansas, Lawrence, Kansas.

Regan, C.T. 1911. The classification of the Teleostean fishes of the order Ostariophysi - 1. Cyprinoidea. Ann. Mag. Nat. Hist. ser. 8, (8): 13–32.

Roberts, T.R. 1973. Interrelationships of ostariophysans, pp. 373–395. In: P.H. Greenwood, R.S. Miles and C. Patterson [eds.]. Interrelationships of Fishes. Academic Press, London.

Rosen, D.E. and P.H. Greenwood. 1970. Origin of the Weberian apparatus and the relationships of the ostariophysan and gonorhynchiform fishes. Am. Mus. Novitates 2428: 1–25.

Sagemehl, M. 1885. Beiträgc zur vergleichenden Anatomie der Fische, III. Das Cranium der Characiniden nebst allgemeinen Bemerkungen üben die mit einem Weber'schen Apparat versehenen Physostomen-familien. Morph. Jahrb. Leipzig (10): 1–119.

Saitoh, K., M. Miya, J.G. Inoue, N.B. Ishiguro and M. Nishida. 2003. Mitochondrial genomics of ostariophysan fishes: perspectives on phylogeny and biogeography. J. Mol. Evol. 56: 464–472.

Schmitt, R. 2005. Filogeografia de Hypopygus lepturus Hoedeman, 1962 (Gymnotiformes: Rhamphicthyidae) ao longo do médio rio Negro, Amazônia. M.Sc. thesis, INPA, Manaus, Brasil.

Smith, A.G. and A. Hallam. 1970. The fit of southern continents. Nature 225: 139–144.
Sullivan, J.P. 1997. A phylogenetic study of the Neotropical hypopomid electric fishes (Gymnotiformes: Rhamphichthyoidea). Ph.D. thesis, Duke University, Durham, U.S.A.
Sullivan, J., J. Lundberg and M. Hardman. 2006. A phylogenetic analysis of the major groups of catfishes (Teleostei: Siluriformes) using rag1 and rag2 nuclear gene sequences. Mol. Phylogenetics and Evolution 412 (2006): 636–662.
Swofford, D.L. 2002. PAUP*. Phylogenetic Analysis Using Parsimony (*and Other Methods). Version 4. Sinauer Associates, Sunderland, Massachusetts.
Szabo, T. 1965. Sense organs of the lateral line system in some electric fish of the gymnotidae, mormyridae and gymnarchidae. J. Morphol. 117: 229–250.
Szabo, T., A.J. Kalmijn, P.S. Enger and T.H. Bullock. 1972. Microampullary organs and a submandibular sense organ in the fresh water ray Potamotrygon. J. Comp. Physiol. 79: 15–27.
Triques, M.L. 1993. Filogenia dos gêneros de Gynotiformes (Actinopterygii, Ostariophysi), com base em caracteres esqueléticos. Comunicações do Museu de Ciências da PUCRS, sér. zool., 6: 85–130.
Triques, M.L. 2005. Análise cladística dos caracteres de anatomia externa e esquelética de Apteronotidae (Teleostei: Gymnotiformes). Lundiana 6(2): 121–149.
Valencio, D.A. and J.F. Vilas. 1969. Age of separation of South America and Africa. Nature 223(5213): 1353–1354.
Vari, R.P. 1979. Anatomy, relationships and classification of the families Citharinidae and Distichodontidae (Pisces, Characoidea). Bull. Br. Mus. Nat. Hist. (Zool) 36(2): 261–344.
Vilas, J.F. and D.A. Valencio. 1970a. Paleogeographic reconstructions of the Gondwana continents based on paleomagnetic and sea-floor spreading data. Earth Planet Sci. Lett. 7: 397–405.
Vilas, J.F. and D.A. Valencio. 1970b. The recurrent Mesozoic drift of South America and Africa. Earth Planet Sci. Lett. 7: 441–444.
Windley, B.F. 1984. The Evolving Continents. John Wiley and Sons, Chichester.
Wu, X., Y. Chen, X. Chen and J. Chen. 1981. A taxonomical system and phylogenetic relationship of the families of the suborder Cyprinoidei (Pisces). Sci. Sinica 24(4): 563–572.
Zakon, H.H. 1986. The electroreceptive periphery, pp. 103–156. In: T.H. Bullock and W. Heiligenberg [eds.]. Electroreception. Wiley Series in Neurobiology. John Wiley and Sons, New York.
Zakon, H.H. 1987. The electroreceptors: diversity in structure and function, pp. 813–850. In: J. Atema, R.R. Fay, A.N. Popper and W.N. Tavolga [eds.]. Sensory Biology of Aquatic Animals. Springer-Verlag, Berlin.
Zaragüeta-Bagils, R., S. Lavoué, A. Tillier, C. Bonillo and G. Lecointre. 2002. Assessment of otocephalan and protacanthopterygian concepts in the light of multiple molecular phulogenies. C.R. Biologies 325 (2002): 1191–1207.

14

A Nomenclatural Analysis of Gonorynchiform Taxa

William N. Eschmeyer[1], Terry Grande[2] and Lance Grande[3]

Abstract

The taxonomic history among nominal gonorynchiform taxa is complex. Until recently, most gonorynchiform taxa were aligned with non-gonorynchiform groups, many of which have no close relationship to the order in its present form. Because of this complexity, researchers working on the group, especially those investigating species-level questions, have spent countless hours tracking down type specimens and trying to sort out nomenclatorial problems. The goal of this chapter, therefore, is to provide a current list of all nominal (fossil and Recent) gonorynchiform taxa along with synonomies, and the location and catalogue numbers of type material. The arrangement of taxa reflects the phylogenetic organization presented in this volume. Three families are recognized: Chanidae, Gonorynchidae and Kneriidae. The genus *Phractolaemus*, unlike in the more traditional classifications, is contained within the Kneriidae. Within each gonorynchiform family, all nominal genera and species are listed with synonomies.

[1] California Academy of Sciences, Department of Ichthyology, Golden Gate Park, 55 Music Concourse Drive, San Francisco, California 94118, USA.
[2] Loyola University Chicago, 1032 West Sheridan Road, Chicago, Illinois 60626, USA.
[3] Field Museum of Natural History, Roosevelt Road at Lake Shore Drive, Chicago, Illinois 60305, USA.

Introduction

The history of the order Gonorynchiformes is long and complex. Taxa (e.g., *Chanos*) have often been aligned with the clupeomorphs, salmonids, or cyprinids based on overall similarities. As a result, some taxa carry a labyrinth of synonomies dating back to Forsskål (1775) or Lacepède (1803). Tracking these synonomies through the vast taxonomic literature is no small task, but synonomies provide important information about nomenclatural history. Synonomies also reflect our changing views of evolution, diversity of taxa, and the development of systematics as a science.

Research on Recent gonorynchiform taxa has revealed that species diversity is higher than previously thought (Moritz *et al.* 2006). Also, a significant body of work based on fossil forms has been published of late. Several new fossil taxa have been described or redescribed. In this chapter we provide a list of all the taxa in the Order Gonorynchiformes including their respective synonomies, type material and collection status. It is our goal that this chapter will provide important historical information about gonorynchiform fishes, and that it will serve as a practical tool to those interested in the nomenclatural history of this group.

Materials and Methods

For each genus group, the valid name is given first, followed by synonyms if any. The original citation to the name, the type species, and the method by which the type species was established are provided. For each species, we provide the original citation, the type locality, and the location of primary type specimen(s). Synonyms are grouped under the current valid species name. Summary lists of genera and species names are given in alphabetical order at the end of each family account. Unavailable species names, ones that for technical reasons cannot be used, are also provided.

Primary type specimens are listed for species. Secondary types, such as paratypes, can be found in Eschmeyer *et al.* 1998 and online at http://research.calacademy.org/research/ichthyology/catalog/fishcatmain.asp

Miscellaneous abbreviations are R. (River), c and s (cleared and stained). Institutional Abbreviations are as follows:

ANSP	Academy of Natural Sciences, Philadelphia, USA
BMNH	Natural History Museum, London. Formerly British Museum of Natural History
CAS	California Academy of Sciences, San Francisco, USA
CM	Carnegie Museum of Natural History, Pittsburgh, USA
DGM-DNPM	Divisão de Geologia e Mineralogia, Departamento Nacional de Producãno Mineral, Rio de Janeiro, Brazil
GIN	Geological Institute of Munster, Germany

HUJ	Hebrew University of Jerusalem, Israel
IPS	Instituto de Paleontologia Sabadell
IRSNB	Institut Royal des Sciences Naturelles de Belgique, Brussels
KUVP	Division of Vertebrate Paleontology of the University of Kansas Museum of Natural History
MD	Museo do Dundo, Dundo, Angola
NMNZ	National Museum of New Zealand
MNHN	Muséum National d'Histoire Naturel, Paris
MOR	Museum of the Rockies, Montana, USA
MPC	Museo Provincial de Cuenca, Spain
MRAC	Musée Royal de l'Afrique Centrale, Tervuren, Belgium
Na	Museum de Nardo, Italy
NMW	Naturhistorisches Museum, Wien, Germany
PIN	Paleontological Institute, Russian Academy of Sciences, Moscow, Russia
QVMS	Queen Victoria Museum, Salisbury, Zimbabwe
R	Mr. Armando Díaz Romeral (private collection)
RMNH	Rijksmusem van Natuurlijke Historie, Leiden
ROM	Royal Ontario Museum, Toronto
SU	Stanford University. Collection now at CAS, California, USA
UAM	Universidad Autónoma de Madrid, Spain
USNM	National Museum of Natural History, Washington, D.C. Formerly United States National Museum, Washington, D.C.
ZMB	Universität Humboldt, Museum für Naturkunde, Berlin, Germany
ZMH	Universität Hamburg, Zoologisches Institut und Museum, Hamburg, Germany
ZMUC	Københavns Universitet Zoologisk Museum [Zoological Museum, University of Copenhagen)

Order Gonorynchiformes

The use of ordinal names is not governed by the International Code of Zoological Nomenclature; therefore there are no priority issues. The order Gonorynchiformes has come into use since Greenwood *et al.* 1966, although some taxa have been variously placed in other groups (see T. Grande 1996: 299). There is currently general agreement on the composition of this order (T. Grande and Poyato-Ariza 1999; this book).

Family Chanidae Jordan, 1887

The family Chanidae is probably the oldest among gonorynchiform families, with a fossil record dating to the Early Cretaceous. It is know by one extant genus, *Chanos*, and about seven fossil genera. *Chanos* is a marine Indo-Pacific form, while the fossil taxa have been collected from marine and freshwater localities in Brazil, Belgium, Equatorial Guinea, Gabon, Italy and Spain.

Nominal Genera of Family Chanidae

†*Aethalinopsis* Traquair 1911 = †*Aethalinopsis* Traquair 1911
†*Apulichthys* Taverne 1997 = †*Apulichthys* Taverne 1997
Chanos Lacepède 1803 = *Chanos* Lacepède 1803
†*Dastilbe* Jordan 1910 = †*Dastilbe* Jordan 1910
†*Gordichthys* Poyato-Ariza 1994 = †*Gordichthys* Poyato-Ariza 1994
†*Halecopsis* Delvaux and Ortlieb 1887 = †*Halecopsis* Delvaux and Ortlieb 1887
Lutodeira van Hasselt 1823 = *Chanos* Lacepède 1803
†*Neohalecopsis* Weiler 1920 = †*Neohalecopsis* Weiler 1920
†*Parachanos* Weiler 1922 = †*Parachanos* Weiler 1922
Ptycholepis Richardson 1843 = *Chanos* Lacepède 1803
Scoliostomus Rüppell 1828 = *Chanos* Lacepède 1803
†*Rubiesichthys* Wenz 1984 = †*Rubiesichthys* Wenz 1984
†*Tharrhias* Jordan and Branner 1908 = †*Tharrhias* Jordan and Branner 1908

Genus *Chanos* Lacepède 1803

Chanos Lacepède 1803: 395. Type species *Chanos arabicus* Lacepède 1803. Type by monotypy.

Lutodeira van Hasselt 1823: 330. Type species *Lutodeira indica* van Hasselt 1823. Type by monotypy.

Scoliostomus Rüppell 1828: 17. Type species *Lutodeira indica* van Hasselt 1823. Type by subsequent designation.

Ptycholepis Richardson in Richardson and Gray 1843: 218. Type species *Mugil salmoneus* Forster 1801. Type by monotypy.

Chanos chanos (Forsskål 1775)

Mugil chanos Forsskål 1775: 74, xiv (Jidda, Saudi Arabia, Red Sea). Syntypes: ZMUC P17154 (dry skin, primary specimen), ZMUC P17751 [no. 110 b] (1).

Mugil salmoneus Forster in Bloch and Schneider 1801: 121, xxxii (Pacific Ocean). No types known.

Chanos arabicus Lacepède 1803: 395, 396 (Arabian Sea). No types known.

Lutodeira indica van Hasselt 1823: 330 (Vishakhapatnam [Vizagapatam], India). No types known.

Cyprinus tolo Cuvier 1829: 276 (Vishakhapatnam [Vizagapatam], India). No types known.

Cyprinus pala Cuvier 1829: 276 (Vishakhapatnam [Vizagapatam], India). No types known.

Leuciscus zeylonicus Bennett 1833: 184 (Sri Lanka). Holotype (unique): BMNH 1855.12.26.288.

Chanos aldrovandi Risso in Cuvier and Valenciennes 1836: 176. No types known.

Chanos orientalis Valenciennes (ex Kuhl) in Cuvier and Valenciennes 1847: 197 (No locality). Holotype (unique): Type material missing but never housed in MNHN as indicated in Eschmeyer 1998 (online update 2008).

Chanos mento Valenciennes in Cuvier and Valenciennes 1847: 194 (Mauritius). Syntypes: (5) MNHN 3627 to 3629 (3).

Chanos chloropterus Valenciennes in Cuvier and Valenciennes 1847: 195 (Madeoplan, India). No types known.

Chanos nuchalis Valenciennes in Cuvier and Valenciennes 1847: 196. No types known.

Chanos lubina Valenciennes in Cuvier and Valenciennes 1847: 199, Pl. 567 [not 533] (Buru I., Moluccas Is., Indonesia). Lectotype: MNHN A 9827.

Chanos cyprinella Valenciennes in Cuvier and Valenciennes 1847: 198 (Hawaiian Is., USA). Holotype (unique): MNHN 3624.

Butirinus argenteus Jerdon 1849: 343 (Coondapoor, n. Canara, India). No types known.

Lutodeira (Chanos) mossambicus Peters 1852: 684 (Quisanga [Querimba], Mozambique). Holotype (unique): ZMB 6614.

Lutodira (Chanos) elongata Peters 1859: 412 (Sandwich I. [Hawaiian Is. or Vanuatu [New Hebrides]]). No types known.

Chanos gardineri Regan 1902: 280 (North pool of Hulule I., Male Atoll, Maldives, Indian Ocean). Syntypes: (3) BMNH 1901.12.21.141–142 (2).

Remarks: *Chanos chanos* apparently is a worldwide species with many synonyms as listed above. Treatment as one species follows Nelson (2006). To our knowledge, no molecular studies have attempted to look at population structure.

Fossils assigned to the genus *Chanos* as per Poyato-Ariza (1996). All taxa are poorly known, fragmentary and in need of revision and confirmation if they indeed belong to the genus *Chanos* or are chanids.

†*"Chanos brevis"* Heckel 1854, Early Miocene deposits of Italy

†*"Chanos zignoi"* Kner and Steindachner 1863, Early Miocene deposits of Italy

†*"Chanos forcipatus"* Heckel 1853, Eocene beds of Monte Bolca, Italy

†*"Chanos compressus"* Stinton 1977, Lower Eocene deposits of Southern England, otoliths only

†*"Chanos torosus"* Danil'Chenko 1968, Upper Paleocene of Turkmenia

†*"Chanos leopoldi"* Costa 1915, Lower Cretaceous of Pietraroia, Benevento, Italy

Genus †*Aethalinopsis* Traquair 1911

†*Aethalionopsis robustus* Traquair 1911 (Lower Cretaceous of Belgium). Lectotype: IRSNB P.1244a, b. Distribution: Germany.

Genus †*Apulichthys* Taverne 1997

†*Apulichthys gayeti* Taverne 1997 (Upper Cretaceous beds of Porto Selvaggiro, near Nardo, Italy). Holotype: Na 244. Paratypes: Na 220, 439.

Genus †Dastilbe Jordan 1910

†*Dastilbe crandelli* Jordan 1910.Type Species (Early Cretaceous of Brazil, Santana Formation, Crato member). Holotype: CMNH 5247/91, type material missing. Distribution: Brazil.
†*Dastilbe elongates* Silva-Santos 1947 (Early Cretaceous of Brazil, Santana Formation). Lectotype: DGM-DNPM 176-P. Distribution: Brazil.
†*Dastilbe moraesi* Silva-Santos 1955 (Early Cretaceous of Brazil, Santana Formation). Holotype: 593-P.DGM-DNPM. Distribution: Brazil.
†*Dastilbe batai* Gayet 1968 (Early Cretaceous deposits of Rio San Benito, Equatorial Guinea). Holotype. MNHN AFE-4a, b. Distribution: Gabon.

Remarks: Considerable debate continues concerning the species composition of †*Dastilbe*. Taverne (1981) and Dietze (2007) propose synonomyzing all species into †*D. crandelli*. Dietze (2007) further argues that †*D. batai* is synonymous with the genus †*Parachanos*.

Genus †*Gordichthys* Poyato-Ariza 1994

†*Gordichthys conquensis* Poyato-Ariza 1994 (Lower Cretaceous deposits of Las Hoyas, Spain). Holotype: MPC LH: 1228R (housed at UAM). Paratypes: MPC LH: 477, 480, 509, 550, 598, LH 602, 631, 643, 700, 735, 739, 818, 843, 928, 929, 951, 960, 985, 951, 985, 1125, 1279, 1305, 1407, 1414, 1415, 1417, 1421, 1585, 1613, 1708, 1625. R: 2179, 4986, 4989. KUVP: LH 123102–03, 12305. Distribution: Spain.

Genus †*Halecopsis* Delvaux and Ortlieb 1887

†*Halecopsis insignis* Delvaux and Ortlieb 1887 (Eocene, London Clay, London; Argiles des Flandres, Belgium and Nord of France). Syntypes: E.F.P. 143–145, I.R.Sc.N.B., IG no. 6852. Distribution: Europe.

Genus †*Neohalecopsis* Weiler 1920

†*Neohalecopsis striatus* Weiler 1920 (Oligocene of Florsheim, Rheinhessen, Germany). Holotype: Mainzer Museum, Mainz, Germany, number not provided. Distribution: Germany.

Genus †*Parachanos* Weiler 1922

†*Parachanos aethiopicus* Weiler 1922 (Cretaceous of Equatorial Guinea). Holotype: MNHN GAB-1. Distribution: Equatorial Guinea.

Genus †*Rubiesichthys* Wenz 1984

†*Rubiesichthys gregalis* Wenz 1984 (Cretaceous of Montsec, Spain). Holotype: IPS PR-4. Distribution: Spain.

Genus †*Tharrhias* Jordan and Branner 1908

†*Tharrhias araripis* Jordan and Branner 1908. Type Species (Upper Cretaceous deposits of Santana Formation, Romualdo member, Brazil). Holotype: Museu Rocha collection no. 4, part missing. Counterpart CAS 58318.

†*Tharrhias rochae* Jordan and Branner 1908 (Upper Cretaceous deposits of Santana Formation, Romualdo member, Brazil). Holotype: Museu Rocha collection no. 5, part missing. Counterpart CAS 58297.

Nominal Species of Family Chanidae (Alphabetized Summary)

aldrovandi, Chanos Risso 1836 = *Chanos chanos* (Forsskål 1775)
arabicus, Chanos Lacepède 1803 = *Chanos chanos* (Forsskål 1775)
argenteus, Butirinus Jerdon 1849 = *Chanos chanos* (Forsskål 1775)
chanos, Mugil Forsskål 1775 = *Chanos chanos* (Forsskål 1775)
chloropterus, Chanos Valenciennes 1847 = *Chanos chanos* (Forsskål 1775)
cyprinella, Chanos Valenciennes 1847 = *Chanos chanos* (Forsskål 1775)
elongata, Lutodeiro (Chanos) Peters 1859 = *Chanos chanos* (Forsskål 1775)
gardineri, Chanos Regan 1902 = *Chanos chanos* (Forsskål 1775)
indica, Lutodeira van Hasselt 1823 = *Chanos chanos* (Forsskål 1775)
lubina, Chanos Valenciennes 1847 = *Chanos chanos* (Forsskål 1775)
mento, Chanos Valenciennes 1847 = *Chanos chanos* (Forsskål 1775)
mossambicus, Lutodeira (Chanos) Peters 1852 = *Chanos chanos* (Forsskål 1775)
nuchalis, Chanos Valenciennes 1847 = *Chanos chanos* (Forsskål 1775)
orientalis, Chanos Valenciennes 1847 = *Chanos chanos* (Forsskål 1775)
pala, Cyprinus Cuvier 1829 = *Chanos chanos* (Forsskål 1775)
salmoneus, Mugil Forster 1801 = *Chanos chanos* (Forsskål 1775)
tolo, Cyprinus Cuvier 1829 = *Chanos chanos* (Forsskål 1775)
zeylonicus, Lesuciscus Bennett 1833 = *Chanos chanos* (Forsskål 1775)

Unavailable Chanid Species Names

oriental, Chanos Eydoux and Souleyet, 1850: 196, Pl. 7 (Fig. 1). *Nomen nudum*. In the synonymy of *Chanos chanos* (Forsskål 1775).
orientalis, Lutodeira Kuhl, 1847: 197. *Nomen nudum*. In the synonymy of *Chanos chanos* (Forsskål 1775).

salmonoides, Chanos Günther, 1879: 471. Misspelling. In the synonymy of
Chanos chanos (Forsskål 1775) — (Fricke 1999:81).

Family Gonorynchidae Scopoli, 1777

†Notogoneidae Jordan 1923
†Judeichthyidae Gayet 1985
†Charitosomidae Gayet 1993

This family (including all fossil and living species) is the most geographically widespread of all families in the order. The marine genus *Gonorynchus* contains five valid species (T. Grande 1999); and occurs in the Indian and Pacific Oceans. Four fossil genera are described. † *Notogoneus* Cope is a widespread freshwater genus known from the United States (Wyoming and Montana), France, Germany, Ukraine, and Australia. †*Charitosomus* van der Marck contains four species from marine deposits in Germany (1) and Lebanon (3). †*Ramallichthys* Gayet is known from a single marine species from Ramallah. †*Charitopsis* Gayet is known from one fossil species from Lebanon. †*Hakelensis* Gayet is known from one species from Lebanon. †*Judeichthys* Gayet is known from one specimen from Ramallah. According to Grande and Grande (2008), only four genera are valid: †*Notogoneus*, †*Charitosomus*, †*Ramallichthys* and †*Charitopsis*. In that publication, †*Hakeliosomus*, †*Judeichthys* and †*Ramallichthys* were synonomized. The name †*Ramallichthys* has priority.

Nominal Genera of the Family Gonorynchidae
†*Charitisomus* Marck 1885 = +*Charitisomus* Marck 1885
 Gonorynchus Scopoli 1777 = *Gonorynchus* Scopoli 1777
†*Hakeliosomus* Gayet 1993 = †*Charitisomus* Marck 1885
†*Judeichthys* Gayet 1985 = †*Judeichthys* Gayet 1985
†*Notogoneus* Cope 1885 = †*Notogoneus* Cope 1885
†*Protocatostomus* Whitfield 1890 = †*Notogoneus* Cope 1885
†*Ramallichthys* Gayet 1982 = †*Ramallichthys* Gayet 1982
Rynchana Richardson 1845 = *Gonorynchus* Scopoli 1777
†*Solenognathus* Pictet and Humbert 1866 = †*Charitisomus* Marck 1885
†*Sphenolepis* Agassiz 1844 = †*Notogoneus* Cope 1885

Unavailable Generic Names
Gonorhynchus Gronow 1763: 56. Suppressed

Genus †*Charitosomus* von der Marck 1885

†*Charitosomus* von der Marck 1885: 257. Type species †*Charitosomus formosus* Marck 1885 by monotypy.

†*Solenognathus* Pictet and Humbert, 1866: 54. Type species †*Solenognathus lineolatus* Pictet and Humbert 1866.
†*Hakeliosomus* Gayet 1993. Type species †*Spinodon hakelensis* Davis 1887.
†*Spinodon* Davis 1887
†*Charitosomus* (Davis 1898)

†*Charitosomus formosus* von der Marck 1885: 257, Pl. 24 (Fig. 1) (Upper Cretaceous. Baumberg, Westphalia; skeleton, much of head missing). Holotype: IGM 8541.

†*Charitosomus hakelensis* (Davis 1887)
[=†*Spaniodon hakelensis* Davis 1887: 591, Pl. 34 (no. 4)] (Upper Cretaceous (Cenomanian), marine. Hakel, Mt. Lebanon, Lebanon). Holotype without number, Edimburg Museum of Science and Art.

†*Charitosomus lineolatus* (Pictet and Humbert 1866)
[=†*Solenognathus lineolatus* Picket and Humbert 1866,
†*Charitosomus lineolatus* Pictet and Humbert 1901: A. S. Woodward, p. 274, pl. 15. Fig. 4] (Upper Cretaceous marine deposits of Sahel Alma, Lebanon). Holotype: Musée d'Histoire naturelle de Ville de Genève, Switzerland.

†*Charitisomus major* Woodward 1901: 272, Pl. 15 (Fig. 3) (Upper Cretaceous, Sahel Alma, Mt. Lebanon, Lebanon). Holotype (unique): BMNH P.9173 (complete fish, part of Lewis Collection).

Remarks: The phylogenetic relationships among †*Charitosomus* species are yet to be determined.

Genus †*Charitopsis* Gayet 1993

†*Charitopsis spinosus* Gayet 1993. Type species by original designation (also monotypic) (Upper Cretaceous, Hakel, Lebanon), Holotype: AMNH 3895.

Genus *Gonorynchus* Scopoli 1777

Gonorynchus Scopoli (ex Gronow) 1777: 450. Type species *Cyprinus gonorynchus* Linnaeus 1766.
Rynchana Richardson 1845: 44. Type species *Rynchana greyi* Richardson 1845. Type by monotypy.

Gonorynchus gonorynchus (Linnaeus 1766)
[=*Cyprinus gonorynchus* Linnaeus 1766: 528 (Cape of Good Hope, South Africa).
Holotype (unique): BMNH 1853.11.12.120 [Gronovius coll.] (skin).

Gonorhynchus gronovii Valenciennes in Cuvier and Valenciennes 1847: 207, Pl. 568 [not 534] (Cape of Good Hope, South Africa; Réunion I.). Syntypes: MNHN 3617 (1), 3618 (1, poor condition), 3619 (1).

Gonorhynchus brevis Kner 1867: 342, Pl. 16 (Fig. 1) Types lost. (Island of St. Paul or Cape of Good Hope, South Africa)]. Distribution: Southeastern Atlantic, Indian Ocean, w. Pacific.

Gonorynchus abbreviatus Temminck and Schlegel 1846: 217, Pl. 103 (Figs. 5, 5a–b) (Japan). Lectotype: RMNH 2686a. Paralectotype: RMNH 2686b. Distribution: Western North Pacific.

Gonorynchus forsteri Ogilby 1911: 34 (New Zealand).Original type series lost. Neotype: NMNZ P 33020. Distribution: Southwestern Pacific.

Gonorynchus greyi (Richardson 1845)

[=*Rynchana greyi* Richardson 1845:45, Pl. 29 (Figs. 1–6) (W. Australia and Port Nicholson, Cook's Strait, New Zealand). Syntypes: BMNH 1855.9.19. 967 (1).

Gonorhynchus parvimanus Ogilby 1911:34 (Moreton Bay, Queensland, Australia). Holotype: J. T. Jameson priv. coll. (considered lost), Woody Point, Australia]. Distribution: Southwestern Pacific.

Gonorhynchus moseleyi Jordan and Snyder 1923: 347, Fig. 1 (Honolulu, Oahu I., Hawaiian Is., USA). Holotype (unique): SU 23239. Distribution: Hawaiian Is.

Remarks: The genus *Gonorynchus* was reviewed by T. Grande 1999. Five valid species were diagnosed. All species are known from the Indian and Pacific Oceans, although two specimens were collected off the coast of Chile (Island de San Fèlix), thus expanding the range of the genus.

Genus †*Judeichthys* Gayet 1985

†*Judeichthys haasi* Gayet 1985: (Upper Cretaceous, marine deposits of Ramallah near Jerusalem). Holotype (unique): HUJ AJ-432 (nearly complete skeleton). Distribution: Ramallah.

Genus †*Notogoneus* Cope 1885

†*Notogoneus* Cope 1885: 1091. Type species *Notogoneus osculus* Cope 1885 by monotypy.

†*Sphenolepis* Agassiz 1844: 87. Type species *Cyprinis squamosus* Blainville 1818. Type by subsequent designation of Woodward 1901:276.

†*Protocatostomus* Whitfield 1890: 120. Type species *Protocatostomus constablei* Whitfield 1890 by monotype.

†*Notogoneus osculus* Cope 1885: 1091 (Early Eocene, Fossil Butte Formation of Green River Formation, Wyoming, USA). Syntypes AMNH 2503 and 2504. Subsequent designations as per Grande and Grande 2008): Lectotype AMNH 2503, Paralectotype 2504.
[=†*Protocatostomus constablei* Whitfield 1890. Holotype: AMNH 3900 (Eocene, Green River Formation, USA)]. Distribution: Wyoming, USA.

†*Notogoneus cuvieri* (Agassiz 1843)
[=†*Sphenolepis cuvieri* Agassiz 1843–44: Pt. 1, 13, pt. 2, 89, pl. 44 (Upper Oligocene of Montmartre, near Paris, France). Holotype (unique): MNHN P.11308 (nealy complete skeleton)]. Distribution: France.

†*Notogoneus gracilis* Sytchevskaya 1986 (Upper Paleocene/Lower Eocene drill core sample from Boltyshka, Ukraine). Holotype: PIN 3119/739. Distribution: Ukraine.

†*Notogoneus janeti* Priem 1908 (Upper Eocene or Lower Oligocene deposits of Paris Basin, France). According to Priem (1908) the type specimens are in M. Janet's private collection, not MNHN, and now considered lost. Distribution: France.

†*Notogeneus longiceps* (von Meyer 1848)
[=†*Cobitis longiceps* von Meyer 1848 [Fowler says Mayer]: 1, 51, Pl 20 (Fig. 2) (Upper Oligocene, Mayence basin, Germany). Partial fish]. Distribution: Germany.

†*Notogoneus montanensis* Grande and Grande 1999: 614, Figs. 2–4. (Late Cretaceous (Campanian), Two Medicine Formation, ne. Montana, USA). Holotype (unique): MOR 1065 (skeleton, missing much of head). Distribution: Montana, USA.

†*Notogoneus parvus* Hills 1934: 164, Pl. 20 (Oligocene, South Queensland, Australia). Distribution: Australia. Holotype: QMC 4.

†*Notogoneus squamosseus* (Blainville 1818)
[=†*Cyprinus squamosseus* Blainville 1818: 371 (Lower Oligocene, Aix-en-Provence, France)]. Distribution: France.

Remarks: Additional †*Notogoneus* material known from fragmentary material and thus not assigned to species are: (1) subopercle and scales from late Paleocene freshwater deposits of Alberta, Canada (Wilson 1980); (2) isolated scale fragments listed as †*Notogoneus* sp. cf. †*N. osculus* from Eocene freshwater deposits of the Coalmont Formation of northern Colorado (Wilson 1981); and (3) fragments listed by Gaudant (1981) as †*Notogoneus* sp. from various Tertiary localities of France, Germany and England.

 As discussed in Grande and Grande (1999, 2008), the widespread distribution of †*Notogoneus* suggests that the genus is much older than

Cretaceous in origin, or that it is a taxon that lived in both freshwater and marine environments during its lifetime. This lends support to the hypothesis of Grande and Bucheim (1994) that †*Notogoneus osculus* was possibly migratory, very much like *Gonorynchus*.

Genus †*Ramallichthys* Gayet 1982

†*Ramallichthys orientalis* Gayet 1982: 405. (Upper Cretaceous, marine deposits of Ramallah, near Jerusalem, in the Middle East). Holotype (unique): HUJ EY-386 (nearly complete skeleton). Distribution: Ramallah.

Nominal Species of the Family Gonorynchidae (Alphabetized Summary)

abbreviatus, *Gonorynchus* Temminck and Schlegel 1846 = *Gonorynchus abbreviatus* Temminck and Schlegel 1846

brevis, *Gonorhynchus* Kner 1867 = *Gonorynchus gonorynchus* (Linnaeus 1758)

†*constablei*, *Protocatostomus* Whitfield 1890 = †*Notogoneus osculus* (Cope 1885)

†*cuvieri*, *Sphenolepis* Agassiz 1843 = †*Notogoneus cuvieri* (Agassiz 1843)

†*formosus*, *Charitisomus* Marck 1885 = †*Charitisomus formosus* Marck 1885

forsteri, *Gonorrynchus* Ogilby 1911 = *Gonorynchus forsteri* Ogilby 1911

gonorynchus, *Cyprinus* Linnaeus 1766 = *Gonorynchus gonorynchus* (Linnaeus 1766)

†*gracilis*, *Notogoneus* Sytchevskaya 1986 = †*Notogoneus gracilis* Sytchevskaya 1986

greyi, *Rynchana* Richardson 1845 = *Gonorynchus greyi* (Richardson 1845)

gronovii, *Gonorhynchus* Valenciennes 1847 = *Gonorynchus gonorynchus* (Linnaeus 1766)

†*haasi*, *Judeichthys* Gayet 1985 = †*Judeichthys haasi* Gayet 1985

†*hakelensis*, *Spaniodon* = †*Charitisomus hakelensis* (Davis 1887)

†*janeti*, *Notogoneus* Priem 1908 = †*Notogoneus janeti* Priem 1908

†*longiceps*, *Cobitis* von Meyer 1848 = †*Notogoneus longiceps* (von Meyer 1848)

†*major*, *Charitisomus* Woodward 1901 = †*Charitisomus major* Woodward 1901

†*montanensis*, *Notogoneus* Grande and Grande 1999 = †*Notogoneus montanensis* Grande and Grande 1999

moseleyi, *Gonorhynchus* Jordan and Snyder 1923 = *Gonorynchus moseleyi* Jordan and Snyder 1923

†*orientalis*, *Ramallichthys* Gayet 1982 = †*Ramallichthys orientalis* Gayet 1982

†*osculus*, *Notogoneus* Cope 1885 = †*Notogoneus osculus* Cope 1885

parvimanus, *Gonorynchus* Ogilby 1911 = *Gonorynchus greyi* (Richardson 1845)

†*parvus*, *Notogoneus* Hills 1934 = †*Notogoneus parvus* Hills 1934

†*squamosseus*, *Cyprinus* Blainville 1818 = †*Notogoneus squamosseus* (Blainville 1818)

Unavailable Species Names
conorynchus, Cyprinius Bloch and Schneider, 1801: 443. In the synonymy of
   Gonorynchus gonorynchus (Linneaus 1758)—(Grande 1999: 456) *gayi*,
   Gonorynchus Benham, 1919: 8, Misspelling of greyi.

Family Kneriidae Steindachner, 1866
This family is restricted to the freshwaters of Africa today, and has no
published fossil record. Phractolaemus is included within Kneriidae based
on the work of Grande and Poyato-Ariza (1999) and Poyato-Ariza, Grande,
Diogo (this volume).

Nominal Genera of Family Kneriidae
Angola Myers 1928 = *Kneria* Steindachner 1866
Cromeria Boulenger 1901 = *Cromeria* Boulenger 1901
Grasseichthys Géry 1964 = Grasseichthys Géry 1964
Kneria Steindachner 1866 = *Kneria* Steindachner 1866
Parakneria Poll 1965 = *Parakneria* Poll 1965
Phractolaemus Boulenger 1901 = *Phractolaemus* Boulenger 1901
Xenopomichthys Pellegrin 1905 = *Kneria* Steindachner 1866

Genus *Cromeria* Boulenger 1901

Cromeria Boulenger 1901: 445. Type species *Cromeria nilotica* Boulenger 1901.
   Type by monotypy.

Cromeria nilotica Boulenger 1901: 445 (White Nile R., Sudan). Loctotype
   BMNH 1907.12.24.461–466 (6) Fashoda, BMNH 1907.12.2. 467–473 (7)
   Lake No.

Cromeria occidentalis Daget 1954
Cromeria nilotica occidentalis Daget 1954: 65, Fig. 11 (Upper Niger R.). Syntypes:
   (25 + 8 + 62) MNHN 1954-009 (6) Faranah, 1960-0398 (17) Faranah,
   1960-0399 (8) Bissikrima. Distribution: Eastern Africa.

Remarks: Until most recently, *Cromeria* was represented by one genus and
two subspecies. Differences in skull and postcranial morophology point to
additional species (Moritz *et al.* 2006).

Genus *Grasseichthys* Géry 1964

Grasseichthys Géry 1964: 4805. Type species *Grasseichthys gabonensis*
   Géry 1964.
Type by original designation (also monotypic).

Grasseichthys gabonensis Géry 1964: 4806 [2], Fig. (Trib. of Ntsimy R., near
Nzingmeyong, Ivindo basin, Gabon). Holotype: MNHN 1967-0442.
   Distribution: Gabon.

Genus *Kneria* Steindachner 1866

Kneria Steindachner 1866: 769. Type species *Kneria angolensis* Steindachner 1866. Type by monotypy.
Xenopomichthys Pellegrin 1905: 145. Type species *Xenopomichthys auriculatus* Pellegrin 1905. Type by monotypy.
Angola Myers 1928: 7. Type species *Xenopomatichthys ansorgii* Boulenger 1910. Type by original designation (also monotypic).

Kneria angolensis Steindachner 1866: 770, Pl. 17 (Fig. 1) (Angola). Holotype (unique): NMW 46059. Distribution: Southern Africa.

Kneria ansorgii (Boulenger 1910)
[=*Xenopomatichthys ansorgii* Boulenger 1910: 542 (Lucala R. at Lucala, Angola). Syntypes: BMNH 1910.11.28.58–59 (2)]. Distribution: Angola.

Kneria auriculata (Pellegrin 1905)
[=*Xenopomichthys auriculatus* Pellegrin 1905: 146 (Muza R., Mozambique). Lectotype: MNHN 1905-0119]. Distribution: Southern Africa, Mozambique.

Kneria katangae Poll 1976:40 (Upemba National Park, Mubale, Zaire). Holotype: MRAC 79-1-P. 1274. Distribution: Central Africa.

Kneria maydelli Ladiges and Voelker 1961: 134 (Rua Cana, Cunene R. system, Angola). Holotype: ZMH H1329. Distribution: Africa: Angola and Namibia.

Kneria paucisquamata Poll and Stewart 1975: 155, Fig. 2 (Headwaters of Luongo R., 30 km west of Luwingu, Zambia). Holotype: ROM 28043. Distribution: Africa.

Kneria polli Trewavas 1936: 64, Pl. 1 (Figs. 1–2) (Mt. Moco, Cuvo R. system, Angola). Syntypes: (17) BMNH 1935.3.20.52–56 (5), 1935.3.20.57–63 (7), 1935.3.20.64 (1, c and s). Distribution: Angola and Zambia.

Kneria ruaha Seegers 1995: 105, Figs. 10–13 (Ruaha drainage, Kisasa R., a brook west of Ifunda, ca. 63 km west of Iringa, 35°25'E, 8°07'S, sw. Tanzania). Holotype: ZMB 32347. Distribution: Tanzania.

Kneria rukwaensis Seegers 1995: 109, Figs. 2, 15–17 (Western Lake Rukwa drainage, Chiwanda R. at Chiwanda, affluent of Momba R., 9°10'S, 32°34'E, Mbeya region, w. Tanzania). Holotype: ZMB 32352. Distribution: Africa (Tanganyika basin).

Kneria sjolandersi Poll 1967: 34, Fig. 3 (Cerilo, Cubal R., Moçâmedes dist., Huila Prov., s. Angola). Holotype: GNHM 1575. Distribution: Angola.

Kneria stappersii Boulenger 1915: 163 (Lumbumbashi R., 5–8 km downstream from Elisabethville [Lubumbashi], Zaire). Syntypes: BMNH 1920.5.26.13–14 (2), MRAC 11836-38 (3). Distribution: Central Africa.

Kneria uluguru Seegers 1995: 101, Figs. 4–6 (Upper Ruvu drainage, Sombesi R., e. slopes of Uluguru Mountains, 6°54'S, 37°51'E, e. Tanzania). Holotype: ZMB 32343. Distribution: Tanzania.

Kneria wittei Poll 1944: 1, Fig. 1 (Makala, near Albertville, Zaire). Holotype (unique): IRSNB 70. Distribution: Eastern Africa.

Genus *Parakneria* Poll 1965

Parakneria Poll 1965: 7. Type species *Parakneria damasi* Poll 1965. Type by original designation.

Parakneria abbreviata (Pellegrin 1931)
[=*Kneria cameronensis* var. *abbreviata* Pellegrin 1931: 206 (French Equatorial Africa). Lectotype: MNHN 1930-0247]. Distribution: Central Africa.

Parakneria cameronensis (Boulenger 1909)
[=*Kneria cameronensis* Boulenger 1909: 171, Fig. 136 (Ja R., Congo system, s. Cameroon). Lectotype: BMNH 1909.4.29.34]. Distribution: Central Africa.

Parakneria damasi Poll 1965: 9, Figs. 1–2 (Lualaba, N'Zilo, Congo R. basin, Zaire). Holotype (unique): MRAC 119211. Distribution: Central Africa.

Parakneria fortuita Penrith 1973: 132, Pl. 1 (Cutato R., at the bridge between Chitembo and Chimbangombe, Angola, 13°31'S, 16°25'E). Holotype: State Museum, Windhoek. Distribution: Africa: Angola and Namibia (?).

Parakneria kissi Poll 1969: 365, Fig. 2 (Hombo R., tributary of Luhoho River, Zaire). Holotype: MRAC 164696. Distribution: Central Africa.

Parakneria ladigesi Poll 1967: 325, Fig. 158 (Cuango River, Cafunfo, Angola, 8°47'S, 18°01'E). Holotype: MD 6599. Distribution: Central Africa.

Parakneria lufirae Poll 1965: 17, Figs. 9–11 (Muyé River, tributary to Lufira River, Upemba National Park, Zaire). Holotype: MRAC 79-1-P. 1336. Distribution: Central Africa.

Parakneria malaissei Poll 1969: 361, Fig. 1 (Luanza River, 3 km west of Kabiashia, Zaire). Holotype: MRAC 164681. Distribution: Central Africa.

Parakneria marmorata (Norman 1923)
[=*Kneria marmorata* Norman 1923: 695 (Kokema River, tributary of Quanza R., Angola, elev. 4000 ft). Holotype (unique): BMNH 1923.8.15.2]. Distribution: Angola.

Parakneria mossambica Jubb and Bell-Cross 1974: 2, Figs. 2a–b (Mutsambidzi River, Chiringoma Plateau, Gorongoza National Park, Mozambique). Holotype: QVMS 3108. Distribution: Mozambique.

Parakneria spekii (Günther 1868)
[=*Kneria spekii* Günther 1868: 372 (Uzaramo, between coast and Usagara, Tanzania). Lectotype: BMNH 1863.8.11.19].
[=*Kneria taeniata* Pellegrin 1922: 349 (Oukami, Tanganyika Territory). Holotype (unique): MNHN 1922-0015]. Distribution: Eastern Africa.

Parakneria tanzaniae Poll 1984: 3, Fig. 1 (Kimani River waterfalls, Tanzania). Holotype: BMNH 1976.10.21.158. Distribution: Tanzania.

Parakneria thysi Poll 1965: 22, Figs. 12–13 (Kiubo rapids, Lufira River, Congo River basin, Zaire). Holotype: MRAC 154367. Distribution: Central Africa.

Parakneria vilhenae Poll 1965: 24, Fig. (Kasai River, Kazumba Territory, Congo River basin, Zaire). Holotype: MRAC 101977. Distribution: Central Africa.

Genus Phractolaemus Boulenger 1901

Phractolaemus Boulenger 1901: 6. Type species *Phractolaemus ansorgii* Boulenger 1901. Type by monotypy.

Phractolaemus ansorgii Boulenger 1901: 6, Pl. 2 (Junction of Ethiop River and Jamieson River, Niger Delta). Syntypes: BMNH 1901.1.28.1–3 (3) and .4 (1, skeleton).
Phractolaemus spinosus Pellegrin 1925: 550 (Sangha River, French Equatorial Africa). Syntypes: BMNH 1930.3.4.1 (1), MNHN 1925-0122 to 0127 (orig. 6), USNM 92978 [ex MNHN 1929-250] (1).
Phractolaemus spinosus carpenteri Fowler 1949: 242, Figs. 9–14 (Small brook in a swamp, 8 mi. south of Oka, Congo system, French Equatorial Africa). Holotype: ANSP 71873. Distribution: Western Africa.

Remarks: The classification presented here differs from the traditional classification of Nelson (2006) in that *Phractolaemus* does not represent its own monotypic family, but is phylogenetically contained within Kneriidae.

Nominal Species of the Family Kneriidae (Alphabetized Summary)
abbreviata, *Kneria cameronensis* Pellegrin 1931 = *Parakneria abbreviata* (Pellegrin 1931)
angolensis, *Kneria* Steindachner 1866 = *Kneria angolensis* Steindachner 1866
ansorgii, *Phractolaemus* Boulenger 1901 = *Phractolaemus ansorgii* Boulenger 1901

ansorgii, Xenopomatichthys Boulenger 1910 = *Kneria ansorgii* (Boulenger 1910)
auriculatus, Xenopomichthys Pellegrin 1905 = *Kneria auriculata* (Pellegrin 1905)
cameronensis, Kneria Boulenger 1909 = *Parakneria cameronensis* (Boulenger 1909)
carpenteri, Phractolaemus spinosus Fowler 1949 = *Phractolaemus ansorgii* Boulenger 1901
damasi, Parakneria Poll 1965 = *Parakneria damasi* Poll 1965
fortuita, Parakneria Penrith 1973 = *Parakneria fortuita* Penrith 1973
gabonensis, Grasseichthys Géry 1964 = *Grasseichthys gabonensis* Géry 1964
katangae, Kneria Poll 1976 = *Kneria katangae* Poll 1976
kissi, Parakneria Poll 1969 = *Parakneria kissi* Poll 1969
ladigesi, Parakneria Poll 1967 = *Parakneria ladigesi* Poll 1967
lufirae, Parakneria Poll 1965 = *Parakneria lufirae* Poll 1965
malaissei, Parakneria Poll 1969 = *Parakneria malaissei* Poll 1969
marmorata, Kneria Norman 1923 = *Parakneria marmorata* (Norman 1923)
maydelli, Kneria Ladiges and Voelker 1961 = *Kneria maydelli* Ladiges and Voelker 1961
mossambica, Parakneria Jubb and Bell-Cross 1974 = *Parakneria mossambica* Jubb and Bell-Cross 1974
nilotica, Cromeria Boulenger 1901 = *Cromeria nilotica* Boulenger 1901
occidentalis, Cromeria nilotica Daget 1954 = *Cromeria nilotica* Boulenger 1901
paucisquamata, Kneria Poll and Stewart 1975 = *Kneria paucisquamata* Poll and Stewart 1975
polli, Kneria Trewavas 1936 = *Kneria polli* Trewavas 1936
ruaha, Kneria Seegers 1995 = *Kneria ruaha* Seegers 1995
rukwaensis, Kneria Seegers 1995 = *Kneria rukwaensis* Seegers 1995
sjolandersi, Kneria Poll 1967 = *Kneria sjolandersi* Poll 1967
spekii, Kneria Günther 1868 = *Parakneria spekii* (Günther 1868)
spinosus, Phractolaemus Pellegrin 1925 = *Phractolaemus ansorgii* Boulenger 1901
stappersii, Kneria Boulenger 1915 = *Kneria stappersii* Boulenger 1915
taeniata, Kneria Pellegrin 1922 = *Parakneria spekii* (Günther 1868)
tanzaniae, Parakneria Poll 1984 = *Parakneria tanzaniae* Poll 1984
thysi, Parakneria Poll 1965 = *Parakneria thysi* Poll 1965
uluguru, Kneria Seegers 1995 = *Kneria uluguru* Seegers 1995
vilhenae, Parakneria Poll 1965 = *Parakneria vilhenae* Poll 1965
wittei, Kneria Poll 1944 = *Kneria wittei* Poll 1944

Acknowledgements

We greatly appreciate the help from the "Catalog of Fishes" staff for data entry, proofing, programming, formatting outputs, and general assistance with this account, especially from C. Ferraris, J. Fong, M. Hoang, W. Poly and F. Ruis. Many thanks for the helpful comments and suggestions of two anonymous reviewers. This research was supported in part by a grant from the National Science Foundation to TG and LG (DEB 0128794).

References

Agassiz, L. 1843–44. Recherces sur les Poissons Fossiles.

Benham, W.B. 1919. Annual report for the year 1818. Otago Univ. Mus. 1–9.

Bennett, E.T. 1833. Characters of new species of fishes from Ceylon. Proc. Zool. Soc. Lond. 1832 (pt 2): 182–184.

Blainville, H. [Grande use de Blainville] 1818. Sur les ichthyolites ou les poisons fossils. Nouv. Dist. Hist. Nat. 37: 310–395. Deterville, Paris.

Bloch, M.E. and J.G. Schneider. 1801. M. E. Blochii, Systema Ichthyologiae iconibus cx illustratum. Post obitum auctoris opus inchoatum absolvit, correxit, interpolavit Jo. Gottlob Schneider, Saxo. Berolini. Sumtibus Auctoris Impressum et Bibliopolio Sanderiano Commissum: i–lx + 1–584, pls. 1–110.

Boulenger, G.A. 1901. On the fishes collected by Dr. W.J. Ansorge in the Niger Delta. Proc. Zool. Soc. Lond. 1901, 1 (pt 1): 4–10, pls. 2–4.

Boulenger, G.A. 1901. Diagnoses of new fishes discovered by Mr. W.L.S. Loat in the Nile. Ann. Mag. Nat. Hist. (Ser. 7), 8 (no. 47): 444–446.

Boulenger, G.A. 1909. Catalogue of the fresh-water fishes of Africa in the British Museum (Natural History), 1: i–xi + 1–373.

Boulenger, G.A. 1910. On a large collection of fishes made by Dr. W.J. Ansorge in the Quanza and Bengo rivers, Angola. Ann. Mag. Nat. Hist. (Ser. 8), 6 (36): 537–561.

Boulenger, G.A. 1915. Diagnoses de poissons nouveaux. II. Mormyrides, Kneriides, Characinides, Cyprinides, Silurides. (Mission Stappers au Tanganika-Moero.). Rev. Zool. Afr. 4 (fasc. 2): 162–171.

Cope, E.D. 1885. Eocene paddle-fish and Gonorhynchidae. Am. Nat. 19: 1090–1091.

Cuvier, G. 1829. Le Règne Animal, distribué d'après son organisation, pour servir de base à l'histoire naturelle des animaux et d'introduction à l'anatomie comparée. Edition 2. v. 2: i–xv + 1–406.

Cuvier, G. and A. Valenciennes. 1836. Histoire naturelle des poissons. Tome onzième. Livre treizième. De la famille des Mugiloïdes. Livre quatorzième. De la famille des Gobioïdes 11: i–xx + 1–506 + 2 pp., pls. 307–343.

Cuvier, G. and A. Valenciennes. 1847. Histoire naturelle des poissons. Tome dix-neuvième. Suite du livre dix-neuvième. Brochets ou Lucioïdes. Livre vingtième. De quelques familles de Malacoptérygiens, intermédiaires entre les Brochets et les Clupes 19: i–xix + 1–544 + 6 pp., pls. 554–590 [not 520–556].

Daget, J. 1954. Les poissons du Niger Supérieur. Mem. Inst. Franc. Afr. Noire (36): 1–391.

Dietze, C. 2007. Redescription of Dastilbe crandalli (Chanidae, Euteleostei) from the Early Cretaceous Crato Fromation of North-Eastern Brazil. J. Vert. Paleontol. 27(10): 8–16.

Eschmeyer, W.N. 1998. Catalogue of Fishes. California Academy of Sciences, San Francisco.

Eydoux, J.F.T. and F. Souleyet. 1850. Poissons. Pp. 155–216. In: Voyage autour du monde exécuté pendant les années 1836 et 1837 sur la corvette La Bonite, commandée par M. Vaillant. Zoologie, 1 (pt 2). Paris. i–iv, i–xxxix, 1–334, pls. 1–10.

Forsskål, P. 1775. Descriptiones animalium avium, amphibiorum, piscium, insectorum, vermium; quae in itinere orientali observavit.... Post mortem auctoris edidit Carsten Niebuhr. Hauniae: 1–20 + i–xxxiv + 1–164, map.

Fowler, H.W. 1949. Results of the two Carpenter African expeditions, 1946–1948. Pt. II—The fishes. Proc. Acad. Nat. Sci. Philadelphia 101: 233–275.

Gaudant, J. 1981. Contribution de la paléoichthyologie continentale à la reconstitution des paléoenvironnements cénozoigues d'Europe occidentale: approche systématique paléoécologique, paléogéographique et paléoclimatologique. Thèse de doctorate d'état, Univ. Pierre et Marie Curie, Paris 6.

Gayet, M. 1982. Cypriniforme ou Gonorhynchiforme? †*Ramallichthys*, Nouveau genre du Cenomanien inferieur de Ramallah (monts de Judee). C.R. Hebd. Séances Acad. Sci. Paris (II) 295: 405–457.

Gayet, M. 1993. Nouveau genre de Gonorhynchidae du Cénomanien inféruer maein de Hakel (Liban). Implications phylogénétiques. C.R. Acad. Sci. Paris 316 (II): 257–263.

Gayet, M. 1993. Gonorhynchoidei du Cretacé superieur marin du Liban et relations phylogénétiques des Charitosomidae nov. fam. Documents Lab Geol., Lyons 126: 1–131.

Géry, J. 1964. Une nouvelle famille de Poissons dulcaquicoles africaines: les Grasseichthyidae. C.R. Hebd. Seances Acad. Sci. 259: 4805–4807.

Grande, L. and T. Grande. 1999. A new species of †*Notogoneus* (Teleostei: Gonorynchidae) from the Upper Cretaceus Two Medicine Formation of Montana, and the poor Cretaceus record of freshwater fishes from North America. J. Vert. Paleontol. 19(4): 612–622.

Grande, L. and T. Grande. 2008. Redescription of the type species for the genus †*Notogoneus* (Teoeostei: Gonorynchidae) based on new, well-preserved material. J. Paleontol. supplement to 82(5): 1–31.

Grande, T. 1996. The interrelationships of fossil and Recent gonorynchiform fishes with comments on two Cretaceous taxa from Israel, pp. 299–318. *In*: G. Arratia and H. Viohl [eds.]. Mesozoic Fishes—Systematics and Paleoecology. Pfeil, Munich.

Grande, T. 1999. Revision of the genus *Gonorynchus* Scopoli 1777 (Teleostei: Ostariophysi). Copeia 1999(2): 453–469.

Grande, T. and F.J. Poyato-Ariza 1999. Phylogenetic relationships of fossil and Recent gonorynchiform fishes (Teleostei: Ostariophysi). Zool. J. Linn. Soc. 125: 197–238.

Gronow, L.T. 1763. Zoophylacii Gronoviani fasciculus primus exhibens animalia quadrupeda, amphibia atque pisces, quae in museo suo adservat, rite examinavit, systematice disposuit, descripsit atque iconibus illustravit Laur. Theod. Gronovius, J.U.D.... Lugduni Batavorum: 1–136, 14 pls.

Günther, A. 1868. Catalogue of the fishes in the British Museum. Catalogue of the Physostomi, containing the families Heteropygii, Cyprinidae, Gonorhynchidae, Hyodontidae, Osteoglossidae, Clupeidae,... [thru]... Halosauridae, in the collection of the British Museum 7: i–xx + 1–512.

Günther, A. 1879. [Observations on the marine fish fauna of Rodriguez made during the transit of Venus expeditions in the years 1874–75.]. Philos. Trans. Roy. Soc. Lond. 168 (extra vol.): 470–472.

Hills, E.S. 1934. Tertiary fresh water fishes from Southern Queensland. Mem. Queensl. Mus. 10(4): 163–172. Fowler says Mills.

Jerdon, T.C. 1849. On the fresh-water fishes of southern India. (Continued from p. 149.). Madras J. Lit. Sci. 15 (pt 2): 302–346.

Jordan, D.S. Stanf. Univ. Publ. Biol. Sci. 3 (no. 2).

Jordan, D.S. 1923. A classification of fishes including families and genera as far as known. Stanf. Univ. Publ. Biol. Sci. 3 (no. 2): 77–243 + i–x.

Jordan, D.S. and J.O. Snyder. 1923. *Gonorhynchus moseleyi*, a new species of herring-like fish from Honolulu. J. Wash. Acad. Sci. 13(15): 347–350.

Jubb, R.A. and G. Bell-Cross. 1974 . A new species of *Parakneria* Poll 1965 (Pisces, Kneriidae) from Moçambique. Arnoldia (Rhod.) 6(29): 1–4.

Kner, R. 1867. Fische. Reise der österreichischen Fregatte "Novara" um die Erde in den Jahren 1857–1859, unter den Befehlen des Commodore B. von Wüllerstorf-Urbain. Wien. Zool. Theil. 1 (3): 275–433, pls. 12–16.

Lacepède, B.G.E. 1803. Histoire naturelle des poisons, 5: i–lxviii + 1–803 + index, pls. 1–21.

Ladiges, W. and J. Voelker. 1961. Untersuchungen über die Fischfauna in Gebirgsgewässern des Wasserscheidenhochlands in Angola. Mitt. Hamb. Zool. Mus. Inst. 59: 117–140, pls. 3–7.

Linnaeus, C. 1766. Systema naturae sive regna tria naturae, secundum classes, ordines, genera, species, cum characteribus, differentiis, synonymis, locis. Laurentii Salvii, Holmiae. 12th ed. 1 (pt 1): 1–532.

Moritz, T., R. Britz and K.E. Linsenmair. 2006. *Cromeria nilotica*, two valid species of the African freshwater fish family Kneriidae (Teleostei: Gonorynchiformes). Ichthyol. Explor. Freshwaters 17(1): 65–72.

Myers, G.S. 1928. Two new genera of fishes. Copeia (166): 7–8.

Nelson, J.S. 2006. Fishes of the World (4th ed.). John Wiley and Sons, Inc., Hoboken, New Jersey.

Norman, J.R. 1923. A new cyprinoid fish from Tanjanyika Territory, and two new fishes from Angola. Ann. Mag. Nat. Hist. (Ser. 9), 12 (no. 72): 694–696.

Ogilby, J.D. 1911. On the genus *Gonorrynchus* (Gronovius). Ann. Queensl. Mus. 10: 30–35.

Pellegrin, J. 1905. Poisson nouveau du Mozambique. Bull. Mus. Natl. Hist. Nat. (Sér. 1), 11 (3): 145–146.

Pellegrin, J. 1922. Poissons nouveaux de l'Afrique orientale. Bull. Mus. Natl. Hist. Nat. (Sér. 1) 28(5): 349–351.

Pellegrin, J. 1925. Sur les poissons africains de la famille des Phractolaemidés. C.R. Hebd. Seances Acad. Sci. 180: 549–551.

Pellegrin, J. 1931. Poissons du Kouilou et de la Nyanga recueillis par M. A. Baudon. Bull. Soc. Zool. Fr. 56: 205–211.

Penrith, M.J. 1973. A new species of *Parakneria* from Angola (Pisces: Kneriidae). Cimbebasia (Ser. A), 2 (11): 131–135, pl. 1.

Peters, W. (C.H.) 1852. Diagnosen von neuen Flussfischen aus Mossambique. Monatsb. Akad. Wiss. Berlin 1852: 275–276, 681–685.

Peters, W. (C.H.) 1859. Eine neue vom Herrn Jagor im atlantischen Meere gefangene Art der Gattung *Leptocephalus*, und über einige andere neue Fische des Zoologischen Museums. Monatsb. Akad. Wiss. Berlin 1859: 411–413.

Poll, M. 1944. Descriptions de poissons nouveaux recueillis dans la région d'Albertville (Congo belge) par le Dr. G. Pojer. Bull. Mus. Roy. Hist. Nat. Belg. 20 (3): 1–12.

Poll, M. 1965. Contribution à l'étude des Kneriidae et description d'un nouveau genre, le genre *Parakneria* (Pisces, Kneriidae). Mem. Acad. Roy. Belg. Cl. Sci. (Ser. 8), 36 (fasc. 4): 1–28, pls. 1–16.

Poll, M. 1967. Contribution à la faune ichthyologique de l'Angola. Publ. Cult. Cia. Diamantes Angola (75): 1–381, pls. 1–20.

Poll, M. 1969. Contribution à la connaissance des *Parakneria*. Rev. Zool. Bot. Afr. 80 (3–4): 359–368.

Poll, M. 1976. Poissons. Exploration Parc National Upemba Mission G. F. de Witte (73): 1–127, 43 pls.

Poll, M. 1984. *Parakneria tanzaniae*, espèce nouvelle des chutes de la rivière Kimani, Tanzanie (Pisces, Kneriidae). Rev. Zool. Afr. 98 (1): 1–8.

Poll, M. and D.J. Stewart. 1975. Un Mochocidae et un Kneriidae nouveaux de la rivière Luongo (Zambia), affluent du bassin du Congo (Pisces). Rev. Zool. Afr. 89 (1): 151–158.

Priem, M.F. 1908. Etude des poisons fossils du bassini Parisien. Publ. Annal. Paléontol., Paris 144 pp., 5 pls.

Regan, C.T. 1902. On the fishes from the Maldive Islands. Faun. Geog. Mald. Lacc. Archip. 1 (pt. 3): 272–281.

Richardson, J. 1844–48. Ichthyology of the voyage of H.M.S. Erebus and Terror,... *In*: J. Richardson and J.E. Gray. The Zoology of the Voyage of H.M.S. "Erebus and Terror," under the command of Captain Sir J.C. Ross ... during ... 1839–43. London 2 (2): i–viii + 1–139, pls. 1–60.

Richardson, J. and J.E. Gray. 1843. List of fish hitherto detected on the coasts of New Zealand, ...; with the description, by J.E. Gray, Esq., and Dr. Richardson, of the new species brought home by Dr. Dieffenbach, pp. 206–228. *In*: E. Dieffenbach. Travels in New Zealand vol. 2.

Rüppell, W.P.E.S. 1828–30. Atlas zu der Reise im nördlichen Africa. Fische des Rothen Meeres. Frankfurt-am-Main 1–141 + 3 pp., col. pls. 1–35.

Scopoli, J.A. 1777. Introductio ad historiam naturalem, sistens genera lapidum, plantarum et animalium hactenus detecta, caracteribus essentialibus donata, in tribus divisa, subinde ad leges naturae. Prague: i–x + 1–506.

Seegers, L. 1995. Revision of the Kneriidae of Tanzania with description of three new *Kneria*-species (Teleostei: Gonorhynchiformes). Ichthyol. Explor. Freshwaters 6(2): 97–128.

Steindachner, F. 1866. Ichthyologische Mittheilungen. (IX.) [With subtitles I–VI.]. Verh. K.-K. Zool.-Bot. Ges. Wien 16: 761–796, pls. 13–18.

Sytchevskaya, E.K. 1986. Presnovodnaya paleogenovaya ikhtiofauna SSSR i Mongolii. Paleogene freshwater fish fauna of the USSR and Mongolia]. Trudy-Sovmestnaya Sovetsko-Mongol'skaya Nauchno-Issledovatel'skaya Nauchno-Issledovatel'stov "Nauka" 29:1–154 [Russian with English summary].

Temminck, C.J. and H. Schlegel. 1843. Pisces. *In*: Fauna Japonica, sive descriptio animalium quae in itinere per Japoniam suscepto annis 1823–30 collegit, notis observationibus et adumbrationibus illustravit P. F. de Siebold. Part 1: 1–20.

Temminck, C.J. and H. Schlegel. 1846. Fauna Japonica, sive descriptio animalium quae in itinere per Japoniam suscepto annis 1823–30 collegit, notis observationibus et adumbrationibus illustravit P. F. de Siebold. Parts 10–14: 173–269.

Trewavas, E. 1936. Dr. Karl Jordan's expedition to South-West Africa and Angola: The fresh-water fishes. Novit. Zool. (Tring) 40: 63–74, pls. 1–2.

van Hasselt, J.C. 1823. Uittreksel uit een' brief van Dr. J. C. van Hasselt, aan den Heer C. J. Temminck. Algem. Konst Letter-bode I Deel (no. 21): 329–331.

Wilson, M.V.H. 1980. Oldest known *Esox* (Pisces: Esocidae), paet of a new Paleocene teleost fauna from western Canada. Canad. J. Earth Sci. 17: 307–312.

Wilson, M.V.H. 1981. Eocene freshwater fishes from the Coalmont Formation Colorado. J. Paleontol. 55:671–674.

Woodward , A.S. 1901. Catalogue of the fossil fishes in the British Museum, 4. London, Taylor and Francis 1–636.

Index

A

Abductor superficialis 118, 119, 135, 136, 137, 138, 139, 140
Adductor arcus palatini 110, 114, 115, 118, 119, 123, 124, 125, 127, 280, 329
Adductor hyomandibulae-1 126, 127, 128
Adductor hyomandibulae-2 126, 127, 128
Adductor mandibulae 108, 110, 111, 112, 113, 114, 115, 116, 117, 118, 119, 120, 121, 122, 123, 124, 139, 140, 141, 143, 278, 279, 280, 300, 303, 304, 314, 315, 329
Adductor operculi 110, 114, 115, 119, 123, 125, 126, 127, 129
Adductor profundus 108, 135, 136, 137, 138, 139, 140, 277, 278, 329
Adductor superficialis 118, 119, 134, 135, 136, 138, 139, 140
Anguloarticular 348
Antorbital bone 112, 262, 473
Arcocentrum 40
Arcualia 40, 81, 82
Arrector dorsalis 108, 118, 119, 133, 134, 135, 136, 137, 138, 139, 140
Arrector ventralis 118, 119, 134, 135, 137, 138
Articular head (of Hyomandibula) 256, 273, 274, 300, 314, 325, 330, 337, 345, 346, 347, 348, 358
Articular process (of Maxilla) 249, 251, 297, 299, 301, 313, 324, 329, 357
Ascending process (of Interopercular bone) 249, 261, 295, 303, 312, 315, 326, 359
Ascending process (of Premaxilla) 249, 261, 295, 303, 312, 314, 316
Autocentrum 40, 42, 50
Autopalatine 249, 251, 254, 255, 295, 312, 305, 313, 314, 316, 324, 325, 357, 479, 480, 488, 489
Autosphenotic 77, 83, 87, 88, 95, 480

B

Basibranchial 150, 151, 162, 254, 257, 288, 300, 303, 304, 314, 315, 325, 339, 341, 346, 347, 349, 352, 353, 354, 358, 359, 430
Basihyal 170
Basioccipital 242, 243, 297, 323, 356
Basipterygium 78, 82, 83, 89, 91, 93, 101, 102, 481
Basipterygoid process 413
Basisphenoid 241, 323
Branchial arches 2, 27, 30–34, 74, 78, 87, 92, 93, 96, 97, 102, 145, 147, 149, 154, 160, 164, 165, 257, 413
Branchiostegal ray 277

C

Caudal endoskeleton 33, 234, 268, 269, 270, 271, 273, 274, 275, 276, 287, 288, 289, 294, 299, 300, 301, 309, 328
Caudal scute 268, 328
Cephalic (= Cranial) rib 1, 2, 5, 7, 18, 32, 54, 65, 66, 243, 266, 286, 295, 312, 323, 356, 369
Ceratobranchial 78, 79, 80, 96, 257, 295, 312, 325, 411, 412, 416, 428, 430, 431, 434
Ceratohyal, anterior 78
Ceratohyal, posterior 78
Chondral bone 27, 77, 94, 97, 101
Chondrocranium (= Endocranium) 2, 73, 103, 104, 106
Chordacentrum 40, 42
Cleithrum 43, 53, 64, 65, 66, 67, 268
Compound centrum 60
Coracoid 64, 66, 67
Coronoid process 277, 328, 362
Coronomeckelian 76, 77, 83, 87, 95, 111, 112, 113, 417, 488
Cranial intermuscular bones 243, 300, 314, 323, 356

D

Dentary 77, 83, 86, 92, 94, 95, 97, 246, 250, 276, 279, 281, 282, 296, 298, 300, 301, 312, 314, 324, 328, 329, 342, 344, 345, 348, 349, 352, 357, 362, 501

Dentary notch (="Leptolepid" notch) 250, 296

Dermal bone 3, 26, 77, 79, 94

Dermal tooth plates 78

Dermatocranium (=Dermocranium) 2, 3

Dermopalatine 254, 286, 325, 358

Dilatator operculi 279, 280, 329, 362

E

Ectopterygoid 254, 255, 287, 304, 315, 325, 358, 411

Endopterygoid (=Entopterygoid) 21, 25, 26, 27, 188, 253, 254, 256, 280, 288, 301, 342, 344, 346, 347, 349, 354

Endopterygoid teeth 301, 347, 349, 354

Epibranchial bone 78, 83, 89, 93, 96, 155, 388, 400, 466

Epibranchial organ 30, 31, 102, 145–165

Epicentral intermuscular 53, 54

Epiotic (=Epioccipital) 77, 83, 88, 89, 90, 95

Epipleural intermuscular 55

Epural 58, 59, 60, 61, 62, 64

Ethmoid 77, 83, 88, 89, 95, 96, 101, 110, 118, 119, 123, 243, 252, 343, 344, 349, 466, 480, 483, 488, 489, 498

Exoccipital 45, 52, 54, 77, 83, 86, 88, 95, 101, 242, 246

Extrascapular 77, 86, 89, 110, 118, 119, 123, 349, 350, 383, 384

F

Fin rays (=Lepidotrichia) 55, 56, 74, 93, 135, 138, 182, 183, 273, 472, 482

Foramen magnum 246, 303, 314, 324, 357

Frontal bone 90

Fulcra (fringing) 268, 298, 327, 361

G

Gas bladder (=Swimbladder) 2, 44, 164, 369, 386, 425, 427, 466, 472

Gill rakers 79, 80, 102

Gingival teeth 247, 324, 357

H

Haemal arch 276, 303, 315, 328, 361

Historical biogeography 213, 215, 307, 308, 443, 509, 517, 519, 520, 547, 562

Hyohyoidei adductores 126, 127, 131, 132, 133

Hyohyoideus abductor 126, 127, 131, 132, 133, 277, 304, 316, 328, 362

Hyohyoideus inferioris 277, 328, 362

Hyoid arch 277

Hyomandibula (=Hyomandibular bone) 256, 280, 346

Hyosymplectic 74

Hypobranchial 31, 33, 78, 86, 87, 93, 96, 257, 482

Hypohyal, dorsal 78

Hypohyal, ventral 78

Hypural 58, 60, 63, 80, 273, 274, 275, 276, 286, 328, 361, 417, 472

I

Infraorbital bone 88

Infraorbital sensory canal 77

Infrapharyngobranchial 79

Intercalar 77, 88

Interfrontal fontanelle 244, 304, 31, 323, 356

Interhyal 126, 127, 257, 304, 315, 325, 358

Intermandibularis 276, 328, 362

Interopercle (=Interopercular bone) 15, 19, 21, 23, 25, 28, 78, 83, 87, 88, 96, 110, 114, 116, 117, 118, 119, 123, 126, 127, 132, 261, 288, 303, 304, 315, 326, 348, 474, 503

Interopercular spine 288

L

Lacrimal (=Infraorbital 1) 21, 83, 87, 88, 95, 111, 262

Lateral ethmoid 252, 466, 480, 483, 488, 489, 498

Lateral line 268, 276, 327, 328, 482, 565

Lateroparietal (skull condition) 244, 323, 356

Leptolepid notch (=Dentary notch) 222, 226, 250, 295

Levator arcus palatini 110, 114, 115, 118, 119, 122, 123, 124, 154, 280, 329

Levator operculi 110, 114, 115, 116, 117, 118, 119, 121, 123, 128, 129, 154

M

Mandibular arch 2, 3, 25, 28
Mandibular sensory canal 251, 300, 313, 329
Maxilla 3, 21, 23, 25, 26, 29, 32, 243, 246, 247, 248, 249, 250, 251, 278, 279, 281, 290, 295, 296, 312, 324, 329
Maxillary process 249, 251, 281, 297, 301, 304, 313, 315, 324
Medioparietal (skull condition) 244, 323, 356
Mesethmoid 2, 3, 5, 7, 8, 9, 14, 15, 28, 246, 255, 256, 298, 303, 304, 314, 316, 325, 326
Mesocoracoid 64, 66, 78, 83, 90, 94, 100, 134, 135, 136, 138, 139, 140, 488, 498
Mesoparietal (skull condition) 244, 323, 356
Metapterygiod bone 3, 27, 77, 87, 88, 122, 124, 125, 159, 253, 256, 280, 296, 325, 358, 479, 501, 504
Metapterygoid process (of Hyomandibula) 3, 21, 25, 27, 252, 253, 256, 280, 286, 296, 303, 312, 314, 325

N

Nasal 3, 5, 7, 243, 297, 304, 313, 315, 323, 339, 341, 343, 350, 352, 353, 356, 477
Neural arch 263, 264, 268, 269, 286, 295, 301, 304, 312, 315, 327, 328, 370
Neural spine 264, 478
Neurocranium (=Endocranium) 2, 280, 329
Notochord 39, 40, 42, 76, 79, 80, 81, 82, 86
Notochordal flexion 79, 80

O

Olfactory nerve 244, 323, 356
Opercle (=Opercular bone) 45, 64, 78, 83, 82, 92, 96, 97
Opercular apparatus 51, 53, 70, 92, 97, 258, 318, 326, 359
Opercular spine 258, 288, 326, 359
Oral process (of Premaxilla) 247, 295, 312, 324, 357
Orbitosphenoid 77, 241, 289, 295, 311, 323, 356, 369, 417, 419
Osteocranium 73, 103

P

Paedomorphosis 69, 167, 318
Palatoquadrate 74, 77, 95, 96, 103
Parapophysis 44, 49
Parasphenoid 77, 83, 86, 94, 95, 96, 97, 101, 171, 342, 343, 344, 345, 347, 349, 358
Parhypural (=Parahypural) 58, 59, 60, 61, 62, 63, 64, 80, 273, 328, 361
Parietal bone 76, 77, 78, 83, 89, 90, 95
Pectoral axillary process 78, 83, 91
Pelvic axillary process 78
Pelvic splint, 78, 83, 90
Peritoneal tunic 370, 371, 386, 387
Pharyngeal teeth 417, 426, 430
Pharyngobranchial 32, 78, 79, 87, 93, 96, 155, 159, 160, 162, 257, 300, 314, 325, 359, 411, 466, 480
Pleural rib 369, 371, 386, 387
Pleurostyle 370, 381, 382, 389
Postcleithrum 268
Posttemporal 10, 14, 16, 64
Premaxilla 3, 21, 23, 25, 29, 32, 110, 111, 114, 116, 117, 118, 119, 121, 122, 123, 243, 246, 247, 248, 249, 251, 278, 287, 289, 295, 312, 324, 329
Premaxillary process (of Maxilla) 25, 247, 248, 295, 312, 324
Preopercle (=Preopercular bone) 19, 21, 22, 23, 25, 28, 110, 111, 112, 114, 118, 119, 122, 123, 124, 125, 126, 127, 129, 132, 133, 260, 280, 329
Preopercular sensory canal 21, 259
Preural centrum 43, 54, 57, 58, 60, 61, 62, 63, 64, 268, 269, 271, 272, 273, 276, 303, 309, 315, 328
Prootic 2, 7, 10, 14, 15, 16, 124, 125, 127, 128
Propterygium 78, 83, 91, 102
Protractor hyoidei 113, 115, 126, 130, 131, 277, 328
Protractor pectoralis 135, 138, 139, 142
Proximal radial 57, 65, 66, 67, 135, 136, 138
Pterosphenoid 1, 5, 7, 10, 14, 16, 32, 110, 118, 119, 123, 241, 289, 295, 300, 311, 313, 323
Pterotic 2, 3, 5, 7, 10, 12, 13, 14, 16, 17, 28, 118, 119, 125, 127, 128, 129, 138, 245

Q

Quadrate 3, 5, 15, 21, 23, 25, 26, 28, 29, 30, 32, 34, 110, 111, 112, 114, 123, 124, 125
Quadrate-mandibular articulation 28

R

Radials 55, 67, 135, 136, 138
Retroarticular bone 29
Retroarticular process 29
Rib 5, 7, 18, 42, 43, 44, 45, 49, 50, 53, 54, 65, 66

S

Scale 577
Scapula 64, 65, 66, 67, 78, 83, 81, 89, 91, 94, 102, 134, 135, 136, 138, 342, 349
Scapular foramen 67
Sclerotic 77, 91
Sphenotic 418
Splanchnocranium 97
Sternohyoideus 154, 158, 159, 171
Subopercle (=Subopercular bone) 177, 180, 186, 193, 342, 344, 348, 349
Subopercular clefts 261, 326, 359
Subopercular spine 359
Suborbital 427
Supracleithrum 64
Supradorsal 46, 49, 50
Supramaxilla 348, 357

Supraneural 44, 45, 49, 50, 51, 52, 180, 181, 186, 267, 297, 313, 327, 479
Supraoccipital bone 246, 281, 301
Supraoccipital crest 14
Supraorbital bone 5, 83, 90, 95, 246, 281, 301
Supraorbital sensory canal 77, 344, 349
Suprapreopercle (=Suprapreopercular bone) 19, 21, 78, 83, 91, 96
Supratemporal commissure 10, 11, 12, 13, 14, 87
Symplectic 3, 21, 25, 27, 28, 78, 83, 86, 87, 92, 95, 96

T

Tendon bone 77
Terminal centrum 43, 44, 57, 58, 60, 63, 64, 68

U

Ural centrum 43, 59, 64
Urodermal 382, 388, 390, 399
Urohyal 74, 76, 83, 86, 95, 103
Uroneural 43, 57, 58, 60, 61, 64

V

Vicariance 215, 307, 443, 457
Vomer 77, 88, 94, 96, 101

W

Weberian apparatus 39, 40, 44, 47, 69, 70

Milton Keynes UK
Ingram Content Group UK Ltd.
UKHW020004071024
449327UK00031B/2651